Environmental Science and Engineering

Series Editors

The ultimate goal of this series is to contribute to the protection of our environment, which calls for both profound research and the ongoing development of solutions and measurements by experts in the field. Accordingly, the series promotes not only a deeper understanding of environmental processes and the evaluation of management strategies, but also design and technology aimed at improving environmental quality. Books focusing on the former are published in the subseries Environmental Science, those focusing on the latter in the subseries Environmental Engineering.

Pravat Kumar Shit · Partha Pratim Adhikary ·
Gouri Sankar Bhunia · Debashish Sengupta
Editors

Soil Health and Environmental Sustainability

Application of Geospatial Technology

Springer

Editors
Pravat Kumar Shit
Geography
Raja N. L. Khan Womens College
(Autonomous)
Midnapore, West Bengal, India

Gouri Sankar Bhunia
GIS Scientist
TPF Getinsa Euroestudios S.l
Gurgaon, Haryana, India

Partha Pratim Adhikary
Soil Science
ICAR Indian Institute of Water
Management
Bhubaneswar, Odisha, India

Debashish Sengupta
Department of Geology and Geophysics
Indian Institute of Technology Kharagpur
Kharagpur, West Bengal, India

ISSN 1863-5520 ISSN 1863-5539 (electronic)
Environmental Science and Engineering
ISBN 978-3-031-09269-5 ISBN 978-3-031-09270-1 (eBook)
https://doi.org/10.1007/978-3-031-09270-1

This Springer imprint is published by the registered company Springer Nature Switzerland AG
The registered company address is: Gewerbestrasse 11, 6330 Cham, Switzerland

Dedicated to
Young Scholars in the field of Soil Science
and Environment

Preface

Soil and agricultural systems are an integral part of the global environment and human well-being, providing multiple goods and services essential for people worldwide and crucial for sustainable development. Soil contamination is an environmental hazard, and has become a big issue related to environmental health. The challenge of twenty-first century is to reduce the contaminant load and bring it to below permissible level. The contamination is not only a problem affecting local environments at the place of occurrence but also spreading to other regions because of easy transportation of pollutants. This leads to direct and indirect contamination of land and aquatic systems, surface water and groundwater, inducing significant risks for natural ecosystems. The liquid waste and leachates generated from the solid waste percolate into the ground causing problems like groundwater contamination, degradation of vegetation, modification of soil properties, etc. Soil is a "universal sink" which bears the greatest burden of environmental pollution. It is getting polluted in a number of ways. However, there is an urgent need for proper management and geospatial modelling to control soil pollution in order to preserve soil fertility and increase productivity as well as reduce the ecological risk.

GIS data management and spatial modelling provide the possibility to examine large amount of data simultaneously by incorporating layers of different data into one, which expands the field of application to environmental modelling and scenario simulation. Spatial tools and geostatistical applications are useful techniques for assessment of values at un-sampled locations and account the spatial correlation between projected and sampled points and minimizing the discrepancy of assessment error, trends and patterns in large amounts of data and implementation costs.

This book is composed of 31 chapters associated with spatial modelling in soil and sediments pollution, and remediation, radioactive wastes, microbiology of soil and sediments, soil salinity and sodicity, pollution from landfill sites, soil erosion and contamination from agricultural activities, heavy metal pollution and health risk, Environmental impact and risk assessment, sustainable land use, landscape management and governance, soil degradation and risk assessment, agricultural soil pollution, pollution due to urban activities, soil pollution by industrial effluents and solid wastes, pollution control and mitigation in extreme environments. The content of this book will be of interest to researchers, professionals and policymakers who work in the soil science and agriculture practices. The book equips with the knowledge and skills to tackle a wide range of issues manifested in geographic data, including those with scientific, societal and environmental implications.

We are very much thankful to all the authors who have meticulously completed their documents on a short announcement and played a vital role in building this edifying and beneficial publication. We do believe that this will be a very convenient book for soil science, geographers, ecologists, environmental scientists and others working in the field of soil-water resources management including research scholars, environmentalists and policymakers. We also acknowledge our deep gratitude to the Springer Publishing House for contracting with us for such timely publication.

Midnapore, India — Pravat Kumar Shit
Bhubaneswar, India — Partha Pratim Adhikary
Gurgaon, India — Gouri Sankar Bhunia
Kharagpur, India — Debashish Sengupta

Acknowledgments

The preparation of this book has been guided by several soil pioneers. We are obliged to these experts for providing their time to evaluate the chapters published in this book. We thank the anonymous reviewers for their constructive comments that led to substantial improvement to the quality of this book. Because this book was a long time in the making, we want to thank our family and friends for their continued support. This work would not have been possible without constant inspiration from our students, knowledge from our teachers, enthusiasm from our colleagues and collaborators, and support from our family. Finally, we also thank our publisher and its publishing editor, Springer, for their continuous support in the publication of this book.

Contents

Editors and Contributors

About the Editors

Pravat Kumar Shit, Ph.D. He is an Assistant Professor at the PG Department of Geography, Raja N. L. Khan Women's College (Autonomous), West Bengal, India. He received his M.Sc. and Ph.D. degrees in Geography from Vidyasagar University and PG Diploma in Remote Sensing and GIS from Sambalpur University. His research interests include applied geomorphology, soil erosion, groundwater, forest resources, wetland ecosystem, environmental contaminants and pollution, and natural resources mapping and modelling. He has published 16 books (Springer—13, Elsevier—02, CRC Press—01) and more than 75 papers in peer-reviewed journals and 74 book chapters. He is also the Guest Editor of "Environmental Science and Pollution Research" and "Applied Water Science". He is currently the Editor of the GIScience and Geo-environmental Modelling (GGM) Book Series, Springer-Nature.

Partha Pratim Adhikary, Ph.D. He is a Senior Scientist at ICAR-Indian Institute of Water Management, Bhubaneswar, India. He obtained his Ph.D. in Agricultural Physics from ICAR-Indian Agricultural Research Institute, New Delhi, India. His research interests include solute transport, soil and water conservation and management, pedotransfer functions and geospatial modelling of natural resources. He has published more than 60 research papers in peer-reviewed journals and 8 books. His other publications include book chapters, popular articles, technology brochures, technical bulletins and scientific reports. He is the Associate Editor of Indian Journal of Soil Conservation. He is also the Guest Editor of "Environmental Science and Pollution Research" and "Applied Water Science". Currently, he is the Editor of Springer-Nature book series GIScience and Geo-environmental Modelling.

Gouri Sankar Bhunia, Ph.D. He received his Ph.D. from the University of Calcutta, India in 2015. His Ph.D. dissertation work focused on environmental control measures of infectious disease using geospatial technology. His research interests include environmental modelling, risk assessment, natural resources mapping and modelling, data mining and information retrieval using geospatial technology. He is Associate Editor and is on the editorial boards of three international journals in Health GIS and Geosciences. He has published more than 75 articles in various journals in Scopus indexed. He is currently the Editor of the GIScience and Geo-environmental Modelling (GGM) Book Series, Springer-Nature.

Debashish Sengupta, Ph.D. He is working as Professor, Higher Administrative Grade and Former Head of the Department of Geology and Geophysics in Indian Institute of Technology (IIT) Kharagpur, West Bengal, India. He has more than 30 years of teaching and research experience. He has completed his Ph.D. in 1987 in Applied Geophysics. His areas of interests are nuclear geophysics including petroleum logging using subsurface nuclear data, radioactive methods and geochronology, radon emanometry and its applications, applications of isotopes and radionuclides in earth and environmental geosciences, heat flow and geothermics.

He has 110 research publications in international journals and more than 60 papers in Conference Proceedings apart from a large number of Invited Talks delivered both in India and abroad. The research work has been seminal and resulted in the formulation of Environmental Regulation policies in various countries both in India and countries like USA, South America and the European Union. He had also been a Visiting Professor at the University of Sao Paulo, Brazil and as Senior Visiting Professor at the University of Salamanca, Spain, earlier, while on a sabbatical leave from the institute. He has received Society of Geoscientists and Allied Technologists (SGAT) Award of Excellence in Earth Sciences for the year 2003. He has published four (04) books.

Contributors

Adhikary Partha Pratim ICAR-Indian Institute of Water Management, Bhubaneswar, Odisha, India

Amato Mariana Scuola di Scienze Agrarie, Forestali, Alimentari ed Ambientali, Università degli Studi della Basilicata, Potenza, Italy

Arvind Department of Soil Science of Agricultural Chemistry, Institute of Agricultural Sciences, Banaras Hindu University, Varanasi, U.P, India

Bahuguna Ayush Department of Soil Science of Agricultural Chemistry, Institute of Agricultural Sciences, Banaras Hindu University, Varanasi, U.P, India

Bajracharya Roshan M. Kathmandu University, Dhulikhel, Nepal;
Tribhuvan University, Kathmandu, Nepal

Banerjee S. K. Ecology & Climate Change Division, TFRI, Jabalpur, India

Banerjee Saikat Ecology & Climate Change Division, TFRI, Jabalpur, India

Barik Shuli Genetics and Cell Biology Laboratory, Department of Zoology, Sidho-Kanho-Birsha University, Purulia, West Bengal, India

Barman Arijit ICAR-Central Soil Salinity Research Institute, Karnal, Haryana, India

Barthwal Akshita Department of Soil Science of Agricultural Chemistry, Naini Agriculture Institute Sam Higginbottom University of Agriculture, Technology and Sciences, Prayagraj, U.P, India

Basak Nirmalendu ICAR-Central Soil Salinity Research Institute, Karnal, Haryana, India

Bedwal Sandeep ICAR-Central Soil Salinity Research Institute, Karnal, Haryana, India

Behera Bhaskar Department of Biosciences and Biotechnology, Fakir Mohan University, Balasore, Odisha, India

Behera Jiban Kumar Department of Zoology, Fakir Mohan University, Balasore, Odisha, India

Bera Biswajit Department of Geography, Sidho Kanho Birsha University, Purulia, West Bengal, India

Bhattacharya Manojit Department of Zoology, Fakir Mohan University, Balasore, Odisha, India

Bhattacharyya Sourav Genetics and Cell Biology Laboratory, Department of Zoology, Sidho-Kanho-Birsha University, Purulia, West Bengal, India

Bhunia Gouri Sankar Paschim Medinipur, West Bengal, India;
Kharagpur, West Bengal, India

Biswas Surjyo Jyoti Genetics and Cell Biology Laboratory, Department of Zoology, Sidho-Kanho-Birsha University, Purulia, West Bengal, India

Chakraborty Baisakhi PG Department of Geography, Raja N. L. Khan Women's College (Autonomous), Midnapore, West Bengal, India

Chandola Dinesh GB Pant National Institute of Himalayan Environment, Kosi-Katarmal, Almora, Uttarakhand, India

Chandra Priyanka ICAR-Central Soil Salinity Research Institute, Karnal, Haryana, India

Chandrakala M. National Bureau of Soil Survey and Land Use Planning, Hebbal, Bangalore, Karnataka, India;
ICAR - National Bureau of Soil Survey and Land Use Planning, Regional Centre, Bangalore, India

Chatterjee Sabyasachi Department of Botany (PG), Ramananda College, Bishnupur, Bankura, West Bengal, India

Chatterjee Soumik PG Department of Botany, Ramananda College, Bishnupur, Bankura, West Bengal, India

Chattopadhyay T. ICAR-National Bureau of Soil Survey and Land Use Planning, Kolkata, West Bengal, India

Dadarwal Basant Kumar Department of Soil Science of Agricultural Chemistry, Institute of Agricultural Sciences, Banaras Hindu University, Varanasi, U.P, India

David Raj Anu Agriculture & Soils Department, Indian Institute of Remote Sensing, Indian Space Research Organization (ISRO), Dehradun, India

Deb Sutapa Department of Civil Engineering, Indian Institute of Technology, Kharagpur, West Bengal, India

Dharumarajan S. ICAR - National Bureau of Soil Survey and Land Use Planning, Regional Centre, Bangalore, India

Dhiman Varun Department of Environmental Sciences, Central University of Himachal Pradesh, Dharamshala, India

Dutta Subrata Department of Plant Pathology, Nadia, Mohanpur, West Bengal, India

Dwivedi B. S. National Bureau of Soil Survey and Land Use Planning, Nagpur, Maharashtra, India;
ICAR-National Bureau of Soil Survey and Land Use Planning, Maharashtra, Nagpur, India

Ertunç Güneş Department of Mining Engineering, Hacettepe University, Beytepe, Ankara, Turkey

Fernández Manuel Pulido GeoEnvironmental Research Group, University of Extremadura, Cáceres, Spain

Giri Anand Department of Environmental Sciences, Central University of Himachal Pradesh, Dharamshala, India

Giri Santosh Kumar Genetics and Cell Biology Laboratory, Department of Zoology, Sidho-Kanho-Birsha University, Purulia, West Bengal, India

Gope Dinesh Genetics and Cell Biology Laboratory, Department of Zoology, Sidho-Kanho-Birsha University, Purulia, West Bengal, India

Gorain Sanjib Genetics and Cell Biology Laboratory, Department of Zoology, Sidho-Kanho-Birsha University, Purulia, West Bengal, India

Gumiere Silvio José Faculty of Agriculture and Food Sciences, Pavillon Paul-Comtois, Rue de L'Agriculture, Université Laval, Québec, QC, Canada

Habib Md. Ahosan Geological Survey of Bangladesh, Segunbaghicha, Dhaka, Bangladesh

Haydar Md. Abu Institute of Nuclear Science & Technology, Bangladesh Atomic Energy Commission, Savar, Dhaka, Bangladesh

Hegde Rajendra National Bureau of Soil Survey and Land Use Planning, Hebbal, Bangalore, Karnataka, India;
ICAR - National Bureau of Soil Survey and Land Use Planning, Regional Centre, Bangalore, India

Hombegowda H. C. ICAR-Indian Institute of Soil and Water Conservation, Research Centre, Ooty, Tamil Nadu, India

Idris Abubakr M. Research Center for Advanced Materials Science (RCAMS), King Khalid University, Abha, Saudi Arabia;
Department of Chemistry, College of Science, King Khalid University, Abha, Saudi Arabia

Islam Md. Saiful Department of Soil Science, Patuakhali Science and Technology University, Dumki, Patuakhali, Bangladesh

Islam Sk Saruk Department of Zoology, Raja N.L. Khan Women's College, Midnapore, West Bengal, India

Jakhar Praveen ICAR-Central Institute of Women in Agriculture, Bhubaneswar, Odisha, India

Jena Roomesh Kumar ICAR-Indian Institute of Water Management, Bhubaneswar, India

Kabir Md. Humayun Department of Environmental Science and Resource Management, Mawlana Bhashani Science and Technology University, Tangail, Bangladesh

Kanwar Nidhi Center for Environment Assessment and Climate Change, G.B. Pant, National Institute of Himalayan Environment, Kosi-Katarmal, Almora, Uttarakhand, India

Kar Niladri Bhusan Department of Zoology, Fakir Mohan University, Balasore, Odisha, India

Karim Md. Masud Institute of Nuclear Science & Technology, Bangladesh Atomic Energy Commission, Savar, Dhaka, Bangladesh

Kaur Harshpreet ICAR-Central Soil Salinity Research Institute, Karnal, Haryana, India

Keshavarzi Ali Laboratory of Remote Sensing and GIS, Department of Soil Science, University of Tehran, Karaj, Iran

Khan Rahat Institute of Nuclear Science and Technology, Bangladesh Atomic Energy Commission, Savar, Dhaka, Bangladesh

Khatana Raghu Nandan Singh Department of Soil Science of Agricultural Chemistry, Naini Agriculture Institute Sam Higginbottom University of Agriculture, Technology and Sciences, Prayagraj, U.P, India

Khatri P. K. TFRI, Jabalpur, India

Kormoker Tapos Department of Emergency Management, Patuakhali Science and Technology University, Dumki, Patuakhali, Bangladesh

Kumar Sanjay ICAR-Central Soil Salinity Research Institute, Karnal, Haryana, India

Kumar Suresh Agriculture & Soils Department, Indian Institute of Remote Sensing, Indian Space Research Organization (ISRO), Dehradun, India

Kumar Sushil ICAR-Central Agroforestry Research Institute, Jhansi, Uttar Pradesh, India

Kuniyal Jagdish Chandra Center for Environment Assessment and Climate Change, G.B. Pant, National Institute of Himalayan Environment, Kosi-Katarmal, Almora, Uttarakhand, India

Lakkad A. P. College of Agricultural Engineering and Technology, NAU, Dediapada, India

Lalitha M. ICAR - National Bureau of Soil Survey and Land Use Planning, Regional Centre, Bangalore, India

Machiwal Deepesh Division of Natural Resources, ICAR-Central Arid Zone Research Institute, Jodhpur, Rajasthan, India

Madhu M. ICAR-Indian Institute of Soil and Water Conservation, Dehradun, Uttarakhand, India

Mahato Madhumita Genetics and Cell Biology Laboratory, Department of Zoology, Sidho-Kanho-Birsha University, Purulia, West Bengal, India

Mandal Subashis ICAR-Central Soil Salinity Research Institute, Karnal, Haryana, India

Maske Sunil P. National Bureau of Soil Survey and Land Use Planning, Hebbal, Bangalore, Karnataka, India

Meena Ram Swaroop ICAR-National Bureau of Soil Survey and Land Use Planning, Regional Centre, Udaipur, India

Midya Sujoy Department of Zoology, Raja N.L. Khan Women's College, Midnapore, West Bengal, India

Mishra Akansha Department of Zoology, Fakir Mohan University, Balasore, Odisha, India

Mishra Alka Department of Rural Technology, Guru Ghasidas University, Bilaspur, Chhattisgarh, India

Mishra Pabitra Department of Zoology, Fakir Mohan University, Balasore, Odisha, India

Moharana Pravash Chandra ICAR-National Bureau of Soil Survey and Land Use Planning, Regional Centre, Udaipur, India

Mondal Krishna Chandra Department of Microbiology, The University of Burdwan, Burdwan, West Bengal, India

Mukhopadhyay S. ICAR-National Bureau of Soil Survey and Land Use Planning, Kolkata, West Bengal, India

Naorem Anandkumar Regional Research Station, ICAR-Central Arid Zone Research Institute, Bhuj, Gujarat, India

Nayak D. C. ICAR-National Bureau of Soil Survey and Land Use Planning, Kolkata, West Bengal, India

Pal Jayeeta Genetics and Cell Biology Laboratory, Department of Zoology, Sidho-Kanho-Birsha University, Purulia, West Bengal, India

Pandey Astha Department of Soil Science of Agricultural Chemistry, Institute of Agricultural Sciences, Banaras Hindu University, Varanasi, U.P, India

Pandey Supriya GB Pant National Institute of Himalayan Environment, Kosi-Katarmal, Almora, Uttarakhand, India

Pant Deepak Department of Environmental Sciences, Central University of Himachal Pradesh, Dharamshala, India

Patel Abhishek Regional Research Station, ICAR-Central Arid Zone Research Institute, Bhuj, Gujarat, India

Patra Monoj Genetics and Cell Biology Laboratory, Department of Zoology, Sidho-Kanho-Birsha University, Purulia, West Bengal, India

Paul Debasish Institute of Nuclear Science & Technology, Bangladesh Atomic Energy Commission, Savar, Dhaka, Bangladesh

Pollice Alessio Dipartimento di Economia e Finanza, Università degli Studi di Bari Aldo Moro, Bari, Italy

Pradhan Upendra Kumar ICAR-Indian Agricultural Statistics Research Institute, New Delhi, India

Proshad Ram Key Laboratory of Mountain Surface Processes and Ecological Regulation, Institute of Mountain Hazards and Environment, Chinese Academy of Sciences, Chengdu, Sichuan, China;
University of Chinese Academy of Sciences, Beijing, China

Rai Arvind Kumar ICAR-Central Soil Salinity Research Institute, Karnal, Haryana, India

Rai Ashish Krishi Vigyan Kendra, Rajendra Prasad Central Agricultural University, Pusa, Bihar, India

Rai Sumit Center for Environment Assessment and Climate Change, G.B. Pant, National Institute of Himalayan Environment, Kosi-Katarmal, Almora, Uttarakhand, India

Rajwar Anup Kumar Genetics and Cell Biology Laboratory, Department of Zoology, Sidho-Kanho-Birsha University, Purulia, West Bengal, India

Rashid Md. Bazlar Geological Survey of Bangladesh, Segunbaghicha, Dhaka, Bangladesh

Reza S. K. ICAR-National Bureau of Soil Survey and Land Use Planning, Kolkata, West Bengal, India

Rodrigo-Comino Jesús Departamento de Análisis Geográfico Regional y Geografía Física, Facultad de Filosofía y Letras, University of Granada, Granada, Spain

Rossi Roberta Consiglio per la Ricerca in Agricoltura e l'Analisi dell'Economia Agraria, Centro Zootecnia e Acquacoltura, Rome, Italy

Roy Sambhunath PG Department of Geography, Raja N. L. Khan Women's College (Autonomous), Midnapore, West Bengal, India

Sadeghi-Chahardeh Alireza Mineral, Metallurgical, and Materials Engineering, Faculty of Science and Engineering, Pavillon Adrien-Pouliot, Avenue de la Médecine, Université Laval, Québec, QC, Canada

Saha Sudipta Institute of Nuclear Science & Technology, Bangladesh Atomic Energy Commission, Savar, Dhaka, Bangladesh

Sahoo Gyanaranjan Krishi Vigyan Kendra, Odisha University of Agriculture and Technology, Angul, Odisha, India

Sahoo Sonalika ICAR-National Bureau of Soil Survey and Land Use Planning, Nagpur, India

Sen Krishnendu Department of Microbiology, Vidyasagar University, Midnapore, West Bengal, India;
Department of Plant Pathology, Nadia, Mohanpur, West Bengal, India

Sengupta Debashish Department of Geology and Geophysics, Indian Institute of Technology (IIT), Kharagpur, West Bengal, India

Sharma Parbodh Chander ICAR-Central Soil Salinity Research Institute, Karnal, Haryana, India

Sharma Sachin Department of Soil Science of Agricultural Chemistry, Institute of Agricultural Sciences, Banaras Hindu University, Varanasi, U.P, India

Shit Pravat Kumar PG Department of Geography, Raja N. L. Khan Women's College (Autonomous), Midnapore, West Bengal, India

Shrivastava P. K. ASPEE College of Horticulture and Forestry, NAU, Navsari, India

Singha Sourav Department of Microbiology, Bankura Sammilani College, Bankura, W.B, India

Singh S. K. ICAR-National Bureau of Soil Survey and Land Use Planning, Maharashtra, Nagpur, India

Sondarva K. N. College of Agricultural Engineering and Technology, NAU, Dediapada, India

Srinivasan R. ICAR - National Bureau of Soil Survey and Land Use Planning, Regional Centre, Hebbal, Bangalore, Karnataka, India

Sujatha K. National Bureau of Soil Survey and Land Use Planning, Hebbal, Bangalore, Karnataka, India

Sundha Parul ICAR-Central Soil Salinity Research Institute, Karnal, Haryana, India

Swamy Singam Laxmana Thakur Chedilal Barrister College of Agriculture and Research, Indira Gandhi Agricultural University, Bilaspur, Chhattisgarh, India

Thathola Pooja GB Pant National Institute of Himalayan Environment, Kosi-Katarmal, Almora, Uttarakhand, India

Tuffour Henry Oppong Department of Crop and Soil Sciences, Kwame Nkrumah University of Science and Technology, Kumasi, Ghana

Wani Afaq Majid Department of Forest Biology and Tree Improvement, College of Forestry, Sam Higginbottom University of Agriculture and Sciences, Uttar Pradesh, Prayagraj, India

Yadav Bharti Department of Soil Science of Agricultural Chemistry, Institute of Agricultural Sciences, Banaras Hindu University, Varanasi, U.P, India

Yadav Rajender Kumar ICAR-Central Soil Salinity Research Institute, Karnal, Haryana, India

Zeraatpisheh Mojtaba Henan Key Laboratory of Earth System Observation and Modeling, Henan University, Kaifeng, China;
College of Geography and Environmental Science, Henan University, Kaifeng, China

Part I
Measurement, Monitoring and Mapping of Soil and Land Resources

Chapter 1
Open-Source Satellite Repository and Geographic Information System (GIS) for Soil Resource Mapping

Nidhi Kanwar, Sumit Rai, and Jagdish Chandra Kuniyal

Abstract To get the most out of geographical data, people are being encouraged to use open-source data and software. Freely available Remote Sensing (RS) and Geographic Information Systems (GIS) software is making rapid progress. These software's can be used to look at various properties that a satellite-based data, and other geographic data from different sources provides. Overall, satellite based remote sensing applications for soil resource mapping and modelling has gained a huge success all over the world. Conventional soil analysis is time-consuming and labor-intensive, resulting in high costs. Interpolation and its derivatives have gained widespread acceptance as an important spatial analysis technique. In this context, prognostic soil mapping approaches have been created as a result of advances in RS and GIS technology. With thorough satellite-derived soil quality indices, in-situ soil quality measurements may be transformed, which can then be scaled up to cover wider geographical areas. The spatial maps can also be used as an ideal input for models that are spatially distributed. It was possible to distinguish four major land degradation categories based on data derived from remote sensing-based satellite data, namely undegraded, moderately degraded, degraded and severely degraded based on the amount of vegetation cover, slope and erosion. Likewise, different researchers have used application of RS & GIS in mapping natural resources, studying soil taxonomy, soil resource assessment and mapping and crop growth indicator integrated with crop models. In addition, RS & GIS application can also be used to track change in forest cover density, Land use land cover, coastal morphology, status of reef and biodiversity of islands. Realizing this, the chapter highlights the utilization of free open satellite dataset and GIS software's as well as the presentation of open-source soil information and use of the application of RS & GIS in soil resource mapping and monitoring in India.

Keywords Open-source · RS & GIS · Soil · Software · Satellite

N. Kanwar · S. Rai (✉) · J. C. Kuniyal
Center for Environment Assessment and Climate Change, G.B. Pant, National Institute of Himalayan Environment, Kosi-Katarmal, Almora 263643, Uttarakhand, India
e-mail: rai.sumit@gbpihed.nic.in

P. K. Shit et al. (eds.), *Soil Health and Environmental Sustainability*, Environmental Science and Engineering,
https://doi.org/10.1007/978-3-031-09270-1_1

1.1 Introduction

The advancements in RS & GIS applications have increased the efficiency of preparing a comprehensive inventory associated to natural resource monitoring and mapping (Sahu et al. 2015; Mulder et al. 2011; Tripathi 2017). Satellite data provides a worldwide time series dataset ranging from days to years, with a synoptic perspective and greater coverage of the region (Sivakumar and Hinsman 2004; Dehni and Lounis 2012). Satellite data is useful in generating consistent observational information on current conditions of the earth surface on large and small scales (Turner et al. 2015; Dowman and Reuter 2016).

Remote sensing for soil uses the optical spectrum (200–1400 nm). Optical spectrum includes three major categories of electromagnetic spectrum, Ultraviolet (10–400 nm), Visible (380–750 nm) and Infra-red wavelengths (700 nm–1 mm) (Fig. 1.1). Satellite system are distinguished by variety of characteristics. Among the most important is the satellite orbit which includes altitude, period, inclination and equatorial crossing time. Another important characteristic is its resolution. Resolution of a system refers to its ability to record and display fine details (Lillesand et al. 2011). In remote sensing we need four diverse types of information such as spatial resolution, spectral resolution, radiometric (intensity) and temporal resolution information (Lillesand and Kiefer 2004). The orbital height and instantaneous field of vision defined a spatial resolution. The spectral resolution comprises the spectral

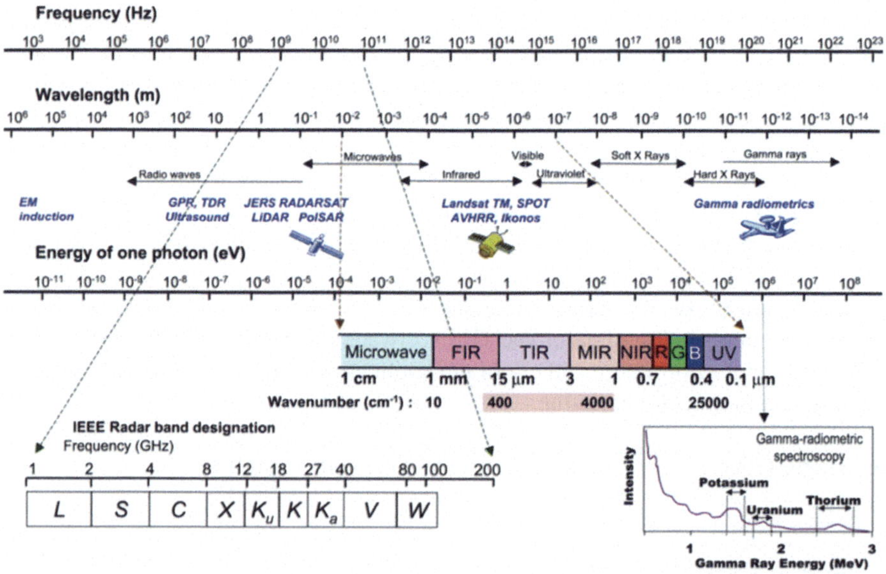

Fig. 1.1 Electromagnetic spectral component highlighting the relevant components, to acquire soil and environmental variables information through RS and PS (after McBratney et al. 2003)

sensor bands' number, width and location. The resolution of the radiometric is based on the number of bits utilized for recording the detected radiation.

Three variables (data continuity, cost access and data) influence the use of RS data disproportionately (Turner et al. 2015; Kumar et al. 2018). Data continuity means maintenance of long-term archive of remote sensing satellite dataset. Long time series of continuous data enabled changes both spatially and temporally about the earth surface features. Cost of data determines the data affordability. Open access to government satellite images will have impact on societal benefits, as expensive images will not be used widely as expected. The abilities of end-users is to discover, gather, modify and extract value from satellite images, and integrate it with other information, is addressed by data access.

GIS capacity to handle geographic data has been an essential tool for a broad range of applications since the 1970s. In recent years, integration of the GIS with other geospatial technologies (GPS), remote sensing and mobile devices has found applications in locational services, interactive mapping, and precise agriculture (Chang 2010; Worboys and Duckham 2004; Liu and Philippa 2009; Kumar 2013). There is an abundance of medium to low resolution satellite data that is open and freely accessible for download these days (Harris and Baumann 2015; Bakillah and Liang 2016; Kumar et al. 2018; Dowman and Reuter 2016; Turner et al. 2015).

Spatio-temporal datasets from satellites is becoming more popular, thanks to fast advancements in the capabilities of GIS systems. Open-source GIS software's are free program for anyone to use, modify, and distribute anywhere. Whereas proprietary GIS software, on the other hand, is very expensive, with extra costs for certain modules. Geographic information systems (GIS) have grown tremendously over the last few decades, and it currently encompasses a great variety of paid and freely open-source software (Maurya et al. 2015; Steiniger and Bocher 2009). Most of proprietary GIS licenced software packages go unused at different institutions and research centres because they need either significant training or easy access in numerous places. When compared to proprietary software in general, freely open-source software are more versatile but requires more effort to use. The source code of open-source software can be viewed and modified by the user, and it can also be redistributed (Maurya et al. 2015). The information provided in this chapter presented the accessibility to free satellite data and GIS software's, open-source soil information and application of RS & GIS in Soil resource mapping and monitoring.

1.2 Spectral Reflectance of the Soil

An in-depth understanding of soil spectral behavior is essential for the identification and characterization of soils utilizing remote sensing methods (Ravisankar and Sreenivas 2011). In the optical spectrum (between 200 and 1400 nm), most of the shortwave solar energy is either absorbed or reflected, with only a small amount of it being transferred. With the help of a spectroradiometer equipment, it is possible to collect data on soil reflectance in the laboratory or on the ground. There are many

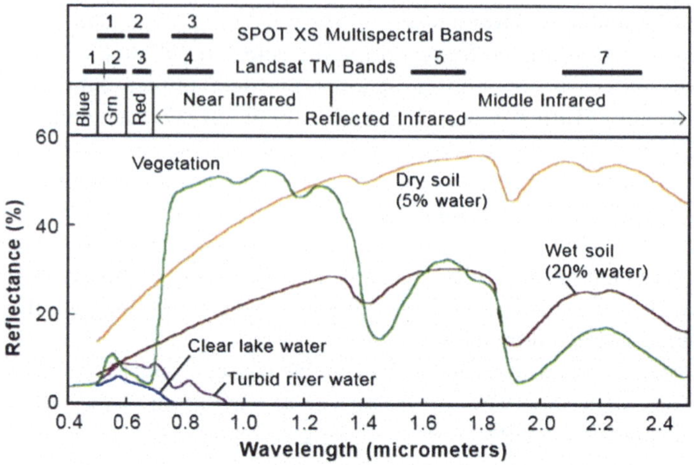

Fig. 1.2 Spectral signatures of soil, water and vegetation (after Mondal 2018)

spectroscopic methods such as visible-near and mid infrared spectroscopy mainly used to forecast and analyze different soil characteristics (Dwivedi 2017). Color, texture, structure, as well as its moisture content and surface conditions/roughness are all significant physical characteristics of soil. Among the chemical characteristics of soils that influence reflectance include soil mineralogy, organic matter, salinity, carbonates, and other compounds. Texture of the soil and amount of organic content therein have a considerable impact on the soil's spectral properties (i.e., sand, silt, and clay) (Baumgardner et al. 1985). In India, Spectral libraries consisting pedons, surface and sub-surface soils types of major physiographic units were developed using a specto-radiometer range of 350–2500 nm (NBSS & LUP 2005). Hyper-spectral libraries of this kind will be very useful for the characterization of soils (Fig. 1.2).

1.3 Commonly Used Open Satellite Data

Satellites are designed and launched in different ways depending on their intended application and orbit. More than thousands remote sensing satellites have been launched so far (Turner et al. 2003). These spacecrafts were upgraded with a higher resolution of new satellites. With hyperspectral sensors (hundreds of spectral bands) the few sensors of the early flights were upgraded. The duration between visits has been reduced from months to daily. Even resolution (spatial, spectral and radiometric) has increased. Furthermore, an increasing amount of remote sensing satellite data is now available as open data sources. Table 1.1 summarises the most regularly utilised remote sensing satellites as well as their specifications.

Table 1.1 Details of freely open satellite dataset

Sr. No.	Satellite	Sensor	Resolution			Scene size	Accessibility
			Spatial	Spectral	Temporal		
1	Landsat-1	MSS/RBV	Band 4 to 7 (60 m)	4 bands (VNIR)	18 days		July 1972 to October 1992, June 2012 to January 2013
2	Landsat-2						
3	Landsat-3						
4	Landsat-4	TM/MSS	Band 1 to 5 and band 7 (30 m) Band 6 (120 m)	7 band (VNIR, SWIR, TIR)	16 days		
5	Landsat-5					170 km × 183 km	July 1982 to May 2012
6	Landsat-7	ETM +	Band 1 to 5 and band 7 (30 m) Band 6 (60 m) Band 8 (15 m)	8 band (VNIR, SWIR, TIR, PAN)	16 days		July 1999 to till date
7	Landsat-8	OLI, TIRS	Band 1 to 7 and band 9 (30 m) Band 10 and 11 (100 m) Band 8 (15 m)	11 band (VNIR, SWIR, TIR, PAN)	16 days		March 2013 till date
8	Landsat-9	OLI-2; TIRS-2	Band 1 to 7 and band 9 (30 m) Band 10 and 11 (100 m) Band 8 (15 m)	11 band (VNIR, SWIR, TIR, PAN)	16 days		
	EO 1	ALI	Band 1 to 7 (30 m) Band 8 (10 m)	10 bands, VNIR, SWIR, PAN	16 days	37 km ×	May 2001 to March 2017
7	NOAA	AVHRR	1.1 km	6 bands VNIR, SWIR, TIR		2400 km × 6400 km	1981 till date

(continued)

Table 1.1 (continued)

Sr. No.	Satellite	Sensor	Resolution			Scene size	Accessibility
			Spatial	Spectral	Temporal		
8	Resourcesat 1	LISS-III	24 m	4 bands	24 days	140 km	Selected dates
		AWiFS	56 m	4 bands	5 days	740 km	
9	MODIS	Terra and Aqua	Band 1 and 2 (250 m) Band 3 to 7 (500 m) Band 8 to 36 (1 km)	36 bands VNIR, SWIR, TIR	1–2 days	2330 km × 10 km	February 2000 till date
10	Sentinel 2A	MIS	Band 2 to 4 and 8 (10 m) Band 5 to 7 and 8A, 11, 12 (20 m) Band 1, 9, 10 (60 m)	13 bands VNIR, SWIR	10 days	100 × 100 km	From October 2015
11	EO 1	Hyperion	30 m	220 bands VNIR, SWIR		7.7 km × 42 km	May 2001 to March 2017
12	IMS1	HySI	500 m	64 bands VNIR	24 days	128 km	Selected dates
13	ALOS	ALOS	30 m	L-band SAR	–	1° × 1°	
14	Cartosat	Cartosat	30 m	Optical	–	1° × 1°	
15	SRTM	Shuttle Radar Topography Mission	30 m	C-band SAR	–	1° × 1°	
16	ASTER	ASTER	30 m	Optical	–	1° × 1°	
17	GeoEye	OrbView 3	1 m	PAN	3 day		Selected number of data

1.3.1 Low Resolution Satellite Data

High repetition rates are associated with lower resolution, which implies that the satellite will analyze the same region more than once in a short duration. Perhaps with a coarse resolution, it is possible to capture large or global region. Advanced Very High Resolution Radiometer (AVHRR) and Moderate Resolution Imaging Spectroradiometer (MODIS) provide the coarse resolution data. AVHRR is generated by National Oceanic and Atmospheric Administration (NOAA). AVHRR provides multispectral data in the four to six band range with continuous (morning and afternoon) global coverage since June 1979 (Table 1.1). The pixel size of one km is approximately equal to the scale range of 1:500000 to 1:1000000 (Dobos et al. 2001). Many studies on small scale have been carried out using low resolution satellite datasets for identification of pattern in the soil (Dobos et al. 1998; Vettorazzi et al. 1995; Odeh and McBratney 1998). On the other hand, Terra and Aqua satellites carry MODIS sensor, that provide detailed earth surface simultaneously. Terra and Aqua provides 2-day repeat global coverage in 36 spectral bands with a radiometric sensitivity of 12 bits (see Table 1.1). Spatial resolution for MODIS sensor is 250, 500 and 1000 m (Justice et al. 2002; Masuoka et al. 1988; Lillesand et al. 2011). MODIS sensor has a 110° field of view (FOV) and 2330 km cross track swath. All of these data sets are publicly accessible for downloading via the websites www.earthexplorer.usgs.gov/, http://glovis.usgs.gov/, and http://reverb.echo.nasa.gov/reverb, respectively. In addition, Hou et al. (2011) and Wang et al. (2015) conducted research to get direct retrieval of soil characteristics, which included the soil texture using MODIS data (Wang et al. 2015).

1.3.2 Moderate Resolution Satellite Data

Global land surface observation is carried out by medium resolution satellites. Medium resolution satellites include Landsat, Resourcesat and Sentinel Satellite data. The Landsat-1 earlier named as ERTS, was the first earth observing satellite launched to study and monitor the earth surface. Eight satellites system have been launched since 1972 to till date which provides free extensive data pertaining to earth surface. These datasets can be downloaded through Earth Explorer or USGS Global Visualization Viewer (GloVis). Landsat 9 satellite which is launched in September, 2021, data availability will be publicly made available by 2022. The Landsat 1, 2, 3 satellite system carries two sensor system, Return Beam Vidicon (RBV) and Multispectral Scanner (MSS) system. Landsat 4 and 5 carries a payload MSS and Thematic Mapper (TM) instruments. Landsat 7 includes the Enhanced Thematic Mapper Plus (ETz+) sensor, Landsat 8 payload consists of two sensors—the Operational Land Imager (OLI) and the Thermal Infrared Sensor (TIRS) and Landsat 9 launched in 2021 carries payload Operational Land Imager-2 (OLI-2) (Shroder 2016; USGS 2014).

Another satellite system is IRS Resourcesat-1 launched by ISRO in 2003 with the data being restricted to India. It includes multispectral sensor Linear Imaging Self Scanning (LISS-III) and Advanced Wide Field Sensor (AWiFS). Both the sensors consist four spectral bands, but difference is only in terms of resolution. The LISS-III data has a spatial resolution of 23.5 m, while the AWiFS sensor has a spatial resolution of 56 m and a coverage of 740 km (see Table 1.1).

Sentinel-2 satellite is the European Space Agency's second constellation. Sentinel missions are equipped with multispectral scanners on board. Sentinel-2 has a higher spatial 10 m and temporal resolution of 5 days. Sentinel-2 consists 13 spectral bands ranging from visible and near-infrared (VNIR) to shortwave infrared (SWIR) wavelengths along a 290-km orbital swath. Out of the 13 bands, 4 bands are having spatial resolution of 10 m, 06 having a resolution of 20 m and 03 bands has 60 m resolution. Bands 4, 8, 9, 11 and 12 are used to generate the soil and vegetation indices (Phiri et al. 2020). Merging of only Sentinel-2 data with any other satellite data improves the overall accuracy.

1.3.3 High Resolution Satellite Data

High-resolution satellites typically have a per pixel size of less than 1 m. The advantage of higher spatial resolution makes them ideal for monitoring changes, detect objects, or track human activities and remotely located natural areas (see Table 1.1). However, high-resolution satellite data have not been made open for reasons like financial viability, proprietary interests, defence and security, national legislations, etc. OrbView-3 satellite collected images between 2003 and 2007 by Orbital Imaging Corporation (now GeoEye) can now be downloaded for free through USGS Earth Explorer at a resolution of up to one metre in panchromatic (black and white) mode or four metres in multispectral (colour) mode.

1.4 Other Earth Resource Satellites

Several earth observing satellite systems are in operation or many others are scheduled to be launched in the near future. These are hyperspectral satellite system; Digital Elevation Models (DEM) and Advanced land Imager (ALI) (see Table 1.1).

1.4.1 Hyperspectral Satellite Systems

In airborne hyperspectral scanning technique, data is collected in a variety of small spectral bands. Hyperspectral remote sensing is a new technique. In hyperspectral sensors, each pixel in the picture is covered by a continuous spectrum of radiation from 200 or more bands. These images are used for the mapping of minerals and highlighting the soil parameters such as moisture, organic content and salinity. Hyperspectral datasets can be downloaded freely from http://earthexplorer.usgs.gov.

1.4.2 Digital Elevation Model (DEM)

DEM is a three-dimensional representation of the earth surface elevation in relation to a datum, such as a reference point. DEM stands for Digital Elevation Models, a type of topographic representation. DEMs are used to determine the elevation at each point, slope, and aspect of the landscape. Using a stereo pair or a radar interferometer, one can construct DEM. DEMs can be generated by optical sensors such as ASTER and Cartosat, which have in-track stereovision, respectively. ASTER is developed by NASA Terra spacecraft and Japan Ministry of Economy, Trade, and Industry (METI). ASTER and Cartosat DEM has a resolution of 30 m and can be downloaded from a variety of websites, including http://earthexplorer.usgs.gov/, http://glovis.usgs.gov/, and http://reverb.echo.nasa.gov/reverb, http://bhuvan-noeda.nrsc.gov.in. Another topographic data produced by NASA's is Shuttle Radar Topographic Mission (SRTM). SRTM provides the data for almost entire world at 1 arc second (approx. 30 m resolution) and 3 arc second (approx. 90 m resolution), respectively (see Table 1.1). On May 2016 the Japan Aerospace Exploration Agency (JAXA) released the digital surface model (DSM) dataset for global terrestrial region with a horizontal resolution of approx. 30 m (1 arc sec) free of charge. The dataset has been compiled using the Advanced Land Observing Satellite "DAICHI" (ALOS)—PALSAR's L-band. The dataset is published based on the DSM dataset (5-m mesh version) of the "World 3D Topographic Data," which is the most precise global-scale elevation data at this time, and its elevation precision is also at a world-leading level as a 30-m mesh version (http://www.eorc.jaxa.jp/ALOS/en/aw3d30/). DEM data based Digital Soil Mapping (DSM) are generally set on some metrics that are supposed to reflect the terrain surface (Costa et al. 2018).

1.5 Open Sources for World Soil Information

Soil information is required for a variety of purposes, including addressing important global issues, giving a global context for local policy makers and supplying basic soil data that helps in the study of earth processes (Table 1.2). Global soil information is provided to a variety of communities, including the international science community, agricultural engineers and researchers and development organizations, soil survey agencies, experts in soil science, farmer, and other stakeholders who require quality-assessed spatial information on soil for studying the earth-atmosphere interactions,

Table 1.2 List of links for free and open soil datasets

Sr. No.	Web directory	Website	Scope	Type	Characteristics features
1	Harmonized Soil Database (HWSD)	http://webarchive.iiasa.ac.at/Research/LUC/External-World-soil-database	Global and local	Soil	It provides regional and national soil datasets and maps from several countries. It contains soil characteristics (such as soil pH, soil depth, and soil texture) having resolution of 30 arc-seconds
2	Global Soil Information System	www.fao.org/global-soil-partnership	Global	Soil	Provide soil information collected by national institutions
3	CGIAR-CSI	www.cgiarcsi.community/data/global-high-resolution-soil-water-balance	Global	Actual evapotranspiration and soil water deficit	Provides data on actual evapotranspiration and soil water deficit
4	SoilGrids1km	www.worldsoilprofiles and www.worldgrids.org	Global	Soil	Provide global soil properties, soil classes and soil mapping using automated mapping
5	ISRIC Soil Information	www.isric.org/explore/soil-geographic-databases	Global	Soil	Provides georeferenced soil related data in point, polygon and grid format
6	ESA CCI SM	esa-soilmoisture-cci.org	Global	Soil moisture	Provides soil moisture product worldwide, as well as the annual

climate change modelling, hydrological process modelling, and crop growth model modelling.

1.5.1 Open Source GIS Packages and Software

In the early 1980s, various free open-source/non-proprietary GIS software packages emerged (Coetzee et al. 2020; Akbari and Rajabi 2013). For a variety of reasons including protection, security protocols and supplier independence, the adoption of this kind of software is growing rapidly. This is because free open-source/non-proprietary software does not have a particular sponsor and is a kind of issues that open-source software deals with. In order to address this issue, academic groups are tasked with assessing and rating this particular kind of software. Technology has advanced at an accelerated pace in recent years, which has made it easier to integrate software and make use of them in assessing spatial information in a variety of activities, particularly in the field of GIS.

In the field of GIS a significant increase has been seen in both desktop and mobile applications in terms of Geospatial libraries, database management systems and web GIS applications (Steiniger and Hunter 2013; Kumar et al. 2018). These applications are developed using various non-proprietary free GIS software. In order to safeguard the rights of end users and ensure freedom of the software uses, the procedure for implementation of 'free software licenses' must be required. There are many organisations, such as the General Public License (GPL) and Berkeley Software Distribution (BSD), which may offer free licence software (Tsou and Smith 2011). Some GIS desktops like GRASS include remote sensing features such as raster processing and image rectification, upgrade, transformation, classification and extraction of geographic objects (Blaschke et al. 2008) and vectorization (i.e., raster to vector conversion). Various desktop systems may also be found on Web or mobile platforms (e.g., QGIS, gvSIG mobile and gvSIG Online) and/or are being utilized as a backend for geospatial processing for Web or Cloud applications. Open-source GIS software such as GRASS, QGIS, SAGA and ILWIS, DIVA-GIS are competitive with commercial/proprietary GIS software such as the Environmental Systems Research Institute (ESRI) ArcGIS desktop when it comes to general-purpose GIS capabilities (Steiniger and Hay 2009; Chen et al. 2010; Kumar et al. 2018). The general-purpose GIS capabilities are data access in raster and vector format, vector editing and manipulations, vector and raster analysis, and presentation capabilities in the form of maps, graphs, plots, and tables (Kumar et al. 2018). Detail of open-source geographic information system (OSGIS) projects are listed on http://www.opensourcegis.org/. The acceptance of free open-source software for GIS depends on its usability (Akbari and Rajabi 2013). List of various open-source GIS software is provided in Table 1.3 with details.

1.6 Application of RS and GIS in Soil Resource Mapping

For instance, in countries like India and other nations, remote sensing methods are regularly employed in the inventorying and monitoring of soil resources, and this

Table 1.3 Details about popular open-source GIS software

Type	Software	Operating system	Platform/Language	License	Application of its use	Home page
Desktop GIS	GRASS GIS 7	Crossplatform	C, Shell, Tcl/Tk, Python	GPL GNU	Geospatial data analysis and management, image processing	http://grass.itc.it/
	Map Window GIS	Windows	C++	MPL	Raster data visualization and analysis	www.mapwindow.org
	QGIS	Linux, Windows, Android	C++, Python, Qt4	GNU GPL	Viewing, editing, analysis, grass-GUI, SAGA-GUI,	http://www.qgis.org
	SAGA GIS	Crossplatform	C++ (MS Visual C++)	LGPL (API), GPL	Analysis, modeling, scientific visualisation	www.saga-gis.uni-goettingen.de/
	uDig	Windows, Linux, Mac OSX	Java (Eclipse RCP)	EPL+BSD	Viewing, editing, analysis,	http://udig.refractions
	Open JUMP	Crossplatform	Java	GNU GPL	Viewing, editing, analysis,	www.openjump.org
	ILWIS	Windows	C++	GNU GPL	Raster analysis	http://52north.org/index.php?
	gvSIG Desktop	Windows, Linux, Mac OSX	Java	GNU GPL	Viewing, editing, analysis (mobile applications)	www.gvsig.com/en
	DivaGIS, 2003	Windows	Delphi/Kylix,	GPL	Biodiversity analysis	DIVA-GIS|free, simple and effective
Geospatial libraries	GDAL/ORG	Cross platform	C, C++	X/MIT	Geospatial data formats (Raster and Vector)	www.gdal.org/
	Geo Tools	Cross platform	Java	GNU LGPL	Provide geospatial data	www.geotools.org/

(continued)

Table 1.3 (continued)

Type	Software	Operating system	Platform/Language	License	Application of its use	Home page
	GOES	Linux, Mac OSX, Windows, Android	C++	GNU LGPL	Spatial predicate functions and spatial operators,	https://trac.osgeo.org/geos/
	libLAS	Windows, Mac OSX, and Linux	C/C++	GNU LGPL	Inspecting, manipulating, transforming, and processing LAS LiDAR data	https://liblas.org/start.html
	JTS	Cross platform	Java	GNU LGPL	Viewing, editing, analysis, (mobile applications)	github.com/locationtech/jts
	Proj4	Cross platform	C	MIT	It transforms geospatial coordinates system (CRS)	http://proj4.org/
Data stores	Post GIS	Linux, Windows, Mac OS X, POSIX-compliant systems	C++	GNU GPL	It allows location queries to be run in SQL	www.postgis.net/
	SpatiaLite	Linux, Windows, Mac	C++	MPL, GPL,	Open source command line interface tool	www.gaia-gis.it/
	Rasdaman	Unix-like	C++	GNU LGPL	For the management and analytics of multi-dimensional arrays	http://rasdaman.org/
	pgRouting	Linux, Windows, Mac OS X, POSIX-	C++	GPL	Provide geospatial routing functionality	https://pgrouting.org/documentation.html

(continued)

Table 1.3 (continued)

Type	Software	Operating system	Platform/Language	License	Application of its use	Home page
Web GIS application	MapServer	Cross platform	C/C++	X/MIT	Its purpose is for publishing spatial data and web mapping applications	www.mapserver.org
	GeoServer	Linux, Windows, MacOS X, POSIX-compliant systems	Java	GNU GPL	Sharing geospatial data	http://geoserver.org
	Mapbender	Cross-platform	PHP, HTML, JavaScript, JSON	MIT	Register, view, navigate, monitor spatial data	Mapbender—OSGeo
	Mapnik	Cross platform	C++	GNU LGPL	Geospatial visualization and processing	http://mapnik.org/
	MapGuide Open Source	Windows; Linux	C++	GNU LGPL	It purpose is to develops web based mapping applications and geospatial services	http://mapguide.osgeo.org/
	Deegree	Cross platform	Java	GNU LGPL	Its purpose is to develop spatial data infrastructures and the geospatial web	http://www.deegree.org/

is becoming more common (Table 1.4). In the soil mapping process, a variety of soils are identified, described, based upon physio-climatic and vegetation, and then depicting those findings on a standard base map based on the information gathered in the field and laboratories. In this case, the RS approaches have significantly decreased the amount of fieldwork required, and soil boundaries have been defined with more precision than using traditional methods (Rao et al. 2004). RS & GIS techniques are useful in soil and land degradation mapping, soil moisture assessment, soil fertility assessment, soil water conservation measures, and soil suitability studies (Ravisankar and Sreenivas 2011).

The soil mapping procedure can be broken down into several steps. These includes, satellite data interpretation, delineation of the land boundary, collection of soil samples, physio-chemical analysis of soil samples, soil correlation, categorization and final step is its mapping (Rao et al. 2004; Wadodkar and Ravisankar 2011). State agencies in India used different resolution satellite data to create soil maps for the whole nation at a size of 1:250000 and 1:500000. In addition, scale of 1:50000 has been used for soil mapping by different agencies for district level planning. As the soil information greater than 1:50000 scale is only accessible for a few parts of the India.

Successful completion of several case studies and operational projects in India has allowed researchers to derive information on various aspects of degraded lands using satellite data, including salt affected soils, eroded soils (wind and rain), water logging, ravinous lands, shifting cultivation, the impact of a soil conservation program on a watershed, and the impact of aquaculture on the coastal zone. For a variety of land surface and atmospheric processes, soil moisture is an essential input parameter to be considered. As a result of their wide area coverage, frequent return capabilities, and the ability to make repeated estimates on a regular basis, remote sensing methods (both optical and microwave) are particularly well suited for soil moisture assessment. Another important advantage of satellite derived remote sensing data is mapping of soil fertility, which may reduce the number of field measurements required by as much as 50%.

1.7 Conclusion

Comparative to proprietary software, free and open-source software provides users with more freedom but necessitates a larger amount of effort in order to make use of them. On the other hand, many established open-source applications also provide good usability. Now a day's soil studies are being done at different spatial image resolution. In India, for example, it is being used operationally for soil and land degradation mapping and monitoring, as well as other purposes. Its use in soil fertility studies, on the other hand, is not well known or practiced. In light of its importance, high-resolution data should be used more often and on a regular basis for mapping and monitoring soil fertility at the village level, in addition to ground observation. Using RS in conjunction with GIS may also provide a very helpful tool for analyzing the

Table 1.4 Different satellite data used for study of soil resource mapping in India

Sr. No.	Satellite	Period	Resolution	Study area	Purpose	Findings	References
1	Indian Remote Sensing (IRS)-1C LISS-III instrument and IRS-1B LISS-II instrument	20 and 21 January, 1996	23.5 m 36.25 m	Rae Bareli and Unnao districts, Uttar Pradesh	Assessment and identification of salt-affected soils and waterlogged area	Results indicates that the observed trend is insignificant, showing that IRS-1B and IC satellite data are equally useful at conveying information	Dwivedi and Sreenivas (1988)
2	Shuttle Radar Topography Mission (SRTM) and Indian Remote Sensing (IRS)-P6 LISS IV instrument	30 October, 2005	90 m 5.8 m	Borgaon Manju watershed, Akola district, Maharashtra	Characterization of landforms and soils	The inventory of soil resources listed fourteen soil series, which are categorized into Lithic Ustorthents, Vertic Haplustepts, Calcic Haplustepts, Typic Haplustepts, Typic Haplusterts, and Sodic Calciusterts	Reddy et al. (2013)
3	Landsat 4–5 TM data	12 June, 2013	30 m	Medinipur Block, Chota Nagpur Plateau, India	Soil organic carbon mapping	71% variance is found in of Soil organic carbon (SOC) distribution and Normalized Difference Vegetation Index values	Bhunia et al. (2017)
4	Landsat-7 ETM+	2014	30 m	Gauriganj block, Amethi district, Uttar Pradesh	Micro level soil resource mapping	83 soil interpretation units, grouped into seven series	Tripathi (2017)

(continued)

Table 1.4 (continued)

Sr. No.	Satellite	Period	Resolution	Study area	Purpose	Findings	References
5	IRS-1A LISS-I1	13 June, 1988	36.25 m	Kandi tract, Chandigarh	Soil resource mapping	Ustorthents are the most common type of soil found in many soilscape units. Besides certain sub-group levels such as Typic/Udic Ustochrepts; Typic Haplustalfs; Typic Ustipsamments and Typic Ustifluvents were also identified	Kudrat et al. (1992)
6	SRTM and Landsat-7 ETM+	2003	90 m 30 m	Eastern part of the Nile Delta	Soil mapping and capability assessment	Study area dominated by flood plain (33.48%), lacustrine plain (21.52%) and the marine plain (3.13%)	Ali and Kotb (2010)
7	Indian Remote Sensing (IRS)—P6, LISS III instrument, LANDSAT TM, and SRTM	April and October, 2006 November, 2000	23.5 m 30 m 90 m	Solani watershed, Haridwar district, Uttarakhand and some part of the Saharanpur district, Uttar Pradesh	Soil resource assessment and mapping	Entisols and Inceptisols are two of the main soil orders found in the majority	Velmurugan and Carlos (2009)
8	IRS 1D LISS III and IRS-I C Panchromatic	March, 2003	23.5 m 5.8	Vellamadai village, Coimbatore district, Tamil Nadu	Soil resource mapping	Seven mapping units belonging to four soil series, Soil reaction ranges slightly acidic to saline along with low soils organic carbon content	Arunkumar et al. (2004)

(continued)

Table 1.4 (continued)

Sr. No.	Satellite	Period	Resolution	Study area	Purpose	Findings	References
9	IRS-IB LISS-II and Landsat-TM	14 May and 23 September,1994 and 23 March, 1995 16th February,1995	36.25 m 30 m	Shankargarh block, Bara Tehsil, Allahabad	Soil resources mapping	Results revealed that soil formation is influenced by the terrain's physiography and lithology	Dwivedi et al. (2000a, b)
10	IRS-1B LISS-II and Landsat TM	25 September, 1994 23 February, 1995	36.25 m 30 m	CharkhariMahoba district, Uttar Pradesh	Mapping of soil resources	There are numerous pediments-inselbergs/ridges/dykes scattered over the alluvial plain in the southern periphery	Dwivedi et al. (2000a, b)
11	IRS 1D LISS III and IRS-I C Panchromatic	–	23.5 m 5.8 m	Junewani village, Hingna Tehsil, Nagpur district, Maharashtra	Large-scale soil mapping	Landform, slope and land use characteristics are all taken into account in a three-tiered strategy for each parcel of property	Srivastava and Saxena (2004)
12	(IRS)-ID LISS-III and ASTER (DEM)	September, 2009	23.5 m 30 m	Simana sub-watershed, Chhotanagpur plateau	Characterization of landforms and soils in complex topo sequence	• six drainage soil class grouped • land capability groupings show moderately good cultivated soils (64.6%) area and (26.3%) has moderate limitation of erosion and soil	Gangopadhyay et al. (2015)
13	Cartosat-2 DEM and IRS-P6 LISS_IV	April, 2018	1 m 5.8 m	Inagalur panchayat, Kadiri, Anantapuramu district, Andhra Pradesh	Mapping of Soil erosion and probability Zones using RUSLE, remote sensing and GIS technique	• Annual soil loss found >40 t/ha/yr in Kharif season and 20–40 t/ha/yr in rabi season • Soil erosion probability zones showed high (8.90%), medium (55%), and low (35%) erosion risk area	Srinivasan et al. (2021)

(continued)

Table 1.4 (continued)

Sr. No.	Satellite	Period	Resolution	Study area	Purpose	Findings	References
14	IRS-1D LISS III	16 November, 2001 and 14 December, 2001	–	Dikrong Basin, Arunachal Pradesh	Assessment of soil erosion using RUSLE, GIS and remote sensing	Avg annual soil loss was found 51 t ha^{-1} $year^{-1}$	Dabral et al. (2008)
15	IRS-1D, LISS-III and PAN data	–	23.5 m	Khulgad watershed, Almora district, Uttarakhand	Soil suitability evaluation	Results found that moderately deep soils with high clay content are suitable for wheat, rice, finger millet and mustard crops, potato, mustard and finger millets cultivation. While degraded hill tops/slopes presently are not suitable for crops cultivation	Surya et al. (2020)
16	IRS-P6, LISS IV and Cartosat-I DEM	2009	5.8 m 2.5 m	Daltonganj watershed, Palamu district, Jharkhand	Estimation of soil loss using RUSLE model, RS & GIS	Avg annual soil Loss was found up to 69 tons ha^{-1} yr^{-1}	Tirkey et al. (2013)
17	IRS-P6, LISS-IV and Cartosat-1	April, 2011	5.8 m 2.5 m	Amethi district, Uttar Pradesh	Soil mapping	Twelve soil series have been mapped	Yadav et al. (2016)
18	IRS-P6, LISS-IV and Cartosat-1	5 October, 2012 and 15 April, 2013	5.8 m 2.5 m	Miniwada Panchayat, Nagpur, Maharashtra	Large-scale soil resource mapping	Total 37 physiography-land use units and six soil series have been discovered	Sahu et al. (2016)

spatial soil nutrient distribution. Thus, combination of soil maps with geo-referenced cadastral, and natural resource maps in a GIS domain, it will be easier to develop appropriate action plans and deliver soil health cards to farmers, allowing for more efficient soil management and maintenance.

Acknowledgements We gratefully acknowledge Director, GBPNIHE, Almora for the kind support and guidance during the manuscript preparation.

References

Akbari M, Rajabi MA (2013) Evaluation of desktop free/open source GIS software based on functional and non-functional capabilities. Tehničkivjesnik 20:755–764

Ali RR, Kotb MM (2010) Use of satellite data and GIS for soil mapping and capability assessment. Nat Sci 8(8):104–115

Arunkumar V, Natarajan S, Sivasamy R (2004) Soil resource mapping of Vellamadai Village Coimbatore district fused (IRS 1C LISS III and PAN merged) space borne multispectral data. Madras Agric J 91(7–12):399–405

Bakillah M, Liang S (2016) Open geospatial data, software and standards. Open Geospat Data Softw Stand 1:1–2

Baumgardner MF, Silva M, Biehl L, Stoner ER (1985) Reflectance properties of soils. Adv Argonomy 38:1–44

Bhunia GS, Shit PK, Pourghasemi HR (2017) Soil organic carbon mapping using remote sensing techniques and multivariate regression model. Geocarto Int 34(2):215–226. https://doi.org/10.1080/10106049.2017.1381179

Blaschke T, Land S, Hay GJ (2008) (eds) Object-based image analysis—spatial concepts for knowledge-driven remote sensing applications. Springer, Berlin

Chang KT (2010) Introduction to geographic information system, 4th edn

Chen D, Shams S, Carmona-Moreno C, Leone A (2010) Assessment of open source GIS software for water resources management in developing countries. J Hydro-Environ Res 4:253–264

Coetzee S, Ivanova I, Mitasova H, Brovelli MA (2020) Open geospatial software and data: a review of the current state and a perspective into the future. ISPRS Int J Geo Inf 9(2):90

Costa EM, Rosa AS, Anjos LHC (2018) Digital elevation model quality on digital soil mapping prediction accuracy. Ciência e Agrotecnologia 42(6):608–622

Dabral PP, Baithuri N, Pandey A (2008) Soil erosion assessment in a hilly catchment of North Eastern India using USLE, GIS and remote sensing. Water Resour Manage 22(12):1783–1798. https://doi.org/10.1007/s11269-008-9253-9

Dehni A, Lounis M (2012) Remote Sensing techniques for salt affected soil mapping: application to the Oran region of Algeria. Procedia Eng 33:188–198

Dwivedi RS (2017) Spectral reflectance of soils. In: Remote sensing of soils. Springer, Berlin, Heidelberg. https://doi.org/10.1007/978-3-662-53740-4_6

Dwivedi RS, Sreenivas K (1998) Delineation of salt affected soils and waterlogged areas in the Indo-Gangetic plains using IRS-1C LISS-III data. Int J Remote Sens 19(14):2739–2751. https://doi.org/10.1080/014311698214488

Dwivedi RS, Ramana KV, Sreeniva K (2000a) Mapping soil resources in part of Northern India using spaceborne multispectral data. Geocarto Int 15(1):78–83

Dwivedi RS, Sreenivas K, Ramana KV (2000b) Detecting soil information in a predominantly black soil region using Indian Remote Sensing Satellite (IRS-1B) Linear Imaging Self-scanning Sensor (LISS-II) data. Int J Remote Sens 21(17):3293–3302. https://doi.org/10.1080/014311600750019903

Dobos E, Micheli E, Baumgardner MF (1998) Statistical analysis of advanced very high resolution radiometer data (AVHRR) soil relationship. Agrokemia es Talajtan, Budapest 47:49–62

Dobos E, Montanarella L, Negre T, Micheli E (2001) A regional scale soil mapping approach using integrated AVHRR and DEM data. Int J Appl Earth Obs Geoinf 3(1):30–42. https://doi.org/10.1016/S0303-2434(01)85019-4

Dowman I, Reuter I (2016) Global geospatial data from Earth observation: status and issues. Int J Digital Earth 10:328–341

Gangopadhyay GK, Obi Reddy GP, Mukhopadhyay S, Singh SK (2015) Characterization of landforms and soils in complex toposequence of Subarnarekha catchment, Chhotanagpur plateau using remote sensing and GIS. Agropedology 25(01):95–109

Hou S, Wang T, Tang J (2011) Soil types extraction based on MODIS image. Procedia Environ Sci 10:2207–2212

Harris R, Baumann I (2015) Open data policies and satellite Earth observation. Space Policy 32:44–53

Justice CO, Townshend JRG, Vermote EFE, Masuoka E, Wolfe RE, Saleous N, Roy DP, Morisette JT (2002) An overview of MODIS land data processing and product status. Remote Sens Environ 83:3–15

Kudrat M, Tiwari AK, Saha SK, Bhan SK (1992) Soil resource mapping using IRS-1A-LISS II digital data—a case study of Kandi area adjacent to Chandigarh-India. Int J Remote Sens 13(17):3287–3302. https://doi.org/10.1080/01431169208904119

Kumar N (2013) Remote sensing and GIS applications in land use/land cover analysis. In: Reddy GPO, Sarkar D (eds) Remote sensing and GIS in digital terrain analysis and soil-landscape modeling, NBSS & LUP pub. No. 152, pp 254–263

Kumar N, Singh SK, Mishra VN, Reddy GPO, Bajpai RK (2018) Open-source satellite data and GIS for land resource mapping. In: Reddy G, Singh S (eds) Geospatial technologies in land resources mapping, monitoring and management. Geotechnologies and the environment, vol 21. Springer, Cham. https://doi.org/10.1007/978-3-319-78711-4_10

Lillesand TM, Kiefer RW, Chipman A (2011) Remote sensing and image interpretation, 6th edn. Wiley, New York

Lillesand TM, Keifer RW (2004) Remote sensing and image interpretation, 3rd edn. John Wiley & Sons, NY

Liu JG, Philippa JM (2009) Essential image processing and GIS for remote sensing. Wiley

Maurya SP, Ohri A, Mishra S (2015) Open source GIS: a review. In: Proceedings of national conference on open source GIS: opportunities and challenges. Varanasi (India), pp 150–155

Masuoka E, Fleig A, Wolfe RE, Patt F (1988) Key characteristics of MODIS data products. IEEE Trans Geosci Remote Sens 36(4):1313–1323

McBratney AB, Santos MLM, Minasny B (2003) On digital soil mapping. Geoderma 117:3–52

Mondal BP (2018) Hyper-spectral analysis of soil properties for soil management. In: Advances in agriculture for sustainable development. Srijan Samiti H. No. 498, Kakarmatta (South), Post-D.L.W., Varanasi-221004 (U.P.), pp 59–65

Mulder VL, De Bruin S, Schaepman ME, Mayr TR (2011) The use of remote sensing in soil and terrain mapping—a review. Geoderma 162(1–2):1–19

NBSS & LUP (2005) Reflectance libraries for development of soil sensor for periodic assessment of state of soil resources, Report No. 835, National Bureau of Soil Survey and Land Use Planning, Nagpur, 40 pp

Odeh IO, McBratney AB (1998) Using NOAA advanced very high resolution radiometric imageries for regional soil inventory. In: Proceedings 16th world congress of soil science, August 20–26 1998, Montpellier France

Phiri D, Simwanda M, Salekin S, Nyirenda VR, Murayama Y, Ranagalage M (2020) Sentinel-2 data for land cover/use mapping: a review. Remote Sens 12(14):2291. https://doi.org/10.3390/rs12142291

Ravisankar T, Sreenivas K (2011) Soils and land degradation in remote sensing applications. In: Roy PS, Dwivedi RS, Vijayan D (eds), NRSC Publication, NRSC, Hyderabad

Rao BRM, Fyzee MA, Wadodkar MR (2004) Utility of remote sensing data for mapping soils at various scales and levels. In: Venkatratnam L, Ravisankar T, Sudarshana R (eds) Soils and crops. NRSC Publication, Hyderabad

Reddy GPO, Nagaraju MSS, Ramteke IK et al (2013) Terrain characterization for soil resource mapping using IRS-P6 data and GIS—a case study from Basaltic Terrain of Central India. J Indian Soc Remote Sens 41:331–343. https://doi.org/10.1007/s12524-012-0240-5

Surya JN, Walia CS, Singh H, Yadav RP, Singh SK (2020) Soil suitability evaluation using remotely sensed data and GIS: a case study from Kumaon Himalayas. J Indian Soc Remote Sens. https://doi.org/10.1007/s12524-020-01143-2

Srivastava R, Saxena RK (2004) Technique of large-scale soil mapping in basaltic terrain using satellite remote sensing data. Int J Remote Sens 25(4):679–688. https://doi.org/10.1080/0143116031000068448

Srinivasan R, Karthika KS, Suputhra SA et al (2021) Mapping of soil erosion and probability zones using remote sensing and GIS in Arid part of South Deccan Plateau. J Indian Soc Remote Sens, India. https://doi.org/10.1007/s12524-021-01396-5

Sivakumar MVK, Hinsman DE (2004) Satellite remote sensing and gis applications in agricultural meteorology and WMO satellite activities. In: A paper from proceedings of the training workshop 7–11 July, 2003, Dehra Dun, India, pp 1–21

Sahu N, Reddy GP, Kumar N, Nagaraju MSS (2015) High resolution remote sensing, GPS and GIS in soil resource mapping and characterization-a review. Agric Rev 36:14–25

Sahu N, Singh SK, Obi Reddy GP, Kumar N, Nagaraju MSS, Srivastava R (2016) Large-scale soil resource mapping using IRS-P6 LISS-IV and Cartosat-1 DEM in Basaltic Terrain of Central India. J Indian Soc Remote Sens 44(5):811–819. https://doi.org/10.1007/s12524-015-0540-7

Shroder J (2016) Issues of hydrologic data collection by remote sensing in Afghanistan and surrounding countries. Transboundary Water Res Afghanistan 289–309. https://doi.org/10.1016/b978-0-12-801886-6.00011-2

Steiniger S, Bocher E (2009) An overview on current free and open source desktop GIS developments. Int J Geogr Inf Sci 23:1345–1370

Steiniger S, Hay GJ (2009) Free and open source geographic information tools for landscape ecology. Eco Inform 4:183–195

Steiniger S, Hunter AJS (2013) The 2012 free and open source GIS software map—a guide to facilitate research, development, and adoption. Comput Environ Urban Syst 39:136–150

Tirkey A S, Pandey AC, Nathawat MS (2013) Use of satellite data, GIS and RUSLE for estimation of average annual soil loss in Daltonganj watershed of Jharkhand (India) of Jharkhand (India). J Remote Sensing Tech 1(1):20–30

Tripathi DK (2017) Landsat ETM plus data processing for micro level soil resource mapping: a case study. Int J Innov Sci Eng Technol 4(6):350–355

Tsou MS, Smith J (2011) Free and Open Source Software for GIS education. White paper, GeoTech Center. http://www.geotechcenter.org/

Turner W, Spector S, Gardiner N, Fladeland M, Sterling E et al (2003) Remote sensing for biodiversity science and conservation. Trends Ecol Evol 18:306–314

Turner W, Rondinini C, Pettorelli N, Mora B, Leidner AK, Szantoi Z, Buchanan G, Dech S, Dwyer J, Herold M, Koh LP, Leimgruber P, Taubenboeck H, Wegmann M, Wikelski M, Woodcock C (2015) Free and open-access satellite data are key to biodiversity conservation. Biol Cons 182:173–176

USGS (2014) Landsat missions timeline. http://landsat.usgs.gov/about_mission_history.php. Accessed 03 Apr 2015

Velmurugan A, Carlos GG (2009) Soil resource assessment and mapping using remote sensing and GIS. J Indian Soc Remote Sens 37:511–525. https://doi.org/10.1007/s12524

Vettorazzi CAI, Bayramin, Baumgardner MF (1995) Evaluation of AVHRR data for delineating regional soil patterns. Post-doctoral research report. Agronomy Department, Purdue University ID

Wadodkar MR, Ravisankar T (2011) Soil resource database at village level for developmental planning. J Indian Soc Remote Sens 39(4)
Wang DC, Zhang GL, Zhao MS, Pan XZ, Zhao YG, Li DC et al (2015) Retrieval and mapping of soil texture based on land surface Diurnal temperature range data from MODIS. PLoS ONE 10(6):e0129977. https://doi.org/10.1371/journal.pone.0129977
Worboys M, Duckham M (2004) GIS: a computing perspective. CRC Press, Boca Raton
Yadav PPS, Singh A, Rajput G, Singh K (2016) Soil mapping at village level in a part of Amethi district, U.P. using IRS LISS-IV and cartosat-1 merged data for sustainable land and crop management. Agropedology 26(02):149–164

Website Accessed

https://seos-project.eu/remotesensing/remotesensing-c03-p02.html. Accessed 23 July 2021
https://www.spatialsource.com.au/gis-data/orbview-3-satellite-data-available-for-free. Accessed 23 July 2021
http://bhuvan-noeda.nrsc.gov.in/download/download/download.php. Accessed 15 July 2021
http://earthexplorer.usgs.gov/. Accessed 20 July 2021
http://glovis.usgs.gov. Accessed 15 July 2021
http://reverb.echo.nasa.gov/reverb. Accessed 15 July 2021
http://srtm.csi.cgiar.org/. Accessed 23 July 2021
http://www.eorc.jaxa.jp/ALOS/en/aw3d30/. Accessed 23 July 2021
http://www.itc.nl/ilwis/downloads/ilwis33.asp. Accessed 23 July 2021

Chapter 2
Applicability of Open Source Satellite Data and GIS for Soil Resources Inventorying and Monitoring

M. Chandrakala, R. Srinivasan, K. Sujatha, Sunil P. Maske, Rajendra Hegde, and B. S. Dwivedi

Abstract The open source satellite data and geographic information system (GIS) have broader capability in soil resources mapping and monitoring. Soil resource mapping, basic requirement is base maps such as satellite imagery, google imagery, toposheets, cadastral maps and GIS software's. The cost of purchasing satellite imagery and GIS soft ware's is too high and not economical, mainly in developing countries like India, hence its very necessary to use freely available open source satellite data and GIS software's for soil resource mapping. Monitoring and mapping of soil resource is a prerequisite to know the soil characteristics as which influences agricultural land use planning through identifying their potentials and problems of the soils/land in a particular region. A study was conducted at a selected mandal, Rayachoty, YSR Kadapa district, Andhra Pradesh, India to map the soil resources using open source satellite imagery such as sentinel-2 and google imagery and DEM data. By using these imagery and DEM, land use land cover, slope, landform were mapped in GIS. With the help of these maps, soil survey was conducted in detail (1:10000 scale) at farm level in Rayachoty mandal. Representative model pedons were selected and soils were brought to laboratory and analysed for physico-chemical properties. Soils were classified in to 10 soil series and 53 mapping units based on physical, chemical and morphological properties. And then soil map was prepared in GIS environment. Using soil map as base, different soil site characteristics viz., soil depth, slope, texture, gravelliness, drainage were mapped. Hence, the study highlights the importance and applicability of freely available open source satellite data and GIS and their greater benefit in soil resource mapping which are cost effective, timely available and reliable thereby have huge impact on agricultural land use planning.

Keywords GIS · Open source · Soil resource · Satellite data · Mapping

M. Chandrakala (✉) · R. Srinivasan · K. Sujatha · S. P. Maske · R. Hegde
National Bureau of Soil Survey and Land Use Planning, Hebbal, Bangalore 560024, Karnataka, India
e-mail: chandra.ssac@gmail.com; Chandrakala.M@icar.gov.in

B. S. Dwivedi
National Bureau of Soil Survey and Land Use Planning, Nagpur, Maharashtra, India

P. K. Shit et al. (eds.), *Soil Health and Environmental Sustainability*,
Environmental Science and Engineering,
https://doi.org/10.1007/978-3-031-09270-1_2

2.1 Introduction

Soil resource is scarce and fundamental unit for any material production. Soils are highly heterogeneous and have diverse physical, chemical, morphological and biological properties. Characterization and classification of soils have therefore foremost important in using those resources based on their land and fertility capability and to use them in sustainable manner. Agricultural land use planning mainly depends on soil potentials and constraints for specific crop planning in a particular region or unit land. The basic requirement for these crop planning/land use planning is detailed land resource inventory at farm level at 1:10000 or less than that scale. Soil resource inventory done through soil survey. Fundamental requirement of soil survey is base maps such as satellite imagery, DEM, google imagery, survey of India toposheets and cadastral maps. In the present situation these base maps are not economical as which are too costly impacts huge cost of investment to R&D departments in long run. Hence its very necessary to use open source satellite imagery and GIS software's for soil resource inventory and mapping. With these backgrounds a detailed soil survey was undertaken at a selected mandal, Rayachoty, YSR Kadapa district, Andhra Pradesh to map the soil resource of the mandal using freely available open source satellite data and GIS. The objectives of the study is to (i) to map the land use land cover at 1:10000 scale, (ii) to map the land form at 1:10000 scale, (iii) to map the soils of the Rayachoty mandal at 1:10000 scale and (iv) to map the other soil resources such as depth, slope, texture, drainage and rock fragments/gravelliness at 1:10000 scale. Govt. of India identified YSR Kadapa district as aspirational district. Rayachoty mandal belongs to Agro-ecological Sub Region 7.1 i.e., South Telangana Plateau (Rayalseema) and Eastern Ghat, hot, dry semi-arid eco-sub region. In this AESR water scarcity and climatic conditions limits the agriculture and also there is no large scale soil resource data available, hence priority has been given to take up mapping of soil resources in the selected mandal.

2.2 Materials and Methods

2.2.1 Location and extent

YSR Kadapa District is one of the known chronically drought prone districts of South Telangana Plateau (Rayalseema) region of Andhra Pradesh. It has a total geographical area of 15,379 km^2 with 3 revenue divisions, 51 mandals, 831 g panchayats and 965 revenue villages. Rayachoty mandal with a total geographical area of 23,240.7 ha (cultivated land is 16768.55 ha), belongs to YSR Kadapa district, lying between north latitudes 13° 59′ 45.28" and 14° 7′ 12.263" and east longitudes 78° 35′ 24.85" and 78° 54′ 5.608". It is divided in to seventeen villages namely Abbavaram (rural), Botlacheruvu, Chennamukkapalle, Cherlopalle, Dullavaripalle, Gorlamudiveedu, Indukurupalle, Masapet, Katimayakunta, Madhavaram, Peddakalavapalle, Sibyala,

Pemmadapalle, Rayachoty (CT), Syamalavaripalle, Yandapalle and Yerranagupalle (Fig. 2.1). This study area belongs to Agro Ecological Sub Region is 7.1 i.e., South Telangana Plateau (Rayalseema) and Eastern Ghat, hot, dry semi-arid eco-sub region.

2.2.2 Climate

Andhra Pradesh receives rainfall in the autumn as well as south west and north east monsoon season. Kadapa characterised by year round high temperatures has a tropical wet and dry climate. Previously recorded more than 50 °C. Summers are having hot and humid climate. During this period temperatures ranges from 34 °C and up to a maximum of 40 °C. Southwest and northeast monsoon contributes 55% and 30% rainfall, respectively. Winters are milder and the temperatures will be low after the onset of the monsoons. During this period, the temperatures ranges from 25 to 35 °C. The annual mean rainfall of Rayachoty mandal is 638 mm with ustic soil moisture regime and annual mean temperature varies between 23 and 34 °C with iso hyperthermic soil temperature regime. Length of growing period (LGP) is 145 days. Water balance diagram of YSR Kadapa district presented in Fig. 2.2.

2.2.3 Geomorphology

YSR Kadapa District has been divided into three units geomorphologically based on the factors relief, slope and soil which are (i) Structural landforms (ii) Denudational landforms and (iii) fluvial landforms (Central Ground Water Board 2013).

(i) The structural landforms: which consists of structural hills and valleys, cuesta hills, Mesa/Buttee, linear ridges, intermontane valleys etc. These landforms occurs in major part in the eastern region of the mandal.
(ii) Denudational landforms: These include pediplain, pediment-inselberg complex.
(iii) Fluvial landforms: which consists of alluvial plains along the major rivers (Pennar river and its main tributaries are Papaghni, Cheyyair, Sagileru, Chitravati, and Kunderu. Mandavi and Pincha are minor streams) and Bazada zones. Flood plains are highly productive regions whereas Bazada forms along foot hills.

2.2.4 Geology

The YSR Kadapa District comprises of various rock types (Central Ground Water Board 2013) of Late Archaean or Early Proterozoic era which are succeeded by rocks of Dharwarian Age and both are traversed by dolerite dykes. The rocks of older

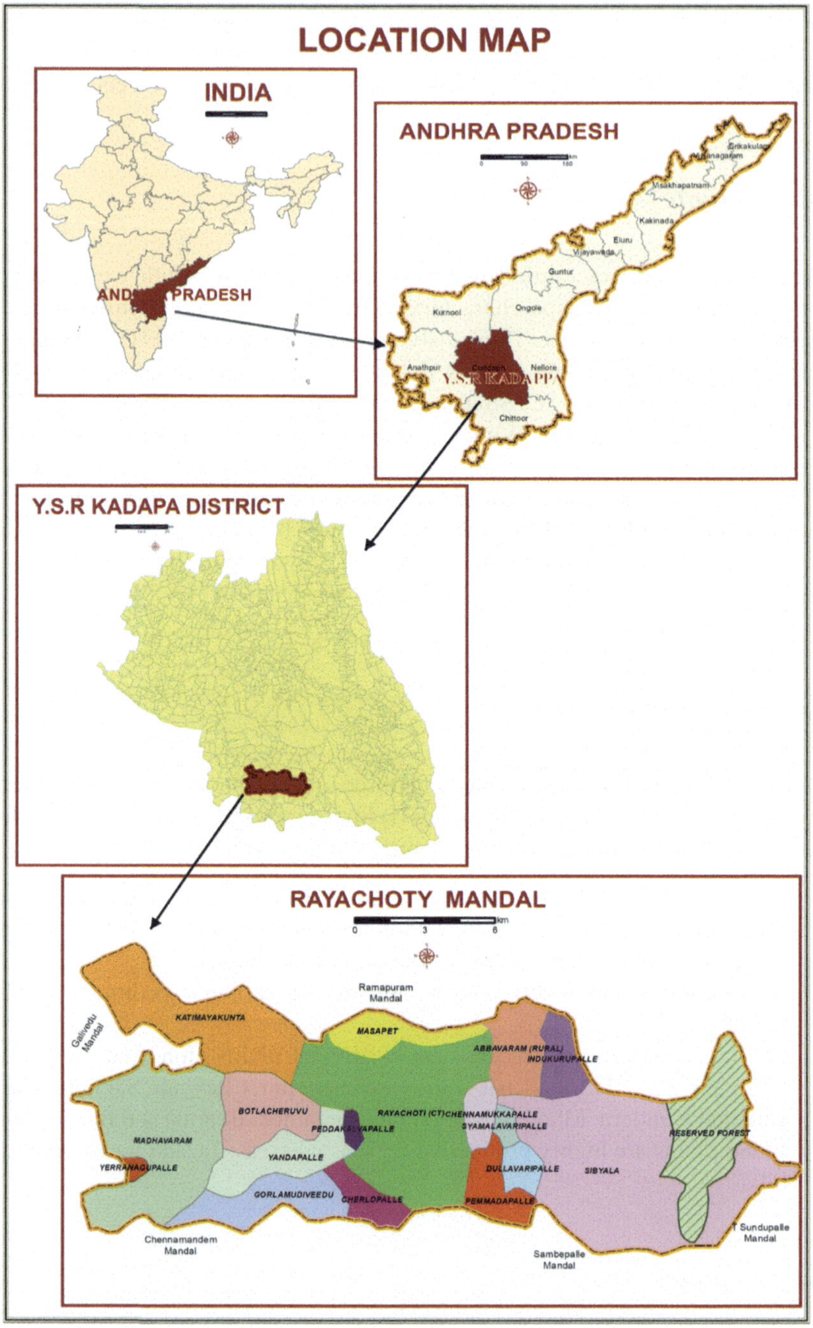

Fig. 2.1 Location map of Rayachoty mandal

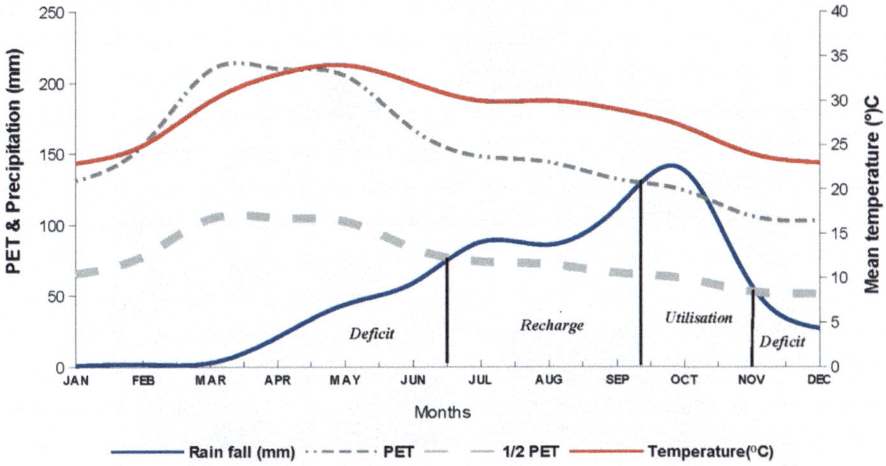

Fig. 2.2 Water balance diagram of Kadapa district

time are overlain by rocks of Kadapa Super group and Kurnool Group belonging to Middle and Upper Proterozoic Age. The Kadapa Sedimentary Basin, falls under depression over the denuded surfaces of older rocks. The important rock types are limestones, quartzites, shales, granites, phyllites, granite gneiss and granodiorites. The Archaean consists of Peninsular Gneissic Complex, represented by granite-gneiss, granite, migmatite and granodiorite. These rocks found in the south western part of the district. The Archaean and Dharwar are both traversed by quartz reefs and dolerite dykes. Alluvium comprises of gravel, sand, silt and clay found along the river courses in the district. Geology of Rayachoty mandal is granite-gneiss (Plate 2.1).

2.2.5 Drainage

In the YSR Kadapa, Pennar river is perennial and flows in NW–SE and its tributaries are Cheyyair, Chitravati, Papaghni, Kunderu and Sagileru. Mandavi and Pincha are minor intermittent streams (Central Ground Water Board 2013). The drainage is parallel to sub parallel indicating structural control.

2.2.6 Base Maps and Image Interpretation

The major land use, habitation, water bodies, drainage and landform were delineated and prepared base map for field survey using sentinel-2 (Fig. 2.3) imagery, google

Plate 2.1 Granite gneiss in Rayachoty mandal

Fig. 2.3 Sentinel-2 imagery of Rayachoty mandal used for soil resource mapping

imagery (Fig. 2.4) and 1:50,000 toposheet (57 J/12, 57 J/16, 57 K/9, 57 K/13), as base available for the mandal.

Fig. 2.4 Google imagery of Rayachoty mandal used for soil resource mapping

2.2.7 Field Investigations

Detailed soil survey (1:10000 scale) was carried out by extensive traverse of the area and also random check to identify permanent features, natural boundaries, water bodies, rock outcrops and habitations. Seven transects were selected along the direction of slopes at different locations in all the villages of the block based on the slope and contour. 103 soil profiles were excavated at close intervals along transects and studied.

Soils were categorised into different soil series based on correlation. Parent material, physiography, horizon sequence, soil depth, soil texture, coarse fragments, soil colour, calcareousness, mottles and any other soil characteristics were the major differentiating characteristics of the soil series. Totally 10 soil series were identified. Horizon-wise soil samples were collected from typifying model pedons of the soil series for laboratory soil characterization. Observations were made and also recorded on variations in surface erosion, soil texture, slope, gravelliness and stoniness (Chandrakala et al. 2019).

2.2.8 Laboratory Characterization of Soils

Soil samples from 19 model soil profiles were brought to the laboratory, shade dried, ground and passed through 2-mm sieve. The samples were used for determining their particle size class, electrical conductivity, soil reaction, cation exchange capacity, organic carbon, exchangeable bases, permanent wilting point, field capacity, and available nutrients status following standard protocols.

2.2.9 *Finalization of Soil Map and Other Soil Characteristics Mapping Using GIS*

Sentinel-2 imagery, survey of India toposheet (1:50,000 scale) and google imagery were used for digitising and drawing habitation, roads, drainage lines and water-body. Using these data, land use land cover and landform map has been prepared in GIS environment. Soil boundaries were drawn using soil phases as mapping units using soil profiles location map and landform map. Soil map was then prepared in the GIS environment. With the soil map as base, soil depth, slope, texture, rock fragments/gravelliness and soil drainage mapping were also done.

2.3 Results and Discussion

2.3.1 *Land Utilization/Land Use Land Cover Mapping*

Hills and ridges and isolated hillocks have exposed rocks. Major land use in upland is rice, cowpea, redgram, groundnut, sesamum, sunflower, coconut and mulberry and mango plantation. Valley plain or lowlands are occupied by rice and also used for cultivating irrigated crops such as sunflower, tomato and chillies. Land use land cover map of Rayachoty mandal is given in the Map 2.1.

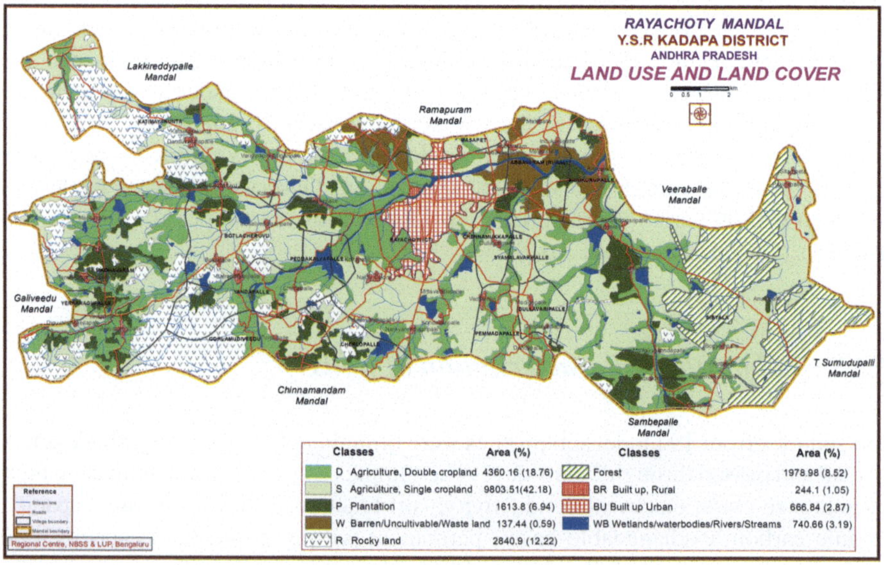

Map 2.1 Land use land cover map of Rayachoty mandal

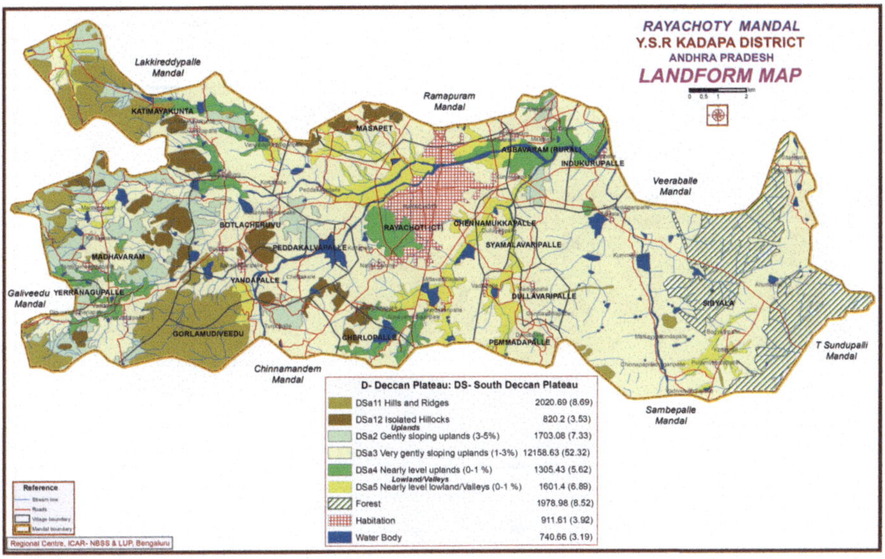

Map 2.2 Landform map of Rayachoty mandal

2.3.2 *Landform Mapping*

The area of the Rayachoty mandal is divided in to 6 landforms (Map 2.2) based on geology and slope with the help of contour measurement in toposheet as well as satellite imagery characteristics and DEM data. They are hills and ridges, isolated hillocks, gently sloping uplands, very gently sloping uplands, nearly level uplands and nearly level lowland/valleys. Among the landforms, very gently sloping uplands occupied the larger area (52.32%).

2.3.3 *Soil and Soil Site Characteristics Mapping*

Soil map (Map 2.3) shows 10 soil series and 53 mapping units or soil phases in the Rayachoty mandal (Table 2.1). All the series formed from granites and gneisses and or its colluvium and alluvium in lowlands. Soils of uplands together occupy the highest area of 15,167.14 ha (65.27%) followed by rock outcrop (2840.9 ha) and forest (1978.98 ha). Lowland soils occupy 1601.41 ha (6.88%).

Soil depth (Map 2.4) is an important soil site characteristics used to differentiate the various soils occurring in the area into different soil series. It determines the depth of effective rooting zone and provides capacity to hold water and nutrients in soil. Moderately deep (23.14%) soils occupy the dominant position followed by moderately shallow (21.38%), deep (19.56%) and shallow (8.07%).

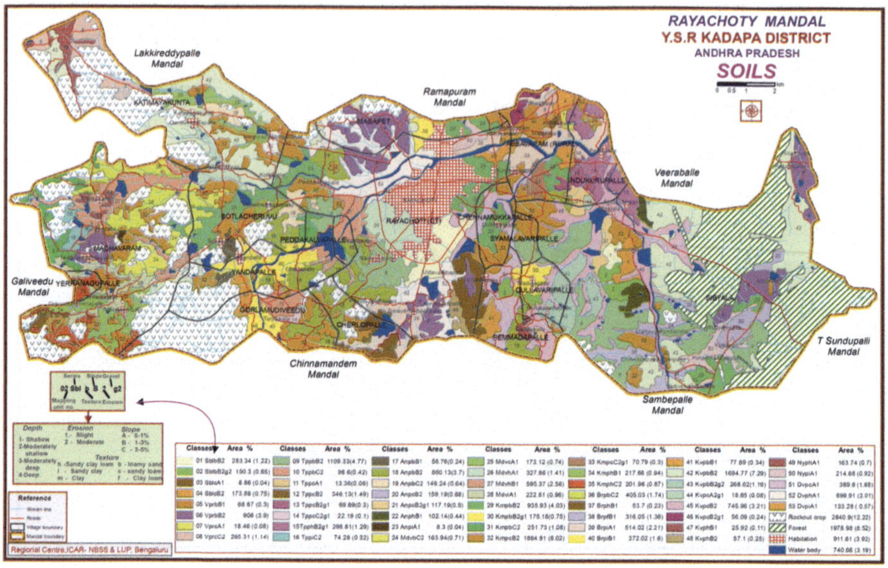

Map 2.3 Soils of Rayachoty mandal

Table 2.1 Identified soil series of Rayachoty mandal, mapping units and area

Series name	Mapping units	Area (ha)	Percentage (%)
Sibyala	1, 2, 3, 4	616.38	2.66
Variyapapireddypalli	5, 6, 7, 8	1258.44	5.40
Turpupalli	9, 10, 11, 12, 13, 14, 16	2032.39	8.75
Anumpalli	17, 18, 19, 20, 21, 22, 23	1452.95	6.20
Madhavaram	24, 25, 26, 27, 28	1483.1	6.38
Kumarapalli	29, 30, 31, 32, 33, 34, 35	3718.06	15.99
Balreddigaripalli	36, 37, 38, 39, 40	1660.82	7.10
Kondavandlapalli	41, 42, 43, 44, 45, 46, 47, 48	2945	12.68
Nayanurpalli	49, 50	378.42	1.16
Duganvandlapalli	51, 52, 53	1222.99	5.26
Soil total		**16768.55**	**72.15**
Rock outcrop		2840.9	12.22
Forest		1978.98	8.52
Habitation		911.61	3.92
Waterbody		740.66	3.19
Total		23240.7	100.00

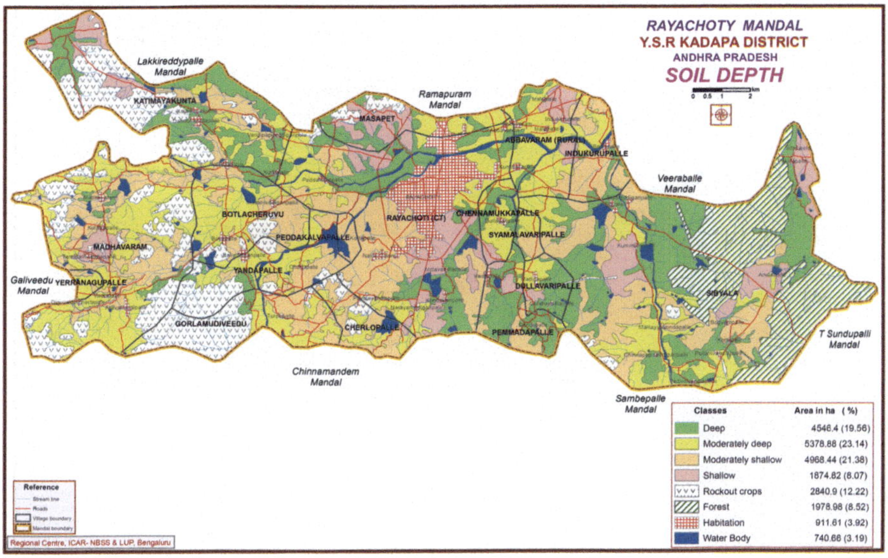

Map 2.4 Soil depth in Rayachoty mandal

Slope (Map 2.5) plays a key role in the formation and development of soils, controls the process of soil erosion and determines the land use. The slope range

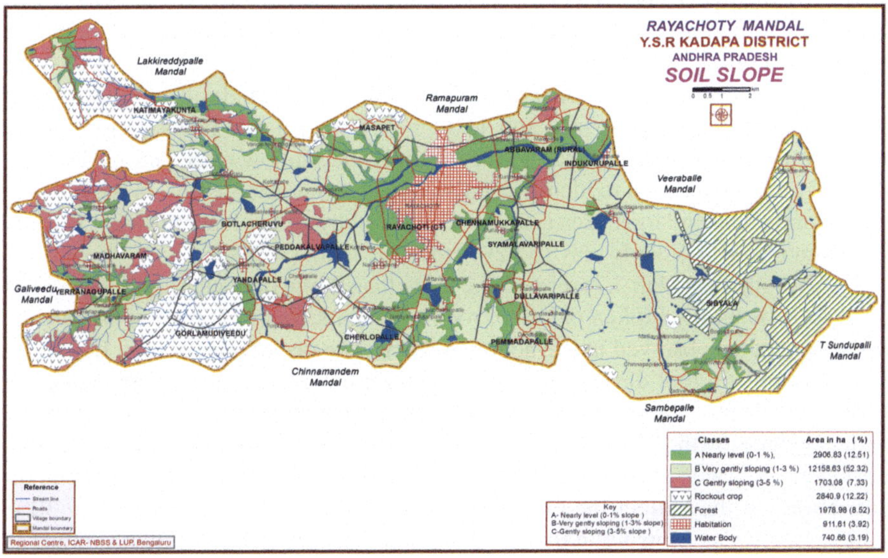

Map 2.5 Slope classes in Rayachoty mandal

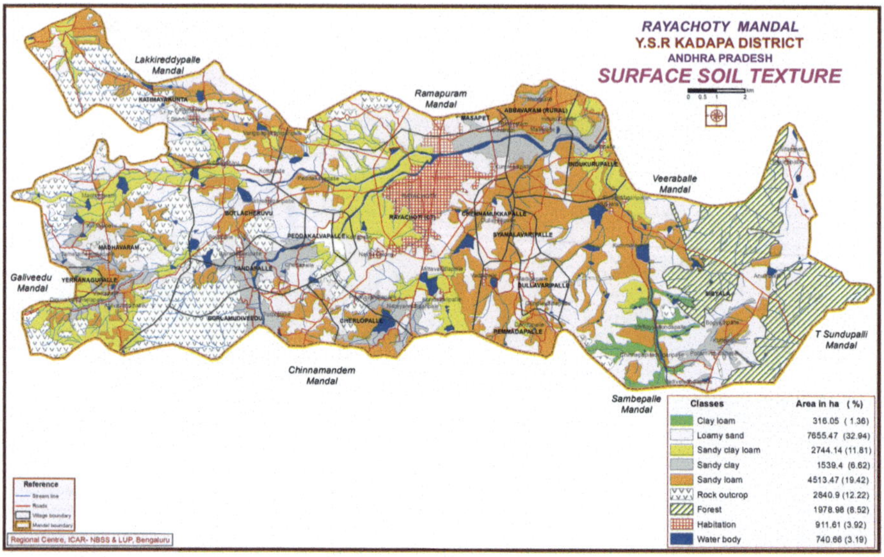

Map 2.6 Surface soil texture in Rayachoty mandal

used to differentiate the various slope classes are presented in Map 2.5. There are three classes of slope occurring in the mandal. 1–3% slope class i.e., Very gently sloping lands distributed in large extent (52.32%) followed by nearly level (12.51%) and gently sloping lands (7.33%). These slopes are good for cultivation of crops along with sufficient suitable soil and water conservation measures are taken before crop production.

Soil texture (Map 2.6) represents the relative fractions of sand, silt and clay content present in the soil. The surface soil texture or plough layer plays major role in influencing the growth and yield of crops, mainly the shallow rooted crops. It is one of the key parameter used to identify phases of soil series established. In the mandal, loamy sand occur in larger area (32.94%) fallowed by sandy loam (19.42%), sandy clay loam (11.81%), sandy clay (6.62%) and clay loam (1.36%).

Rock fragments (Map 2.7) refer to the more than 2 mm diameter particles present in the soil. Generally broken pieces of rock fragments, quartz gravels constitute the coarse fragments observed in Rayachoty mandal. The occurrence of rock fragments in the soil affects root development and seedling emergence through reducing the volume of soil that can be drawn upon by plants for water and nutrients there by hinders plant growth. In clayey soils, presence of coarse fragments helps in the free movement of water and air and upto some amount they are not a constraint for cultivation but they affect the crop performance when it exceeds more than 30%. Some of the soils occurring in the mandal has coarse fragments distributed either in the whole profile or confined to some depth in the solum. The amount and depth of gravel occurrence were used as an major characteristic in differentiating soils into

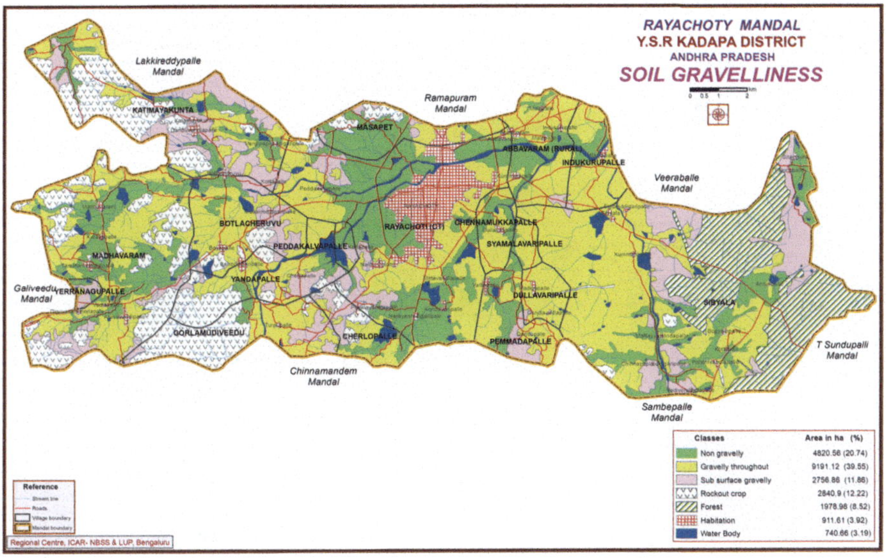

Map 2.7 Surface coarse fragments in Rayachoty mandal

various soil series. Different gravelliness classes seen in the mandal studied. In the area surveyed, gravelly throughout profile (39.55%) occur in larger area followed by non-gravelly (20.74% area) and sub surface gravelly occupied by 11.86% area.

The term drainage indicates how fast rain water after reaching the soil will get infiltrated into the soil and get percolated later to ground water. The rate at which added water (rainfall) removes from the soil is by both run-off and flow through the soil to underground storage. Drainage affects both soil management and crop productivity. All most arable crops excluding paddy prefer well drained soils and productivity gets affected if there is deviation in the drainage. Soil site characteristics, mainly slope of the land area, soil texture and ground water depth affects soil drainage. Drainage classes are identified based on soil morphology and terrain features in the field. The soil drainage classes given by the Soil survey staff (2017) and IARI (19711971) were applied to the soil-site data to derive the drainage class for the mapping units. The various drainage classes identified and their extent of occurrences in 17 villages are given in the map. Well drained soils occur in large extent (53.43%) followed by moderately well drained (10.38) and somewhat poorly drained soils occur in 7.21% area (Map 2.8).

Similar studies on soil resources mapping were done by Chandrakala et al. (2017) using open source satellite data and GIS.

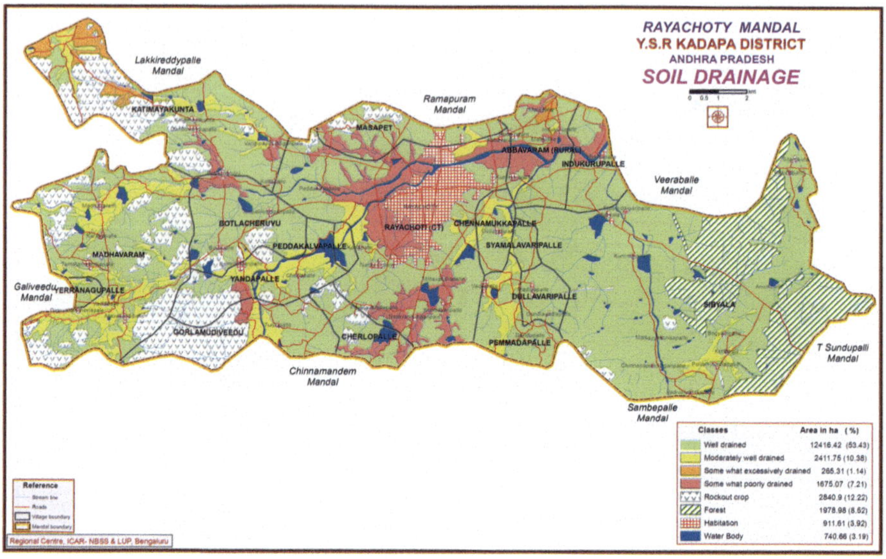

Map 2.8 Soil drainage in Rayachoty mandal

2.4 Conclusion

Applicability of open source satellite data and GIS are greater in soil resource mapping which are user friendly, timely available and reliable as which are freely available and economical to the public domain and saves lot of money to R&D departments. In the present study soil mapping and also mapping of land use land cover, landform and also different soil site characteristics were mapped at 1:10000 scale through by utilizing open source satellite data and GIS. These maps have greater impact to identify the potentials and constraints of the soil resources in specific unit land. There by helps to agricultural land use planning at farm level through addressing soil and water conservation measures and soil constraints to achieve higher production and productivity.

References

Central Ground Water Board (2013) Ground water brochure, YSR Kadapa district, Andhra Pradesh, Ministry of Water Resources, Government of India

Chandrakala M, Srinivasan R, Anil Kumar KS, Nair KM, Maske SP, Ramesh Kumar SC, Srinivas S, Hegde R, Singh SK (2017) Land resource inventory of Elamdesom block, Idukki district, Kerala on 1:10000 scale for optimal agricultural land use planning, using geo-spatial techniques. ICAR-NBSS Publ. No. 1049, NBSS&LUP, Nagpur

Chandrakala M, Srinivasan R, Bhaskar BP, Maske SP, Ramesh Kumar SC, Srinivas S, Hegde R (2019) Land resource inventory of Rayachoty mandal, YSR Kadapa district, Andhra Pradesh on

1:10000 scale for optimal agricultural land use planning, using geo-spatial techniques. NBSS Publ. No.1120, ICAR-NBSS&LUP, Nagpur
Indian Agricultural Research Institute (IARI) (1971) Soil Survey Manual, All India Soil and Land Use Survey Organisation, IARI, New Delhi, 121 pp
Soil Survey Staff (2017) Soil survey manual. In: Agriculture handbook No. 18, USDA, Washington, DC. https://nrcspad.sc.egov.usda.gov/DistributionCenter/. Accessed 5 08 2019

Chapter 3
Application of Discrete Element Method Simulation in Environmental Modeling

Alireza Sadeghi-Chahardeh and Silvio José Gumiere

Abstract The compaction of arable soil, on the one hand, changes the geometry of the soil pores and reduces cavities between soil particles, consequently reducing the ability to retain fluids in these pores. On the other hand, the compaction alters the soil constituents, which has a negative effect on the soil ecological efficiency. Although many efforts have been made to gain a deep and quantitative understanding of the stress transfer and deformation process in arable soils, using a realistic approach can better predict the effects of soil management, such as those from farmland traffic on soil yield. One way to obtain information about the behavior of granular media is to perform simulations with the discrete element method (DEM), which provides the opportunity to track the motion of every single particle in the soil, and as a consequence, it can discern how microstructures affect the macroscopic properties of the soil. In fact, DEM modeling is a virtual laboratory in which the physical and mechanical behavior of granular materials can be predicted with respect to their smaller components. It would be difficult to investigate the effects of these smaller components with other experimental methods. In this chapter, the fundamentals of DEM are reviewed, and two applications of this method for modeling of soil compaction in agricultural field traffic are presented, along with a discussion of the results. The ability to visualize the results of DEM modeling is a great advantage, compared with experimental methods that require high-tech devices, such as X-rays and high-resolution cameras.

Keywords Soil ecological efficiency · Discrete element method simulation · Compaction of soil

---

A. Sadeghi-Chahardeh (✉)
Mineral, Metallurgical, and Materials Engineering, Faculty of Science and Engineering, Pavillon Adrien-Pouliot, Avenue de la Médecine, Université Laval, Québec, QC G1V 06, Canada
e-mail: alireza.sadeghi-chahardeh.1@ulaval.ca

S. J. Gumiere
Faculty of Agriculture and Food Sciences, Pavillon Paul-Comtois, Rue de L'Agriculture, Université Laval, Québec, QC G1V 06, Canada
e-mail: silvio-jose.gumiere@fsaa.ulaval.ca

P. K. Shit et al. (eds.), *Soil Health and Environmental Sustainability*,
Environmental Science and Engineering,
https://doi.org/10.1007/978-3-031-09270-1_3

3.1 Introduction

Soil compaction is the process by which soil particles are rearranged to decrease void space (Glossary of soil science terms 1997). The effects of soil compaction depend on many factors, including compaction effort, soil texture and structure, soil moisture, landscape position, and agricultural practice (Radford et al. 2000; Miller et al. 2002; Sillon et al. 2003; Zhang et al. 2006). For example, compaction in wet clay soils and loamy soils has caused a significant increase in soil bulk density and decreased unsaturated hydraulic conductivity and porosity (Radford et al. 2000; Sillon et al. 2003). However, in calcareous soils, unsaturated hydraulic conductivity is observed to increase under the influence of compaction (Sillon et al. 2003). In addition, other studies have shown that compaction alters soil water retention curves through its effect on the pore size distribution and pore connectivity (Miller et al. 2002; Perrier et al. 1996; McGarry 2003; Horn and Albrechts 2002).

The decrease in micro porosity and increase in micro porosity induced by compaction lead to a generally higher water content under a wide range of pore pressures. Because macro pores are the main paths for infiltration into soils (Hillel 1998), the breaking apart of soil aggregates (e.g., by compaction) may increase surface runoff and the risk of water erosion (Hamza and Anderson 2005). The infiltration rate of Australian heavy clay soil was reduced by four to five times under compaction (Li et al. 2001). Many studies conducted in different parts of the world have shown that compaction is an essential source of runoff, including Hortonian and saturation excess runoff and subsurface flow (Fullen 1985; Burt and Slattery 2006; Boiffin and Monnier 1994; Hillel 1998). Subsoil compaction favors a perched water table near the surface, possibly increasing saturation excess runoff and subsurface flow, thus enhancing soil erosion and modifying water flow paths throughout the watershed.

Soil degradation (e.g., soil compaction) modifies water, sediment, and contaminant flow paths (watershed connectivity) at the watershed scale (Gumiere et al. 2011). Studies have shown a high degree of spatial variability in runoff-generating mechanisms due to agricultural activity (which is prone to compaction) at the watershed scale (Croke et al. 2005; Luce and Cundy 1994; Ziegler and Giambelluca 1997; Croke and Croke 1999; Verbist et al. 2007). Disturbed areas may be characterized by highly variable saturated hydraulic conductivity, leading to non-uniform overland flow and variable runoff response times (Croke et al. 2005). The deterioration of soil structure affecting surface sealing may enhance the ponding of water at the soil surface, which increases hydrological and sediment logical connectivity of the watershed (Appels et al. 2011; Perrier et al. 1996).

Notwithstanding the many efforts to understand soil compaction at different scales, few studies have attempted to address the effects of soil compaction on hydrological connectivity at the watershed scale, either by modeling or by experimental study. In a world where food security is a challenging commodity, the agricultural machinery will become heavier, and monoculture agricultural systems' choices will increase. As a result, agricultural soil compaction and degradation will increase, changing the response of agricultural watersheds to rainfall, modifying the pathway of water and sediments, and making existing BMPs less and less effective (by saturation or bypass flow). However, the process of soil compaction at all scales is still complex. Numerical modeling can help researchers and agronomists understand the parameters and conditions of soil compaction and its interactions with soil hydrology. At a tiny scale, the discrete element method (DEM) (Cundall and Strack 1979) can be used to simulate particle–particle and particle–fluid interaction in compaction processes.

The DEM method allows the modeling of granular materials, soils, and rocks in a discontinuous way instead of the classical continuum mechanics that have been used to examine soils and rocks and to develop numerical solution techniques, such as the finite element method (FEM). FEM involves the constitutive relation of materials, while no reliable constitutive laws can accurately predict the behavior of granular materials (Liang and Zhao 2019). It should be noted that the constitutive laws derived from classical continuum mechanics do not take into account the dimensions of granular elements (Alshibli et al. 2006; Andrade et al. 2011).

Consequently, these constitutive laws suffer from mesh dependency when they are employed for a large deformation (De Borst 1991; Tang et al. 2019). However, DEM can provide applicable equipment for considering the internal length scale of a granular material without involving the sophisticated mathematics of non-classical continuum mechanics (Scholtès and Donzé 2013). Therefore, using the DEM method to model complex processes, such as soil compaction, rock milling, or crushing, is more realistic than FEM-based models. DEM is more appropriate for problems with multiple discontinuities and particulate materials and has been gaining popularity in the analysis of granular materials, soils, and rocks. Many aspects of this method still need to be examined in more depth.

This chapter aims to present the bases of DEM and presents some cases of study applied to soil compaction and land degradation. In Sect. 3.2, the concept of DEM with a rolling resistant effect is reviewed, and the setup for further simulations is discussed. In Sect. 3.3, the first case study involves an investigation of the tractor tire effect on the compaction of soil. The second case concerning the compaction of soil under cyclic loading is also explored. Section 3.4 offers a summary and a discussion of the most salient results of this work.

3.2 Materials and Methods

3.2.1 Discrete Element Method

DEM was introduced by Cundall and Strack (1979) to simulate the mechanical behavior of granular materials in which the particles are considered essential components. Although the modeling of the soil compaction involves complexities such as different size distributions, particle shapes, and solid–fluid interactions, it has been shown that DEM can successfully simulate some properties of granular materials, such as the mechanical and failure behaviors (Sadeghi-Chahardeh et al. 2021b, Sadeghi-Chahardeh et al. 2021c). Therefore, DEM has shown that it has many capabilities in providing a realistic model of soil samples.

Let us consider two particles $\boldsymbol{i}$ and $\boldsymbol{j}$ with arbitrary shapes (see Fig. 3.1a) that are in contact with each other on a surface, and force interactions between them $(\boldsymbol{f}_1, \boldsymbol{f}_2, \cdots, \boldsymbol{f}_n)$. Now, assume these two particles are replaced by two spheres with radii $\boldsymbol{R}_i$ and $\boldsymbol{R}_j$ (see Fig. 3.1b); to be able to consider the effects of these force interactions, we can apply a rolling moment $(\boldsymbol{M}_{r,ij})$ at the point of contact of the two spheres in addition to the normal $(\boldsymbol{F}_{n,ij})$ and tangential $(\boldsymbol{F}_{t,ij})$ contact forces. This rolling moment, in addition to being an alternative to the force interactions between particles, can also be considered as a parameter of the shape of the particles that prevents their free rotation (Sadeghi-Chahardeh et al. 2021c). Therefore, the motion of particle $\boldsymbol{i}$ is governed by the Newton–Euler equations as follows:

$$\begin{aligned} m_i g + \sum_{j=1}^{N_i} \boldsymbol{F}_{n,ij} + \sum_{j=1}^{N} \boldsymbol{F}_{t,ij} &= m_i \frac{d\boldsymbol{v}_i}{dt}, , \\ \sum_{j=1}^{N_i} \boldsymbol{M}_{r,ij} &= I_i \frac{d\omega_i}{dt}, , \end{aligned} \tag{3.1}$$

where N_i is the number of contacts, m_i is the particle mass, and I_i is the principal moment of inertia. In addition, the normal $(\boldsymbol{F}_{n,ij})$ and tangential $(\boldsymbol{F}_{t,ij})$ contact forces, as well as the rolling resistance moment $(\boldsymbol{M}_{r,ij})$, are defined.

by

$$\begin{aligned} &\boldsymbol{F}_n = K_n \delta_n, \\ &\Delta \boldsymbol{F}_t = -K_t \Delta U_t with \|\boldsymbol{F}_t\| \leq \|\boldsymbol{F}_n\| tan\phi_c, \\ &\Delta \boldsymbol{M}_r = -K_r \Delta\theta_r with \|\boldsymbol{M}_r\| \leq \|\boldsymbol{F}_n\| \eta_r min(R_i, R_j), \end{aligned} \tag{3.2}$$

where K_n, K_t, and K_r represent the constant normal stiffness, the constant tangential stiffness, and the constant of rolling stiffness, respectively; ϕ_c is the contact friction angle; η_r is the coefficient of rolling friction; δ_n is the overlapping distance between spheres; U_t is the relative tangential displacement at the contact point; and θ_r is the relative rolling rotation of particles. The constants of stiffness are defined from an elastic modulus, E, and dimensionless tangential and rolling coefficients, α_t and α_r, respectively:

$$\begin{aligned} K_n &= 2E \frac{R_i R_j}{R_i + R_j}; \\ K_t &= \alpha_t K_n; \\ K_r &= \alpha_r R_i R_j K_t. \end{aligned} \tag{3.3}$$

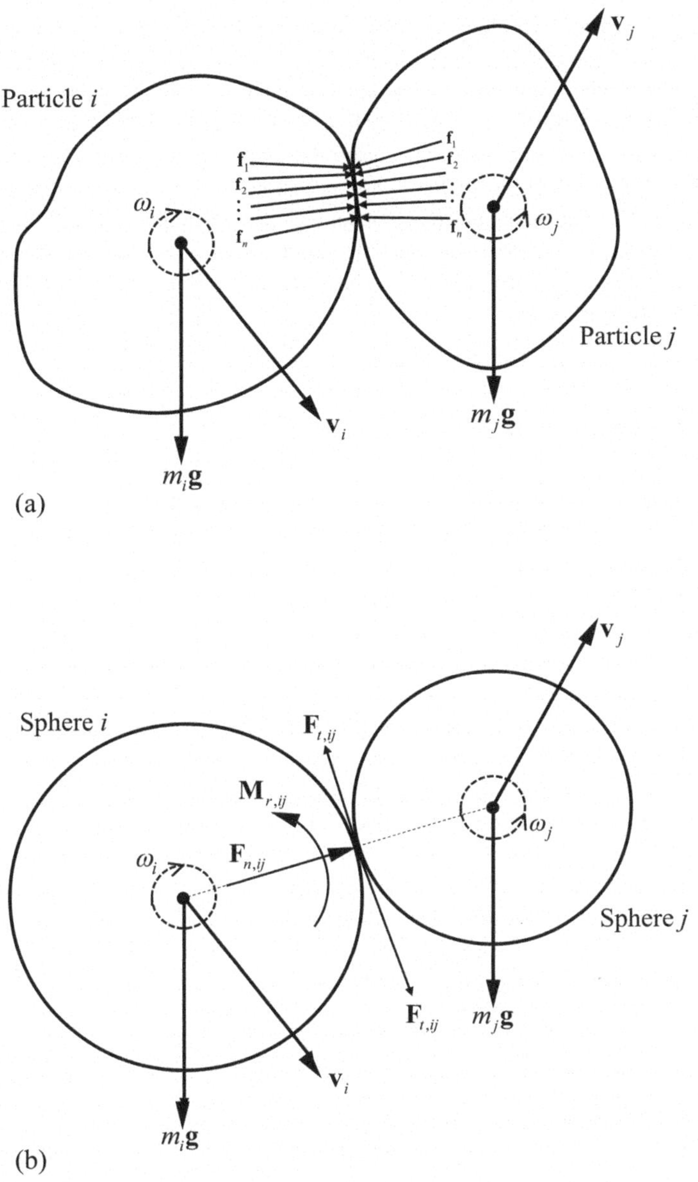

Fig. 3.1 The interaction between (**a**) two real particles and (**b**) two spheres in DEM (m_i is the mass of particle i, $\mathbf{g}$ is the gravitational acceleration, $\mathbf{F}_{t,ij}$ is the tangential contact force between the particles, $\mathbf{F}_{n,ij}$ is the normal contact force between the particles, $\mathbf{M}_{r,ij}$ is the rolling moment between the particles, $\mathbf{f}_n$ is the force interaction between the particles, $\mathbf{v}_i$ is the linear velocity, and ω_i is the angular velocity)

The rolling resistance model is phenomenologically developed to produce a larger simulated shear strength (Zhao et al. 2018). In other words, by considering this model, it is possible to simulate a phenomenon in which the particles have less rotation due to the effect of their shape. Hosn et al. (2017) numerically showed that the plastic macroscopic behaviorof the granular material is a function of the plastic parameters at the microscopic scale (ϕ_c and η_r) and mainly corresponds to the plastic rolling moment ($\| F_n \| \eta_r min(R_i, R_j)$), reflecting the particle's shape. Therefore, the main parameter for considering the particle shape effect is η_r, and the dimensionless rolling coefficient α_r does not affect the plastic macroscopicbehavior of the granular materials (Hosn et al. 2017).

3.2.2 Simulation Setup

In this paper, the DEM computations are realized using the open-source software YADE (Šmilauer et al. 2015). The interactions between the particles are simulated in the normal direction to the contact by a linear elastic spring with a stiffness, $\boldsymbol{K}_n$, and in the tangential direction by a linear elastic spring with a stiffness, $\boldsymbol{K}_t$, and the tangential perfect plasticity with a friction angle of $\boldsymbol{\phi} = 18°$. The properties of thematerials for the DEM simulations are also given in Table 3.1. At the beginning of a computational time step, the position of all the elements and the boundaries are known. The contacts are detected by the algorithm according to the known position of the elements, and the magnitude of the possible overlaps between the elements are discovered. The propagated contact forces and momentum on each sphere are then calculated by the interaction law (Eq. 3.1). After that, the forces are measured according to the law of motion for each particle, and the velocity and acceleration of the particles are calculated. Then, the new sphere positions are calculated by applying Newton's second law of motion. The integration time in Newton's second law and the interaction contact law are both carried out by way of an explicit scheme. The positions of all the particles and the boundaries in the current time step are determined

Table 3.1 The characteristics of the soil particles used in the DEM model (De Pue and Cornelis 2019)

Property	Soil particle
Density (kg/m^3)	2500
Elastic modulus (MPa)	250
Poisson ratio	0.4
Friction angle (rad)	0.31
Damping ratio	0.4
Dimensionless of tangential coefficient	0.385
Dimensionless of rolling coefficient	2
Coefficient of rolling friction	0.1

by the obtained values. This cycle of the calculations is repeated and solved at each time step, and thus the flow or the deformation of the material is simulated.

3.2.3 Soil Sample Preparation

The soil samples in this paper are generated by randomly inserting the particles within a rectangular domain with dimension of ($\boldsymbol{L} \times \boldsymbol{W} \times \boldsymbol{H} = 50 \times 200 \times 200$ cm) and applying gravity to the particles to naturally compact them. The boundaries of the cube are assumed to be displacement-controlled and frictionless. Therefore, the interaction of the particles and the boundaries of the cube will be in the normal direction of their contacts. The boundaries of the cube in the lateral directions (X and Y axes) and its bottom are fixed. The number of particles in the soil sample is 120,000 particles, with a particle size of 0.5 to 1.5 cm with a normally distributed size. The soil sample is then compacted by a frictionless movable wall in the Z-direction until the sample reaches the desired porosity which is equal to 0.457. It should be noticed that the porosity is defined as $\boldsymbol{\phi} = \frac{\boldsymbol{V_T} - \boldsymbol{V_s}}{\boldsymbol{V_T}}$, in which $\boldsymbol{V_s}$ is the volume of soil spheres and $\boldsymbol{V_T}$ is the total volume that is occupied by the soil spheres.

The pre-compression response of the sample is depicted in Fig. 3.2. The velocity of the wall in the pre-compression process, both in the loading and the unloading, is equal to 10 cm/s. This figure shows how to reduce the sample height (increase the sample density) at the end of the pre-compression process, when the external load is removed. It should be noted that the increase in soil sample density is due to the rearrangement of the soil particles. The final soil sample is shown in Fig. 3.3. The final height of the soil sample after the gravitational deposition and the pre-compression process is equal to 111.1 cm.

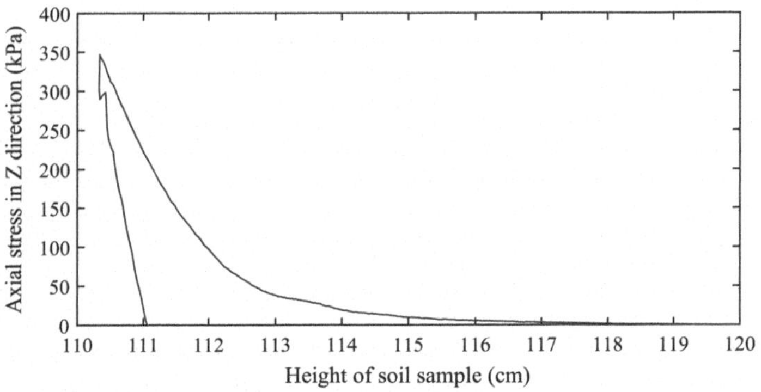

Fig. 3.2 The axial stress response of the soil sample during the pre-compression process (the velocity of loading and unloading in the process is 10 cm/s)

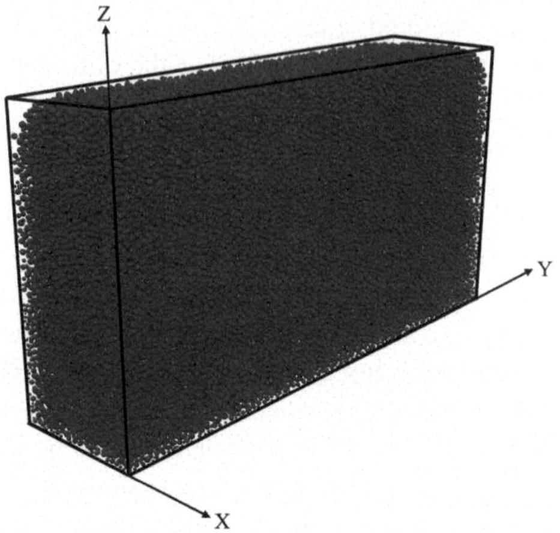

Fig. 3.3 The soil sample after the gravitational deposition and the pre-compression process

The soil particles are assumed to be cohesion-less during the gravitational deposition and the pre-compression process when a homogeneous soil sample is achieved.

For subsequent studies, the normal and shear cohesion with amounts of 1.5×10^2 and 1×10^3 kPa, respectively, will be added between the particles to model the cohesion of the real soil particles (De Pue and Cornelis 2019).

3.3 Case Study

3.3.1 Effect of Tire Compaction on Soil

As the population grows and agriculture develops, so does the need for heavier and faster agricultural implements. However, soil compaction due to the traffic of agricultural implements is one of the main challenges in soil management. The soil compaction, on the one hand, reduces porosity (or, conversely, increases soil compaction), which reduces the soil's ability to hold air and water. On the other-hand, it changes the mechanical stiffness of the soil matrix and rearranges the soil grains (Keller et al. 2013). Predictions of stress distributions in the soil due to tractor tires have been investigated through the experimental (Reaves et al. 1960; Soane et al. 1980), theoretical (Sohne 1958), and numerical (De Pue and Cornelis 2019; Perumpral et al. 1971) methods. The results of these investigations have demonstrated that due to non-uniform stress distribution, the maximum stress can be several times higher than the average stress, which represents the weight of the tractor divided by the contact surface of the track (Zhao et al. 2018). In this work, the DEM simulation will be employed to give a better understanding of the effect of tractor tires on soil compaction.

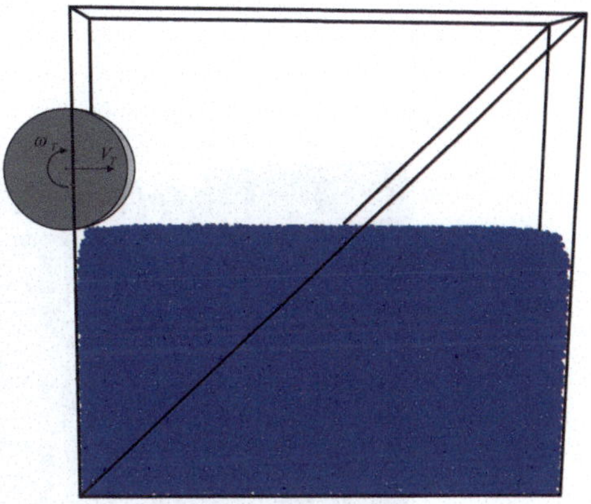

Fig. 3.4 The position of the tractor wheel on the soil sample, in which ω_T and V_T are the rotational and the linear velocities of the tire, respectively

Therefore, a cylinder with a height of 20 cm and a diameter of 50 cm is added to the top of the soil sample to represent a tire (see Fig. 3.4). The linear and angular velocities of the tireis 1 m/s and 4 rad/s, respectively. The linear and angular velocities of the tire will be chosen as the tire rolls perfectly on top of the soil sample. It is worth knowing that the initial position of the tire is determined to simulate the tire of a tractor with a weight of 3000 kg, in which the weight of each tire handled will be 750 kg. Therefore, the average tire pressure on top of the soil sample will be about 280 kPa. Hence, from Fig. 3.4, it can be concluded that the center of the tire will be at a height of 135.5 cm.

The distribution of normal and tangential forces in the soil sample due to the rolling of the tire on top of the sample are shown in Fig. 3.5. For more visualization, the soil sample is cut from the middle to make clearer the effects of the tire on the soil sample. Both the normal and tangential forces between theparticles meet their maximum values in the front of the tire. The amount of the normal force is one order magnitude higher than the amount of the tangential force. Figures 3.5a, b show that the diffusion of the normal forces inside the soil sample are higher than the tangential forces. As the normal force is a sign of a reduction of soil porosity and the holes between the soil particles, Fig. 3.5a indicates that the most affected area of normal soil due to the tire is about 50 cm below the tire. In contrast, the tangential force is represented by the deformation and rotation of the soil particles, which are then rearranged. Figure 3.5b represents that the most affected area of tangential force due to the tire movement is at a height of about 30 cm. It should be noted that, due to the granular nature of these materials and their complex mechanism of energy dissipation, not all particles will equally participate in the deformation.

Figure 3.6 depicts the normal and shear stress distributions due to the tire movement on the soil sample. Both the normal and tangential stresses reach their maximum just below the tire. Despite the distribution of the normal force, the normal stress is more likely to penetrate the soil sample. Although the normal force does not penetrate deeply into the soil sample, the smaller particles show a high amount of normal stress due to their small areas (Fig. 3.6).

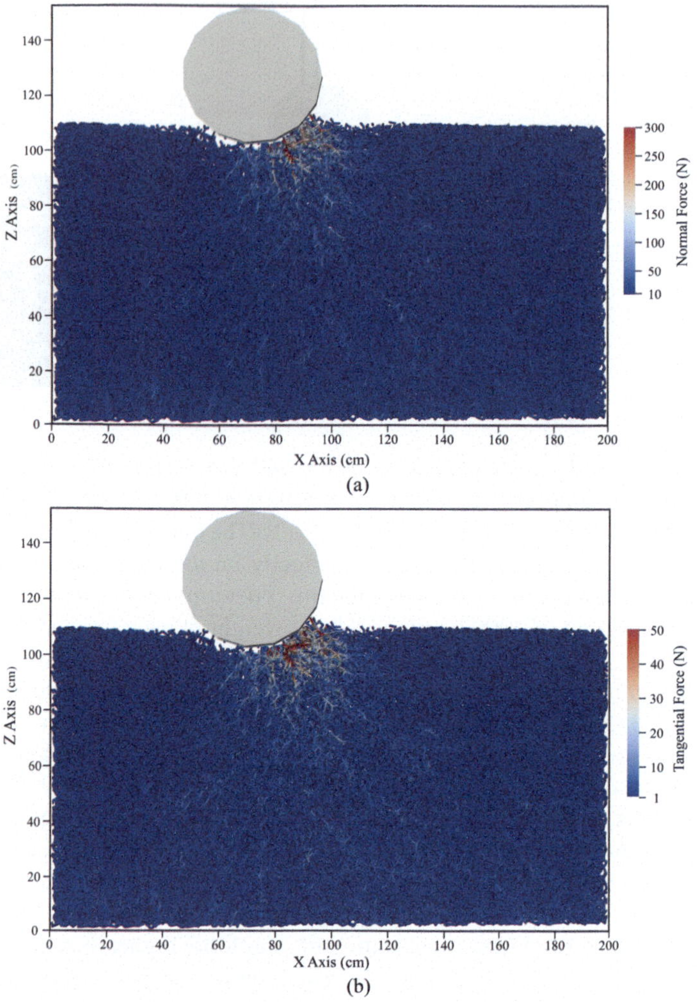

Fig. 3.5 The distribution of (**a**) the normal and (**b**) the tangential force chains of the soil sample due to the tractor tire movement (the soil sample is cut from the middle to visualize the force chains below the tire)

The DEM modeling of soil samples under tractor tire loading shows that the normal forces between soil particles are greater than their tangential forces. As a result, the prevailing phenomenon due to the passage ofagricultural implements over the soil is the reduction of soil porosity and cavities between soil particles. However, in the vicinity of thetire, shear stresses reach their maximum, and as a result, a number of changes in soilgeometry also occur. As a result, like a virtual laboratory, DEM can model the effects of different tire sizes and devices with different weights, without spending significant time and cost, and provide a visualization of the results in color.

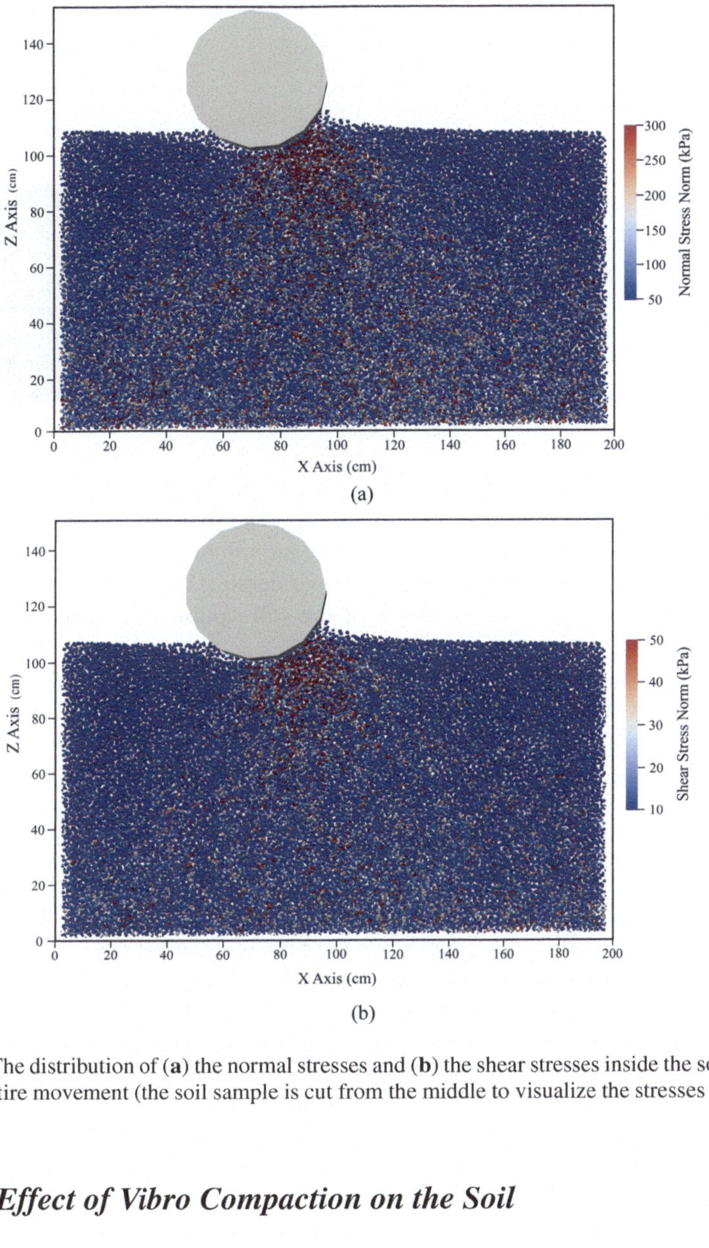

Fig. 3.6 The distribution of (**a**) the normal stresses and (**b**) the shear stresses inside the soil sample due to the tire movement (the soil sample is cut from the middle to visualize the stresses below the tire)

3.3.2 *Effect of Vibro Compaction on the Soil*

Cyclic loading results in surface displacement patterns that exhibit a permanent deformation in the soil. Therefore, the soil response to cyclic loading is always interesting forresearchers. The permanent deformation in the soil affects the presence of fluid between the soil cavities (Keller et al. 2013). Cyclic loading also results in

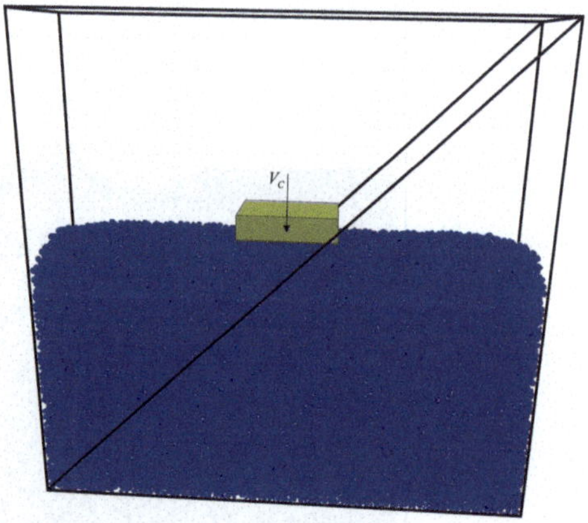

Fig. 3.7 The position of the vibro compactor on the soil sample, in which V_C is the vibro compactor linear velocity

cumulative effects that can be defined as time-dependent strain effects (Sadeghi-Chahardeh et al. 2021a; Wiermann et al. 2000). The first loading cycle causes the most probable soil deformation with a significant volume loss and a change in soil functions that can be detected down to a deeper depth (Hamza & Anderson, 2005). Therefore, in this section, the behavior of the soil sample during its first cycle is investigated.

DEM modeling enables us to study the behavior of a soil sample after cyclic loading due to a vibro compactor. The vibro compactor is modeled with a cube that moves along the Z direction. For applying cyclic loading on top of the soil sample, the velocity of the cube is defined as follows:

$$V_c = 0.5sin\left(\frac{\pi}{2}5t\right)(m/s). \tag{3.4}$$

The cube is placed on top of the soil sample and on its free surface. The maximum deformation in the first cycle occurs when the time is equal to 0.4 s. The distribution of the force chain networks and the stress inside the soil sample are shown in Figs. 3.8 and 3.9, respectively. The force chain networks associated with the maximum deformation of the first cycle are shown in Fig. 3.8. The soil sample is cut from the middle for better visualization. Similar to the tractor tire effect, the normal forces in the vibro compaction response are much greater than the tangential forces. In addition, the penetration of the normal force is more than the tangential force. The soil particles up to 50 cm from the cube can sense the normal forces. However, thepenetration depth of the tangential force is about 32 cm inside the soil sample. As a result, the changes in the volume of soil and its cavities are predominant in the vibro compaction process. Conversely, the changes in the geometry of the soil are limited to the depth of 32 cm of the soil sample. It should be noted that in granular materials, not all particles feel the external force at the same time. This phenomenon, which is intensified during dynamic loading, can be easily seen in Fig. 3.8.

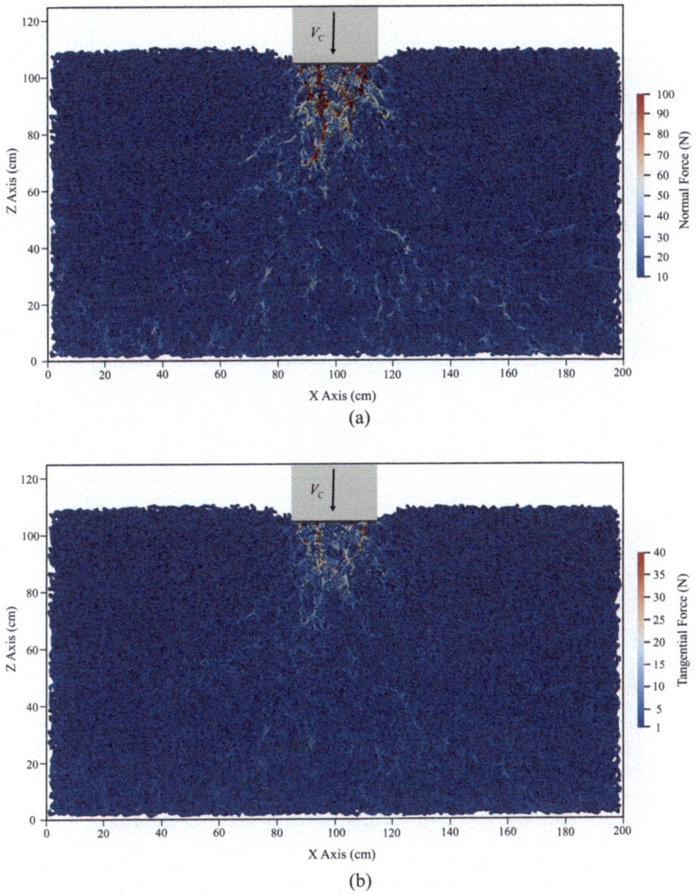

Fig. 3.8 The distribution of (**a**) the normal and (**b**) the tangential force chains of the soil sample due to the vibro compactor (the soil sample is cut from the middle to visualize the force chains below the cube)

The normal and shear stresses inside the soil sample, which is cut from the middle, are depicted in Fig. 3.9. The normal stress due to the vibro compactor at the maximum deformation of its first cycle penetrates more than the shear stress inside the soil sample. The penetration of normal stress is not limited to the area under the cube, but includes a v-shaped area in the soil sample. Therefore, the reduction in porosity of the soil sample under cyclic loading involves a larger area than the vibro compactor.

However, the shear stress is limited only to the area under the cube, and its dispersion appears as a narrower rectangle than the vibro compactor. Figure 3.9b shows the distribution of the shear stress under the vibro compactor.

As a result, the area that undergoes geometric deformation during cyclic loading is smaller than the area in contact with the vibro compactor.

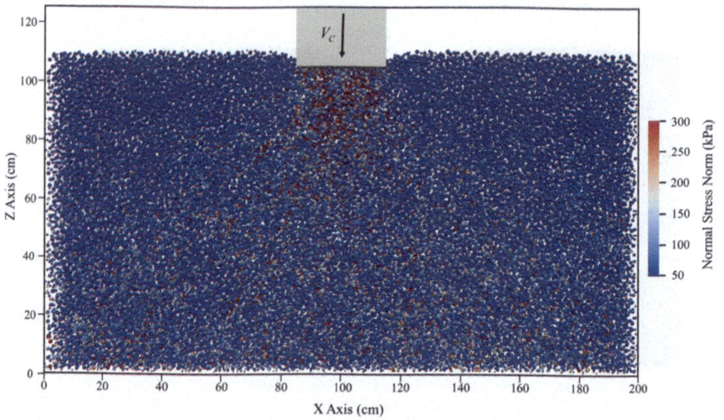

(a)

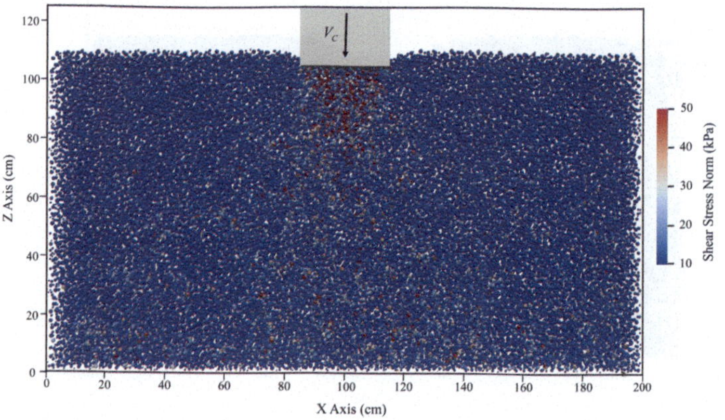

(b)

Fig. 3.9 The distribution of (**a**) the normal and (**b**) the shear stresses inside the soil sample due to the vibro compactor (the soil sample is cut from the middle to visualize the stresses below the cube)

3.4 Conclusion

In this work, DEM modeling and its applications in simulations related to soil mechanics and agricultural problems were investigated. The main conclusions can be summarized as follows:

- DEM, similar to a virtual laboratory, is able to investigate the mechanical behavior of soil during various agricultural processes, and its results are visually displayed. Agricultural issues, such as the effect of tractor wheels and compressors on soil compaction and the depth of penetration of the resulting forces, can be modeled, and the results can be easily compared with experimental results.

- The simulation results of the effect of a tractor wheel on soil compaction show that the amount of normal force between the soil particles is one order of magnitude greater than their tangential force. As a result, normal stress is more prominent than shear stress. It should be noted that the normal stress changes the volume of soil and thus changes the degree of water permeability, while shear stress indicates soil deformation and changes its geometry.
- The DEM modeling of the soil sample under the vibro compaction process showed that the amount and penetration of normal stress in the soil sample was greater than the amount and penetration of the shear stress in it. In addition, normal stress penetrates the soil sample as a V-shaped region. In contrast, shear stress in this type of loading penetrates as a rectangular area that is smaller than the occupied area of the vibro compactor.

The main focus of this article was to introduce DEM and its applications in agriculture. However, the soil behavior in agriculture is influenced by many environmental factors, including the presence of water between soil particles that cause soil particles to stick together, as well as the presence of fine soil particles, which were not addressed in this article. Therefore, in the future, the effects of water between particles and its influences on the force transfer between particles should be discussed.

Acknowledgements The authors gratefully acknowledge the financial support provided by the Natural Sciences and Engineering Research Council of Canada (NSERC). The research presented in this article was financed by NSERC discovery grant.

References

Alshibli KA, Alsaleh MI, Voyiadjis GZ (2006) Modelling strain localization in granular materials using micropolar theory: numerical implementation and verification. Int J Numer Anal. Methods Geomech 30:1525–1544

Andrade JE, Avila CF, Hall SA, Lenoir N, Viggiani G (2011) Multiscale modeling and characterization of granular matter: from grain kinematics to continuum mechanics. J Mech Phys Solids 59:237–250

Appels WM, Bogaart PW, van der Zee SE (2011) Influence of spatial variations of microtopography and infiltration on surface runoff and field scale hydrological connectivity. Adv Water Resour 34:303–313

Boiffin J, Monnier G (1994) Suppression du labour et érosion hydrique dans le contexte agricole français: bilan et possibilité d'application des références disponible. Les Colloques de l'INRA 85–103

Burt TP, Slattery MC (2006) Land use and land cover effects on runoff processes: agricultural effects. Encycl Hydrol Sci

Croke J, Croke J (1999) Managing sediment sources and movement in forests: the forest industry and water quality. Coop Res Cent Catchment Hydrol

Croke J, Mockler S, Fogarty P, Takken I (2005) Sediment concentration changes in runoff pathways from a forest road network and the resultant spatial pattern of catchment connectivity. Geomorphology 68:257–268

Cundall PA, Strack OD (1979) A discrete numerical model for granular assemblies. Geotechnique 29(1):47–65

De Borst RE (1991) Simulation of strain localization: a reappraisal of the Cosserat continuum. Eng Comput

De Pue J, Cornelis WM (2019) DEM simulation of stress transmission under agricultural traffic Part 1: comparison with continuum model and parametric study. Soil Tillage Res 195:104408

Fleige H, Horn R, others (2000) Field experiments on the effect of soil compaction on soil properties, runoff, interflow and erosion. Adv Geoecol 258–268

Flowers MD, Lal R (1998) Axle load and tillage effects on soil physical properties and soybean grain yield on a mollic ochraqualf in northwest Ohio. Soil Till Res 48:21–35

Fullen MA (1985) Compaction, hydrological processes and soil erosion on loamy sands in east Shropshire England. Soil Till Res 6:17–29

Glossary of Soil Science Terms (1997)

Gumiere SJ, Le Bissonnais Y, Raclot D, Cheviron B (2011) Vegetated filter effects on sedimentological connectivity of agricultural catchments in erosion modelling: a review. Earth Surf Process Landf 36:3–19

Hamza MA, Anderson WK (2005) Soil compaction in cropping systems: a review of the nature, causes and possible solutions. Soil Till Res 82:121–145

He X, Wu W, Cai G, Qi J, Kim JR, Zhang D, Jiang M (2020) Work–energy analysis of granular assemblies validates and calibrates a constitutive model. Granul Matter 22:28

Hillel D (1998) Environmental soil physics: fundamentals, applications, and environmental considerations. Elsevier

Horn R and Albrechts C (2002) Stress strain effects in structured unsaturated soils on coupled mechanical and hydraulic processes. In: 2002 ASAE annual meeting, p 1

Hosn RA, Sibille L, Benahmed N, Chareyre B (2017) Discrete numerical modeling of loose soil with spherical particles and interparticle rolling friction. Granul Matter 19:1–12

Keller T, Berli M, Ruiz S, Lamandé M, Arvidsson J, Schjønning P, Selvadurai AP (2014) Transmission of vertical soil stress under agricultural tyres: comparing measurements with simulations. Soil Till Res 140:106–117

Keller T, Lamandé M, Peth S, Berli M, Delenne J-Y, Baumgarten W, … others (2013) An interdisciplinary approach towards improved understanding of soil deformation during compaction. Soil Till Res 128:61–80

Li Y, Tullberg JN, Freebairn DM (2001) Traffic and residue cover effects on infiltration. Soil Res 39:239–247

Liang W, Zhao J (2019) Multiscale modeling of large deformation in geomechanics. Int J Numer Anal. Methods Geomech 43:1080–1114

Lipiec J, Hatano R (2003) Quantification of compaction effects on soil physical properties and crop growth. Geoderma 116:107–136

Luce CH, Cundy TW (1994) Parameter identification for a runoff model for forest roads. Water Resour Res 30:1057–1069

Martínez E, Fuentes J-P, Silva P, Valle S, Acevedo E (2008) Soil physical properties and wheat root growth as affected by no-tillage and conventional tillage systems in a Mediterranean environment of Chile. Soil Till Res 99:232–244

Mathier L, Roy AG (1993) Temporal and spatial variations of runoff and rainwash erosion on an agricultural field. Hydrol Process 7:1–18

Matthews GP, Laudone GM, Gregory AS, Bird NR, de G Matthews AG, Whalley WR (2010) Measurement and simulation of the effect of compaction on the pore structure and saturated hydraulic conductivity of grassland and arable soil. Water Resour Res 46

McGarry D (2003) Tillage and soil compaction. In: Conservation agriculture. Springer, pp 307–316

Miller CJ, Yesiller N, Yaldo K, Merayyan S (2002) Impact of soil type and compaction conditions on soil water characteristic. J Geotech Geoenviron Eng 128:733–742

Morgan RP (2009) Soil erosion and conservation. Wiley

Nicot F, Lerbet J, Darve F (2017) Second-order work criterion: from material point to boundary value problems. Acta Mech 228:2483–2498

Perrier E, Rieu M, Sposito G, de Marsily G (1996) Models of the water retention curve for soils with a fractal pore size distribution. Water Resour Res 3025–3031
Perumpral JV, Liljedahl JB, Perloff WH (1971) A numerical method for predicting the stress distribution and soil deformation under a tractor wheel. J Terrramech 8:9–22
Radford BJ, Bridge BJ, Davis RJ, McGarry D, Pillai UP, Rickman JF, … Yule DF (2000) Changes in the properties of a Vertisol and responses of wheat after compaction with harvester traffic. Soil Till Res 54:155–170
Reaves CA, Cooper AW et al (1960) Stress distribution in soils under tractor loads. Agric Engng 41:20–21
Sadeghi-Chahardeh A, Mollaabbasi R, Picard D, Taghavi S, Alamdari H (2021a) A numerical analysis of the mechanical behavior of coke aggregates under monotonic and cyclic loading. In: The 39th international conference and exhibition of ICSOBA. Manama, Bahrain
Sadeghi-Chahardeh A, Mollaabbasi R, Picard D, Taghavi S, Alamdari H (2021b) Discrete element method modeling for the failure analysis of dry mono-size coke aggregates. Materials 14(9):2174
Sadeghi-Chahardeh A, Mollaabbasi R, Picard D, Taghavi S, Alamdari H (2021c) Effect of particle size distributions and shapes on the failure behavior of dry coke aggregates. Materials 14(19):5558
Scholtès L, Donzé F-V (2013) A DEM model for soft and hard rocks: role of grain interlocking on strength. J Mech Phys Solids 61:352–369
Sillon JF, Richard G, Cousin I (2003) Tillage and traffic effects on soil hydraulic properties and evaporation. Geoderma 116:29–46
Šmilauer V, Catalano E, Chareyre B, Dorofeenko S, Duriez J, Gladky A, … others (2015) Yade documentation. In: The Yade project
Soane BD, Blackwell PS, Dickson JW, Painter DJ (1980) Compaction by agricultural vehicles: a review II. Compaction under tyres and other running gear. Soil Till Res 1:373–400
Sohne W (1958) Fundamentals of pressure distribution and soil compaction under tractor tires. Agric Eng 39:290
Tang H, Dong Y, Wang T, Dong Y (2019) Simulation of strain localization with discrete element-Cosserat continuum finite element two scale method for granular materials. J Mech Phys Solids 122:450–471
Verbist K, Cornelis WM, Schiettecatte W, Oltenfreiter G, Van Meirvenne M, Gabriels D (2007) The influence of a compacted plow sole on saturation excess runoff. Soil Till Res 96:292–302
Vervoort RW, Dabney SM, Römkens MJ (2001) Tillage and row position effects on water and solute infiltration characteristics
Wiermann C, Werner D, Horn R, Rostek J, Werner B (2000) Stress/strain processes in a structured unsaturated silty loam Luvisol under different tillage treatments in Germany. Soil Till Res 53:117–128
Zhang XY, Cruse RM, Sui YY, Jhao Z (2006) Soil compaction induced by small tractor traffic in Northeast China. Soil Sci Soc Am J 70:613–619
Zhao S, Evans TM, Zhou X (2018) Shear-induced anisotropy of granular materials with rolling resistance and particle shape effects. Int J Solids Struct 150:268–281
Ziegler AD, Giambelluca TW (1997) Importance of rural roads as source areas for runoff in mountainous areas of northern Thailand. J Hydrol 196:204–229

Chapter 4
Geospatial Techniques and Methods for Sustainability in Agricultural Management

Mariana Amato, Alessio Pollice, and Roberta Rossi

Abstract The role of high resolution geospatial techniques is increasingly recognized as key for managing inputs and making technical decisions in agriculture in view of the overall sustainability of the soil-crop-farming system. The task is challenging due to complex interactions between permanent and variable properties of soils, crop behavior, weather, climate, and anthropic dynamics. This implies that detection techniques are combined with approaches to data treatment and analysis and with support tools for decision making. Major issues are related to:

- choosing strategic or tactic approaches to site-specific management
- integrating plant and soil information from geospatial sensor data for decision making and identification of Management zones
- producing spatially- and agronomically-sound instructions for management.

Applications are many including research, farm and environmental operations, and range from supporting soil sampling and experimental design, to monitoring environmental sustainability or providing spatial information for interpreting biometric, productive and qualitative data. This chapter reviews open issues and presents approaches to data treatment and interpretation based on the joint analysis of soil and plant spatial behavior.

Keywords Crop modelling · Decision making · Management zone · Precision agriculture · Spatial technique

---

M. Amato (✉)
Scuola di Scienze Agrarie, Forestali, Alimentari ed Ambientali, Università degli Studi della Basilicata, Potenza, Italy
e-mail: mariana.amato@unibas.it

A. Pollice
Dipartimento di Economia e Finanza, Università degli Studi di Bari Aldo Moro, Bari, Italy
e-mail: alessio.pollice@uniba.it

R. Rossi
Consiglio per la Ricerca in Agricoltura e l'Analisi dell'Economia Agraria, Centro Zootecnia e Acquacoltura, Rome, Italy
e-mail: roberta.rossi@crea.gov.it

P. K. Shit et al. (eds.), *Soil Health and Environmental Sustainability*,
Environmental Science and Engineering,
https://doi.org/10.1007/978-3-031-09270-1_4

4.1 Introduction

Increasing efficiency in agricultural production is instrumental to developing both intensive and subsistence agriculture towards the goals of sustainability set by future challenges to primary production for human and animal consumption.

The 2030 Agenda for Sustainable Development of the United Nations issued 17 Goals regarding Sustainable Development (United Nations General Assembly 2015), among which the reduction of hunger and poverty in a world of growing population, while protecting the environment and pursuing social equity and innovation. The ensuing specific general target for agriculture is summarized as "sustainable intensification" of crop production, as set since 2009 among the strategic objectives of the United Nations Food and Agriculture Organization (FAO 2009).

The definition of sustainable intensification implies unit yield increases without an increase—or with a decrease—of adverse environmental impact of cropping techniques (Royal Society 2009). The compatibility of agricultural intensification with sustainability has been questioned (see discussion in Pretty and Bharucha 2014; Struik and Kuyper 2017) to the point of defining sustainable intensification an oxymoron. Pretty and Bharicha (2014) have highlighted the many facets and ambiguities of even the single terms intensification and sustainability. Nevertheless, evidence of shifts to a lower input—higher yield combination exist (Alromeed et al. 2015). Struik and Kuyper (2017) propose that sustainable intensification may be pursued through the "de-intensification" of intensive systems in order to make them more sustainable and the intensification of systems where yield gaps are found so they become more productive.

The issue is more complex than this framework, though. For instance, the increase in inputs in low-yielding conditions may or may not result in a higher yield and therefore may increase or decrease sustainability according to the ability to identify causes of poor yield and apply the appropriate means (Oliver et al. 2010).

In all cases the key to sustainable intensification is increasing resource efficiency and therefore reducing environmental impact per unit surface or per unit yield. To this end, Precision agriculture (PA) is one of the most promising approaches. The core of PA is to detect and manage the spatial variation of crop performance within fields, as opposed to uniform management. Sound agronomy and geospatial technology play a key role in PA, for within-field resource analysis and optimization. Oliver et al. (2010) make a good analysis of the issue by presenting nine case studies related to Precision agriculture, and namely to reduction of inefficiencies linked to the simultaneous presence of high- and low-yield zones in the same field. Authors remark that farmer's knowledge of their own fields allow them to map and rank areas of different yield performance in a way comparable to technological means; the reasons of poor performance, though, are not correctly identified in many cases and therefore farmers may apply additional fertilizer or other inputs in poor yielding areas and not obtain improvements which justify the extra amount of resources. This way inefficiencies increase rather than decreasing. Authors show how only finding the soil spatial constraints allows to diagnose the causes of poor yield and to plan alternatives for the correct management decisions.

The role of geospatial techniques in such a framework is increasingly recognized and can be key in discriminating cases where poor yielding potential needs to be addressed with low inputs to reduce waste of resources, from other instances where constraints to yield may be removed and an increase in inputs will result in higher yields. Geophysical techniques are especially useful to this end since they allow to map soil features such as texture and impeding layers in depth, while spectroscopic methods are limited to soil surface exploration but allow to detect additional chemical soil properties.

The task is challenging, though, due to complex interactions between permanent and variable properties of soils, crop behavior, weather, climate, and anthropic dynamics. This implies that techniques for the detection of the spatial variability of relevant properties in the system need to be combined with approaches to data treatment and analysis, and with support tools for decision making.

This chapter addresses some open issues and presents approaches to data treatment and interpretation based on the joint analysis of soil and plant spatial behavior.

4.2 Using Geospatial Techniques for Decision Making in Agriculture

While farmers' knowledge of field variability has always empirically guided choices to some extent, the wide availability of positioning systems for farm machines has made it possible to implement a spatially-aware form of agriculture which has two main broad fields of application: tractor guidance systems and Precision agriculture.

Tractor guidance systems include the early satellite-assisted driving and the latest automated or self-driving devices, all aimed at reducing overlapping and gaps in farming operations and therefore reduce time, soil compaction and the consumption of fuel, seed, fertilizers and other inputs. Such savings may improve resource efficiency (e.g. 20% in Kharel et al. 2020) and reduce costs and impacts on the environment.

Precision agriculture is a system where the management of crops is not uniform within a field, but site-specific according to the different needs of areas with different characteristics relevant to plant production and to the impact of agronomic techniques on the environment.

Spatial techniques for PA are aimed at two main objectives:

- detecting and mapping the spatial variation of crop behavior and environmental factors relevant to crop production and impact
- guiding the differential application in space of agricultural inputs (fertilizers, water, crop protection, …). This is also called variable-rate application.

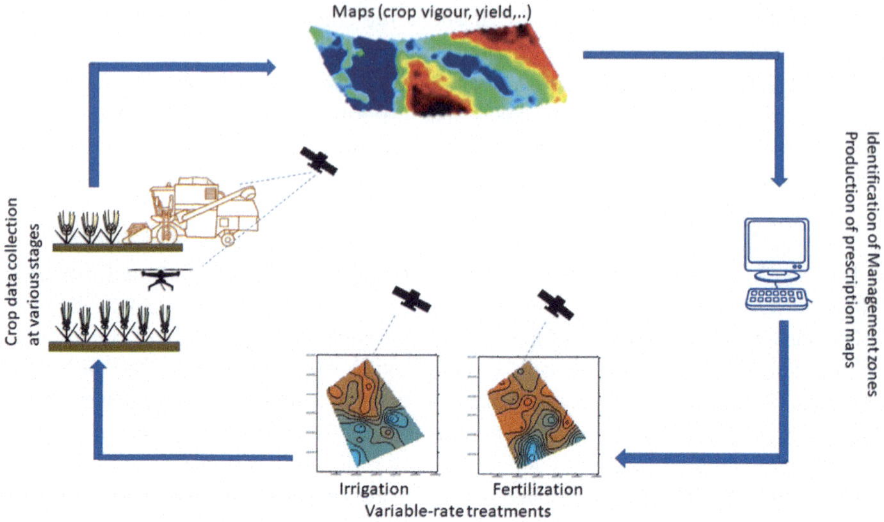

Fig. 4.1 Architecture of the early Precision agriculture cycle

The earliest schemes of spatialization in Precision agriculture were such as depicted in Fig. 4.1. Classic spatial data collection consists in acquiring georeferenced plant data during the vegetative growth of crops and/or harvest, and produce maps trough spatial statistics methods. Data analysis and application of agronomic know-how are then integrated in a further step to produce geo-referenced instructions for the site-specific management of crops in subsequent years.

Crop and soil spatial data, and possibly spatial data on meteorological information are used alone or in combination to make decisions for the differential management of areas within a field. Choices may be tactic or strategic.

Tactic decisions are taken at or slightly before the time of input application ("on the go"), and are typically based on one single layer of spatial data—e.g. from a sensor of crop status—which drives choices about the amount of inputs needed to correct a deficit or a stress of plants in a given field area. An example may be irrigation or nitrogen fertilization given in amounts which vary as the machine proceeds in the field with a variable-rate irrigation or fertilization device driven by an electromagnetic sensor of plant water or nitrogen status. A pro of this approach is that in principle it accounts for the actual space and time variability of crops (Casa et al. 2017). A big problem, tough, is that translating information on plant conditions in a management decision is hardly appropriate with on-the-go decisions based on a single sensor.

This depends on three main reasons:

- one indicator is often not enough to characterize the status of plants.
- even when the condition of a crop is correctly assessed the question is whether poor performance needs to be addressed with higher or lower inputs. As mentioned in the introduction this depends on what the reasons of bad plant growth, status or

yield are. If soil depth or texture in a given area is insufficient for holding enough water, for instance, more irrigation will not improve water stress but it will result in more water deep percolation and/or runoff and therefore more waste of this precious resource without improving yields. The combination of previous spatial knowledge (e.g. soil texture map) with instant measurements may be a way of improving information relevant for management choices. In this case, though, decisions may be less immediate and less continuous (e.g. they will depend on previous zonation to some extent).
- choices in agriculture are made with a target of amount and quality of produced goods, and responses to inputs are often the result of interactions with other factors. For example full nitrogen nutrition may induce excessive early growth of plants, with over-use of water during vegetative stages and this may leave insufficient water in the soil for the reproductive stage. Therefore decisions mediated by agronomic knowledge may be better than automatic choices.

The latter remark introduces a common issue to tactic and strategic approaches: translating sensor data into agronomic prescriptions remains a bottleneck, and future improvements may include a stronger incorporation of agronomic know-how into spatial approaches or vice-versa. At present, spatial data are increasingly used within agronomic tools:

- as an added dimension for decisions based on classical agronomic approaches like the water or nutrient balance.
- jointly with simulation models of plant growth and production in order to predict the spatial variability of crop performance under different conditions.
- within "decision support systems" where spatial information is used with other tools such as weather analyzers or generators, crop models, market analysis, simulators of environmental impact and more.

This way agronomists produce various possible scenarios of uniform or spatially-aware management in different meteorological conditions, price or regulations framework and thus help decisions makers—from farmers to politicians—analyze pros and cons of each choice (Ritchie and Amato 1990; Alromeed et al. 2015).

In strategic approaches choices are based on a spatial tool called "***prescription map***": a set of geo-referenced instructions for the site-specific management of crops. Prescription maps may be issued for one or several inputs (such as irrigation, fertilization, pest control, …) and they may be different for each type and time of application. Prescription maps are based on the structure of variation of relevant plant and terrain attributes and may follow their continuous variation or adopt a division in areas. Such areas are called "***uniform management zones***" (MZ) and spatial techniques for identifying them are the object of research, regarding both data acquisition methods and data treatment and criteria.

4.3 Spatial Techniques in Agriculture: Data Acquisition

Although farmers' knowledge appear to be effective in identifying zones of different crop behavior, the causes of such differences need to be identified before MZs can be established, and they often reside in soil attributes. Among properties to be mapped for Precision agriculture, Nawar et al. (2017) list farmers' knowledge, morphology, pedology, soil chemistry, yield and vegetation indices, soil properties from proximal or remote sensors.

4.3.1 Crop Spatial Data

Sensors for crop behavior may be classified based on several criteria. The most important destructive methods are those used for yield mapping, based on weight measurement upon harvest. Many spatial sensors, though, provide non-destructive mapping of crop behavior, where calibration and ground-truthing may be necessary and spatial resolutions vary widely. Technologies range from proximal to remote and span across a wide array of scales.

The most widespread methods are based on transmittance or reflectance of radiation from crop canopies at different wavelengths used alone or in combination:

- In the visible (VIS 0.4–0.7 μm) and near-infrared (NIR, 0.7–1.3 μm)
- In other infrared regions such as SWIR, (1.3–2.5 μm) or thermal (TIR 7.0–20.0 μm).

Fluorescence spectroscopy (at 0.68 and 0.74 μm wavelengths) is also increasingly used.

Such methods rely on different physical phenomena interfering with reflection and transmission of radiation. Some are based on known interactions of crop physiology with physical processes or on behaviors under study. The two most exploited are based on distinctive features of plants, and specifically the regulation of chlorophyll concentration and stomata opening in response to stress:

a. Plant water status or nitrogen nutrition status are linked to the concentration of a pigment unique to healthy plants: the chlorophyll. The turnover of chlorophyll in leaves is fast therefore environmental or management problems are reflected in changes of chlorophyll levels within a short time. Chlorophyll exhibits a distinctive behavior with regard to solar radiation, and spectroscopy uses the unique feature of chlorophyll of absorbing at the wavelength corresponding to the VIS red color while exhibiting a low absorption (therefore a high reflection and transmission) at the near infrared wavelengths. Measuring reflectance or transmittance of solar radiation in the VIS red and in the NIR has given rise to a few spatial vegetation indices used for the mapping of plant cover and stress and of nitrogen status. The most used are: NDVI = (NIR − VIS-RED)/(NIR + VIS-RED) and Chlorophyll Content Index = NIR/VIS RED. Where NIR

= reflectance or transmittance of radiation in the near infrared; VIS RED = reflectance or transmittance of radiation at wavelength/s corresponding to the red color. Each sensor may vary in the specific wavelength selected.

b. Stomata are openings on leaf surfaces which allow the exchange of water and carbon dioxide between plants and the atmosphere. Plants are able to finely regulate the status of such openings: a short term response of plants to water stress (or few other stress conditions) is the partial or total closure of stomata which causes a drastic reduction of water loss. Under such circumstances crop systems cannot rely on their main thermoregulation system, that is water passing from liquid to vapour at the plant-atmosphere interface employing latent heat. The consequent increase in temperature of the crop surfaces is therefore an index of stress.

Other methods include x-ray, laser or ultrasound imaging aimed at mapping crop height or biomass cover.

Sensors of crop status are the object of a large body of literature and were recently reviewed by Galieni et al. (2021).

4.3.2 Detection and Mapping Techniques for Agricultural Soils

4.3.2.1 Destructive Soil Sampling

Earlier approaches to soil measurement for addressing the spatial variability of crop performance were based on direct measurements of soil properties such as texture, pH, salinity, content of nutrients and possibly hydrologic constants. With this method sample collection is destructive and provides point information. Sampling density is typically low therefore interpolation accuracy is necessary. A great effort in improving interpolation has been devoted to this task in the past years (Nawar et al. 2017).

If methods of interpolation are aimed at assessing the spatial dependency of data with geostatistics, at least 100 data points are needed (Webster and Oliver 2007). This number of samples is small for recent sensor data, but large for traditional methods which are destructive and labor intensive. Traditional methods of point sampling of soil coupled with geostatistics have therefore been applied in research-oriented field campaigns (e.g. Castrignanò et al. 2020) but may not be proposed to farmers in regular practice.

An alternative approach is to apply stratified sampling (also called "targeted sampling" or "surface-response sampling") where a small number of field regions is identified based on plant behavior, and only one or few samples per region are collected. Ritchie and Amato (1990) identified zones of different crop performance in a maize field on the basis of aerial photographs and plant height, and were able to characterize differences in soil properties by sampling five soil profiles only. Oliver et al. (2010) used targeted soil sampling on the basis of zones of different crop performance as established from farmers' knowledge and vegetation indices in nine different case studies in Australia. They identified different types of soil constraints in areas of poor plant performance, and this led to envisage different farming strategies on the basis of crop modelling scenarios: where it was possible to remove constraints then yields could be increased and a matching high input of resources was beneficial. Where limiting soil factors could not be overcome, though, potential yields stayed low and the best strategy was to reduce inputs accordingly.

4.3.2.2 Geospatial Techniques for Agricultural Soils

Geospatial sensor techniques result in a revolution in approaches to the variability of agricultural soils. They provide non-destructive and almost continuous high spatial sampling resolution (e.g. >1500–2000 readings per ha in Mouazen et al. 2007) of electromagnetic soil behavior or other physical properties which are proxies of soil properties relevant to crop behavior. This allows fast zonation but the relations between sensor data and soil properties required for field management need to be established.

As a consequence, after geospatial sensor techniques were introduced in soil science research has focused somewhat less on interpolation methods and more on calibration (Casa et al. 2013).

Based on contact, methods may be remote or proximal, whereas based on sensor type techniques may be classified in two broad categories: spectroscopic: based on radiation just like plant sensors, and geophysical, based on electromagnetic behavior of soil components (Amato and Priori 2020).

The most used spectroscopic methods for agriculture include gamma and VIS–NIR sensors, increasingly used in multispectral and hyperspectral mode to detect different features of the soil surface. Spectroscopes may be remote or proximal. Pros of spectroscopy used in remote mode are the possibility to map large soil surface areas rapidly and at low cost. Problems include the need to measure when/where the soil is bare and the fact that measurements refer to the surface soil layer/s only.

Geophysical methods cannot be used remotely since they rely on proximal or contact sensors, but they have the great advantage of allowing measurements at surface and deep soil layers. This is an important feature for applications in agriculture, since the soil conditions beyond the surface horizons have been identified as crucial for identifying the causes of different crop behavior at the field scale and detecting constraints to agricultural production (Oliver et al. 2010). Therefore geophysical mapping improve the feasibility of precision farming and represent a revolution in the Precision agriculture cycle, where soil mapping can be effectively incorporated in the decision process. Geophysical surveying methods may be broadly classified as those making use of natural Earth fields and methods which need the application of artificially generated energy (Bitella et al. 2015).

Spectroscopic and geophysical methods are able to detect different soil properties, though, and the most informative strategy is a combined use of techniques in order to exploit the characteristics of each sensor.

Example of spectroscopic methods in agriculture-related uses (Selige et al. 2006; Casa et al. 2019; Priori et al. 2016) include gamma-ray spectroscopy measuring the spectrum of gamma-rays emitted by the surface 20–30 cm of soil, which are useful to map soil texture, mineralogy, stoniness and carbonate content. Spectroscopes measuring reflectance in VIS–NIR are used in multispectral or hyperspectral mode to detect texture, organic carbon, calcium carbonates, nitrogen, and water in soils. This kind of fast and low-cost mapping can help delineate Management zones based on soil fertility as reviewed by Nawar et al. (2017) and be of great use to quantify and monitor the effect of agricultural management on environmental variables. As an example, Priori et al. (2016) propose a combination of spectroscopic methods to map soil surface (0–30 cm) carbon stock with a limited number of destructive measurements. They used passive gamma-ray ("The Mole sensor"—Medusa Systems-The Netherlands) in proximal mode to map topsoil spatial variability, and lab VIS–NIR to estimate the carbon stock of fine earth at several points, based on existing spectral libraries. Calibration was then performed on few destructive measurements, at the average density of one sample per hectare. Gamma-ray data and elevation were used for interpolation through geostatistical methods and corrections of VIS–NIR carbon estimation based on gamma-ray mapped stoniness.

Geophysical methods in agricultural production and related environmental issues have a wide range of uses from texture mapping to monitoring soil water and studying plant roots. Many works point at them as an invaluable tool for precision farming (King et al. 2005; Alromeed et al. 2015; Rossi et al. 2015). Methods under study range from radar to seismic (Bitella et al. 2015). But the most widespread technology for agriculture is related to the measurement of electric resistivity (ER) or its inverse bulk electric conductivity (Ec) with galvanic or electro-magnetic induction devices. Principles, advantages and disadvantages have been reviewed in recent years (Samouelian et al. 2005; Bitella et al. 2015; Romero Ruiz et al. 2018).

Their success in agriculture is based:

- on the sensitivity of electrical resistivity/conductivity to the electrical properties of soil materials, which allow not only to discriminate minerals, but also to detect and quantify water, salt concentration, porosity, and resistive materials such as plant structures (Bitella et al. 2015; Amato et al. 2012).
- on the possibility to use them statically (galvanic) or on-the-go (galvanic and electro-magnetic induction) on bare or vegetated surfaces, and thus achieve a fast coverage of agricultural and natural fields.

One of the first and classical uses is the mapping of field soil salinity, an important constraint for crops (e.g. Corwin et al. 2003).

Regarding more specific applications, Basso et al. (2010) show a static application of geo-electrical methods for imaging and quantification of the effects of soil tillage methods on soil porosity in layers relevant for plant root growth. A series of works show field and container measurements based on electric resistivity tomography where plant root mass density could be mapped under trees (Amato et al. 2008; Rossi et al. 2010) and to a lower extent under herbaceous crops (Amato et al. 2009). Some examples are shown in Fig. 4.2.

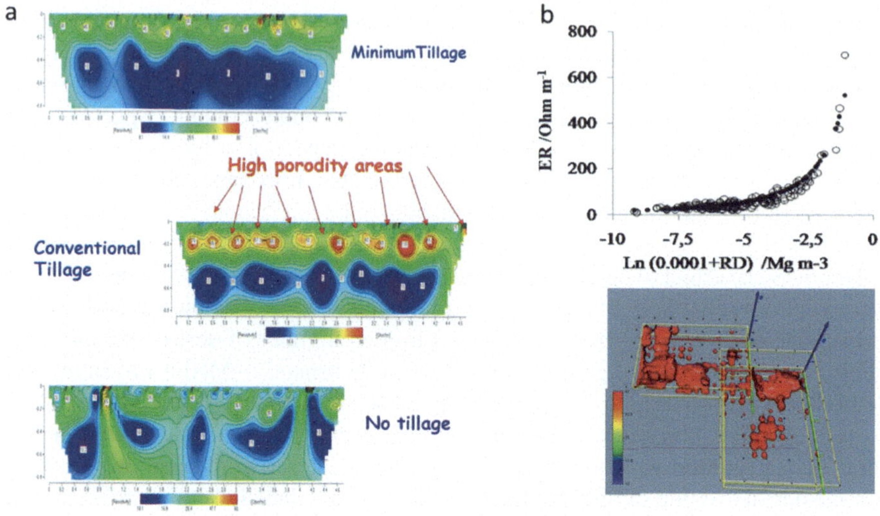

Fig. 4.2 Examples of applications of static geo-electrical methods in plant-soil systems: **a** vertical ER tomography in three different tillage systems (redrawn from data in Basso et al. 2010); **b** results of ER tomography measurements for plant root studies. Top: relation between ER and root biomass density under *Alnus glutinosa* (L.) trees. Bottom: volume reconstruction of root in a *Citrus* orchard with 3-D ER tomography (redrawn from data in Rossi et al. 2010)

4.4 Geospatial Techniques in Agriculture: Data Treatment and Management zones

Identifying Management zones is still one of the major open issues in precision farming. In many instances, the Precision agriculture cycle of Fig. 4.1 still the prevalent approach, where plant sensor data are the only zonation criterion. Problems with this scheme include that:

- the spatial pattern of plant behavior is variable in time, due to interactions of different soil and terrain features with weather (e.g. Machado et al. 2002). Therefore proper assessment of stable Management zones needs a multi-year analysis (McBratney et al. 2005).
- crop behavior is the resultant of many factors. A scheme where the spatial pattern of plant behavior is established but causes are not inquired is based on the stability in time of zones of good and poor yield and not on the identification of limiting factors. This may result in wrong management decisions (Oliver et al. 2010).

What changes the perspective is to use terrain attributes relevant to plant production as well, and namely soil properties and topography. This allows to address soil- and morphology-based sources of variability in the field such as soil constraints or different soil texture in selected areas and therefore to identify many of the causes of yield variability and address them appropriately. Also, such sources of variability are often permanent, and this allows to identify stable Management zones in the field in one go or coupled with a reduced number of years of crop performance observation compared with plant-based observations alone.

4.4.1 Classical Criteria for Identifying Management zones

Management zones are ideally regions where it is appropriate to apply a specific crop input uniformly, but with a rate that is different from that of a different neighboring region; MZs need to be stable in time in order to be used for planning. This definition implies that within MZs the combination of yield-limiting factors is relatively homogeneous, so that optimal use of resources may be pursued with a uniform application of each crop input (Vrindts et al. 2005) but different from that needed in surrounding MZs. In a simpler way Haghverdi et al. (2015) define MZs as subregions of a field that are homogeneous with respect to soil-landscape attributes.

Management zones may be identified based on one or more layers of spatial data and may include logistic criteria linked to farm operation, such as the dimension of machines and input application devices or time and economic constraints.

Nawar et al. (2017) review common approaches to MZ and his conceptual framework of classical MZ research, after the collection of spatial data lists the following steps:

a. identifying homogeneous areas
b. finding the optimal number of classes
c. establishing MZs and evaluating the effectiveness of classification.

The individuation of homogeneous areas is considered the most challenging step since it requires choices as to the definition of zone boundary (Nawar et al. 2017). Also, given the many interactions of factors determining crop yield and environmental impact, multiple data layers are often collected and the relevance of each data layer needs to be assessed, in addition to using techniques for data merging, fusion and multivariate analysis.

Research focuses on:

- statistical techniques to identify the most relevant data layers and their association, such as principal component analysis
- methods to group data layers, like the calculation of indices where single soil or crop properties may bear different weights
- clustering techniques including machine learning with parameters set by users or found through fuzzy/neural network methods.

The choice of technique for this step will then imply methods and indicators for finding the optimal number of classes.

The use of multiple data layers often results in improved effectiveness of MZ individuation compared to one layer only (Nawaret al. 2017), although costs are higher. More specifically, it is the joint use of soil and crop data that makes a difference rather than multiple information on crop only or soil only.

Assessing the effectiveness of classification may be performed on the basis of statistical criteria (e.g. the comparison of variability within and between MZs) or by analyzing the performance of uniform management versus variable-rate management with MZs obtained with one or more criteria. The most interesting methods include the production of scenarios through joint use of spatial data and crop modelling, and cost–benefit analyses.

4.4.1.1 Joint Use of Spatial Data and Crop Modelling for Scenarios

Among pioneers of this approach Ritchie and Amato (1990) used aerial photos of a maize field during water shortage in Michigan to identify areas of different sensitivity to water stress. Five different zones were thus mapped, where plant biometrics and yield as well as soil properties were measured, and differential crop behavior was explained in terms of spatial variability in soil available water in the profile. Management options for this field were then compared based on scenarios produced with the CERES-MAIZE crop model (Jones and Kinry 1986) and a weather generator: yield and use of irrigation water were simulated for 30 years of weather for southern Michigan. Irrigation strategies ranged from uniform watering of the whole field with different criteria, to PA with differential irrigation where timing and amounts of irrigation events were scheduled according to water retention characteristics of each of the five zones. Differential irrigation was the only strategy which allowed to reach the maximum yield and use the lowest amount of irrigation water in all of the five zones.

Oliver et al. (2010) propose the use of crop sensor data or farmers' knowledge to identify areas of different crop behavior, followed by soil sampling to identify soil constraints, and crop modelling for comparing scenarios as a basis for management decisions.

Alromeed et al. (2015, 2019): used electrical soil mapping and irrigation-oriented modelling for precision irrigation planning. An Automatic Resistivity Profiler (ARP © Geocarta—Paris) was used to obtain resistivity maps at the depths of 50, 100, 200 cm in Southern Italy. The map of electrical resistivity was used for sampling soil at a limited number of sites where soil texture was measured and then translated into total available water (TAW) calculated with the Saxton and Rawls (2006) pedotransfer corrected for gravel. Values of TAW were then applied to the ER map and used as an input for the ISAREG model (Teixeira and Pereira 1992) applied to each of six soil areas within the field. Increasing ER corresponded to increasing coarse soil fraction content and decreasing TAW, which ranged between 216 and 121 mm with respect to the whole 200 cm profile and between 120 and 66 mm over 100 cm. Daily weather data for 15 years (1999–2013) were used for simulations comparing uniform and differential irrigation with different criteria on 6 crop types. Differences in irrigation requirements between soil zones identified with ER mapping were 10–44% and varied with irrigation strategy. Differential irrigation of each area according to its own TAW up to 100 cm allowed to save an average of 20% of water without yield losses compared to uniform irrigation using average TAW.

4.4.1.2 Cost–Benefit Analysis

Nawar et al. (2017) review the use of economic criteria for farm-scale comparison of uniform and variable-rate agriculture, and show that precision farming allows to increase yield and/or allow savings of inputs with an overall favourable economic balance. The profitability of precision versus uniform management depends, of course, on the degree of variability within a field, but reviewed yield increases range from 1 to 10% and resource savings from 4 to 46% due to precision application of fertilizers. The overall net return is always higher with precision farming. A case study is also presented on nitrogen fertilization of cereals in the UK where different approaches to Management zones delineation for variable-rate farming are compared. The study shows that MZ delineation based on soil data only is less profitable than MZ identification obtained through soil and crop data.

Farmer's cost–benefit analyses, though, cannot be considered a complete account of economic benefits of a given agricultural management approach, since costs or revenues linked to environmental impacts or ecosystem services should also be taken into account. They would show that Precision agriculture is even more economically viable than uniform management since a lower use of agricultural inputs implies a lower waste of resources and pollution potential.

4.4.2 Using Soil–Plant Spatial Relations to Identify Management zones

In classical clustering procedures different data layers may have different weights in statistical treatment but their relations in view of agronomic criteria are not usually studied. A different approach is found in Rossi et al. (2015, 2018), Pollice et al. (2019), who used the relation between soil and crop data for preliminary zonation before applying other criteria as summarized in the following paragraphs.

In a study on differential irrigation in an alfalfa field in southern Italy (Rossi et al. 2015) electrical resistivity mapping was performed with automatic profiling (ARP © Geocarta Paris) at three layers: 50, 100 and 200 cm of depth from the soil surface (Fig. 4.3).

Values ranged between 3.7 and 64 Ω m with definite spatial variability. Surface-response sampling was used in order to choose 6 positions in the field corresponding to different ER values spanning across the whole range of values. Traditional soil profile studies and soil lab analyses were performed at the 6 positions, and ER was shown to be a proxy of soil texture, being sensitive to clay and sand, just as shown in many other instances in the literature.

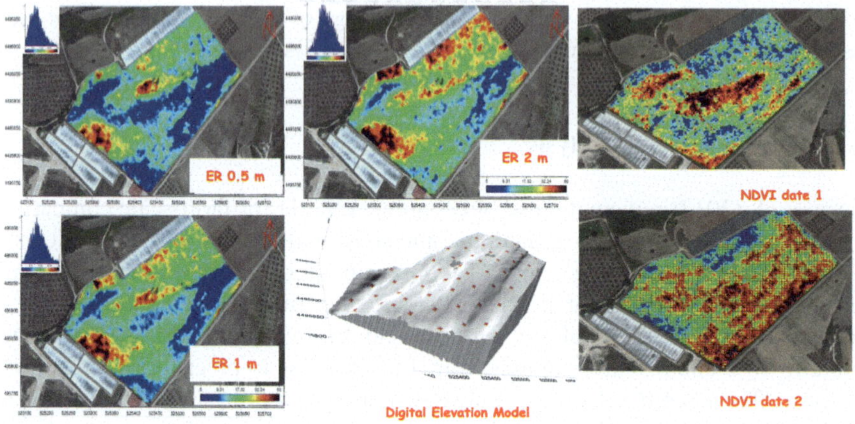

Fig. 4.3 Electrical resistivity maps up to three depths, elevation model and NDVI measured at 2 dates in a 7-ha alfalfa field in southern Italy. Redrawn from data in Rossi et al. (2015)

The crop biomass was also sampled at the same sites while vegetation cover and vigor were mapped with proximal VIS–NIR methods for the calculation of NDVI at 4 dates corresponding to different phenological stages of alfalfa (Rossi et al. 2018). Examples of NDVI at 2 dates are mapped in Fig. 4.3.

Both biomass and the NDVI index were strongly correlated with ER. Alfalfa is a forage crop and the whole plant is fed to livestock, hence biomass coincides with crop yield. Therefore in this study the spatial variation in yield was well predicted by ER maps. The best relationship (statistically strongest) was found between vegetation indices and ER at the deepest measured layer (up to 200 cm) and this was explained by pointing out that alfalfa is a perennial crop with a deep root, therefore sensitive to soil changes at depth (Rossi et al. 2015).

The soil-crop relationship from such data was studied using generalized additive models (Rossi et al. 2015, 2018; Pollice et al. 2019) and showed a complex behavior, which is depicted in Fig. 4.4 where a function of the crop index NDVI and field slope is plotted against ER of the deepest layer.

This function may be divided in three distinct regions, each with a different soil-crop relation:

Region I: low ER (<12–15 Ω m). This region of the relationship corresponds to the dark blue areas on the soil ER map. Here the crop behavior is not a strong function of ER: the wide gray zone around the function line corresponds to a wide confidence interval due to erratic crop response. Also the inversion of the slope in the soil-crop relation points to problem areas. The study of the soil profile in one of these zones shows that the landscape position and a high clay content result in a hydromorphic soil profile (A1 in Fig. 4.4) indicating waterlogging. This explains the erratic crop behavior: the area may be productive at times of low precipitation but less productive in periods of high rainfall when the soil will hold too much water and impair root physiology and production.

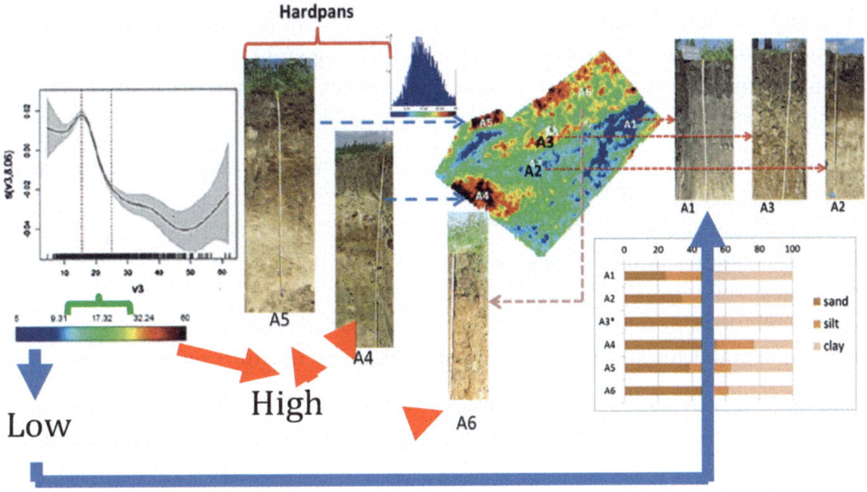

Fig. 4.4 Soil-crop relationship function, soil profiles and fine earth granulometric composition in different regions of the ER map up to 200 cm. Redrawn from data in Rossi et al. (2015, 2018)

Region II Intermediate ER (approximately between 12 and 25 Ω m) Most of the field falls in this relatively narrow range of ER values (areas in green and yellow on the ER map). Here the crop biomass is very sensitive to ER: the vegetation function increases steeply as ER decreases and confidence intervals are very narrow, therefore the ER-crop relation is statistically strong and vegetation will reliably behave in response to ER. Management can usefully be planned according to ER in this area, with continuously variable application of agricultural inputs or through ER-based zonation of the field using common clustering criteria.

Region III High ER (>25 Ω m). This corresponds to red–black areas in the field. In this area the soil-vegetation relationship is again erratic, and has a lower, variable and even inverted slope. This is a strong suggestion of problem areas, and the study of soil profiles in such spots revealed a high stone percentage or the presence of impeding soil layers such as hardpans.

Authors (Rossi et al. 2018; Pollice et al. 2019) therefore propose a two-step procedure for identifying Management zones where step 1 consists of studying the soil-crop relation and separate zones with different relationships and step 2 consists of applying common clustering criteria for MZ to the separate zones or only to the field areas where the soil-crop relation is strong and reliable. Also, finding areas where the soil-crop relationship changes may help identify zones with soil constraints and provides a spatial indication to study their properties and suggest different management options than what effective in other areas.

Rossi et al. (2018) compared this approach with common clustering criteria used in the literature such as fuzzy clustering applied to the whole field, and point out that fuzzy clustering alone, based on similarity criteria and not on the functional relation of crop and soil data, would result in wrong management decisions in areas I and III of the field and not provide indications for identifying soil constraints.

Datasets in this study showed several features common to spatial data in agricultural research. Among them:

- different resolution and density or misalignment of data from different sensors (e.g. resistivity and NDVI maps)
- many sources of data errors
- lack of normality in the frequency distribution of data
- non-linear relations between soil and crop data
- spatial relations of data (autocorrelation, covariance, …)
- need to account for soil-plant relationship in a way useful for applications.

The treatment of such and other features was addressed in the papers (Rossi et al. 2015, 2018; Pollice et al. 2019) and is the object of the following paragraph of this chapter.

4.4.2.1 Data Filtering, Spatial Interpolation, Statistical Modeling and Management Zone Delineation

In agriculture, techniques using proximal soil and crop sensors allow to acquire data fast and at extremely high spatial resolutions, and therefore to obtain a large number of data points per unit surface. Depending on the sensor and the terrain some datasets contain a variable amount of systematic and random errors that include georeferencing errors, operator error, few extreme values due to local crop failure, to poor establishment or planter skip. This is very typical of yield maps that require data filtering prior to interpolation (Simbahan et al. 2004), or soil geophysical surveys that can be affected by different sources of noise contamination. Sometimes the goal is to analyse the functional relationships between variables (i.e. yield responsiveness to soil and terrain attributes). To this regard it is worth noticing that different techniques exhibit different data density and spatial distribution, but also that data acquired at different dates are commonly misaligned in space even if measured with the same sensor. Very often, finding a functional relation between yield and soil or sensor surrogates of soil variables requires appropriate treatment of problems related to spatial misalignment (or change of support Problem—COSP, see Gelfand et al. 2010, Chap. 29) and to the large data size. Also sensor data can show a relatively large amount of outliers due to accidental sensor flaws or (this is the case of proximal soil sensors) to rough terrains which prevent optimal coupling between sensor and soil surface. Pre-processing is usually necessary for many sensors to remove outliers, reduce background noise, correct positioning errors and align data acquired at different locations. Data filtering and spatial interpolation on a common lattice covering the field allow to upscale the data to a common support. Here we address three steps frequently taken in sensor data processing: data filtering, spatial interpolation and statistical modeling. We refer to methods by far more computationally efficient than those traditionally used in this field. Indeed, standard variogram modeling and kriging would hardly be feasible even with few thousands data points.

In the following paragraph we will give a brief overview of:

i. the median filter as an example of simple but efficient filter for crop and soil sensor data,
ii. deterministic interpolators (inverse distance, spline and nearest neighbors) as rapid tools for large sensor datasets
iii. statistical modeling for high density misaligned sensor data aimed at facilitating yield map interpretation or as a tool for field zonation.

Data Filtering: The Median Filter

Proximal sensing can yield massive amounts of finely spaced data. Depending on the technology and the terrain, some datasets can be affected by a variable amount of unwanted noise. Conductivity meters, for instance, are sensitive to the electrical interference from nearby metal objects. For galvanic sensors a poor contact between soil and electrodes can result in a large number of null values, while hitting rock fragment clusters can spike up resistivity values of several orders of magnitude. This unwanted clutter can be filtered out to facilitate the recognition of a broader pattern. The median filter is a simple yet effective tool to remove extreme values without altering the general trend. A median filter operates by calculating the middle value of an ascending-ordered sequence of numbers within a moving window of a given dimension. Every time an observation, within this windows, departs from the median (above or below a user's defined threshold) it is replaced by the window's median. Tabbagh (1988) gives the first example of the use of median filtering for improving on-the-go resistivity survey data quality. The author showed how median filtering de-spikes data without altering the broad pattern. Median filtering algorithms are now available at no cost in many open-source software items such as "R" (https://CRAN.R-project.org). Specifically designed for sub-surface geophysical survey data processing, WuMapPy is an open source python package which provides a median filtering routine (Marty et al. 2015). Median filtering has also been used for plant-based data image processing to improve the accuracy of discrimination and mapping of weed patches in sunflower fields (Peña-Barragán et al. 2007). A data filtering protocol for management zone delineation, which includes median filtering, was described by Córdoba et al. (2016). Median filtering was also used by Mavridou et al. (2019) for the morphological analysis of fruits based on image analysis.

Spatial Interpolation

Spatial interpolation can be considered a special case of statistical inference because it implies prediction over a spatial process. Observed values at certain geographical locations (sampling sites) are used to predict the unobserved values at unsampled sites. Interpolation techniques can be classified into two comprehensive categories: deterministic versus stochastic. Deterministic approaches rely on mathematical functions to derive surfaces from sample data points either on the basis of similarity between points or on the degree of smoothing (Adhikary and Dash 2017). Stochastic methods are founded on the statistical properties of sample points, the field is regarded as a random process and the optimality of the smoothing method is established in terms of minimizing some specific criterion (Babak and Deutsch 2009). Deterministic techniques include: inverse distance weighting (IDW) and the use of spline functions. To stochastic interpolators belongs the large family of geostatistical kriging with its many variations such as simple, universal, ordinary kriging which rely on variogram estimation. A number of works have compared deterministic versus stochastic interpolators but mixed outcomes were obtained (Gong et al. 2014; Gotway et al. 1996; Kravchenko 2003; Mueller et al. 2001, 2004). Very often these techniques were used to interpolate sparse data (e.g. soil survey data). A typical soil survey datasets for variogram estimation comprises few hundreds data points over hectares (Webster and Oliver 1992). This data density is greatly overridden with proximal soil sensing. As an example, Rossi et al. (2013) measured soil resistivity with a continuous resistivity profiling equipment over a 3.5 ha. Tempranillo vineyard. Data were acquired along parallel rows spaced approximately 5.60 m. Data density along the rows was very high, with a measure every 20 cm, yielding over 115,000 data points. In this case the computational effort required to estimate variogram parameters might not be compensated by a hypothetical gain in precision. "Light" deterministic techniques are very efficient and allow to process large geophysical datasets within minutes (Rossi et al. 2013). Inverse Distance Weighing is a spatially-weighted average of the sample values within a search neighbourhood (Shepard 1968; Diodato and Ceccarelli 2005). Unlike kriging, no prior information is needed for spatial prediction. IDW exhibits sensitivity to properties of data and data-bases (e.g. skeweness, anisotropy, samples spatial distribution) (Babak and Deutsch 2009). For instance, with IDW the choice of the exponent value (which greatly affects map accuracy) needs to be based on the data skewness coefficient (Kravchenko and Bullock 1999; Weber and Englund 1994). Examples of IDW potential use in Precision agriculture can be found in several works (Souza et al. 2016; Usowicz and Lipiec 2017; Robinson and Metternicht 2006). Other popular deterministic interpolators include spline functions. A nice definition of spline interpolation can be found in McKinley and Levine (1998, p. 1): "[The fundamental idea behind cubic spline interpolation is based on the engineer's tool used to draw smooth curves through a number of points. This spline consists of weights attached to a flat surface at the points to be connected…The weights are the coefficients on the cubic polynomials used to interpolate the data. These coefficients 'bend' the line so that it passes through each of the data points without any erratic behaviour or breaks in continuity]". Spline interpolation (e.g. cubic spline) has been used in crop science to interpolate climate time-series such as seasonal evapotranspiration (Sadler et al. 2000). Boer et al. (2001) applied thin plate splines to estimate temperatures and precipitations in Jalisco (Mexico-Boer et al. 2001). Rossi et al. (2013) interpolated high density multi-depth soil apparent resistivity spatial data using a cubic spline interpolation. Different soil variables might require specific interpolators. Robinson and Metternicht (2006) showed that while kriging was the best interpolator for electrical conductivity, pH required IDW, organic matter content was best estimated using a cubic spline.

Upon reviewing the literature it is clear that there is no such thing as a "one fits all" interpolation method. As a general rule of thumb for sparse data, uncertainty measures may help increase map accuracy. If a valid sample variogram can be computed, geostatistical kriging is likely to give good predictions in many circumstances. For dense proximal sensing dataset, users may benefit from non-computational-intensive deterministic interpolators to process large datasets in a reasonable time. Using these simple tools, however, always requires a close look to the data. Exploratory analysis is absolutely needed to select the most appropriate interpolator.

One of the difficulties in establishing a functional relationship between variables observed over a common spatial domain comes from the possible difference between the spatial support of the observations (i.e. each variable having its own data points). In Pollice et al. (2019), COSP has been treated through a non-standard approach: multiple spatial data were upscaled by interpolating data points to a square lattice overlaying the studied field. Because of differences in number and location of sampled spots corresponding to each spatial variable, a proportional nearest neighbors neighborhood structure was used to calculate the upscaled values. For each spatial variable the number of neighbors was considered proportional to the samples size at all grid points, where the neighbors' values of mean, variance and covariance between spatial variables were obtained. Although such statistics were not the main intended product of the upscaling procedure, they were useful as inputs of the model likelihood for model fitting.

This latter example of multiple sensor data interpolation leads to the last topic covered by this paragraph: the statistical modeling aimed at estimating functional relationships between variables.

Spatial Modeling

The key concept of Precision agriculture is to deliver crop inputs when and where crops require. In most applications this implies the need of subdividing the field into two or more MZs: management units or zones where crop inputs (i.e. seed/fertilizer rates) are applied at different levels but uniformly within a given unit (Moral et al. 2010). Unsupervised classification techniques such as the fuzzy k-means algorithm are routinely employed in multivariate clustering for automatically identifying homogeneous field zones, also thanks to the availability of freeware user-friendly softwares that perform the task (Fridgen et al. 2004; Paccioretti et al. 2020). Spatial clustering techniques are an invaluable instrument to delineate functional spatial units for variable rate-equipment but do not answer key questions regarding the causes of variation in crop performance and give no clues on crop requirements. Establishing functional relations between plant production and individual soil characteristics can help identify which soil characteristics limit yield in different zones of the field. Shatar and McBratney (1999) compared different empirical methods such as neural networks, projection pursuit regression, generalized additive models and regression trees to model sorghum yield as a function of soil properties. Field-scale yield responsiveness to local field conditions can also be modelled through the boundary-line analysis (Shatar and McBratney 2004). This work features a comparison between individual yield-response functions and a single predictor variable (among soil properties) was chosen at each location on the basis of prediction of the smallest yield value. Simple implements such as correlation analysis can still give some indications about influential variables and subsequent management options albeit more sophisticated modelling techniques such as geographically weighted regression (GWR), specifically designed to handle spatial non-stationarity, may be more suitable if local variation of yield-response function exists (Fotheringham et al. 2003). GWR alone or in combination with temporally weighted regression predicted plant production with a relatively high accuracy (Feng et al. 2021). To select an optimal linear regression model, the shape of the relationship between the target variables should be inspected: a spatially constant relation requires a linear regression model, while GWR is well suited for a spatially varying relationship or even semi-varying (as in mixed geographically weighted regression) when some regression coefficients are globally constant while the others are geographically varying (Yang et al. 2019). Soil–plant relationships often exhibit non-linear features (Shatar and McBratney 1999). Non-linear relationships are conveniently modelled using smooth functional effects in generalized additive models (GAM) (Rossi et al. 2018). When nonlinear spatial patterns emerge in the map of regression residuals, it is possible to add a smooth function of sample coordinates to the model predictor, for instance specifically an anisotropic bivariate smooth represented using tensor product splines can be used. This model term accounts for the nonlinear spatial pattern of the vegetation which is not explained by the predictor variables. GAMs are fitted by penalized likelihood and automatic choice of smoothing parameters may be used minimizing an internal Generalized Cross Validation (GCV) criterion (Wood 2006). The first step is variable selection; starting from a set of alternative candidate models, a stepwise process of variable selection can be followed. The final model selection is performed through comparison of the proportion of the null deviance explained by the model, the Akaike information criterion (AIC), and the Bayesian information criterion (BIC). Such criteria quantify the model goodness of fit penalizing for model complexity to control over-fitting (Zuur et al. 2009). This kind of models apply however to a very simple scenario: a one-time measurement of two continuous variables. But what if things are more complex: say the variables are temporally and spatially misaligned and sensor data have been acquired at different spatial resolution? What if there are repeated measures that need to be jointly analyzed and the dataset involves several thousand observations? We will need a more sophisticated yet very computationally efficient approach. In the next paragraph we will review a case study reported in details in Pollice et al. (2019) in which the nature of the dataset required to address all these issues together.

In Pollice et al. (2019) the objective was to represent the nonlinear relationship between plant production (indirectly estimated by proximal sensing NDVI measurements) and soil information (represented by continuous electrical resistivity measures at three soil depths) through a smooth function. The specific characteristics of the dataset required to address the lack of correspondence of data sampling location and scale between soil and plant sensor data (spatial misalignement), the problem of the repeated measurements in time and along a depth gradient and the issue of the residual variation of unsampled spatial features. As is usually the case, a model-based solution for data integration is set within the Bayesian framework that allows to consider different sources of uncertainty and to rely on Markov chain Monte Carlo (MCMC) simulations to generate observations from the joint posterior distribution (Gelfand 2019). Hierarchical models are generally denoted through the structure: [data|process, parameters] * [process|parameters] * [parameters]. In the notation the brackets denote the probability distributions while the vertical bar symbolizes a conditional specification. The data driven by the underlying natural and anthropic processes is represented through a stochastic model at the top level of the hierarchy. Likewise a stochastic model will be specified for all the other processes under study. Uncertainty in both levels of the modelling will lead to unknowns/parameters. Models for the parameters are deferred to the third stage of the hierarchical specification. This specification which may appear very simple at a first sight is in fact quite rich: it allows multiple data sources, spatial and dynamic structure for both the data and the processes, measurement error in explanatory variables and much more. A structured distributional regression model was implemented to account for the position and the scale heterogeneity of the fodder yield surrogate variable. This model relies on the assumption that response conditional probability distribution belongs to a parametric family and that a regression specification underlies the estimate of each parameter. Specifically an additive composition of potentially nonlinear effects and an additional overall intercept forms the predictor of each model parameter. The nonlinear part of the predictor is composed of different effects such as spatial fields, interaction surfaces but also nonlinear effects of some continuous covariate (e.g. soil information). All the mentioned effects can be approximated through a linear combination of basis functions (Klein et al. 2015). Standard regression model are inadequate to treat cases in which an explanatory variable (our sensor data) has multiple measurements (e.g. repeated measures in time or space) that need to be jointly matched to a single-valued response variable While standard regression theory assumes that explanatory variables are deterministic or error-free in most cases biological processes do not follow this rule. If repeated observations of explanatory variables are available, they can be used to quantify the variation due to measurement error. This latter is actually a problem for inferences based on regression models. Bayesian measurement error correction addresses the issue by including the unknown true covariate values as additional unknowns to be imputed by MCMC simulations accompanying the estimation of all other parameters in the structured distributional regression model. Replicates are considered as contaminated observations of the unknown true covariate. The lack of independence between nearby observations determined by all sources of spatial variability, including erratic and deterministic components, is accounted for including a nonlinear trend surface within the framework of a structured distributional regression model. In this way the functional relationship under analysis (in our case the effect of soil on vegetation) is cleared of any source of spatial variability.

All inferences on the structured additive distributional regression model are based on MCMC simulations performed within the open source software BayesX (Belitz et al. 2015).

Information retrieved from the analysis of yield—response functions, not only facilitate yield pattern interpretation, but can be used as a field zoning criteria itself, as shown in Sect. 4.4.2.1 (Rossi et al. 2018) where a field zonation is proposed, based on the functional relationship between alfalfa NDVI and soil electrical resistivity.

The study illustrated in Sect. 4.4.2 (Rossi et al. 2015, 2018; Pollice et al. 2019) estimates a nonlinear effect of soil (i.e. the surrogate variable soil resistivity) on fodder biomass (i.e. the surrogate variable NDVI), which increases monotonically as resistivity increases, then at certain values of resistivity the shape and sign of the relationship changes with a subsequent decline in NDVI which is steep at first and moderate subsequently. Cut-offs for the smooth function illustrated in Sect. 4.4.2 were defined to classify the field into zones corresponding to a different soil–plant relationship (Rossi et al. 2018). The field zonation illustrated in Sect. 4.4.2 can be considered an informed-clustering that splits the fields into areas of high, low, or season-driven yield responsiveness. Namely, the three zones the relationship was divided in can be considered: zones where fodder biomass is affected by even little changes in soil properties (zone II); zones where soil properties cannot be changed (zone III) or areas when evaluations are needed throughout the crop cycle (zone I). Each part of the smooth function then carries information on the extent and direction of the relations between soil and plant variation. The consequent identification of field zones therefore corresponds to areas where soil constraints are different from the point of view of management. This information is also invaluable as a basis to increase the representativeness and efficiency of ground truth validation destructive sampling which remains necessary for identifying the actual nature of soil constraints in each area. Zones where plant responsiveness to soil variation is high are identified as preferential areas for precision management, because the cause of variation is known and in some cases can be managed (i.e. precision irrigation, precision planting). In such areas a map of soil variation easily translates into a prescription map.

4.5 Conclusions

Geospatial techniques provide an invaluable tool for making decisions in agriculture, and especially for the spatial monitoring of farming impact on the environment and for spatially-aware management methods such as Precision agriculture with its variable rate application of inputs. They are also a basis for stratification, variation and co-variation criteria for the design and interpretation of soil and agronomic surveys and experiments. One of the most fruitful applications is the identification of uniform zones for management and sampling, and the best use of geospatial techniques for agriculture is a combination of data from different sensors. In order to overcome drawbacks of simple data fusion of different information layers, though, advances in data treatment should rely on agronomic interpretation of data, appropriate statistical treatment and management-oriented modelling of soil-vegetation relationships.

References

Alromeed AA, Rossi R, Bitella G, Bochicchio R, Amato M (2015) Irrigation scenarios for artichokes and dry bean as a result of soil variability on the basis of resistivity mapping in south west Italy. Ital J Agron 10(3):151–154

Adhikary PP, Dash AP (2017) Comparison of deterministic and stochastic methods to predict spatial variation of groundwater depth. Applied Water Science 7(1):339–348

Alromeed A, Bitella G, Rossi R, Bochicchio R, Amato M, Perniola M (2019) On-the-go automatic resistivity profiler (ARP©) mapping as a basis for modelling crop precision irrigation strategies. Irrigation Matera 2019. In: IX International symposium on irrigation of horticultural crops book of abstracts, p 149

Amato M, Priori S (2020) Innovative technologies of soil mapping as a tool for precision agriculture. XLIX Convegno SIA 16–18 settembre 2020. XLIX Convegno Nazionale Società Italiana di Agronomia. Consiglio per la ricerca in agricoltura e l'analisi dell'economia agraria Centro di ricerca Agricoltura e Ambiente, sede di Bari. In: Dalla Marta A, Ventrella D (eds) Proceedings of the 49th national conference of the Italian Society for Agronomy "Sustainable management of croppingsystems", Bari, Italy, 16th–18th September 2020. SIA, pp 32–33

Amato M, Basso B, Celano G, Bitella G, Morelli G, Rossi R (2008) In situ detection of tree root distribution and biomass by multielectrode resistivity imaging. Tree Physiol 28(10):1441–1448

Amato M, Bitella G, Rossi R, Gómez JA, Lovelli S, Ferreira Gomes JJ (2009) Multi-electrode 3-D resistivity imaging of alfalfa root zone. Eur J Agron 31:216–222

Amato M, Lapenna V, Rossi R, Bitella G (2012) Chapter 11—Multi electrode resistivity imaging. In: Mancuso S (ed) Measuring roots—an updated approach. Springer, Berlin-Heidelberg-New York, pp 189–212

Babak O, Deutsch CV (2009) Statistical approach to inverse distance interpolation. Stoch Env Res Risk Assess 23(5):543–553

Basso B, Amato M, Bitella G, Rossi R, Kravchenko A, Sartori L, Carvahlo LM, Gomes JA (2010) Two-dimensional spatial and temporal variation of soil physical properties in tillage systems using electrical resistivity tomography. Agron J 102(2):440–449

Belitz C, Brezger A, Kneib T, Lang S, Umlauf N (2015) BayesX: software for Bayesian inference in structured additive regression models Version 3:2. http://wwwBayesXorg

Bitella G, Rossi R, Loperte A, Satriani A, Lapenna V, Perniola M, Amato M (2015) Geophysical techniques for plant, soil, and root research related to sustainability. In: Vastola A (ed) The sustainability of agro-food and natural resource systems in the Mediterranean Basin 2015. Springer, Berlin and Heidelberg, pp 353–372

Boer EP, de Beurs KM, Hartkamp AD (2001) Kriging and thin plate splines for mapping climate variables. Int J Appl Earth Obs Geoinf 3(2):146–154

Casa R, Castaldi F, Pascucci S, Alombo A, Pignatti S (2013) A comparison of sensor resolution and calibration strategies for soil texture estimation from hyperspectral remote sensing. Geoderma 197–198:17–26

Casa R, Basso B, Morari F (2017) Agricoltura di precision. In: Ceccon P (ed) Agronomia. Edises, pp 453–466

Casa R, Castaldi F, Pascucci S, Pignatti S (2019) Potential of hyperspectral remote sensing for field scale soil mapping and precision agriculture applications. Remote Sens 11(3):309

Castrignanò A, Giugliarini L, Risaliti R, Martinelli N (2020) Study of spatial relationships among some soil physico-chemical properties of a field in central Italy using multivariate geostatistics. Geoderma 97(1–2):39–60

Córdoba MA, Bruno CI, Costa JL, Peralta NR, Balzarini MG (2016) Protocol for multivariate homogeneous zone delineation in precision agriculture. Biosys Eng 143:95–107
Corwin DL, Kaffka SR, Hopmans JW, Mori Y, Lesch SM, Oster JD (2003) Assessment and field-scale mapping of soil quality properties of a saline-sodic soil. Geoderma 114(3–4):231–259
Diodato N, Ceccarelli M (2005) Interpolation processes using multivariate geostatistics for mapping of climatological precipitation mean in the Sannio Mountains (southern Italy). Earth Surf Process Landf J Br Geomorphol Res Group 30(3):259–268
FAO (2009) Strategic framework 2009–2019 conference report Rome, 18–23 November 2009, 36 pp. http://www.fao.org/3/k5864e01/k5864e01.pdf
Feng L, Wang Y, Zhang Z, Du Q (2021) Geographically and temporally weighted neural network for winter wheat yield prediction. Remote Sens Environ 262:112514
Fotheringham AS, Brunsdon C, Charlton M (2003) Geographically weighted regression: the analysis of spatially varying relationships. Wiley
Fridgen JJ, Kitchen NR, Sudduth KA, Drummond ST, Wiebold WJ, Fraisse CW (2004) Management zone analyst (MZA) software for subfield management zone delineation. Agron J 96(1):100–108
Galieni A, D'Ascenzo N, Stagnari F, Pagnani G, Xie Q, Pisante M (2021) Past and future of plant stress detection: an overview from remote sensing to positron emission tomography. Front Plant Sci 11:609155
Gelfand AE (2019) Modeling for environmental and ecological processes In: Alan Gelfand AE, Fuentes M, Hoeting JA, Smith RL (eds) Handbook of environmental and ecological statistics. CRC Press
Gelfand AE, Diggle P, Guttorp P, Fuentes M (2010) Handbook of spatial statistics. Chapman & Hall
Gong G, Mattevada S, O'Bryant SE (2014) Comparison of the accuracy of kriging and IDW interpolations in estimating groundwater arsenic concentrations in Texas. Environ Res 130:59–69
Gotway CA, Ferguson RB, Hergert GW, Peterson TA (1996) Comparison of kriging and inverse-distance methods for mapping soil parameters. Soil Sci Soc Am J 60(4):1237–1247
Haghverdi A, Leib BG, Washington-Allen RA, Ayers PD, Buschermohle MJ (2015) Perspectives on delineating management zones for variable rate irrigation. Comput Electron Agric 117:154–167
Jones CA, Kinry JR (1986) CERES-maize. A simulation model of maize growth and development. Texas University Press, p 198
Kharel TP, Ashworth AJ, Shew A, Popp MP, Owens PR (2020) Tractor guidance improves production efficiency by reducing overlaps and gaps. Agric Environ Lett 5:e20012
King JA, Lark RM, Wheeler HC, Park W, Bradley RI, Mayr TR (2005) Mapping potential crop management zones within fields: use of yield-map series and patterns of soil physical properties identified by electromagnetic induction sensing. Precis Agric 6:167–181
Klein N, Kneib T, Klasen S, Lang S (2015) Bayesian structured additive distributional regression for multivariate responses. J R Stat Soc C Appl Stat 64(4):569–591
Kravchenko AN (2003) Influence of spatial structure on accuracy of interpolation methods. Soil Sci Soc Am J 67(5):1564–1571
Kravchenko A, Bullock DG (1999) A comparative study of interpolation methods for mapping soil properties. Agron J 91(3):393–400
Machado SE, Bynum D, Archer TL, Bordovsky J, Rosenow DT, Peeterson C, Bronson K, Nesmith DM, Lascano RJ, Wilson LT, Segarra E (2002) Spatial and temporal variability of sorghum grain yield: influence of soil, water, pests and diseases relationships. Precis Agric 3:389–406
Marty P, Darras L, Tabbagh J, Benech C, Simon FX, Thiesson J (2015) WuMapPy-an open-source software for geophysical prospection data processing. Archaeologia Polona 53:563–566
Mavridou E, Vrochidou E, Papakostas GA, Pachidis T, Kaburlasos VG (2019) Machine vision systems in precision agriculture for crop farming. J Imaging 5(12):89
McBratney A, Whelan B, Ancev T (2005) Future directions of precision agriculture. Precis Agric 6:7–23
McKinley S, Levine M (1998) Cubic spline interpolation. Coll Redw 45(1):1049–1060

Moral FJ, Terrón JM, Da Silva JM (2010) Delineation of management zones using mobile measurements of soil apparent electrical conductivity and multivariate geostatistical techniques. Soil Tillage Res 106(2):335–343

Mouazen AM, Maleki MR, De Baerdemaeker J, Ramon H (2007) On-line measurement of some selected soil properties using a VIS–NIR sensor. Soil Tillage Res 93:13–27

Mueller TG, Pierce FJ, Schabenberger O, Warncke DD (2001) Map quality for site-specific fertility management. Soil Sci Soc Am J 65(5):1547–1558

Mueller TG, Pusuluri NB Mathias, KK Cornelius, Barnhisel RI, Shearer SA (2004) Map quality for ordinary kriging and inverse distance weighted interpolation. Soil Sci Soc Am J 68(6):2042–2047

Nawar S, Corstanje R, Halcro G, Mulla D, Mouazen AM (2017) Chapter four: delineation of soil management zones for variable-rate fertilization: a review. Adv Agron 143:175–245

Oliver YM, Robertson MJ, Wong MTF (2010) Integrating farmer knowledge, precision agriculture tools, and crop simulation modelling to evaluate management options for poor-performing patches in cropping fields. Eur J Agron 32(2010):40–50

Paccioretti P, Córdoba M, Balzarini M (2020) Fast mapping: software to create field maps and identify management zones in precision agriculture. Comput Electron Agric 175:105556

Peña-Barragán JM, López-Granados F, Jurado-Expósito M, García-Torres L (2007) Mapping Ridolfia segetum patches in sunflower crop using remote sensing. Weed Res 47(2):164–172

Pollice A, Jona Lasinio G, Rossi R, Amato M, Kneib T, Lang S (2019) Bayesian measurement error correction in structured additive distributional regression with an application to the analysis of sensor data on soil–plant variability. Stoch Env Res Risk Assess 33:747–763

Pretty J, Bharucha ZP (2014) Sustainable intensification in agricultural systems. Ann Bot 114(8):1571–1596

Priori S, Fantappié M, Bianconi N, Ferrigno G, Pellegrini S, Costantini EAC (2016) Field-scale mapping of soil carbon stock with limited sampling by coupling gamma-ray and Vis-NIR spectroscopy. Soil Sci Soc Am J 80(4):954–964

Ritchie JT, Amato M (1990) Field evaluation of plant extractable soil water for irrigation scheduling. Acta Hort 278:595–615

Robinson TP, Metternicht G (2006) Testing the performance of spatial interpolation techniques for mapping soil properties. Comput Electron Agric 50(2):97–108

Romero Ruiz A, Linde N, Keller T, Or D (2018) A review of geophysical methods for soil structure. Rev Geophys 56(4):672–697

Rossi R, Amato M, Bitella G, Bochicchio R, Ferreira Gomes JJ, Lovelli S, Martorella E, Favale P (2010) Electrical resistivity tomography as a non-destructive method for mapping root biomass in an orchard. Eur J Soil Sci 62(2):206–215

Rossi R, Pollice A, Diago MP, Oliveira M, Millan B, Bitella G, Amato M, Tardaguila J (2013) Using an automatic resistivity profiler soil sensor on-the-go in precision viticulture. Sensors 13(1):1121–1136

Rossi R, Pollice A, Bitella G, Bochicchio R, D'Antonio A, Alromeed AA, Stellacci AM, Labella R, Amato M (2015) Soil bulk electrical resistivity and forage ground cover: nonlinear models in an alfalfa (Medicago sativa L.) case study. Ital J Agron 10(4):215–219

Rossi R, Pollice A, Bitella G, Labella R, Bochicchio R, Amato M (2018) Modelling the non-linear relationship between soil resistivity and alfalfa NDVI: a basis for management zone delineation. J Appl Geophys 159:146–156

Royal Society (2009) Reaping the benefits: science and the sustainable intensification of global agriculture. The Royal Society, London

Sadler EJ, Bauer PJ, Busscher WJ, Millen JA (2000) Site-specific analysis of a drought corn crop: II water use and stress. Agron J 92(3):403–410

Samouelian A, Cousin I, Tabbagh A, Bruand A, Richard G (2005) Electrical resistivity survey in soil science: a review. Soil Tillage Res 83(2):173–193

Saxton KE, Rawls WJ (2006) Soil water characteristic estimates by texture and organic matter for hydrologic solutions. Soil Sci Soc Am J 70:1569–1578

Selige T, Boehner J, Schmidhalter U (2006) High resolution topsoil mapping using hyperspectral image and field data in multivariate regression modeling procedures. Geoderma 136(1–2):235–244

Shatar TM, McBratney AB (2004) Boundary-line analysis of field-scale yield response to soil properties. J Agric Sci 142(5):553–560

Shepard D (1968) A two-dimensional interpolation function for irregularly-spaced data. In: Proceedings of the 1968 23rd ACM national conference, pp 517–524

Simbahan G, Dobermann A, Ping JL (2004) Screening yield monitor data improves grain yield maps. Agron J 96(4):1091–1102

Souza EG, Bazzi CL, Khosla R, Uribe-Opazo MA, Reich RM (2016) Interpolation type and data computation of crop yield maps is important for precision crop production. J Plant Nutr 39(4):531–538

Struik PC, Kuyper TW (2017) Sustainable intensification in agriculture: the richer shade of green. A review, Agron Sustain Dev 37, 39

Tabbagh J (1988) Traitement des données et élimination des valeurs erronées en prospection électrique en continu. Archéo Sciences Revue D'archéométrie 12(1):1–9

Teixeira JL, Pereira LS (1992) ISAREG, an irrigation scheduling model. ICID Bull 41(2):29–48

United Nations General Assembly (2015) Transforming our world: the 2030 agenda for sustainable development. Resolution adopted by the General Assembly on 25 September 2015 n. 70.1 15-16301 (E) Pp 35. https://undocs.org/A/RES/70/1

Usowicz B, Lipiec J (2017) Spatial variability of soil properties and cereal yield in a cultivated field on sandy soil. Soil Tillage Res 174:241–250

Vrindts E, Mouazen AM, Reyniers M, Maertens K, Maleki MR, Ramon H, Baerdemaeker J (2005) Management zones based on correlation between soil compaction, yield and crop data. Biosyst. Eng., 92: 419–428.

Webster R, Oliver MA (1992) Sample adequately to estimate variograms of soil properties. J Soil Sci 43(1):177–192

Webster R, Oliver MA (2007) Geostatistics for environmental scientists statistics in practice. Wiley, The Atrium

Weber DD, Englund EJ (1994) Evaluation and comparison of spatial interpolators II. Math Geol 26(5):589–603

Wood SN (2006) Generalized additive models: an introduction with R. CRC–Chapman and Hall, Boca Raton

Yang SH, LiuF SXD, Lu YY, Li DC, Zhao YG, Zhang GL (2019) Mapping topsoil electrical conductivity by a mixed geographically weighted regression kriging: a case study in the Heihe River Basin northwest China. Ecol Indicators 102:252–264

Zuur AF, Ieno EN, Walker NJ, Saveliev AA, Smith GM (2009) Mixed effects models and extensions in ecology with R New York: Springer

Chapter 5
Soil and Vegetation in Pachmarhi Biosphere Reserve and Their Correlation

Saikat Banerjee, P. K. Khatri, and S. K. Banerjee

Abstract Pachmarhi Biosphere Reserve (PBR) comprises three protected sites: Bori Sanctuary, Satpura National Park and Pachmarhi Sanctuary. Forests represent approximately 63% of the area of the PBR. The natural forest of the area is a natural ecotone of two important timber species i.e. Teak (*Tectona grandis*) and Sal (*Shorea robusta*). However, bulk of the area carries mixed forest where neither Teak nor Sal predominates. The vegetation diversity in the area is basically dependent on the climate and topography, particularly varying degrees of slope, elevation and aspect influencing the macro and micro vegetation by creating local climate and edaphic conditions. Geology of the area obviously envisages the relationship between soil type and diverse vegetation. In the present review we have presented the characteristics of the soils and their impact on vegetation and vice-versa.

Keywords Biosphere reserve · Ecology · Land use pattern · Soil · Vegetation · Pachmarhi

5.1 Introduction

Biosphere Reserve is an area of terrestrial or coastal ecosystem promoting solutions to reconcile the conservation of biodiversity with its sustainable use. At the initiative of the United Nations Educational, Scientific and Cultural Organization (UNESCO) Man and Biosphere Programme to establish an international network of protected areas, the Indian Government has established 18 biosphere reserves (BR) all over India to protect larger areas of natural habitat than a typical natural park or animal sanctuary, and that include one or more national parks or animal sanctuaries or reserves, along with buffer zones that are open to some economic uses. Protection

S. Banerjee · S. K. Banerjee (✉)
Ecology & Climate Change Division, TFRI, Jabalpur, India
e-mail: banerjeeskdr@gmail.com

P. K. Khatri
TFRI, Jabalpur, India

P. K. Shit et al. (eds.), *Soil Health and Environmental Sustainability*, Environmental Science and Engineering,
https://doi.org/10.1007/978-3-031-09270-1_5

Fig. 5.1 Pachmarhi biosphere reserve

is granted not only to the flora and fauna of the protected region but also to the human communities who inhabit these regions and their ways of life. All the BRs are internationally recognized, nominated by national governments and remain under sovereign jurisdiction of the states where they are located. Biosphere reserves contain virgin vegetation plus various kinds of cultural landscape, in the whole of which conservation is practiced (Rai 2000).

The Pachmarhi Biosphere Reserve (PBR) is located in the bio-geographical region of the Decan Peninsula and the biotic province of Central India (22.4685 °N and 78.4412 °E) and was established on 3rd March 1999. It ranks 11th of the country. UNESCO designated it as biosphere reserve in 2009. The Satpura mountain ranges cross India from west to east and Pachmarhi lies in its centre. The highest peak is the Dhoopgarh which reaches 1,362 m above sea level while the Pachmarhi hills are characterized by steep slope in the northern region. The boundary of the biosphere reserve lies along a road with cultivation farms close to Dudhi river while the southern boundary borders the Tawa plateau (Mehta 2018). Geographically, PBR falls in three civil districts of Hoshangabad, Betul and Chhindwara in Madhya Pradesh (Fig. 5.1).

The climate of Pachmarhi is mild, generally warm and temperate, with an average annual temperature of 21.7 °C. The summers here have much more rainfall than winter. The rainfall averages at 2012 mm. In summer, the average temperature is 30.3 °C and May is the hottest month and in winter December is the coldest month and the average temperature is 15.5 °C.

PBR comprises three protected sites: the Bori sanctuary, Satpura National Park and Pachmarhi sanctuary—otherwise known as Satpura Tiger Reserve. The PBR is also known as "Queen of Satpuras" because it contains valleys, marshes, streams and waterfalls, all of which have led to the development of a unique and varied biodiversity (Pandey et al. 1993).

Forests represent approximately 63% of the area of the biosphere reserve, while agricultural lands (30%), wastelands (2.46%), water bodies (5%) and human settlement areas (0.54%) account for the remainder.

Table 5.1 Position of villages in buffer zone of PBR

Sr. No	Name of division	Total No. of villages	Revenue villages	Forest villages
1	Hoshangabad	304	274	30
2	West Chhindwara	158	158	0
3	North Betul	49	44	5

A biosphere reserve usually comprises of three zones—the core zone, the buffer zone and the transition zone. At the core area of PBR is the Satpura National Park, covering approximately 524 km^2. The buffer and the transitions zone are comprised of Bori Wildlife Sanctuary covering 518 km^2 area and the Pachmarhi Sanctuary covering 461 km^2.

The core zone contains suitable habitat for numerous plant and animal species, including higher order predators and may contain centres of endemism. In the buffer zone, which adjoins or surrounds core zone, uses and activities are managed in ways that protect the core zone. These uses and activities include restoration, demonstration sites for enhancing value addition to the resources, limited recreation, tourism, fishing, grazing, etc. which are permitted to reduce its effect on core zone. In buffer zone, sustainable use of natural resources by local inhabitants is allowed and need-based sustainable management interventions are made by the managers. The transition area is the outermost part of a biosphere reserve. This is usually not delimited one and is a zone of cooperation where conservation knowledge and management skills are applied.

In the buffer zone, there are 511 villages with a population of 2,17,820. Most of the villages are revenue villages (Table 5.1).

According to the guidelines for Biosphere Reserves, the buffer zone of PBR is considered as a manipulative or utilizable zone and emphasis is given on the sustainable use of natural resources by the local inhabitants, considering them as an integral part of the ecosystem.

Land use pattern

About 60.3% of the total area under PBR falls in Hoshangabad district followed by that in Chhindwara (28.2%) and Betul (11.5%) districts. Among the total forest cover, 85.3% area is categorized as dense forest. The remaining 8.2%, 4.2% and 2.3% areas are categorized as open, degraded and blank forests respectively. As per the landsat imageries of April 1990 and November 1991–January 1992, the Tawa reservoir and Denwa river are the main water bodies in the area.

Configuration

The altitude varies from 320 to 1352 m at different localities in the area. In Hoshangabad district, the lowest elevation is 352 m. However, the general elevation in Betul district varies from 380 to 1005 m and that in Chhindwara district from 380 to 1211 m.

The general configuration of the country is hilly, undulating terrain and at places precipitous having deep narrow gorges around Mahadeo hill in Pachmarhi plateau (Khatri et al. 2004).

Geology

The Satpura hills consist of a series of parallel ridges running east to west between the rivers Narmada and Tapti. In the western portion, they are composed of the Deccan traps further in east they comprise, in succession, the Mahadeo or Pachmarhi hills of Gondwana formation. The rocks of Mahadeo series are exposed in Pachmarhi. This series is named after Mahadeo hills on which a celebrated Mahadeo shrine is situated near Pachmarhi. The rocks of Pachmarhi stage form the magnificent scrap above which the town of Pachmarhi is situated having a huge lenticular mass of sandstone between the Denwa and Bijori beds, and consists of red and buff sandstones with some red clay near the base and top. There are layers of haematitic clay and platy veins of hard dense ferruginous matter which, on weathering, resemble broken pieces of pottery (Pandey et al. 1993).

The climate of the Pachmarhi Biosphere Reserve (PBR) is typically monsoon type, with three distinct seasons. The Tawa Reservoir is the major constituent of the reserve's water bodies. The area is large enough to be effective as a conservation unit. The PBR contains 4 of the 21 preservation plots identified in the state (M.P) that cover various representative forest types. The reserve also contains 3 of the 26 endemic centres, identified by the Government of India all over the country.

There are two things which make PBR unique: (i) occurrence of Sal in the Teak predominant area and (ii) this area is the upper limit for growth of Sal as well as western limit of growth of Sal. In this area lots of studies have been carried out by a number of scientists and foresters on the distribution of flora and fauna, distribution of medicinal plants and their uses, wildlife and habitat etc. but the characteristics of the soils and their impact on vegetation and vice-versa have not been well documented. In the present review we have presented the same.

Vegetation spectrum

The state of Madhya Pradesh, as a geographical heart of the Indian sub-continent is bestowed with plenty of natural heritage. The natural forest of the area is a natural junction (ecotone) of two most important timber species i.e. Teak (*Tectona grandis*) and Sal (*Shorea robusta*). The dividing line which segregates the isolated patches of Pachmarhi Sal from the Bori teak, the hills of Betul and part of Chhindwara forest, runs from Binora on Denwa river in south to Rorighat, Kajri and Nagdwari river then going up to Neemghan and finally touching the Denwa river in the north. It then reaches Pagara and Kanjighat villages and again joins the Denwa river.

The forests of Madhya Pradesh constitute is a union of biological diversity in the form of diverse forest types. Teak being hardy and vigorous coppicer has gradually gained predominance in areas where the underlying rock is trap and the soil is well drained. Similarly sal, which is characteristically mesophyllous species, has colonized the better rainfall areas in the east of the state. However, bulk of the area carries mixed forest, where neither teak nor sal predominates. This is more akin to the original mixed forests types, somewhat modified due to biotic factors (Buch 1991). In the north, where conditions are xerophytic, it is mostly comprised of thorn forests and the growth is very stunted. Bamboos as undergrowth are found in dry and moist forests of the greater parts of the state.

Pachmarhi Biosphere Reserve is a true natural junction of representative forest types prevailing in the state. The vegetation associated in the forest is representative of Sub Himalayan, Central Hill Forests, and South Indian Dry and Moist Deciduous Forests ecosystem.

. On the basis of geology, topography, climatic conditions and dominating tree vegetation, the entire forests of the Biosphere Reserve have broadly been categorized as: 1. Tropical Moist Deciduous, 2. Tropical Dry Deciduous and 3. Central Indian Sub-Tropical Hill forests. These three major types are further divided into sub types (Champion and Sett 1968). The vegetation diversity of the area is basically dependent on the climate, soil and topographical features particularly varying degree of elevation, slopes and aspects, which influence the macro as well as the micro vegetation by creating local climatic and edaphic conditions. Geology of the area obviously envisaged the relationship between soil types and diverse vegetation spectrum of the study area (Banerjee and Jain 2011).

Central part, including western and eastern boundaries of the biosphere reserve is occupied by rich vegetation representing all the three (Teak, Sal and Mixed) forest types. South–east, south–west and northern boundaries of the area are subjected to biotic influence by occupation of habitations of aboriginal tribes e.g. Gond, Korkoo and Bharia. However, in southernmost boundaries, there are few dense patches of moist mixed deciduous forests with some pockets having teak. With the changing micro climatic conditions and topographical features, association of plant species including ground flora obviously indicates the marked variation in different plant communities.

In the miscellaneous forests, there are few specific pockets, which are dominated by a particular tree species in almost pure form which may also be highlighted as diverse forest patches of the area. These specific pockets are (1) *Hardwickia* dominating, (2) *Gardinia* dominating and (3) *Chloroxylon* dominating forests. At Pachmarhi plateau there is an isolated patch of high hill Sal forests which also represent the western limit of Sal forest. This is localized in a small area having stunted growth of vegetation.

The reserve also supports more xerophytic vegetation that closely resembles the tropical dry deciduous forests. The rich plant diversity and gene pool of the reserve area include 71 species of thallophytes; 83 species of bryophytes belonging to 34 families and 56 genera; 71 species of pteridophytes belonging to 16 families; 7 species of gymnosperms belonging to 3 families; and also 1190 species of angiosperms (flowering plants) belonging to 127 families and 633 genera. The flora of the reserve are distributed in 180 families, out of which 54 are represented by just one genus each, and 29 are represented by two or three genera each. Many of the thallophytes, bryophytes and pteridophytes can be seen exclusively in this reserve because of the special topographical and climatic features of the locality (Pandey et al. 1993).

Among the 71 species of non flowering plants found in the reserve, 48 species belong to ferns and the rest belong to fern allies. Some of the most notable species of ferns found here include *Psilotum triquetrum, Isoetes panchanail, Selaginella exigua*and *Ophioglossum nudicaule.* The tree ferns like *Cyathea gigantea* and *C. spinulosa* are also found in the reserve. A few clumps of rare and endemic species like *Melastoma melabarthicum, Murraya paniculata, Holmskioldia sanguinea, Blumea lanceolaria* and *Sophora interrupta* can be found exclusively in the moist teak forest of this reserve.

Soil

Khatri (2000) studied the morphological, physical and chemical characteristics of soils under three plant communities viz. *Chloroxylon–Terminalia, Shorea–Terminalia* and *Syzygium–Terminalia* corresponding to <900 m, 900–1100 m and 1100–1350 m elevation respectively in the forest of Pachmarhi Biosphere Reserve. The profile characteristics of each representative site are given below.

The profile 1 was under plant community *Chloroxylon-Terminalia* at the elevation of 900 m. Physiography is rolling to undulating at lower hills of the area with a slope of 1–3%. Land surface shows scattered stoniness. Exposed rocks are common feature. Erosion is moderate with moderately to well drain soils. This soil is deep to moderately deep and is developed on sandstone.

Texture of the soil is clay loam in the surface to sandy clay loam down below. Silt suddenly decreases with depth. Soil does not contain much gravel. It increases with depth from 5.56 to 28.0%. Soil reaction varies from slightly acidic to medium acidic and electrical conductivity from 0.06 to 0.01 m mhos/cm. Organic matter per cent is maximum (6.28%) in upper layer and decreases to 0.41% after 25 cm depth. Cation exchange capacity of this profile ranges from 30.0 to 36.2 me%. Exchangeable Mg increases steadily with depth. In case of exchangeable Ca, however, the increasing trend is not very regular. Exchangeable K increases with depth with more accumulation in 40–58 cm layer. Per cent base saturation ranges from 59.47 to 95.13 in this soil profile. The soil profile has Ochric epipedon with cambic horizon down below. It is, therefore, put under Order Inceptisol and Sub-order Ochrept. Since the profile enjoys ustic moisture regime, it is included in the Great Group Ustochrept and Sub-group Udic Ustochrept. Thus, these soils belong to the member of fine loamy mixed, hyperthermic family of Udic Ustochrept (Soil Survey Staff 1975) (Table 5.2).

Table 5.2 Physicochemical characteristics of the soil at an elevation of 900 m

Depth (cm)	pH	Org. C (%)	CEC (me%)	Exch. Ca^{2+} (me%)	Exch. Mg^{2+} (me%)	Clay (%)
0–10	6.2	4.0	37.3	15.5	8.4	40.7
10–25	6.3	3.1	33.0	13.3	6.3	41.2
25–40	6.7	2.0	19.6	10.8	5.8	40.2
40–58	6.8	0.8	11.7	6.7	4.0	39.2
58–68	6.8	0.6	6.8	5.6	4.2	39.7

Table 5.3 Physicochemical characteristics of the soil at an elevation of 1050 m

Depth (cm)	pH	Org. C (%)	CEC (me%)	Exch. Ca^{2+} (me%)	Exch. Mg^{2+} (me%)	Clay (%)
0–8	6.1	3.9	30.3	14.2	7.6	46.6
8–24	5.9	2.6	22.5	9.7	4.1	48.2
24–40	6.8	1.2	16.4	7.5	3.1	44.6
40–58	6.7	0.8	8.6	3.3	0.9	47.6

The soil profile 2 was under the plant community *Shorea–Terminalia* which occurs at the elevation of 1050 m. The site is located at mid hill position on north–east aspect. The area is covered with lot of stones and rock out crops are scattered throughout. Soil is deep to moderately deep.

Top horizon of this profile is gravelly sandy clay loam, followed by gravelly clay loam. Further lower horizons have texture of similar nature where clay varies from 30 to 37%. Silt percent is minimum of 4% in the lowest horizon.

Top layer consists of 66% gravels which decrease to 34% in the last layer. Soil reaction is slightly acidic to neutral. Cation exchange capacity of this profile is 24.6 to 49.6 me% with dominance of exchangeable Ca and Mg which vary from 10.8 to 24.9 me% and from 9.0 to 24.0 me%, respectively. Per cent base saturation is more than 80% in this profile.

This soil profile under *Shorea–Terminalia* community has Ochric epipedon because the surface horizon is not darker than the lowermost horizon in either colour value or chroma by one or two units, respectively, both in moist and dry conditions. This is followed by cambic horizon. Hence, this soil profile belongs to Order Inceptisol and Sub-order Ochrept. Since there exists ustic moisture regime, the soil is put under Great Group Ustochrept and Sub-group Udic Ustochrept. These soils are the member of loamy skeletal, mixed hyperthermic family of Udic Ustochrept (Table 5.3).

The soil profile 3 was at an elevation of 1280 m under *Syzygium-Terminalia* community which lies on the elevation range of 1100–1350 m. Since this profile lies at the hill top position, rock outcrops is the common feature. The site is hilly and undulating. Slope at this site is 3–6% facing to north-east aspect. Soils are well drained with moderate to severe erosion. Colour of the soil in general varies from dark yellowish brown to dark reddish brown. Soil is shallow to moderately deep with paralithic contact occurring after 24 cm. The morphological features of the soil profile are given below.

Top layer of the profile is sandy clay loam, followed by loam in the second horizon. Below this, there occurs a horizon with weathered material which consists of plenty of gravels. Organic matter is 6.26% in the surface horizon which reduces to 4.23% in the second horizon. Available N, P and K are rated medium in range. Per cent base saturation in the surface soil is 96% and CEC 38.8 me%. Exchangeable Ca, Mg and K in the surface layer are found to be 22.8 me%, 14.0 me% and 0.38 me%, respectively.

Table 5.4 Physicochemical characteristics of the soil at an elevation of 1280 m

Depth (cm)	pH	Org. C (%)	CEC (me%)	Exch. Ca^{2+} (me%)	Exch. Mg^{2+} (me%)	Clay (%)
0–8	5.7	3.2	28.6	13.7	6.2	44.8
8–24	5.6	2.1	21.7	10.6	4.7	35.8
24–40	5.5	1.4	12.3	5.0	2.7	26.2

This soil profile is put under Order Entisol as there is no distinct diagnostic horizon and is placed into Sub-order Ustorthent. Soil is further put into the sub-group Lithic Ustorthent as there is a paralithic contact within 50 cm soil depth i.e. 24 cm. The soils are the member of loamy skeletal mixed hyperthermic family of Lithic Ustorthent (Soil Survey Staff 1975) (Table 5.4).

The morphological characteristics of soils of the three elevations reveal that they have, in general, hues of 5YR, chroma 3–4 and values of 2–4.

Site and Soil Characters Related to Tree Growth

In modern ecosystem approach of ecology, environment is divided into biotic and abiotic components. The edaphic, climatic and biotic factors which constitute the environment, affect plants, their population and community growth and dynamics in a holistic manner. In order to understand the mechanism of environmental influences, each component/factor of environment needs to be studied separately. Soil, as an ecological factor, is of great significance for it affords a medium for anchorage of plants. It is a store house of minerals and water which is very much influenced by parent material, topography, drainage; land form, organic matter content, etc. Thus, physical properties of soil monitor the growth of plants in association with other ecological processes occurring in forest ecosystem.

Topography influences soil development in many ways since drainage pattern of soil water is affected by it. This brings a lot of difference in soil quality in terms of its morphological, physical and chemical characteristics. Soil at 1100–1350 m elevation (profile No. 3) is under *Syzygium–Terminalia* community. Champion and Sett (1968) put this forest into 8A/C3, Central Indian Sub-tropical Hill Forest type. Due to hilly tract, surface soil is subject to acute soil water runoff activities and nutrients are washed down to lower elevation resulting into shallow soil depth where A horizon is 7–8 cm deep placed directly on C horizon. Soil is acidic in nature (pH 5.4–5.7) with low insoluble salts and also poor in available nutrients. Rocky land surface and bouldery soil matrix influence vegetation growth in terms of its height and girth. Only hardy species like *Syzygium cumini, Terminalia tomentosa, Terminalia chebula, Mangifera indica, Buchanania lanzan, Madhuca indica, Anogeissus latifolia, Cassia fistula, Mimusops hexandra, Bahinia retusa, Saccopetalum tomentosum,* etc. can survive with stunted growth on different terrain existing in 1100–1350 m elevation of the area. Their density, frequency of occurrence and IVI vary from 15 to 80/ha, 30 to 90% and 8.61 to 29.24 respectively indicating that they are most common ones to flourish even in these difficult edaphic conditions having the following site and soil characteristics.

At an altitudinal range of 900–1100 m, the community identified is *Shorea-Terminalia.* Champion and Sett (1968) put this forest type in 5 B/C1(IV), Northern Tropical Dry Peninsular Low Level Sal. Activity of settling of transported and run off material from top hilly areas is more pronounced in this zone due to which soils here are moderately deep to deep (70 cm) with thickness of A horizon varying from 24–25 cm overlying on B horizon. Soil belonging to Udic Ustochrept/Udic Haplustalf contains more gravels varying from 60 to 66% in upper horizon to 34–48% in lower layers. Soil is slightly acidic to near neutral in reaction (pH 6.2–6.8). Organic carbon at the surface ranges from 4.4 to 2.2% decreasing down the profile to 3.3–0.6%. Available nutrients like N and K are in the medium range, while P is in the low range. Species like *Shorea robusta, Terminalia tomentosa, Buchanania lanzan, Diospyros melanoxylon, Syzygium cumini, Lagerstroemia parviflora, Chloroxylon swietenia, Anogeissus latifolia, Caseaeria graveolens, Bauhinia variegata, Madhuca indica, Phyllanthus emblica* are some of the dominant tree species whose density, frequency of occurrence and IVI vary from 17.5 to 305.0/ha, 30–100% and 7.19–88.5, respectively. Since these species are showing good growth even in the hostile edaphic conditions occurring at 900–1100 m elevation of the Biosphere Reserve, it may be concluded that the following soil characteristics are in tune with the growth of these tree species existing at this elevation.

Shorea robusta is found to grow luxuriantly on soils derived from geological formation consisting of sand stone, granite or gneiss where soils are acidic in reaction and contain less calcium and magnesium (Bhatnagar 1961, Totey et al. 1986). However, in PBR the soils contain more calcium and magnesium and pH of soils is mild acidic to neutral. Growth of sal in this tract is, therefore, stunted. Totey et al. (1986) observed that soils containing more exchangeable calcium and magnesium and more cation exchange capacity having neutral pH support stunted growth of sal.

At the elevation of 900–1100 m, the second dominant species is *Terminalia tomentosa* which has attained somewhat better growth mainly due to higher depth of soil (Khatri et al. 2004). Its height is about 18 m in this soil as against shallow soil ats elevation of 1100–1350 m where the height of *Terminalia* varies from 11 to 12 m. Another important species which growsat this elevation fairly well is *Buchanania lanzan* whose height varies from 8 to 12 m and girth 39.0–51.6 cm. Neutral pH and high degree of soil fertility combined with low electrical conductivity have been found to be responsible for its better growth (Luna 1996). *Buchanania lanzan* normally prefers undulating hills than the plains.

At lower elevation <900 m, there occurs a community recognized as *Chloroxylon-Terminalia.* Champion and Sett (1968) put this community in the forest type 5 A/C3, Dry Mixed Deciduous Forest (Mixed Forest). Since this community is at the lowest elevation as compared to that of earlier two, it is the main site for accumulation of colluvial material brought by the gravitational force from the top elevation. Intense activity of factors of weathering and soil formation gives rise to deeper soils with thick surface horizon. This soil belonging to sub-group Udic Ustochrept is moderately deep to deep (68–75 cm) with 25–27 cm thick surface horizon. Soil reaction is acidic to medium acidic and organic matter raging from 2.2 to 6.3% in upper layer and 0.4–1.4 in lower horizons. Available nitrogen, phosphorus and potassium are rated to be in the medium range. In this hilly zone too, there occur lot of big stones and also rock outcrops. Tree growth at this site, in general, is stunted like that at the earlier two sites. Even in such hostile soil conditions species like *Chloroxylon swietenia, Terminalia tomentosa, Buchanania lanzan, Terminalia chebula, Madhuca indica, Diospyros melanoxylon* are found growing luxuriantly. Their density, frequency and IVI vary from 42.5 to 175/ha, 50–90% and 9.95–42.1 respectively. *Shorea robusta* and also *Syzygium cumini* are found in this mixed forest and are members of *Chloroxylon-Terminalia* community. Other dominant trees at this elevation which are fairly high in the density, frequency and also IVI are *Lagerstroemia parviflora, Mimusops hexandra, Terminalia bellerica, Saccopetalum tomentosum, Anogeissus latifolia, Phyllanthus emblica, Cassia fistula* and *Anogeissus pendula.* Existence of these species with fair abundance at this <900 m elevation could be possible under the following site and soil characteristics which are in harmony with the growth of above species.

Chloroxylon swietenia is the most dominant species on soils at <900 m elevation. This species, although not dominant, is also present in the soil occurring at elevation 900–1100 m. However, growth-wise its performance is better at <900 m elevation suggesting that site conditions and soil characteristics are congenial for the growth of this species. The second dominant tree at this elevation is *Terminalia tomentosa* which has also performed better at <900 m elevation as compared to other two elevations in terms of its number of individuals, average diameter growth and to some extent average height growth, although this species is co-dominant in upper two elevations taking part in community formation and the best suited species in the whole landscape. At this elevation, *Buchanania lanzan* also did well.

Thus, soil site characteristics like land form, erosion, slope, soil depth, soil drainage, stoniness, rockiness, parent material etc. are important since their combined integration results in creating soil environment suitable for plant growth. Similar observations on slope soil relationship were also reported by Jyotyprakash (1968), Bhattacharjee et al. (1971) and Gaikwad et al. (1974). In an undulated landscape where the parent material and climatic conditions are more or less the same, topography and consequently the external and internal drainage conditions may largely change the properties of soils so also the growth and distribution of vegetation.

There is more similarity of about 68% in tree species occurring on lower hills (<900 m) and middle (900–1100 m) hills as compared to tree species on middle hills and upper hills where similarity is of the order of 63%. Shrub species on lower hills and middle hills showed more similarity of 50% due to similarity in microclimate. On the other hand, there is only 31% similarity between shrub layer species occupying at mid hills and upper hills. Herb species corresponding to <900 m and 900–1100 m elevation are more similar to each other with 43% similarity.

In the class of angiosperms, total 147 plant species have been recorded in different communities. Out of 147 plant species, 52 represent tree species, 32 shrubs and 63 herbs and grasses. All angiospermic species present in the park are put into 51 families of which topmost position is occupied by the family Papilionaceae, followed by Asteraceae and Poaceae. Family Asteraceae which has occupied 2nd position in the Satpura National Park occurs normally more in the temperate region and is the most important characteristic of temperate climate (Khatri 2000). It is, therefore, assumed that the nature of vegetation in the Satpura National Park is semi temperate. In the flora, out of 51 families, 22 families are represented by one species, 9 by two species, and 6 by three species.

Soil Characteristics under Teak Forests

Though teak is a very good timber species and its durability is the gift of nature, its other qualities are due to matchless properties, such as termite, fungus and weather resistance, lightness with strength, workability and seasoning capacity without splitting, cracking and materially altering shape (Tewari 1992). It is a tall tree species having long straight cylindrical bole and mellow colour (Banerjee and Prakasham 2013). M.P. teak has been famous for these qualities and fetches a high price in the international market. The teak forest is put in four quality classes. Teak of Bori forest is of quality I (AIG-I). Bhowmik and Totey (1990) characterized some of the soils under teak in Bori and classified them.

The natural teak forest of Bori, Hoshangabad district, M.P. is situated between 78°0′ to 78°30′ E longitude and 22°0′ to 22°30′ N latitude at an altitude up to 429 m on gently sloping (1–5%) upland. The parent material consists of numerous intrusions of 20–30 m wide dolerite sills of Deccan trap which are formed in the Bijori formation of lower Gondwana. The annual precipitation is 1370 mm (10 years average) and the area enjoys hyperthermic soil temperature of 23.5 °C. Mean monthly temperature is the highest in May (34 °C) and the lowest in January (18 °C). Bhowmik and Totey (1990) studied four representative soil profiles and their descriptions are given below.

Profile No. 1

The profile is on sloping land (3–5% NE) of deccan trap, the erosion is slight, B horizon is partially weathered parent material. In the second and third horizons many clay skins and clay cutans are located which is a clear indication of the formation of argillic sub-surface diagnostic horizon. Gravel (>2 mm) percentage increases with depth. With the increase of the depth of the profile, pH also increases (5.8–7.2).

Profile No. 2

The profile is situated on sloping upland (1–3% NE). From the second horizon to 4th horizon, clay skins and clay cutans are located forming the argillic horizon, The distribution of gravels is not uniform. The most characteristic feature is the colour which is the same in all the horizons. The soil is very deep and slightly acidic to near neutral.

Profile No. 3

The profile 3 is situated in sloping upland (5% NE) on Deccan trap and well drained. Erosion is slight. The lowest horizon is partially weathered loose material. From 2nd horizon, gravel percentage is in decreasing order. The soil reaction is acidic throughout the horizon and pH varies from 5.1 to 5.7. The profile is very deep.

Profile No. 4

The profile 4 is situated on sloping upland (3–5%NE) of Deccan trap, well drained, slight erosion. The soil is very deep and from the depth of 70 cm and below the horizons comprise partially weathered trap boulders and partially weathered trap/sand stone. Soils of all the horizons are acidic.

The morphological characteristics of these soils reveal that they have, in general, hues of 5YR, chromas of 3–4 and values of 2–4. The reddish brown colour is attributable to the presence of non hydrated iron oxides released due to the weathering of ferro-magnesium minerals (Saxena and Singh 1982; Challa and Gaikawad 1986).

Soil-vegetation correlation

pH of soils of profile 1 and 2 is slightly acidic to neutral. Higher acidity in profiles 3 and 4 is attributed to the formation of more organic acid owing to the higher organic matter content. In most cases, pH increases down the profile. Translocation of clay in profiles 1 and 2 is very distinct as evidenced by the formation of well developed argillic subsurface diagnostic horizons. The clay owes its origin to the doleritic sills. Mechanical composition of the soils is clear evidence of good infiltration and drainage condition of the soils.

The soil organic carbon in the surface is comparatively much higher due to the accumulation of a large proportion of inputs in the form of litter on the surface and their decomposition. With increasing depth, there is a sharp fall in the contents of organic carbon. This may be related to the number of trees in the top storey per unit area. At the site of profile 1, there exists a natural teak forest having 777 teak trees per hectare of height varying from 35 to 37 m with average GBH of 104.7 cm. In the middle storey, there are species like *Lagerstroemia parviflora, Syzygium cumini, Saccopetalum tomentosa, Cassia fistula* etc. The site of profile 2 bears Davidson plantation of 1869 with 699 teak trees per hectare of the average height of 35 m and average GBH of 119.9 cm. In the middle storey, there are species like *Dendrocalamus strictus, Lagerstroemia parviflora, Cassia fistula, Ougeinia dalbergioides. Dalbergia latifolia, Bauhinia racemosa, Saccopentalum tomentosa, Flacourtia ramontchi* etc. Profile 3 is under a natural teak forest (Preservation Plot) where teak exists in association with miscellaneous tree species like *Lagerstroemia parviflora, Syzygium cumini, Terminalia tomentosa, Terminalia chebula, Diospyros melanoxylon, Scheleichera oleosa, Madhuca indica, Ougeinia dalbergioides, Dalbergia paniculata* in the top storey. The density of teak trees is thus reduced to 155 trees per hectare with height varying from 37 to 40 m and average GBH is 160.3 cm. In the middle storey, there are plants like *Dendrocalamus srtictus, Syzygium cumini, Ougeinia dalbergioides, Ficus glomerata, Diospyros melanoxylon* etc. Profile 4 also carries teak plantation of 1944 with 865 teak trees per hectare having height of 35 m and average GBH 69.6 cm.

Cation exchange capacity does not bear any relationship either with the distribution pattern of clay or organic matter. However, significant amount of CEC of soils in all these four profiles is suggestive of contribution by non-clay fraction to CEC

(Karale et al. 1969). Profile 3 has the highest exchangeable Ca^{2+} which increases with the depth. The exchangeable Ca^{2+} may be related to the distribution of total CaO which is more influenced by the presence of calcite in profiles 2 and 3. Leaching of the products of decomposition of leaf litter of teak, which are rich in calcium and magnesium produces higher cation content in surface as well as sub-surface horizons. Calcium accumulates in greater quantities in the foliage than the other cations and also does not go out of the foliage just before leaf fall as large portions of other elements do. As such, the percentage of calcium and magnesium is higher in these profiles (Banerjee et al. 2020). The content of exchangeable K in all the profiles is higher in the surface layer with decreasing trend down the profile and is closely associated with the distribution of organic matter and total K_2O.

Exch. Ca^{2+} is the dominant cation which forms about 70–80% of CEC and the remaining is mostly Mg^{2+}. More of calcium as compared to magnesium is recycled from lower to A horizon in profiles 1 and 4, which affects Ca:Mg ratio. High Ca:Mg ratio (Table 5.5) in the upper layer relative to lower ones in profiles 2 and 3 indicates lower order interaction of vegetation on soil development (Singhal and Sharma 1985). In the B and C horizons Mg^{2+} tends to persist larger in an exchangeable form than Ca^{2+}. Ca:Mg ratios generally decrease with increased weathering and leaching as in profiles 1 and 2.

On the basis of morphological characteristics of the soil epipedon and structure, organic matter and base saturation (Soil Survey Staff 1975), all the four profiles come in the order Mollisol. Due to the presence of ustic moisture regime, they are included in the order Ustall. Profiles 1 and 2 have argillic horizons and as such, qualify for the great group Argiustoll. The other two profiles, however, qualify for the great group Haplustoll. Profile 1 belongs to fine loamy, hyperthermic family of Udic Argiustoll, while profile 2 comes in fine clayey, mixed hyperthermic family of Udic Argiustoll. Profiles 3 and 4 belong to fine loamy, mixed, hyperthermic family of Udic Haplustoll (Soil Survey Staff 1975).

Among the 71 species of non flowering plants found in the Pachmarhi Biosphere Reserve, 48 species belong to ferns and the rest belong to fern allies. Apart from that, many angiospermic plants are also observed to be rare in the reserve area. Some of the most notable species of ferns found here include *Psilotum, triquetum, Isoetes panchanani, Selaginella exigua, Ophioglossum* spp. (Pandey et al. 1993).

Teak bearing forests cover about 800 km^2 in central as well as west–southern part of the reserve (now wildlife sanctuary). Altitude varies from 520 to 650 m with precipitous slopes to undulating terrain. Underlying rock is Deccan trap in major portion of this type of forests. Annual average rainfall varies from 1500 to 2000 mm. Slopes of hillocks and valleys are occupied by good formation of bamboo as an understorey crop, though plains are devoid of it. On the basis of presence and absence of bamboo the entire forest of this category can also be divided into two groups i.e. with bamboo and without bamboo forests. However, there are certain pure bamboo patches in Nagdwari valley.

Due to variation in rainfall and topographical features in different localities the composition of teak and their associates shows marked changes in community formations (Pandey et al. 1993).

Table 5.5 Physicochemical characteristics of the soils under Teak

Horizon	Sand	Silt	Clay	Org. C	CEC	Ca^{2+}	Mg^{2+}	K^{+}	Na^{+}	Ca/Mg ratio
	(%)				me%					
Profile I										
A1	50.6	24.8	24.6	2.61	18.7	11.4	4.5	0.4	0.2	2.5
B21t	45.4	24.2	30.4	1.45	19.5	11.4	6.4	0.1	0.1	1.7
B22t	50.2	17.8	32.0	0.81	20.7	11.4	6.4	0.1	0.1	1.7
B3	58.2	15.2	24.6	0.69	18.6	9.5	6.4	0.1	0.1	1.4
Profile II										
A1	50.2	23.6	25.4	1.97	27.7	15.2	8.3	0.5	0.2	1.8
B21t	44.4	8.6	47.0	1.27	27.8	15.2	8.3	0.5	0.2	1.8
B22t	39.8	15.8	41.4	1.04	25.5	15.2	8.3	0.1	0.1	1.8
B23t	49.0	14.0	37.0	0.75	24.6	13.3	8.3	0.1	0.1	1.6
IIB3	48.0	25.0	27.0	0.63	23.2	13.3	6.4	0.1	0.1	2.0
IIIC	64.2	16.0	19.8	0.44	23.6	13.3	6.4	0.1	0.1	2.0
Profile III										
A11	45.8	25.6	28.6	3.22	32.8	22.8	6.4	0.3	0.3	3.5
A12	44.0	36.0	20.0	1.79	33.7	22.8	8.3	0.3	0.2	2.7
B21t	45.8	22.8	31.4	0.72	37.1	22.8	12.1	0.2	0.3	1.5
IIB22	53.6	22.0	24.4	0.38	41.4	26.6	12.1	0.1	0.4	2.1
IIC1	66.1	18.0	15.8	0.20	44.4	26.6	14.0	0.1	0.4	1.9
IIIC2	44.6	32.6	22.8	0.06	49.4	36.1	8.3	0.1	0.6	4.3
Profile IV										
A1	45.6	21.2	33.3	3.21	18.1	13.3	2.6	0.7	0.2	5.0
A3	39.4	26.4	34.2	1.83	19.4	13.3	2.6	0.6	0.3	5.0
B2	43.6	31.4	25.0	0.69	17.7	11.4	2.6	0.1	0.2	4.2
IIB3	47.6	29.2	23.2	0.59	18.4	11.4	4.5	0.1	0.3	2.5
IIC1	53.6	24.4	22.0	0.49	20.3	11.4	6.4	0.1	0.4	1.7
IIC2	63.6	18.4	18.0	0.35	25.9	13.3	8.3	0.1	0.3	1.5
IIIC3	63.5	26.8	9.7	0.29	23.8	11.4	8.3	0.1	0.4	1.8

In the moist deciduous teak forests, teak is the most characteristic species. In the top and second canopy, the moist teak forest is associated with *Terminalia tomentosa, Diospyros melonoxylon, Buchanania lanzon, Adina cordifolia, Pterocarpus marsupium, Madhuca indica, Schleichera oleosa, Careya arborea, Lagerstroemia parviflora, Kydia calycina, Butea monosperma, Syzigium cumini, Ougenia oojenensis, Saccopetalum tomentosum, Cordia myxa, Cassia fistula, Mallotus phillippensis, Casearia tomentosa, Gardinia latifolia, Bridelia retusa, Bauhinia retusa, Albizzia odoratissima, Casearia graveolensi, Ficus cunia, F. glomerata,* etc. However, *Terminalia tomentosa, Pterocrpus marsupium* and *Lagerstroemia parviflora* are frequently distributed in the area.

In the ground vegetation,*Rungia pactinata, Blepharis maderaspatensis, Euphorbia hirta, Asparagus recemosus, Achyranthus aspera, Sida acuta, Atylosia scarabaeoides, Gymnena sylvestre, Leucas lanata, Ichnocarpus frutescence,, Lavendula bipinnata, Chelianthus teunifolia (Fern), Zingiber zerumbet* etc. are the prominent herbs and shrubs found in the area. Some important grasses and sedges i.e. *Apluda varia, Dichanthium annulatum, Eragrostis viscosa, Cynodon dactylon, Cyperus iria, Themeda quadrivolvia, Heteropogon contortus* are commonly observed.

In the forest where bamboo is associated as understorey crop, ground vegetation shows poor density but diversity was found to be good in comparison to localities without bamboo. In bamboo areas, the common ground plants are *Ichnocarpus fruitescence, Sida acuta, Apluda varia, Hemidesmes indicus, Elephentopus scaber, Urena repanda, Celosia urgencia, Occimum spp. Adiantum spp., Vernonia cinerea, Vicoa indica, Ageratum conyzoides, Hetropogon contortus, Themeda quadrivalvis* etc. But in localities without bamboo, *Cassia tora, Abrus precctorius, Tridax procumbense, Cympopogon martinii, Heteropogon contortus* etc. are found.

Slightly moist teak forests occur in the western part of Bori reserve. The annual rainfall is comparatively less (1200–1600 mm) than that in the occurrence of moist teak forest and soil is moderately deep loamy. Though, the dominating species is teak, yet the density of teak trees is lesser than that in the previous forest type.

Southern moist mixed deciduous forest is almost similar to moist teak bearing forest except in the composition of teak which is only occasionally present along with other associates. This type of forest is believed to have been developed as a result of those in secondary succession (Champion and Sett 1968) and is localized in small patches in damp valleys with shallow or porous soil. Formation of this type of forest shows the intermittent stage between moist deciduous teak bearing forest in the north and southern tropical dry deciduous mixed forest in the study area. The elevation where this type of forest is found ranges from 500 to 650 m above msl. The climatic conditions of these specific localities are almost similar to those in moist teak bearing forest.

Floristically, the forest represents mixed forest with teak. Teak is occasionally found in scattered form with other associates e.g. *Pterocarpus marsupium, Terminalia bellerica, Anogeissus latifolia, Dalbergia latifolia, Terminalia tomentosa, Lannea coromandelica, Madhuca indica, Garuga pinnata* etc. in the top canopy and *Saccopetalum tomentosum, Lagerstroemia parviflora, Emblica officinalis, Grewia tiliaefolia,* etc. in the middle canopy.

Similar to other forest types, *Dendrocalamus strictus* appears on the slopes of hillocks but it is not uniform in the area. Ground vegetation, however is found to be similar to that in the moist teak forest.

The dry deciduous forest is met with in north –eastern part of the biosphere reserve. The area occupied by this type is undulating having small to medium hillocks and plains. The annual rainfall is 1000–1300 mm. The maximum temperature has been recorded to be 45 °C and minimum 6 °C in December–January. The upper canopy in this forest type in not much dense as found in the moist teak forests. Bamboo is present on slopes as an understorey but the growth is not as luxuriant as in the moist forests. The specific feature at this site is that the Anjan (*Hardwickia binata*) appears in north eastern boundaries in pure form. However, Sal has also appeared contributing to the main associates in the plateau of biosphere by forming a dry peninsular Sal forest in the area.

On the basis of composition of principal species in the prevailing association at different localities, this forest type is further divided into two sub-groups (Champion and Sett 1968) such as 1. Southern Tropical dry Deciduous Forests and 2. Dry Peninsular Sal.

In southern tropical dry deciduous forests, teak is invariably present, though their distribution is not uniform. Thus, on the basis of distribution of teak, it is again divided into two sub types: (i) Dry teak forest and (ii) Moist deciduous forest.

In dry teak forest, teak is the dominant species but its frequency and distribution is sparse. The main species teak, is being associated with *Anogeissus latifolia, Terminalia tomentosa, Diospyros melanoxylon, Pterocarpus marsupium, Cassia fistula, Dalbergia latifolia, Butea monosperma, Adina cordifolia, Bridelia retusa, Aegle marmelos, Lagerstroemia parviflora, Wrightia tinctoria, Bauhinia retusa* and *Hardwickia binata* in the top canopy. In the lower canopy, the common trees and bushes are *Nyctanthus arbortristis, Woodfordia latifolia, Helicteres isora, Grevia hirsuta, Gymnosporia spinosa, Indigofera pulchella, Adhatoda vasica, Carissa spinarum, Holarrhena antdysenterica, Lantana camara* etc. In the ground vegetation, the common herbs and shrubs are *Cassiatora, Euphorbia spp., Xanthium strumarium* (in the vicinity of habitations), *Abrus precatorius, Acacia pinnata, Milletia auric7lata, Vantilago calyculata, Zizyphus oenoplia* etc. The common grasses are *Apluda mutica, Aristida* spp., *Chlorisinftata, Eragrostis tenella, Eragrostiella* spp., *Eragrstis viscosa, Eragrostis unioloides, Iseilemaaxum, Heteropogon contortus, Themeda guadrivolvis, Dichanthium annulatum* etc. Bamboo is also found in the understorey on the slopes towards Pachmarhi plateau.

The forest of subgroup mixed deciduous forest is almost identical to thatof dry teak deciduous forests, but it differs mainly in floristic composition, where some typical species are more conspicuous. Thorny plants have appeared and tend to increase in proportion in the localities subjected to heavy grazing. Bamboo is often absent and usually of poor quality.

Dry Peninsular Sal

The Sal (*Shorea robusta*) is the main constituent dominating in this type of forests and situated in the centre of Pachmarhi plateau, extending towards the southern most part of Delakhari and Tamia Forests.

This sub type has occupied the localities of shallow soils. The soil often rests directly on hard impervious laterite and is sometime calcareous. Sal regeneration is fair but slow. This forest type covers the area of Pachmarhi plateau, extending towards eastern boundaries of the area. At plateau, the growth of Sal is found to be stunted, representing the highest limit for dry Sal. This type of vegetation is undoubtedly edaphically conditioned, occurring on dry sandy pebble soil derived from Pachmahri sand stones and conglomerates, in a climate where the natural vegetation is representing to sub-tropical wet hill forests.

In the vegetational composition of this forest type, trees are sparsely distributed but forms the diverse vegetation constituted by various species of trees, shrubs, herbs and climbers. Ground vegetation also shows the diversity at different localities.

In the tree categories, the important tree species are *Shorea robusta, Terminalia tomentosa, Chloroxylon swietenia, Buchanania lanzan, Anogeissus latifolia, Madhuca latifolia, Emblica officinalis, Miliusa tomentosa, Litsea chinensis, Launea grandis, Bauhinia varigata, Terminalia chebula, Casearia tomentosa, Casearia gravelolence, Cassia fistula, Syzygium cuminii, Cariya arborea, Pterocarpus marsupium* etc.

Sal forest of Pachmahiri plateau has very special features regarding the ground flora. The common herbs and shrubs found in this locality are *Vicoa indica, Vicoa cernua, Vernonia cinerea, Ageratum conyzoides, Laggera alata, Justicia sp., Strobilanthus spp., Andrographis paniculata, Sida acuta* etc. Some common grasses and sedges, e.g. *Apluda varia, Themeda guadrivalvis, Symbopogon martini, Heteropogon contortus, Chloris dolichostachya, Eragrostis astrovirens, Cyperus iria* etc. are commonly found though sparsely distributed in the area.

Moreover, Pachmahri plateau consisting of several deep gorges, perennial and seasonal streams, creating micro-climatic conditions is congenial for the growth of several species of pteridophytes, bryophytes, algae, orchids and several other moisture loving plants. Some of the important ferns are *Psilotum nudum, Lycopodium cerninum, Polybotry appendiculata, Polytrichum amabile, Cyathea gigantiea, Adiantum lunatum, Chaeilanthus farinosa* etc., Similarly, amongst bryophytes, some important species are *Riccia stricta, Riccia sanguinea, Targionia hypophylla, Reboulia hemispharica, Riccardia levieri, Pycanthus stricutus, Brachymenium nepalensis* etc. Unique Drosera patches are also found distributed in various localities in this area.

Central Indian Sub Tropical Hill Forests

The prevailing forests of this category have occupied the hill top of Pachmahri plateau. The site is exposed, having very poor soil and has long been subjected to human settlement and grazing. The forests in such localities are of inferior type, the trees being short boled and branchy. It is doubtful, if any climax forest survives though the vegetation of sheltered glens and steep and narrow gullies probably represents it to some extent (Champion and Sett 1968). The site carries more xerophytic vegetation in degraded form.

The elevation of these localities varies from 400–800 m above msl. The minimum average temperature being 10 °C in December–January and the maximum temperature rising up to 45 °C in the month of May–June. Annual rainfall received is about 1500 mm which is definitely higher than that in other parts, though there is a long dry season after rainy period. Soils are typically shallow where the topography is steep but of fair depth on plains.

The tree associates found in this forest type are *Syzygium cumini, Terminalia tomentosa, Anogeissus pendula, Emblica officinalis, Cassia fistula, Eliodendrum glaucum, Casearia tomentosa, Trema orientalis, Bombax ceiba, Flacourtia indica, Litsea glutinosa, Grewia tilifolia, Kydia calycina, Albizia odoretissima, Ficus glomerata, Bridelia retusa, Miliusa tomentosa, Terminalia chebula, T. bellirica, Mimosops hexandra, Anogeissus alata, Ficus hispida* etc. The common shrubs and woody climbers observed in the area are *Euphorbia royleana, Doclonaea sp, Lantana indica/camera Sophora interrupta, Phoenix acaualis, Gardinia turgida, Zizyphus oenoplea, Murraya exotica, Flamingia bracteata, Vernonia divergens, Colebrookia oppositifolia.*

In the ground vegetation, the common herbs and herbaceous climbers are *Justicia crinata, Rungia* spp., *Ageratum conyzoides, Leucas montanum, Leuas lanata, Sida acuta, Andrographis paniculata, Euphorbia hirta, Plaetranthus incanus, Strobilanthes callosa, Asparagus racemosus, Dioscorea spp., Vicoa indica, Vicoa cernua, Trichodesma zeylanicum, Lavendula bipinnata, Oxalis corniculata, Artemisia parviflora, Vernonia cinerea* etc. Grasses like *Eragrostis unioloides, Eulaliopsis binata, Sympopogon martini, Eragrostis atrovirens, Themeda guadrivalvis* are commonly found in the area.

Conclusion

The concept of Biosphere Reserve is of immense value to conserve the gene pool resources of flora and fauna in potentially rich habitat. Biosphere reserve serves not only to conserve the endangered and rare species of flora and fauna but makes a platform for carrying out research in it. Pachmarhi plateau has several rare and endemic species in it. However, these habitats need urgent conservation measures for in situ protection of such rare and endemic species.

Some of the important and rare species, which are observed may be considered as gene bank of rare species in these localities. Over sixty species of pteridophytes have been reported from the area. Out of these, 48 species belong to ferns and rest to fern allies. Most of the fern species are terrestrial in habit and growing deep inside the gorges and ravine.

Moreover, some plants like *Hymenodictyon excelsum, Alangium salvifolium, Peucedanum dhana, Sopubia delphinifolia, Lobelia nicotianafolia, Leea macrophylla, Andrographis echioides, Psolalea corylifolia, Centella asiatica, Hydrocotyle sibthorpoides, Artemisia nilgirica, Plumbago zeylanica, Swertia minor, Chirita bifoliuta, Arisfolochia bracteoluta, Gloriosa superba, Smilex zeylanica, Curuma aromatica, Zinziber roseum, Curculigo orchiolies, Urginea indica, chlorphytum tuberrosum* have medicinal values and found sparsely distributed. These sites are the points constantly visited by tourists and under severe exploitation of such rare endemic and economically important species in the area. Repeated exploitation of such species for the last many years have put their existence in danger and the species need urgent protection measures for their in situ conservation.

References

Banerjee SK, Jain A (2011) Indian forest and research—past and present. In: Trivedi PC (ed) Plant—environmental interaction. Pointer Publishers, Jaipur, Rajasthan, pp 171–194

Banerjee SK, Prakasham U (2013) Biomass carbon pool and soil organic carbon sequestration in *Tectona grandis* plantation. Indian for 139(9):797–802

Banerjee SK, Saikat B, Sahoo TK, Shukla PK (2020) Changes in soil characteristics and carbon sequestration potential consequent upon differences in forest cover in a plantation area of Kalimpong Division, West Bengal. J Trop for 36(4):55–63

Bhatnagar HP (1961) Factor in the distribution of sal (*Shorea robusta)* forest in India with special reference to U.P. and M.P. J Indian Bot Soc XI 1:104–112

Bhattacharjee JC, Barde NK, Shende NK, Raut BP (1971) Morphology and distribution of soil of lower Ib watershed. J Indian Soc Soil Sci 19:423–439

Bhowmik AK, Totey NG (1990) Characteristics of some soils under teak forests. J Indian Soc Soil Sci 38:481–487

Buch MN (1991) The forest of Madhya Pradesh. Madhya Pradesh Madhyam, Bhopal

Challa O, Gaikawad ST (1986) Soils in a catena from Dadra and Nagar Haveli: their characteristics and classification. J Indian Soc Soil Sci 34(3):543–550

Champion HG, Sett SK (1968) A revised survey of forest types of India. The Manager of Publication, Delhi, p 6

Gaikwad ST, Rao YS, Verma HKG (1974) Characteristics of catenary soil developed on basalt parent rock in Nagpur district, Maharashtra. J Indian Soc Soil Sci 22:181–190

Jyotyprakash (1968) The study of physical and chemical properties of soil of the Couvery Catchment area of Mysore state with reference to their geomorphology. J P G School 6:106–107

Karale RL, Tamhane RV, Das SC (1969) Soil genesis as related to parent material and climate. 1. Morphology, physical, chemical and physicochemical properties. J Indian Soc Soil Sci 17:227–239

Khatri PK (2000) Study on biodiversity in tropical forest ecosystem of Satpura National Park, Madhya Pradesh. PhD thesis, Forest Research Institute University

Khatri PK, Totey NG, Pandey RK (2004) Altitudinal variation in structural composition of vegetation in Satpura National Park. Indian for 130:1141–1154

Luna RK (1996) Plantation trees. International Book Distributors, Dehra Dun

Mehta V (2018) Pachmarhi Biosphere Reserve, Madhya Pradesh—Complete Guide, toppr.com

Pandey RK, Pantane NK, Sharma P (1993) Preliminary project report on flora of Pachmarhi of detailed project formulation to constitute Pachmarhi Biosphere Reserve. State Forest Research Institute, Jabalpur, p 91

Rai RK (2000) Management of Biosphere Reserve in India: present status, future prospects and constraints. In: Proceedings of the review meeting: biosphere and their management, 8–11 Sept, Peechi, Kerala

Saxena SC, Singh KS (1982) Pedo-chemical characterization of soils of Rajasthan (semi-arid to humid agro-climatic region). J Indian Soc Soil Sci 30(4):515–522

Singhal RM, Sharma SD (1985) Study of the soils of Doon Valley forests. J Indian Soc Soil Sci 33:627–634

Soil Survey Staff (1975) Soil taxonomy, Agri. Handb., U.S. Dep. Agric. 436, Indian Reprint, 1979 by NBSS & LUP, Nagpur

Totey NG, Bhowmik AK, Khatri PK (1986) Performance of sal (*Shorea robusta*) on soil derived from different parent materials in Shahdol forest division, Madhya Pradesh. Indian for 112:18–31

Tewari DN (1992) A monograph on teak (*Tectona grandis* Linn). International Book Distributors, Rajpur Road, Dehra Dun, 479 pp

Chapter 6
Salt Affected Soils: Global Perspectives

Nirmalendu Basak, Arvind Kumar Rai, Arijit Barman, Subashis Mandal, Parul Sundha, Sandeep Bedwal, Sanjay Kumar, Rajender Kumar Yadav, and Parbodh Chander Sharma

Abstract Salts are the primary sources of salinity in soil and water. Around one billion ha of global earth land are more or less affected by different kinds of salt and associated threats. The demand for expansion of intensive irrigated agriculture in canal networks, climate change with the temperature rise, the incidence of drought, water scarcity arises the more demand of evapotranspiration of the crops and subsequently import excess salts in the soil under saline irrigation, rise in sea level, limitation of freshwater, ingress of seawater, unpredictable behavior of precipitation and inappropriate drainage facilitate the salinization problems, sodication, and the infestation of high sodium adsorption ratio (SAR), damaging soil chemical environment, and development of sodicity causes a deleterious impact on soil

N. Basak (✉) · A. K. Rai (✉) · A. Barman · S. Mandal · P. Sundha · S. Bedwal · S. Kumar · R. K. Yadav · P. C. Sharma
ICAR-Central Soil Salinity Research Institute, Karnal 132001, Haryana, India
e-mail: nirmalendu.basak@icar.gov.in; basaknirmalendu@gmail.com

A. K. Rai
e-mail: ak.rai@icar.gov.in

A. Barman
e-mail: arijir.barman@icar.gov.in

S. Mandal
e-mail: s.mandal@icar.gov.in

P. Sundha
e-mail: parul.sundha@icar.gov.in

S. Bedwal
e-mail: sbedwal519@gmail.com

S. Kumar
e-mail: sanjaysahu493@gmail.com

R. K. Yadav
e-mail: rk.yadav@icar.gov.in

P. C. Sharma
e-mail: parbodh.chander@icar.gov.in

P. K. Shit et al. (eds.), *Soil Health and Environmental Sustainability*, Environmental Science and Engineering,
https://doi.org/10.1007/978-3-031-09270-1_6

physical health. To meet the food-feed-fibre of the bargaining population rehabilitation of salt-affected soil (SAS) is a main agenda in present policies of countries extended with salinity and irrigation depends on the marginal quality of water. Here, we described the distribution and occurrence of SAS; narrated the causes and drivers for salinization/sodification; characteristics and properties of SAS; estimated the production loses of the crop because of salinity; created the economic importance of SAS, and finally mechanism-based management options are described for rehabilitation of SAS for greening the baran underproductive land, ensure for food security, empowering livelihood and checking the mass migration of peoples in the future.

Keywords Salinity · Sodicity · Climate change · Reclamation technology

6.1 Introduction

Among the abiotic stress, salinity hurts soil properties and its quality and crop growth and production. Countries that reside in arid and semiarid regions are much suffering from the infestation of salt balance that suppresses crop growth, nutrition, subsequently in productivity failure and severely affecting food production and consequently insecurity for livelihood and badly affect the socio-economic status of the habitant resides in SAS. Presently, the development of canal network irrigation and expected changes in climate particularly rise in temperature, unpredictable behavior of rainfall will increase in evapotranspiration (ET) requirement of the crops and thus deposited a more salinity in case of saline water irrigation or assured irrigation in canal command area and climate change will harm agriculture, particularly in arid regions (Hopmans et al. 2021). Therefore, monitoring the expansion of soil salinity and its intensity of severity is crucial to quantify the damaging impact on crop productivity and land degradation (Barman et al. 2021). Characterization and inventory of salt-affected soils (SAS) is a preliminary step towards forecasting and implementations, management, and reclamation programs for sustainable agricultural productivity of SAS and canal command irrigated areas because soils in these ecologies endow with diverse in nature and amounts of salts. The first category of SAS contains a large proportion of soluble salts that are neutral in a chemical reaction and the SAS has a high saline water table. Other categories of soils bear salts that show alkalinity upon water hydrolysis but the concentration of the electrolyte varies due to soil type or external deposition of salt (Rai et al. 2020a). Provisionally occurrence of sodicity arises because of intensive and prolonged irrigation with alkaline water (RSC, residual sodium carbonate) (Choudhary et al. 2011; Murtaza et al. 2021; Sheoran et al. 2021a). The fresh or canal water crisis the farmers to irrigate with marginal water (e.g. saline, treated wastewaters, and desalinated water) to utilize for meeting the growing demands (Soni et al. 2021). Seasonal or unprecedented salt/brackish-water intrusion and shallow water tables increase the development of soil salinity in coastal lines (Dasgupta et al. 2015). Based on adverse effects of salinity, different groups of SAS on soil characteristics, crop production and type and severity of difficulties, and package and practices of cultivation are required special management practices. Therefore, to manage adverse effects, precise and thoughtful reclamation processes to alleviate salinity and associated limits are diverse. Under dry wet, cropped, and uncropped conditions for direct estimation of soil salinity in the field. Further, technological options are addressed to combat salinity issues for crop production.

6.2 Global Distribution and Occurrence

The extent of salt-affected soil (SAS) covers a significant land area of global earth ~932.2 Mha (Rengasamy 2006). This problem is further aggravated with the practice of faulty irrigation practices which affect 34.2 Mha or more than 10% of the total irrigated area (Mateo-Sagasta and Burke 2011; Aquastat 2016). Soil salinity/sodicity is a dynamic process. More than a hundred countries are affected by SAS and no continent is entirely free from salinity/sodicity or twin problem of both (Fig. 6.1). Central and southeastern Asian countries (China, India, Pakistan, Iran, and Iraq, etc.) and other western countries (United States,), a major part of Australia, Argentina and Brazil from southern hemisphere; and Spain and Italy from Europe, etc. are the key hot spots of global soil salinisation (Ghassemi et al. 1995; Aquastat 2016). Some important river basins where the problem of soil salinization extensively reported in the past decades are Central Asia of Aral Sea Basin, Indo-Gangetic Basin in India, Indus Basin in Pakistan, Yellow River Basin in China, Euphrates Basin in Iraq

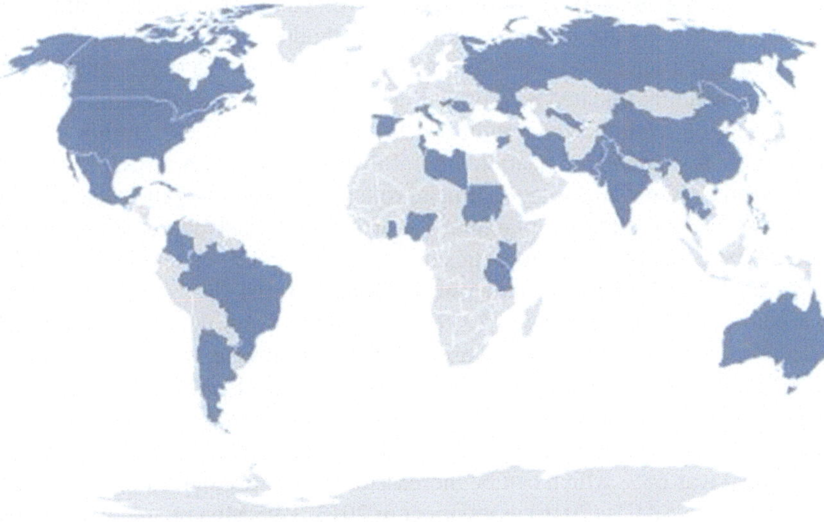

Fig. 6.1 Countries map affected by soil salinity/sodicity problem (Redrawn https://www.researchgate.net/publication/262495450)

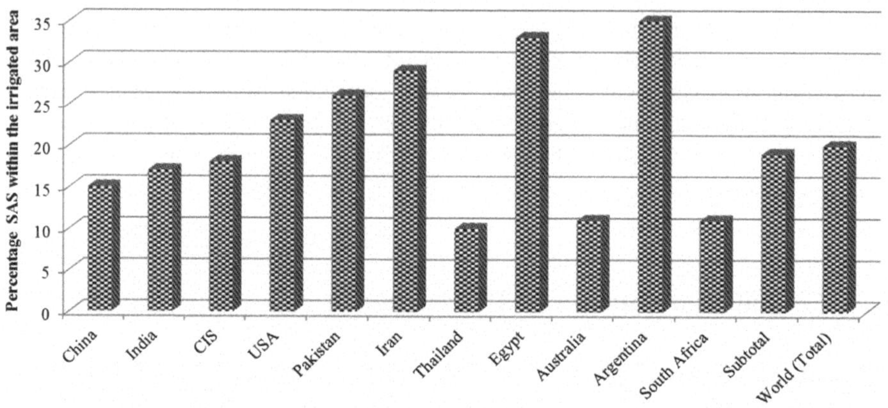

Fig. 6.2 Global estimation of SAS by secondary salinization within the world irrigated area (Modified from Shahid et al. 2018)

and Syria, Murray-Darling Basin in Australia, and San Joaquin Valley in California (Chang and Silva 2014; Qadir et al. 2014) There is an alarming rate of 10% increment in area of salinity of entire Globe (Nachshon 2018).

The global SAS within the irrigated area of 45 mha out of 227 mha infested by secondary salinization (Ghassemi et al. 1995; Mashali 1995; Shahid et al. 2018) is presented in Fig. 6.2. Argentina, Egypt, Iran, Pakistan, and the USA are the most affected country due to faulty irrigation practices.

In India, about 3.67 Mha of geographic land are affected with salinity which is deviated from earlier estimates (6.74 Mha) done by NRSC, CSSRI, and NBSS&LUP based on Landsat imagery of 1996 (SAC 2016). This difference for estimation of SAS area illustrates data accuracy, followed methodology, and scalability problem. The salt-affected land increased from 0.232 Mha in 1996 to 0.315 Mha in 2010–2013 of the recently updated map of Haryana, India by CSSRI (Mandal et al. 2011) due to irrigation from the canal in impeded drained area and use of high RSC/SAR groundwater for irrigation. Spatiotemporal distribution pattern of SAS, its factors, and occurrence knowledge plays a crucial role in understanding the changes of SAS area and necessary effective planning for arresting further degradation process in future climatic and other uncertainties drivers of SAS. The changes of salt-affected area from 1986 to 2016 based on thermal IR imagery (10.4–12.5 μm) from Landsat 5 and 8 satellites by Ivushkin et al. (2019). World map of soil salinity classes for 2016 presented in Fig. 6.3. They have observed that salt-affected areas increased more than 100 mha after 32 years, at 2–5 mha area increased per year. This increment was pronounced for slight salinity class which reflects that the unaffected area is converting to saline soil (Table 6.1). The validation accuracy of the 2016 world soil salinity map was 68% which can be increased by more ground truth.

According to SAC's (2016) assessment in 2013, 1.95 Mha of area was reclaimed between 2003–2005 and 2011–2013, whereas, 3.63 Mha unaffected land was degraded within the same period. Assessment of SAS using remote sensing and

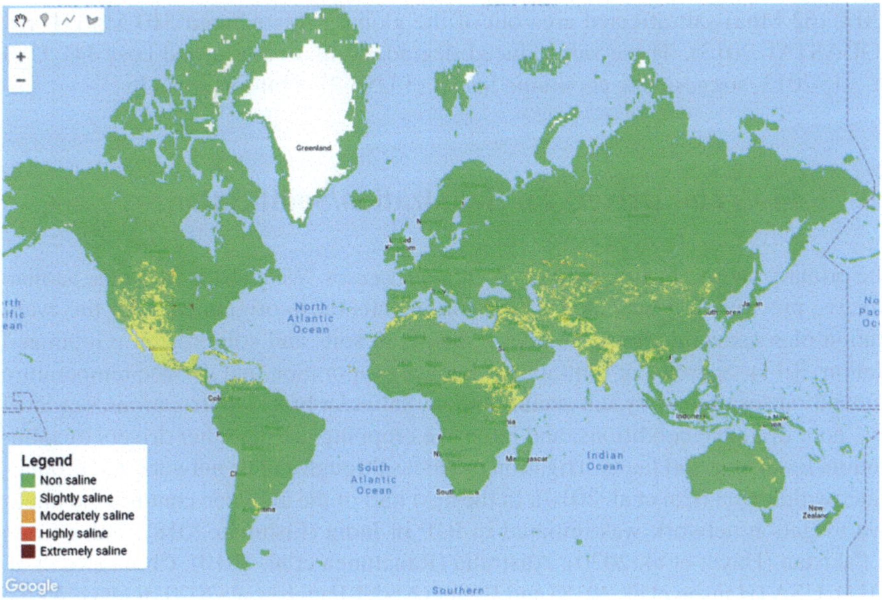

Fig. 6.3 World soil salinity map based on 2016 Thermal IR Landsat imagery (Adopted from http://www.fao.org/3/ca9215en/ca9215en.pdf)

Table 6.1 Changes the predicted area (percentage of total salt-affected class) of soil salinity class from 1986 to 2016

Year	Slightly saline	Moderately saline	Highly saline	Extremely saline
May-86	95.89	3.31	0.23	0.57
May-00	95.38	3.26	0.45	0.91
May-02	97.13	2.11	0.29	0.48
May-05	96.73	2.42	0.22	0.63
May-09	97.21	1.87	0.22	0.70
May-16	96.90	2.32	0.23	0.54
User's accuracy (%), 2016	46.6	91	100	98.1
Producer's accuracy (%), 2016	88	61	47	52

GIS helped to monitor frequently the changes of the spatial distribution of soil salinity/sodicity globally. However, this method limit to the separation of saline soil from sand and other human-induced low productive land. Remote sensing and GIS methods create a problem with low accuracy in separation between slightly and moderately saline soil without proper ground truth. Qadir et al. (2014) reported that

~20% (62 Mha) salt-affected area out of the global irrigated area 310 Mha (FAO-AQUASTAT 2013). These salt-induced degraded inflation-adjusted cost 441 US$ ha^{-1} in 2013, suggests the economic losses of US$ 27.3 billion annually.

6.3 Causes and Drivers for Salinization/Sodification

The primary cause of salinization is natural processes. Weathering of rocks, aeolian process, precipitation mediated deposition of salts those originated from the evaporation of sea, accumulating soluble minerals in soils and soil erosion (Gupta and Mathur 2011). Scarcity of rainfall and intense evaporation and surface temperature demand causes secondary salt accumulation. Saline/or brackish irrigation/, sea-level rise, poor drainage conditions, and intensive cropping are the other drivers of salinization. Extensive and faulty irrigation practices through canals network and groundwater withdrawal (Cai et al. 2010). In the next half of the eighteen century, the intensive irrigation network was initiated in IGP in India (Fishman 2018), Indus basin in Pakistan (Hayat et al. 2020), Australia (Ranatunga et al. 2010), China (Xu et al. 2013), USA (Hansen et al. 2018) and Egypt (Abdel-Fattah et al. 2020) to develop the settlement prospects, assure crop productivity and improving livelihood. The presence of groundwater saline aquifer compounded the effect of salinization. The saline water carries considerable amounts of electrolytes (Na, Ca, and Mg salts). The relative proportions of ions yield the hydraulic conductivity of soils or aggregate failure (Choudhary et al. 2011). The global coastal ecosystems are threatening with rise in mean sea-level rise and surface and sub-surface brackish water intrusion (Dasgupta et al. 2015). Melting of glacial snow and thermal expansion of oceans cause sea-level rise and subsidence of nearby coastland. The torrential storm and/or surges influence the fresh flood and inundation of low-lying coastal areas with brackish water. The recurring problem of seawater intrusion promotes for shrimp or selfish farming instead for paddy cultivation (Blankespoor et al. 2017). The rise in anthropogenic greenhouse gas (GHGs) emissions increases the projected earth surface temperature, rise in seawater temperatures, and shifts in rainfall pattern on latitude. Warming of the Indian Ocean declines annual maximum precipitation and pattern of monsoonal precipitation (Roxy et al. 2015). The rise in temperature accelerates the evapotranspiration demand and causes the increasing risk of extreme rainfall and fresh submergence (Wasko and Sharma 2017).

Weathering of alkaline alumino-silicate minerals supply carbonate (Na_2CO_3 and $NaHCO_3$) in an environment of shallow groundwater table and high evaporation demand or aeolian deposition (Jobbágy et al. 2017). The IGP and Indus basin of India and (Qadir et al. 2018; Sheoran et al. 2021a), the Vertisols (Shirale et al. 2018) of southern India, irrigated agriculture in Australia, Iraq, and Iran with the Aral Sea Basin in Central Asia (Raiesi and Kabiri 2016), river stream of Nile and Niger in Egypt and Africa (Abdel-Fattah et al. 2020) and intensive irrigated basin of Argentina (Sione et al. 2017) showed the deposition of Na salts and prognosis of sodification (Table 6.2). Poor microbial activity in an alkaline environment limits inorganic carbon

Table 6.2 Salinity and sodicity: implication and associated threats

Regions	Projected driving change
Australia	The sodicity threat, landscape modification, urbanization, and sea level rising (Rengasamy 2016)
Argentina	Sodicity threat promote biocrusts formation (Sione et al. 2017)
Central and Eastern Europe	Groundwater recharge impedence, marginal areas threat to salinization, water contamination, and coastal erosion (Falloon and Betts 2010)
Iran, Iraq, and Central Asia	Presence of calcareous saline-sodic soil, saline under groundwater with an appreciable amount of Mg than Ca (Raiesi and Kabiri 2016)
Indo-Gantetic plain India and Pakistan	Aeolian deposits cause sodification, underground waters are saline/sodic, canal command areas are a threat to irrigation induce sodification (Sheoran et al. 2021a; Hayat et al. 2020)
Nile delta, Egypt	Intensely cultivation, soil salinization, inundation, and gradual transgression of seawater and waterlogging (Abdel-Fattah et al. 2020)
Sunderban delta Bangladesh and India	Sea-level rise, saline water intrusion, soil salinization (Dasgupta et al. 2015; Mandal et al. 2019b)
USA	The decline in estuarine, algal blooms, erosion, sediment runoff, sedimentation, eutrophication, dredging, pollution, and biological invasion (Neubauer et al. 2019)
West coast, India	Risk of land degradation, the threat for agro-biodiversity and soil salinization (Mahajan et al. 2021)
Yellow river, China	Fluctuations in tidal levels, flooding, heavy metal contamination in wetland, inundations due to rising sea level threat local ecosystem stability, impaired ecological service function (Liu et al. 2017)

supply and the organic matter remains mobile (Datta et al. 2019). The breakdown of rock is the main origin of inorganic carbon in the soil where the system restricts degassing of CO_2. Chemical weathering and exchange processes laterally supply Ca^{2+} and Mg^{2+} in the soil-water environment and lengthy water trajectory remains saturated with Ca^{2+}/Mg^{2+} over HCO_3^- (Jobbágy et al. 2017). Oppositely in short trajectories, the possibility of HCO_3^- may occur because of chemical weathering of calcareous rocks. Additionally, Na^+ selectivity in a sodic environment reverts Ca and causes precipitation out of $CaCO_3$. An extreme rise in soil pH (~12.0) appears on hydrolysis of sodium carbonate (Bajwa and Swarup 2012); moreover deposition of Ca^{2+} in form of $CaCO_3$ in the soil profile facilitates the occurrence of OH^- ions in soil solution. The spontaneous release of OH^- manifested the further rise in the pH of calcareous sodic soils compared to non-calcareous sodic soil. The high exchangeable Na in the nonexistence of an substantial amount of soluble neutral salts resulted in extreme pH (Basak et al. 2015).

$$2NaHCO_3 \rightarrow Na_2CO_3 + CO_2 + H_2O$$

$$Na_2CO_3 + 2HOH \rightarrow 2Na^+ + 2OH^- + H_2CO_3$$

$$Na - feldpars + HOH \rightarrow H - silicate\ clay + NaOH$$

6.4 Definition and Characteristics of SAS

The definition and characterization of SAS is the primary step towards the execution of a reclamation project for greening the barren or under-productive land. Usually, all types of SAS carry electrolytes of Na^+, K^+, Ca^{2+}, Mg^{2+} and Cl^-, SO_4^{2-}, CO_3^{2-}, HCO_3^- and SiO_x^{-n}, etc. Soluble, quasi soluble, and near to insoluble salts are generated on combinations of two categories of ions (cations and anions). The correspondence solubility of salts generated by cations is: $K^+ > Na^+ > Mg^{2+} > Ca^{2+}$ and anions is $Cl^- > SO_4^{2-} > HCO_3^- > CO_3^{2-}$. The degree of alkaline hydrolysis is $Na^+ > Mg^{2+} > Ca^{2+} > K^+$ and $CO_3^{2-} > HCO_3^-$ and anions Cl^- and SO_4^{2-} produce salts of near neutral to acidic (Tavakkoli et al. 2015). The forms of silicate in soil water solution depend on the pH of the soil environment. Decomposition of primary minerals liberates alkaline salts in the case of extreme equilibrium of hydrological stagnation with short trajectories (Jobbágy et al. 2017). Saline, sodic and saline-sodic soils are the three categories of SAS that are normally available.

6.4.1 *Saline Soil*

Soluble electrolytes concentrations create osmotic effects which define as 'physiological drought'. This generates ion toxicity to plant roots and soil microbes. An encrustation of white salt on the soil surface and poor crop stand is the common syndromes of saline soil. Cl^-, SO_4^{2-}, Na^+, Ca^{2+}, and Mg^{2+} are the predominant electrolytes in this category of soil, and the presence of these electrolytes disorders the soil chemical equilibria and imbalance in plant nutrition but soil physical quality is merely unaffected. The pH of saturation paste (pH_s) of saline soil less than 8.2, ESP (exchangeable sodium percent) less than 15.0% and SAR [sodium adsorption ratio (SAR_e)] less than 13.0 $mmol^{1/2}\ L^{-1/2}$, and electrolytic conductivity of the soil water-saturated paste extract (EC_e) higher than 4.0 dS m^{-1} at 25 °C (Abrol et al. 1988). The origin of saline soil is due to the occurrence of electrolytes in soil water solution, a capillary rise of underground saline water, prolonged irrigation with a saline/or marginal water, and congestion of natural drainage or intrusion of the sea or brackish water at seacoast region (Singh 1998).

6.4.2 Sodic Soil

Sodic soils interchangeably define as 'alkali soils' carry unduly greater quantity of Na^+ than Ca^{2+} and Mg^{2+} in the sites of soil colloid and the soil solution. The ESP of soils has more than 15% and SAR_e more than 13.0 $mmol^{1/2}\ L^{-1/2}$, $pH_s > 8.2$, and EC_e generally vary depending on the presence of soluble salts. The presence of exchangeable Na harms soil physical properties and disperse clay, blocking of pore space, impaired water and air entry (poor soil hydraulic conductivity), extensive mobility and erosion loss of organic matter. In strong alkaline environment, Na appears as toxic a form and Ca precipitated out as $CaCO_3$ and occurring Na induced nutritional deficiency. Deficiency of nitrogen in sodic soil usually occurs due to losses of soil organic matter (SOM) and low symbiotic N fixation and wasteful conversion of supplied N (Sundha et al. 2017; 2022). In an alkaline environment, SOM remains dissolved and procedures black organic clay coatings upon soil aggregates. Carbonate is soluble and its hydrolysis to increase pH upto 12.0 (Bajwa and Swarup 2012). In arid and semi-arid environments, the deposited $CaCO_3$ in the profile and constantly facilitates the release of OH^-. In soil solution. Therefore, the OH^- produce a alkaline pH in calcareous sodic soils than that in non-calcareous alkali soil. A generation of exchangeable Na without a large quantity of neutral soluble salts will always result in alkaline pH.

6.5 Salt Affected Soils and Crop Production

Salinity-induced land degradation and faulty irrigation managements without proper drainage systems aggravated the development of salts in the root zone, severely affecting soil properties and losses of crop productivity. Five percent of the earth land area is salt-affected and 75% of the cultivable area under irrigated agriculture is affected by salinity (Hopmans et al. 2021). For feeding the nine billion people by 2050, it is anticipated that crop production can be meti by cultivating SAS soil with appropriate agro-techniques. The yield reduction of crops of wheat, rice, sugarcane, and cotton reduced by 40, 45, 48, and 63% are grown on SAS in the Indo-Gangetic Plain of India, respectively (Sharma et al. 2015). In the Indus basin in Pakistan, wheat and rice yield reduced 20–43 and 36–69% from SAS (Murtaza 2013). Further, crop yield reduced by 10, 30, 40, and 40% in the USA, Egypt, Uzbekistan, and Turkmenistan (Pitman et al. 2004). Around 2.0 and 1.3 billion US$ monetary loss of produce in India and Australia and other SAS affected countries. Long-time irrigation with high sodic water increased the ESP and soil pH and reduced the productivity wheat and rice by 14 and 16%, respectively (Sheoran et al. 2021a). However, low permeability and waterlogging in occurrence with the concretion of '*kankar* layer of $CaCO_3$' below the rhizosphere of paddy produce desirable grain yield in some areas in India and Pakistan. But, the alkalinity induce submergence along with toxic as the appearance of HCO_3^-, B, Al, Fe, and Mn hampered the productivity of wheat and winter crops in IGP and sodic areas in Australia (Sharma et al. 2018). The cotton yield largely affected in southern Kazakhstan at Aral Sea Basin, the effect of salinity and presence of Mg in soil exchange sites cause a yield loss of cotton (Vyshpolsky et al. 2010).

6.6 Management Options

6.6.1 Agronomic Practices for Saline Soil

Washing of excess salts below the root zone by irrigation with the most available quality water (seasonal rain, canal, or underground water) is accomplished by ponding water in a well-leveled field. The amount of salts washed from soils depends on the quantity of irrigation applied and the presence of soluble salts and soil texture. Flushing with irrigation water is recommended to wash surface deposited salts in low permeable soils. Field-scale salinity management requires appropriate land, water, and crop production strategies to maintain economic productivity and sustainable cultivation in saline soil and minimize the risk of salinity development. Properly leveled land, conservation or minimum tillage, mulching, conjunctive use of saline irrigation, implementation of deficit irrigation (Soni et al. 2021; Rai et al. 2022), cycling and mixing mode (Minhas et al. 2020), avoiding the application of saline irrigation at physiological critical stages (Zwart and Bastiaanssen 2004), the adoption of more productive pressurized irrigation techniques (surface and subsurface drip/sprinkler) (Barakat et al. 2016) allow to leaching of root zone salinity and sustaining crop production seems promising in productive utilization of salinity affected land and use of saline water.

6.6.2 Subsurface Drainage (SSD) for Rehabilitation of Continental Saline Soil with a Shallow Water Table

Surface or sub-surface or drainage is a feasible, reliable, and socially acceptable solution for dropping water table and washing of salts and to deliverable option a favorable salt-balance in surface soil (Plate 6.1). Perforated corrugated PVC pipe enclosed with synthetic filter mechanically fitted inappropriate design beneath the effective rooting depth to lower down saline water table and wash excess salts and water by gravitational action or hydraulic pump (Kamra et al. 2019) (Plate 6.1). The average cost of intervention and output per unit area is US$ 806 for one ha in IGP of northwest India and US$1007 for one ha for fine textured *Vertisols* soils in southern India (Chinchmalatpure et al. 2015). Several countries the USA, Pakisthan (Imran et al. 2021), Egypt, and Gulf countries adopted SSD technology to rehabilitate an appreciable area of waterlogged saline soil. Because of the increase in crop yields, SSD increases to three-fold in farmers' income.

6.6.3 Land Shaping Technology

Land modification/shaping techniques mainly alters the landscape by generating sunken and raised beds by alternatively excavating soil from one strip and placing it another (Plate 6.1). The altered land surface provides the scope for adopting integrated farming with versatile cropping throughout the year, harvesting water, generating irrigation facilities, managing salinity and increasing drainage congestions and improving soil health (Mandal et al. 2019b). The likely cost of intervention is about US$ 1329 per ha for soil excavation (Chinchmalatpure et al. 2015).

Plate 6.1 Reclamation technologies for rehabilitation of salt-affected soils

6.6.4 Bio-Drainage

Waterlogging in shallow saline soils with impeded drainage problems can be rehabilitated by bio-drainage. A shallow water table can be overcome by enabling the physiological transpiration of tree plants. This also generates extra advantage of timber and related ecological services and lowering of the water table (Plate 6.1). Benefits gained by greater cropping intensity up to 300% *vis-à-vis* greater nutrient use efficiency, to cultivate arable crops including pulses and oilseed, those are unable to grow on waterlogged soils, and generate employment (Dagar et al. 2016).

6.6.5 Gypsum and Alternate Reclamation Technology for Sodic Soil and Water

Mine gypsum (Abrol and Bhumbla 1979), FGD (flue gas desulfurization) (Zhao et al. 2020), elemental sulfur, acids, acid-formers (Ganjegunte et al. 2018), press mud (Sheoran et al. 2021a), fly ash (Mishra et al. 2019) phosphogypsum, aluminum sulphate/chloride (Luo et al. 2015), fluoro-gypsum or boro-gypsum, pyrites (Sharma and Swarup 1997), conjunctive use of gypsum with bio augmented material (Gupta et al. 2016)/or city waste compost (Sundha et al. 2018) (Rai et al. 2020b), etc. generally advocated for reclaiming soil sodicity. Sodicity reclamation may be one-time investment if irrigation water is safe for cultivation (Rai et al. 2020a). Rice-based cropping system reports its productivity potential mostly in three years from the application of amendments (Plate 2). Soil incipient sodicity generates as a concern of sodic irrigation, application of soluble Ca^{2+} (gypsum dissolution) or other amendment are advocated at the recurring interval. The application of gypsum or alkalinity neutralization is prescribed when RSC (residual sodium carbonate) of water applied for irrigation is more than 2.5 me L^{-1}. The net present worth (NPW) of gypsum-based sodicity reclamation technology is generally estimated to be US$ 698 per ha with BC (benefit cost ratio) of 1.43 and an IRR (internal rate of return) of technology is 25% (Table 6.3). The technology has been successfully implemented for increasing crop yield, improving health, increasing resouces use efficiency, raising farm income, minimizing flood hazards and waterlogging, and augmenting groundwater recharge. Performance of different alternative amendments alone or in conjunctive application to neutralize the sodicity stress are described in Table 6.4.

6.6.6 Crop Management and Salt-Tolerant Varieties

Cultivation of crops tolerant to salinity and sodicity can be a suitable option for harvesting economic yield and productive utilization of SAS. Less water requiring crops like oilseed, nutrient-dense coarse grain (quinoa), seed spices, biofuel, certain

Table 6.3 Gypsum based technology improve yield (Mg ha^{-1}) in sodic soils

Treatment	Ding et al. (2021)	Hussain et al. (2001)	Rasouli et al. (2013)	Sheoran et al. (2021a)	Zia et al. (2007)		Murtaza et al. (2019)	
Crops	Wheat	Wheat	Wheat	Rice	Rice (g pot^{-1})	Wheat (g pot^{-1})	Rice (g pot^{-1})	Wheat (g pot^{-1})
Initial soil pH	8.1	8.8	9.1	9.1	8.23		8.78	
Unamended control	4.4	2.0	2.0	2.2	46.0	33.9	15.0	20.4
Gypsum	5.6	3.2	3.8	2.8	61.0	40.2	17.5	20.9

fruit trees, forage (*A. lentiformis, D. palmeri*), agroforestry and oilseed halophyte (*S. bigelovii*) another halophyte *Batis maritime, Distichlis spicata, Juncus roemerianus, Paspalum vaginatums* can tolerate irrigation water salinity than the salinity sensitive cultivars. Large quantity biomass produced by *Sesbania aculeata* followed by *Leptochloa fusca, Echinochloa colona, Eleusine coracana*, respectively (Qadir et al. 1996). In coastal ecosystems, paddy is advocated because of receiving heavy rainfall in *kharif* season leaching of soluble electrolytes. Usual irrigation with sodic water is avoided for high water requiring crop rice and sugarcane. But, sodic water having low infiltrability and typically occurrence of '*kankar* layer' behind the plough layer facilitate waterlogging and accomplish noticeable paddy yield in some pockets of Indo-Gangetic Plain affected with alkalinity (Sheoran et al. 2021c). The list of salt-tolerant cultivers along with their alkalinity and salinity tolerance is given in Table 6.5.

6.7 Economic Importance of Salt-Affected Soil World-Wise

Rehabilitation of degraded lands due to salinization will contribute to accomplish the important global Sustainable Development Goals (SDGs) such as poverty alleviation, land and water resources conservation, food security and economic growth, and the preservation of livelihoods in rural areas (Negacz et al. 2021). Salinity is a consequence of natural and man made processes. The major cause of human-induced salinization is attributed to ill-designed, large-scale irrigation projects. It is estimated that the irrigation of cultivated lands shares for ~70% of underground and surface water demands which is leading to severe unsustainable agricultural water uses in many regions. Nearly one billion hectares which is ~7% of the global land area is presently salt-affected. Except the natural geochemical processes, ~30% of irrigated lands in the global are salt-affected because of irrigation led salinization (Hopmans et al. 2021). The groundwater lifting and surface irrigation disorders the natural water cycle, salt balance and cause stresses for downstream ecosystems and

Table 6.4 Alternate amendments of gypsum for sodicity reclamation and recommendation

Amendments	Recommendation
Farmyard manure/press mud	Application of FYM/press mud (3.75–5.0 Mg ha^{-1}) in conjunctive application with gypsum (3.75–5.0 Mg ha^{-1}) is advocated to sustain the productivity in areas having alkali groundwater for irrigation (Yaduvanshi and Swarup 2005; Sheoran et al. 2021a)
Farmyard manure and *Sesbania* green manure (GM)	Organic amendments mobilize precipitated $CaCO_3$ and release Ca. Green manure @ 20 Mg ha^{-1}, and wheat straw @ 6 Mg ha^{-1} are recommended before transplanting rice in sodic water irrigated areas (Choudhary et al. 2011)
Iron pyrite (contain S 22%)	Pyrites with 6 and 8% soluble sulfur are equally effective as gypsum. The Fe- and S-oxidizing bacteria of pyrite increase the solubility of S. Therefore, low S containing pyrites need to store before sodicity reclamation (Sharma and Swarup 1997)
Phosphogypsum	Phosphogypsum neutralizes soil alkalinity and declines soil ESP and therefore improved yield of rice and wheat than mine gypsum. PG @ 10.0 Mg ha^{-1} is advocated to improve aggregation, hydraulic conductivity, and rise soil biological properties (Nayak et al. 2013)
Inorganic polymer: polymeric aluminum ferric sulfate (PAFS)	PAFS generates H^+ through hydrolysis rapidly reduces soil pH and mobilizes $CaCO_3$ and produces Ca^{2+} which replaces exchangeable Na^+ in saline-sodic soils. For reclamation of surface sodicity an application of PAFS @ 15 Mg ha^{-1} is recommended (Luo et al. 2015)
Flue gas desulfurization gypsum (FGDG)	Synthetic gypsum is generated as a by-product of industrial processes. The small particle size of FGDG react between gypsum and sodic soil. FGDG amelioration is a safe, no ecological risk, and effective way to reclaim soil sodicity (Zhao et al. 2018)
Elemental S	By burning of S° produces acidified (dilute sulphuric acid) irrigation water which neutralizes alkali soil under saturated (thin layer of overlying surface water) and aerobic (60% WFPS). S° mobilizes the native $CaCO_3$ that is present in the root zone, resulting in the formation of $CaSO_4$. $CaSO_4$ on dissolution supplies soluble Ca^{2+} (Ganjegunte et al. 2018)
Wood chip biochar, biosolids co-compost (BSC), and green waste compost	Biochar application @ 75 Mg ha^{-1} was most likely due to the action of physico-chemical mechanisms. Biochar, biosolids compost and green waste compost improve the physico-chemical properties of a saline-sodic soil (Chaganti and Crohn 2015)

(continued)

Table 6.4 (continued)

Amendments	Recommendation
Conjunctive use of mineral gypsum (25GR) with MSWC (municipal solid waste compost) @ 10 Mg ha^{-1} with	Purity and quality of agricultural grade gypsum is an concern, therefore gypsum (GR25) and MSWC (10 Mg ha^{-1}) is recommended for neutralising alkalinity and salinity stress of soil underuse with sodic water (Sundha et al. 2020)

Table 6.5 Salt-tolerant cultivars

Crop	Tolerant cultivers	Abiotic stressors		
		Alkaline	Saline	Coastal saline
		$pH_{1:2}$	EC_e (dS m^{-1})	EC_e (dS m^{-1})
Paddy	CSR19, CSR23*, CSR27*, CSR30*, CSR36*, CSR36 and Lunishree, Vytilla 1, Vytilla 2, Vytilla 3, Vytilla 4, Panvel 1, Panvel 2 (India)	9.8–10.2	6–11	–
	BRRI dhan 40, BRRI dhan 41 (from Bangladesh); and OM2717, OM2517, OM3242 (from Vietnam)		6.11	–
	CSR1-3, CSR4*, CST7-1*, SR26B, Sumati*	–	6–9	4
Wheat	KRL 1-4*, WH157, Raj3077, KRL19*, KRL210, KRL213 (India)	<9.3	6–10	–
	S-24, LU-26S (Pakisthan)	<9.3		–
Indian mustard	Pusa Bold, Varuna	8.8–9.2	6–9	–
	Kranti, CS52*, CSTR330-1			
Raya	CST609-B 10, CS54*	8.8–9.3	6–9	–
Gram	Karnal Chana 1	<9.0	<6.0	–
Sugarbeet	Ramonskaaya 06, Maribo Resistapoly	9.5–10	<6.5	–
Sugarcane	Co453, Co1341	<9.0	EC_e–10	–

(Krishnamurthy et al. 2017; Hussain et al. 2021); http://www.knowledgebank.irri.org/ricebreedingcourse/Breeding_for_salt_tolerance.htm

habitats. Excessive irrigation develops soil salinity in dry regions with large quantity of salt content in the subsoil that occur further osmotic stress in the root zone and demands additional irrigation. Therefore, overall increases the cost of irrigation but reduces water productivity. Despite the extent and magnitude of the problem, accurate estimation and present statistics are not available at the global level. The best available estimates propose salinity and sodicity with an area of ~412 and 618 million ha (UNEP 1992), however, this data does not discriminate areas where salinity and

Table 6.6 Salt-affected soils distributed in different continents (Mha) (UNEP 1992)

Continent	Saline soils	Sodic soils	Total
Africa	122.9	86.7	209.6
South Asia	82.3	1.8	84.1
North and Central Asia	91.5	120.2	211.7
Southeast Asia	20.0	–	20.0
South America	69.5	59.8	129.3
North America	6.2	9.6	15.8
Mexico/Central America	2.0	–	2.0
Australasia	17.6	340.0	357.6
World	412.0	618.0	1030.0

sodicity occur together (Table 6.6). However, there is a skewed distribution of salt-affected land in various countries with a range of 9–34%, and a world average of 20% (Ghassemi et al. 1995). Salinization reduces crop yields and beyond certain thresholds causes complete losses of crop yield. Globally, every year, soil salinization decreases the production potential of 46 Mha of land and takes up to 1.5 Mha of farmland (per year) out of production. The annual loss in agricultural productivity affected by salinization is near US$ 31 million (FAO 2015).

Several researchers corroborated the yield penalty because of the widespread problem of salinity and shortage of freshwater irrigation. However, the types and cultivars of crops have major role in tolerance to salinity stress (Watson 2000). Depending on the crop cultivated, category of land degradation and its intensity, irrigation water characteristic, provision and drainage network congestion and on-farm soil, water and crop management, the saline areas of India, Pakisthan and Kazakhstan reported a wide variation in yield penalty from 40–63, 36–69 and 6–71%, respectively. Salt-regulated land degradation estimated a cost value of US$ 264 ha^{-1} in 1990. The global data showed that because of inflation-adjustment the cost of salt-affected land degradation hiked to US$ 441 ha^{-1} in 2013 (Qadir et al. 2014). Further, the aggregated total annual economic loss was US$30 billion at the whole Globe (Shahid et al. 2018). Economic costs for salt-affected lands might be multifaceted firstly because of yield penalty and additionally investment of higher input. Further, type of salinization (salinity, sodicity or water-born salinity/sodicity), the degree of salinization (the current state of salinization), the types of crops grown in the targeted region, the market price of cultivated crops, stressors (e.g. climate change-mediated sea-level rise/brackish water inundation) and farm-level decisions to reclaim the salinity hazard (which may include the adopting of salt-tolerant crops/cultivars). Yield losses are particularly detrimental at the farm level but often not realized at macro level estimation (Mandal et al. 2018). Based on farm level data, a case study on small-holder farmers in India estimated the production loss of 0.16, 0.45 and 1.35 Mg ha^{-1} for the crops, paddy (dry/winter season), tomato and potato, respectively, under coastal salt-affected soils in India. The value of crop production loss was estimated to be US$ 32 ha^{-1}, US$ 50 ha^{-1}, and US$ 116 ha^{-1} for the same paddy, tomato and potato crops (Mandal et al. 2019a). India loses 16.8 Mt of farm production (cereals, pulses, oilseeds, and cash crops) valued at US$ 3.1 billion (₹230.2 billion) per annum because of salinity and associative constraints (moving average data during 2012–2014) (Mandal et al. 2019a). The consideration of other cost components such as the environmental costs certainly hiked the total cost of cultivation at SAS (Negacz et al. 2021).

Food production under saline conditions and innovation in this field can help to create an economic and social perspective for the SAS affected regions and populations. The most important factor for scaling up the innovations is the salinity level. Each salinity level has its economics which should not be compared with conventional agriculture. The scaling up must be done where similar soil, water, and environmental conditions may be existing, similar to where technology is going to be transferred. There must be a reason for scaling up based on market demand (Negacz et al. 2021). It is imperative to capture the socio-economic costs associated to irreversible groundwater withdrawal and concomitantly degrading water resources and showing the true economic value of water. Water quantity and quality are interlinked, so that deterioration of water quality must be understanded when using poor quality water use for irrigation. Policy decisions will vary in special and temporal variability, with changes in economic development and public favourites regarding the impacts of irrigation on its society and the environment (Hopmans et al. 2021). There is a need for a transdisciplinary approach in developing and implementing a "Research to Development Continuum" to ensure a continuous interaction between farmers, researchers, marketers, and investors (Negacz et al. 2021) for the management of SAS and harnessing benefits out of technological interventions. Irrigation-induced soil salinization and sodification processes have been extensively studied globally, however, the magnitude of their impacts on soils as well as ecosystem services are still largely remained unknown or unaccounted. If adequate restoration strategies are not targeted or not managed carefully, salinization further impacts on the prices of produces grown in vulnerable areas, which will increase the risk of food uncertainty in the affected regions and the mass migration of peoples may occur in the coming future (FAO 2021).

6.8 Conclusions and Way Forward

Salt balance is an indispensable part of salt-affected areas and irrigated agriculture. Salt-affected soils are categorized as saline and sodic/saline-sodic. The Land Degradation Neutrality, Sustainable Development Goals, and several other environmental policies of Globe underscores known amendments (gypsum) and searching locally available amendments for sustainably enhancing the crop resilience and affordable solutions for arresting sodicity induced soil and land degradation. Oppositely necessity of pilot-scale base community project subsurface drainage for managing waterlogged saline soil. Agro-forestry-based bio drainage needs a definite time for managing waterlogged soil. Various agronomic practices which mitigate the influence of projected climate change, advancement of conservation agricultural practices (such as zero tillage, bed planting, residue management, the inclusion of legume, crop rotation, and diversification). Screening salt, moisture, and other abiotic stress-tolerant cultivars are advocated for increasing the use efficiency of water, nutrients, and energy conservation and farm inputs. Promoting crop insurance, multi-enterprise, mechanical and precision farming can cope up the impact of weather aberrations due to climate change.

References

Abdel-Fattah MK, Abd-Elmabod SK, Aldosari AA, et al (2020) Multivariate analysis for assessing irrigation water quality: a case study of the Bahr Mouise Canal, Eastern Nile Delta. Water 12

Abrol IP, Bhumbla DR (1979) Crop responses to differential gypsum applications in a highly sodic soil and the tolerance of several crops to exchangeable sodium under field conditions. Soil Sci 127

Abrol IP, Yadav JSP, Massoud FI (1988) Salt-affected soils and their management. FAO Soils Bulletin No. 39, Food and Agriculture Organization of the United Nations, Rome

Aquastat (2016) FAO's information system on water and agriculture [WWW Document]

Bajwa MS, Swarup A (2012) Soil salinity and alkalinity. In: Goswami NN, Rattan RK, Dev G, et al (eds) Fundamentals of soil science, 2nd edn. New Delhi, pp 329–339

Barakat M, Cheviron B, Angulo-Jaramillo R (2016) Influence of the irrigation technique and strategies on the nitrogen cycle and budget: a review. Agric Water Manag 178:225–238. https://doi.org/10.1016/j.agwat.2016.09.027

Barman A, Basak N, Narjary B, Mitran T (2021) Land degradation assessment using geospatial techniques BT—geospatial technologies for crops and soils. In: Mitran T, Meena RS, Chakraborty A (eds) Springer, Singapore, pp 421–453

Basak N, Chaudhuri SK, Sharma DK (2015) Influence of water quality on exchange phase-solution phase behavior of texturally different salt-affected soils. J Indian Soc Soil Sci 64:365–372. https://doi.org/10.5958/0974-0228.2015.00048.1

Blankespoor B, Dasgupta S, Lange G-M (2017) Mangroves as a protection from storm surges in a changing climate. Ambio 46:478–491. https://doi.org/10.1007/s13280-016-0838-x

Cai X, Sharma BR, Matin MA, et al (2010) An assessment of crop water productivity in the Indus and Ganges river basins: current status and scope for improvement. IMWI Research Report 140. Columbo

Chaganti VN, Crohn DM (2015) Evaluating the relative contribution of physiochemical and biological factors in ameliorating a saline–sodic soil amended with composts and biochar and leached with reclaimed water. Geoderma 259–260:45–55. https://doi.org/10.1016/j.geoderma.2015.05.005

Chang AC, Silva DB (2014) Prologue BT—salinity and drainage in San Joaquin Valley, California: science, technology, and policy. In: Chang AC, Brawer Silva D (eds) Springer Netherlands, Dordrecht, pp 1–6

Chinchmalatpure A, Ali S, Kulshrestra N, et al (2015) Intellectual property management and commercialization of ICAR-CSSRI technologies for management of salt-affected and waterlogged soils of India. ICAR-Central Soil Salinity Research Institute, Haryana

Choudhary OP, Ghuman BS, Bijay-Singh, et al (2011) Effects of long-term use of sodic water irrigation, amendments and crop residues on soil properties and crop yields in rice–wheat cropping system in a calcareous soil. Field Crop Res 121:363–372. https://doi.org/10.1016/j.fcr.2011.01.004

Dagar JC, Lal K, Ram J, et al (2016) Eucalyptus geometry in agroforestry on waterlogged saline soils influences plant and soil traits in North-West India. Agric Ecosyst Environ 233:33–42. https://doi.org/10.1016/j.agee.2016.08.025

Dasgupta S, Hossain MM, Huq M, Wheeler D (2015) Climate change and soil salinity: the case of coastal Bangladesh. Ambio 44:815–826. https://doi.org/10.1007/s13280-015-0681-5

Datta A, Setia R, Barman A, et al (2019) Carbondynamics in salt-affected soils. In: Dagar J, Yadav R, Sharma P (eds) Research developments in saline agriculture. Springer, Singapore

Ding Z, Kheir AMS, Ali OAM, et al (2021) A vermicompost and deep tillage system to improve saline-sodic soil quality and wheat productivity. J Environ Manag 277:111388. https://doi.org/10.1016/j.jenvman.2020.111388

Falloon P, Betts R (2010) Climate impacts on European agriculture and water management in the context of adaptation and mitigation—the importance of an integrated approach. Sci Total Environ 408:5667–5687. https://doi.org/10.1016/j.scitotenv.2009.05.002

FAO-AQUASTAT (2013) Area equipped for irrigation and percentage of cultivated land. http://www.fao.org/nr/water/aquastat/globalmaps/index.stm. Accessed 16 Sep 2013

FAO (2021) Salt-affected soils are global issue, ITPS Intergovernmental Technical Panel on Soils, Soil Letters, No 3

FAO I (2015) Status of the world's soil resources—main report. Food and Agriculture Organization of the United Nations and Intergovernmental Technical Panel on Soils, Rome, Italy

Fishman R (2018) Groundwater depletion limits the scope for adaptation to increased rainfall variability in India. Clim Change 147:195–209. https://doi.org/10.1007/s10584-018-2146-x

Ganjegunte GK, Clark JA, Parajulee MN, et al (2018) Salinity management in pima cotton fields using sulfur burner. Agrosyst Geosci Environ 1:180006. https://doi.org/10.2134/age2018.04.0006

Ghassemi F, Jakeman AJ, Nix HA (1995) Stalinization of land and water resources: human causes, extent, management and case studies. CABI, Wallingford, UK

Gupta M, Srivastava PK, Shikha, et al (2016) Use of a bioaugmented organic soil amendment in combination with gypsum for withania somnifera growth on sodic soil. Pedosphere 26:299–309. https://doi.org/10.1016/S1002-0160(15)60044-3

Gupta RN, Mathur A (2011) Origin, distribution and classification of the salt-affected soils of India. In: Rattan R, Singh AK (eds) Bulletin of the Indian society of soil science. Indian Society of Soil Science, New Delhi, India, pp 1–28

Hansen JA, Jurgens BC, Fram MS (2018) Quantifying anthropogenic contributions to century-scale groundwater salinity changes, San Joaquin Valley, California, USA. Sci Total Environ 642:125–136. https://doi.org/10.1016/j.scitotenv.2018.05.333

Hayat K, Bundschuh J, Jan F et al (2020) Combating soil salinity with combining saline agriculture and phytomanagement with salt-accumulating plants. Crit Rev Environ Sci Technol 50:1085–1115. https://doi.org/10.1080/10643389.2019.1646087

Hopmans JW, Qureshi AS, Kisekka I, et al (2021) Chapter one—Critical knowledge gaps and research priorities in global soil salinity. In: Sparks DLBT-A in A (ed) Academic Press, pp 1–191

Hussain N, Ghaffar A, Zafar ZU et al (2021) Identification of novel source of salt tolerance in local bread wheat germplasm using morpho-physiological and biochemical attributes. Sci Rep 11:10854. https://doi.org/10.1038/s41598-021-90280-w

Hussain N, Hamdy G, Arshadullah M, Mujeeb F (2001) Evaluation of amendments for the improvement of physical properties of sodic soil. Int J Agric Biol 3:319–322

Imran MA, Xu J, Sultan M, et al (2021) Free discharge of subsurface drainage effluent: an alternate design of the surface drain system in Pakistan. Sustain 13

Ivushkin K, Bartholomeus H, Bregt AK, et al (2019) Global mapping of soil salinity change. Remote Sens Environ 231:111260. https://doi.org/10.1016/j.rse.2019.111260

Jobbágy EG, Tóth T, Nosetto MD, Earman S (2017) On the fundamental causes of high environmental alkalinity (ph $\geq$ 9): an assessment of its drivers and global distribution. L Degrad Dev 28:1973–1981. https://doi.org/10.1002/ldr.2718

Kamra SK, Kumar S, Kumar N, Dagar JC (2019) Engineering and biological approaches for drainage of irrigated lands BT—research developments in saline agriculture. In: Yadav RK, Sharma PC (eds) Dagar JC. Springer, Singapore, pp 537–577

Krishnamurthy SL, Sharma PC, Sharma DK et al (2017) Identification of mega-environments and rice genotypes for general and specific adaptation to saline and alkaline stresses in India. Sci Rep 7:7968. https://doi.org/10.1038/s41598-017-08532-7

Liu J, Rong Q, Zhao Y (2017) Variations in soil nutrients and salinity caused by tamarisk in the coastal wetland of the Laizhou Bay, China. Ecosphere 8:e01672. https://doi.org/10.1002/ecs2.1672

Luo J-Q, Wang L-L, Li Q-S, et al (2015) Improvement of hard saline–sodic soils using polymeric aluminum ferric sulfate (PAFS). Soil Tillage Res 149:12–20. https://doi.org/10.1016/j.still.2014.12.014

Mahajan GR, Das B, Manivannan S, et al (2021) Soil and water conservation measures improve soil carbon sequestration and soil quality under cashews. Int J Sediment Res 36:190–206. https://doi.org/10.1016/j.ijsrc.2020.07.009

Mandal AK, Obi Reddy GP, Ravisankar T (2011) Digital database of salt affected soils in India using geographic information system. J Soil Salin Water Qual 3:16–29

Mandal S, Raju R, Kumar A et al (2018) Current status of research, technology response and policy needs of salt-affected soils in India—a review. J Indian Soc Coast Agric Res 36:40–53

Mandal S, Mandal UK, Lama TD et al (2019a) Economic analysis of farm-level agricultural risks in coastal region of West Bengal in India. J Soil Salin Water Qual 11:269–279

Mandal UK, Burman D, Bhardwaj AK, et al (2019b) Waterlogging and coastal salinity management through land shaping and cropping intensification in climatically vulnerable Indian Sundarbans. Agric Water Manag 216:12–26. https://doi.org/10.1016/j.agwat.2019b.01.012

Mashali AM (1995) Integrated soil management for sustainable use of salt-affected soils and network activities. In: Proceedings of the international workshop on integrated soil management for sustainable use of salt-affected soils. Bureau of soils and water. Manila, Philippines, pp 55–75

Mateo-Sagasta J, Burke J (2011) Agriculture and water quality interactions: a global overview (SOLAW Background Thematic Report-T R08, FAO)

Minhas PS, Ramos TB, Ben-Gal A, Pereira LS (2020) Coping with salinity in irrigated agriculture: crop evapotranspiration and water management issues. Agric Water Manag 227:105832. https://doi.org/10.1016/j.agwat.2019.105832

Mishra VK, Jha SK, Damodaran T, et al (2019) Feasibility of coal combustion fly ash alone and in combination with gypsum and green manure for reclamation of degraded sodic soils of the Indo-Gangetic plains: a mechanism evaluation. L Degrad Dev 30:1300–1312. https://doi.org/10.1002/ldr.3308

Murtaza G (2013) Economic aspects of growing rice and wheat crops on salt-affected soils in the Indus Basin of Pakistan (unpublished data). Institute of Soil and Environmental Sciences, University of Agriculture, Faisalabad, Pakistan

Murtaza B, Zaman G, Imran M et al (2019) Municipal solid waste compost improves crop productivity in saline-sodic soil: a multivariate analysis of soil chemical properties and yield response. Commun Soil Sci Plant Anal 50:1013–1029. https://doi.org/10.1080/00103624.2019.1603305

Murtaza G, Rehman MZ, Qadir M et al (2021) High residual sodium carbonate water in the Indian subcontinent: concerns, challenges and remediation. Int J Environ Sci Technol. https://doi.org/10.1007/s13762-020-03066-4

Nachshon U (2018) Cropland soil salinization and associated hydrology: trends, processes and examples. Water 10

Nayak AK, Mishra VK, Sharma DK et al (2013) Efficiency of phosphogypsum and mined gypsum in reclamation and productivity of rice-wheat cropping system in sodic soil. Commun Soil Sci Plant Anal 44:909–921. https://doi.org/10.1080/00103624.2012.747601

Negacz K, Vellinga P, Barrett-Lennard E, et al (eds) (2021) Future of sustainable agriculture in saline environmentsnull. Taylor & Francis

Neubauer SC, Piehler MF, Smyth AR, Franklin RB (2019) Saltwater intrusion modifies microbial community structure and decreases denitrification in tidal freshwater marshes. Ecosystems 22:912–928. https://doi.org/10.1007/s10021-018-0312-7

Pitman AJ, Narisma GT, Pielke Sr. RA, Holbrook NJ (2004) Impact of land cover change on the climate of southwest Western Australia. J Geophys Res Atmos 109. https://doi.org/10.1029/2003JD004347

Qadir M, Qureshi RH, Ahmad N, Ilyas M (1996) Salt-tolerant forage cultivation on a saline-sodic field for biomass production and soil reclamation. L Degrad Dev 7:11–18. https://doi.org/10.1002/(SICI)1099-145X(199603)7:1<11::AID-LDR211>3.0.CO;2-C

Qadir M, Quillérou E, Nangia V, et al (2014) Economics of salt-induced land degradation and restoration. Nat Resour Forum 38:282–295. https://doi.org/10.1111/1477-8947.12054

Qadir M, Schubert S, Oster JD, et al (2018) High-magnesium waters and soils: emerging environmental and food security constraints. Sci Total Environ 642:1108–1117. https://doi.org/10.1016/j.scitotenv.2018.06.090

Rai AK, Basak N, Sundha P (2020a) Chemistry of salt-affected soils. In: Minhas PS, Yadav RK, Sharma PC (eds) Managing Salt-Affected Soils for Sustainable Agriculture. Directorate of Knowledge Management in Agriculture, Indian Council of Agricultural Research (ICAR), New Delhi, India

Rai AK, Basak N, Sundha P (2020b) Alternate reclamation sources of sodic soils. In: Minhas PS, Yadav RK, Sharma PC (eds) Managing Salt-Affected Soils for Sustainable Agriculture. Directorate of Knowledge Management in Agriculture, Indian Councilof Agricultural Research (ICAR), New Delhi

Rai AK, Basak N, Soni PG, Kumar S, Sundha P, Narjary B, Yadav G, Patel S, Kaur H, Yadav RK, Sharma PC (2022) Bioenergy sorghum as balancing feedback loop for intensification of cropping system in salt-affected soils of the semi–arid region: energetics biomass quality and soil properties. Eur J Agron 134:126452 https://doi.org/10.1016/j.eja.2021.126452

Raiesi F, Kabiri V (2016) Identification of soil quality indicators for assessing the effect of different tillage practices through a soil quality index in a semi-arid environment. Ecol Indic 71:198–207. https://doi.org/10.1016/j.ecolind.2016.06.061

Ranatunga K, Nation ER, Barodien G (2010) Potential use of saline groundwater for irrigation in the Murray hydrogeological basin of Australia. Environ Model Softw 25:1188–1196. https://doi.org/10.1016/j.envsoft.2010.03.028

Rasouli F, Kiani Pouya A, Karimian N (2013) Wheat yield and physico-chemical properties of a sodic soil from semi-arid area of Iran as affected by applied gypsum. Geoderma 193–194:246–255. https://doi.org/10.1016/j.geoderma.2012.10.001

Rengasamy P (2006) World salinization with emphasis on Australia. J Exp Bot 57:1017–1023. https://doi.org/10.1093/jxb/erj108

Rengasamy P (2016) Soil chemistry factors confounding crop salinity tolerance—a review. Agronomy 6:53

Roxy MK, Ritika K, Terray P et al (2015) Drying of Indian subcontinent by rapid Indian Ocean warming and a weakening land-sea thermal gradient. Nat Commun 6:7423. https://doi.org/10.1038/ncomms8423

SAC (2016) Desertification and Land Degradation Atlas of India (Based on IRS AWiFS data of 2011–13 and 2003–05). Ahmedabad, India

Shahid SA, Zaman M, Heng L (2018) Soil salinity: historical perspectives and a world overview of the problem BT—guideline for salinity assessment, mitigation and adaptation using nuclear and related techniques. In: Shahid SA, Heng L (eds) Zaman M. Springer International Publishing, Cham, pp 43–53

Sharma P, Swarup A (1997) Comparison of pyrites varying in water-soluble sulfur with gypsum for the reclamation of alkali soils under a rice-wheat rotation. Biol Fertil Soils 24:96–101. https://doi.org/10.1007/BF01420227

Sharma DK, Thimmappa K, Chinchmalatpure AR, Mandal AK, et al (2015) Assessment of production and monetary losses from salt-affected soils in India. Technical Bulletin. Karnal, Haryana, India

Sharma SK, Kulshreshtha N, Kumar A et al (2018) Waterlogging effects on elemental composition of wheat genotypes in sodic soils. J Plant Nutr 41:1252–1262. https://doi.org/10.1080/01904167.2018.1434541

Sheoran P, Basak N, Kumar A, et al (2021a) Ameliorants and salt tolerant varieties improve rice-wheat production in soils undergoing sodification with alkali water irrigation in Indo–Gangetic Plains of India. Agric Water Manag 243:106492. https://doi.org/10.1016/j.agwat.2020.106492

Sheoran P, Kumar A, Sharma R, et al (2021b) Managing sodic soils for better productivity and farmers' income by integrating use of salt tolerant rice varieties and matching agronomic practices. Field Crop Res 270:108192. https://doi.org/10.1016/j.fcr.2021c.108192

Shirale AO, Kharche VK, Wakode RR et al (2018) Influence of gypsum and organic amendments on soil properties and crop productivity in degraded black soils of central India. Commun Soil Sci Plant Anal 49:2418–2428. https://doi.org/10.1080/00103624.2018.1510952

Sione SMJ, Wilson MG, Lado M, González AP (2017) Evaluation of soil degradation produced by rice crop systems in a Vertisol, using a soil quality index. CATENA 150:79–86. https://doi.org/10.1016/j.catena.2016.11.011

Soni PG, Basak N, Rai AK et al (2021) Deficit saline water irrigation under reduced tillage and residue mulch improves soil health in sorghum-wheat cropping system in semi-arid region. Sci Rep 11:1880. https://doi.org/10.1038/s41598-020-80364-4

Sundha P, Basak N, Rai A, et al (2017) N and P release pattern in saline-sodic soil amended with gypsum and municipal solid waste compost. J Soil Salinity Water Qual 9:145–155

Sundha P, Basak N, Rai AK et al (2018) Utilization of municipal solid waste compost in reclamation of saline-sodic soil irrigated with poor quality water. J Indian Soc Soil Sci 66:28–39

Sundha P, Basak N, Rai AK, et al (2020) Can conjunctive use of gypsum, city waste composts and marginal quality water rehabilitate saline-sodic soils? Soil Tillage Res 200:104608. https://doi.org/10.1016/j.still.2020.104608

Sundha P, Rai AK, Basak N, et al (2022) P solubility and release kinetics in the leachate of saline–sodic soil: effect of reclamation strategies and water quality. Soil Tillage Res 222:105440. https://doi.org/10.1016/j.still.2022.105440

Tavakkoli E, Rengasamy P, Smith E, McDonald GK (2015) The effect of cation–anion interactions on soil pH and solubility of organic carbon. Eur J Soil Sci 66:1054–1062. https://doi.org/10.1111/ejss.12294

UNEP (1992) Proceedings of the Ad-hoc expert group meeting to discuss global soil databases and appraisal of GLASOD/SOTER, February 24–28. Nairobi, UNEP

Vyshpolsky F, Mukhamedjanov K, Bekbaev U, et al (2010) Optimizing the rate and timing of phosphogypsum application to magnesium-affected soils for crop yield and water productivity enhancement. Agric Water Manag 97:1277–1286. https://doi.org/10.1016/j.agwat.2010.02.020

Wasko C, Sharma A (2017) Global assessment of flood and storm extremes with increased temperatures. Sci Rep 7:7945. https://doi.org/10.1038/s41598-017-08481-1

Watson W, Hall NH, Hamblin A (2000) Economic aspects of sodic, acid and saline soils in Australia. Tatura, VIC, Australia, p 3

Xu X, Huang G, Sun C, et al (2013) Assessing the effects of water table depth on water use, soil salinity and wheat yield: searching for a target depth for irrigated areas in the upper Yellow River basin. Agric Water Manag 125:46–60. https://doi.org/10.1016/j.agwat.2013.04.004

Yaduvanshi NPS, Swarup A (2005) Effect of continuous use of sodic irrigation water with and without gypsum, farmyard manure, pressmud and fertilizer on soil properties and yields of rice and wheat in a long term experiment. Nutr Cycl Agroecosyst 73:111–118. https://doi.org/10.1007/s10705-005-3361-1

Zhao Y, Wang S, Li Y, et al (2018) Extensive reclamation of saline-sodic soils with flue gas desulfurization gypsum on the Songnen Plain, Northeast China. Geoderma 321:52–60. https://doi.org/10.1016/j.geoderma.2018.01.033

Zhao Y, Zhang W, Wang S, et al (2020) Effects of soil moisture on the reclamation of sodic soil by flue gas desulfurization gypsum. Geoderma 375:114485. https://doi.org/10.1016/j.geoderma.2020.114485

Zia MH, Sabir M, Ghafoor A, Murtaza G (2007) Effectiveness of Sulphuric Acid and Gypsum for the Reclamation of a Calcareous Saline-Sodic Soil Under Four Crop Rotations. J Agron Crop Sci 193:262–269. https://doi.org/10.1111/j.1439-037X.2007.00262.x

Zwart SJ, Bastiaanssen WGM (2004) Review of measured crop water productivity values for irrigated wheat, rice, cotton and maize. Agric Water Manag 69:115–133. https://doi.org/10.1016/j.agwat.2004.04.007

Chapter 7
Application of Remote Sensing and GIS Techniques in Assessment of Salt Affected Soils for Management in Large Scale Soil Survey

R. Srinivasan, M. Lalitha, M. Chandrakala, S. Dharumarajan, and Rajendra Hegde

Abstract Salt-affected soils are most common land degradation processes in arid and semi-arid regions, where evaporation exceeds precipitation. Under such climatic conditions, the soluble salts are accumulates in the soil influence the soil properties and crop productivity as well. Therefore, mapping of saline and sodic soils is essential for understanding the soil resource for sustainable use and management. The extent of primary salt-affected soils in the world is about 955 Mha, as secondary salinization affects some 77 Mha out of which 58% are in irrigated areas. Estimates reveal that nearly 20% of all irrigated land is salt-affected, and the proportion tends to increase in spite of reclamation. Hence careful monitoring of the soil salinity and sodicity status is required for change detection and identification of hotspot areas for arresting the soil degradation. Remote sensing has surpassed the traditional method of assessing the soil salinity and sodicity areas offering more rapid and informative assessment techniques for monitoring and mapping. Multi-temporal, optical and microwave remote sensing has significant role in detecting temporal and spatial changes of salt-related surface features. Airborne geophysics and ground based electromagnetic induction meters, combined with ground truth data, have shown potential for mapping depth of salinity occurrence. Recent satellite sensors (e.g., Resourcesat-1, Cartosat-1, IKONOS I, and RISAT-2), along with improved image processing techniques integrated with terrain and other spatial data using a geographic information system, are enabling mapping at large scale. The variations in salt encrustation at the surface imposed by soil moisture variation, water logging, vegetative barrier, and dynamics of subsurface salts present constraints could be overcome through recent satellite sensors and mapped by using spatial techniques for better management.

Keywords Soil salinity · Alkalinity · Remote sensing · GIS · Assessment and monitoring

R. Srinivasan (✉) · M. Lalitha · M. Chandrakala · S. Dharumarajan · R. Hegde
ICAR - National Bureau of Soil Survey and Land Use Planning, Regional Centre, Bangalore 560024, India
e-mail: R.Srinivasan@icar.gov.in

P. K. Shit et al. (eds.), *Soil Health and Environmental Sustainability*, Environmental Science and Engineering,
https://doi.org/10.1007/978-3-031-09270-1_7

7.1 Introduction

Soil Salinity and Sodicity are found in all continents under different climatic conditions and are a major threat to agriculture. These soils are present most extensively in arid and semiarid regions of the world and cover approximately 7% of the total land area of the Earth (Ghassemi et al. 1995; Singh 2005; Hillel 2000). The United Nations Food and Agriculture Organization (FAO) has estimated that saline soil covered 397 million hectares of the total land area of the world (Koohafkan 2012). These soils occur as large stretches of salt-encrusted lands and isolated patches interspersed with normal soils, forming no contiguous pattern and decreasing soil quality. Soil Salinity and Sodicity are considered major salt affected soils (SAS) in global due to its adverse impact on the ecosystems, environment, agricultural productivity and sustainability. Thus adversely affects plant growth, crop production, soil and water quality, and it eventually results in soil erosion and land degradation (Zhu 2001; Corwin and Lesch 2003; Srinivasan et al. 2017a). Salt affected soils impacts are not limited only to the environment but also extend to the national economy.

Salt affected soils, the second major cause of land degradation after soil erosion, has been a cause of decline in agricultural societies for 10,000 years. Globally about 2000 ha of arable land is lost to production every day due to salinity and sodicity. Salt affected soils can cause yield decreases of 10–25% for many crops and may prevent cropping altogether when it is severe and lead to desertification. Addressing problem soils through improved soil, water and crop management practices is important for achieving sustainable food security and to avoid desertification (Srinivasan et al. 2018a). Generally, Salt affected soils present in arid and semi-arid regions where rainfall is poor or insufficient to meet the water requirements of the crops, and leaching the mineral salts out of the root-zone. The association between humans and salinity has existed for centuries and historical records show that many civilizations have failed due to increases in the salinity of agricultural fields. Problem soils undermine the resource base by decreasing the soil quality and can occur due to natural causes or from misuse and mismanagement.

Effective soil resource use and different management requires scientific based understanding of soil salinization and sodification. It is particularly important in the regions where salinity occurs, to determine extent and risk of salinity, of which salinity mapping and regular monitoring has a great role to play. Salinity and sodicity information at local, regional and national levels, as well as in irrigated fields, therefore, becomes very important for decision making and resources management. Managing salt affected soils (SAS) is highly site specific and depends on factors such as nature of soils, content soluble salts and hydrological conditions (Singh et al. 2010). Therefore, for greater development and implementation of sufficient soil reclamation programs and preventing any further SAS to sustain agricultural lands and natural systems, information on the spatial extent, nature and distribution of SAS is becoming very essential. Thus, timely detection of soil salinity and sodicity, monitoring and assessment of its severity level and extent become very important at local and regional scales.

7.2 Development of Salt-Affected Soils

The development of Salt-Affected Soils (SAS) depends on climate, topography, geology, soil mineral weathering, drainage, irrigation source, hydrology, ground water depth and quality, and management practices (Ghassemi et al. 1995). Accumulation of sodium or neutral salts in soils over a period leading to the formation of saline, saline-alkali, or alkali soils may be compounded by natural or irrigation-induced factors, such as weathering of natural salt bearing soil minerals; irrigation with salt rich waters; and water logging due to a rising ground water table (Zaman et al. 2018).

The distribution pattern of rainfall in India varies with two monsoons season; southwest (SWM) and northeast (NEM) monsoon are prevalent. The SWM influences soil salinity or sodicity in arid and semiarid regions. The rainy season (mid-June to September), referred as "*kharif*" or crop period, when about 80% of the rainfall occurs. The second phase is the cool and dry season (October to March), referred to as the "*rabi*" (winter) crop period. The third phase is the hot and dry weather (April to mid-June) is called summer. The winter and summer periods are dry, water-deficit, whereas the *kharif* crop season has surplus water. The build-up of salts in soils is significantly influenced by wet and dry cycles set in by the monsoon and prevailing water or irrigation practices. Summer period, the land remains fallow, and an upward moisture flux or movement is dominant due to high evaporation demand (5–10 mm d^{-1}), which resulted a buildup salts. The maximum possible quantity concentration of salts in croplands (up to 12 dS m^{-1}) and non-arable lands (>12 dS m^{-1}) is observed in the pre-monsoonal period (may–June) in waterlogged saline areas (Tyagi 2003; Srinivasan et al. 2018b). When the monsoon and the planting of crop start in October, the desalinization takes place in the soils, and salt levels reaches their minimum. From November to February, the evaporative demands are low, but the upward flux begins to increase. This favors irrigation with saline, alkali, and saline-alkali ground waters in areas of deficit canal water supply, leading to increase in soil salinity/alkalinity. Salt-affected soils, including waterlogged, contain excess concentrations of soluble neutral salts and exchangeable sodium (Na^{+}), or both, impairs the seed germination and plant growth, leads to poor crop permanence and yield (Srinivasan et al. 2017b) (Fig. 7.1).

7.3 Characterization and Identification of Salt-Affected Soils

Salt-affected soils are classified into saline, saline-alkali, and alkali soils on the basis of soil reaction of saturation paste (pHs), electrical conductivity of saturation paste (ECe), exchangeable sodium percentage (ESP), and the sodium adsorption ratio (SAR) (Table 7.1). The Indian system of classification for characterization of SAS is essentially the same as that of the USDA (U.S. Salinity Laboratory Staff

During rainy season water logging condition | During cropping period poor crop growth | During summer salt encrustation on surface of the soils

Fig. 7.1 Development of salt affected soils in Coastal West Bengal (Canning II block)

Table 7.1 Classification of SAS according to USDA system (adapted from USSL (1954) and Eynard et al. (2006)

Soil class	ECe†	pHs	ESP	SAR
	$dS\ m^{-1}$			$(m\ molc\ L^{-1})^{1/2}$
Non-saline, non-alkali	<4.0	<8.5	<15	<13
Saline	>4.0	<8.5	<15	<13
Alkali	<4.0	>8.5	>15	>13
Saline-alkali	>4.0	<8.5	>15	>13

† ECe, electrical conductivity of saturation extract; ESP, exchangeable sodium percentage; pHs, pH of saturation paste; SAR, sodium adsorption ratio

1954), except the pH criteria was reconsidered from 8.5 to 8.2 because this value of pH initiates the sodification process and is associated with an ESP of 15–20 (Abrol et al. 1980). Unlike the USDA classification, the Indian classification system for reclamation has classified SAS into two main categories: saline or alkali. The saline-alkali soil category is reconsidered to be saline or alkali based on a ratio of $(CO_2{}^{3-} + HCO^{3-})/(Cl^- + SO_2{}^{4-})$ or $Na^+/(Cl^- + SO_2{}^{4-})$. If the ratio is >1, the saline-alkali soil is treated as alkali; if the ratio is <1, the soil is treated as saline.

Under the shallow water table (within 2 m) conditions, saline soils can be identified by the presence of a light gray to dull white crust of chlorides of Na, Ca and K salts on the surface, good physical conditions, high permeability, and patchy, stunted and wilted plant growth that is often deep green to bluish color even when the soil contains enough moisture. Natural halophytic grasses (Cyperus and Chloris) are grown on such soils (Singh 2005). Alkali soils are identified by the presence of white or dull white crust of $NaHCO_3$ (sodium bicarbonate), Na_2CO_3 (sodium carbonate) or both salts on the soil surface with low permeability and poor physical conditions caused by deflocculation of the Na^+ ion. The black color brought in alkali soils is due to the dispersion of organic matter and clay at high soil pH. These soils turn black, slippery, and soft when wet and very hard when dry. Characterization

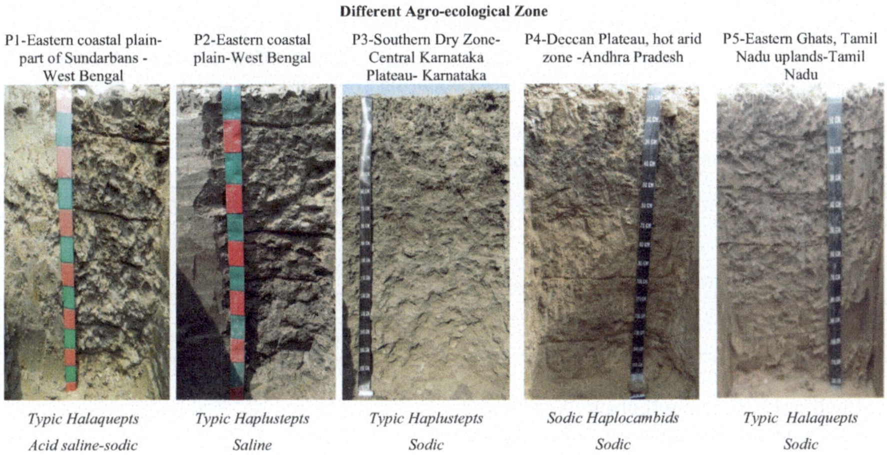

Fig. 7.2 Different salt affected soils (saline, sodic and saline-sodic) distributed in different agro-ecological zone of India—characterization and classification (Srinivasan et al. 2015, 2017a, b, 2019)

and classification of Different salt affected soils (saline, sodic and saline-sodic) in different Agro-ecological Zone of India given in Fig. 7.2 and Table 7.2.

7.4 Classification of Salt-Affected Soils

Soil contains soluble salts in the root-zone which are sufficiently high enough to impair the crop growth is defined as saline. However, because salt injury depends on crop species, variety, stage of plant growth, environmental factors, and nature of the salts, it is very difficult to define a saline soil precisely. Having said, the most widely accepted definition of a saline soil is one that has ECe more than 4 dS m^{-1} at 25 °C.

7.4.1 Saline Soils

Saline soils are defined as the soils which have pH usually less than 8.5, ECe is more than 4 dS m^{-1} and exchangeable sodium percentage (ESP) <15. A high ECe with a low ESP tends to flocculate the soil particles into aggregates. The soils are usually recognized by the presence of white salt crust in surface during some part of the year particularly during summer when the evapo-transpiration exceeds precipitation. Permeability is either greater or equal to 'normal' soils.

Table 7.2 Soil physical and chemical characteristics of different agro-ecological zone of India

Depth (cm)	pH	ECe (dS m^{-1})	ESP	Sand	Silt	Clay	CEC (Cmol (p^{+}) kg^{-1})
			%				
P1-Chandipur series, Gosaba block, West Bengal (*Fine, mixed, hyperthermic Typic Halaquepts*)							
0–15	4.9	15.2	17.7	0.9	61.3	37.8	15.8
15–42	6.3	4.42	20.7	0.6	56.5	42.9	14.5
42–62	5.5	5.02	18.4	0.6	65.3	34.1	15.2
62–95	4.6	5.55	14.8	3.9	53.8	42.3	14.9
95–114	6.3	5.63	17.9	1.6	66.7	31.7	14.0
114–150+	7.1	4.71	21.3	0.8	70.3	28.9	14.1
P2-Deuli series, Canning II block, West Bengal (*Fine-silty, mixed, hyperthermic, Typic Haplustepts*)							
0–18	7.5	4.11	3.4	0.7	74.9	24.4	11.6
18–48	8.2	2.46	1.8	0.1	66.2	33.7	15.8
48–82	8.2	6.88	3.4	0.6	73.5	25.9	11.5
82–115	8.2	8.20	7.0	4.6	82.4	13.0	7.1
115–150	8.1	9.72	5.7	2.7	83.7	13.6	6.9
P3-Thumakuru series, Balichakra hobli, Yadgir taluk, Karnataka (*Fine, mixed, isohyperthermic Typic Haplustepts*)							
0–12	9.6	0.35	6.6	62.9	15.7	21.3	21.8
12–29	9.7	1.27	27.3	45.9	18.5	35.5	30.5
29–74	9.1	3.44	36.0	48.4	16.2	35.2	28.6
74–132	9.3	2.52	23.1	38.2	20.5	41.1	34.9
132–158	9.2	2.07	24.5	36.8	19.9	43.1	34.2
P4-Ingaluru series, Obuladevaracheruvu Mandal, Ananthapur district of Andhra Pradesh (*Fine-loamy, mixed, isohyperthermic, Sodic Haplocambids*)							
0–10	8.5	0.09	3.6	77.7	14.0	8.1	14.0
10–29	9.7	0.80	43.9	50.4	21.7	27.8	21.7
29–55	9.6	1.15	57.2	40.3	25.2	34.4	25.2
55–90	9.2	1.08	53.8	61.1	15.2	23.5	15.2
90–127	9.4	0.58	41.6	77.7	8.6	13.6	8.6
P5-Sowttahalli series, Kaveripattinam block, Krishnagiri district of Tamil Nadu (*Coarse-loamy, mixed, isohyperthermic Typic Halaquepts*)							
0–16	8.8	0.56	15.3	55.8	9.0	35.2	15.4
16–25	9.1	0.33	17.2	58.4	9.3	32.2	9.3
25–42	8.9	0.46	9.4	82.9	6.3	10.7	8.9
42–67	9.0	0.4	13.4	62.1	9.5	28.4	11.8
67–90	7.3	0.075	21.7	65.1	12.0	21.9	16.9

7.4.2 *Saline-Sodic Soils*

Saline-sodic soils contain sufficient soluble salts (ECe > 4 dS m^{-1}) to interfere the growth of most crop plants with sufficient soil ESP (>15) to affect the soil properties and plant growth adversely primarily by degrading the soil structure. The pHs may be less or more than 8.5.

7.4.3 *Sodic Soils*

Sodic soils exhibit an ESP more than 15 and show an ECe of <4 dS m^{-1}. The pHs generally ranged from 8.5 to 10 and may be even as high as 11. The low ECe and high ESP tend to deflocculate soil aggregates and hence, lower the permeability of soil water.

7.4.4 *Distribution of SAS*

In India, northern part, these soils are occurs in Indo-Gangetic plains (IGP), spread over major states of Uttar Pradesh, Delhi, Haryana, Punjab, and Bihar. In the western part, the states viz., Gujarat, Rajasthan and Maharashtra states have sizable areas of SAS. In the center and southern part of the nation, such as the states of Madhya Pradesh, Andhra Pradesh, Karnataka, and Tamil Nadu have extensive areas of SAS. In the eastern part, SAS are found in the coastal and deltaic parts of West Bengal and Odisha (Singh 2005). The information of nine benchmark profiles of SAS out of 64 benchmark soils from all over the country were synthesized and classified into 12 associations of great groups (Murthy et al. 1980). Natrustalf, Natraqualf, Haplaquept, and saline phases of Calciorthids, Haplargid, Camborthid, Ustochrept, Fluvaquent, and Haplaquept occur in the northern Indian plains. Salorthids, Natrargid, Haplaquept, and saline phases of Ustochrept form the major units in the western region of the country. Haplaquept and saline phases of Haplaquept occur in the eastern region. The saline and alkali phases of Pellustert, Chromustert, Ustifluvent, and Haplaquept are found in southern Peninsular India (Murthy et al. 1980). The surface view of different salt affected soils in Different Agro-ecological Zone of India shown in Fig. 7.3.

7.5 Soil Salinization

One billion of the 13 billion ha land on earth covered with saline and sodic soils in that, between 25 and 30% of irrigated lands are salt-affected and commercially

Fig. 7.3 A view of different salt affected soils in different agro-ecological zone of India (Srinivasan et al. 2015, 2017a, b, 2018a, b, 2019)

unproductive. In India 20% cultivable lands are salt affected and distributed mainly in Rajasthan, coastal Gujarat and Indo-Gangetic Plains. Soil salinization is a global issue and affects almost all continents; it is not static but dynamic. Salinization can affects ecosystem to an extent where it will not be able to provide environmental services to its full potential. Many factors contribute to the development of saline soil conditions.

7.5.1 *Types of Soil Salinity*

7.5.1.1 Dry Land Soil Salinity

Soils salinity in dry land develops through a rising water table and the subsequent evaporation of soil water. There are many reasons for the rising water table, e.g. restricted drainage due to an impermeable soil layer and when deep-rooted trees are replaced with shallow-rooted crops. Under such situations, the groundwater may dissolves salts embedded in rocks and in the soil. The salty water eventually reaches the surface of the soils through capillary rise due to high evaporational rate causing soil salinity. Dry land salinity can also present in un-irrigated landscapes.

7.5.1.2 Secondary Soil Salinity

In opposite to dry land salinity, secondary salinity refers to the salinization of soil due to human induced practices or activities such as irrigation. Water scarcity in arid and desert environments necessitates the use of salt water or saline and brackish water to meet a part of the water requirement of crops. The improper use of poor quality irrigation waters which has high soluble salts, especially in soils having drainage problem, results in the encrustation of salt on the soil surface. These results the development of surface and sub-surface salinity and, thereby reducing the productivity soil resource and ecosystem services (Zaman et al. 2018).

7.5.2 *Damage Caused by Soil Salinity*

Some of the damages caused by increasing the soil salinity (Shahid 2013) are listed below:

- Declines in crop yields
- Loss of biodiversity and ecosystem disruption
- Abandonment or desertification of previously productive farm land
- Increasing numbers of dead and dying plants
- Increased risk of soil erosion due to loss of vegetation

- Contamination of drinking water
- Roads and building foundations are weakened by an accumulation of salts within the natural soil structure
- Lower soil biological activity due to rising saline water table.

7.5.3 Socio-economic Impacts of Salinity

- Reduced crop productivity on saline land leads to poverty due to income loss
- Farmers abandon their land and migrate into urban, which leads to unemployment
- High costs for soil reclamation if taken
- Loss of good quality soils, requires more inputs and caused financial pressure to farmers
- Saline agriculture system give lower cash returns than conventional crop production systems.

7.5.4 Visual Indicators of Soil Salinity

Once soil salinity develops in irrigated agriculture fields, it is easy to see the effects on soil properties and plant growth Visual indicators of soil salinization (Shahid and Rahman 2011) (Fig. 7.4) include:

- A white salt crust during summer
- Fluffy soil surface
- Salt stains on the dry soil surface
- Reduced or no seed germination
- Patchy crop establishment
- Reduced plant vigor
- Foliar damage—leaf burn
- Marked changes in leaf color and shape occur

Fig. 7.4 Some visual indicators of Soil Salinity in Southern India

- The occurrence of naturally growing halophytes such as *Prosophis* spp.
- Trees are either dead or dying
- Affected area worsens after a rainfall
- Water logging.

7.5.5 *Field Assessment of Soil Salinity*

Visual appraisal of salinity in the field only provides a qualitative indication but it does not give a quantitative measure on the level and severity of soil salinity. That is possible only through laboratory electrical conductivity (EC) measurement of the soil.

7.5.6 *Classes of Soil Salinity and Plant Growth*

Electrical conductivity of the soil saturation extract (ECe) is the standard measure of salinity. USSL Staff (1954) has described general relationship of ECe and plant growth (Table 7.3).

7.6 Soil Sodicity

Sodicity is a measure of sodium (Na^+) ions in soil water, relative to calcium and magnesium ions. It is expressed either as exchangeable sodium percentage (ESP) or sodium adsorption ratio (SAR). If the SAR of the soil equals or is greater than 13 (milli equivalents/liter) or the ESP equals or is greater than 15, the soil is called sodic (USSL Staff 1954).

Table 7.3 Relationship between ECe and plant growth

Salinity classes	ECe (dS m^{-1})	Plant growth
Non-saline	2	Negligible effect to crop yields
Very slightly saline	2–4	Yields of very sensitive crops may be restricted
Slightly saline	4–8	Yields are restricted
Moderately saline	8–16	Only salt tolerant crops can get good yields
Strongly saline	>16	Only few very salt tolerant crops get good yields

Fig. 7.5 Some visual indicators of soil sodicity in DAM catchment of Tamil Nadu

7.6.1 Visual Indicators of Soil Sodicity

Soil sodicity can be predicted visually in the field (Fig. 7.5) in the following ways.

- Poorer vegetative growth compared than normal soils (stunted plants or trees)
- Heights of the plants is variable
- Poor soil hydraulic conductivity
- Surface sealing and crusting
- Shallow plant root growth and depth
- Soil looks black in color due to the formation of a Na-humic substances complex
- High force required for tillage (fine textured soils)
- Poor pore space (blockage with dispersed clay).

7.6.2 Field Assessment of Soil Sodicity

Field assessment of relative level of soil sodicity can be determined through the use of a turbidity test on soil:water (1:5) suspensions, with ratings:

- Clear suspension—non sodic
- Partly turbid or cloudy—medium sodicity
- Very turbid cloudy—high sodicity.

7.6.3 Laboratory Assessment of Soil Sodicity

Soil sodicity diagnostics can be made by analyzing soil samples in the laboratory. The standard presentation of soil sodicity is the exchangeable sodium percentage (ESP) or using sodium adsorption ratio (SAR). ESP can be determined through measurement of exchangeable sodium (ES) and cation exchange capacity (CEC), as below.

$$\text{ESP} = \frac{\text{Exchangeable Na} + \text{ions}}{\text{CEC}} \times 100$$

$$SAR = \frac{[Na^+]}{\sqrt{\frac{Ca^{2+}+Mg^{2+}}{2}}}$$

where, ES and CEC are represented as meq/100 g soil. An ESP of 15 is the threshold for designating soil as being sodic soil (USSL Staff 1954). At this ESP level, the soil structure starts degrading and negative effects on plant growth appear.

7.6.4 Sodicity and Soil Structure

Lack of fresh water used for irrigation in arid and semi-arid regions often results in a increased level of soluble salts and sodium ion concentration in soil. Sodicity recognized as one of the main problem which affects the soil permeability, swelling and dispersion of soil fine clays and destroys the original soil structure likely physical property finally affecting the plant growth (Fig. 7.6). The soil bulk density, hydraulic conductivity and porosity are mainly used as parameters for the soil structure evaluation. The effect of the sodicity of soil has shown significant impact on surface and sub-surface soil crusting or sealing (Shahid et al. 1992). In surface sealing, the soil sodicity causes a breakdown and slaking of soil aggregates due to wetting and when the soil surface dries, a surface crust is formed. In sub-surface soil sealing, the clay particles dispersed and translocated into lower or sub-surface layers, where they are deposited on the surface of the voids, thereby reducing void volume and blocking the pores, thus restricting further water movement. The surface crusting and sealing due to either or combined effects of sodicity and raindrop splash action, have both positive and negative effects (Srinivasan et al. 2017b).

Surface characteristics of sodic soils Subsurface soil columnar structure

Fig. 7.6 Development of sodicity in surface and sub-soils in part of paddy growing soils in coastal West Bengal (Ramnagar block)

7.7 Remote Sensing for Soil Affected Soil Mapping

7.7.1 Remote Sensing Data

Remote sensing is a technique which acquires information about the surface of Earth's without having physical any contact. This is done by sensing and recording the reflected or emitted energy using sensors and processing, analyzing and applying the information using image processing techniques. In much of remote sensing, the process involves an interaction between incident radiation and targets of interest. Different features on the Earth's surface will absorb and reflect different parts of the electromagnetic (EM) spectrum depending on their chemical make-up. In this way, different parts of the electromagnetic spectrum provide information about the Earth's surface features that may be useful for the detection of salinisation and sodification (Allbed and Kumar 2013). The majority of the images are supplied from satellites known as Landsat. Thematic Mapper provides observations in bands ranging from visible to thermal on each area of the Earth's surface and sending information back to Earth's observation and processing unit.

7.7.1.1 Digital Image

Digital image is a two dimensional array or otherwise known as grids of small areas called pixels. Each pixel corresponds spatially to an area on the earth's surface which otherwise represented as digital number (DN). The two dimensional array or pixel grid structure is also known as raster, so image data is often referred as raster data. Band is a set of data file values for a specific portion of the electromagnetic spectrum of reflected light also called as channel.

7.7.1.2 Satellite Image Resolution

The smallest object that can be detected by the sensor is known as spatial resolution. The area on the ground is represented by pixel and finer the resolution means lower the pixel size (Fig. 7.7). Spectral Resolution is the specific wavelength intervals in the electromagnetic spectrum which a sensor can record. For example, band 1 of Landsat Thematic Mapper sensor records energy between 0.45 and 0.52 μm in the visible part of the electromagnetic spectrum.

7.7.1.3 Geographical Information System (GIS)

The advancements in computer technology, image processing, global position system and geo statistics have resulted in development of Geographical Information System (GIS) to store, retrieve and management of spatial data (maps derived from remotely

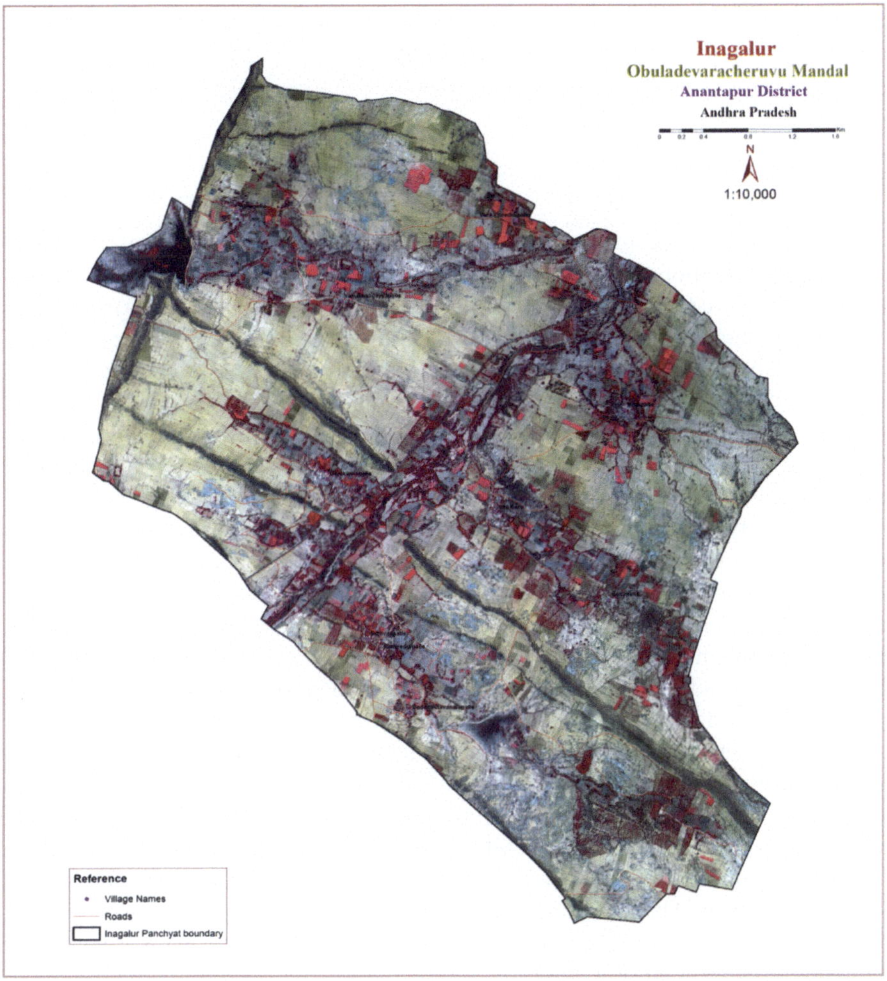

Fig. 7.7 The Cartosat-2 (<1 m resolution) + LISS IV merged image with village cadastral map and SOI topographical sheets of 1:10,000 scale for detailed soil survey purpose

sensed data etc.), attribute data (soil properties, climatic parameters, and topography etc.) and other related ancillary data more efficiently. GIS proved to be effective tool for handling spatial data at different scales and also high quantity point data such as rainfall, temperature, soil information, and socioeconomic data etc. It also helps in performing integrated analysis for any region and arriving different management and optimum solutions for various problems. In India, GIS is used for various fields such as optimum land use planning, land resources management, assessment of water requirements for different crops, development of wastelands etc. The efforts of GIS use in crop yield modelling, developing reclamation measures/salt-affected soils management, quantification of soil loss and suitable conservation measures, evaluation of soils for various purposes like agriculture, horticulture, agroforestry, forestry and aquaculture development are immense.

7.7.2 Methodology

The two methods used in interpretation and analysis of remotely sensed data to derive the information on problem soils are visual interpretation and digital image analysis. Case study was carried out in Inagalur Panchayat (2938 ha) Obuladevaracheruvu Mandal, Ananthapuramu District of Andhra Pradesh to develop detailed Land Resource Inventory information for farm level planning using remote sensing and GIS techniques (Hegde et al. 2019). Visual interpretation involves identification and delineation of different salt affected soils that are manifested on False Colour Composite (FCC) or black and white prints in different shape, size, tone, texture, pattern etc. (Fig. 7.7).

The remotely sensed data in GIS are analyzed with the help of different image analysis software mostly in ArcGIS software. The spectral reflectance of SAS forms the basis in the digital analysis. Both visual and digital techniques are used in extracting information on SAS from remotely sensed data. The False Color Composites are analyzed with the help of SOI topographical maps, published reports and other available ancillary data to delineate broad categories of degraded lands. Again each unit will be divided into subunits on the basis of erosion status, drainage density, vegetation cover and land use. These delineated units will be transferred on to base maps prepared from Survey of India topographical maps. Representative sample areas will be selected for various SAS for ground truth collection (Fig. 7.8). During field visits, features of topography and soil profiles will be studied and marked in base map (Figs. 7.9 and 7.10). Site characteristics and soil samples will be collected for laboratory analysis. The preliminary interpreted maps will be modified in the light of field data (Fig. 7.11) and soil chemical analytical data and the final maps are prepared with appropriate legend (Fig. 7.12).

7.7.3 Detection of Soil Salinity by Remote Sensing

7.7.3.1 Delineation of Salt Affected Area

Soil salinity can be detected from remotely sensed (RS) data through salt features visible at the soil surface in bare soil with white salt crusts on the surface (Matinfar 2013) or indirectly from presence of halophytic plant and performance of salt-tolerant crops (Iqbal 2011; Aldakheel 2011).

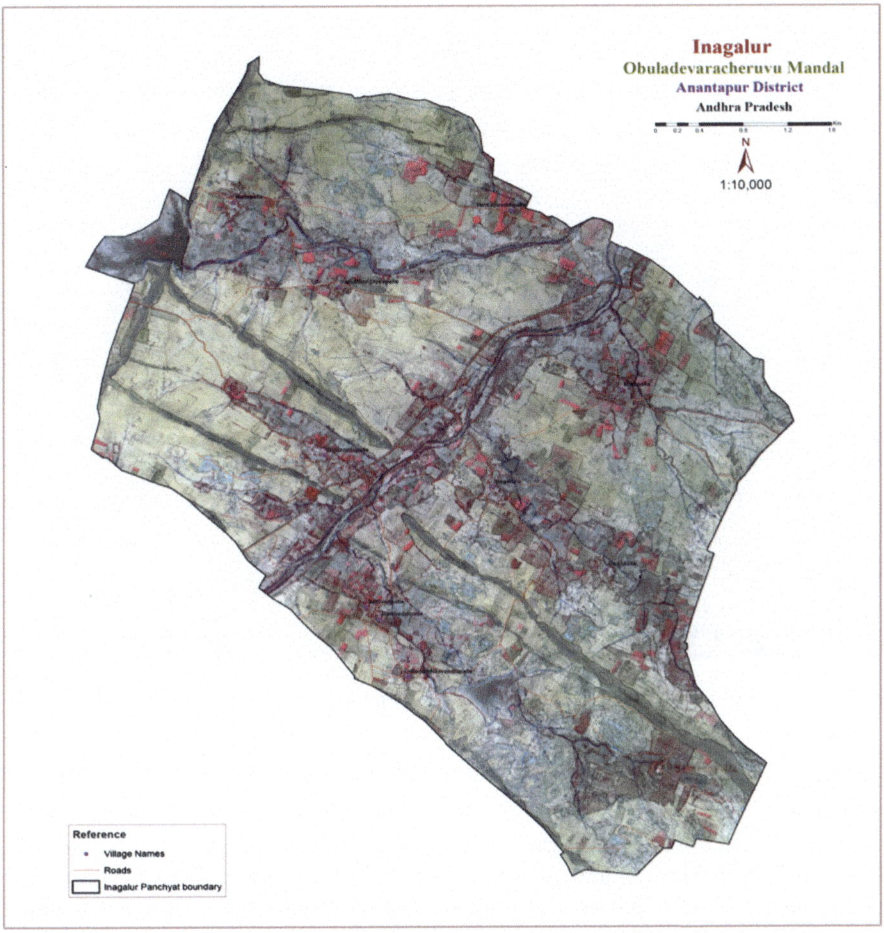

Fig. 7.8 Visual and digital interpretation of RS imagery (FCC) based on colour, tone and texture

Fig. 7.9 Field verification and base map correction and locate the soil profiles

Fig. 7.10 Soil profile observation cum study observation marked on base map

Salt Features at the Soil Surface

Formation of salt affected soils are the dynamic processes operate at the soil surface often limit the monitoring and assessment of the salinization process because they are influenced by the spectral, spatial and temporal behavior of the salt features (Metternicht and Zinck 2008). Physico-chemical properties of soil such as soil moisture content, organic matter, soil texture, types of clay content, soil color and surface roughness soil which affects the spectral reflectance is determined (Demattê et al. 2004; Brown et al. 2006; Shrestha et al. 2005). Due to salinity, change in these soil properties affect the spectral reflectance of features that occur at the soil surface, including salt crusts and efflorescence besides variations in surface texture and structure (Schmid et al. 2008; Thomas 2011). Similarly, Singh and Sirohi (1994) noted that a crusted saline soil surface is generally smoother than a non-saline surface and exhibits high reflectance in the visible and NIR bands, which has been confirmed by Rao et al. (1995). Despite the effects of salt features on the soil surface reflected in the electromagnetic spectrum, they have been considered as good direct indicators of soil salinity as well.

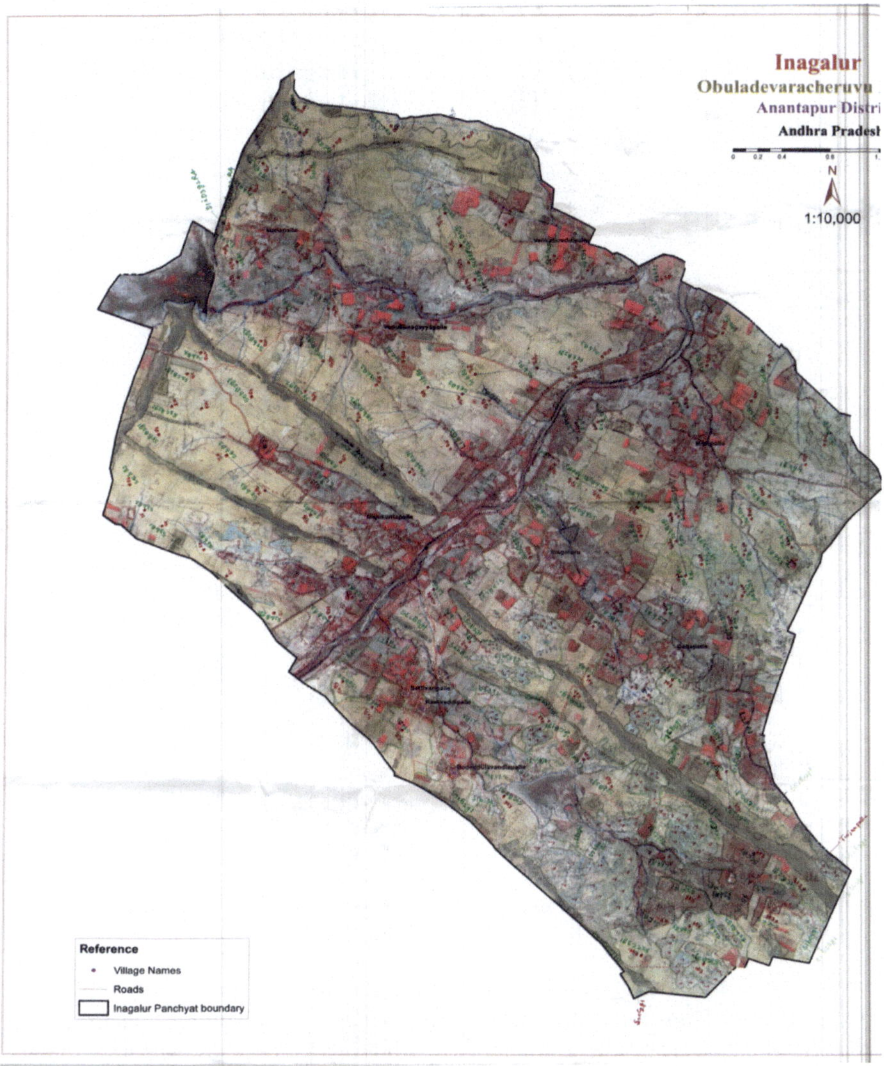

Fig. 7.11 Tentative soil map of Inagalur panchayat after filed work

Presence of Halophytic Plants

Halophytic plants (salt-tolerant plants) are tolerant to high salt concentrations of the soil and can be grown in salt affected land (Glenn et al. 1999). Although halophytic plants are common in salt affected area, not all have been found to be good remote sensing indicators of soil salinity. For instance, due to lower chlorophyll content the spectral reflectance curve of *C. Dactylon* a halophyte grown in salt affected soil, increased continuously in the visible and NIR bands. Same way *P. Juliflora*, an

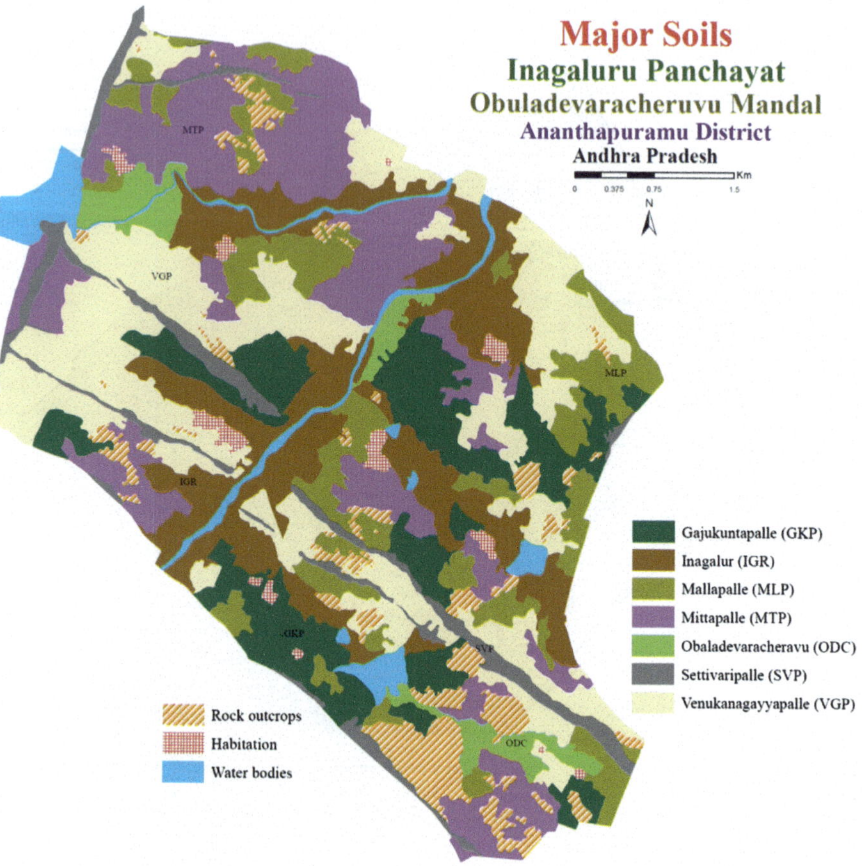

Fig. 7.12 Mapping of major soils in Inagalur panchayat using remote sensing and GIS

invasive sp. occupied in most of the salt affected soils also shown very reflectance in the NIR region but low in the visible reflectance. These studies concluded that halophytic plants were promise indicator to distinguish saline areas from non-affected ones. Some cultivated SAS land surfaces are occupied by different vegetation and barren in shown in Fig. 7.13.

Crop Performance

The performance of crops grown on saline soils, such as paddy, wheat, barley and cotton, reflect the severity of soil salinity. Cotton or paddy is largely cultivated on irrigated land, is therefore considered an ideal indirect indicator for soil salinity, so it has been used as salinity indicators in a different studies (Metternicht and Zinck 2008). For example, based on the high correlations between the Normalized

Fig. 7.13 Presence of different vegetation and other halophytic plants in surface of salinity soils

Fig. 7.14 Poor crop performance in saline and sodic soils

Difference Vegetation Index (NDVI) values of cotton, sugar-cane crops and the EC, Wiegand et al. (1994, 1996) successfully assessed the severity and extent of soil salinity in terms of the economic impact on crop production and also distinguished saline soils from non-affected soils (Fig. 7.14).

7.7.4 *Salinity Mapping and Monitoring*

A reliable method for salinity and sodicity assessment and mapping requires delineating the areas into soil salinity and sodicity status zones or maps (Fig. 7.15). At the national level, salinity mapping information will helps stakeholders in land use planning and to address the reverse causes of salinization. In agricultural farms salinity maps help farmers to understand subtle difference in soil properties across their fields, allowing them to develop more precise management zones and selection of salt tolerant crops and, ultimately for potentially higher yields (Srinivasan et al. 2019). It sounds complicated, but salinity mapping at the farm level is one of the simplest and least expensive salinity measurement tools. Soil salinity mapping using airborne remote sensing and spectroscopy, salinity assessment by combined use of RS and GIS (Brena et al. 1995; Casas 1995), selection of best possible Landsat TM band combination for the delineation of salt-affected soils (Dwivedi and Rao 1992), salinity monitoring using RS and GIS are the proved methods which can be used for

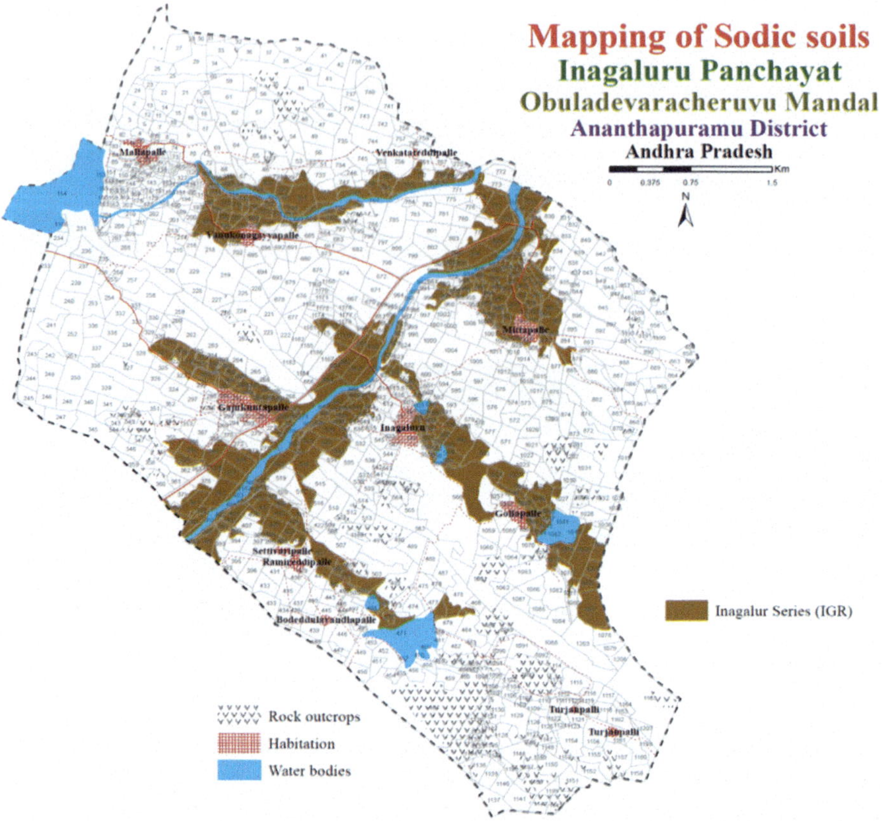

Fig. 7.15 Mapping of sodic soils (Inagalur series) in Inagalur panchayat of Andhra Pradesh

salinity status mapping and monitoring. In addition, studies developed successful image interpretation keys such as, the soils remain devoid of natural vegetation, except for the presence of very hardy grasses (*Sporobolus, Leptochloa, Cynodon, and Sueada*) and the salt encrustation at soil surfaces in saline and alkali soils varies greatly over time and space and is influenced by the prevailing irrigation practices and climatic conditions of India (Singh 2005).

7.7.5 Delineation of Salt-Affected Soils in India

7.7.5.1 Conventional Approach

The first survey of SAS in India using a conventional approach was conducted in 1902 to delineate several patches of alkali soil in the Etah and Mainpuri districts of Uttar Pradesh state in the IGP (Leather 1902). After the nation's independence in 1947,

Table 7.4 Estimates of salt-affected soils in India by different organizations/researchers (Singh et al. 2010)

Individual/organization	Area under SAS	Individual/organization	Area under SAS
	Mha		Mha
		Bhargava (1989)	10.0
National commission on Agriculture (1976)	7.17	Ministry of Agriculture (1990)	9.4
Ponnamperuma and Bandyopadhyay (1980)	26.1	Dent et al. (1992)	7.02
Ministry of Agriculture (1980)	7.0	Abrol (1994)	10.9
Bhumbla and Khare (1984)	7.2	Chauhan (1996)	7.2
Ministry of Agriculture (1985)	9.08	Singh and Bandyopadhya (1996)	8.6

the occurrence of SAS using a conventional approach including laboratory analysis was reported in other parts of IGP, the Peninsula, Rajasthan, and Gujarat states and other parts of the country (Agarwal et al. 1979). The national estimates of SAS based on a conventional approach from various sources varied from 6.0 to 26.1 Mha. The wide variations in the national estimates reflected the variation in methodology and class definition of SAS adopted by different researchers/organizations (Table 7.4).

The first estimate from the Central Soil Salinity Research Institute (CSSRI) put the area under SAS at 7 Mha in the country (Abrol and Bhumbla 1971). Such a small-scale map provided a synoptic view of the problem at national scale but did not help in locating the area under SAS at the district level required for reclamation. Consequently, the area under SAS in India reported by researchers from other countries in the late 1970s and early 1980s was more than 20 Mha. The most detailed and comprehensive surveys and classification were performed by Singh (1994b), who reported 8.57 Mha of SAS in the country. There was large variation in the national estimates by the conventional approach, and it was difficult to obtain precise spatial distribution at a national scale. Therefore, it was imperative to resort to remote sensing technology to provide reliable and rapid spatiotemporal information of SAS for planning effective reclamation strategies.

7.7.5.2 Remote Sensing Approach

Remote sensing is able to detect SAS directly from salt-encrusted surfaces of varying salt mineralogy or indirectly from vegetation/crop conditions and has been widely used in India for mapping and monitoring of SAS. The earliest systematic mapping of land degradation that included SAS in India was performed by visual interpretation of aerial photographs in the late 1950s. In the 1960s and early 1970s, the mapping and monitoring of SAS for categorization were performed at the local and regional scales for assessment of extent and spatial distribution (Hilwig and Karale 1973; Iyer et al. 1975). Visual interpretation techniques developed using aerial photography were even extended to satellite multispectral data for delineating and mapping of SAS at the regional and national scales (Hilwig 1980; Karale et al. 1983; Manchanda and Iyer 1983; Singh 1994a). The appearance of SAS with salt encrustation at the surface is generally smoother than normal soil surfaces and has higher reflectance in the visible and near-infrared bands (Singh and Sirohi 1994; Rao et al. 1995). Salt-affected soils on standard FCCs of satellite data are expressed as bright white to dull white patches within light reddish-brown background of normal soils (Singh et al. 1977). The slightly or low-encrusted salt affected soils initially occur in small patches within the cultivated normal soils, and these extend to large patches with time.

Accuracy of SAS mapping on satellite FCC has been found superior by the adoption of on-screen visual interpretation techniques. This is most extensively used in national and state soil survey organizations. Subsequently, digital image processing is being applied in the country for local and regional scale mapping due to the enhanced availability of image processing hardware and software in national and state remote sensing application centers. In digital image processing, statistical pattern recognition techniques based on inherent spectral reflectance properties have aided in the differentiation of SAS classes. Standard per pixel classifiers (e.g., as maximum likelihood) and advanced algorithms (e.g., fuzzy logic, decision trees, and artificial neural networks) have been used for inventorying and monitoring (Dwivedi et al. 2008) the SAS over different regions of the globe. Space-borne data for mapping and monitoring provide greater accuracy and economy than the conventional approach at the district scale.

It cannot be denied that medium or low spatial resolution of the satellite images can limit the mapping and detection of SAS regions, particularly when the affected areas are smaller than the pixel size. Thus, high resolution multispectral sensors with pixel size of less than 5 m are becoming an essential for soil salinity studies (Dwivedi et al. 2008). However, only limited attempts have been made to identify and map soil salinity problems using fine spatial resolution (0.6–4 m) images that are available from IKONOS (4 bands) and Quickbird (4 bands) satellites, as well as WorldView-2, which has 8 multispectral bands at 1.84 m spatial resolution and one panchromatic band at 0.5 m spatial resolution (Navulur 2006). This is most likely due to the higher cost of this higher-resolution imagery and these sensors being more recently developed systems. Dwivedi et al. (2008) conducted a comparative study on the performance of IKONOS imagery and imagery from the IRS-ID LISS-III sensor for mapping salt-affected soils. Different image classification and transformation techniques were used in their study and an overall accuracy of 92.4% was gained when using IKONOS data compared to an overall accuracy of 78.4 and 84.3% obtained when using the IRS-ID LISS-III multispectral sensor, which indicates the great potential of high spatial resolution IKONOS images for soil salinity mapping and detection.

Hyperspectral Remote Sensing Data

The development of airborne and satellite-based hyperspectral sensors has overcome some of the spatial and spectral limitations of multispectral satellite imagery for monitoring and mapping soil salinity, both regional and local. Hyperspectral sensors offer a large number of spectral bands with high spatial resolution that allow the discrimination of halophyte plants from non-halophyte plants as well as the identification of surface salt features in more detail than the multispectral sensors (Dutkiewicz 2006). The potential of the HyMap airborne hyperspectral sensor, which captures images within a spectral range of 450–2500 nm in 128 bands, for soil salinity studies has been tested by Dehaan and Taylor (2002, 2003). The study concluded that HyMap has significant potential in mapping saline areas and also characterize the severity of salinity levels and scattered halophyte plants.

Vegetation and Soil Indices

As mentioned previously, halophytic plants grow naturally in SAS, and can be adapted to high salinity level. Therefore, vegetation or plants has been used as an indirect indicator for predicting and mapping of SAS. Accordingly, several researchers have conducted studies on the mapping and delineation of soil salinity using different Spectral Vegetation Indices (SVI). Among the vegetation indices, NDVI, SAVI, Ratio Vegetation Index (RVI) and Tasseled Cap Transformation that consists of Soil Brightness Index (SBI), the Green Vegetation Index (GVI), and the Wetness Index (WI) have been widely used in SAS studies (Matinfar 2013; Aldakheel 2011; Lobell et al. 2010). Due to absorption in the visible range and high reflectance in the NIR range of the electromagnetic spectrum, the NDVI has been extensively used to map soil salinity by monitoring halophytic plants (Fernandez-Buces et al. 2006; Elnaggar and Noller 2009). The difference in reflectance between the visible and NIR bands is divided by the sum of the two bands' reflectance. This normalizes differences in the amount of incoming light and produces a number from -1 to 1; the range of actual values is approximately 0.1 for bare soils to 0.9 for healthy vegetation (Deering and Rouse 1975).

Geo-statistics

Geo-statistics is a spatial tool used for mapping the land surface features from limited sample data. It is widely used in fields where spatial data is studied. Geo-statistical estimation is a two-stage process. First step is studying the gathered data to establish a predictability of values from place-to-place within the study area. This results in a graph known as a semi-variogram, which models the difference between a value at one location and the value at another location according to the distance and direction between them. The second step is estimating values at those locations which have not been sampled. This process is known as kriging. The basic technique of ordinary

Table 7.5 Descriptive statistics parameters of pH of the study area

Soil properties	Minimum	Maximum	Mean	SD	CV (%)	Skewness	Kurtosis	Shapiro–Wilk's p
pH	5.18	9.10	7.05	0.90	12.83	0.21	−0.96	0.0004

Table 7.6 Values of model parameters used to find the best semivariogram of the study area

Soil properties	Model	Nugget	Partial sill	Nugget/sill (%)	Range (m)	RMSE	Spatial dependency
pH	Spherical	0.008	0.008	1.00	0.019	0.810	Weak

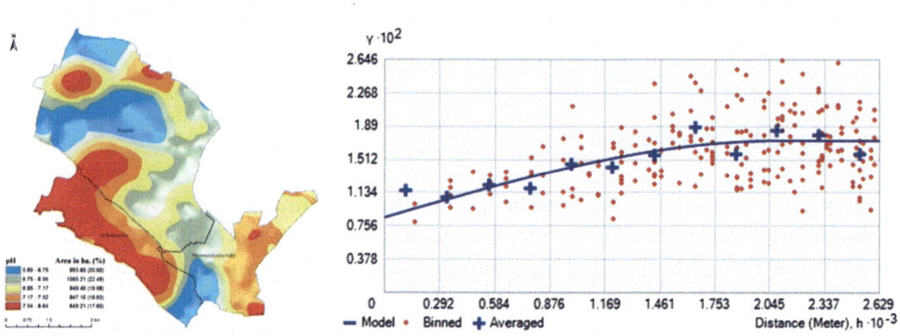

Fig. 7.16 Spatial distribution of pH in Inagalur panchayat using geostatistics

kriging uses a weighted average of neighboring samples to estimate the 'unknown' value at a given location. The weights are optimized using the semi-variogram model based on the locations of the samples, and all the relevant interrelationships between known and unknown values. The method also yields a standard error which can be used to calculate the confidence levels. The application of geo-statistical techniques, such as ordinary and co-kriging, have been applied to salinity survey data in an effort to represent more accurately the spatial distribution of soil salinity (Vaughan et al. 1995; Reza et al. 2015). A case study attempted to predict spatial map of soil pH in Inagalur panchayat is given in Tables 7.5, 7.6 and Fig. 7.16.

7.7.6 Constraints in Remote Sensing of SAS Mapping

Satellite images can help to assessing the extent of SAS areas and monitoring the changes in real time. SAS fields are often identified by the presence of spotty white patches of precipitated salts. Such precipitates usually occur in unvegetated areas, where water evaporates and leaves salt left behind. Such salt crusts, can be detected on satellite images, are not reliable evidence of high salinity in the root zone. Inadequate resolution of low cost remote sensing (RS) data in optical range limited to surface salt encrustation, therefore, identification of subsurface salinity and waterlogging using optical RS data becomes difficult. Other limitation in SAS mapping with multispectral imagery is where problem soils support productive plant growth (Furby et al. 1995) such as biosaline agriculture, where plant cover obscured direct sensing of the soil, while salt tolerant plants could not be differentiated from other cover.

7.8 Management of Salt Affected Soils Using Remote Sensing

A conventional field survey approach including maps was used to assess the nature, magnitude, and areal extent and distribution of SAS. However, these methods are expensive, time consuming and labour intensive (Dwivedi and Rao 1992). Area estimates using such methods were subject to sampling error; therefore, reliable estimates were often not available. Furthermore, surveying of inaccessible areas and inhospitable terrain was difficult using a conventional approach. Remote sensing technology, with its unique characteristics of systematic, synoptic, rapid, and repetitive coverage, has emerged as a cost-effective approach for studying and mapping salt affected soils and other degraded lands in space and time domains (Navalgund et al. 2007; Metternicht and Zinck 2008). Aerial photographs with limited field surveys were visually interpreted in the 1960s and early 1970s. Subsequently, visual interpretation of satellite multi-spectral data was widely adopted for mapping and monitoring of salinity and sodicity at regional and national scales. The use of remotely sensed data has improved the mapping accuracy because different types of SAS exhibited distinctive patterns on standard false color composites (FCCs). Multi-temporal imageries and other ancillary data integrated using a geographic information system (GIS) have been used to create spatiotemporal databases to monitor the status and trends of salinisation and sodification and the impact of various reclamation of soils (Srinivasan et al. 2017a). Calibration of image classification algorithms and quantification of classification accuracy were supported with ground-truth data through an appropriate sampling method. With the advancement in multi-spectral sensor and image processing technologies, it is now possible to generate and update information at moderate and severe levels of salts at the farm scale in a cost-effective manner.

7.9 Conclusion

Soil affected soils are naturally occur or induced by human which is a serious environmental problem, especially in arid and semi-arid soils. Saline and sodic soils are complex and dynamic process with serious problems for environment as well as, climatic, geochemical, hydrological, agricultural, and economic impacts. Being a serious problem, the regular detection and assessment is very important at local and regional scales of soil affected soils and its extent and severity. Remote sensing and GIS are become valuable tools in studying the spatial extent of salt affected soils and monitoring the changes that have taken place over a period time. Multispectral RS satellite sensors are the better option for mapping and monitoring problem soils, largely due to the cost effective manner and operational basis. The usage of GIS is very important tool for handling voluminous data generated on SAS through conventional and remote sensing techniques and for integrated analysis of data to derive plan for reclamation and conservation of natural resources.

References

Abrol IP, Bhumbla DR (1971) Saline and alkali soils of India: their occurrence and management. In: Soil survey and soil fertility research in Asia and the Far East: report on regional seminar, New Delhi, India. World Soil Resources Report No. 41. FAO, Rome, Italy pp 42–52

Abrol IP, Chhabra R, Gupta RK (1980) A fresh look at the diagnostic criteria for sodic soils. In: Proceedings on the international symposium of salt-affected soils, Karnal, India. Central Soil Salinity Research Institute, Karnal, India

Abrol IP (1994) Land degradation: a challenge to sustainability. In: Rao DLN et al (eds) Salinity management for sustainable agriculture. CSSRI, Karnal, India, pp 7–8

Agarwal RR, Yadav JSP, Gupta RN (1979) Saline and alkali soils of India, Revised edn. Indian Council of Agricultural Research, New Delhi, India

Aldakheel YY (2011) Assessing NDVI spatial pattern as related to irrigation and soil salinity management in Al-Hassa Oasis, Saudi Arabia. J Indian Soc Remote Sens 39(2):171–180

Allbed A, Kumar L (2013) Soil salinity mapping and monitoring in arid and semi-arid regions using remote sensing technology: a review. Adv Remote Sens 2:373–385

Bhargava GP (1989) Salt-affected soils of India. Oxford & IBH Publication, New Delhi, India

Bhumbla DR, Khare A (1984) Estimates of wastelands in India. Society for Promotion of Wastelands Development (SPWD), New Delhi, India

Brena J, Sanvicente H, Pulido L (1995) Salinity assessment in Mexico. In: Vidal A, Sagardoy JA (eds) Water Report No. 4: use of remote sensing techniques in irrigation and drainage, Rome, FAO, pp 173–178

Brown DJ, Shepherd KD, Walshb MG, Maysc MD, Reinschc TG (2006) Global soil characterization with VNIR diffuse reflectance spectroscopy. Geoderma 132(3):273–290

Casas S (1995) Salinity assessment based on combined use of remote sensing and GIS. In: Vidal A, Sagardoy JA (eds) Water Report No. 4: use of remote sensing techniques in irrigation and drainage, Rome, FAO, pp 185–197

Chauhan HS (1996) Management of problem area in irrigation commands through conjunctive use and other methods. In: Proceedings of the national workshop reclamation of waterlogged saline and alkali lands and prevention, New Delhi. Ministry of Water Resources, New Delhi, India, pp 79–86

Corwin D, Lesch S (2003) Application of soil electrical conductivity to precision agriculture. Agron J 95(3):455–471

Deering D, Rouse J (1975) Measuring 'Forage Production' of grazing units from Landsat MSS Data. In: 10th international symposium on remote sensing of environment, ERIM, Ann Arbor, pp 1169–1178

Dehaan RL, Taylor GR (2002) Field-derived spectra of salinized soils and vegetation as indicators of irrigation-induced soil salinization. Remote Sens Environ 80(3):406–417

Dehaan RL, Taylor RGR (2003) Image-derived spectral endmembers as indicators of salinisation. Int J Remote Sens 24(4):775–794

Demattê JAM, Campos RC, Marcelo MC, Fiorioa PR, Nanni MR (2004) Visible-NIR reflectance: a new approach on soil evaluation. Geoderma 121(1–2):95–112

Dent FJ, Rao YS, Takeuchi K (1992) Regional strategies for arresting land degradation (Womb of the Earth). FAO/RAPA, Bangkok, Thailand

Dutkiewicz A (2006) Evaluating hyperspectral imagery for mapping the surface symptoms of dry land salinity. The University of Adelaide, Adelaide

Dwivedi RS, Rao BRM (1992) The selection of the best possible Landsat TM band combination for delineating salt-affected soils. Int JRS 13(11):2051–2058

Dwivedi RS, Kothapalli RV, Singh AN (2008) Generation of farm level information on salt-affected soils using IKONOS-II multispectral data. In: Metternicht G, Zinck JA (eds) Remote sensing of soil salinization: impact on land management. CRC Press, Taylor & Francis, Boca Raton, FL, pp 73–90

Elnaggar AA, Noller JS (2009) Application of remote-sensing data and decision-tree analysis to mapping salt-affected soils over large areas. Remote Sens 2(1):151–165

Eynard A, Lal R, Wiebe KD (2006) Salt-affected soils. In Lal R (ed) Encyclopedia of soil science, 2nd edn, vol 2. Taylor & Francis, Abingdon, UK, pp 1538–1541

Fernandez-Buces N, Siebe C, Cram S, Palacio JL (2006) Mapping soil salinity using a combined spectral response index for bare soil and vegetation: a case study in the former lake Texcoco, Mexico. J Arid Environ 65(4):644–667

Furby SL, Vallace JF, Caccetta PA, Wheaton GA (1995) Detecting and monitoring salt-affected land: a report from the LWRRDC project detecting and monitoring changes in land condition through time using remotely sensed data. Remote Sensing and Image Integration Group, CSIRO Division of Mathematics & Statistics, Western Australia

Ghassemi F, Jakeman AJ, Nix HA (1995) Salinization of land and water resources: human causes, extent, management and case studies. CAB International, Wallingford, UK

Glenn EP, Jed Brown J, Blumwald E (1999) Salt tolerance and crop potential of halophytes. Crit Rev Plant Sci 18(2):227–255

Hegde R, Srinivasan R, Srinivas S, Niranjana KV, Dhanorkar BA, Singh SK (2019) Land Resource Inventory (LRI) of Inagalur Panchayat, Ananthapuramu District on 1:10000 scale under Andhra Pradesh Drought Mitigation Project (APDMP), Andhra Pradesh. APDMP LRI Atlas No.1, ICAR – NBSS & LUP, RC, Bangalore, p 70

Hillel D (2000) Salinity management for sustainable irrigation: integrating science, environment, and economics. World Bank Publications, Washington DC

Hilwig FW (1980) Visual interpretation of multi-temporal Landsat data for inventories of natural resources: a north Indian case study. ITC J 2:297–327

Hilwig FW, Karale RL (1973) Physiographic systems and elements of photo-interpretation as applied to soil survey in Ganges plain. J Indian Soc Soil Sci 21:205–212

Iqbal F (2011) Detection of salt affected soil in rice-wheat area using satellite image. Afr J Agric Res 6(21):4973–4982

Iyer HS, Singh AN, Kumar W (1975) Problem area inventory of parts of Hoshiarpur district through photo-interpretation. J Indian Soc Remote Sens 3:79

Karale RL, Bali YP, Seshagiri Rao KV (1983) Soil mapping using remote sensing techniques. Sadhana 6:197–208

Koohafkan P (2012) Water and cereals in dry lands. The Food and Agriculture Organization of the United Nations and Earthscan, Rome

Leather JW (1902) Some excessively saline Indian well waters and soils. J Am Chem Soc 18:887–888

Lobell D, Lesch SM, Corwin DL, Ulmer MG, Anderson KA, Potts DJ, Doolittle JA, Matos MR, Baltes MJ (2010) Regional-scale assessment of soil salinity in the Red River valley using multi-year MODIS EVI and NDVI. J Environ Qual 39(1):35–41

Manchanda M, Iyer H (1983) Use of Landsat imagery and aerial photographs for delineation and categorization of SAS of part of northwest India. J Indian Soc Soil Sci 31:263–271
Matinfar HR (2013) Detection of soil salinity changes and mapping land cover types based upon remotely sensed data. Arab J Geosci 6(3):913–919
Metternicht GI, Zinck JA (2008) Remote sensing of soil salinization: impact on land management. CRC Press, Taylor & Francis, Boca Raton, FL
Ministry of Agriculture (1980) Status of land degradation in India. Directorate of Economics and Statistics. Ministry of Agriculture. New Delhi, India
Ministry of Agriculture (1985) Status of land degradation in India. Directorate of Economics and Statistics. Ministry of Agriculture. New Delhi, India
Ministry of Agriculture (1990) Status of land degradation in India. Directorate of Economics and Statistics. Ministry of Agriculture. New Delhi, India
Murthy RS, Hirekar LR, Bhattacharya JC (1980) The taxonomy of the salt-affected soils of India sub-continent. NBSSLUP, Nagpur, India
National Commission on Agriculture (1976) Report of the national commission on agriculture. Part V and IX, Abridged Report. Ministry of Agriculture and Irrigation, New Delhi, India
Navalgund RR, Jayaraman V, Roy PS (2007) Remote sensing applications: an overview. Curr Sci 93:1747–1766
Navulur K (2006) Multispectral image analysis using the object-oriented paradigm. CRC Press, Boca Raton
Ponnamperuma FN, Bandyopadhyay AK (1980) Extent of salt-affected soils and their management. In: Proceedings of the international symposium on priorities for alleviating soil-related constraints to food production in the Tropics, Los Banos, Philippines, IRRI, Philippines, pp 3–19
Rao BRM, Ravi Sankar T, Dwivedi RS, Thammappa SS, Venkataratnam L, Sharma RC, Das SN (1995) Spectral behavior of salt-affected soils. Int J Remote Sens 16:2125–2136
Reza SK, Nayak DC, Chattopadhyay T, Mukhopadhyay S, Singh SK, Srinivasan R (2015) Spatial distribution of soil physical properties of alluvial soils: a geostatistical approach. Arch Agron Soil Sci. https://doi.org/10.1080/03650340.2015.1107678
Schmid T, Koch M, Gumuzzio J (2008) Application of hyperspectral imagery to soil salinity mapping. In: Metternicht G, Zinck J (eds) Remote sensing of soil salinization: impact on land management. CRC Press, Boca Raton, pp 113–137
Shahid SA (2013) Developments in salinity assessment, modeling, mapping, and monitoring from regional to submicroscopic scales. In: Shahid SA, Abdelfattah MA, Taha FK (eds) Developments in soil salinity assessment and reclamation—innovative thinking and use of marginal soil and water resources in irrigated agriculture. Springer, Dordrecht/Heidelberg/New York/London, pp 3–43
Shahid SA, Rahman KR (2011) Soil salinity development, classification, assessment and management in irrigated agriculture. In: Passarakli M (ed) Handbook of plant and crop stress. CRC Press/Taylor & Francis Group, Boca Raton, pp 23–39
Shahid SA, Jenkins DA, Hussain T (1992) Halite morphologies and genesis in the soil environment of Pakistan. In: Proceedings of the international symposium on strategies for utilizing salt affected lands. Bangkok, Thailand, pp 59–73
Shrestha D, Margateb DE, van der Meera F, Anhc HV (2005) Analysis and classification of hyperspectral data for mapping land degradation: an application in Southern Spain. Int J Appl Earth Obs Geoinf 7(2):85–96
Singh AN (1994a) Monitoring change in the extent of salt-affected soils in northern India. Int J Remote Sens 15:3173–3182
Singh NT (1994b) Land degradation and remedial measures with reference to salinity, alkalinity, waterlogging and acidity. In: Deb DL (ed) Natural resources management for sustainable agriculture and environment. Angkor Publication, New Delhi, India, pp 442–443
Singh NT (2005) Irrigation and soil salinity in the Indian subcontinent: past and present. Lehigh University Press, Bethlehem, PA

Singh RP, Sirohi A (1994) Spectral reflectance properties of different types of soil surfaces. Int Soc Photogram Remote Sens 49:34–40
Singh G, Bundela DS, Sethi M, Lal K, Kamra SK (2010) Remote sensing and geographic information system for appraisal of salt-affected soils in India. J Environ Qual 39:5–15
Singh NT, Bandyopadhya AK (1996) Chemical degradation leading to salt-affected soils and their management for agriculture and alternate uses. In: Biswas TD, Narayanasamy G (ed) Soil management in relation to land degradation and environment. Indian Society of Soil Sciences Bulletin 17, New Delhi, India, pp 89–101
Singh AN, Baumgardner MF, Kristof ST (1977) Delineating salt affected soils in part of Indo-Gangetic plain by digital analysis of Landsat data. Technical Report. LARS, Purdue University, West Lafayette, IN
Srinivasan R, Mukhopadhyay S, Nayak DC, Singh SK (2015) Characterization, classification and evaluation of soil resources in coastal eco-system—a case study of Gosaba block (part), South 24 Parganas, West Bengal. Agropedology 25(02):195–201
Srinivasan R, Singh SK, Chattopadhyay T, Gangopadhyay SK, Nayak DC, Mukhopadhyay S (2017a) Characterization of soils and cropping pattern of Coastal West Bengal—a case study in Canning II block. J Soil Salin Water Qual 9(2):249–256
Srinivasan R, Singh SK, Nayak DC (2017b) Assessment of soil degradations in coastal ecosystem of Sundarbans, West Bengal—a case study. J Soil Salin Water Qual 9(2):257–269
Srinivasan R, Hegde R, Karthika KS, Maddileti N, Singh SK (2019) Effect of different land uses on soil pedogenic properties and sodicity development in Krishnagiri reservoir project dam catchment in Tamil Nadu. J Soil Salin Water Qual 11(1):10–17
Srinivasan R, Singh SK, Nayak DC, Dharumarajan S (2018a) Assessment of soil and water salinity and alkalinity in Coastal Odisha—a case study. J Soil Salin Water Qual 10(1):14–23
Srinivasan R, Gangopadhyay SK, Mukhopadhyay S, Das K, Nayak DC, Singh SK (2018b) Land resource inventory at 1:10000 scale in Ganjam block of Ganjam district, Odisha for optimal agricultural land use planning using geospatial technique. NBSS&LUP Technical Bulletin No. 1114, p 170
Thomas DSG (2011) Arid zone geomorphology: process, form and change in dry lands. Wiley, Chichester
Tyagi NK (2003) Managing saline and alkali water for higher productivity. In: Kijne JW et al (ed) Water productivity in agriculture: limits and opportunities for improvement. CAWMA Series, No 1. CABI Publication, London, UK, pp 69–87
U.S. Salinity Laboratory Staff (1954) Diagnosis and improvement of saline and alkali soils. USDA Handbook No. 60. U.S. Department of Agriculture, Washington, DC
Vaughan PJ, Lesch SM, Crown DL, Cone DG (1995) Water content effect on soil salinity prediction. A geostatistical study using co-kriging. Soil Sci Soc Am J 59(4):1146–1156
Wiegand C, Rhoades JD, Escobar DE, Everitt JH (1994) Photographic and videographic observations for determining and mapping the response of cotton to soil salinity. Remote Sens Environ 49(3):212–223. https://doi.org/10.1016/0034-4257(94)90017-5
Wiegand C, Anderson G, Lingle S, Escobar D (1996) Soil salinity effects on crop growth and yield-illustration of an analysis and mapping methodology for sugarcane. J Plant Physiol 148(3–4):418–424
Zaman M, Shahid SA, Heng L (2018) Guideline for salinity assessment, mitigation and adaptation using nuclear and related techniques. Springer open. ISBN 978-3-319-96189-7; ISBN 978-3-319-96190-3 (eBook) https://doi.org/10.1007/978-3-319-96190-3
Zhu JK (2001) Plant salt tolerance. Trends Plant Sci 6(2):66–71

Chapter 8
Status and Challenges of Monitoring Soil Erosion in Croplands of Arid Regions

Deepesh Machiwal, Abhishek Patel, Sushil Kumar, and Anandkumar Naorem

Abstract Soil erosion is the greatest threat to soil health and soil ecosystem services globally. Several studies are reported in literature on monitoring and assessment of soil erosion in semi-arid and humid regions both at catchment scale and field level. However, soil erosion studies are rare from arid regions especially Indian arid region. It is learnt that a variety of methods have been used for monitoring of soil erosion and rates of soil erosion vary considerably for regional and global estimates according to the method used to derive them. This chapter aims at providing an overview of methods used for monitoring runoff and soil erosion from agricultural lands. Then it describes different scales ranging from micro-plot to field and catchment scale assessment of soil erosion from agricultural/arable land. Thereafter, different devices and methods used for measurement or estimation of soil erosion in the literature are explained. Furthermore, a case study is presented to demonstrate a step-by-step methodology for measurement of runoff and soil erosion from agricultural fields of an arid region of Gujarat, India and results are discussed. The case study revealed that the highest soil loss occurred from the field plots of cultivated fallow (108.03 kg ha^{-1} yr^{-1}) and unploughed fallow (78.95 kg ha^{-1} yr^{-1}). The best intercropping practice in reducing field-level soil erosion is found as green gram intercropped with sorghum and pearl millet, which checked erosion of fertile soil by 69–79% more effectively than the cultivated and unploughed fallow plots. Moreover, challenges and issues faced in regular monitoring of soil erosion in arid climate are discussed.

Keywords Arid zone · Erosion modelling · Nutrient loss · Runoff

---

D. Machiwal (✉)
Division of Natural Resources, ICAR-Central Arid Zone Research Institute, Jodhpur, Rajasthan 342003, India
e-mail: dmachiwal@rediffmail.com; Deepesh.Machiwal@icar.gov.in

A. Patel · A. Naorem
Regional Research Station, ICAR-Central Arid Zone Research Institute, Bhuj, Gujarat 370105, India

S. Kumar
ICAR-Central Agroforestry Research Institute, Jhansi, Uttar Pradesh 284003, India

P. K. Shit et al. (eds.), *Soil Health and Environmental Sustainability*, Environmental Science and Engineering,
https://doi.org/10.1007/978-3-031-09270-1_8

8.1 Introduction

Soil erosion is continually-occurring natural geomorphic process where soil is destroyed and eroded through human activities, transporting it from one place and depositing to other place (Pennock 2019). The external forces enhance the rates of soil detachment and transport and disturb its balance with naturally occurring rates (Wang et al. 2018). Recently, a report of the Food and Agriculture Organization of the United Nations, Rome highlighted that research and extension efforts made over a century could not check soil erosion by water, wind and tillage to be the greatest threat to soil health and soil ecosystem services throughout the world (Pennock 2019). Furthermore, it is observed that the rates of soil erosion considerably vary for regional and global estimates according to the method of measurement or estimation. Also, amounts of the mean annual soil loss from field plots are reported to be considerably higher (8–50 t ha^{-1} yr^{-1}) than those estimated using regional and global models (2–4 t ha^{-1} yr^{-1}).

It is reported for the first time in 1993 that fertile soil is eroded at a rate of about 75 billion tons yr^{-1} from agricultural systems globally, which is much higher than the erosion rates occurring in natural ecosystems (Myers 1993). The same global estimates of the soil erosion causing a monetary loss of 400 billion USD annually are still reported by the Food and Agriculture Organization (FAO) led Global Soil Partnership states (GSP 2017). As the reliable worldwide soil erosion quantities are generally lacking, it is mentioned in the recent literature that the scientific community is forced to opt to the first studies performed in early 1990s (Borrelli et al. 2013; Machiwal et al. 2021). Annual loss of fertile soil from the world's two largest land occupying countries, together occupying 13% of the total area of the world, is 6.6 billion tons (India) and 5.5 billion tons (China) (Lal and Stewart 1990; Pimentel 2006). Likewise, about 3 billion tons of soil is lost in the United States annually (Pimentel 2006). Croplands or agricultural land is more conducive to occurrence of soil erosion than other types of land use. This is mainly due to difficulty for the crop cover in achieving long-term stability as the cover in an agricultural field may change considerably in a year. In agricultural lands, soil erosion redistributes and destructs soil, reduces organic carbon and nutrient contents, and decreases availability of water. Soil erosion in croplands makes the tillage layer thinner, declines soil fertility, restricts crop cultivation, escalates fertilizer cost, and diminishes productivity, and ultimately, erosion makes the crop more susceptible to drought occurrence (Fenton 2012; Wang et al. 2018). Soil erosion has a devastating effect on crop productivity and detrimental effects on soil quality and crop yield, and therefore, it ultimately affects socio-economics of farming community. In an intensive cropping system, soil erosion reduces crop yields significantly due to the restricted soil water storage and reduced water availability to plants (Bakker et al. 2004). In addition, nutrient supply gets reduced and rooting space becomes limited. If sufficient soil depth is available, losses of yield may be diminished by enhancing fertilizer application in order to compensate for the nutrient losses (Bakker et al. 2004, 2007). However, declines in yield may be substantially high under the conditions of limiting soil thickness.

Plentiful studies on soil erosion monitoring and assessment from croplands are available in literature for humid and semi-arid areas worldwide. However, it is observed that studies on soil erosion measurement, estimation and prediction from agricultural lands are relatively less for the arid or extremely arid regions especially in the Indian context (e.g., Machiwal et al. 2021). Very few numbers of rainy storms, less quantities of rainwater and little amount of rainfall-generated runoff in arid regions pose hurdles in accurate and precise monitoring of runoff and soil erosion. Some studies existing in literature provides soil erosion estimates from micro-watershed (e.g., Moharana et al. 2016) though such studies for different crop covers in the Indian arid regions are not found in the literature. Hence, the limitations and challenges experienced in soil erosion monitoring in arid regions are somewhat lacking to be known through the literature. It is felt that a gap exists in literature in finding the status of soil erosion measurement in arid regions and future scope of advancing the studies related to soil erosion measurement from croplands. This chapter is an attempt to bridge this knowledge gap by providing current status and challenges faced in soil erosion monitoring studies in arid regions. Furthermore, this chapter provides an overview of available methods of measuring/monitoring soil erosion from agricultural lands reported in literature both at local level (runoff plot study) as well as regional level (catchment study). Afterwards, it illustrates a case study of an Indian arid region where quantities of soil erosion from different sole crops and intercrop covers were measured and analyzed. Then, it discusses challenges and issues related to soil erosion monitoring in arid regions, and finally, future needs are briefly explained and concluding remarks are provided.

8.2 Overview of Methods Used for Monitoring Runoff and Soil Erosion from Arable/Agricultural Lands

It is revealed from the literature that soil erosion studies have vast domain and are being performed in relation to several spatial and temporal scales, on different climatic conditions and land use types, amounting the eroded soil on volumetric and weight (mass) basis, either with direct measurement/observation or estimation/prediction, and using empirical or physical process-based models. The results of these studies differ from each other in terms of amount and rate of erosion, however, are capable to serve towards new advances in research, methodology and model development, and finally for planning and decision making in order to archive sustainable agriculture and environmental functionalities.

Soil erosion assessment at spatial scale is classified in many ways by different researchers such as a point or site (plot or field) and areal extent (catchment or watershed) (Boardman and Evans 2020), micro (laboratory setup), meso (plot) and macro (catchment, field) scale (Millington 1981). The studies for understanding the on-site and off-site effects of soil erosion are generally connected with smaller and larger spatial scales, respectively (Ciesiolka and Rose 1998). The temporal scale varies among the studies based on consideration of spatial scale, for example, the raindrop splash impact in plot studies can be understood within short period, whereas the sediment delivery studies from catchment scales are long-time dependent and may take larger time frame (Van Rompaey et al. 2005). Climatic conditions of a region, such as arid, semi-arid, sub-humid and humid, have their own set of factors contributing to soil erosion. The diverse crop-soil conditions and rainfall quantities of different climatic regions are the main causes behind consideration of dissimilar processes for soil erosion estimation. Plot based study to understand the drivers of erosion generally considers the mass basis assessment of soil erosion from a plot, however, this approach has been criticized (Boardman 2006; Evans 1993; Evans et al. 2016). The field based volumetric assessment has been extensively done in number of studies (Boardman 1988, 1991, 2003; Boardman et al. 1996, 2009). Volumetric estimation, useful for rill, inter-rill and gully erosion, uses the change detection methods in cross-section profile (before and after rainfall and runoff generation) throughout considered length of rill or gully. Plot based volumetric assessment has been considered as the easiest way for rate assessment (Evans and Boardman 1994). However, the pin or stack based volumetric methods generally accompany with uncertainties (Haigh 1977). The measurement of soil erosion generally relates the rainfall, soil, land use/land cover, slope, and their attributes to the soil erosion. These plot/field level procedures have been established for different time scales. Empirical relationships of soil erosion with other relevant factors are continually being improved with more and more availability of reliable soil erosion database worldwide (Renard et al. 1997; Kinnell and Risse 1998), and this relationship is a key factor in planning and design of appropriate soil conservation measures (Cao et al. 2015). One of the widely adopted empirical relationships in soil erosion studies is the Universal Soil Loss Equation (USLE) (Wischmeier and Smith 1978). Boardman (2006) indicated that application of the original USLE model may lead to reduction in event-based erosion studies. However, Nearing (2013) pointed out its usefulness for the specific purposes. In empirical assessment of soil erosion, the inclusion of empirical relationship must be in line with the actual physical processes of erosion at the scale of consideration. However, empirical models are associated with limitations of spatial and temporal scales leading towards development of physical or process based models. The physical models may be considered as a substitute to the empirical relationships. However, the physical models are usually based on generalized assumptions (de Oliveira Salumbo 2020) and require a large number of parameters.

There is no exact and well-defined boundary/border between the different methods, and thus, there is also not a clear demarcation among methods of soil erosion measurement/estimation. In general, all soil erosion models overlay and remain complementary to each other. Methods are interchangeably used for different scales both spatial and temporal and climatic conditions for erosion assessment. Several studies have been performed globally on soil erosion monitoring from agricultural/arable lands/fields using a variety of measurement/estimation methods. The spatial scales and methods used for soil erosion monitoring are discussed in the

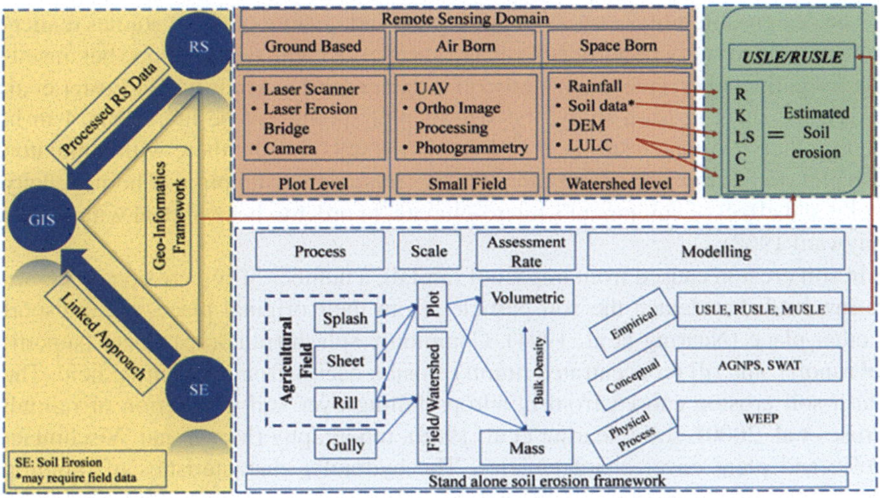

Fig. 8.1 Schematic showing interactions and relations among soil erosion processes, scales, measurement basis, modeling approach and coupling with remote sensing and GIS

following sub-sections. Interactions and relations among scales, methods, erosion-types, modeling approach, and their coupling with the modern tools of remote sensing (RS) and geographic information system (GIS) are illustrated in Fig. 8.1.

8.2.1 Assessment of Soil Erosion at Different Scales from Agricultural/Arable Land

8.2.1.1 Plot or Micro-scale Studies

Runoff plots are often used in soil erosion studies from which runoff and eroded soil are sampled, analyzed and measured. The plot based studies have been conducted under both actual (natural) and simulated (artificial) rainfall conditions. The rainfall simulator helps in reducing time involvement in soil erosion studies by creating an artificial condition of rainfall occurrences. The runoff plots generally have a defined boundary within which the runoff gets generated and is collected at end of each simulation (Mwango et al. 2016; Sadeghi et al. 2015). Sometimes, compacted bunds are constructed around the plot boundary or metallic sheets are inserted at edges of the plot to avoid escape of the generated runoff from the plot. It was the runoff plot where a series of plot studies on soil erosion were conducted at about 50 locations for more than 20 years in the United States (Gilley and Flanagan 2007), outcome of which resulted in development of the USLE and its further refinements and modifications (Renard et al. 1997; Wischmeier and Smith 1978).

The design suitability, easy replication, and randomization of plot studies renders them suitable to study raindrop/splash and sheet effect with desired slope but unsuitable for gully and creep erosion study (cf. Seginer 1966). According to Foster et al. (1981), for sheet erosion study, the size of plot should not be more than 4 m in length, otherwise underestimated soil erosion results will be there with simulator. Also plot studies must be taking care of biases related to assumption of homogeneity of plot and effect of equipment's micro-climate, in order to be confident with results (Hayward 1968).

In soil erosion caused from impact of rainfall, a number of soil erosion processes are involved that detach the soil particles from their original place and transport to other place (Nearing et al. 1994). Commonly splash (raindrop), sheet (superficial runoff), and rill (concentrated runoff) erosion occurs from agriculture field. The splash soil erosion caused from raindrops falling over soil is function of rainfall (Cruse et al. 2000), soil (Nearing et al. 1988), topography (Meyer and Wischmeier 1969), and plant cover characteristics. The hydraulic characteristics of raindrop (mass, velocity, kinetic energy, and impact angle of drop) greatly affect splash process (Cruse et al. 2000). Catching the splashed soil particle using trap container were used for splash erosion assessment (Morgan 1978; Bolline 1980). Parlak and Parlak (2010) performed plot scale splash experiment with four crop covers (control, vetch, barley, ryegrass) at 4 and 9% slope-combination during the years of 2006 and 2007. They studied the impact of rainfall in terms of its kinetic energy, cover crop, and slope over splash erosion. The experiment was replicated three times using splash cup of 10 cm as inner (small wall) and 30 cm as outer (high wall) diameter cylindrical cups with one diametric partition of inter-cylindrical area (Morgan 1978). The splashed amount of soil was calculated in terms of loss rate (g m^{-2}) occurred from inner cylinders' open area (diameter 10 cm) after each rainfall event. The rainfall was highest and lowest during third and second replications, respectively. The corresponding splash erosion was found to be highest (3173.7 g m^{-2}) and lowest (1155.2 g m^{-2}). The correlations of splashed soil erosion with 9 and 4% slopes were found to be 0.84 and 0.83. They concluded that splash soil erosion amplifies with increase in kinetic energy of rainfall, and slope and decreases with crop cover. However, the higher amount of rainfall may cause the more or similar erosion from high crop cover condition as compared to the combination of low rainfall and high crop-cover. Hence, further studies were required to address the inter-relationship of rainfall kinetic energy and crop cover adequately. Several splash erosion measurement devices such as cup (Zhou et al. 2013), funnel (Jordán et al. 2016), gutter (Jomaa et al. 2012), tower of funnels, Morgan tray (Beguería et al. 2015), and Tübingen cup (Tcup) (Liu et al. 2015) have been used in measuring splash erosion (Fernández-Raga et al. 2019).

Rainfall simulator is the equipment that mimics the natural rainfall in terms of intensity, duration and amount over experimental erosion plot helping to understand erosion from plot (Martínez-Mena et al. 2001). Though simulated rainfall does not exactly match the natural rainfall, results obtained are having good correlation with actual soil erosion values in field (Hamed et al. 2002). Such facility greatly helps in data generation and comparative studies in shorter time period than natural rainfall-erosion condition. Moreover, generated dataset allow erosion modelling (calibration, validation and prediction) using models such as USLE (Wischmeier and Smith 1978).

Nolan et al. (1997) estimated natural storm (rainfall) erosion using plot-size simulator (1 m^2 size) in barley tillage site of British Columbia. They measured 12 storm erosions, due to natural rainfall, from plots (144 m^2 in size) under three tillage conditions, i.e., zero tillage (ZT), reduced tillage (RT), and conventional tillage (CT), during the years 1987–1991. The rainfall erosivity factor (R) was taken as base factor (Wischmeier and Smith 1978). The simulator was adjusted for similar erosivity that of natural rainfall, and corresponding simulated erosions were measured during May 27–30, 1991. Two simulation runs, i.e., high at 140 mm h^{-1} and low at 60 mm h^{-1} intensity were carried out in dedicated 1 m^2 plots for simulator in each 144 m^2 plots. On an average, six runoff samples from simulator plot at every 3 min were collected. These samples were used to proportionate the total amount of soil loss. Moreover, the cumulative soil loss-curves were generated from dataset. They performed three major adjustments in simulator referring to kinetic energy of rainfall simulator (E), smaller land slope (L), and runoff lag time after simulator began (B). This E-L-B adjustment was found to give simulated erosion 95, 112, and 99% of natural erosion for ZT, RT, and CT, respectively. The low intensity (60 mm h^{-1}) was found to provide the best estimation of storm soil erosion from simulated soil erosion value. The study experienced the limitation related to scaling up of simulator measurement, and if so, there will be potential error in estimated storm soil erosion.

It is seen that most of the plot studies in literature are done for a short duration of time, and thus, these studies provide the fundamental information of soil erosion in a small time frame. The conditions where high magnitude and low frequency rainfall occurs, the short-term plot study does not reflect the long-term erosion process and soil losses. In such areas, the long-term plot measurements are required, which assess long-term erosion process and harmonize short-term plot studies.

In this direction, the efforts were made by Novara et al. (2011), who used a fixed reference to measure soil erosion over 9-year period (long-term). Using fixed-pole, they readily monitored and determined the long-term soil erosion by measuring poles' height over the ground. This study was conducted in vineyards of southwestern Sicily. To quantify erosion, rate a topographical approach was applied in which poles of 220 cm height were installed as erosion markers. At beginning of study (time Ti) poles were installed at 60 cm depth in three rows with 5 m inter-row distance. At this time, the reference height (Hi) of 160 cm was noted for all poles. At the end of study (after 9-year, time Tf) the final heights (Hf) of poles over the ground were recorded insuring that poles are vertical. The topographical change in plot was indicated by the net difference in pole heights (H) at end and beginning of study (H = Hf – Hi). This H is the depth of soil erosion or soil deposition within the plot. The inter-pole distance along a row was interpolated with the polynomial curve. Amount of soil erosion was estimated by multiplying the polynomial area with plot length. In order to convert this volumetric soil erosion into the weight of eroded soil, the bulk density (measured at three points) was used. The inter pole area index (I) was used to detect the soil erosion, deposition, and neutral points in the plot. This index was given as the ration of the differences of final heights of two consecutive poles along the slope to the initial height of pole. The negative, positive and zero value of area index corresponds to the soil erosion, soil deposition and neutral inter-pole area, respectively. The H value was found to be in range from −0.40 to 0.25 m. At the upper and middle parts of the slope, the H values in the upper (0.11) and middle (0.20 m) portions of the slope, indicated that the middle part was under-grown with more erosion than the upper part of plot. Whereas, area index was reported ranging from −0.06 to +0.02. The net volume of erosion over a long period (9-years) was estimated as difference of eroded and deposited volume within the plot. The erosion rate reported in this study was found to be in range of 86–118.5 Mg ha^{-1} yr^{-1} with the mean value of 102.2 Mg ha^{-1} yr^{-1}. The upper, middle and lower parts were reported to be the most eroded, less eroded and deposition zone, respectively. The innovative long-term plot study is inexpensive and determines long-term soil erosion both on volumetric and weight basis with spatial reference in particular, rather than giving erosion estimation for a plot as a whole.

The runoff plot method is used for assessment of soil erosion quantities that go out of the area along with flowing runoff water (McDonald et al. 2003) and is suitable in relative studies among different treatments as experiment (Mitchell and Bubenzer 1980). The runoff plots studies disturb the natural soil conditions due to establishment of plot boundary and do not measure the erosion within the field. Mess Bag (MB) method was initiated in 1992 (Hsieh 1992) to measure in-field soil erosion. The measurement of in-field soil erosion has been carried out using mess bags (MB) with undisturbed soil condition at Florida A&M University (FAMU), Tallahassee, Florida (Hsieh et al. 2009). The MB of three sizes (100, 400, and 900 cm^2) were fabricated using two layers of nylon mess of 4 mm (upper layer) and 0.1 mm (lower layer) opening. The lower layer of mess was chosen to ensure negligible soil loss occurring through it and to achieve a high accuracy of soil loss measurement in the experiment. The experimental plot (35 m^2 with 25% slope) was set up with 18 randomly placed MB (six replicates of each MB size) in grid pattern of 1 m spacing in close contact with bare soil of plot. After rainfall events, the soil from MB was taken out for oven drying and weight measurement. The rate of soil erosion was assessed considering weight of eroded soil particles, collected by MB, having size less than 2 mm per unit area of MB (t ha^{-1}). A relationship was developed between eroded soil in MB and from runoff plot (RP). They found the MB soil as 7.1 times that collected from the runoff plot $\{MB = (7.1 \times RP) + 10.7\}$ with correlation coefficient (r) value of 0.92 and may be attributed to fact that gross quantities of soil transported over small distances may be large, but net quantities moved downslope may be small. This study concluded that MB method has strong positive correlation with RP soil erosion. It is a straightforward quantitative method that is applicable for spatial and temporal soil erosion studies.

8.2.1.2 Field and Catchment Scale Studies

Plot based soil erosion studies were among the initial and controlled attempts for monitoring soil erosion. However, plot studies were having some drawbacks in accurate measurement of soil erosion. Loughran (1989) highlighted glitches of plot-based soil erosion measurement. Plot based studies are subjected to short and linear slopes, which may not be illustrative of actual topography of field (Edwards 1977; Evans 1990). It was observed that small plot studies do not represent rill and gully actions of the soil erosion while monitoring soil erosion quantities (Evans 1993). Therefore, direct extrapolation of findings of plot study to macro or field/catchment level may result inaccuracy in soil erosion assessment (Daniels et al. 1985). A runoff plot having well-defined boundaries stops water to flow naturally outside the plot border and concentrate it along the plot border that would have flown out in actual field condition and resulting more runoff from plot (Jinze 1981). On the other hand, field-scale based soil erosion studies consider the actions of rill and gully erosion adequately in monitored values of soil erosion. Hence, plot scale study considering only splash and sheet erosion, when extrapolated to field and catchment scale conditions, usually overestimates the soil erosion quantities than the real field conditions. This is the reason that field or catchment scale studies are preferred and considered superior to the plot based studies.

8.2.2 Devices and Methods Used for Measurement or Estimation of Soil Erosion

There exists a wide variety of devices and methods used for monitoring and/or estimation of soil erosion worldwide. In this section, widely used devices and salient tools used in soil erosion studies are briefly discussed. Salient methods used for soil erosion monitoring are categorized into suitable classes as shown in a classification chart (Fig. 8.2).

8.2.2.1 Flow Measuring and Runoff Sampling Approach

In general, total collection tanks and flumes are mostly used to monitor runoff and soil loss from small size runoff plots (Toy et al. 2002). Capacity of such tanks must be sufficiently large to hold runoff and soil loss generated at least in 24–48 h period. However, measurement of soil erosion from a field requires handling of huge amount of runoff that cannot be stored. Also, construction of larger-sized permanent structure would cost a lot for this purpose. In this case, the field is usually equipped with devices to measure runoff and soil sampling. Devices used for measuring or estimating water flow from a field include flumes (e.g., H-flume, long-throat flume), drop boxes and notches (e.g., v-notch). Flow measurement from these devices depends on

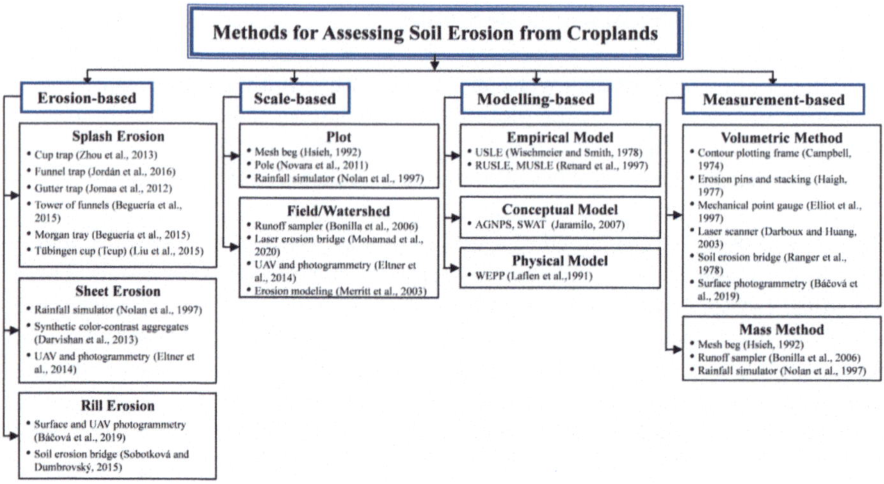

Fig. 8.2 Chart showing classification of methods for monitoring soil erosion

the water level recorded during the runoff flow and runoff volume is computed using rating curve or flow meter. The second component in such soil erosion measurement studies is sampling the sediment from the runoff water. Runoff sample may be collected by using slot samplers (PAP/RAC 1997) installed after the flow measuring device. Most common sediment sampler is multi-slot divisor (stationary slots) and the Coshocton wheel sampler (revolving slots). An automatic sediment sampler is also used with H-flume for regular sampling. The larger field size (catchment scale) measurement adopts weirs where the samples are collected at fixed time interval with integrated samplers and sediment discharge rating curves are used for sediment analysis (Morgan 2005). These methods are associated with uncertainty due to sampling frequency and may underestimate actual erosion assessed from field based on sediment samples. The collected samples may also be assessed for nutrient loss and soil management, which is one of the advantages of such sampling-based soil erosion studies.

Bonilla et al. (2006) performed the instrumentation for runoff and sediment measurement along with chemical loss analysis from agricultural field. They modified multi-slot divisor used by Pinson et al. (2004) to felicitate wide range runoff rate measurements and can be used for larger field area. This runoff sampler has a collector (5–7 m wide), multi-slot divisor and data storage device. Runoff is intercepted by collector and routed to multi-slot divisor. This divisor uses slotted crown and collects discharge-weighted sample to measure runoff and soil loss (Pinson et al. 2004). The discharge-weighted sample from divisor is sent to the storage tank equipped with water sensor, which makes this device capable to measure time varying runoff rate. In their study, eight such devices were installed in agricultural fields of Wisconsin. Since 2003–2006, total of 300 experimental data were collected. After each rainfall storm, generated runoff was measured and runoff samples were collected, which were subsequently transported to laboratory for determining soil erosion quantities. Results of one experiment indicated that the runoff sampler collected about 54.78 L of runoff volume. The total 2.7 m^3 of total runoff volume was estimated from field. The sediment analysis showed a total of 8.10 kg ha^{-1} sediment eroded from field during the particular storm. The study found to be useful for effective collection of sediment and their analysis from large agricultural field.

8.2.2.2 Approach Based on Volumetric Methods

Volumetric methods of soil erosion assessment are an alternative to runoff and sediment sampler-based methods particularly suitable for macro-level large areas. The volumetric methods such as flume and sampler are advantageous over the runoff sampler methods as the former can be conveniently replicated at different sites and do not require any structure to be constructed for monitoring runoff and soil loss. Volumetric assessment of soil erosion include a number of methods such as contour plotting frame (Campbell 1974), erosion pins and stacking (Haigh 1977; Novara et al. 2011), mechanical point gauge (Elliot et al. 1997), laser scanner (Darboux and Huang 2003), soil erosion bridge (Ranger and Frank 1978; De Santisteban et al. 2006), and surface photogrammetry (Rieke-Zapp and Nearing 2005).

In the year 2015, Sobotková and Dumbrovský (2015) developed volumetric method for rill erosion estimation in a heavy rainfall season at four sites in cadastral Šardice, Czech Republic. The soil erosion bridge, developed by Brno University of Technology, Czech Republic, was used in the maize crop fields having 10% land slope. This bridge has three equipment: (i) a square frame (2×2 m^2), (ii) removable profile (support motion of bridge), and (iii) soil erosion bridge. About 1300 number of rill profile measurements were done at four sites, each having two transects. Each transect had 5 square plots (4 m^2) along the slope length of transect. Five surface profiles were measured within each square plot and erosion volume from rill was calculated. This volumetric value was converted into mass value of soil erosion using bulk density of field soil. The high amount of soil erosion (500 t ha^{-1}) confirmed the high erodibility of soil from maize crop field.

Another method, i.e., non-contact photogrammetric method, has also been used for rill erosion estimation in earlier studies (e.g., Castillo et al. 2012; Wells et al. 2016). Recently, Báčová et al. (2019) performed python-liked geographic information system (GIS) based photogrammetry on digital surface model (DSM) prepared for artificially created rill in the catchment of Central Bohemian Region, Czech Republic. The experimental setup consisted of a metal reference frame (1.6×1.1 m^2) and a camera for taking photos. The DSM of the rill was generated using Photo Modeler Scanner software. Methodology for volume calculation of rill was developed using python tool in Arc-GIS software, which do not consider the actual initial surface rather estimate it from manually drawn rill polygon. Hypothesis behind this methodology was that as in actual field condition the initial field surface before rill erosion is not available and must be simulated before volume calculation. The DSM before and after rill formation (using rill polygon) were generated and volume of rill was calculated by differentiating both DSMs. To validate the result of artificial rill volume, the actual rill volume of naturally generated rill was measured using Structure from Motion (SFM) method (Szeliski 2010; Ullman 1979). The results showed 10–15% underestimation of rill volume than the actual volume, which was attributed to the more curved natural surface than the simulated initial surface. Therefore, the results were found satisfactory and the method was found to be advantageous for the fast and effective rill erosion estimation when the prior surface information is not available. In addition, the method has some limitations of poor resolution, uncertainty of rill polygon position, exclusion of undercut rill-wall, and no idea about sheet erosion.

8.2.2.3 Modelling Based Approach

Soil erosion estimation and soil management have received a great attention globally since 1930s'. Consequently, several models for soil erosion assessment have been proposed in the literature, which also have been utilized in many studies over the decades (Lal 2001). Soil erosion models do not directly measure erosion rather these estimate or predict the erosion based on involved parameters and their defined/possible interaction (Anejionu et al. 2013). Models help in prediction of likely events well in advance and allow us to act accordingly. The developed models are grouped into three categories (Merritt et al. 2003): (i) empirical, (ii) semi-empirical, and (iii) physical or process-based models. Empirical soil erosion models, represented by mathematical equations, are developed using observed soil erosion data and are static in nature, whereas physical or process-based models define dynamic mechanism of considered parameters. Semi-empirical models fall in between the empirical and physical models. Three widely used empirical models of soil erosion assessment are USLE, Revised USLE (RUSLE) and Modified USLE (MUSLE) (Tesfahunegn 2011). Similarly, Agricultural Non-Point Source Pollution (AGNPS) (Jaramilo 2007) is the most popular example of conceptual model and Water Erosion Prediction Project (WEPP) is a well-known process-based model (Laflen et al. 1991). Applicability of different models depends on intension of research, required input dataset, type of output from model, among others, as model have their own complexities and conceptualization (Merritt et al. 2003).

The USLE model (Wischmeier and Smith 1978) was developed for estimation of soil erosion from level or gentle croplands of USA (Ganasri and Ramesh 2016). The USLE requires six factors named rainfall erosivity (R, relate rainfall intensity), soil erodibility (K, relate soil properties), slope length (L) and slope steepness (S) (both relate topography), cover-management practices (C, relate crop condition), and supporting conservation practices (P, relate farming/agronomical practices). The USLE is developed based on 10,000 plot-years data of sheet erosion obtained from plots and small-scale watersheds (Wischmeier and Smith 1978). Though, it is simple and easy but do not predict event-based erosion rather gives annual soil loss, hence, it may not provide highest soil loss event of year (Merritt et al. 2003). The USLE cannot be used for estimation of gully erosion (Morgan 2005) and individual rainfall-erosion event. In the past, many attempts were carried out to revise the classic USLE model for improving its accuracy to estimate soil erosion from crop fields (McCool et al. 1995). In the revisions, the classic structure of the USLE model was preserved, however, the latest insights from studies conducted with the USLE after 1978 were incorporated and computerized version in form of RUSLE was developed (Renard and Freimund 1994). The existing USLE model was revised in order to make it applicable on complex watershed level by giving important revisions of topographic factors in the USLE (Renard et al. 1997). This classic LS factor was adjusted and improved to incorporate different terrain conditions (including steep one), their shape (convex or concave) and discharge of convergence effects (relate runoff conversion from upstream to downstream). This revised form of model was named as RUSLE3D (Renard et al. 1997). Use of geographical information system (GIS) is necessary for working with this model. Using GIS tool, the thematic layers for incorporated factors: rainfall erosivity (R-factor), pedological units (K-factor), cumulative flux direction (LS-factor), and land-use (C-factor) are created, and their algebraic operation predicts soil erosion as spatial soil erosion map for watershed scale. Thereafter, RUSLE3D model has been applied in different studies on soil erosion assessment in order to develop soil erosion risk map, identify susceptible area for erosion, and take necessary action for soil conservation (e.g., Beskow et al. 2009; Oliveira et al. 2014). Later on, it was realized that adoption of the USLE and/or RUSLE models for soil erosion estimation from agricultural fields requires computation of R-factor based on long-term (at least 15-years) rainfall data, and hence, both the models cannot simulate event-based soil loss. In order to overcome this limitation, the individual rainfall-runoff events on peak discharge basis were considered in place of R-factor. This modification in the USLE model came up with modified-USLE or MUSLE model.

In contrast to the USLE model and its derivatives (RUSLE and MUSLE), the AGNPS is an event based model having capability of simulating runoff, soil and nutrient losses from a cropland. This conceptual model was proposed collaboratively by the Agricultural Research Service of the United States Department of Agriculture (USDA), the Minnesota Pollution Control Agency and the Soil Conservation Service (SCS). It is noted that the AGNPS model incorporates the concepts of existing models like fusion of RUSLE for soil loss prediction with SCS-Curve Number method for runoff prediction on grid-cell basis (Merritt et al. 2003). The AGNPS model can also be used for suitable planning and management of drainage basins for analysis and identification of land degradation and environmental issues (Nugroho 2003). Higher complexity and data requirement of the AGNPS model than empirical one may be attributed to its capability to simulate complex runoff-erosion-pollution phenomenon (Merritt et al. 2003).

The USLE and its family models are restricted to estimate the long-term mean annual soil loss from agricultural fields, which is not of much use for managing the soil resources in large scale environmental conservation projects. In order to cope with these drawbacks of the local scale USLE model and its derivatives, process-based WEPP model was developed collaboratively by several agenesis of USA to substitute the USLE and its variants (Laflen et al. 1991; Flanagan et al. 2007; Deb and Shukla 2011). This model consists of three computer modules, i.e., hillslope, watershed and gridded modules (Morgan 2005) and requires four types of input database related to soil, climate, topography, and management (Pieri et al. 2007).

8.2.2.4 Remote Sensing and Geographic Information System Based Approach

Applicability of geo-informatics in soil erosion estimation for management of soil resources has been found useful by several researchers worldwide (e.g., Pradhan et al. 2012; Vrieling et al. 2010; Ganasri and Ramesh 2016). Geo-informatics is an advance technology, which combines modern tools of remote sensing (RS) and geographic information system (GIS). Methodology of the geo-informatics, involving fusion of RS and GIS for soil erosion estimation, is found cost-effective, time-saving, dynamic, and accurate (Seutloali et al. 2017). The RS and GIS based soil erosion models offer detailed analysis of spatial and temporal variability of soil erosion process. It enables applicability of soil erosion assessment methods over large areas, sometimes at continental or global level, which are usually remain impractical for conducting survey and physically inaccessible for the most time in a year (Sepuru and Dube 2018). There is a large impact of areal extent or scale on erosion phenomenon and its assessment, and use of RS data enhances capacity to study soil erosion at different landscape extents. Besides, use of RS and GIS techniques does not disturb soil surface and provides fast and repetitive (temporal) assessment of soil losses over an area of interest (Seutloali et al. 2017). Furthermore, geo-informatics is relatively a new approach having a vast potential in soil erosion modeling (Senanayake et al. 2020). The widespread use of GIS technique enabled coupling of the empirical USLE model with GIS tool (Jazouli et al. 2017). Likewise, the RUSLE model has also been successfully incorporated within the GIS environment at watershed and basin scales (Bahrawi et al. 2016; Gaubi et al. 2017).

8.3 Case Study of Soil Erosion Monitoring in an Indian Arid Region

It is seen from the extensive literature search that runoff and soil loss studies are rarely reported from the Indian arid regions. This is the first study reported from the arid region of Gujarat where runoff and soil loss are measured from croplands located in arid climate. In this section, a case study is presented demonstrating a step-by-step methodology and results where runoff and soil erosion from croplands were measured during 2013–2015 in arid Kachchh region of Gujarat, western India.

8.3.1 Overview of Study Area

This field study was performed in research field of a Regional Station of the Central Arid Zone Research Institute, located in Kukma village of Bhuj, Gujarat, India. Longitude of the experimental field covering 72 m × 66 m area varies from 69°47′31″ to 69°47′34″ E and latitude from 23°12′53″ to 23°12′56″ N. In the study area, the mean annual rainfall is 389 mm (1988–2015) received in average 13 rainy days in a year. The area is situated in a hot-arid climate where 38–68% of the total annual rainfall is received in 2–4 consecutive days through southwest monsoon (Machiwal et al. 2016). The mean potential evapotranspiration is very high (1900 mm $year^{-1}$) (Singh and Kar 1996). The 102-year (1901–2002) mean monthly maximum and minimum temperatures vary from 22.1 to 31.9 °C and from 8.8 to 22.7 °C, respectively (Machiwal et al. 2017). During monsoon season, relative humidity over the coastal lands is observed to be more than 80% while it is more than 65% in inland. In afternoon, the relative humidity decreases up to 25% or less both in summer and winter seasons (Singh and Kar 1996). Soil of the area is medium-textured, loamy to sandy loam, mixed hyperthermic, developed from sandstone and shale (Mangalassery et al. 2014a). Depth of the soil is restricted beyond 100 cm depth underlain by weathered rocks with limitations of soil depth, salinity and water erosion. In general, the slope of the field does not exceed 2%.

8.3.2 Methodology

8.3.2.1 Field Setup and Treatment Details

In this study, size of experimental field-plots was kept as 20 m (length) × 5 m (width) following recommendations of the Food and Agriculture Organization (FAO) of the United Nations, Rome (Hudson 1993). Field bunds were constructed around every individual plot to prevent runoff of one plot entering to other plots and also to guide surface runoff towards the outlet. In addition, an alley of 2 m width was maintained between two nearby plots for providing an easy access to the plots and avoiding interference of one plot to another. A schematic of the field setup is shown in Fig. 8.3. Multi-slot divisors were installed at outlet of each field plot in rainy season of year 2013 for monitoring of runoff as well as soil losses from the field plots. A non-recording type of raingauge was installed at the study site to measure daily rainfall.

Soil erosion was monitored from field plots of commonly-grown crops in arid region of Gujarat. A total of four crops, i.e., two cereal (sorghum and pearl millet) and two leguminous (green gram and cluster bean) crops were considered in this study for soil erosion monitoring. A total of 10 treatments consisting of sole (four individual crops), intercropping (legumes with cereals) and control plots were considered. There were two control treatments, where field plots were kept empty without sowing of

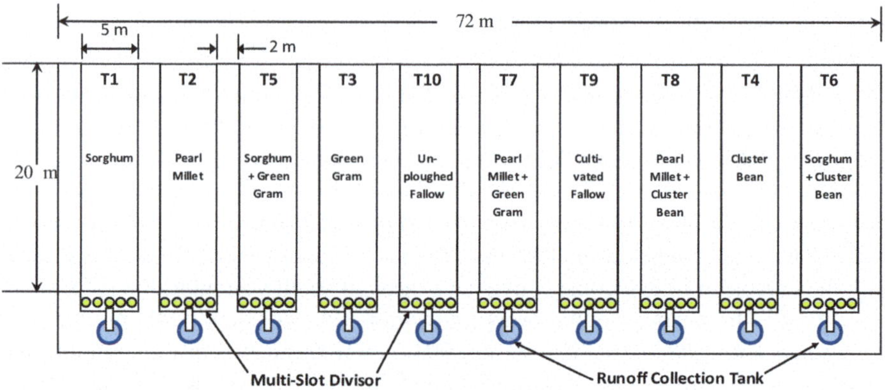

Fig. 8.3 Experimental layout showing 10 field plots of first replication

any crops: first control plot enabled evaluating effect of ploughing and inter-culturing operations, and second control plot offered examining effect of unploughed fallow land on soil erosion. Thus, the ten treatments were: T1-sole sorghum crop, T2-sole pearl millet crop, T3-sole green gram crop, T4-sole cluster bean crop, T5-Inter-cropping of sorghum and green gram crops, T6-Inter-cropping of sorghum and cluster bean crops, T7-Inter-cropping of pearl millet and green gram crops, T8-Inter-cropping of pearl millet and cluster bean crops, T9-cultivated fallow (control), and T10-unploughed fallow (control).

8.3.2.2 Construction and Installation of Soil Erosion Measuring Device

Runoff and soil erosion measuring device, i.e., multi-slot divisor, was designed according to local settings of the study area. Rainfall intensity in Kachchh region is very high, which was a major factor in designing soil loss measuring device. Hence, the dimensions of the device, e.g. size and number of slots along with capacity of the collection tank were decided in such a manner to accurately measure runoff and soil loss for any rainfall storms of expected intensity. Different shapes for the multi-slot divisor were designed and a variety of construction materials were considered, and a final shape and material were chosen based on the cost-effectiveness, easiness in construction, and availability of the facilities locally (Fig. 8.4).

8.3.2.3 Monitoring Rainfall, Runoff and Soil Loss

Daily rainfall during four years (2013–2016) was recorded near the field experiment using a non-recording type of raingauge. During the study period, amount of runoff generated in ten field plots and magnitude of eroded soil carried with flowing runoff water was determined from samples of runoff collected after every storm. Many

Fig. 8.4 Installation of fabricated multi-slot divisor and runoff sampling tank

rainy storms wre found inadequate in generating runoff water to be flown in the field plots. The runoff and eroded soil samples were collected in storage tanks of 50 L capacity, which were connected to three outlets of the multi-slot divisors. The stored runoff water was stirred thoroughly and three samples of the runoff water with suspended soil were collected, mixed together and carried to laboratory for filtration and chemical analyses. Filtration of sampled water (Whatman No.1 filter paper) separated out eroded soil from the runoff water, and the former was then oven-dried to get dry soil. The dried soil was weighed to know the amount of soil erosion from ten field plots.

8.3.2.4 Collecting Soil Samples and Chemical Analysis

Soil samples were collected at 0–10 cm depth from all ten field plots after harvesting of the crops in every year. The samples were collected from three locations of the field plot, i.e., upper, middle and lower parts, from every replication, which were mixed together for every plot. The composite soil samples were then transported to laboratory for determining soil chemical properties, i.e., soil organic carbon (OC), available P_2O_5 and available (exchangeable) K_2O by using Degtjareff method (Walkley and Black 1934), ascorbic acid method (Olsen et al. 1954), and neutral ammonium acetate method (Metson 1957), respectively. The macro-nutrients P_2O_5, K_2O and OC are considered as the necessary nutrients required for optimum growth of plant, which together define the soil fertility. Values of soil nutrients in mg g^{-1} were converted to kg ha^{-1} by considering value of soil bulk density as 1.45 g cm^{-3} and 50-cm soil depth of root-zone (Mangalassery et al. 2014b).

8.3.2.5 Statistical Significance of Crop Covers

Relative significance of sole crop and intercrop covers in checking soil erosion was evaluated statistically by applying analysis of variance (ANOVA). The ANOVA reveals statistical-significance of variations in soil erosion values due to annual and treatment-wise variability as well as due to their interactions.

8.3.3 Results and Discussion

8.3.3.1 Annual Variability of Runoff

Treatment-wise mean annual runoff values for 2013–2015 are shown in Fig. 8.5a. It is apparent that the mean annual runoff for three years as percent of annual rainfall is lowest (26%) for PM + GG (26 mm yr^{-1}) and PM + CB (27 mm yr^{-1}) plots, and the highest runoff percentage (41%) is for unploughed fallow (41 mm yr^{-1}) and cultivated fallow (36%) plots (Fig. 8.5a). This finding indicates that close-growing legume intercropping system reduces the amount of runoff generation over the fallow/bare

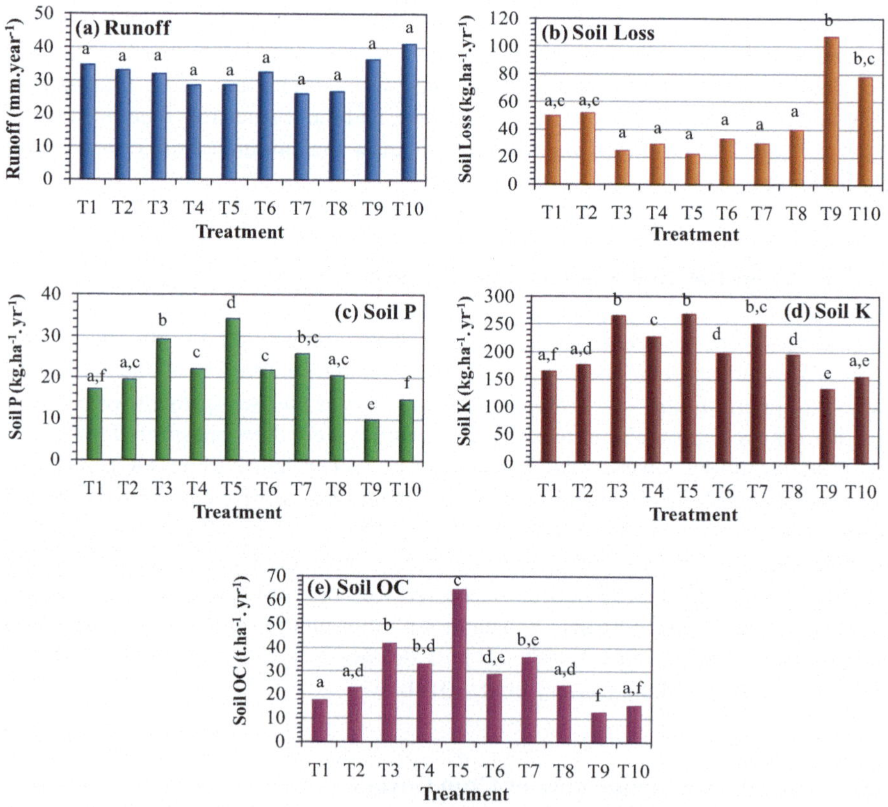

Fig. 8.5 Bar charts of the mean annual values for **a** runoff, **b** soil loss, **c** soil phosphorus, **d** soil potassium, and **d** soil organic carbon content during 2013–2015

and sole cereal-crop plots. The largest runoff quantity generated in the field plots in the year 2013 matches well with the highest annual rainfall received in the year 2013 during 2013–2015, although the lowest runoff is generated in the year 2014 when the rainfall is not the lowest of 3 years. It is found that of the total annual rainfall, only 205, 137 and 89 mm rainfall received in 3, 3 and 1 rainy days in the years 2013, 2014 and 2015 could cause runoff water flowing in the experimental plots. Hence, the runoff generation does not depend entirely on the annual rainfall amounts rather factors such as rainfall intensity and its distribution pattern influences the runoff generation.

Test-statistic values of Tukey's honestly significant difference (HSD) test were obtained from the ANOVA for annual runoff of ten treatments. Results of Tukey's HSD test indicate that the mean annual runoff for sole and intercrops do not differ significantly from that of two fallow plots (cultivated and ploughed) at 0.05 level of significance. This suggests that runoff generated from cropped and fallow plots is not significantly different from each other.

8.3.3.2 Annual Variability of Soil Erosion

Treatment-wise mean annual soil erosion is depicted in Fig. 8.5b. The figure reveals that the mean value of annual soil erosion is lowest (22.81 ± 12.42 kg ha^{-1} yr^{-1}) for S + GG intercrop and sole GG (25.27 ± 20.44 kg ha^{-1} yr^{-1}) and CB (29.71 ± 21.17 kg ha^{-1} yr^{-1}). On the contrary, the soil erosion is highest from field plots of cultivated fallow (108.03 ± 49.95 kg ha^{-1} yr^{-1}) and unplowed fallow (78.95 ± 28.42 kg ha^{-1} yr^{-1}). It is further seen that soil erosion from sole crops of S, PM, GG and CB are respectively 53%, 51%, 77% and 73% lower than that from cultivated fallow and respectively 36%, 33%, 68% and 62% lower than that from unplowed fallow. Similarly, soil erosion from intercrops of S + GG, S + CB, PM + GG and PM + CB are 79%, 69%, 72% and 63%, respectively, lower than that from cultivated fallow and 71%, 57%, 62% and 49%, respectively, lower than that from unplowed fallow (Fig. 8.5b). Hence, reduction in soil erosion is revealed from leguminous crops (GG and CB) and intercrops of green gram with cereals as compared to fallow plots. It is further apparent that the highest annual soil erosion occurred in the year 2013 and lowest in the year 2015 from all plots (Fig. 8.5b). This does not show harmony to the magnitudes of annual rainfall, which are 291, 193 and 346 mm in years 2013, 2014 and 2015, respectively. Hence, it is important to understand that of the total 16, 16 and 10 rainy days in years 2013, 2014 and 2015, respectively, runoff in the field plots could be produced during only 3, 3 and 1 rainy days. The effective annual rainfall generating 205, 136, and 89 mm runoff in years 2013, 2014 and 2015, respectively has a direct correspondence with the amount of total annual soil erosion.

Results of Tukey's HSD test reveals that magnitude of the mean annual soil erosion occurring from sole crops (S, PM, GG and CB) and intercrops differ significantly ($p < 0.05$) from that in cultivated and unploughed fallow plots. Furthermore, quantity of the soil erosion occurring from sole legumes (GG and CB) and all intercrops is found significantly different ($p < 0.05$) from that measured in unploughed fallow plot.

Moreover, difference between the mean soil erosion occurring from sole cereals (S and PM) is not statistically different from that in unploughed fallow plot.

8.3.3.3 Annual Variability of Soil Nutrients

Mean of three soil nutrients (P_2O_5, K_2O and OC) for 10 treatments are given in Fig. 8.5c–e. The mean annual value of soil phosphorous is lowest for cultivated fallow (10.06 ± 1.13 kg ha^{-1} yr^{-1}) and unplowed fallow (14.85 ± 1.01 kg ha^{-1} yr^{-1}); and highest (34.17 ± 4.32 kg ha^{-1} yr^{-1}) is for S + GG and sole GG (29.34 ± 4.61 kg ha^{-1} yr^{-1}). Likewise, the mean value of soil potassium is lowest for cultivated fallow (135.61 ± 13.60 kg ha^{-1} yr^{-1}) and unplowed fallow (156.79 ± 12.22 kg ha^{-1} yr^{-1}); and highest is for S + GG (270.78 ± 0.42 kg ha^{-1} yr^{-1}), followed by GG (266.86 ± 27.09 kg ha^{-1} yr^{-1}). Similarly, the mean value of organic carbon is lowest for cultivated fallow (13.10 ± 2.02 t ha^{-1} yr^{-1}) and unplowed fallow (15.74 ± 4.40 t ha^{-1} yr^{-1}); and highest is for S + GG (64.85 ± 11.64 t ha^{-1} yr^{-1}) and GG (41.95 ± 6.96 t ha^{-1} yr^{-1}). Overall, it is evident that soil nutrients essential for optimum plant growth remains in the highest proportions in treatments of sole GG and S + GG intercrop. This finding emphasizes the important role of legume crops in covering the ground optimally and conserving soil and nutrient resources in arid croplands (Bashagaluke et al. 2018). The quantities of soil nutrients show an increase over the three years including fallow plots, which is attributed to fertilizers applied during the experiment.

Results of the Tukey's HSD test for soil P_2O_5, soil K_2O and soil organic carbon are presented in Fig. 8.5c–e. Sole legumes (GG and CB) and their intercrops conserved soil P_2O_5 significantly in comparison to the cultivated and ploughed fallow plots ($p < 0.05$). Likewise, the mean K_2O of sole legumes and their intercrops are found significantly different from that of cultivated and unploughed fallow plots ($p < 0.05$). Also, the mean of soil organic carbon in case of legumes, both sole crops and intercrops, showed statistical dissimilarity from that obtained for the two fallow plots (Fig. 8.5e). Overall, the amount of soil nutrients present in two fallow plots is found significantly different from that found in sole leguminous crops and cereal-legume intercrops.

8.3.4 Conclusions of the Case Study

In this case study, runoff, soil erosion and nutrients from agricultural fields of an Indian arid region are monitored, and importance of cereal-legume intercrops in checking soil erosion is illustrated. Such estimates of runoff and soil erosion are necessary for adequate planning and appropriate design of soil conservation structures or measures at field scale and are not readily available for the arid regions of Gujarat, India. It is revealed from the results that the highest amount of soil erosion occurred from the field plots of cultivated fallow (108.03 kg ha^{-1} yr^{-1}) and unplowed fallow (78.95 kg ha^{-1} yr^{-1}). The best intercrop practice in reducing field-level soil erosion is found as green gram intercropped with sorghum and pearl millet, which checked erosion of fertile soil by 69–79% more effectively than the cultivated and unploughed fallow plots. All the obtained results are statistically significant ($p < 0.05$).

8.4 Challenges and Issues in Regular Monitoring of Soil Erosion in Arid Climate

Soil erosion due to water occurs over lands situated under all climate regions such as arid, semi-arid, sub-humid, and humid. However, a large difference exists in the process of erosion over different climates, partially supported by soil types that pose different kind of challenges and issues towards studies related to monitoring/measurement of soil erosion. Here, a few of those challenges and issues for the arid climate mostly experienced in the Indian arid regions are discussed.

8.4.1 High Speed Winds

Soil erosion occurs due to both water as well as wind. High velocity wind and extreme air temperatures are common phenomena in arid regions that cause significant reduction in soil moisture and blow a large amount of arid soils from one place to another. It is further learnt that soils of the arid climate regions is mostly sandy or sandy loam in nature as almost all deserts of the world are located under the arid climate. These sandy soils are not capable of holding the infiltrated water for a long time and water is escaped due to either evaporation or free drainage. Hence, arid soils remain dry or moisture deficient for a considerable period of time in a year, which causes easy blowing of soil particles due to high velocity winds. These winds erode the soil of one place, transport the eroded soil to another place and deposit it there. In deserts, shifting of sand dunes over place to place has been reported in the literature. Therefore, wind plays more powerful role in erosion of arid soils in comparison to soil of other climate regions including sub-humid and humid.

8.4.2 Infrequent Rainy Days and Runoff

In arid regions, number of rainy events is less than that occurs in semi-arid and humid regions. Thus, there is little less scope for monitoring soil erosion caused by runoff generated from the infrequent rainy days. Perhaps, this is the reason that estimates of soil erosion are available for relatively less area over the world's arid regions than that exist for areas under other than arid climate. Furthermore, it is seen that soil erosion studies conducted in arid regions are usually based on scanty data used for analyses. Thus, the results obtained from such studies may sometimes be weak or inferior and need further strengthening or confirmation in future when more data of soil erosion are available. In general, the magnitude of rainfall received in arid regions is also less as compared to that obtained in sub-humid or humid climate regions, which offers much less opportunity for the studies performed in arid regions to monitor or record a large soil erosion database. Despite the fact that soil erosion database in any part of the arid regions may be of inadequate length of time, such studies should always be encouraged as results of those studies provide an initial estimates of soil erosion for that area.

8.4.3 Shallow Soil Thickness

Soil erosion is a continuous process that always remains in operation whenever the factors, i.e., wind and water, working on it become active in any part of arid regions. There are certain portions of the arid lands where soil erosion by either or both of water and wind factors have been put into action for the past several decades. Consequently, huge quantities of soils have already been eroded and transported from their original source and deposited towards downstream portions where lowlands are situated. This has resulted in complete loss of topsoil and even the sub-soils from the deeper horizons, which has caused significant decline in soil depths. Currently, many areas of the arid regions are facing limitations of soil thickness where the soil has already been lost due to erosion occurring over a considerably period of time. Besides, restricting the soil thickness, large quantities of the soil eroded from uplands get deposited in depressions of the downstream lowlands mainly creating another problem of reservoir sedimentation. The restricted soil thickness may hamper crop growth especially in croplands due to availability of the limited space for plant root zone. In addition, there will be an inadequate root-zone for extracting nutrients by the plant roots.

8.4.4 Changing Rainfall Patterns Due to Climate Change

Arid regions are generally perceived as the areas that experience a little rainfall over a few rainy days in a year with severe drought years. With this outlook, it seems that measurement of soil erosion may not be a difficult task in an arid region with only a few days' observations, may be less than a week, whenever little rainfall occurs in the area. However, researchers such as hydrologists and soil & water managers working for soil conservation works in the arid regions always look forward to have more and more number of rainy days in order to strengthen the obtained soil erosion database. The four monsoon months (July–October) are a very critical period for the people involved in monitoring of soil erosion in an arid region as every single rainy event is important for them to capture amount of eroded soil looking at rare occurrences of the rainy days. In the recent times, changes in rainfall pattern of arid regions have been reported by some researchers. In arid regions, trends in rainfall are found to be significantly increasing mainly after the year 2000. This increase in the rainfall is attributed to occurrence of high-intensity rainfall storms of short duration in the arid regions, which is also stated in the report of International Panel of Climate Change (IPCC). The highly intense rainfall results in severe soil erosion in arid croplands only in just one or two events. It is experienced that accurate monitoring of soil erosion quantities in response to extreme rainy events in the arid croplands is difficult. Sometimes, it is consecutive rainy days causing severe soil erosion from agricultural field, and there may not be sufficient time to collect runoff and soil sediment samples between two sequential rain events. Thus, collection of runoff samples and monitoring soil erosion in arid lands involve complex and difficult challenge under the changing rainfall patterns.

8.4.5 Unfavorable Soil Workability Conditions

When soil erosion is to be measured in croplands, it is difficult to get enter into an agricultural field just after occurrence of a rainy event of the adequate quantity generating runoff in the field. This is due to enhanced soil water content beyond the soil field capacity and saturation capacity. In spite of the fact that soil workability conditions says 'no-go' day on a field (Simalenga and Have 1992), it is very much imperative to enter the field for measurement of soil erosion. In lands other than agricultural fields, it is quite easy to have an easy access to the spot of soil erosion measurement such as a catchment having little amount of soil.

8.5 Future Needs and Concluding Remarks

Based on overview of the previous studies on soil erosion measurement from croplands of arid regions as reported in literature as well as from authors' personal experience of working on the similar aspect in the Indian arid regions, challenges and issues related to soil erosion monitoring arising from the arid regions are described in the previous section. Here, future needs to resolve the issues and to face the challenges are briefly discussed.

Database on soil erosion both due to water and winds need to be strengthened for the arid regions. This may be achieved by taking long-term experiments on soil erosion measurements from croplands. As the cropland experiments are location-specific, there is a need to carryout such experiments at multi-sites especially on farmers' fields covering entire area in scattered manner. It is noticed that there may always be a deficiency of data on soil erosion from croplands although the studies performed systematically following a standard step-by-step procedure should always be appreciated and researchers be motivated to extend their work further in order to strengthen the database and confirm the earlier findings when adequate quantum of data become available. Currently, some parts of the traditionally water-short arid regions are becoming wetter in response to extreme or heavy rainy storms, which are of high-intensity and occur rarely in the area. It is very important for the hydrologists to capture the process of soil erosion for those extreme rainy events, which may prolong over more than two or three days in the area. There should be a task force ready to deal with measurement of soil erosion occurring during the unusual rainfall and unexpected runoff generated from the croplands of arid regions. Such database might play a crucial role in suitable planning and appropriate design of appropriate soil conservation measures/structures in the future years under the climate change scenario.

Advanced planning for monitoring of soil erosion from crop fields under wet soil may be advantageous. Provision of an alternative path to reach at the soil monitoring site in case of excessive rainfall may be beneficial when soil workability conditions do not favour in providing access to the site. Work of soil erosion measurement should be imagined in mind prior to actual happenings of the worst conditions, which may help in taking necessary precautions and keeping possible options in hand whenever the need arises. Appropriate soil conservation measures may be adopted in arid regions depending upon the major factor (water or wind) to check soil erosion and conserve soil and nutrient resources and protect agricultural production. In wind affected areas, mostly shelter-belts or other agronomic intervention practices such as cereal-legume intercropping are implemented to resist the soil erosion. In areas impacted by the water erosion, agronomic as well as engineering practices such as earthen bund, gabion structure, etc. may be adopted. In checking water erosion, the aim is to break the flow of runoff water through some kind of barrier and reduce runoff velocity. This will result in relatively large opportunity time for the runoff water to get infiltrated and increase the soil moisture, which may also favour crop growth and agricultural production.

It is learnt from the literature that soil erosion from croplands of arid regions is assessed in relatively less number of studies than that performed for soil loss assessment at catchment or basin scale. Thus, more and more studies are required to be undertaken at field level to monitor soil erosion from different types of croplands situated in other arid lands globally.

Acknowledgements Authors gratefully acknowledge the facilities provided by the Director, ICAR-Central Arid Zone Research Institute, Jodhpur, Rajasthan to carry out the case study presented in the chapter. This chapter is based on outcome of institute's research project CAZRI/T1/31.

References

Anejionu OC, Nwilo PC, Ebinne ES (2013) Long term assessment and mapping of erosion hotspots in South East Nigeria: TSO 3B—remote sensing for land use and planning—6448. Environment for sustainability, FIG working week, pp 1–19

Báčová M, Krása J, Devátý J, Kavka P (2019) A GIS method for volumetric assessments of erosion rills from digital surface models. Eur J Remote Sens 52:96–107

Bahrawi JA, Elhag M, Aldhebiani AY, Galal HK, Hegazy AK, Alghailani E (2016) Soil erosion estimation using remote sensing techniques in Wadi Yalamlam Basin, Saudi Arabia. Adv Mater Sci Eng 2016. Article ID 9585962. https://doi.org/10.1155/2016/9585962

Bakker MM, Govers G, Rounsevell MDA (2004) The crop productivity-erosion relationship: an analysis based on experimental work. CATENA 57(1):55–76

Bakker MM, Govers G, Jones RA, Rounsevell MDA (2007) The effect of soil erosion on Europe's crop yields. Ecosystems 10:1209–1219

Beguería S, Angulo-Martínez M, Gaspar L, Navas A (2015) Detachment of soil organic carbon by rainfall splash: experimental assessment on three agricultural soils of Spain. Geoderma 245:21–30

Beskow S, Mello CR, Norton LD, Curi N, Viola MR, Avanzi JC (2009) Soil erosion prediction in the Grande River Basin, Brazil using distributed modeling. CATENA 79:49–59

Boardman J (1988) Severe erosion on agricultural land in East Sussex, UK October 1987. Soil Technol 1:333–348

Boardman J (1991) Land use, rainfall and erosion risk on the South Downs. Soil Use Manag 7:34–37

Boardman J (2003) Soil erosion and flooding on the eastern South Downs, southern England, 1976–2001. Trans Inst Br Geogr 28:176–196

Boardman J (2006) Soil erosion science: Reflections on the limitations of current approaches. CATENA 68:73–86

Boardman J, Evans R (2020) The measurement, estimation and monitoring of soil erosion by runoff at the field scale: Challenges and possibilities with particular reference to Britain. Prog Phys Geogr Earth Environ 44:31–49

Boardman J, Burt TP, Evans R, Slattery MC, Shuttleworth H (1996) Soil erosion and flooding as a result of a summer thunderstorm in Oxfordshire and Berkshire, May 1993. Appl Geogr 16:21–34

Boardman J, Shepheard ML, Walker E, Foster ID (2009) Soil erosion and risk-assessment for on- and off-farm impacts: a test case using the Midhurst area, West Sussex, UK. J Environ Manage 90:2578–2588

Bolline A (1980) Splash measurements in the field. In: De Boodt M, Gabriels D (eds) Assessment of erosion. Wiley, Chichester, pp 441–453

Bonilla CA, Kroll DG, Norman JM, Yoder DC, Molling CC, Miller PS, Panuska JC, Topel JB, Wakeman PL, Karthikeyan KG (2006) Instrumentation for measuring runoff, sediment, and chemical losses from agricultural fields. J Environ Qual 35:216–223

Borrelli P, Robinson DA, Fleischer LR, Lugato E, Ballabio C, Alewell C, Meusburger K, Modugno S, Schutt B, Ferro V et al (2013) An assessment of the global impact of 21st century land use change on soil erosion. Nat Commun 8. https://doi.org/10.1038/s41467-017-02142-7

Campbell IA (1974) Measurements of erosion on badlands surfaces. Z Geomorphologie Supplement Bänd 21:122–137

Cao L, Zhang K, Dai H, Liang Y (2015) Modeling interrill erosion on unpaved roads in the loess plateau of China. Land Degrad Dev 26:825–832

Castillo C, Pérez R, James M, Quinton J, Taguas EV, Gómez JA (2012) Comparing the accuracy of several field methods for measuring gully erosion. Soil Sci Soc Am J 76:1319–1332

Ciesiolka CAA, Rose CW (1998) The measurement of soil erosion. In: Penning de Vries FWT, Agus F, Kerr J (eds) Soil erosion at multiple scales: principles and methods for assessing causes and impact. CAB International, Wallingford, pp 287–301

Cruse RM, Berghoefer BE, Mize CW, Ghaffarzadeh M (2000) Water drop impact angle and soybean protein amendment effects on soil detachment. Soil Sci Soc Am J 64:1474–1478

Daniels RB, Gilliam JW, Cassel DK, Nelson LA (1985) Soil erosion class and landscape position in the North Carolina Piedmont. Soil Sci Soc Am J 49:991–995

Darboux F, Huang C (2003) An instantaneous-profile laser scanner to measure soil surface microtopography. Soil Sci Soc Am J 67:92–99

de Oliveira Salumbo AM (2020) A review of soil erosion estimation methods. Agric Sci 11:667–691

De Santisteban LM, Casalí J, López JJ (2006) Assessing soil erosion rates in cultivated areas of Navarre (Spain). Earth Surf Process Landf J Br Geomorphol Res Group 31:487–506

Deb SK, Shukla MK (2011) An overview of some soil hydrological watershed models. In: Shukla MK (ed) Soil hydrology, land use and agriculture: measurement and modeling. CAB International, Wallingford, U.K, pp 75–116

Edwards KA (1977) Cultural practice and changes in catchment hydrology: a review of hydrological research techniques as aids to development planning in the humid tropics. In: Lal R, Greenland DJ (eds) Soil conservation and management in the humid tropics. Wiley, Chichester, pp 33–48

El Jazouli A, Barakat A, Ghafiri A, El Moutaki S, Ettaqy A, Khellouk R (2017) Soil erosion modeled with USLE, GIS, and remote sensing: a case study of Ikkour watershed in Middle Atlas (Morocco). Geosci Lett 4:1–12

Elliot WJ, Laflen JM, Thomas AW, Kohl KD (1997) Photogrammetric and rillmeter techniques for hydraulic measurement in soil erosion studies. Trans ASAE 40:157–165

Eltner A, Baumgart P, Maas HG, Faust D (2015) Multi-temporal UAV data for automatic measurement of rill and interrill erosion on loess soil. Earth Surf Proc Land 40(6):741–755

Evans R (1990) Water erosion in British farmers' fields-some causes, impacts, predictions. Prog Phys Geogr 14:199–219

Evans R (1993) Extent, frequency and rates of rilling of arable land in localities in England and Wales. In: Wicherek S (ed) International symposium on farm land erosion in temperate plains environment and hills. Elsevier, Amsterdam, pp 177–190

Evans R, Boardman J (1994) Assessment of water erosion in farmers' fields in the UK. In: Conserving soil resources: European perspectives. Selected papers from the first international congress of the European society for soil conservation. Cab International, pp 13–24

Evans R, Collins AL, Foster ID, Rickson RJ, Anthony SG, Brewer T, Deeks L, Newell-Price JP, Truckell IG, Zhang Y (2016) Extent, frequency and rate of water erosion of arable land in Britain–benefits and challenges for modelling. Soil Use Manag 32:149–161

Fenton TE (2012) The impact of erosion on the classification of Mollisols in Iowa. Can J Soil Sci 92(3):413–418

Fernández-Raga M, Campo J, Rodrigo-Comino J, Keesstra SD (2019) Comparative analysis of splash erosion devices for rainfall simulation experiments: a laboratory study. Water 11(6):1228. https://doi.org/10.3390/w11061228

Flanagan DC, Gilley JE, Franti TG (2007) Water erosion prediction project (WEPP): development history, model capabilities, and future enhancements. Trans ASABE 50:1603–1612

Foster GR, Simanton JR, Renard KG, Lane LJ, Osborn HB (1981) Discussion of application of the Universal soil loss equation to rangelands on a per-storm basis by Trieste and Gifford. J Range Manag 33:66–70 (1980). Rangel Ecol Manag/J Range Manag Arch 34:161–165

Ganasri BP, Ramesh H (2016) Assessment of soil erosion by RUSLE model using remote sensing and GIS—a case study of Nethravathi Basin. Geosci Front 7:953–961

Gaubi I, Chaabani A, Mammou AB, Hamza MH (2017) A GIS-based soil erosion prediction using the revised universal soil loss equation (RUSLE) (Lebna watershed, Cap Bon, Tunisia). Nat Hazards 86:219–239

Gilley JE, Flanagan DC (2007) Early investment in soil conservation research continues to provide dividends. Trans ASABE 50:1595–1601

GSP (2017) Global soil partnership endorses guidelines on sustainable soil management. http://www.fao.org/global-soil-partnership/resources/highlights/detail/en/c/416516/. Accessed 14 May 2020

Haigh MJ (1977) The use of erosion pins in the study of slope evolution. Br Geomorphol Res Group Tech Bull 18:31–49

Hamed Y, Albergel J, Pépin Y, Asseline J, Nasri S, Zante P, Berndtsson R, El-Niazy M, Balah M (2002) Comparison between rainfall simulator erosion and observed reservoir sedimentation in an erosion-sensitive semiarid catchment. CATENA 50:1–16

Hayward JA (1968) The measurement of soil loss from fractional acre plots. Lincoln College. A Research Publication of the New Zealand Agricultural Engineering Institute, Lincoln College, Canterbury, New Zealand. Lincoln papers in water resources, 63 pp

Hsieh Y-P (1992) A mesh-bag method for field assessment of soil erosion. J Soil Water Conserv 47:495–499

Hsieh YP, Grant KT, Bugna GC (2009) A field method for soil erosion measurements in agricultural and natural lands. J Soil Water Conserv 64:374–382

Hudson NW (1993) Field measurement of soil erosion and runoff. FAO Soils Bull 68. Food and Agriculture Organization (FAO) of the United Nations, Rome

Jinze M (1981) The establishment of experimental plots for studying runoff and soil loss in the rolling loess regions of China. IAHS Publication

Jomaa S, Barry DA, Brovelli A, Heng BCP, Sander GC, Parlange J-Y, Rose CW (2012) Rain splash soil erosion estimation in the presence of rock fragments. Catena 92:38–48

Jordán A, Zavala LM, Granged AJ, Gordillo-Rivero ÁJ, García-Moreno J, Pereira P, Bárcenas-Moreno G, de Celis R, Jiménez-Compán E, Alanís N (2016) Wettability of ash conditions splash erosion and runoff rates in the post-fire. Sci Total Environ 572:1261–1268

Kinnell PIA, Risse LM (1998) USLE-M: empirical modeling rainfall erosion through runoff and sediment concentration. Soil Sci Soc Am J 62:1667–1672

Laflen JM, Elliot WJ, Simanton JR, Holzhey CS, Kohl KD (1991) WEPP: soil erodibility experiments for rangeland and cropland soils. J Soil Water Conserv 46:39–44

Lal R (2001) Soil degradation by erosion. Land Degrad Dev 12(6):519–539

Lal R, Stewart BA (1990) Soil degradation. Springer, New York

Liu W, Luo Q, Li J, Wang P, Lu H, Liu, Wenyao LH (2015) The effects of conversion of tropical rainforest to rubber plantation on splash erosion in Xishuangbanna, SW China. Hydrol Res 46:168–174

Loughran RJ (1989) The measurement of soil erosion. Prog Phys Geogr 13:216–233

Machiwal D, Kumar S, Dayal D (2016) Characterizing rainfall of hot arid region by using time-series modeling and sustainability approaches: a case study from Gujarat, India. Theoret Appl Climatol 124:593–607

Machiwal D, Dayal D, Kumar S (2017) Long-term rainfall trends and change points in hot and cold arid regions of India. Hydrol Sci J 62(7):1050–1066

Machiwal D, Kumar S, Islam A, Kumar S, Jat SR, Vaishnav M, Dayal D (2021) Evaluating effect of cover crops on runoff, soil loss and soil nutrients in an Indian arid region. Commun Soil Sci Plant Anal 52(14):1669–1688

Mangalassery S, Dayal D, Meena SL, Ram B (2014a) Carbon sequestration in agroforestry and pasture systems in arid northwestern India. Curr Sci 107(8):1290–1293

Mangalassery S, Dayal D, Ram B, Meena SL (2014b) Effect of tillage and soil amendments on soil quality and yield of clusterbean (*Cyamopsis tetragonoloba*) in shallow hardpan soils of arid Gujarat. Indian J Agric Sci 84(3):428–431

McCool DK, Foster GR, Renard KG, Yoder DC, Weesies GA (1995) The revised universal soil loss equation. In: Department of Defense/interagency workshop on technologies to address soil erosion on department of defense lands San Antonio, TX, p 9

McDonald MA, Lawrence A, Shrestha PK (2003) Soil erosion. In: Schroth G, Sinclair FL (eds) Trees, crops and soil fertility: concepts and research methods. Cabi, pp 325–343

Merritt WS, Letcher RA, Jakeman AJ (2003) A review of erosion and sediment transport models. Environ Model Softw 18:761–799

Metson AJ (1957) Methods of chemical analysis for soil survey samples. Department of Scientific and Industrial Research, Bulletin 12 of Soil Bureau, Wellington, New Zealand, 208 pp

Meyer LD, Wischmeier WH (1969) Mathematical simulation of the process of soil erosion by water. Trans ASAE 12:754–758

Millington AC (1981) Relationship between three scales of erosion measurement on two small basins in Sierra Leone [Macroscale, mesoscale, microscale]. Food and Agriculture Organization (FAO) of the United Nations, IAHS Publication, vol 133, pp 485–492
Mitchell JK, Bubenzer GD (1980) Soil loss estimation. In: Kirkby MJ, Morgan RPC (eds) Soil erosion. Wiley, Brisbane, pp 17–61
Moharana PC, Santra P, Singh DV, Kumar S, Goyal RK, Machiwal D, Yadav OP (2016) ICAR-central arid zone research institute, Jodhpur: erosion processes and desertification in the Thar Desert of India. Proc Ind Nat Sci Acad 82(3):1117–1140
Morgan RPC (1978) Field studies of rainsplash erosion. Earth Surf Process 3:295–299
Morgan RPC (2005) Soil erosion and conservation, 3rd edn. Blackwell Publishing Ltd., Oxford, UK, p 304 pp
Mwango SB, Msanya BM, Mtakwa PW, Kimaro DN, Deckers J, Poesen J (2016) Effectiveness of mulching under miraba in controlling soil erosion, fertility restoration and crop yield in the Usambara Mountains, Tanzania. Land Degrad Dev 27:1266–1275
Myers NG (1993) An Atlas of planet management. J Acad Librariansh 19:200
Nearing MA (2013) Soil erosion and conservation. In: Wainwright J, Mulligan M (eds) Environmental modelling: finding simplicity in complexity, 2nd edn. Wiley, pp 365–378
Nearing MA, Lane LJ, Lopes VL (1994) Modeling soil erosion. In: Lal R (ed) Soil erosion: research methods. St. Lucie Press, Delray Beach, p 32 pp
Nearing MA, West LT, Brown LC (1988) A consolidation model for estimating changes in rill erodibility. Trans ASAE 31:696–700
Nolan SC, van Vliet L, Goddard TW, Flesch TK (1997) Estimating storm erosion with a rainfall simulator. Can J Soil Sci 77:669–676
Novara A, Gristina L, Saladino SS, Santoro A, Cerdà A (2011) Soil erosion assessment on tillage and alternative soil managements in a Sicilian vineyard. Soil Tillage Res 117:140–147
Nugroho SP (2003) Application of the agricultural non-point source pollution (AGNPS) model for sediment yield and nutrient loss prediction in the Dumpul sub-watershed, Central Java, Indonesia. In: Boer D de, Froehlich W, Mizuyama T, Pietroniro A (eds) Erosion prediction in ungauged basins: integrating methods and techniques, no (279), pp 125–130
de Oliveira VA, de Mello CR, Durães MF, da Silva AM (2014) Soil erosion vulnerability in the Verde river basin, southern Minas Gerais. Ciência e Agrotecnologia 38:262–269
Olsen SR, Cole CV, Watanabe FS, Dean LA (1954) Estimation of available phosphorous in soils by extraction with sodium bicarbonate. Circular No. 939, United States Department of Agriculture, Washington, DC, 19 pp
PAP/RAC (1997) Guidelines for mapping and measurement of rainfall induced erosion processes in the Mediterranean Coastal Areas. PAP-8/PP/GL.1. Split, Priority Actions Programme Regional Activity Centre (MAP/UNEP), with the cooperation of FAO, 70 pp
Parlak M, Parlak AO (2010) Measurement of splash erosion in different cover crops. Turk J Field Crop 15:169–173
Pennock D (2019) Soil erosion: the greatest challenge to sustainable soil management. Food and Agriculture Organization (FAO) of the United Nations, Rome, p 100 pp
Pieri L, Bittelli M, Wu JQ, Dun S, Flanagan DC, Pisa PR, Ventura F, Salvatorelli F (2007) Using the water erosion prediction project (WEPP) model to simulate field-observed runoff and erosion in the Apennines mountain range, Italy. J Hydrol 336:84–97
Pimentel D (2006) Soil erosion: a food and environmental threat. Environ Dev Sustain 8:119–137
Pinson WT, Yoder DC, Buchanan JR, Wright WC, Wilkerson JB (2004) Design and evaluation of an improved flow divider for sampling runoff plots. Appl Eng Agric ASAE 20:433–437
Pradhan B, Chaudhari A, Adinarayana J, Buchroithner MF (2012) Soil erosion assessment and its correlation with landslide events using remote sensing data and GIS: a case study at Penang Island, Malaysia. Environ Monit Assess 184:715–727
Ranger GE, Frank FF (1978) 3-F erosion bridge—a new tool for measuring soil erosion. Range improvement studies. Publication No. 23. Department of Forestry, California, 7 pp

Renard KG, Foster GR, Weesies GA, McCool DK, Yoder DC (1997) Predicting soil erosion by water: a guide to conservation planning with the Revised Universal Soil Loss Equation (RUSLE). Agric Handb 703:25–28
Renard KG, Freimund JR (1994) Using monthly precipitation data to estimate the R-factor in the revised USLE. J Hydrol 157:287–306
Rieke-Zapp DH, Nearing MA (2005) Digital close range photogrammetry for measurement of soil erosion. Photogram Rec 20:69–87
Sadeghi SHR, Gholami L, Homaee M, Khaledi Darvishan A (2015) Reducing sediment concentration and soil loss using organic and inorganic amendments at plot scale. Solid Earth 6:445–455
Seginer I (1966) Gully development and sediment yield. J Hydrol 4:236–253
Senanayake S, Pradhan B, Huete A, Brennan J (2020) A review on assessing and mapping soil erosion hazard using geo-informatics technology for farming system management. Remote Sens 12:4063. https://doi.org/10.3390/rs12244063
Sepuru TK, Dube T (2018) An appraisal on the progress of remote sensing applications in soil erosion mapping and monitoring. Remote Sens Appl Soc Environ 9:1–9
Seutloali KE, Dube T, Mutanga O (2017) Assessing and mapping the severity of soil erosion using the 30-m Landsat multispectral satellite data in the former South African homelands of Transkei. Phys Chem Earth Parts a/b/c 100:296–304
Simalenga TE, Have H (1992) Estimation of soil tillage workdays in a semi-arid area. J Agric Eng Res 51:81–89
Singh S, Kar A (1996) Integrated natural and human resources appraisal for sustainable development of Kachchh District. Report of the Central Arid Zone Research Institute, Jodhpur, 165 pp
Sobotková V, Dumbrovský M (2015) The new volumetric approach for field measurements of rill erosion. Eurasian J Soil Sci 4:94–99
Szeliski R (2010) Computer vision: algorithms and applications. Springer. Springer, London, pp 303–334
Tesfahunegn GB (2011) Soil erosion modelling and soil quality evaluation for catchment management strategies in Northern Ethiopia. Unpublished PhD thesis, Rheinischen Friedrich-Wihelms University
Toy TJ, Foster GR, Renard KG (2002) Soil erosion: processes, prediction, measurement, and control. Wiley, 352 pp
Ullman S (1979) The interpretation of structure from motion. Proc R Soc London Ser B Biol Sci 203:405–426
Van Rompaey A, Bazzoffi P, Jones RJ, Montanarella L (2005) Modeling sediment yields in Italian catchments. Geomorphology 65:157–169
Vrieling A, Sterk G, de Jong SM (2010) Satellite-based estimation of rainfall erosivity for Africa. J Hydrol 395:235–241
Walkley A, Black IA (1934) An examination of the Degtjareff methods for determining soil organic matter, and a proposed modification of the chromic acid titration method. Soil Sci 37(1):29–38
Wang S, Sun B, Li C, Li Z, Ma B (2018) Runoff and Soil erosion on slope cropland: a review. J Resour Ecol 9(5):461–470
Wells RR, Momm HG, Bennett SJ, Gesch KR, Dabney SM, Cruse R, Wilson GV (2016) A measurement method for rill and ephemeral gully erosion assessments. Soil Sci Soc Am J 80:203–214
Wischmeier WH, Smith DD (1978) Predicting rainfall erosion losses: a guide to conservation planning. US Department of Agriculture, Washington DC, Agriculture Handbook, vol 537, 62 pp
Zhou H, Peng X, Young A, Darboux F (2013) Effect of rainfall kinetic energy on crust formation and interrill erosion of an Ultisol in subtropical China. Vadose Zone J 12(4). https://doi.org/10.2136/vzj2013.01.0010

Bashagaluke JB, Logah V, Opoku A, Sarkodie-Addo J, Quansah C (2018) Soil nutrient loss through erosion: impact of different cropping systems and soil amendments in Ghana. PLoS ONE 13(12):e0208250. https://doi.org/10.1371/journal.pone.0208250

Chapter 9
Application of RUSLE and MUSLE Models to Assess Erosion Sensitivity of a Sub-watershed Using ArcSWAT Interface

A. P. Lakkad, P. K. Shrivastava, and K. N. Sondarva

Abstract Land degradation in term of water erosion at sub-watershed level has been monitored, measured and modelled using remote sensing and GIS techniques. Sub-basin (5D1A5c) was selected as the study area to estimate soil and sediment erosion rates input using the ArcGIS interface and the ArcSWAT model. The different RUSLE parameters were estimated separately to calculate the average annual soil erosion, while the ArcSWAT model was used to estimate the sediment yield using MUSLE. The average annual erosion rate for the study area is estimated at 39.25 tons/ha/year. The soil loss tolerance limit (ie e. 5 tons/ha/year) of the study area used for erosion susceptibility mapping. The results show that 42.53% of the area is in the "safe category" while 22.15% of the area is in the "very high" priority for land conservation. The SWAT model has been run for 30 years (1986–2015). Alluvial production is simulated at 39.42 tons/ha/year with a total alluvium volume of 59.94 tons/ha/year. The model was calibrated using standard USLE_P values for sloping cropland with <25% slope. After correction, the silt production was simulated at 22.78 tons/ha/year while the total sediment production was 34.61 tons/ha/year. Sediment samples for 12 rainfall events in 3 years (201,315) using homemade sediment samplers collected at the outlet of the basin. Four stats viz. R^2, NSE, RSR and PBIAS are used to evaluate the operation and applicability of the SWAT model for the study area. All of them were found within the acceptable range. The average sediment production from the Karjan Reservoir catchment area was estimated at 21.56 tons/ha/year over 29 years (1984–2013), which is close to the simulated SWAT sediment yield of 22.78 tons/ha/year. This estimated 0.58 sediment input ratio indicates that approximately 58% of the eroded soil leaves the basin, which reduces soil depth, depletes the basin's soil fertility, and reduces the annual storage capacity of water from the Karjan Reservoir.

P. K. Shrivastava
ASPEE College of Horticulture and Forestry, NAU, Navsari, India

A. P. Lakkad (✉) · K. N. Sondarva
College of Agricultural Engineering and Technology, NAU, Dediapada, India
e-mail: larunp@nau.in

P. K. Shit et al. (eds.), *Soil Health and Environmental Sustainability*,
Environmental Science and Engineering,
https://doi.org/10.1007/978-3-031-09270-1_9

Keywords ArcSWAT interface · Erosion risk map · MUSLE · RUSLE · Sediment delivery ratio

9.1 Introduction

"Soil is the living outer layer of the Earth, the medium for plant growth, and an important natural, non-renewable resource that provides habitat for animals and other microorganisms" (NAAS 2012). Humans derive more than 99.7% of their food from soils and less than 0.3% from oceans and other aquatic ecosystems (Pimentel 2006). Land degradation due to reckless use and mismanagement is a major national problem that negatively impacts agricultural productivity, environmental quality and ultimately the sustainability of human life. Land degradation is another strong driver of climate change due to current and projected impacts on land use and management changes. The current global estimate of from the Harmonized World Soil Database (HWSD) is about 141.7 billion tonnes of SOC trapped in the first soil and about 716 billion in the top 30 cm soil profile of Caon and Vargas, 2017. Contains tons of SOC. # 41;... Globally, approximately 19.65 million km^2 of land is man-made due to water erosion (55.67%), wind erosion (27.94%), chemical degradation (12.16%) and physical degradation (4.22%). It is exposed to soil degradation... The cost of soil degradation is approximately 6,580 million annually, which poses a major long-term environmental threat to fiber and feed safety (Young and Orsini 2015). Water erosion is a major cause of soil degradation worldwide (Odeman 1992). It is estimated that soil loss due to water erosion transport between 23–42 million tonnes of nitrogen and 15–26 million tonnes of phosphorus from cultivated land around the world. Of the major causes of land loss in India, water erosion is estimated to be the most serious, covering nearly 68.39% of the affected areas and causing about 5.3 billion tonnes of soil loss annually. (Maji et al. 2010). Accelerated soil erosion depletes the SOC supply severely and rapidly. Organic carbon loss due to erosion and spillage can be as high as 0.5–3.0% even on gentle slopes (Banerjee et al. 1991). The process of estimating soil loss using traditional methods is tedious, time consuming and costly. The outcome of basin planning and development depends on the quality and quantity of available information on topographical parameters, climate and socio-economic resources. Both remote sensing and GIS provide relevant, reliable, and timely information geospatial support needed for optimal planning, development, and management of land development activities. ArcSWAT was designed to use the GIS interface to predict the impact of land management practices on water, sediment and pesticide yields in large and complex watersheds over time. In this study, the ArcSWAT interface was adapted to use the RUSLE and MUSLE models to model land degradation due to water erosion of selected sub watershed.

9.2 Materials and Methods

9.2.1 Description of Study Area

9.2.1.1 Study Location

Sub watershed that catches water from main stream of Dediapada region (Dist.-Narmada, Gujarat) was selected for the study purpose. The sub watershed falls within 73° 31′ 52.63″ and 73° 38′ 58.02″ East longitude and 21° 33′ 23.83″ and 21° 40′ 14.18″ North latitude. The selected sub-watershed is indicated as 5D1A5c in watershed Atlas of AIS & LUS, having 7710.64 ha area (Anonymous 2014a, b). The location of the sub-watershed is shown in Fig. 9.1.

9.2.1.2 Meteorological Data

The 30 years weather data i.e. daily rainfall, max. and min. temp., relative humidity, wind speed and sunshine hours were collected from State Water Data Centre, Gandhinagar, Gujarat (India) for the duration of 1986–2015.

9.2.1.3 Topographical Data

SOI 1:50,000 scale Toposheet number F43N10 was collected from Map Sale Office, Gujarat Daman Diu Geo-Spatila Data Centre, Survey of India, Gandhinagar.

9.2.1.4 Remotely Sensing Satellite Data

The 30 m resolution satellite image captured by Landsat OLI-TIRS sensor on 6th Nov., 2014 was collected from www.landsat.usgs.gov website.

9.2.1.5 ArcGIS Desktop

ArcGIS Desktop is a collection of software products for building complete geographic information systems (GIS). It includes GIS applications that support a number of GIS tasks including mapping, data creation, compilation, analysis, geo-database management and geographic information sharing. ArcMap, ArcToolbox and ArcCatalog were used to carry out all the study related analysis work during this study.

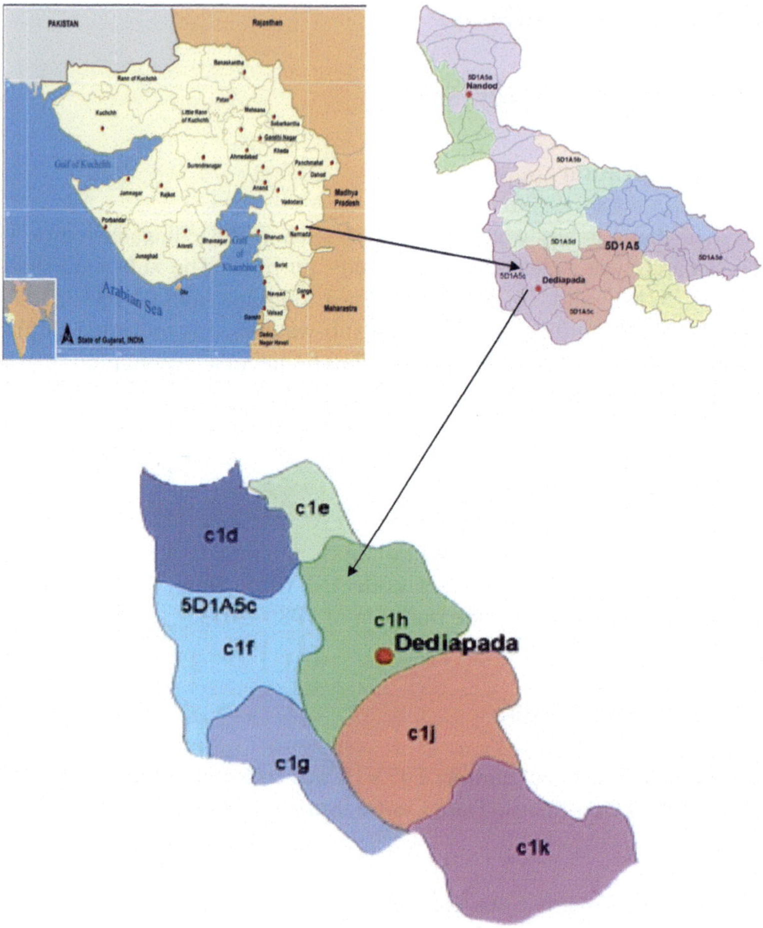

Fig. 9.1 Location map of selected sub-watershed 5D1A5c

9.2.1.6 Arc GIS Extensions

A number of powerful analytical tools are available as add-on in ArcGIS Desktop. Each tool enables user to add specialized analytical techniques such as raster geo-processing, 3D terrain analysis and various spatial analysis. Following extensions were used in our analysis.

1. Spatial Analyst tools: hydrology, interpolation, map algebra, reclass, surface
2. Analysis tools: overlay—intersect
3. Arc Hydro tools: terrain pre-processing
4. Conversion tools: raster to vector and vector to raster tool
5. Data Management Tools: projections and transformations, features

6. Geo-referencing tools: geo-referencing of maps using ground control point (GCP)
7. Editor tools: digitization of polygons, polylines and points.

9.2.1.7 ArcSWAT

The ArcGIS extension i. e. ArcSWAT is a graphical user interface for the SWAT (Soil and Water Assessment Tool) model. In order to estimate sediment yield, Modified Universal Soil Loss Equation (MUSLE) model incorporates by SWAT interface was used. The ArcSWAT version 2012, an Extension of ArcGIS 10.2 was used in this study.

9.2.1.8 Image Processing Software: ERDAS Imagine

For geospatial applications, ERDAS Imagine software was designed by ERDAS. It is an extension enables the user to perform various operations on a remote sensing image and produce a desired result to specific geographical questions. The level of reflectance of EMR from the surfaces from the image could be helpful for numerous analysis operations. ERDAS Professional 2013 was used to prepare land use/land cover map of selected sub-watershed.

9.2.1.9 Survey Instruments

Garmin 5.5 Global Positioning System (GPS), Dumpy Level, Levelling Staff, Homemade Sediment Sampler.

9.3 Models

Two models were used for the study, namely Revised Universal Soil Loss Equation (RUSLE) and Modified Universal Soil Loss Equation (MUSLE). RUSLE was used to estimate the average gross soil erosion rate whereas; MUSLE was used to derive the sediment yield at sub-watershed outlet.

9.3.1 Revised Universal Soil Loss Equation

The Revised Universal Soil Loss Equation (Renard et al. 1997) is a model developed for estimating soil loss in terms of sheet and rill erosion from most undisturbed lands experiencing overland flow, from lands undergoing disturbance, and from newly or established reclaimed lands. It does not estimate gully or stream-channel erosion. RUSLE retains the same structure of its predecessor, the Universal Soil Loss Equation (USLE, Wischmeier and Smith 1978), namely:

$$A = RKLSCP \tag{9.1}$$

where,

A Average gross soil erosion (mt/ha per year)
R Rainfall Erosivity Factor (MJ mm ha^{-1} hr^{-1} per year)
K Soil Erodibility Factor (mt · hr MJ^{-1} mm^{-1} per unit R)
L Slope Length Factor (dimensionless)
S Slope Steepness Factor (dimensionless)
C Crop/Cover Management Practice (dimensionless)
P Conservation/Support Practice Factor (dimensionless)

9.3.2 MUSLE Model

Modified Universal Soil Loss Equation is modified version of Universal Soil Loss Equation. The MUSLE equation is applicable to the points where overland flow enters the streams and then all those points are summed up to give the total amount of sediment delivered to the stream network within a watershed. In general, MUSLE is expressed as follows,

$$SY = 11.8\left(Q \times q_p\right)^{0.56} \times K \times LS \times C \times P \tag{9.2}$$

Where,

SY the sediment yield to the stream network in metric tons,
Q the runoff volume from a given rainfall event in m^3,
qp the peak flow rate in m^3/s,

K, LS, C and P are same as in equation 9.1.

9.4 Methods Used to Estimate Various Model Parameters

9.4.1 Rainfall Erosivity Factor (R)

The soil loss potential of a given rainfall event can estimated by rainfall erosivity. The erosivity factor (R) is a function of the falling drops and the rainfall intensity. Wischmeier and Smith (1965) found that the product of kinetic energy of raindrop and the maximum intensity of rainfall over the duration of 30 min in a storm was the best estimator of soil loss. It requires individual rainfall event intensity which is not available for the study area and hence, relationship between rainfall erosivity index and annual rainfall was developed with the data available from various meteorological observatories of India by Singh et al. in 1981 and presented as Eq. 9.3 was used for the study to estimate rainfall erosivity.

$$Y = 79 + 0.393\,X \tag{9.3}$$

where,

Y Annual Rainfall Erosivity (MJ mm ha^{-1} hr^{-1} per year)

X Average Annual Rainfall (mm)

9.4.2 Soil Erodibility Factor (K)

9.4.2.1 Soil Physico-Chemical Properties

The soil map of study area was collected from Bhaskaracharya Institute for Space Applications and Geo-Informatics (BISAG), Gandhinagar. The map shows that there are mainly two types of soils i.e. clay and clay loam which covers 53.39 and 46.61% of study area respectively. Soil samples from top 15 cm soil layer were collected from several locations that represents various land cover categories of the study area and finally 10 representative soil samples from clay soils and 8 representative soil samples for clay loam soils were prepared to estimate the soil physico-chemical properties of study area. Various soil properties were computed using the methods and techniques described in the Table 9.1.

9.4.3 Soil Erodibility Factor Computation

Rate of susceptibility of soil particles to erosion is indicated by Soil Erodibility factor (K). It can be computed per unit of rainfall erosivity factor (R) for a specified soil on a unit plot having a 9% uniform slope and a slope length of 22.13 m over a continuously clean fallow land with up and down slope farming. The soil erodibility depends on sand, silt, clay contents, organic matter content and rock fragment content of soil. The classical formula (Eq. 9.4) could be used to determine the soil erodibility K factor.

$$K = 2.77 \times 10^{-6} \times M^{1.14}(12 - a) + 0.043(b - 2) + 0.033(c - 3) \tag{9.4}$$

Table 9.1 Analytical methods adopted soil physico-chemical properties determination

Sr. No	Soil physical properties	Method and techniques	References
1	Soil textural class	International pipette method	Jackson (1973)
2	Organic carbon (%)	Wet oxidation followed by Walkley and black's rapid titration method	Jackson (1973)
3	Rock fragment content	Sieve analysis technique	Jackson (1973)
4	Electrical conductivity (dSm^{-1})	Conductometric method (EC meter)	Jackson (1973)
5	Moist bulk density (gm/cm^3)	Soil texture triangle Hydraulic properties calculator	www.pedosphere.com/resources/texture/workable_us.cfm
6	Saturated hydraulic conductivity (cm/hr)	Soil texture triangle Hydraulic properties calculator	http://www.afrc.uamont.edu/ficklinr/soils/soiltexture.htm
7	Available water capacity of soil layer (cm^3 of water/cm^3 of soil)		

where,

K the soil erodibility factor ($mt \cdot hr\ MJ^{-1}\ mm^{-1}$ per unit R),
M particle size parameter (% silt + % very fine sand) × (100-clay),
a Organic matter content (%),
b Soil structure code (1-very fine granular, 2-fine granular, 3-medium or coarse granular, 4-blocky, platy or massive),
c soil permeability class (1-rapid, 2-moderate to rapid, 3-moderate, 4-slow to moderate, 5-slow, 6- very slow).

This equation could be used only for the soils having greater than 70% silt, less than 4% organic content and less than 1.5% rock fragment. Hence, based on silt content, organic matter content and fraction of soil covered with rock fragments, Auerswald et al. (2014) has reclassified and modified this formula as presented in Eqs. 9.5–9.12.

For silt + very fine sand <70%,

$$K_1 = 2.77 \times 10^{-6} \times M^{1.14} \tag{9.5}$$

For silt + very fine sand >70%,

$$K_1 = 1.75 \times 10^{-6} \times M^{1.14} + 0.0024(\%\ \text{silt} + \%\ \text{very fine sand}) + 0.16 \tag{9.6}$$

For organic matter <4%,

$$K_2 = 100 - \text{clay} \tag{9.7}$$

For organic matter >4%,

$$K_2 = 0.8 \tag{9.8}$$

For $K_1 * K_2 > 0.2$,

$$K_3 = 2.77 \times 10^{-6} \times M^{1.14}(12 - a) + 0.043(b - 2) + 0.033(c - 3) \tag{9.9}$$

For $K_1 * K_2 < 0.2$,

$$\begin{aligned} K_3 &= 0.091 - 0.34 * K_1 * K_2 + 1.79 * (K_1 * K_2)^2 \\ &\quad + 0.24 * K_1 * K_2 * b + 0.033(c - 3) \end{aligned} \tag{9.10}$$

For rock fragment <1.5%,

$$\text{Soil Erodibility factor K} = K_3 \tag{9.11}$$

For rock fragment >1.5%,

$$\text{Soil Erodibility factor K} = K_3 * \left[1.1 - \{\exp(-0.24 \times F_{rk}) - 0.06\}\right] \tag{9.12}$$

where, F_{rk} = Rock fragment content (%).

9.4.4 *Slope Length Factor (L)*

The L factor is the ratio of the actual horizontal slope length to the experimentally measured slope length of 22.13 m. Slope length is the distance from the point of origin of overland flow to either the point where the slope decrease to the extent that deposition begins or the point where runoff enters well defined channels (Wischmeier and Smith 1978).

9.4.5 *Unit Stream Power Erosion and Deposition (USPED) Model*

In USLE and RUSLE, L depends on the linear distance λ, which is the horizontal length from the beginning of sediment transport to any point on the slope. So in the USPED model, it is essentially a one-dimensional function. A topographical factor is a change in the transport capacity in the direction of flow, which is positive in areas of potential topographic sedimentation and negative in areas of potential for erosion.

USPED uses elevated areas to promote flow at any point on the stream. The USPED model replaces the previous slope length with area.

The L calculation on a slope is shown in Equation

$$L = (m + 1)\left(\frac{\lambda_A}{22.13}\right)^m \tag{9.13}$$

where,

L	slope length factor.
λ_A	area of upland flow,
22.13	unit plot length.
m	variable exponent calculated from the ratio of rill-to-inter rill erosion, as described in Eq. 9.14.

$$m = \frac{\beta}{1 + \beta} \tag{9.14}$$

in which, β dependent on slope, It was computes using Formula 9.15.

$$\beta = \frac{\sin\theta}{0.0896\left[3(\sin\theta)^{0.8} + 0.56\right]} \tag{9.15}$$

The $m + 1$ comes from the fact that, in order to get a value for $L = \left(\frac{\lambda}{22.13}\right)^m$ that is considerate of the area of contributing upland flow on the slope up to any point i, we must integrate L over the interval $[0..i]$.

The Digital Elevation Model (DEM) is required to analyze the topographic properties of study area in order to estimate the slope length and slope steepness factor.

9.4.6 Digital Elevation Model (DEM)

Topography i.e. Slope steepness and slope length of an area plays major role for the water erosion. Topography has been represents by Digital Elevation Model (DEM) into GIS interface. DEM could be used to derive flow direction, flow accumulation; slope steepness; slope direction; flow length and flow pattern. 20 m DEM for study area was generated using three shape files of contour, elevation points and watershed boundary which were digitized from F43N10 Topo-sheet. Topo to Raster interpolation method was selected as it generates a hydrologically correct DEM. The resolution of DEM was selected as 20 m due to the fact that this is closest to 22.13 m slope length, which is used for the derivation of model relations. Raster calculator was used to derive slope length factor map from 20 m DEM into ArcGIS interface as per Eqs. 9.13–9.15.

9.4.7 Slope Steepness Factor (S)

On steep slopes the flow velocity is high, which causes scouring and cutting of soil. The slope steepness factor expresses the ratio of soil loss from a plot of known slope to soil loss from a unit plot under identical conditions. The Eqs. 9.16 and 9.17 given by Mc Cool et al. (1989) have been used to estimate and prepare the thematic map on slope steepness factor in ArcGIS interface.

$$S = 10.8\sin\theta + 0.03 \quad \text{for slope gradient} \le 9 \tag{9.16}$$

$$S = 16.8\sin\theta - 0.50 \quad \text{for slope gradient} > 9\% \tag{9.17}$$

where,

S is the slope steepness factor

θ is the slope in degrees.

9.4.8 Cover Management Factor/Vegetative Cover Factor (C)

The cover and management factor is the ratio of soil loss from land cropped under specified conditions to the corresponding loss from clean-tilled, continuous fallow land. Vegetation cover protects the soil by dissipating the raindrop energy before reaching the soil surface. The value of C factor depends on land use/ land cover, vegetation type, stage of growth and cover percentage. The C factor ranges from near 0 for a high density of vegetation to 1 for barren land. It is difficult and costly to estimate the value from wide range of vegetations available in nature.

Kurothe et al. (1991–92) had derived the C factor for most of the crops of study area at Indian Institute of Soil and Water Conservation, Vasad. C factor values derived by Kurothe et al. (1991–92) were used to derive crop-wise C factors for agricultural land. The average area covered by different crops of the Dediapada block during last 3 years (Annual Progress Report, 2012–13, 2013–14, 2014–15) was used to derive the weighted C factor value as shown in Table 9.2. In order to derive the C factor map, C factor values estimated by Singh et al. (1981) and Narain et al. (1994) for forest and wasteland (Table 9.3) was used.

9.4.9 Land Use/Land Cover Map

In order to assign the cover management factor C values as per the land use/land cover of study area, Land cover/Land use details for study area was required. Land cover/Land use was classified as per Level II land use classification techniques i.e.

Table 9.2 Weighted C factor value for agricultural land

Sr. No	Crop	Area	C factor value	Area × C factor value
1	Cotton	7124	0.31	2208.44
2	Paddy	7086	0.28	1984.08
3	Pigeon pea	8013	0.43	3445.59
4	Maize	1089	0.50	544.5
5	Moong	779	0.30	233.7
6	Sorghum	475	0.36	171
7	Castor	475	0.79	375.25
8	Groundnut	84	0.42	35.28
9	Total	25,125		8997.84
10	Weighted C factor value			0.358

Table 9.3 C factor values for different land cover classes

Sr. No	Land cover class	C factor value
1	Evergreen forest	0.004
2	Mixed forest	0.08
3	Deciduous forest	0.4
4	Pasture	0.6
5	Wasteland without scrub	1.0

single crop and double crop agricultural land, evergreen, mixed and deciduous forest land and land with scrub/pasture and land without scrub with low density resident (Anderson 1976) for study area. Landsat 8 image dated November 6th, 2014 (Fig. 9.4) was sharpened with PAN band of study area image in order to convert the image resolution from 30 to 15 m. Google image (Fig. 9.2) and Toposheet (Fig. 9.3) of the study area was used as base map for creating polygons of training area from Pan Sharpen Landsat 8 image (Fig. 9.5). Minimum 10 training area were selected for each land use type and merged to form one cluster and saved as signature file in.sig format. Average spectral values of each land use class cluster are given in the Table 9.4. Because the research area was well-known during the reconnaissance survey, the land use/land cover map was created using a supervised classification technique. The Land Use/Land Cover Map was created using the ERDAS IMAGINE 2013 Interface. Total 185 ground truthing sample points were taken using GPS from study area based on weighted area covered by each categories of land use/land cover for accuracy assessment of LU/LC map. The overall efficiency and Kappa co-efficient were derived for estimating accuracy of land use/land cover map.

Fig. 9.2 Google image of study area

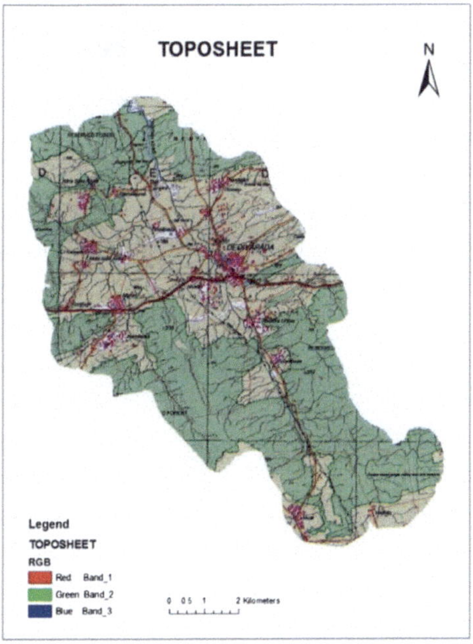

Fig. 9.3 Topo-sheet portion of study area

Fig. 9.4 Landsat 8 image (06-11-14)

Fig. 9.5 PAN sharpen Landsat 8 image

Table 9.4 Average spectral values of selected training area

Sr. No	Land use/land cover	R	G	B
1	Single crop land (yellowish brown soil)	0.624	0.620	0.618
2	Single crop land (black cotton soil)	0.486	0.479	0.457
3	Double crop land	0.652	0.668	0.770
4	Evergreen forest	0.053	0.044	0.021
5	Mixed forest	0.159	0.147	0.156
6	Deciduous forest	0.507	0.507	0.502
7	Pasture	0.576	0.575	0.580
8	Wasteland without scrub/low density Resident	0.581	0.576	0.555

9.4.10 Conservation/Support Practice Factor (P)

The support practise factor (P) is the proportion of soil loss caused by a certain conservation practise compared to the same loss caused by upslope and downslope farming under equal conditions. P factor is always less than one. The conservation practice factor depends on different conservation measures constructed to conserve the eroded soil and runoff water. Information regarding conservation measures was obtained through field observations in the study area. Entire agricultural lands were cultivated using cross slope farming and there were not any other type of agronomical or engineering measures were adopted in the area under study. P factor values (Table 9.5) recommended by Dhruvnarayan (2007) was assigned to prepare conservation practice factor P map.

9.5 Gross Soil Erosion Estimation

Gross soil erosion at each grid cell was computed by using maps of all the 6 derived components of RUSLE through raster calculator in ArcGIS interface.

Table 9.5 P factor values as per land cover class and land slope class

Sr. No	Land use/land cover	Slope (%)	P factor value
1	Agricultural land	0–2	0.6
2		2–7	0.5
3		7–12	0.6
4		12–16	0.7
5		16–20	0.8
6		20–25	0.9
7		>25	1.0
8	Other land use pattern	–	1.0

9.6 Erosion Susceptibility Map

The erosion risks were identified based on gross erosion rate and soil loss tolerance limit of the study area. Erosion risk values were used to identify the priorities of planning and development of watershed management activities of study area. The erosion risk values were derived using Eq. 9.18 (Sharda et al. 2013) while soil loss tolerance limit for the study area was taken as 5 ton/ha/year (Sharda et al. 2013 and Anonymous 2008–09, ICAR Annual Report). The erosion risk map was prepared for study area by adopting following standard procedure.

$$\text{Erosion Risk} = \text{Gross Erosion Rate} - \text{Soil Loss Tolerance Limit} \quad (9.18)$$

9.6.1 ArcSWAT: An ArcGIS Extension

The ArcSWAT ArcGIS extension is a graphical user interface for the SWAT model. SWAT model incorporates the Modified Universal Soil Loss Equation (MUSLE) model for estimation of sediment yield. Sediment yield of selected sub-watershed was estimated by MUSLE using ArcSWAT version 2012 for the period of 30 years (1986–2015).

The procedure for the use of the ArcSWAT model has been divided into two parts. During first part, input data required by SWAT at its various stages of operation were prepared while during second part, the input data were edited and uploaded into the model in order to run and calibrate the model to estimate the desired outputs.

9.7 Preparation of Arcswat Input Data

The procedure adopted for preparation of the required datasets is described below.

(A) **ArcSWAT Spatial Datasets**

(B) Digital Elevation Model (DEM): The derived 20 m DEM was used in the SWAT model to reclassify the slope parameters of the selected sub-watershed.

(C) Land Cover/Land Use Data: The derived land use land cover map was used in the SWAT model to reclassify the selected sub-watershed as per their land use.

(D) Soil Data: Soil map collected from BISAG, Gandhinagar was used in SWAT model to reclassify the selected sub-watershed as per their soil types.

(B) **ArcSWAT Tables and Text Files**

Table 9.6 Land covers database details

Sr. No	Land cover class	CPNM	USLE_C	Manning's n for overland flow	SCS curve number			
					A	B	C	D
1	Single crop agriland	AGSL	0.358	0.06	67	78	85	89
2	Double crop agriland	AGDL	0.358	0.17	67	78	85	89
3	Evergreen forest	FRSE	0.004	0.80	25	55	70	77
4	Mixed forest	FRST	0.08	0.40	36	60	73	79
5	Deciduous forest	FRSD	0.40	0.15	45	66	77	83
6	Wasteland without Scrub/pasture	PAST	0.50	0.13	49	69	79	84
7	Wasteland without Scrub	WLWS	0.50	0.011	59	74	82	86

9.7.1 Land Use/Land Cover Database Input Files

User defined land use/land cover data were prepared to be used in SWAT Model. The details of land use/land cover to generate the hydrologic response unit (HRU) in SWAT Model are given in Table 9.6. The crop factor C derived for RUSLE model were used in SWAT model for different land use pattern. The Manning's n values for overland flow in different land use pattern had been assigned based on the roughness co-efficient (Manning's n) for sheet flow given in the report of Urban Hydrology for Small Watersheds-TR-55 (1986) by USDA.

The Curve Number were selected based on soil hydrologic group and land use pattern as per values given in the report of Urban Hydrology for Small Watersheds—TR-55 (1986) by USDA. The Curve Number values for cultivated land were selected from the table of runoff CNs for cultivated agricultural land for cultivated area and for forest land and wasteland were selected from the table of CNs for other agricultural land.

9.7.2 Soil Database Input Files

Soil database details (Table 9.7) was prepared to edit in SWAT model. Based on the textural examination of the 18 soil samples from the study area, the soil hydrologic group (HYDGRP) was chosen. The main crops of study area was paddy, pigeon pea and cotton (Table 9.2) therefore the maximum rooting depth of soil profile (SOL_ZMX) was considered as 400 mm for clay soil and 500 mm for clay loam soil. The depth of soil surface to bottom of layer (SOL_Z) were collected from details given in the attribute table of soil map of study area collected from BISAG, Gandhinagar. Moist bulk density (SOL_BD), available water capacity of the soil

Table 9.7 Soil database details

No	Field	Definition	Value	Value	Unit
1	SNAM	Soil name	Clay	Fine	_
2	HYDGRP	Soil hydrologic group	D	C	A, B, C, D
3	SOL_ZMX	Maximum rooting depth of soil profile	400	500	mm
4	TEXTURE	[OPTIONAL] Texture of soil layer	Clay	Clay loam	TEX_CODE
5	SOL_Z1	Depth from soil surface to bottom of layer	500	750	mm
6	SOL_BD1	Moist bulk density	1.26	1.33	gm/cm3
7	SOL_AWC1	Available water capacity of the soil layer	0.12	0.11	mm/mm
8	SOL_K1	Saturated hydraulic conductivity	1.5	2.2	mm/hr
9	SOL_CBN1	Organic carbon content	0.85	0.87	% wt
10	CLAY1	Clay content	49.39	35.09	% wt
11	SILT1	Silt content	17.56	20.06	% wt
12	SAND1	Sand content	33.05	44.85	% wt
13	ROCK1	Rock fragment content	12.82	13.69	% wt
14	SOL_ALB1	Moist soil albedo	0.08	0.12	fraction
15	USLE_K1	USLE equation soil erodibility (K) factor	0.177	0.236	unitless
16	SOL_EC1	Electrical conductivity	0.094	0.095	dS/m
17	NLAYERS	Number of layers in the soil	1	1	no
18	NUMLAYER	The layer being displayed	1	1	no

layer (SOL_AWC) and saturated hydraulic conductivity (SOL_K) were computed by using soil texture triangle hydraulic properties calculator using soil textural components and organic matter. The average values of rock fragment content (ROCK) found through sieve analysis from all soil samples was used into soil database. Moist soil albedo (SOL_ALB) values were selected based on the soil texture and average moisture content of soil during monsoon season as given by Ten Berge (1987). The Organic carbon content (SOL_CBN) and soil electrical conductivity (SOL_EC) were computed and used as per the methodology given in Table 9.1.

9.7.3 ArcSWAT Weather Data Input Files

9.7.3.1 Weather Generator Gage Location Table

In order to prepare the data files of various weather parameters, in desired format (Weather Generator Gage Location Table format) for the software, the data has to be

Table 9.8 Statistical parameters of daily rainfall for period 1986–2015

Sr. No	Month	PCP_MM	PCPSTD	PCPSKW	PR_W1	PR_W2	PCPD
1	Jan	0.47	0.40	29.40	0.00	0.50	0.07
2	Feb	0.57	0.58	29.10	0.00	0.00	0.03
3	Mar	0.20	0.20	30.50	0.00	0.00	0.03
4	Apr	0.73	0.73	30.00	0.00	0.00	0.03
5	May	6.17	3.36	21.33	0.00	0.50	0.20
6	Jun	159.05	15.68	4.66	0.16	0.57	8.17
7	Jul	414.40	30.28	6.93	0.37	0.79	20.53
8	Aug	311.97	20.91	4.79	0.40	0.75	20.03
9	Sep	186.73	16.02	4.58	0.20	0.64	11.57
10	Oct	23.50	5.49	10.21	0.03	0.41	1.53
11	Nov	1.87	0.95	20.41	0.01	0.00	0.20
12	Dec	0.77	0.75	30.50	0.00	0.00	0.03

prepared in the standard format. This is done for user specified weather data such as rainfall, maximum and minimum temperature, relative humidity, solar radiation and wind speed are required in the model.

9.7.3.2 Rainfall Statistical Parameters

The rainfall statistical parameters i.e. average monthly precipitation (PCP MM), standard deviation (PCPSTD), skew coefficient (PCPSKEW), probability of a wet day following a dry day (PCP W1), probability of a wet day following a wet day (PCP W2), and average number of days of precipitation in month (PCPD) were computed and used in the SWAT model as shown in Table 9.8. The pcpSTAT software was used to compute various statistical precipitation parameters from daily rainfall data of 30 years (1986–2015).

9.7.3.3 Temperature Statistical Parameters

The statistical parameters of temperature such as average of monthly maximum temperature (TMPMX), average of monthly minimum temperature (TMPMN), standard deviation of monthly maximum temperature (TMPSTDMX), standard deviation of monthly minimum temperature (TMPSTDMN) were computed using pivot table properties of excel sheet are given in the Table 9.9 and used in SWAT model.

Table 9.9 Statistical parameters of daily temperature for period 1986–2015

Sr. No	Month	TMPMX	TMPMN	TMPSTDMX	TMPSTDMN
1	Jan	30.30	14.14	2.42	2.66
2	Feb	32.85	15.38	2.69	2.92
3	Mar	37.34	18.39	2.74	2.45
4	Apr	40.42	21.96	2.23	2.18
5	May	41.05	25.49	2.19	1.50
6	Jun	36.32	26.10	4.49	1.27
7	Jul	29.63	24.26	3.23	0.93
8	Aug	28.07	23.18	2.50	1.44
9	Sep	30.17	22.34	2.97	1.58
10	Oct	33.34	20.04	2.69	2.43
11	Nov	32.61	17.92	1.94	2.54
12	Dec	30.46	15.18	2.26	2.78

9.7.3.4 Relative Humidity Statistical Parameters

The statistical parameter of relative humidity such as average dew point temperature in month (DEWPT) were computed using *dew02.exe* program are given in the Table 9.10 and used in SWAT model.

Table 9.10 Average dew point temperature in month for period 1986–2015

Sr. No	Month	Average daily maximum temperature	Average daily minimum temperature	Average daily relative humidity	Average dew point temperature in month
1	Jan	33.09	15.51	28.93	5.51
2	Feb	36.43	17.58	27.41	6.93
3	Mar	39.68	21.22	31.12	12.25
4	Apr	40.16	24.6	40.9	17.98
5	May	36.96	25.57	58.7	21.94
6	Jun	31.19	24.66	79.21	23.86
7	Jul	28.47	23.53	88.57	24.02
8	Aug	29.58	22.52	85.54	23.62
9	Sep	32.48	20.45	62.57	18.68
10	Oct	32.75	18.43	44.15	12.61
11	Nov	31.14	16	36.44	8.46
12	Dec	30.44	14.6	34.24	6.63

9.7.3.5 Wind Velocity and Solar Radiation Statistical Parameters

The statistical parameters of wind velocity and sunshine hours respectively such as average monthly wind velocity (WNDAV) and average monthly solar radiation (SOLARAV) were computed using pivot table properties of excel sheet as given in the Table 9.11 and used in SWAT model.

9.8 ArcSWAT Model Operation

Methods and procedures used by ArcSWAT model using ArcGIS platform to operate, run and estimate the sediment yield was done in 6 steps.

9.8.1 SWAT Project Set-Up

This is the first step of SWAT model to set up and save the model in ArcGIS interface.

9.8.2 Watershed Delineator

In this step, the process of watershed delineation of selected study area was carried out by using watershed delineator menu. The raster layer of 20 m DEM was uploaded in SWAT model and used to generate the stepwise thematic maps of flow direction, flow

Table 9.11 Monthlies average wind velocity and solar radiation for 1986–2015

Sr. No	Month	Average monthly wind velocity	Average monthly solar radiation
1	January	2.66	20.43
2	February	2.73	23.63
3	March	3.13	25.51
4	April	4.03	25.83
5	May	4.15	21.75
6	June	3.88	15.43
7	July	3.42	13.64
8	August	2.63	16.96
9	September	2.12	19.92
10	October	2.28	18.29
11	November	2.57	17.15
12	December	2.77	17.77

accumulation, stream network, basin delineation, sub-basin delineation and longest flow path delineation for each sub-basin and study area.

9.8.3 Hydrologic Response Unit (HRU) Analysis

The study area's land use, soil, and slope characterization were done with commands from the ArcSWAT Toolbar's HRU Analysis menu. Users can load land use and soil layers into the current project, analyze slope characteristics, and determine the land use/soil/slope class combinations and distributions for the entire watershed and each sub-watershed using these tools. The HRU distribution was determined using user-specified parameters after the land use and soil datasets were imported and matched to the SWAT databases.

9.8.4 Write Input Tables

In this step, prepared weather parameter gage location table text files and daily weather data table text files were uploaded during first section of "weather stations". In second section of "Write SWAT Input Tables", the model writes all the uploaded information and generates geo database to store input values in the SWAT model.

9.8.5 Edit SWAT Input

The Edit SWAT Input menu enables the user to edit the details of different SWAT model database such as user soils database, land cover/plant growth database, tillage database, user weather stations database immediately after the step of project set up.

9.8.6 SWAT Simulation

SWAT Simulation is the last and principle step of SWAT model. During this step, SWAT model was set up for run and the model was run for the desired period of 1986–2015. The output files of sediment yield for each sub-basin reach and study area outlets were analyzed to draw out the results.

9.8.7 SWAT Model Calibration

The simulated SWAT model was calibrated using Manual Calibration helper command of SWAT Simulation menu. The conservation practices P value of 1.0 has been used by SWAT model for all land use/land cover categories. However, in the study, the USLE_P values were replaced for the agricultural land of the study area during calibration of the model. As per conservation practices P values based on slope classes given in the Table 9.5, USLE_P value for 0–2, 2–7, 7–12 and 12–25% slope classes were replaced with 0.6, 0.5, 0.6 and 0.8 respectively for agricultural land.

9.8.8 Collection of Suspended Sediment Samples

The sediment samples for 12 rainfall event during 3 years (2013–15) were collected at study area outlet in order to validate the calibrated SWAT model. Sediment samplers were used to collect the sediment samples during the events of rainfall which generate the runoff at the watershed outlet. Sediment sampler was locally manufactured adopting dimensions (Fig. 9.6) as per standard procedure described by Bartram et al. (1996) on behalf of United Nations Environment Programme and the World Health Organization (UNEP/WHO).

The cross-section of study area outlet was computed using dumpy level and levelling staff. The reduced level was measured at 5 m interval at cross section of outlet. Maximum width and maximum depth of selected sub-watershed outlet is 75 m and

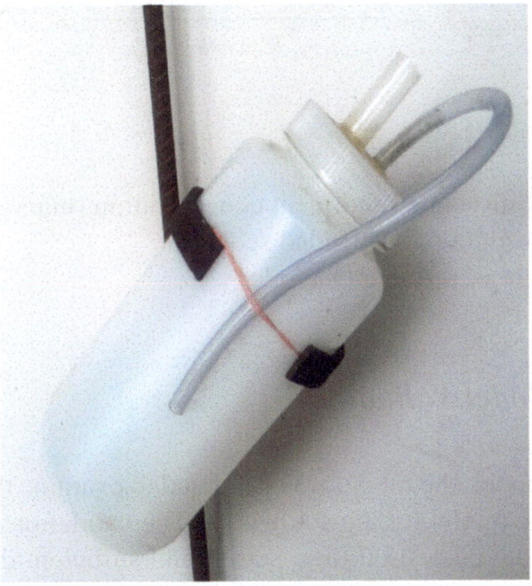

Fig. 9.6 Manufactured sediment sampler as per standard design

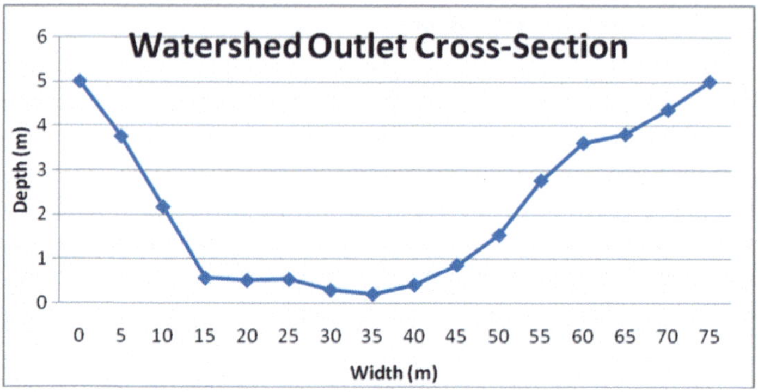

Fig. 9.7 Cross-section of study area outlet

4.80 m respectively (Fig. 9.7). The sediment samples are collected at 15 m from both sides of stream bank and at middle point of stream at approximately mid depth as shown in outlet cross-section.

9.8.9 Sediment Concentration

The collected sediment sample were weighted and oven dried at 105 °C up to complete removal of water and again weighted in order to estimate the event wise suspended sediment concentration by using the Eq. 9.19.

$$C_{estimated} = \frac{w(sand + silt + clay)}{w_{sample}} \times 10^6 \tag{9.19}$$

where

$C_{estimated}$	total suspended sediment concentration (mg/kg).
$W_{(sand+silt+clay)}$	dry weight of suspended sediment.
W_{sample}	weight of sediment sample.

9.8.10 SWAT Model Validation

After calibration of model, the model was validated to examine the performance and applicability of SWAT model for the study area. The validation of calibrated model needs the observed data of desired time period and sufficient duration. The runoff and sediment data are not recorded by any agency for the study area as the area is at sub-watershed level. The 12 samples of sediment concentration collected were used to examine the performance and validate SWAT model particularly for sediment yield.

9.8.11 Model Evaluation Statistics

The performance and applicability of SWAT model for study area was evaluated. The sediment concentrations of collected samples have been compared with simulated sediment concentration from SWAT for 12 events of 3 years (2013–2015). Four model evaluation criteria for sediment yield namely Co-efficient of Determination (R^2), Nash–Sutcliffe efficiency (NSE), RMSE-observations standard deviation ratio (RSR) and Percent bias (PBIAS) recommended by Moriasi et al. (2007) were computed to judge the performance and applicability of SWAT Model for the study area.

9.8.12 Sediment Delivery Ratio

The sediment yield from a region divided by the gross erosion of the same area is known as the sediment delivery ratio (SDR). The efficiency of the watershed in transferring soil particles from erosion areas to the point where sediment yield is recorded is represented by SDR, which is stated as a percentage. The Sediment delivery ratio was calculated using Eq. 9.20.

$$Sediment\ Delivery\ Ratio = \frac{Sediment\ yield}{Gross\ Erosion} \tag{9.20}$$

9.9 Results and Discussion

9.9.1 Components of RUSLE Model

9.9.1.1 Rainfall Erosivity Factor (R)

Average annual R factor from 30 years average rainfall (1986–2015) was calculated as 480.63 MJ mm ha^{-1} hr^{-1} for the study area. This R factor was used to estimate the average annual gross erosion for the study area. As per the graph of annual rainfall verses rainfall erosivity (Fig. 9.8), it could be inferred that rainfall erosivity is directly related with annual rainfall, when there was increase in annual rainfall, erosivity was found to be increased and vice-versa. Maximum (736.76 MJ mm ha^{-1} hr^{-1}) and minimum (241.99 MJ mm ha^{-1} hr^{-1}) rainfall erosivity was observed in the year

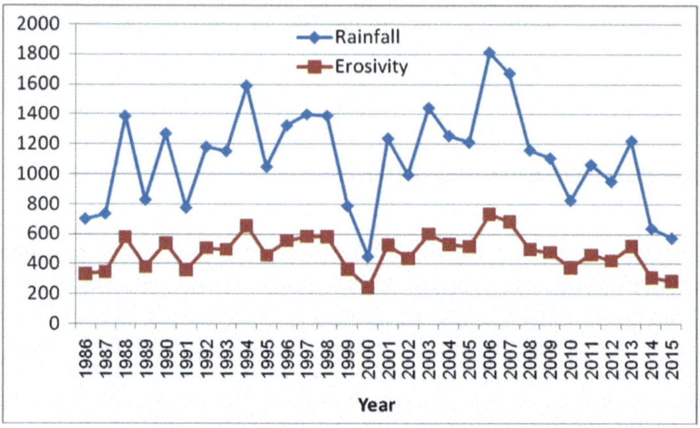

Fig. 9.8 Time series of annual rainfall and erosivity of Dediapada region

2006 and 2000 respectively as annual rainfall was maximum in the year 2006 and minimum in 2000. With an average annual rainfall of 1106.42 mm, the average rainfall erosivity over these 30 years was 480.63 MJ mm ha^{-1} hr^{-1}.

9.9.2 *Soil Erodibility Factor (K)*

9.9.2.1 Soil Physico-chemical Properties

A standard approach was used to examine the textural qualities of 18 soil samples. Course particles range from 8.05 to 28.73%, fine particles from 20.81 to 34.03%, silt particles from 13.33 to 27.58%, and clay particles from 29.00 to 39.66%, with average values of 16.77, 28.08, 20.06, and 35.09%, respectively, in clay loam soils, while course particles range from 3.57 to 19.83%, fine particles from 15.06 to 27.38%, silt particles from 13.20 to 23.72 For clay and clay loam soils, the average value of rock fragments was 13.69% and 12.82%, respectively. It is not a heavy clay soil because the clay content in both soils is less than 50%.

For soil samples, the Walky and Black method was employed to assess organic carbon and organic matter. In clay soil and clay loam soil, the average organic matter was 1.50% and 1.47%, respectively. The erodibility factor has a significant influence in water erosion since the organic matter level in both soils was less than 4%.

9.9.2.2 Soil Erodibility Factor Map

The amounts of silt, organic matter, and rock fragments in the soil have a significant impact on its erodibility. Because all of the soil samples have >70% silt and less

Table 9.12 Sample wise soil erodibility factors

Soil class	Soil sample	K_1	K_2	K_1 x K_2	K_3	Rock fragment (%)	K	Average K
Clay loam soil	1	0.029	8.95	0.261	0.29	11.30	0.23	0.236
	2	0.029	10.53	0.303	0.34	23.55	0.19	
	3	0.030	10.31	0.306	0.34	2.65	0.33	
	4	0.033	10.22	0.333	0.37	11.07	0.28	
	5	0.030	10.97	0.332	0.37	10.41	0.29	
	6	0.017	11.30	0.197	0.22	19.57	0.14	
	7	0.024	11.15	0.265	0.30	11.37	0.23	
	8	0.023	10.62	0.241	0.27	12.62	0.20	
Clay soil	9	0.023	9.96	0.233	0.30	31.02	0.14	0.177
	10	0.018	11.08	0.196	0.25	8.99	0.21	
	11	0.020	10.37	0.208	0.27	12.79	0.20	
	12	0.018	10.68	0.193	0.25	8.72	0.21	
	13	0.020	10.45	0.208	0.27	14.18	0.20	
	14	0.020	10.84	0.215	0.28	17.96	0.18	
	15	0.017	10.10	0.175	0.24	18.58	0.15	
	16	0.011	10.91	0.124	0.20	4.41	0.19	
	17	0.009	10.09	0.089	0.18	7.59	0.16	
	18	0.011	10.81	0.114	0.20	12.61	0.15	

than 4% organic matter, K1 and K2 were calculated using Eqs. 9.5 and 9.7. Because K1•K2 0.2 in seven soil samples (samples no. 6, 10, 12, 15, 16, 17, and 18), Eq. 9.9 was used to estimate the K factor, whereas Eq. 9.8 was utilised for the remaining soil samples (Table 9.12). For clay and clay loam soil types, the average soil erodibility factors were 0.236 and 0.177, respectively.

The average erodibility of clay loam and clay soil was estimated at 0.236 and 0.177 respectively (Table 9.12). Higher value in clay loam soils makes it more susceptible for erosion. The soil map (Fig. 9.9) shows that clay soil is more (53.39%) while erodibility value is lower than clay loam (46.61%) soil, which is also reflected by the area of both soil classes in erodibility map (Fig. 9.10).

9.9.2.3 Digital Elevation Model

Figure 9.11 shows the Toposheet with digitized contour lines of 20 m interval and elevation points of the sub-watershed 5D1A5c under study, whereas, Fig. 9.12 depicts the reclassified DEM of the study area derived from shape files of contour lines, elevation points and watershed boundary. The reclassified DEM indicates that 98.67 ha (1.28%) area is coved by more than 275 m altitude and 47.81 ha (0.62%) area is

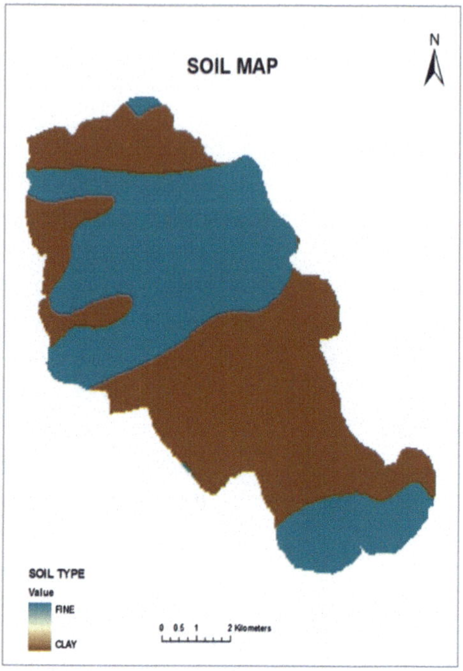

Fig. 9.9 Soils map

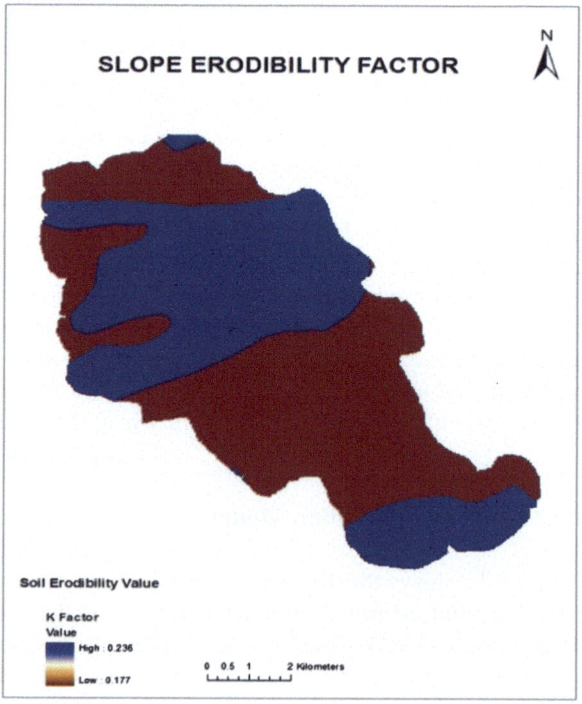

Fig. 9.10 Soil erodibility factor map

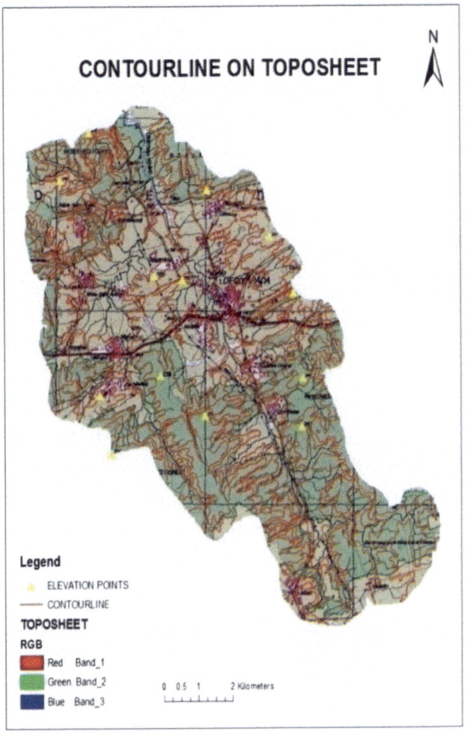

Fig. 9.11 Contour digitization on toposheet

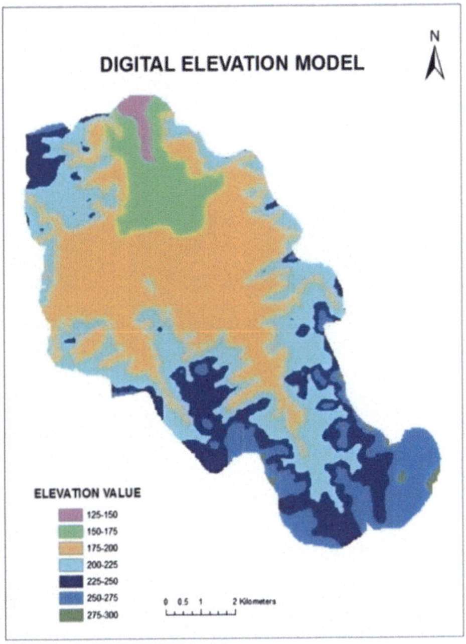

Fig. 9.12 Digital elevation model

covered by less than 150 m altitude while 7563.78 ha (98.1%) area of study area falls between 150 to 275 m altitudes. Lowest and highest altitude values of study area were 139.39 m and 288.35 m respectively. It could be concluded that major part of study area are highly susceptible to erosion due to overland flow, because of major area have elevation difference of 125 m.

9.9.2.4 Slope Length Factor

Rill to inter rill erosion ratio is used as exponent 'm' to derive the slope length factor. The exponent m value was derived by using Eq. 9.14 as shown in Fig. 9.13. The maximum value of exponent m is 0.44. The reclassified exponent m value map (Fig. 9.14) shows that 51.68% area has value less than 0.10 while 25.01% area has exponent m values between from 0.10 to 0.20. Though highest value of exponent is 0.44, but it is of very small area, whereas most of the area (77%) has less than 0.20 values, therefore it shows that erosion susceptible covers less area (23%).

Figure 9.15 describes the slope length factor at each grid cell of study area. The slope length factor value ranges from 0 to 15.58. Reclassified slope length factor (Fig. 9.16) indicates that 95.50% (7672.28 ha) area of the sub-watershed having slope length value of less than 4, while only 0.50% (38.35 ha) area having slope

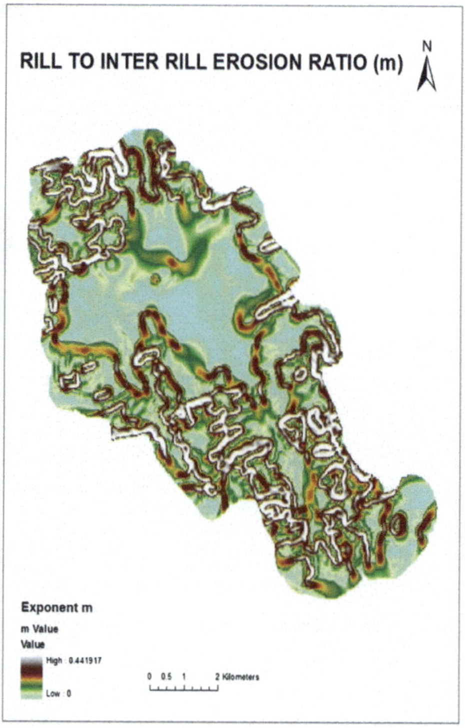

Fig. 9.13 Exponent m map

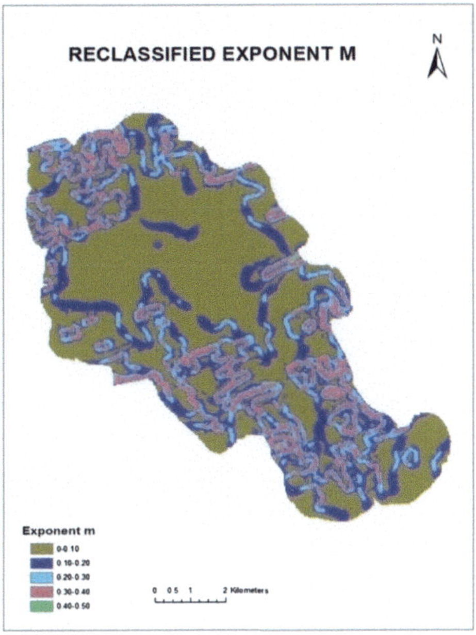

Fig. 9.14 Reclassified exponent m map

length factor values of more than 4 which falls only on high altitudes hilly terrain. It could be inferred from the above results that when value of L was more than erosion was more, in steep areas, where as when it was less, in plain topography, erosion was less. Also, that exponent 'm' plays a major role in affecting L factor.

9.9.2.5 Slop Steepness Factor

Figure 9.17 depicts the slope in percent, as shown in the figure, 5924.27 ha (76.83%) area has slope $\leq 9\%$, so equation no. 16 was used to compute the slope steepness for this area while 1786.37 ha (23.17%) area has slope greater than 9% so Eq. 9.17 was used to compute the slope steepness for that area.

Equations 9.16 and 9.17 was used to derive slope steepness for study area using raster layer (Fig. 9.18). In order to get the final slope steepness map (Fig. 9.19), attributes of slope steepness for the resulting raster layer was transferred to the raster layer of Fig. 9.18. Reclassified slope steepness map (Fig. 9.20) indicates that 76.83% of study area has slope steepness value less than 1.0 while it is greater than 1.0 only for 23.17% of study area therefore average gross erosion value of study area was less.

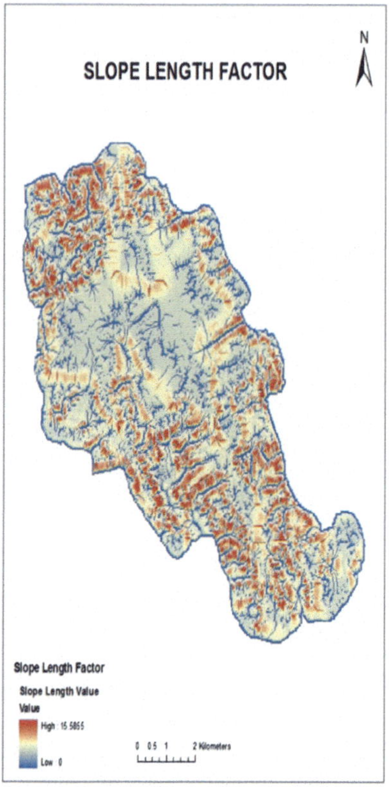

Fig. 9.15 Slope length factor map

9.9.3 *Crop/Cover Management Factor (C)*

9.9.3.1 Land Use/Land Cover Map

Cover management factor values as per the land use categories of the study area were assigned using Land use/land cover map. The area covered by different types of land cover of the selected sub-watershed is shown in land use/land cover map (Fig. 9.21) and presented in the Table 9.13. The map shows that about 41.55% (3203.55 ha) land is used for cultivation while 31.95% area is covered by forest land and 2043.34 ha (26.50%) area is under both wasteland with scrub/pasture and wasteland without scrub with low density residential area.

9.9.3.2 Land Use/Land Cover Accuracy Assessment

The GPS locations of collected ground truthing points from different land cover categories were used to locate points into land use/land cover map using ArcGIS interface. The confusion matrix was prepared using collected ground truthing sample points

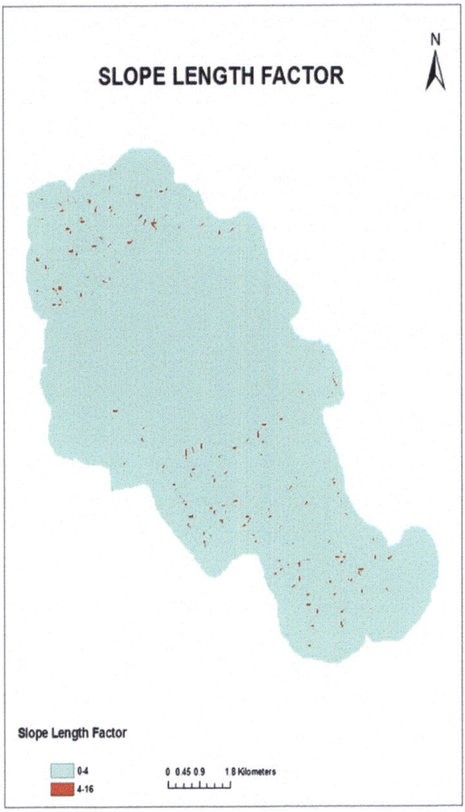

Fig. 9.16 Reclassified slope length map

and number of points fall into different classes of land use/land cover map. The overall efficiency and Kappa co-efficient were computes as 0.87 and 0.83 respectively.

9.9.3.3 Crop/Cover Management Factor

The crop management factor map (Fig. 9.22) depicts that maximum area of sub-watershed has cultivated land and the C value or crop management factor of which was found to be 0.358, followed by deciduous forest (0.4); pasture (0.6); wasteland without scrub (1.0); mixed forest (0.08); evergreen forest (0.004) in descending order of area under each type of land cover. C value of 1 has the highest susceptibility of erosion whereas less value indicates lesser erodibility. The lowest value of c factor 0.004 was for evergreen forest, but, it covers a very small area in the watershed and maximum area comes under cultivated land, which is prone to erosion.

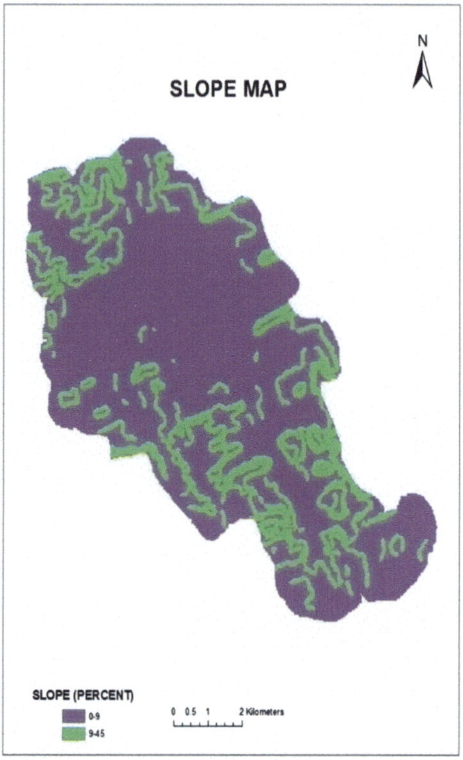

Fig. 9.17 Slope map in percent

9.9.3.4 Conservation/Support Practice Factor (P)

Land use/land cover and slope maps were used to assign the conservation practice value as per the land use and slope categories of the study area to derive P factor map. The area covered by different slope classes is presented into Table 9.14 and Fig. 9.23. The highest (38.83%) area is covered under 0–2% class followed by 2–7; 7–12; 12–16; 16–20 and 20–25% class. This shows that there is lesser area of higher slope, which has higher erosion susceptibility and large part of watershed has less slope having lesser susceptibility to erosion.

Figure 9.24 shows the intersection map of land use land cover polygon map with slope polygon map. The P value for agricultural land having less than 25% slope as per given in Table 9.14 was assign to the attribute table of the intersection map in order to prepare the P factor map (Fig. 9.25).

The P factor map indicates that 3192.98 ha (41.41%) land has conservation practice factor values of less than 1.0 while 4517.66 ha (58.59%) land has conservation practice factor value of 1.0. As per Table 9.13, total are under cultivation was 3203.51 so that 99.67% cultivated areas having conservation practice factor values of less than 1.0 and less than 25% slope while only 0.33% cultivated areas having more than 25% slope.

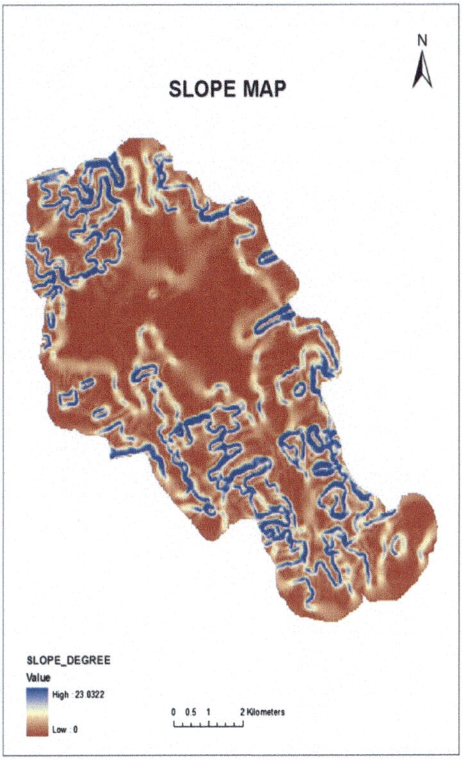

Fig. 9.18 Slope map in degree

9.9.4 Gross Soil Erosion Using RUSLE Model

9.9.4.1 Gross Soil Erosion

Gross soil erosion using Revised Universal Soil Loss Equation is computed using raster layer of all the RUSLE parameters. The raster layer of all the 5 parameters K, L, S, C and P and computed R value were multiplied using raster calculator in order to prepare gross soil erosion map. The map was reclassified as per erosion class as shown in Fig. 9.26. The Fig. 9.26 and Table 9.15 indicates that 3278.99 ha (42.53%) area having average annual erosion rate less than 5 ton/ha/year while about 1015.16 ha (13.17%) area having average annual erosion rate of more than 80 ton/ha/yr. About 44.30% (3416.49 ha) areas having gross soil erosion between 5–80 tons/ha/yr. This shows that, as erosion rate increases, percent area under particular erosion class decreases. Average annual erosion rate for study area was estimated at 39.25 tons/ha/yr.

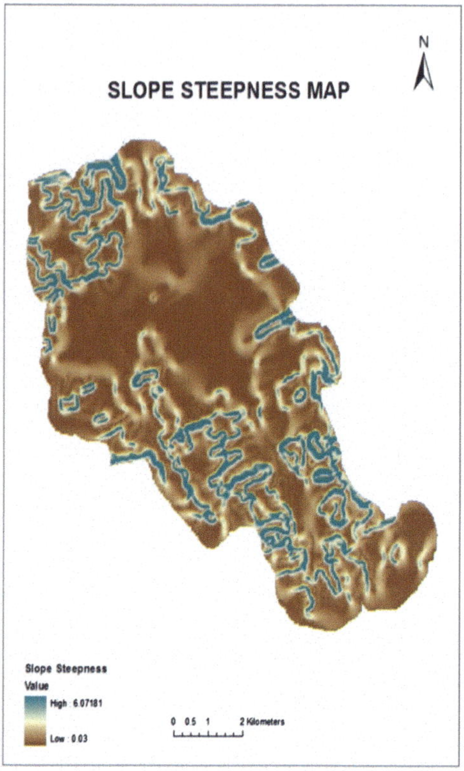

Fig. 9.19 Slope steepness factor map

9.9.4.2 Erosion Susceptibility Map

Figure 9.27 and Table 9.16 shows that highest area (42.53%) of sub-watershed comes under safe zone followed by very high priority zone (22.15%); very less priority zone (12.23%); less priority (11.85%); medium priority (6.51%) and high priority (4.73%), therefore, the area under very high priority should be treated first in priority to cover 66.68% area of sub-watershed under safe zone for sustainable watershed development and protection from further deterioration.

9.9.5 SWAT Model Simulation Results

9.9.5.1 SWAT Model Simulation

SWAT Simulation is the principle and final step of SWAT model. In this step, SWAT model was set up and the model was run for the desired period of 1986–2015. The

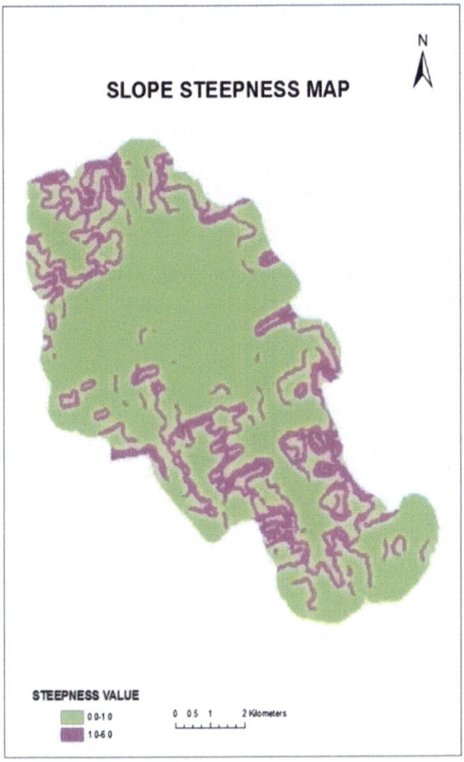

Fig. 9.20 Reclassified slope steepness map

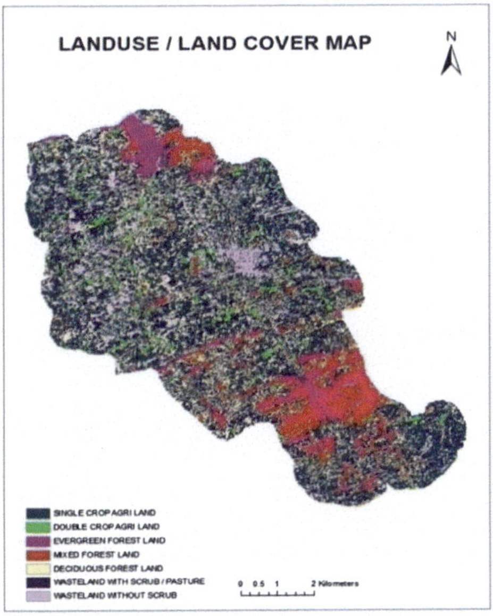

Fig. 9.21 Land use/land cover map

Table 9.13 Areas under different land use land cover of selected sub-watershed

Sr. No	Land use/land cover	Area (ha)	Percent of total area
1	Single crop agriculture land	2823.10	36.61
2	Double crop agriculture land	380.45	4.93
3	Evergreen forest	312.37	4.05
4	Mixed forest	875.93	11.36
5	Deciduous forest	1275.64	16.54
6	Pasture/wasteland with scrub	1038.98	13.47
7	Wasteland without scrub/low density resident	1004.36	13.03
8	Total	7710.64	100.00

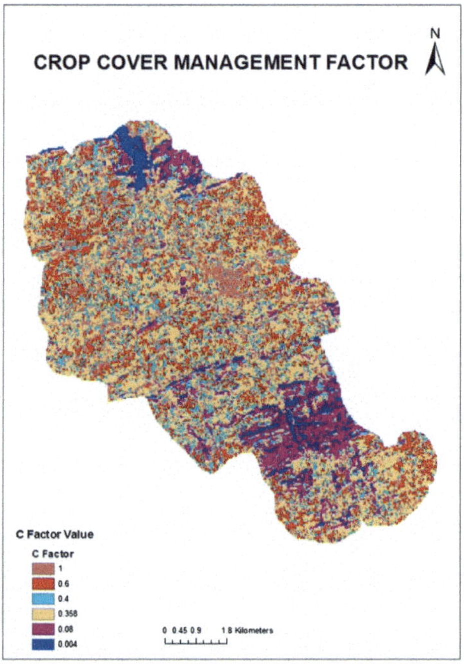

Fig. 9.22 Cover management factor map

Table 9.14 Areas under different slope classes of selected sub-watershed

Sr. No	Slope class (%)	Area (ha)	Percent of total area
1	0–2	2994.28	38.83
2	2–7	2479.10	32.15
3	7–12	962.65	12.48
4	12–16	479.61	6.22
5	16–20	362.48	4.70
6	20–25	304.14	3.94
7	>25	128.39	1.67
8	Total	7710.64	100.00

Fig. 9.23 Reclassified slope map

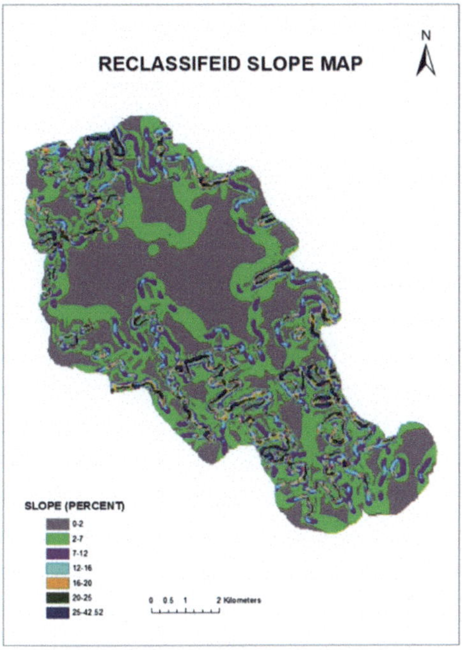

Fig. 9.24 Intersection of LU/LC with slope

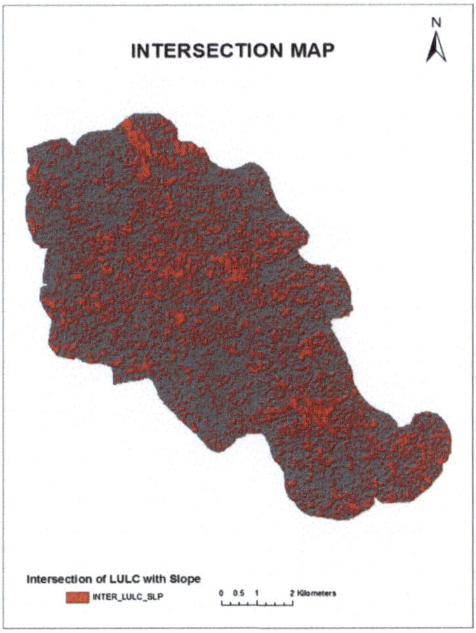

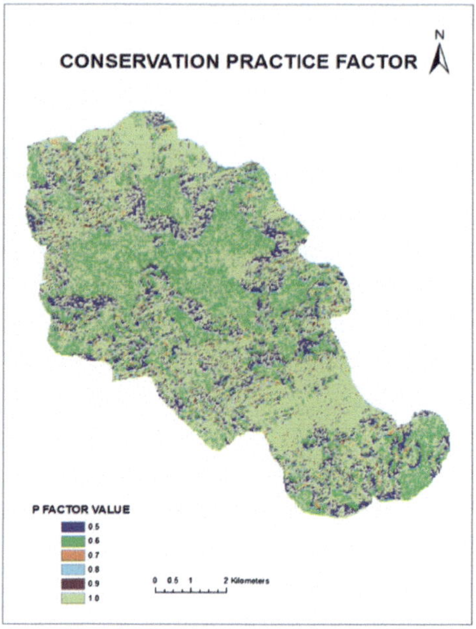

Fig. 9.25 Conservation factor P map

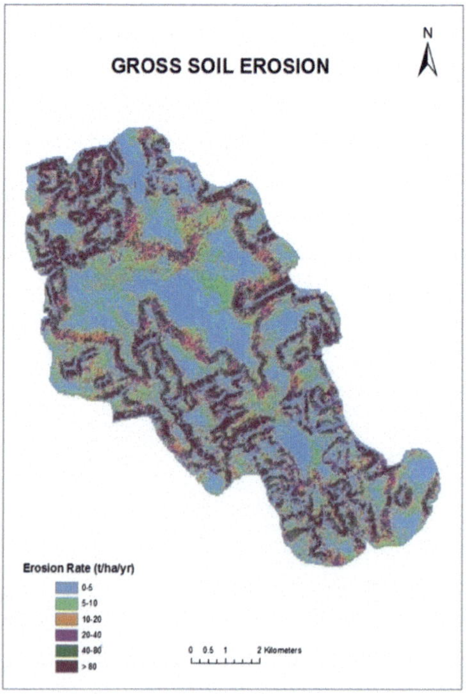

Fig. 9.26 Gross soil erosion map

Table 9.15 Gross soil erosion classes of selected sub-watershed

Sr. No	Erosion Class	Range (tons/ha/yr)	Area (ha)	Percent area
1	Slight	0–5	3278.99	42.53
2	Moderate	5–10	943.12	12.23
3	High	10–20	913.65	11.85
4	Very High	20–40	866.91	11.24
5	Severe	40–80	692.81	8.99
6	Very Severe	>80	1015.16	13.17

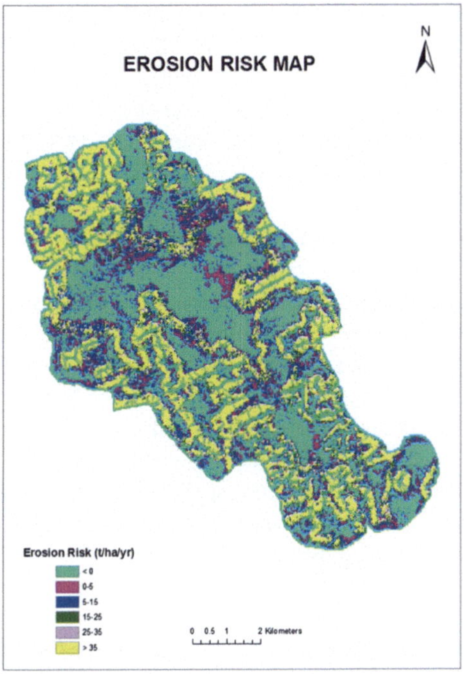

Fig. 9.27 Erosion risk map

Table 9.16 Erosion risk area of selected sub-watershed

Sr. No	Erosion risk (ton/ha/year)	Area (ha)	Percent area (%)	Erosion risk criteria
1	<0	3278.99	42.53	Safe
2	0–5	943.12	12.23	Very less priority
3	5–15	913.65	11.85	Less priority
4	15–25	501.89	6.51	Medium priority
5	25–35	365.03	4.73	High priority
6	>35	1707.97	22.15	Very high priority

gross sediment loading (gross erosion) of 59.94 tons/ha/year and sediment yield (at the outlet) of 39.42 tons/ha/year was estimated by the SWAT model.

9.9.5.2 SWAT Model Calibration

The SWAT model was calibrated using Manual Calibration helper command of SWAT Simulation menu. The USLE_P values were replaced for the agricultural land of the study area. As per conservation practices P values based on slope classes given in the Table 9.5, USLE_P values for 0–2, 2–7, 7–12 and 12–25% slope classes were replaced with 0.6, 0.5, 0.6 and 0.8 respectively.

The model was re-run after the calibration. The gross sediment loading were estimated to be 34.61 tons/ha/yr by SWAT model while the sediment yield was estimated at 22.78 tons/ha/yr (Fig. 9.28).The surface runoff was remaining constant before and after the calibration of SWAT Model using USLE_P values.

9.9.5.3 SWAT Model Validation

The catchment area of Karjan reservoir is 1404 km^2. The study area covers 5.49% of Karjan reservoir catchment area. According to the Compendium on Silting of Reservoir in India (2015) of Watershed & Reservoir Sedimentation Directorate, Central Water Commission, Government of India, siltation rate of 29 years (1984–2013) of Karjan reservoir was 2.241 thousand $m^3/km^2/yr$. Therefore, from density and siltation rate of Karjan reservoir, the average sediment yield for 29 years was estimated to be 21.56 tons/ha/year. Whereas, the sediment yield by SWAT was estimated to be 22.78 tons/ha/year. The gross sediment loading estimated by SWAT model was 34.61 tons/ha/year.

9.9.5.4 Performance Evaluation of SWAT Model

The computed sediment concentration from collected sediment samples and SWAT model simulated sediment concentration is given in Table 9.17. The model performance was evaluated based on these sediment concentration values by using recommended statistical parameters.

The performance evaluation depends on the value of Co-efficient of Determination (R^2), Nash–Sutcliffe efficiency (NSE), RMSE-observations standard deviation ratio (RSR) and Percent bias (PBIAS). The values obtained in Table 9.18 indicate that the SWAT model performs satisfactorily and could be applicable for the study area for the purpose of estimating gross soil erosion and sediment yield at watershed outlet for planning, execution and management of soil and water conservation programmes at sub-watershed level.

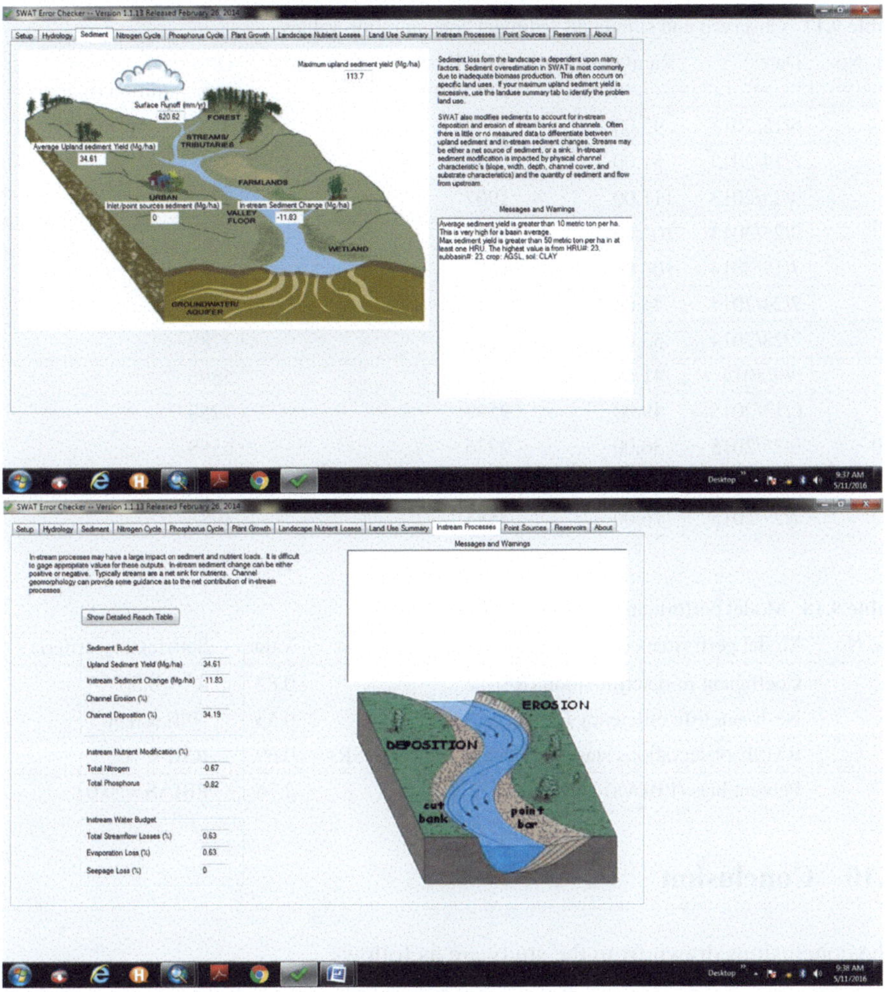

Fig. 9.28 SWAT checkup results for sediment yield

9.9.5.5 Sediment Delivery Ratio

Sediment delivery ratio was estimated by taking ratio of gross erosion estimated from RUSLE (39.25 tons/ha/yr) with sediment yield estimated from SWAT simulation (22.78 ton/ha/year). The sediment delivery ratio was computed as 0.58. This ratio indicates that about 58% of eroded soils go out of the watershed which reduces the soil depth and fertility of land on site and reaches Karjan reservoir thus reducing the capacity of water storage each year.

Table 9.17 Observed and simulated sediment concentration values

Sr. No	Date	Rainfall (mm)	SWAT simulated sediment concentration (mg/kg)	Observed sediment concentration (mg/kg)
1	6/15/2013	92.00	2380	3856
2	8/14/2013	51.00	10,430	9852
3	9/23/2013	111.00	1907	2745
4	9/25/2013	102.00	1268	1988
5	7/19/2014	104.00	1357	1863
6	7/24/2014	44.00	9726	8590
7	7/28/2014	53.00	12,240	7789
8	9/9/2014	41.00	5138	5896
9	6/13/2015	49.00	9599	7758
10	6/25/2015	36.00	2936	6588
11	7/25/2015	74.00	1446	2807
12	7/27/2015	76.00	1584	1985

Table 9.18 Model performance evaluation criteria values

Sr. No	Model performance evaluation criteria	Value	Satisfactory criteria
1	Coefficient of determination (R^2)	0.82	$R^2 > 0.50$
2	Nash-sutcliffe efficiency (NSE)	0.53	NSE > 0.50
3	RMSE-observations standard deviation ratio (RSR)	0.69	RSR < 70
4	Percent bias (PBIAS)	2.76	PBIAS = 0.0

9.10 Conclusion

The conclusions drawn from the study are as follows:

1. The average annual rainfall erosivity for study area was estimated at 480.63. MJ mm/ha hr.
2. The soil erodibility for clay loam soil was estimated at 0.236 t/ha per unit R while for clay soil it was 0.177 t/ha per unit R.
3. The high resolution of 20 m DEM performed well to generate stream network and it perfectly matched with actual drainage network.
4. Reclassified slope length factor indicates that 95.50% area of the sub-watershed has slope length factor value of <4, while only 0.50% area has slope length factor values of >4 which falls only at high altitudes hilly terrain.
5. As per Reclassified slope steepness map, 76.83% area having <9% slope has slope steepness <1.0 value of while 23.17% area having >9% slope has 1.0–6.0 slope steepness.

6. Supervised classification techniques for land use/land cover classification found to suit well as overall efficiency and Kappa co-efficient was determined as 0.87 and 0.83 respectively.
7. Cover management factor values was derived as 0.358, 0.004, 0.08, 0.4, 0.6 and 1.0 for agricultural land, evergreen forest, mixed forest, deciduous forest, pasture and low density resident respectively.
8. 41.41% land has P value of less than 1.0 while 58.59% land has P value of 1.0.
9. Average annual erosion rate for study area was estimated at 39.25 tons/ha/yr.
10. Area under average gross soil erosion rate under different classes of slight, moderate, high, very high, severe and very severe were 42.53, 12.23, 11.85, 11.25, 8.99 and 13.17% respectively.
11. Sediment yield simulated though SWAT model was 22.78 tons/ha/year for the study area
12. The SWAT model was validated as values of co-efficient of determination (R^2), Nash–Sutcliffe efficiency (NSE), RMSE-observations standard deviation ratio (RSR) and percent bias (PBIAS) were computed as 0.82, 0.53, 0.69 and 2.76 respectively and these values were within acceptable limits.
13. The sediment delivery ratio was computed as 0.58.
14. SWAT model performs satisfactorily and could be applicable for the study area for the purpose of estimating gross soil erosion and sediment yield at sub-watershed level.

References

Anonymous (2009) Soil resource inventory and management. Soil Water Prod. DARE/ICAR Annual Report, pp 7–11

Anonymous (2012–13, 2013–14 & 2014–15) Annual progress report, Krushi Vigyan Kendra, Navsari Agricultural University, Dediapada

Anonymous (2008–09) Soil resource inventory and management. Soil and Water Prod. DARE/ICAR Annual Report, pp 7–11

Anonymous (2014a) Watershed Atlas of India, office of the chief soil survey, soil and land use survey of India. Department of Agriculture and Co-operation, Ministry of Agriculture, Government of India, Aravali Printers & Publishers Pvt. Ltd., New Delhi, 2nd edn, pp 107

Anonymous (2014b) Watershed Atlas of India, soil and land use survey of India, Ministry of Agriculture, Government of India, Aravali Printers & Publishers Pvt. Ltd., New Delhi, pp 107

Auerswald K, Fiener P, Martin W, Elhaus D (2014) Use and misuse of the K factor equation in soil erosion modeling: an alternative equation for determining USLE nomograph soil erodibility values. CATENA 118:220–225

Banerjee SK, Chinnamani S, Jha MN (1991) Forest soils of north and northeast Himalayas and constraints in crop production. Indian Soc Soil Sci, New Delhi, India, Bull 15:164–176

Bartram J, Richard B (1996) Water quality monitoring: a practical guide to the design and implementation of freshwater quality studies and monitoring programmes. CRC Press, pp 1–15

Caon L, Vargas R (2017) Threats to soils: global trends and perspectives. Global soil partnership food and agriculture organization of the United Nations. In: Brajendra (eds) A contribution from the intergovernmental technical panel on soils. Global land outlook working paper

Dhruvnarayan VV (2007) Soil and water conservation research in India, ICAR, New Delhi, pp 40–41

Jackson ML (1973) Soil chemical analysis. Prentice-Hall of India Pvt. Ltd., New Delhi, India, pp 39–415

Kumar S, Kushwaha SPS (2013) Modelling soil erosion risk based on RUSLE-3D using GIS in a Shivalik sub-watershed. J Earth Syst Sci 122(2):389–398

Kurothe RS (1991–92) Determination of 'C' and 'P' factors of USLE for important crops and management practices in vasad region. Annual Report, Central Soil and Water Conservation Research and Training Institute, Dehradun, 100–101

Maji, A.K., Obi Reddy, G.P. and Sarkar, D. (2010). Degrad-ed and Wastelands of India: Status and Spatial Distribu-tion. Indian Council of Agricultural Research, New Delhi, India, pp 23, 159.

McCool DK, Brown LC, Foster GR, Mutchler CK, Meyer LD (1987) Revised slope steepness factor for the Universal Soil Loss Equation. Trans ASAE 30(5):1387–1396

Mitasova H, Hofierka J, Zlocha M, Iverson L (1996) Modelling topographic potential for erosion and deposition using GIS. Int J Geogr Inf Syst 10(5):629–641

Moriasi DN, Arnold JG, Van Liew MW, Bingner RL, Harmel RD, Veith TL (2007) Model evaluation guidelines for systematic quantification of accuracy in watershed simulations. Trans ASABE 50(3):885–900

NAAS (2012) Sustaining agricultural productivity through Integrated Soil Management. Policy Paper No. 56, National Academy of Agricultural Sciences, New Delhi, pp 24

Narain P, Khybri ML, Tomar HPS, Sindhwal NS (1994) Estimation of runoff, soil loss and USLE parameters for doon valley. Ind J Soil Conser 22(3):129–132

Odeman LR, Hakkeling RTA, Sombroek WG (1992) World map of the status of human induced soil degradation. ISRIC, Wageningen and UNEP, Nairobi 26

Pimentel D (2006) Soil erosion: a food and environ-mental threat. Environ Dev Sustain-Abil 8:119–137

Renard KG, Weesies GA, McCool DK, Yoder DC (1997) Predicting soil erosion by water: a guide to conservation planning with the Revised Universal Soil Loss Equation (RUSLE). USDA Handbook, 703, U.S. Dep. Agric., Washington, D.C 404

Sharda VN, Mandal D, Ojasvi PR (2013) Identification od soil erosion risk areas for conservation planning in different states of India. J Envi Boil 34:219–226

Singh G, Babu R, Chandra S (1981) Soil loss prediction research in India, Tech Bull T-12/D-9. Central Soil and Water Conservation Research and Training Institute, Dehradun, India.

Singh G, Babu R, Narain P, Bhushan LS, Abrol IP (1992) Soil erosion rates in India. J Soil Water Conserv 47(1):97–99

Wischmeier WH, Smith DD (1965) Predicting rainfall-erosion losses from cropland east of the rocky mountains. Agriculture Handbook, 282. U.S. Dept. Agric., Washington D.C

Wischmeier WH, Smith DD (1978) Predicting rain-fall erosion losses—a guide to conservation planning. USDA Handbook, 537, U.S. Dep. Agric., Washington, D.C

Young R, Orsini S (2015) Soil degradation as big a threat to humanity as climate change. Sustain Food Trust, UK, pp 1–55

Chapter 10
Delineation of Irrigation Management Zones Using Geographical Weighted Principal Component Analysis and Possibilistic Fuzzy C-Means Clustering Approach

Pravash Chandra Moharana, Upendra Kumar Pradhan, Roomesh Kumar Jena, Sonalika Sahoo, and Ram Swaroop Meena

Abstract Delineation of irrigation management zones (IMZs) depend on spatial variability of soil hydro-physical properties like soil texture, bulk density (BD), field capacity (FC), permanent wilting point (PWP) and available water content (AWC). This work presents a method for delineation of irrigation zones under such constraints. A total of 67 geo-referenced soil profiles were collected from the study area covering an area of 4206 ha. The spatial variability and correlations of hydro-physical properties were firstly characterized using geostatistics and principal component analysis. Their spatial variability was analyzed and geostatistical analysis showed that Gaussian, spherical and circular models were the best-fit models. Then, IMZs were delineated by geographical weighted principal component analysis (GWPCA) and possibilistic fuzzy C-means (PFCM) clustering algorithm. Optimum clusters were identified using fuzzy performance index (FPI) and normalized classification entropy (NCE). The study area was divided into two IMZs by PFCM clustering, and soil hydro-physical properties had high uniformity in each subzone. The IMZs can provide the basis for decision making of precision irrigation practices. The IMZ-based crop water requirement reduces the application quantity of water significantly at a large extent and maximizes crop production.

P. C. Moharana (✉) · R. S. Meena
ICAR-National Bureau of Soil Survey and Land Use Planning, Regional Centre, Udaipur 313 001, India
e-mail: pravashiari@gmail.com

U. K. Pradhan
ICAR-Indian Agricultural Statistics Research Institute, New Delhi 110 012, India

R. K. Jena
ICAR-Indian Institute of Water Management, Bhubaneswar 751 023, India

S. Sahoo
ICAR-National Bureau of Soil Survey and Land Use Planning, Nagpur 440 033, India

P. K. Shit et al. (eds.), *Soil Health and Environmental Sustainability*,
Environmental Science and Engineering,
https://doi.org/10.1007/978-3-031-09270-1_10

Keywords Irrigation management zone · Geostatistics · Geographical weighted principal component analysis · Possibilistic fuzzy c-means · Hot arid ecosystem

10.1 Introduction

Water resource is under tremendous pressure to meet the growing demand for agriculture, urbanization and industrial sectors. In the future also the need of water will increase to sustain the rapidly growing population. Hence, water has become a precious input for the agricultural production system as climate change driven rainfall uncertainty has aggravated the challenges for freshwater allocation towards agricultural sector (Falloon and Betts 2010). Considering the current pressure on freshwater resources and future demand of irrigation water, enhancing water use efficiency in agricultural production system is very much essential in arid and semi arid region (Daccache et al. 2014). In arid regions of India, agriculture is adversely affected by low and erratic rainfall coupled with high evaporative demand and low moisture retention by light textured soils. On the other hand indiscriminate use of scarce water through conventional irrigation management practices led to exhaustion of ground water resources and development of water logging in canal command area. Hence, efficient management of limited water is the need of the hour for achieving sustainable production for longer period on light textured soils of arid regions.

Precision irrigation, an important component of precision agriculture includes precise and optimal application of water as per the requirement of each irrigation management zone (IMZ) to maximize water use efficiency, crop production and economic profitability along with minimal adverse environmental impact (Jiang et al. 2011). Management zone delineation in an agricultural field is generally done on the basis of relatively homogeneous soil-landscape attributes. It assists in site specific application of inputs across the management zones (Haghverdi et al. 2015) which results in efficient resources use and yield optimization (Schepers et al. 2004). Thus, management zones delineation has emerged as popular approach for site specific input management or precision irrigation management.

The IMZ concept advocates the identification of regions (management zones) within the area delimited with similar soil characteristics. Informations related to either soil and landscape properties or crop yield map or combining both the information are used for delineation of IMZs. Soil-landscape attributes includes soil survey maps, evaluation of soil physical and chemical properties and remote sensing images (Fraisse et al. 2001; Johnson et al. 2003; Vitharana et al. 2008). To reduce the sampling cost, errors associated with estimation and interpolation to unsampled location, geostatistical tools and simulation with cluster and kriging simulations are being used in delineation of management zone (Saito et al. 2005; Brevik et al. 2016; Verma et al. 2018). Several methods of cluster analysis have been widely used to classify management areas (Anderberg 1973). Possibilistic fuzzy C-means (PFCM) is a better clustering algorithm as it has the potential to give more value to membership or typicality values (Pal et al. 2005). PFCM inherits both the properties of possibilistic c-means (PCM) and fuzzy C-means (FCM) that often avoids various problems like cluster coincidence and noisy sensitivity.

Considering the high soil hydro-physical variability, it was hypothesized that soil properties that were studied in the same area differed in spatial distribution. In view of the above facts, the present study was done to (i) characterize the spatial variability of the soil hydro-physical attributes using geostatistical analysis, (ii) identify irrigation management zones (IMZ) by using robust geographical weighted principal component analysis (GWPCA) and possibilistic fuzzy C-means (PFCM) cluster algorithms, and (iii) evaluate the potential of defined IMZ for site-specific irrigation management in the arid region of India.

10.2 Materials and Methods

10.2.1 Site Description

This study was conducted in Central State Farm, Sardargarh (29° 20′ 53″–29° 24′ 47″ N, 73° 30′ 00″–73° 37′ 38″ E), located in the western plains of Rajasthan, India (Fig. 10.1). The farm is ~4206 ha which is intensively cultivated. The area has very scanty rainfall with average of 286 mm. The dominant soils are deep to very deep, either calcareous or non-calcareous and sandy clay loam in nature (Soil Survey Staff 1999).

10.2.2 Soil Sampling and Analysis

Grid-sampling scheme was imposed on the field to collect 67 soil samples with an interval of 500 m. Eight soil hydro-physical properties were measured by standard procedure. Soil texture was determined with the international pipette method (Jackson 1973). Soil organic carbon (SOC) content of the soil samples was determined by the Walkley and Black (1934) method. Soil moisture retention at field capacity (FC) and permanent wilting point (PWP) was measured by pressure plate apparatus (Klute 1986). Soil bulk density was determined by collecting undisturbed soil samples using core sampler of known volume (Veihmeyer and Hendrickson 1948). Available water content (AWC, mm mm^{-1}) was obtained by measuring the difference in the water content between FC and PWP.

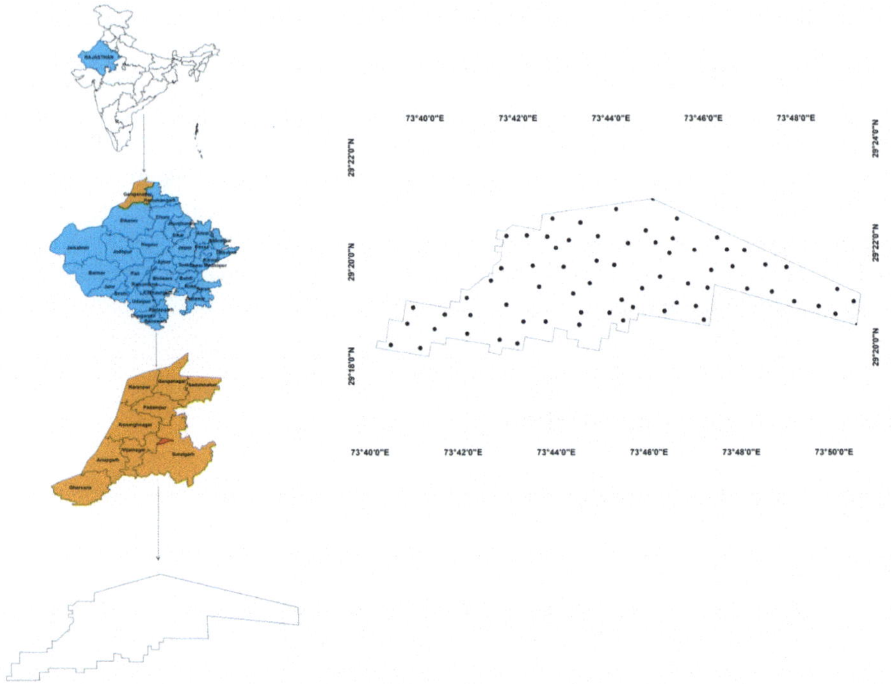

Fig. 10.1 Study site and soil-sampling points in irrigated hot arid environment of India

10.2.3 Descriptive and Geostatistical Analysis

Before conducting the geographically weighted principal component analysis (GWPCA), an exploratory analysis of the data was carried out in order to find relationships between soil hydro-physical properties by using R-statistics software (R-Core-Team 2019). In geostatistics, spatial variability of soil properties is expressed by semivariogram $\gamma(h)$, which measures the average dissimilarity between the data separated by a vector h (Goovaerts 1998). It was computed as half of the average squared difference between the components of data pairs:

$$\widehat{\gamma}(\mathrm{h}) = \frac{1}{2N(h)} \sum_{i=1}^{N(h)} [z(x_i) - z(x_i + h)]^2 \quad (10.1)$$

where, $N(h)$ is the number of data pairs within a given class of distance and direction, $z(x_i)$ is the value of the variable at the location x_i, $z(x_i + h)$ is the value of the variable at a lag of h from the location x_i. Experimental semivariograms [$\widehat{\gamma}$(h)] as obtained from the above equation were fitted with standard models using weighted least square technique Semivariogram parameters were estimated using the gstat R package (Pebesma 2004; Gräler et al. 2016) with R version 3.5.3 package (R-Core-Team 2019).Arc GIS 10.3.1 software (ESRI Inc., Redlands, California, USA) was used for soil hydro-physical properties mapping.

10.2.4 Principal Components Analysis and Fuzzy Clustering

GWPCA which is an extension of the global principal component analysis (PCA) was performed by procedure outline by Harris et al. (2011, 2014). The data are converted into a spatial data frame to run the robust GWPCA using GW model R package (Gollini et al. 2015). GWPCA score for each covariate at each data point was computed and used in the input for the possibilistic fuzzy C-means (PFCM) algorithm. Figure 10.2 is a schematic representation of the proposed methodology. In this study, the PFCM was performed using the "ppclust" R-package (Cebeci et al. 2018). Fuzzy performance index (FPI) (McBratney and Moore 1985) and normalized classification entropy (NCE) were used as indicators of optimum cluster number (Bezdek 1981) as follows:

$$\text{FPI} = 1 - \frac{c}{c-1}\left[1 - \frac{\sum_{i=1}^{c}\sum_{k=1}^{n}(\mu_{ik})^2}{n}\right] \tag{10.2}$$

$$\text{NCE} = \frac{n}{n-c}\left[-\frac{\sum_{k=1}^{n}\sum_{i=1}^{c}\mu_{ik}\log_a(\mu_{ik})}{n}\right] \tag{10.3}$$

where c is the number of clusters and n is the number of observations, μ_{ik} is the fuzzy membership and $\log_a$ is the natural logarithm. The FPI measures the degree of

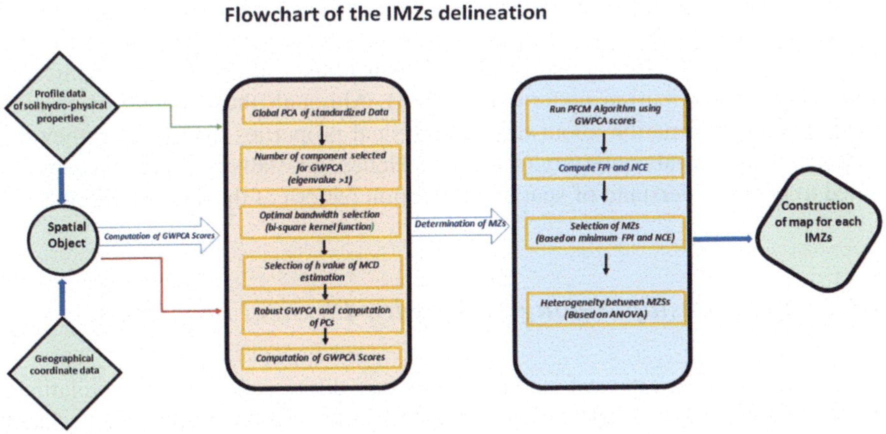

Fig. 10.2 Schematic representation of the proposed methodology for IMZ delineation

fuzziness created by a specified number of classes. Values of FPI may range from 0 to 1. Values approaching 0 indicate distinct classes with little membership sharing while values near 1 indicate no distinct classes with a large degree of membership sharing. The NCE is an estimate of the amount of disorganization created by a specified number of classes. The optimal number of clusters for each computed index (FPI and NCE) is obtained when the index is at the minimum, representing the least membership sharing (FPI) and greatest amount of organization (NCE) as a result of the clustering process (Fridgen et al. 2004). Furthermore, analysis of variance was used to indicate heterogeneity among different IMZs. Descriptive statistics were obtained by R software. R software was used in implementing the GWPCA and PFCM clustering algorithm.

10.3 Results and Discussion

10.3.1 Descriptive Statistics of Soil Hydro-Physical Properties

The statistical implication of sand, silt, clay, SOC, BD, FC, PWP and AWC are listed in Table 10.1. The study adopted the classification of Wilding and Dress (1983) for evaluating the variations of soil properties based on their coefficient of variation (CV) (%): (1) if CV is less than 15, then the parameter has low variations, (2) if CV is between 15 and 35, the variable has medium variations and (3) otherwise, the parameter has high variations. Accordingly, the BD and FC had low variability (CV < 15%) while sand, silt, clay, SOC, PWP and AWC exhibited medium variability (CV of 15–35%). The values of CV for hydro-physical properties ranged from 6.31 to 25.01%. A similar result was reported by Jiang et al. (2011). The mean values of sand, silt, clay, SOC, BD, FC, PWP and AWC in this region were 35.80%, 34.17%, 30.03%, 0.37%, 1.37 Mg m^{-3}, 26.27%, 14.48% and 0.16 mm mm^{-1}, respectively. Skewness values of –0.34 to 1.02 for soil properties revealed that data were not normally distributed. Histograms of all soil hydro-physical properties graphically depict the data distribution pattern (Fig. 10.3). Although these statistical studies provide useful information about the soil physical properties distribution, they do not describe the spatial continuity of the data. Hence, geostatistical techniques were applied to better understand of spatial distribution pattern of the studied variables.

10.3.2 Relationship Among Soil Hydro-Physical Properties

By correlation analysis, it was found that there were different degrees of correlations among the eight soil physical properties (Fig. 10.4). FC was significantly correlated to BD ($r = -0.443$), and the correlation coefficients between FC with sand and SOC were –0.327 and 0.271, respectively. Pachepsky et al. (2001) found that water

Table 10.1 Descriptive statistics of soil hydro-physical properties in the hot arid regions of India

	Minimum	Maximum	Mean	SE	SD	CV (%)	Skewness	Kurtosis	$P_{K\text{-}S}$	Distribution pattern
Sand (%)	24.48	57.91	35.80	0.829	6.79	18.96	0.70	0.67	0.007	Log
Silt (%)	14.77	49.76	34.17	1.007	8.24	24.12	-0.34	-0.67	0.001	Log
Clay (%)	19.08	48.53	30.03	0.761	6.23	20.75	0.73	0.23	0.000	Log
SOC (%)	0.19	0.58	0.37	0.009	0.07	18.98	0.00	0.78	0.001	Log
BD (Mg m^{-3})	1.22	1.63	1.37	0.011	0.09	6.31	1.02	0.67	0.000	Log
FC (%)	16.61	38.11	26.27	0.528	4.32	16.46	0.15	0.13	0.200	Normal
PWP (%)	8.27	22.08	14.48	0.443	3.62	25.01	0.24	-1.06	0.027	Log
AWC (mm mm^{-1})	0.07	0.27	0.16	0.005	0.04	23.66	0.30	0.31	0.200	Normal

Note $p^{K\text{–}S}$ is the significance level of Kolmogorov–Smirnov test and $p^{K\text{–}S} > 0.05$ indicates normal distribution

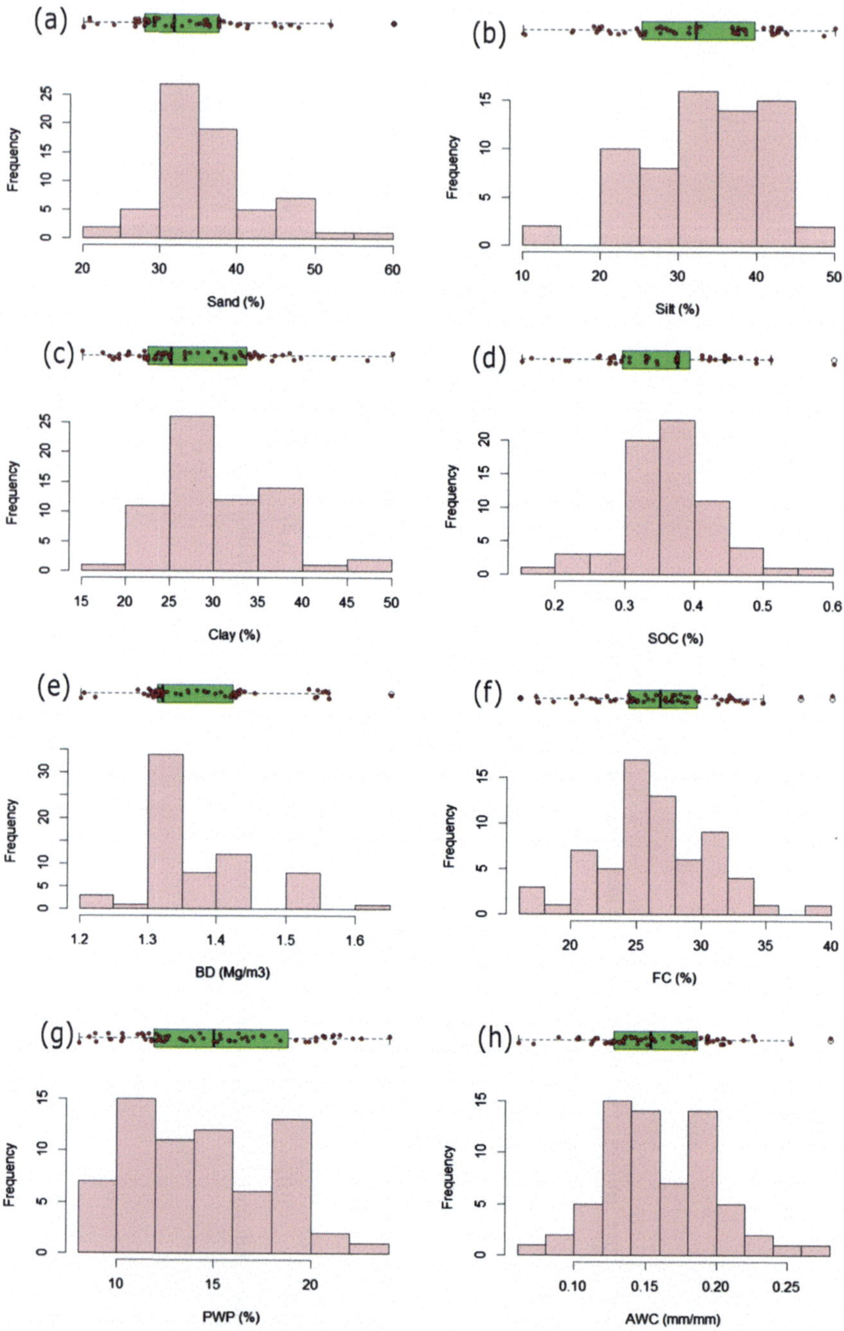

Fig. 10.3 Histogram of soil hydro-physical properties in the hot arid regions of India *Cross mark indicates the insignificant coefficients according to the specified p-value significance level*

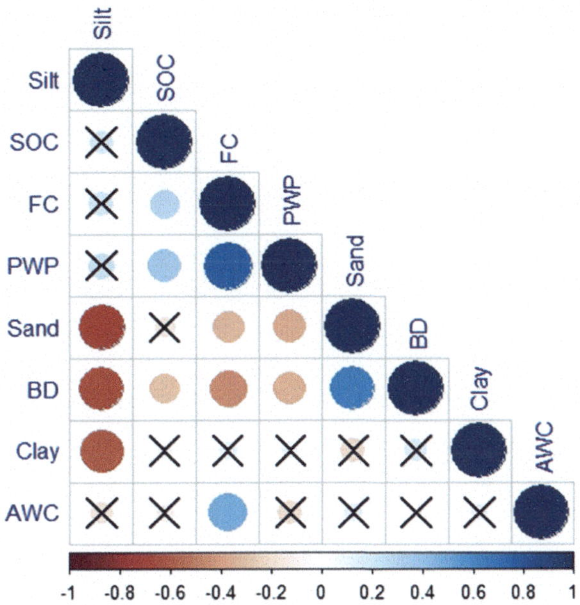

Fig. 10.4 Correlation matrix of soil hydro-physical properties in the hot arid regions of India

retention at some capillary pressures exhibited a strong dependence on soil texture, but similar significant correlation was found for water-saturated soil at PWP. Wilting point had negative correlation with soil BD, which was also confirmed by Jiang et al. (2011). The sand and silt contents positively and negatively correlated with FC, PWP and AWC, but particle size distribution could not significantly affect AWC. This might be due to the limited range of soil textural classes of the study area. These results are in agreement with the findings of Jiang et al. (2011) and Reyes et al. (2019).

10.3.3 Geostatistical Interpolation

The semivariogram parameters of the soil properties were determined by running R software and the results are listed in Table 10.2 and Fig. 10.5. The best models were identified by semivariogram analysis. Root-mean-squared residuals were found minimally corresponding to the spherical model for silt, clay and BD. The circular model was found best for the sand and PWP, respectively. Similarly, a Gaussian model was found best for the FC and AWC. The nugget/sill ratios of the soil properties were all less than 25%, suggesting that the variables were strongly spatially dependent (Reza et al. 2016). Reyes et al. (2019) found that soil water retention properties had strong relationships with soil texture. In a small research area, the possible reasons for spatial variability of water retention properties were the local heterogeneity of soil texture. The spatial correlation scale was varied from 1024 to

Table 10.2 Semivariogram parameters of soil hydro-physical properties in the hot arid regions of India

Variables	Model	Nugget	Partial sill	Sill	Nugget/sill	Nugget/Sill (%)	Range (m)	RMSE	Spatial dependence
Sand (%)	Circular	0.0091	0.0247	0.0338	0.2696	26.96	1269	0.183	Moderate
Silt (%)	Spherical	0.0047	0.0674	0.0721	0.0650	6.50	1041	0.269	Strong
Clay (%)	Spherical	0.0162	0.0282	0.0444	0.3652	36.52	1586	0.205	Moderate
SOC (%)	Circular	0.0292	0.0108	0.0400	0.7307	73.07	1721	0.199	Moderate
BD (Mg m^{-3})	Spherical	0.0014	0.0016	0.0030	0.4605	46.05	1183	0.061	Moderate
FC (%)	Gaussian	17.3757	2.1110	19.4867	0.8917	89.17	3533	4.351	Weak
PWP (%)	Circular	0.0048	0.0622	0.0670	0.0715	7.15	1024	0.251	Strong
AWC (mm mm^{-1})	Gaussian	0.0013	0.0004	0.0018	0.7557	75.57	3026	0.037	Weak

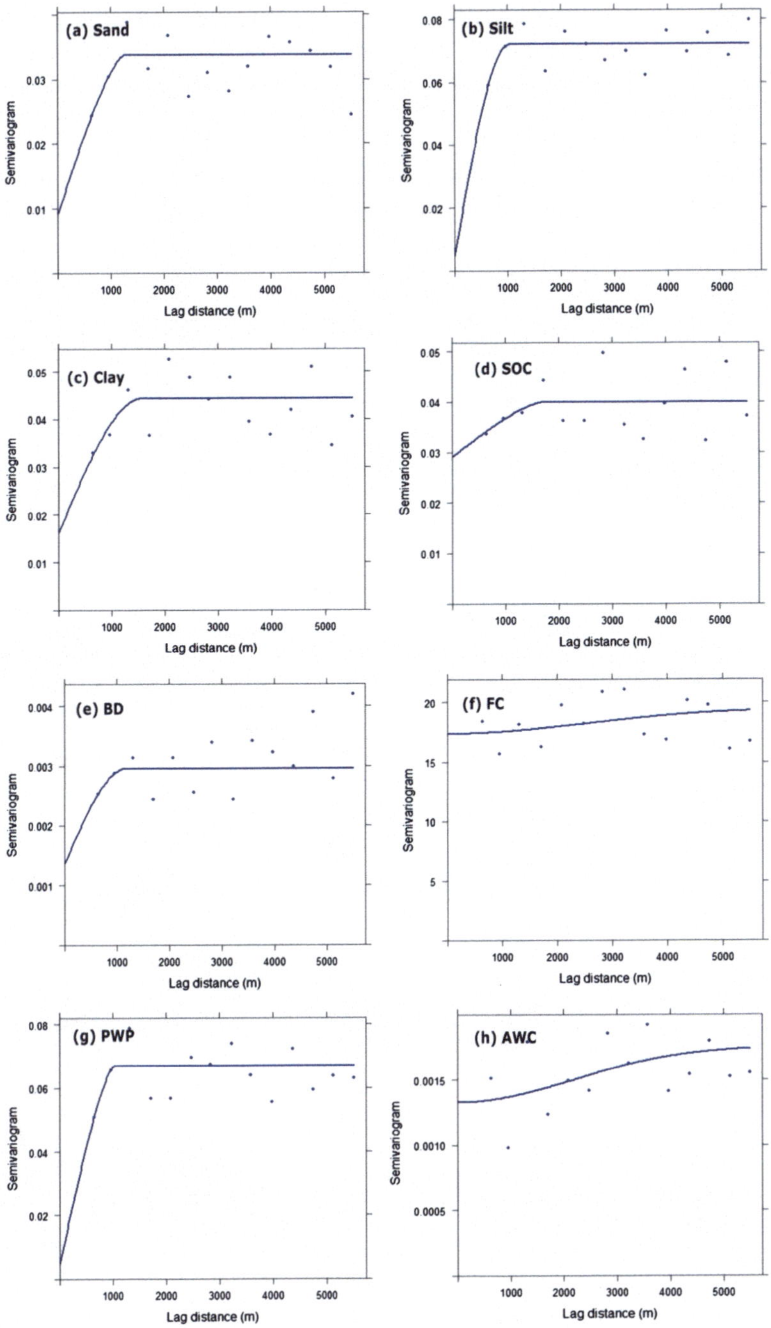

Fig. 10.5 Semivariogram model for spatial distribution of soil hydro-physical properties in the hot arid regions of India

3533 m for soil properties. Thus, the grid spacing (500 m) was adequate for characterizing the spatial variability of the soil properties. The sampling interval, soil type and inherent variability have a significant influence on the variance structure and the spatial correlation scale. FC, AWC and SOC had similar spatial distributions.

An ordinary kriging was used to estimate the values of the soil hydro-physical properties at non-sampling positions and the spatial distribution maps are shown in Fig. 10.6. Field capacity and bulk density had similar spatial distributions. The

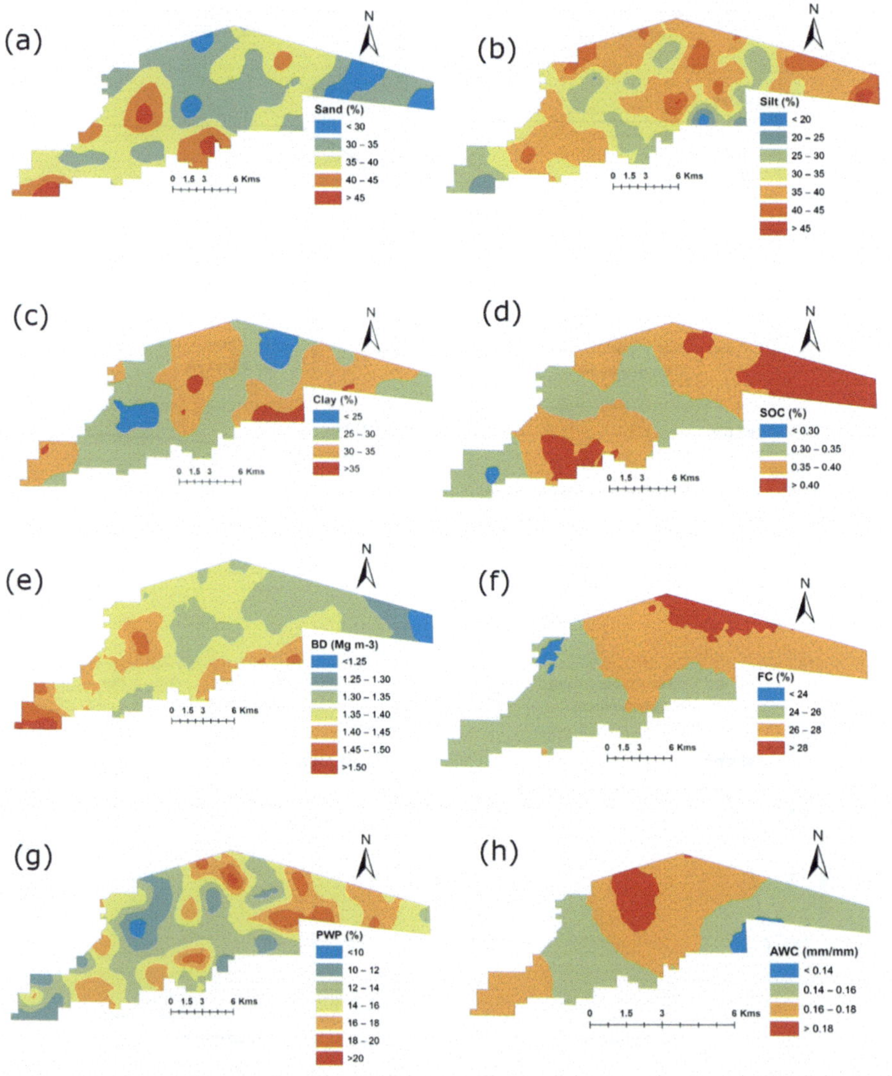

Fig. 10.6 Spatial distribution map of soil hydro-physical properties generated based on measured data and fitted variogram in the hot arid regions of India

spatial distribution of wilting point was patchy and relatively homogeneous. Spatial distribution maps of a data source can characterize its spatial heterogeneity more intuitively than statistic analysis.

10.3.4 Determining Clustering Variables for Irrigation Management Zones

The results of correlation analysis showed that strong correlation existed among soil hydro-physical properties. The correlated data should be removed to minimize the interaction among data and to reduce the effects of multi-correlation among data on clustering results before delineating management zones. To removing correlated data from the properties was analyzed by geographically weighted principal component (Table 10.3). The first four PCs have eigenvalues greater than unity, and for these four PCs, the cumulative percentage of total variance (PTV) exceeds 85%. The four PCs (PC1, PC2, PC3 and PC4) explained 28.1, 24.7, 19.1 and 13.4% of total variance, respectively. By calculating the load matrices of PC1, PC2, PC3 and PC4, they can be expressed as

$$PC1 = 0.259 \times \text{Sand} + 0.039 \times \text{Silt} - 0.334 \times \text{Clay}$$

Table 10.3 Global principal component analysis of soil properties and loading coefficient for the first four principal components

Principal component	Eigen values		Proportion of the total variance (PTV)			Cumulative PTV	
PC1	1.501		0.281			0.28	
PC2	1.408		0.247			0.53	
PC3	1.237		0.191			0.72	
PC4	1.036		0.134			0.85	
PC5	0.894		0.100			0.95	
PC6	0.613		0.047			1.00	
PC7	0.015		0.000			1.00	
PC8	0.009		0.000			1.00	
	PC loading for each variable						
	Sand	Silt	Clay	SOC	BD	FC	PWP	AWC
PC1	0.259	0.039	−0.334	−0.162	0.122	−0.616	−0.619	−0.128
PC2	0.469	−0.659	0.361	−0.155	0.287	0.083	−0.089	0.307
PC3	0.315	−0.202	−0.076	0.618	0.194	−0.151	0.303	−0.565
PC4	0.397	0.168	−0.654	0.212	0.104	0.336	0.065	0.462

SOC soil organic carbon; *BD* bulk density; *FC* Field capacity; *PWP* permanent wilting point; *AWC* Available water content

$$- 0.162 \times \text{SOC} + 0.122 \times \text{BD} - 0.616 \times \text{FC} - 0.619 \times \text{PWP} - 0.128 \times \text{AWC} \quad (10.4)$$

$$\text{PC2} = 0.469 \times \text{Sand} - 0.659 \times \text{Silt} + 0.361 \times \text{Clay}$$
$$- 0.155 \times \text{SOC} + 0.287 \times \text{BD} - 0.083 \times \text{FC} - 0.089 \times \text{WP} + 0.307 \times \text{AWC} \quad (10.5)$$

$$\text{PC3} = 0.315 \times \text{Sand} - 0.202 \times \text{Silt} - 0.076 \times \text{Clay}$$
$$+ 0.618 \times \text{SOC} + 0.194 \times \text{BD} - 0.151 \times \text{FC} + 0.303 \times \text{PWP} - 0.565 \times \text{AWC} \quad (10.6)$$

$$\text{PC3} = 0.397 \times \text{Sand} - 0.168 \times \text{Silt} - 0.654 \times \text{Clay}$$
$$+ 0.212 \times \text{SOC} + 0.104 \times \text{BD} + 0.336 \times \text{FC} + 0.065 \times \text{WP} + 0.462 \times \text{AWC} \quad (10.7)$$

In the first principal component PC1, the coefficients of FC and PWP were relatively large, indicating that PC1 was the comprehensive index affected jointly by field capacity and wilting point. For PC2, silt and sand produced a large contribution relative to the other variables, suggesting that soil texture played a decisive role on PC2. The PC3 was dominated by SOC and AWC. Similarly, clay had the most significant influence on PC4. Thus the four principal components were used as the comprehensive clustering variables for identifying IMZs. The scores map of PC1 was similar to the FC and PWP distribution maps. The scores map of PC2 was similar to the map of silt and sand. Distribution maps of PC3 and SOC and AWC were similar. The spatial concentrations of large negative scores are in the north-eastern part of the farm while large positive scores are in the eastern part (Fig. 10.7), although, there is no clear geographical trend. GWPCA provides additional information which is obscured by PCA, and the former comprises the major focus in this paper. To compare GWPCA and PCA, only the first four components (PC1–PC4) from each calibration are considered. PCs scores from GWPCA and the global PCA are mapped in Fig. 10.7. The majority of the soil account for between 85.9 and 93.8% of the variance in the data with an average of 89.9%, which explains 23% more variability than Global PCA.

10.3.5 Delineating Irrigation Management Zones

GWPCA score for the first four PCs were imported into "ppclust" package of R software for management zone analysis. PFCM cluster algorithm was performed to classify the four PCs into IMZs. PFCM algorithm repeated multiple times to get cluster validity indices for 2 to 8 IMZs. The values of cluster validity indices FPI and NCE are plotted against the number of classes in Fig. 10.8. The optimum number of clusters is determined when each index is at the minimum representing

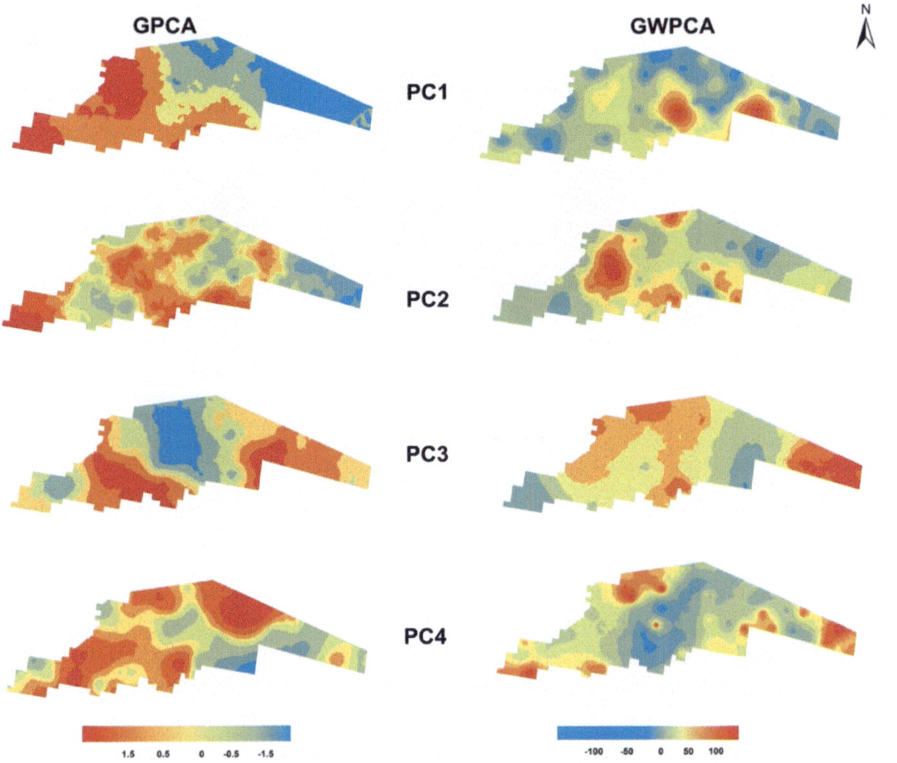

Fig. 10.7 PC scores maps for GPCA and GWPCA. GPCA, global principal component analysis; GWPCA, geographically weighted principal component analysis; PC, principal component

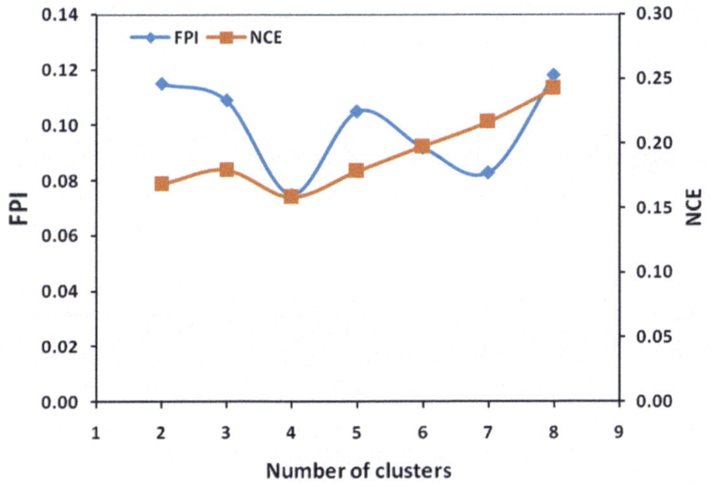

Fig. 10.8 Fuzzy performance index (FPI) and normalized classification entropy (NCE) calculated for identifying the optimum clusters for the study area

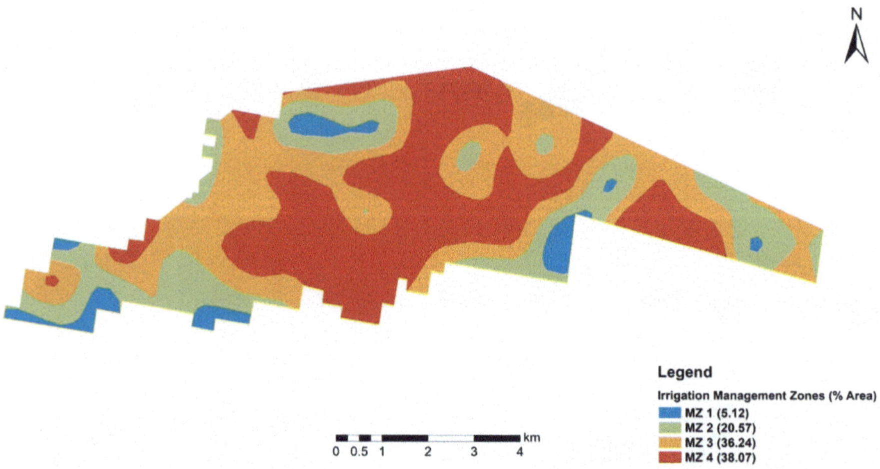

Fig. 10.9 Irrigation management zone (IMZ) map for four clusters in the hot arid regions of India

the least membership. FPI and NCE values were minimum for four clusters and the membership for all observations in different clusters were calculated. The IMZ map was developed by performing zonal statistics in ArcGIS 10.3.1 (Fig. 10.9). Tukey multiple comparisons test was performed to assess the effectiveness of different of IMZs. Analysis of variance indicated that significant statistical difference ($P <$ 0.01) existed among the four IMZs (Table 10.4). There was no significant difference for SOC, BD and PWP among the IMZs. IMZ 4 is covering the major portion of the study area (38.07%) followed by IMZ 3 (36.24%), IMZ2 (20.57%) and IMZ1 (5.12%). This zonation concept will be helpful in effective and efficient scientific management of irrigation. Hence, the information regarding IMZs could be used by farmers and other stakeholders for site specific irrigation management.

Table 10.4 Variance and Tukey's multiple comparisons test for soil hydro-physical properties of in irrigation management zones in the hot arid regions of India

IMZs	Sand (%)	Silt (%)	Clay (%)	SOC (%)	BD (Mg m^{-3})	FC (%)	PWP (%)	AWC (mm mm^{-1})
1	38.898^{a}	29.933^{b}	31.169^{a}	0.347^{a}	1.405^{a}	23.731^{b}	13.304^{a}	0.145^{b}
2	34.376^{b}	36.062^{a}	29.563^{b}	0.393^{a}	1.357^{a}	26.750^{ab}	14.539^{a}	0.166^{a}
3	35.408^{b}	36.443^{a}	28.149^{b}	0.363^{a}	1.366^{a}	26.381^{ab}	14.450^{a}	0.162^{a}
4	35.035^{b}	32.217^{ab}	32.748^{a}	0.387^{a}	1.362^{a}	28.155^{a}	15.742^{a}	0.169^{a}

Values in a column followed by different letters are significantly different at $P < 0.01$
SOC soil organic carbon; *BD* bulk density; *FC* Field capacity; *PWP* permanent wilting point; *AWC* Available water content

10.3.6 Application of IMZ Results

The management zone maps and hydro-physical properties in each management zone (Table 10.3) will help the planning of monitoring water balance and the adoption of a differentiated irrigation management. By the delineated IMZs, precision irrigation procedure can be designed for the implementation with the existing soil moisture monitoring technology and variable irrigation technology. First of all, soil moisture monitoring devices can be installed on the basis of the optimal sampling number and spatial dependence distance of AWC. Then, a decision on irrigation can be made based on information on the real-time soil moisture, crop type, and knowledge of crop water requirements at different growth stages. Results from case studies provided estimates of the potential for water conservation using precision irrigation using geostatistical and fuzzy clustering approach.

10.4 Conclusions

Research precision irrigation has focused on dividing a field into a few relatively uniform homogeneous zones as a practical, environmentally sustainable and cost-effective approach for managing water resources in hot air environment. The soil hydro-physical properties were used as the data source to identify irrigation management zones. In this study, the spatial variability of eight soil properties was quantified using geostatistical methods and was aggregated into IMZs using GWPCA and fuzzy clustering algorithms. According to geostatistical analysis, spherical, circular and gaussian models were best fit models for estimated soil parameters. Results from fuzzy clustering indicated that the optimal number of IMZs was four. Statistical analysis and one-way ANOVA analysis showed that delineation of IMZs was reasonable and could provide references for the zoning operation of site-specific irrigation in the study area. The IMZ-based precision irrigation can be useful in effective use of water resources in different stages of crop growth and improving overall productivity.

References

Anderberg MR (1973) Cluster analysis for applications. Academic Press Inc., New York, USA

Bezdek JC (1981) Pattern recognition with fuzzy objective function algorithms. Plenum, New York, USA

Brevik EC, Calzolari C, Miller BA, Pereira P, Kabala C, Baumgarten A, Jordan A (2016) Soil mapping, classification and pedologic modeling: History and future directions. Geoderma 264:256–274

Cebeci Z, Yildiz F, Kavlak AT, Cebeci C, Onder H (2018) Ppclust: probabilistic and possibilistic cluster analysis. R package version 0.1.1, https://CRAN.R-project.org/package=ppclust. Accessed 22 Aug 2019

Daccache A, Ciurana JS, Rodriguez Diaz JA, Knox JW (2014) Water and energy footprint of irrigated agriculture in the Mediterranean region. Environ Res Lett 9:124014 (12 pp)
Falloon P, Betts R (2010) Climate impacts on European agriculture and water management in the context of adaptation and mitigation—the importance of an integrated approach. Sci Total Environ 408:5667–5687
Fraisse CW, Sudduth KA, Kitchen NR (2001) Delineation of site-specificmanagement zones by unsupervised classification of topographic attributesand soil electrical conductivity. Trans ASAE 44(1):155–166
Fridgen JI, Kitchen NR, Sudduth KA, Drummond ST, Wiebold WJ, Fraisse CW (2004) Management zone analyst (MZA): software for subfield management zone delineation. Agron J 96:100–108
Gollini I, Lu B, Charlton M, Brunsdon C, Harris P (2015) GWmodel: an R package for exploring spatial heterogeneity using geographically weighted models. J Stat Softw 63:1–50
Goovaerts P (1998) Geostatistical tools for characterizing the spatial variability of microbiologycal and physico-chemical soil properties. Biol Fertil Soils 27:315–334
Gräler B, Pebesma E, Heuvelink G (2016) Spatio-temporal interpolation using gstat. R J 8(1):204–218
Haghverdi A, Leib BG, Washington-Allen RA, Ayers PD, Buschermohle MJ (2015) Perspectives on delineating management zones for variable rate irrigation. Comput Electron Agric 117:154–167
Harris P, Brunsdon C, Charlton M (2011) Geographically weighted principal components analysis. Int J Geogr Inf Sci 25:1717–1736
Harris P, Brunsdon C, Charlton M, Juggins S, Clarke A (2014) Multivariate spatial outlier detection using robust geographically weighted methods. Math Geosci 46:1–31
Jackson ML (1973) Soil chemical analysis. Prentice Hall of India Private Limited, New Delhi
Jiang Q, Fu Q, Wang Z (2011) Delineating site-specific irrigation management zones. Irrig Drain 60:464–472
Johnson CK, Wienhold BJ, Shanahan JF, Doran JW (2003) Site-specific management zones based on soil electrical conductivity in a semiarid cropping system. Agron J 95:303–315
Klute A (1986) Water retention: laboratory methods; In: Klute A (Ed) Methods of soil analysis, Part I—physical and mineralogical methods. Agronomy Series 9, 2nd edn, Am Soc of Agronomy, Madison, WI, USA, pp 635–662
Mc Bratney AB, Moore AW (1985) Application of fuzzy sets to climatic classification. Agric Forest Meteorol 35:165–185
Pachepsky YA, Rawls WJ, Gimenez D (2001) Comparison of soil water retention at field and laboratory scales. Soil Sci Soc Am J 65:460–462
Pal NR, Pal K, Keller JM, Bezdek JC (2005) A possibilistic fuzzy c-means clustering algorithm. IEEE Trans Fuzzy Syst 13:517–530
Pebesma EJ (2004) Multivariable geostatistics in S: the gstat package. Comput Geosci 30:683–691
R Core Team (2019) R: a language and environment for statistical computing. R foundation for statistical computing, Vienna, Austria. https://www.R-project.org/
Reyes J, Wendroth O, Matocha C, Zhu J (2019) Delineating site-specific management zones and evaluating soil water temporal dynamics in a farmer's field in Kentucky. Vadose Zone J. 18:180143. https://doi.org/10.2136/vzj2018.07.0143
Reza SK, Nayak DC, Chattopadhyay T, Mukhopadhyay S, Singh SK, Srinivasan R (2016) Spatial distribution of soil physical properties of alluvial soils: a geostatistical approach. Arch Agron Soil Sci 62:972–981
Saito H, McKenna A, Zimmerman DA, Coburn TC (2005) Geostatistical interpolation of object counts collected from multiple strip transects: ordinary kriging versus finite domain kriging. Stoch Env Res Risk A 19:71–85
Schepers AR, Shanaham JF, Liebig MA, Schepers JS, Johnson SH, Luchiari JA (2004) Appropriateness of management zones for characterizing spatial variability of soil properties and irrigated corn yields across years. Agron J 96:195–203
Soil Survey Staff (1999) Soil taxonomy. A basic system of soil classification for making and interpreting soil surveys. 2nd edn. USDA, U.S. Gov. Print Office, Washington, DC

Veihmeyer FY, Hendrickson AH (1948) Soil density and root penetration. Soil Sci 65:487–493
Verma RR, Manjunath BL, Singh NP, Kumar A, Asolkar T, Chavan V, Srivastava TK, Singh P (2018) Soil mapping and delineation of management zones in the Western Ghats of coastal India. Land Degrad Dev. https://doi.org/10.1002/ldr.3183
Vitharana UWA, Meirvenne MV, Simpson D, Cockx L, Josse De Baerdemaeker JD (2008) Key soil and topographic properties to delineate potential management classes for precision agriculture in the European loess area. Geoderma 143:206–215
Walkley AJ, Black IA (1934) An examination of the Degtjareff method for determining soil organic matter and a proposed modification of the chromic acid titration method. Soil Sci 37:29–38
Wilding LP, Dress LR (1983) Spatial variability and pedology. In: Wilding LP, Smeckand NE, Hall GF (eds) Pedogenesis and soil taxonomy. I. Concepts and interactions. Elsevier Science Pub, New York (NY), pp 83–116

Part II
Geospatial Modeling and Risk Assessment

Chapter 11
Soil Quality Assessment: Integrated Study on Standard Scoring Functions and Geospatial Approach

Ali Keshavarzi, Manuel Pulido Fernández, Mojtaba Zeraatpisheh, Henry Oppong Tuffour, Gouri Sankar Bhunia, Pravat Kumar Shit, and Jesús Rodrigo-Comino

Abstract Assessment of soil quality indices is highly influential in sustainable agriculture. A myriad of methods is currently used to select the most relevant soil quality indicators. However, information about the most accurate and precise methods for agricultural areas at the catchment scales is lacking. Therefore, the main aim of the present study was to assess ten soil quality indicators from a factor analysis (FA) to obtain the most suitable soil quality indicators in combination with an indicator selection method (standard scoring functions). The study was conducted in an irrigated agriculture area in the Mashhad Plain in Northeast Iran. Results of FA by maximum likelihood method showed that four factors were the most significant in explaining

A. Keshavarzi (✉)
Laboratory of Remote Sensing and GIS, Department of Soil Science, University of Tehran, P.O. Box: 4111, 31587-77871 Karaj, Iran
e-mail: alikeshavarzi@ut.ac.ir

M. P. Fernández
GeoEnvironmental Research Group, University of Extremadura, 10071 Cáceres, Spain

M. Zeraatpisheh
Henan Key Laboratory of Earth System Observation and Modeling, Henan University, Kaifeng 475004, China

College of Geography and Environmental Science, Henan University, Kaifeng 475004, China

H. O. Tuffour
Department of Crop and Soil Sciences, Kwame Nkrumah University of Science and Technology, Kumasi, Ghana

G. S. Bhunia
Paschim Medinipur, West Bengal 721424, India

P. K. Shit
Department of Geography, Raja NL Khan Women's College, Gope Palace, Medinipur, West Bengal 721102, India

J. Rodrigo-Comino
Departamento de Análisis Geográfico Regional y Geografía Física, Facultad de Filosofía y Letras, University of Granada, Campus Universitario de Cartuja, Granada, Spain

P. K. Shit et al. (eds.), *Soil Health and Environmental Sustainability*, Environmental Science and Engineering,
https://doi.org/10.1007/978-3-031-09270-1_11

the system variance and collectively accounted for 78.9% of the total. The magnitude of the loadings, which explains a great part of the variance in each factor, was used for naming the factors. On the surface, nitrogen (0.12), electrical conductivity (0.11), exchangeable sodium percentage (0.11), and sodium adsorption rate (0.11) had the highest scores. In the subsoil, however, the scoring was sodium adsorption ratio (0.12), exchangeable sodium percentage (0.12), calcium carbonate equivalent (0.12), pH (0.11), and electrical conductivity (0.11). The lowest scores were obtained for soil nutrients Olsen-P and exchangeable K. Overall, higher soil quality was observed in the subsoil relative to the surface, which is a strong confirmation of the rapid land degradation processes developed in the area.

Keywords Factor analysis · Soil properties · Soil quality indicator · Spatial variability · Standard scoring function

11.1 Introduction

The soil is a three-dimensional resource, living and has complex configurations that undergo constant and dynamic changes (Grant 2017; Smith 2018). Soils are also involved in several different vital processes for wildlife and human health, and a good understanding of the factors that explain their quality is vital for sustainable land management (Alexander et al. 2015; Soil Survey Staff 2014).

During the last decade, interest in soil quality assessment has grown among researchers, and in recent times, it is considered a key theme in soil and agronomic sciences (Calleja-Cervantes et al. 2015; Zeraatpisheh et al. 2022). Measurable soil properties that affect soil capacity for the ability to produce specific agricultural products are called soil quality characteristics (Qi et al. 2009; Zeraatpisheh et al. 2022). Fundamentally, soil quality indicators comprise physical, chemical, and biological attributes. These allow for the measurement of the capacity of the soil to perform its important functions in respect of sustainable soil management (Mukherjee and Lal 2014; Zeraatpisheh et al. 2020). Soil quality is described in two aspects, firstly, the inherent quality (Bastida et al. 2008), and one related to the dynamics or variations, which indicate the state of soil health (Kraaijvanger and Veldkamp 2015; Stockdale et al. 2013).

Understanding soil quality and the key properties that can function as main quality indices are very economical in both time and monetary terms and also enhances the effectiveness of the land use management plans (Buchholz et al. 2017; McGrath and Zhang 2003). Additionally, this knowledge allows for identifying key problematic areas or regions with adverse trends (Bindraban et al. 2000; de Paul Obade and Lal 2016a; Zeraatpisheh et al. 2022) and provides valuable basic information for future investigations (Mukherjee and Lal 2014). For several semiarid areas in rapidly developing regions, soil quality indicators are not readily accessible to farmers and local technicians (Sulieman et al. 2018; Maleki et al. 2021). As a result, soil quality status is usually inferred from some basic soil parameters (Mairura et al. 2007;

Bakhshandeh et al. 2019; Zeraatpisheh et al. 2020), which is a high limitation since the use of a single soil parameter is not a reliable and ideal indicator of soil quality (Hosseini et al. 2017).

Studies on soil quality are essential for evaluating agricultural land status (Bogunovic et al. 2017) since soil quality reflects the environmental stewardship of land use management strategies (Pulido et al. 2017; Zeraatpisheh et al. 2022). The selection of a suitable number of soil quality indices from different land uses is the first and most important step in evaluating soil quality (Mishra et al. 2017, 2018; Zeraatpisheh et al. 2020). Thus, information on soil quality status is essential for policymakers and stakeholders in agriculture, especially in areas with severe losses in soil fertility (Hosseini et al. 2017).

In this regard, soil quality characteristics have to be introduced as evaluation criteria. Soil quality characteristics such as bulk density (BD), soil texture, aggregate stability, and organic carbon content (OC) are the most commonly used due to their easy interrelationships and low cost of determination (Lado et al. 2004; Shukla et al. 2006). However, sampling, analyses, and finally, preparation of a detailed distribution map on large scales could be highly costly and may impose a big limitation on soil quality studies (Khaledian et al. 2016).

Considering the spatial distribution patterns of soil quality indicators is required to enhance understanding of soil processes, optimal use of resources, energy, and other inputs for sustainable agriculture and development to meet global demands (Karlenet al. 1998). As a result, geostatistical methods have, in recent years, gained widespread usage for interpolations or accurate estimations of soil properties in non-studied areas based on some sampling points (Gray et al. 2016; Tuffour et al. 2016; Keshavarzi et al. 2018). As it is well-known, soil properties can be similar in the adjoining points; however, the spatial variability, especially in agricultural fields, can increase drastically, being the intra-plot variability extremely difficult to extract as common patterns (Taguas et al. 2015). In contrast to classical statistical methods, geostatistics computes the location of the variable and allows for the calculation of estimation errors (Omran 2012; Kamali et al. 2013; Zeraatpisheh et al. 2022). Therefore, spatial statistics combined with highly accurate geoinformation in geographic information systems (GISs) could be ideal for providing soil quality indicator maps (de Paul Obade and Lal 2016b).

The intricacy of mixing multiple variables into an overall soil quality index (SQI) could not identify by specific soil quality parameters. An essential concern in soil quality evaluation is the calculation of SQI, which is usually based on a combined estimation of parameters and their loads with an indirect approach. Several quantitative models are available for the spatial assessment of soil quality; among them are the integrated quality index (IQI) (Doran and Parkin 1994), geostatistical methods (Sun et al. 2003), factor analysis (FA) using total porosity, stable aggregates in water, BD, soil particle size distribution, electrical conductivity (EC), soil moisture content, saturated hydraulic conductivity (K_S), OC, total N, and pH values (Kang et al. 2005; Shukla et al. 2006; Qi et al. 2009). However, the use of minimum parameters in calculating SQIs is another point to be discussed (Mukherjee and Lal 2014, 2015; Zhang et al. 2016).

For example, Swanepoel et al. (2014) confirmed the value of the SQI concerning the agricultural systems' sustainability and verified low SQI values of improperly managed soil. Also, Lima et al. (2013), Nesbitt and Adl (2014), and Zeraatpisheh et al. (2020) have proposed the evaluation of soil physical, chemical, and biological attributes as soil quality indicators. On the other hand, based on Hazarika et al. (2014), soil quality could also be assessed throughout the soil deterioration index as well, which the deviations of soil physicochemical attributes of an area under anthropic action that are in comparison to the baseline of a contiguous natural area or an area with the same situation of soil and climate.

Some researchers have pointed out that plowing methods and the types of plants cultivated in a crop rotation also affect soil quality characteristics and soil quality indicators (Kang et al. 2005; Mukherjee and Lal 2015). Currently, the study of SQI using a standard scoring function in fast-developing countries such as Iran is limited. Therefore, the main aims of this study were to (i) present a SQI using a standard scoring function; and (ii) investigate the spatial distribution of this one for two different soil depths. To demonstrate the validity of the SQI, an area in the Mashhad Plain, Khorasan-e-Razavi Province, Northeast Iran, which is adegraded area with no information on soil properties and soil quality, was selected for the study.

11.2 Materials and Methods

11.2.1 Site Description

The study was conducted in a catchment in the Mashhad plain of Khorasan-e-Razavi Province (35° 59′ and 37° 04′ N, and 58° 22′ and 60° 07′ E), Northeast Iran, (Fig. 11.1). The geological formation, mainly the alluvial plain, developed into a thick sediment-dominated environment belonging to the Quaternary period. Loam, sandy clay loam, and sandy loam are the predominant soil textures, described as *Calcaric Cambisols*, *Calcaric Fluvisols*, *Gypsic Regosols*, and *Calcaric Regosols* (IUSS Working Group WRB 2014). The horizons contain low soil organic carbon contents and high calcium carbonate concentrations, covering pediment plains, gravelly colluvial fans, plateau, and upper terraces, respectively (Fig. 11.1). The plain is extended along an NW–SE direction surrounded by Kopetdagh and Binaloud mountains. The mean altitude is 1200 m a.s.l., and the slope varies from 0 to 8.2%. The maximum and minimum rainfall events occur during March (44.8 mm) and September (1.2 mm), respectively. Mean annual precipitation and temperature are 222.1 mm $yr^{-1}$and 15.8 °C, respectively (Keshavarzi et al. 2016). The irrigated and dryland farming around the Kashfrod River is the main land uses system in the area.

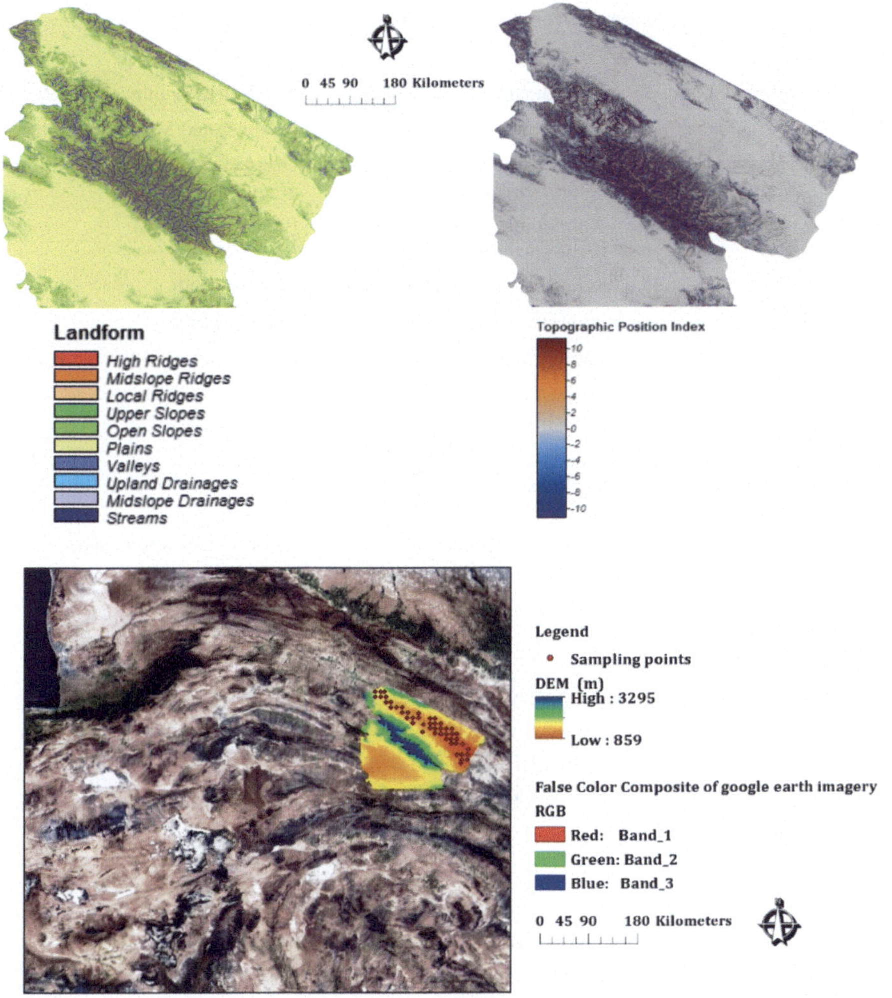

Fig. 11.1 Location of the study area with 48 soil profiles

11.2.2 Field and Laboratory Procedures

Forty-eight representative soil profiles were described by a stratified random sampling technique(IUSS Working Group WRB 2007), and soil samples were collected from two different depths, topsoil (from 0 to 30 cm) and subsoil (from 30 to 60 cm). Plant residues and gravels were removed, and the soil samples were transported to the laboratory, air-dried, sieved through 2 mm mesh, and stored in polyethylene bags under ambient temperature. The analytical protocols used are presented in Table 11.1.

Table 11.1 Protocol measurements for indicators selected in the study

Indicator	Protocol	References
Organic Carbon (OC)	Dichromate wet combustion	Nelson and Sommers (1982)
Total N (N)	Kjeldahl	Bremner and Mulvaney (1982)
Available phosphorus (P)	Sodium bicarbonate extraction, colorimetric detection	Olsen et al. (1954)
Available potassium (K)	Ammonium acetate extraction, flame spectrometry detection	Bond et al. (2006)
pH_e	Saturated paste extract	Thomas (1996)
Electrical conductivity (EC_e)	Saturated paste extract	Thomas (1996)
CEC (Cation Exchange Capacity)	Sodium saturation	Bower et al. (1952)
CCE (Calcium Carbonate Equivalent)	Back titration	Nelson (1982)
ESP (Exchangeable Sodium Percentage)	Exchangeable sodium	Sumner and Miller (1996)
SAR (Sodium Adsorption Ratio)	Sodium, calcium, and magnesium concentrations	Oster and Sposito (1980)
Cations and anions	Soluble ions	Sparks et al. (1996)

Statistical analysis was performed using descriptive statistics, including minimum, maximum, arithmetic mean, standard deviation (SD), and coefficient of variation (CV). Characterization of CV (Wilding and Dress 1983) was employed, where CV values from 0 to 15% were classed as low, 16–35% as moderate, and greater than 35% were high.

11.2.3 Integrated Quality Index (IQI) and Weight Assignment

In this study, the Qi et al. (2009) equation was used to compute the IQI:

$$IQI = \sum_{i=1}^{n} W_i \cdot N_i \tag{11.1}$$

where W_i, N_i and n represent the selected weight, the indicator score, and the number of indicators, respectively. The weight of each soil quality characteristic was determined by two multivariate techniques, namely, factor analysis (FA) and principal component analysis (PCA). These were used to reduce the number of the most appropriate soil quality indicators for the selected territory from the list of all indicators.

Further, due to differences in indicator units, the standard scoring function (SSF) (Andrews et al. 2002; Qi et al. 2009; Mukherjee and Lal 2014) was used to score soil indicators. Depending on the function of the indicator on soil quality, four types of indicators were selected, namely, descriptive function, the upper limit, lower limit, and peak limit. Since there was no prior information about the upper and lower thresholds in Iran, the minimum and maximum observational values of the region variables were considered lower and upper thresholds, respectively. Table 11.2 presents the SSF equations in both depths for the indicators. Regarding the fact that the range of soil reaction improves soil quality, the values of the range have a score equal to 1, and increasing the distance from this range (more or less), the score decreases. For example, the optimal pH value of 7 (Marzaioli et al. 2010); however, since the pH of the soil in all measured points was higher than 7, and they were classified as "less is better". The weight of each indicator was assigned by communality using mathematical statistics of standardized FA (Sun et al. 2003; Shukla et al. 2006).

11.2.4 Spatial Variability of an Integrated Quality Index (IQI)

In this study, trend analysis and anisotropy were performed before using the geospatial interpolation method in ArcGIS 10.2.1 (ESRI, USA). For precise modeling of experimental variograms, which are the inputs of different kriging interpolation methods, the direction of the spatial continuity of the data should be specified (Fu et al. 2010). To draw an anisotropy ellipse, the variogram was drawn in different directions, and the level of the range was obtained in all directions. Anisotropy ratio, which is equal to the ratio of the largest range (large ellipse diameter) to the smallest (small ellipse diameter), was used as a criterion. In this study, the value of this ratio was greater than 1, which indicated partial anisotropy.

Following Lark (2000) and Robinson and Metternicht (2006), we considered the value of the experimental variogram for a separation distance of h (referred to as the lag), which is half the average squared difference between the value at $z(x_i)$ and the value at $z(x_i + h)$:

$$\gamma(h) = \frac{1}{2N(h)} \sum_{i=1}^{N(h)} [Z(X_i + h) - Z(X_i)]^2 \tag{11.2}$$

where the number of paired data within a specific class of distance and direction is $N(h)$. If the values at $z(x_i)$ and $z(x_i + h)$ register an autocorrelation, the result of Eq. (11.2) will reach lower values relative to uncorrelated paired points.

In assessing the experimental variogram, a suitable model was fitted with the weighted least squares, and the parameters (e.g., range, nugget, and sill) were used to generate the distribution maps using ordinary kriging (OK). By this method,

Table 11.2 Standard scoring functions (SSFs) and parameters for soils (0–30 and 30–60 cm, $n = 48$)

Scoring function	Upper limit (UL)		Lower limit (LL)		Function	Unit	Indicator
	0–30	30–60	0–30	30–60			
$f(x) = \begin{cases} 1 & x < L \\ 1 - 0.9\frac{x-L}{U-L} & L \leq X \leq U \\ 0.1 & x > U \end{cases}$	8.30	8.40	7.70	7.90	L	–	pH_e
	9.00	16.00	0.01	0.01	L	dSm^{-1}	EC_e
	21.83	23.08	0.29	0.29	L	%	ESP
	18.70	20.10	0.20	0.20	L	$(l/mmol)^{0.5}$	SAR
	39.70	39.70	4.90	5.10	L	%	CCE
$f(x) = \begin{cases} 1 & x < L \\ 0.9\frac{x-L}{U-L} & L \leq X \leq U \\ 0.1 & x > U \end{cases}$	1.61	1.58	0.13	0.11	M	%	OC
	0.20	0.20	0.01	0.00	M	%	N
	30.40	32.80	1.60	1.20	M	$mg\ kg^{-1}$	P
	525.49	525.49	92.63	53.98	M	$mg\ kg^{-1}$	K
	37.11	39.08	16.05	11.93	M	$C\ mol\ (+)\ kg^{-1}$	CEC

the type of model, sill, range, nugget, lag, and minimum and maximum neighborhoods were optimized to obtain the least error and the highest correlation. Since the normal distribution of data is a fundamental condition for the use of many geostatistical estimators, the data of the integrated SQI were subjected to a non-parametric Kolmogorov–Smirnov test. Finally, the maps of distribution and spatial distribution of the integrated SQI in the study area were generated.

11.3 Results and Discussion

11.3.1 Indicators Among Different Depths

The values of the soil indicators comprising maximum, minimum, coefficient of variation (CV), standard deviation, and arithmetic mean are presented in Table 11.3.

Table 11.3 Summary statistics of soil indicators

Indicator[a]	Unit	Depth (cm)	Min	Max	Mean	Std.	Skewness	CV (%)
pH_e	–	0–30	7.70	8.30	8.08	0.128	−0.963	1.58
		30–60	7.90	8.40	8.11	0.119	0.372	1.46
EC_e	dSm^{-1}	0–30	0.01	9.00	1.94	1.841	1.714	94.89
		30–60	0.01	16.00	2.31	1.871	2.897	80.95
ESP	%	0–30	0.3	21.8	4.7	4.5	1.645	95.74
		30–60	0.3	23.1	5.5	5.3	1.2	96.36
SAR	mmol $lit^{0.5}$	0–30	0.20	18.70	3.75	3.32	2.000	88.53
		30–60	0.20	20.10	4.33	4.12	1.560	95.15
CEC	cmol (+) kg^{-1}	0–30	16.05	37.11	30.46	4.561	−1.106	14.97
		30–60	11.93	39.08	29.45	4.920	−1.427	16.70
CCE	%	0–30	4.90	39.70	21.4	8.4	0.080	39.25
		30–60	5.1	39.7	22	9.1	−0.037	41.36
OC	%	0–30	0.13	1.61	0.6	0.3	1.772	50
		30–60	0.1	1.6	0.4	0.2	2.413	50
N	%	0–30	0.01	0.20	0.1	0.03	0.612	30
		30–60	0.00	0.2	0.1	0.04	1.118	40
P	mg kg^{-1}	0–30	1.60	30.40	10.15	7.471	1.360	73.59
		30–60	1.20	32.80	5.62	5.163	3.676	91.81
K	mg kg^{-1}	0–30	92.63	525.49	239.59	91.317	0.684	38.11
		30–60	53.98	525.49	194.30	89.644	0.909	46.13

[a]EC: Electrical Conductivity; ESP: Exchangeable Sodium Percentage; SAR: Sodium Adsorption Ratio; CEC: Cation Exchange Capacity; CCE: Calcium Carbonate Equivalent; OC: Organic Carbon; N: Total Nitrogen; P: Available Phosphorus; K: Available Potassium

Among the indicators, exchangeable sodium percentage (ESP) and soil pH showed the highest and the lowest variability as evidenced by their CVs, respectively. Since the study area is often under cultivation, soil salinity, organic carbon, total nitrogen, phosphorus, and potassium were high. This could be due to fertilizer application and other soil management practices, such as tillage, irrigation, etc., in other land uses (Andrews et al. 2002; Costa et al. 2015; Dai et al. 2015). With similar results, in Iran, Fard and Harchagani (2009) reported high CVs for these soil characteristics due to fertilizer application and soil amendment. Also, Kavian et al. (2018) and Maleki et al. (2021) confirmed a high variability of total nitrogen in an Iranian irrigated area with different land uses in the last 30 years. One of the main reasons for the high coefficient of variation in this area could also be attributed to the landform changes, such as the plateau physiography and slope dynamics. The topographic effects on the distribution of soil particles, organic matter, and nutrients are due to erosion and sedimentation, which result in the physical and chemical characteristics of the soil varying between the up and down slopes (Wilson and Gallant 2000; Manandhar and Odeh 2014) and connectivity processes (Turnbull et al. 2018).

11.3.2 Minimum Data Set Selection

The results of PCA of the soil quality indicators in both top-and sub-soils are summarized in Table 11.4. With these results, the facts-based model was applied to generate

Table 11.4 Results of principal component analysis (PCA) of soil quality indicators in soils (0–30 and 30–60 cm, $n = 48$)

PCs[a]	PC1		PC2		PC3		PC4	
	0–30	30–60	0–30	30–60	0–30	30–60	0–30	30–60
Eigenvalue	3.191	3.37	2.568	2.00	1.091	1.23	1.042	1.06
Cumulative percent	31.9	33.72	57.6	53.70	68.5	65.99	78.9	76.60
Eigenvectors								
pH_e	**0.54**[b]	0.03	0.25	−0.52	0.3	−0.28	−0.53	**0.31**[b]
EC_e	−0.03	0.12	−0.05	−0.18	−0.36	0.15	**0.89**	**−0.48**
ESP	0.28	0.10	0.09	−0.27	−0.29	0.07	**0.89**	**−0.50**
SAR	0.27	0.10	0.10	−0.29	−0.33	0.08	**0.88**	**−0.50**
CEC	**−0.48**	0.07	−0.43	**0.51**	0.45	−0.14	0.35	−0.28
CCE	0.28	**0.84**	**−0.84**	0.24	−0.17	0.12	−0.14	0.11
OC	0.25	0.17	0.03	−0.28	**0.78**	**−0.58**	0.29	−0.05
N	0.12	0.29	−0.08	0.05	**0.85**	**−0.57**	0.34	−0.05
P	0.23	−0.10	−0.14	−0.31	**0.61**	**0.32**	0.29	−0.14
K	−0.40	**−0.36**	0.31	0.22	**0.46**	−0.32	0.34	−0.24

Note [a]PC, principal component. [b]Bolded factor loadings correspond to the highly weighted the SQI and establish the minimal data set within the model to avoid data redundancy. In the topsoil, the PCA explained 78.9% of the total variability using 4 factors. Thus, the bold values in Table 11.4 (i.e., pH and CEC for PC-1, CCE for PC-2, OC, N, P and K for PC-3, and EC, ESP, and SAR for PC-4) were considered highly weighted eigenvectors and, therefore, were selected first. For the subsoil, 76.6% was obtained; the bold-face values as presented in Table 11.4 (i.e., CCE and K for PC-1, CEC for PC-2, OC, N and P for PC-3, pH, EC, ESP, and SAR for PC-4) were considered highly weighted eigenvectors, and hence, were used.

The results of the loadings were used for naming the factors. All the unique observations (untransformed data) of each soil had been included in the PCA model. The PCs with excessive eigen values represented the most variant in the dataset. For a given PC, each variable had a unique corresponding eigen vector weight value or factor loading (Table 11.4).

For situations where two or more variables were retained under a specific component, a multivariate correlation matrix was conducted to determine the correlation coefficients among the parameters (Yang et al. 2016), as presented in Table 11.5. When the parameters showed a correlation coefficient (r) greater than 0.6, the value with the highest loading factor was used in the model, and all others were removed to avoid possible redundancy. The non-correlated parameters under a particular PC were also considered relevant and saved in the model to give new insights into the final interpretation. Table 11.5 presents the Pearson linear correlation coefficient. The highest correlations were found between SAR with ESP (0.99). The synergistic relationships between pH and EC, OC and nitrogen, and SAR and Exchangeable Sodium Percentage are key to understanding the growth of plants (Recena et al. 2017; Roberts et al. 2017), which can also be influenced by the calcareous parent material (Dai et al. 2015). Several authors (e.g., Brimhall et al. 1991; Yavitt 2000;

Table 11.5 Pearson linear correlation coefficient. Significant differences are indicated as $p < 0.05^{*}$ and $p < 0.01^{**}$. n.s, not significant

	pH_e	EC_e	ESP	SAR	CEC	CCE	OC	N	P	K
pH_e	1									
EC_e	-0.50^{**}	1								
ESP	-0.33^{*}	0.83^{**}	1							
SAR	-0.32^{*}	0.85^{**}	0.99^{**}	1						
CEC	-0.33^{**}	0.24^{*}	$0.18^{n.s}$	$0.16^{n.s}$	1					
CCE	0.01^{ns}	$-0.07^{n.s}$	$-0.10^{'}$	$-0.10^{n.s}$	$0.01^{n.s}$	1				
OC	0.22^{*}	$-0.02^{n.s}$	$0.05^{n.s}$	$0.04^{n.s}$	0.24^{*}	$-0.12^{n.s}$	1			
N	0.06^{ns}	$-0.04^{n.s}$	$0.01^{n.s}$	$-0.01^{n.s}$	0.31^{**}	$-0.03^{n.s}$	0.69^{**}	1		
P	-0.06^{ns}	$0.02^{n.s}$	$0.06^{n.s}$	$0.05^{n.s}$	0.25^{*}	$-0.08^{n.s}$	0.37^{**}	0.48^{**}	1	
K	-0.13^{ns}	$0.14^{n.s}$	$0.15^{n.s}$	$0.15^{n.s}$	0.32^{**}	-0.25^{*}	0.30^{**}	0.36^{**}	0.27^{**}	1

EC: Electrical conductivity; ESP: Exchangeable Sodium Percentage; SAR: Sodium adsorption ratio; CEC: Cation-exchange capacity; CCE: Calcium Carbonate Equivalent; OC: Organic carbon; N: Total Nitrogen; P: available Phosphorus; K: available Potassium

Table 11.6 Estimated communality and the weight value of each soil quality indicator

Indicator	Communality		Weight	
	Topsoil	Subsoil	Topsoil	Subsoil
pH_e	0.73	0.80	0.10	0.11
EC_e	0.79	0.78	0.11	0.11
ESP	0.80	0.84	0.11	0.12
SAR	0.78	0.84	0.11	0.12
CEC	0.74	0.63	0.10	0.09
CCE	0.75	0.89	0.10	0.12
OC	0.70	0.68	0.10	0.10
N	0.84	0.65	0.12	0.09
P	0.46	0.39	0.06	0.05
K	0.58	0.59	0.08	0.08

EC: Electrical conductivity; ESP: Exchangeable Sodium Percentage; SAR: Sodium adsorption ratio; CEC: Cation-exchange capacity; CCE: Calcium Carbonate Equivalent; OC: Organic carbon; N: Total Nitrogen; P: Available Phosphorus; K: Available Potassium

Mohammadkhan et al. 2011) have also reported on the vital influence of the parent material on nutrient transfers and bio-geocycles, which significantly influence on soil quality.

11.3.3 Weight Assignment Values of Every Soil Quality Indicator

Within the topsoil, nitrogen (0.12), electrical conductivity (0.11), exchangeable sodium percentage (0.11), and sodium adsorption rate (0.11) had the highest scores (Table 11.6). In the subsoil, sodium adsorption ratio (0.12), exchangeable sodium percentage (0.12), calcium carbonate equivalent (0.12), pH (0.11) and electrical conductivity (0.11) had the highest scores. On the other hand, the lowest scores were obtained for Olsen-P and exchangeable K. Therefore, it is demonstrated that soil properties, which used to be considered limiting factors for the plants, and subsequently, for agricultural activities such as nitrogen (Al-Kaisi et al. 2005), sodium adsorption (Gorji et al. 2017), carbonates (Dai et al. 2015) and water retention (Pérez-de-los-Reyes et al. 2015) are the most important soil quality indicators. Thus, the clear relationship between soil and plant will determine the final aptitude of the pedological quality (Fischer et al. 2014; Gabarrón-Galeote et al. 2013).

11.3.4 Integrated Quality Index (IQI) Calculation

Soil quality indices were calculated after the indicators were weighted using the IQI (Eq. 11.1). Table 11.7 and Fig. 11.2 present the summary statistics of the IQI and Box-and-Whisker plot of IQIs for both depths. The potential benefits of soil quality assessment herein would be its flexibility in choosing selected soil characteristics and measured indicators, which ensured that assessments were suitable for specific management objectives (Mishra et al. 2017, 2018). The results of regression analysis for SQIs were shown in Fig. 11.3.

Table 11.7 Summary statistics of the integrated quality index (IQI)

Statistics	Mean	Max	Min	CV (%)	Skewness	Std.
Topsoil	27.41	48.70	14.70	28.20	0.33	7.72
Subsoil	24.24	49.66	10.17	32.80	0.47	7.95

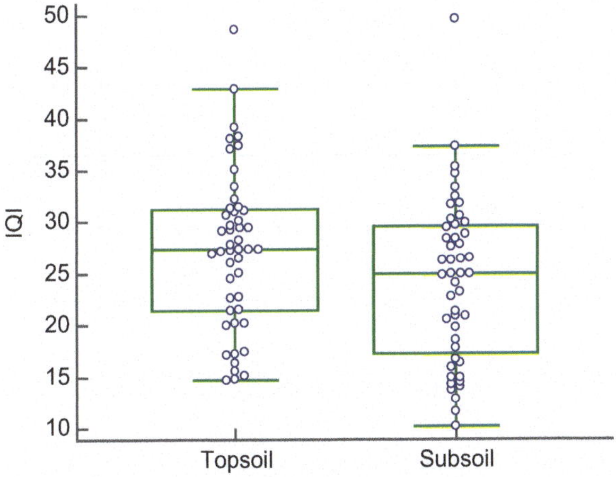

Fig. 11.2 Box-and-Whisker plot of IQIs in 0–30 and 30–60 cm depths

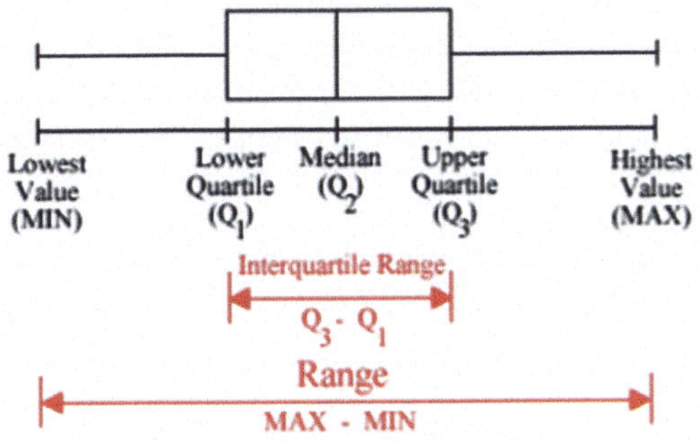

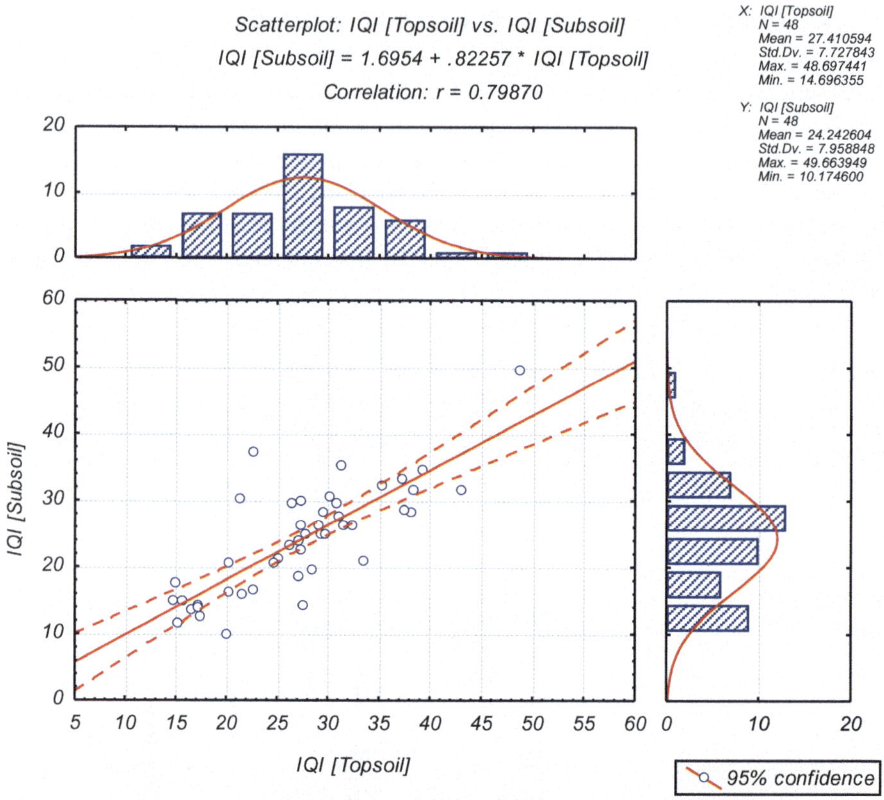

Fig. 11.3 Linear regression and variance analysis between IQIs in depths of 0–30 and 30–60 cm

11.3.5 Spatial Analyses of Soil Quality Index (SQI)

The variogram model parameters for the integrated soil quality index are shown in Table 11.8. The experimental variograms were omnidirectional since there was no obvious anisotropy in the sample data. The variables included 10 soil properties, SQI, and its residual data. Gaussian and exponential models were fitted to the experimental variograms of soil in the 0–30 cm and 30–60 cm depths, respectively (Fig. 11.4). The Gaussian model had R^2 of 0.46 while the exponential had 0.28 and a higher range

Table 11.8 Variogram parameters of integrated soil quality index

Soil depth (cm)	Model	Nugget (C_0)	Sill ($C+C_0$)	(Nugget/Sill) * 100	R^2	Range (m)
0–30	Gaussian	0.1	55.50	0.18	0.46	6400
30–60	Exponential	0.4	55.28	0.70	0.28	4600

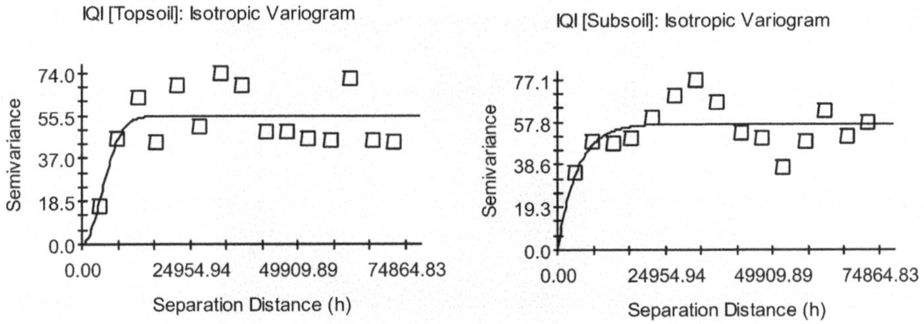

Fig. 11.4 Omnidirectional semi-variograms related IQIs using the ordinary kriging method

effect (6400 vs. 4600 m). The ratio $C_o/(C_o + C)$ was used to describe the spatial self-dependency of soil variables (Omran 2012; Melenya et al. 2015; Tuffour et al. 2015; Bakhshandeh et al. 2019). In this study, the ratio was less than 25% in both depths, which showed strong spatial self-dependency. Further, the spatial distribution of the soil quality indices based on the optical prediction map is presented in Fig. 11.5. Close observation of the maps revealed that soil quality is higher in the NW part of the study area and lower in the middle and the southern parts. In addition, soil quality is higher in the subsoil than in the topsoil.

This result demonstrates that intensive agricultural practices affect soil quality and degrade the surface horizons as reported in other areas in Iran (Kavian et al. 2017; Sadeghi et al. 2017; Samani et al. 2018). This requires strict measured by policymakers and stakeholders to avoid future permanent loss of soil fertility (Grant 2017; Kraaijvanger and Veldkamp 2015) and/or water contamination (Kumar et al. 2018).

For the future, we determine that it would be important to assess and integrate the influence of other parameters such as parent material, topography, crop production, and land use type in order to consider specific management for each study area.

11.4 Conclusions

This research argues the parameters that could be used to monitor the soil quality in the Mashhad Plain, Khorasan-e-Razavi Province, one of the most important agricultural areas in Northeast Iran. This research determined the weight of each soil quality characteristic by principal component analysis (PCA) and factor analysis (FA). In the soil surface, nitrogen, electrical conductivity, exchangeable sodium percentage, and sodium adsorption ratio reached the highest scores. In the subsoil, also sodium adsorption ratio (0.12), exchangeable sodium percentage (0.12), calcium carbonate equivalent (0.12), pH (0.11) and electrical conductivity (0.11). On the other hand, the lowest scores were obtained for soil nutrients (Olsen-P and available K). Also, higher soil quality was detected in the subsoil than in the surface, confirming the

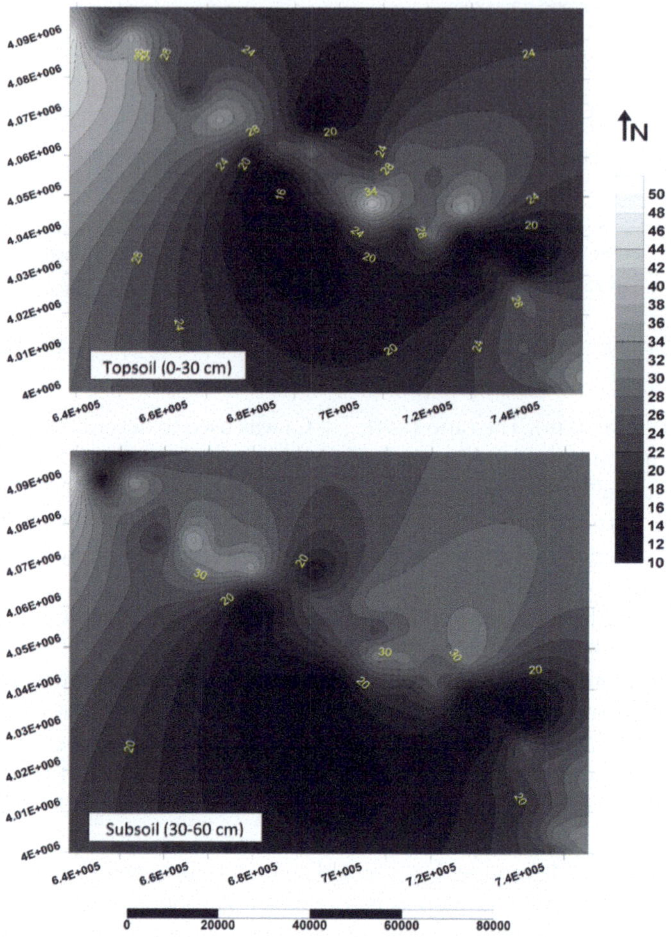

Fig. 11.5 The spatial distribution of soil quality in Mashhad Plain, Khorasan-e-Razavi Province

rapid land degradation processes. Moreover, the clear relationship between soil and plant was confirmed, affecting the selected number of soil quality indicators, which coincided with the most limiting factors for plant growth. The data presented can be useful for sustainable land management in the region as there with a serious lack of studies of soil characteristics that can serve as soil quality indicators.

Acknowledgements This study was partially supported by the Department of Soil Science, University of Tehran, Iran.

Declaration of Interest None.

References

Alexander P, Paustian K, Smith P, Moran D (2015) The economics of soil C sequestration and agricultural emissions abatement. SOIL 1:331–339. https://doi.org/10.5194/soil-1-331-2015

Al-Kaisi MM, Yin X, Licht MA (2005) Soil carbon and nitrogen changes as influenced by tillage and cropping systems in some Iowa soils. Agric Ecosyst Environ 105:635–647

Andrews SS, Mitchell JP, Mancinelli R, Karlen KL, Hartz TK, Horwath WR, Pettygrove GS, Scow KM, Munk DS (2002) On-farm assessment of soil quality in California's central valley. Agron J 94:12–23

Bakhshandeh E, Hossieni M, Zeraatpisheh M, Francaviglia R (2019) Land use change effects on soil quality and biological fertility: a case study in northern Iran. Eur J Soil Biol 95:103119. https://doi.org/10.1016/j.ejsobi.2019.103119

Bastida F, Zsolnay A, Hernández T, García C (2008) Past, present and future of soil quality indices: a biological perspective. Geoderma 147:159–171

Bindraban PS, Stoorvogel JJ, Jansen DM, Vlaming J, Groot JJR (2000) Land quality indicators for sustainable land management: proposed method for yield gap and soil nutrient balance. Agric Ecosyst Environ 81:103–112

Bogunovic I, Pereira P, Brevik EC (2017) Spatial distribution of soil chemical properties in an organic farm in Croatia. Sci Total Environ 584–585: 535–545

Bond CR, Maguire RO, Havlin JL (2006) Change in soluble phosphorus in soils following fertilization is dependent on initial Mehlich-3 phosphorus. J Environ Qual 35:1818–1824

Bower CA, Reitmeir RF, Fireman M (1952) Exchangeable cation analysis of saline and alkali soils. Soil Sci 73:251–261

Bremner JM, Mulvaney CS (1982) Nitrogen—total. In: Page AL, (ed) Methods of soil analysis, part 2. Chemical and microbiological properties, 2nd edn. American Society of Agronomy, Inc., Madison, Wisconsin, USA

Brimhall GH, Lewis CJ, Ford C, Bratt J, Taylor G, Warin O (1991) Quantitative geochemical approach to pedogenesis: importance of parent material reduction, volumetric expansion, and eolian influx in lateritization. Geoderma 51: 51–91

Buchholz J, Querner P, Paredes D, Bauer T, Strauss P, Guernion M, Scimia J, Cluzeau D, Burel F, Kratschmer S, Winter S, Potthoff M, Zaller JG (2017) Soil biota in vineyards are more influenced by plants and soil quality than by tillage intensity or the surrounding landscape. Sci Rep 7:17445

Calleja-Cervantes ME, Fernández-González AJ, Irigoyen I, Fernández-López M, Aparicio-Tejo PM, Menéndez S (2015) Thirteen years of continued application of composted organic wastes in a vineyard modify soil quality characteristics. Soil Bio Biochem 90:241–254

Costa JL, Aparicio V, Cerdà A (2015) Soil physical quality changes under different management systems after 10 years in the Argentine humid pampa. Solid Earth 6:361–371

Dai Q, Liu Z, Shao H, Yang Z (2015) Karst bare slope soil erosion and soil quality: a simulation case study. Solid Earth 6:985–995

de Paul Obade V, Lal R (2016a) Towards a standard technique for soil quality assessment. Geoderma 265:96–102

de Paul Obade V, Lal R (2016b) A standardized soil quality index for diverse field conditions. Sci Total Environ 541:424–434

Doran JW, Parkin BT (1994) Defining and assessing soil quality. In: Doran JW, Coleman DC, Bezdicek DF, Stewart BA (eds) Defining soil quality for a sustainable environment. Soil Science Society of America Inc., Madison, WI, USA, pp 3–21

Fard MM, Harchagani HB (2009) Comparison of artificial neural network and regression pedo transfer functions models for prediction of soil cation exchange capacity in Chaharmahal-e-Bakhtiari province. J Water Soil 23(4):90–99

Fischer C, Roscher C, Jensen B, Eisenhauer N, Baade J, Attinger S, Scheu S, Weisser WW, Schumacher J, Hildebrandt A (2014) How do earthworms, soil texture and plant composition affect infiltration along an experimental plant diversity gradient in grassland? PLoS ONE 9:e98987

Fu W, Tunney H, Zhang C (2010) Spatial variation of soil nutrients in a dairy farm and its implications for site-specific fertilizer application. Soil Tillage Res 106:185–193

Gabarrón-Galeote MA, Ruiz-Sinoga JD, Quesada MA (2013) Influence of aspect in soil and vegetation water dynamics in dry Mediterranean conditions: functional adjustment of evergreen and semi-deciduous growth forms. Ecohydrology 6:241–255

Gorji T, Sertel E, Tanik A (2017) Monitoring soil salinity via remote sensing technology under data scarce conditions: a case study from Turkey. Ecol Indic 74:384–391

Grant CA (2017) Soil fertility and management. In: International encyclopedia of geography. American Cancer Society, pp 1–10

Gray J, Bishop T, Wilford J (2016) Lithology and soil relationships for soil modelling and mapping. CATENA 147:429–440

Hazarika S, Thakuria D, Ganeshamurthy AN, Sakthivel T (2014) Soil quality as influenced by land use history of orchards in humid subtropics. CATENA 123:37–44

Hosseini M, Rajabi Agereh S, Khaledian Y, Jafarzadeh Zoghalchali H, Brevik EC, Movahedi Naeini SAR (2017) Comparison of multiple statistical techniques to predict soil phosphorus. Appl Soil Ecol 114:123–131

IUSS Working Group WRB (2007) Land evaluation. Towards a revised framework, 2nd edn. Land and Water discussion paper. FAO, Rome

IUSS Working Group WRB (2014) World reference base for soil resources 2014. World soil resources report. FAO, Rome

Kamali MR, Omidvar A, Kazemzadeh E (2013) 3D geostatistical modeling and uncertainty analysis in a carbonate reservoir, SW Iran. J Geol Res

Kang GS, Beri V, Sidhu BS, Rupela OP (2005) A new index to assess soil quality and sustainability of wheat-based cropping systems. Biol Fertil Soils 41:389–398

Karlen DL, Gardner JC, Rosek MJ (1998) A soil quality framework for evaluating the impact of CRP. J Prod Agric 11:56–60

Kavian A, Golshan M, Abdollahi Z (2017) Flow discharge simulation based on land use change predictions. Environ. Earth Sci 76:588

Kavian A, Mohammadi M, Gholami L, Rodrigo-Comino J (2018) Assessment of the spatiotemporal effects of land use changes on runoff and nitrate Loads in the Talar River. Water 10:445

Keshavarzi A, Omran E-SE, Bateni SM, Pradhan B, Vasu D, Bagherzadeh A (2016) Modeling of available soil phosphorus (ASP) using multi-objective group method of data handling. Model Earth Syst Environ 2:157

Keshavarzi A, Tuffour H, Bagherzadeh A, Duraisamy V (2018) Spatial and fractal characterization of soil properties across soil depth in an agricultural field, Northeast Iran. Eurasian J Soil Sci 7:35–45

Khaledian Y, Kiani F, Ebrahimi S, Brevik EC, Aitkenhead-Peterson J (2016) Assessment and monitoring of soil degradation during land use change using multivariate analysis. Land Degrad Dev 28(1):128–141

Kraaijvanger R, Veldkamp T (2015) Grain productivity, fertilizer response and nutrient balance of farming systems in Tigray, Ethiopia: a multi-perspective view in relation to soil fertility degradation. Land Degrad Develop 26:701–710

Kumar V, Sharma A, Minakshi, Bhardwaj R, Thukral AK (2018) Temporal distribution, source apportionment, and pollution assessment of metals in the sediments of Beas river, India. Hum Ecol Risk Assessment Int J 24:2162–2181

Lado M, Paz A, Ben-Hur M (2004) Organic matter and aggregate-size interactions in saturated hydraulic conductivity. Soil Sci Soc Am J 68:234

Lark RM (2000) Estimating variograms of soil properties by the method of moments and maximum likelihood. Eur J Soil Sci 51:717–728

Lima ACR, Brussaard L, Totola MR, Hoogmoed WB, de Goede RGM (2013) A functional evaluation of three indicator sets for assessing soil quality. Appl Soil Ecol 64:194–200

Maleki S, Karimi A, Zeraatpisheh M, Poozeshi R (2021) Long-term cultivation effects on soil properties variations in different landforms in an arid region of eastern Iran. CATENA 206:105465. https://doi.org/10.1016/j.catena.2021.105465
Mairura FS, Mugendi DN, Mwanje JI, Ramisch JJ, Mbugua PK, Chianu JN (2007) Integrating scientific and farmers' evaluation of soil quality indicators in Central Kenya. Geoderma 139:134–143
Manandhar R, Odeh I (2014) Interrelationships of land use/cover change and topography with soil acidity and salinity as indicators of land degradation. Land 3:282–299
Marzaioli R, D'Ascoli R, De Pascale RA, Rutigliano FA (2010) Soil quality in a Mediterranean area of Southern Italy as related to different land use type. Appl Soil Ecol 44:205–212
McGrath D, Zhang CS (2003) Spatial distribution of soil organic carbon concentrations in grassland of Ireland. Appl Geochem 18:1629–1639
Melenya C, Logah V, Aryee D, Abubakari A, Tuffour HO, Yeboah IB (2015) Sorption of phosphorus in soils in the semi deciduous forest zone of Ghana. Appl Res J 1:169–175
Mishra G, Marzaioli R, Giri K, Borah R, Dutta A, Jayaraj RSC (2017) Soil quality assessment under shifting cultivation and forests in Northeastern Himalaya of India. Arch Agron Soil Sci 63:1355–1368
Mishra, G., Marzaioli, R., Giri, K.and Pandey, S. 2018. Soil quality assessment across different stands in tropical moist deciduous forests of Nagaland, India. *J. For. Res.,* 1–7.
Mohammadkhan S, Ahmadi H, Jafari M (2011) Relationship between soil erosion, slope, parent material, and distance to road (Case study: Latian Watershed, Iran). Arab J Geosci 4:331–338
Mukherjee A, Lal R (2014) Comparison of soil quality index using three methods. PLoS ONE 9(8):1–15
Mukherjee A, Lal R (2015) Short-term effects of cover cropping on the quality of a Typic Argiaquolls in Central Ohio. CATENA 131:125–129
Nelson DW, Sommers LE (1982) Total carbon, organic carbon, and organic matter. In: Page AL, Miller RH, Keeney DR (eds) Methods of soil analysis, part 2—chemical and microbiological properties. ASA-SSSA, Madison, WI, pp 539–594
Nelson RE (1982) Carbonate and gypsum. In: Page AL, Miller RH, Keeny R (eds) Methods of soil analysis, part II-chemical and microbiological properties, Madison, WI, pp 181–196
Nesbitt JE, Adl SM (2014) Differences in soil quality indicators between organic and sustainably managed potato fields in Eastern Canada. Ecol Indicat 37:119–130
Olsen SR, Cole CV, Watanable FS, Dean LA (1954) Estimation of available phosphorus in soils by extraction with sodium bicarbonate. USDA Circular 939, Washington
Omran E-SE (2012) Improving the prediction accuracy of soil mapping through geostatistics. Int J Geosci 3(3):574
Oster J, Sposito G (1980) The Gapon coefficient and the exchangeable sodium percentage-sodium adsorption ratio relation 1. Soil Sci Soc Am J 44(2):258–260
Pérez-de-los-Reyes C, Ortíz-Villajos JAA, Navarro FJG, Martín-Consuegra SB, Ballesta RJ (2015) Effects of sugar foam liming on the water-retention properties of soil. Commun Soil Sci Plant Ana 46:1299–1308
Pulido M, Schnabel S, Lavado Contador JF, Lozano-Parra J, Gómez-Gutiérrez Á, Brevik EC, Cerdà A (2017) Reduction of the frequency of herbaceous roots as an effect of soil compaction induced by heavy grazing in rangelands of SW Spain. Catena 158:381–389
Qi Y, Darilek JL, Huang B, Zhao Y, Sun W, Gu Z (2009) Evaluating soil quality indices in an agricultural region of Jiangsu Province, China. Geoderma 149:325–334
Recena R, Díaz I, Delgado A (2017) Estimation of total plant available phosphorus in representative soils from Mediterranean areas. Geoderma 297:10–18
Roberts WM, Gonzalez-Jimenez JL, Doody DG, Jordan P, Daly K (2017) Assessing the risk of phosphorus transfer to high ecological status rivers: integration of nutrient management with soil geochemical and hydrological conditions. Sci Total Environ 589:25–35
Robinson TP, Metternicht GM (2006) Testing the performance of spatial interpolation techniques for mapping soil properties. Comput Electron Agri 50:97–108

Sadeghi SH, Kiani Harchegani M, Asadi H (2017) Variability of particle size distributions of upward/downward splashed materials in different rainfall intensities and slopes. Geoderma 290:100–106

Samani AN, Rad FT, Azarakhshi M, Rahdari MR, Rodrigo-Comino J (2018) Assessment of the Sustainability of the Territories affected by gully head advancements through aerial photography and modeling estimations: a case study on Samal Watershed, Iran. Sustainability 10:2909

Shukla MK, Lal R, Ebinger M (2006) Determining soil quality indicators by factor analysis. Soil Tillage Res 87:194–204

Smith P (2018) Managing the global land resource. Proc R Soc B 285:20172798

Soil Survey Staff (2014) Keys to soil taxonomy, 12th edn. USDA-Natural Resources Conservation Service, Washington DC

Sparks DL, Page AL, Helmke PA, Leoppert RH, Soltanpour PN, Tabatabai MA, Johnston GT, Summer ME (1996) Methods of soil analysis. Soil Sci Soc Am J Book Series No. 5. ASA and SSSA, Madison, Wisconsin

Stockdale EA, Goulding KWT, George TS, Murphy DV (2013) Soil fertility. In: Soil conditions and plant growth. Wiley-Blackwell, pp 49–85

Sulieman M, Saeed I, Hassaballa A, Rodrigo-Comino J (2018) Modeling cation exchange capacity in multi geochronological-derived alluvium soils: an approach based on soil depth intervals. CATENA 167:327–339

Sumner ME, Miller WP (1996) Cation exchange capacity and exchange coefficients. Methods of soil analysis. Part 3. Chemical Methods Soil Science Society of America Inc., Madison, pp 1201–1229

Sun B, Zhou SL, Zhao QG (2003) Evaluation of spatial and temporal changes of soil quality based on geostatistical analysis in the hill region of subtropical China. Geoderma 115:85–99

Swanepoel PA, Preez CC du, Botha PR, Snyman HA, Habig J (2014) Soil quality characteristics of kikuyu-ryegrass pastures in South Africa. Geoderma 232–234:589–599

Taguas EV, Guzmán E, Guzmán G, Vanwalleghem T, Gómez JA (2015) Characteristics and importance of rill and gully erosion: a case study in a small catchment of a marginal olive grove. Cuad Investig Geográfica 41:107–126

Thomas GW (1996) Soil pH and soil acidity. In: Sparks DL (ed) Methods of soil analysis. Part 3. Chemical methods. No. 5. ASA and SSSA, Madison, WI, pp 475–490

Tuffour HO, Abubakari A, Bashagaluke JB, Djagbletey D (2016) Mapping spatial variability of soil physical properties for site-specific management. Internet Res J Eng Tec 3:149–163

Tuffour HO, Adjei-Gyapong T, Abubakari A, Melenya C, Aryee D, Khalid AA (2015) Assessment of changes in soilhydro-physicalpropertiesresultingfrominfiltrationofmuddywater. Appl Res J 1:137–140

Turnbull L, Hütt M-T, Ioannides AA, Kininmonth S, Poeppl R, Tockner K, Bracken LJ, Keesstra S, Liu L, Masselink R, Parsons AJ (2018) Connectivity and complex systems: learning from a multi-disciplinary perspective. Appl Netw Sci 3:11

Wilding LP, Dress LR (1983) Spatial variability and pedology. In: Wilding LP, Smeckand NE, Hall GF (eds) Pedogenesis and soil taxonomy. I. Concepts and interactions. Elsevier Science Pub., pp 83–116

Wilson, J.P.and Gallant, J.C. 2000. Terrain Analysis, Principle and Applications. John Wiley and Sons, Inc., NY.

Yang W, Zheng F, Han Y, Wang Z, Yi Y, Feng Z (2016) Investigating spatial distribution of soil quality index and its impacts on corn yield in a cultivated catchment of the Chinese Mollisol Region. Soil Sci Soc Am J 80:317–327

Yavitt JB (2000) Nutrient dynamics of soil derived from different parent material on Barro Colorado Island, Panama. Biotropica 32:198–207

Zeraatpisheh M, Bakhshandeh E, Hosseini M, Alavi SM (2020) Assessing the effects of deforestation and intensive agriculture on the soil quality through digital soil mapping. Geoderma 363:114139. https://doi.org/10.1016/j.geoderma.2019.114139

Zeraatpisheh M, Leonel E, Bakhshandeh E, Owliaie HR, Taghizadeh-Mehrjardi R, Kerry R, Scholten T, Xu M (2022) Spatial variability of soil quality within management zones: homogeneity and purity of delineated zones. CATENA 209:105835. https://doi.org/10.1016/j.catena.2021.105835

Zhang G, Bai J, Xi M, Zhao Q, Lu Q, Jia J (2016) Soil quality assessment of coastal wetlands in the Yellow River Delta of China based on the minimum data set. Ecol Indicat 66:458–466

Chapter 12
Spatial Pattern Analysis and Identifying Soil Pollution Hotspots Using Local Moran's I and GIS at a Regional Scale in Northeast of Iran

Ali Keshavarzi, Gouri Sankar Bhunia, Pravat Kumar Shit, Güneş Ertunç, and Mojtaba Zeraatpisheh

Abstract The spatial distribution of soil physicochemical characteristics and four heavy metals (Mn, Fe, Zn, and Cu) in the semi-arid climatic region of Neyshabur plain in Northeast of Iran was investigated and identified soil pollution hotspots zone using Moran's I and GIS techniques. The geostatistical techniques, Pearson's correlation matrix, and spatial autocorrelation were used to locate the pollution sources and concentration. Geostatistical interpolation techniques determined the spatial distribution of heavy metals. The mean values of Iron (Fe), Manganese (Mn), Zink (Zn), Copper (Cu) were 2.31, 7.18, 2.84, 1.16 mg/kg, respectively. The routs comes of the spatial statistical method have established the gravity of pollutions and their anthropogenic impact based on spatial changes in contamination levels. The genesis of the pollution process was influenced by natural factors (e.g., the high soil shale, the sandstone, the calcareous and the metamorphic parents and the background values) as well as by anthropogenic factors (e.g., waste disposal, extraction from mines of distinct mineral ores and high, unmanaged practices of fertilizer). Although nearly

---

A. Keshavarzi (✉)
Laboratory of Remote Sensing and GIS, Department of Soil Science, University of Tehran, P.O.Box: 4111, 31587-77871 Karaj, Iran
e-mail: alikeshavarzi@ut.ac.ir

G. S. Bhunia
Paschim Medinipur, Kharagpur, West Bengal 721424, India

P. K. Shit
Department of Geography, Raja NL Khan Women's College, Gope Palace, Medinipur, West Bengal 721102, India

G. Ertunç
Department of Mining Engineering, Hacettepe University, 06800 Beytepe, Ankara, Turkey
e-mail: gertunc@hacettepe.edu.tr

M. Zeraatpisheh
Henan Key Laboratory of Earth System Observation and Modeling, Henan University, Kaifeng 475004, China

College of Geography and Environmental Science, Henan University, Kaifeng 475004, China

P. K. Shit et al. (eds.), *Soil Health and Environmental Sustainability*, Environmental Science and Engineering,
https://doi.org/10.1007/978-3-031-09270-1_12

all the monitoring classes of land use suffered from contamination by heavy metals, farmland was the most contaminated. This evidence will help land use planners and environmental menace administrators to promote environmentally sound economic expansion policies.

Keywords Environmental pollution · Geostatistics · GIS · Heavy metals · Hotspots zone · Iran

12.1 Introduction

The natural components of the earth's crust are heavy metals. A number of these constituents are of biological importance and play an important part in human life when they trace the water, the air, dust, soils, and sediments. The soil is the most polluting habitat as a "universal trap." In a variety of cases, it gets tainted. Okrent (1999) stated that soil pollution is demarcated as the growth of obstinate toxic soils, chemicals, salt, or disease-causing substances that adversely affect crop growth and animal health. Soil pollution must be monitored urgently to protect soil fertility and increase productivity. One main source of heavy metals in the soil, and is accountable for an improved pervasiveness and incidence of heavy metal pollution on the Earth's surface, is anthropogenic activity such as mining and metal smelting (Bhattacharya et al. 2006). Generally, water, sewage, improper dumping or by-products, or contamination from the processing of something of value absorbs much of the pollutants into the ecosystem (Soffianian et al. 2014). Opencast mining operations, which produced millions of tons of sulfide-rich waste, have a significant environmental effect on soils and water sources (Parizanganeh et al. 2010). By accelerating erosion, we somewhat lose this important natural resource. Besides that, the enormity of man-made waste, sludge, and other products' from new waste treatment plants also cause or lead to polluted soil. To sustain the fertility and productivity of the soil, rigorous control measures must be implemented, hence increasing the health of all living things.

Evaluating the ecological menace of polluted soil, pesticide application, sewage sludge, and other anthropogenic activities resulting in exposure to hazardous substances in the terrestrial environment is a complex task with manyalliedglitches. In the present way that we evaluate the menace and the effect of anthropogenic agents on the terrestrial climate, even though those factors were ignored, there are a variety of unanswered issues. An assessment of the bioavailable percentage of radioactive metals may be carried out to assess soil contamination of heavy metals. Soil metal mobility has commonly been evaluated by a chemical method based on selective withdrawals.

Iran has experienced broad developments in the last four decades, including rapid urbanization, industrial development, and intensive cultivation in many regions. Sometimes these variations have been escorted by neglected environmental devaluation (Moghtaderi et al. 2018, 2019; Khamesi et al. 2020). This is also an imperative zone for agriculture where crops like maize, barley, and sugar beet are grown. Soils can be polluted by industrial and urban contaminations in agricultural areas, posing a danger to humans, as showed by Doabi et al. (2018, 2019) in other areas of Iran, by consuming food grown in these countries.

Geospatial analytical techniques are key tools for soil parameter characterization (Hou et al. 2017). In previous soil pollution studies, classical statistical methods have been commonly used, but these approaches are affluent and time-consuming and do not quantify assessment errors. Soil contamination can be well known by Geographic Information System (GIS) and geostatistical methods at present (Soffianian et al. 2015). In order to assess the spatial structure of heavy metals and soil physico chemical characteristics, GIS are essential for the implementation through geostatistical and multivariate analyses (Santos-Francés et al. 2017). In fact, it is not possible to arrange adequate samples from the subject areas. Therefore, spatial statistical approaches haves wapped traditional statistics, as they can precisely detect pollutant changes in time and space and calculate estimation errors (Soffianian et al. 2014). Several studies have examined spatial distribution in industrial areas worldwide of heavy metal pollution in the surface ground. For instance, Wang et al. (2017) reported a less national standard but less than the natural baseline values for the geographical dispersal of Cu, Zn, Cr, Cd, As, and Hg concentrations in the industrial area of Sichuan, China. In the industrial city of Aran-o-Bidgol, Iran, Ravankhah et al. (2016) carried out the assessment of the ecological menace of heavy metals from surface soil. The Cd, Pb, Ni, and Cu levels were recorded above the background values.

The study was showed in order to classify the area where heavy metals are tainted. More specifically, first the spatial distribution of some main soil properties and heavy metals such as pH, OC, Sand, Silt, Clay, Phosphorus, Fe, Mn, Zn, and Cu were determnined and then the spatial distribution of soil properties and heavy metals were applied to find toxic hotspots and to detect potential causes of contaminants in surface soils in the Neyshabur plain, Khorasan-e-Razavi Province, Northeast Iran. Moreover, in order to reduce the uncertainties associated with parameters, the datasets were further statistically analyzed using statistical approaches such as the correlation matrix, spatial autocorrelation, and spatial modeling.

12.2 Study Area

The research was carried out in a catchment in the part of Neyshabur plain of Khorasan-e-Razavi Province (36°2′–36°10′ N, and 58°52′–59°07′ E) of Northeast Iran (Fig. 12.1). The study area covers by an area of almost 170 km^2 with an elevation of 1256 m above mean sea level. The region is considered by the semi-arid climate with mean temperature of 14.5 °C and annual precipitation of 233.7 mm. The primary land use structure of the area is irrigated farming (Bagherzadeh et al. 2016). The general slope of the plain extends in NW–SE direction. The major land type is described as piedmont plain and Qft2 unit is the key geological unit, which

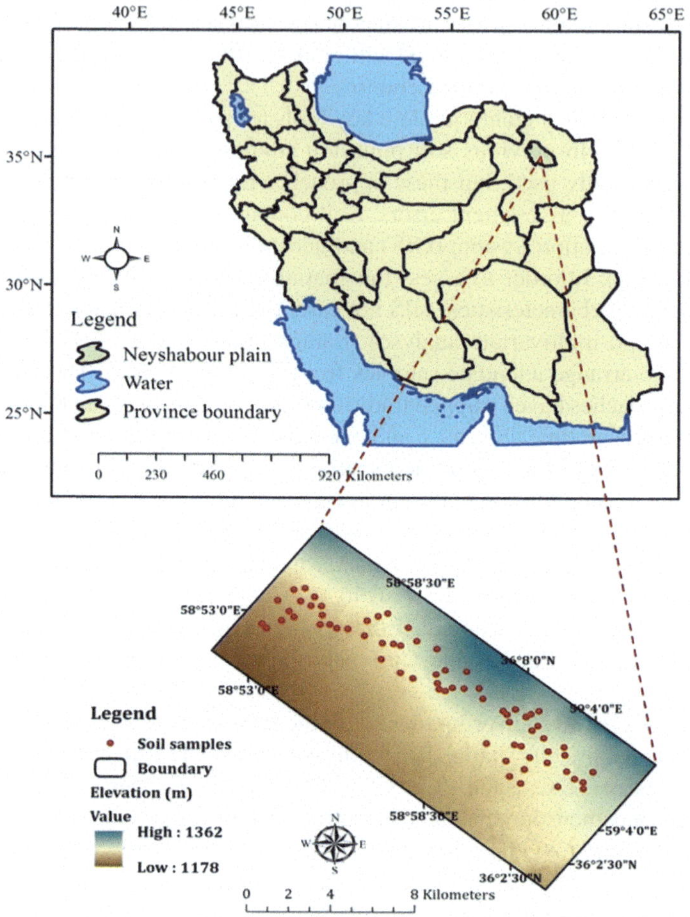

Fig. 12.1 Soil sampling locations in parts of Neyshabur Plain, Khorasan-e-Razavi region, Iran

indicates low levels of piedmont fan and valley terrace deposits. Aridisols and Entisols are the most common soil types in the area, according to Bagherzadeh et al. (2016).

12.3 Methods

12.3.1 Sampling and Analysis

Sixty-eight representative soil samples (during the period between 2018 and 2019) were collected using a random sampling technique for an suitable demarcation of soil sampling areas to reflect the geographical distribution of the parameters distressing the soils and heavy metals. A portable Global Positioning System (GPS) was applied during soil sampling to find the sampling locations. Soil samples were taken from 0 to 20 cm because our goal was to focus on the topsoil, or the part of the soil influenced by crop roots and water infiltration. Large plant materials and pebbles in the samples were parted by hand and discarded. Bulk samples of the soil were spread on trays in the laboratory and were air-dried for two weeks under ambient conditions. The samples were subsequently tiled in a 2 mm mesh and dried in an oven at 50 °C for approximately 48 h with mortar and pestle (Lu et al. 2010). The samples were then homogenated and placed in polyethylene containers. The hydrometer method was used to determine the textural fractions of sand (0.05–2 mm), silt (0.002–0.05 mm), and clay (<0.002 mm) in soil (Gee and Bauder 1986). The approach of Walkley and Black was used to quantify the number of organic carbon (OC) in the soil (Walkley and Black 1934). The method of Olsen et al. (1954) was used to determine the amount of available phosphorus (P). The process of extraction with 1 M ammonium acetate (NH_4OAC) at pH = 7 was used to quantify available potassium (K) (Thomas 1996). A digital EC-pH metre was used to measure pH in saturated paste extract (Thomas 1996). An atomic absorption spectrometer was used to analyse heavy metals like Mn, Fe, Zn, and Cu. Following the procedure of Heidari et al. (2019) the soil samples were digested using the aqua-regia process (HNO_3:HCl in a ratio of 1:3). The digested samples were filtered and diluted in 20 mL double steam distilled water before being utilised in the experiment. After every five samples, the standards and blanks were run for quality assurance and quality control to ensure the machine's 95% accuracy (Arora et al. 2008). The 95–100% recovery rates for samples spiked with standards confirmed the accuracy of the results (Xiao et al. 2013).

12.3.2 Statistical Analysis

The use of statistical methods helps us understand the dynamic soil quality data matrices, classify potential causes that affect soil resources, and provide useful knowledge for effective soil management (Simeonov et al. 2004; Reghunath et al. 2002). The correlation matrix was performed using Microsoft Excel version 2013. *Pearson's* correlation was done by the correlation matrix of soil heavy metal parameters. A correlation coefficient (near +1 or −1) means that a strong relationship between two variables is established and '*0*' indicates that no relationship occurs between them. All the statistical analyses were performed at a significance level of $P < 0.05$.

12.3.2.1 Incremental Spatial Autocorrelation

The Incremental Spatial Autocorrelation (ISA) method uses the *Global Moran's I* function, which calculates the strength of spatial clustering for each of the distances, to construct a sequence of that distances, and can be intended based on the following equation:

$$I = \frac{n}{S_0} \frac{\sum_{i=1}^{n} \sum_{j=1}^{n} \omega_{i,j}, Z_i Z_j}{\sum_{i=1}^{n} Z_i^2}$$

where, Z_i is the deviation of heavy metallic soil parameters for sample location i from its mean $(x_i - \overline{x})$, $\omega_{i,j}$ is the spatial weight between sample location i and j, n is equal to the total number of sampling sites, S_o is the cumulative of all the spatial weights:

$$S_0 = \sum_{i=1}^{n} \sum_{j=1}^{n} w_{i,j}$$

The *Moran 's Index* is positive because the data collection continues to cluster geographicalluy (high values cluster next to other high values, low values cluster near other small ones). The index would be negative if high values repel certain interests and appear to be close to low. If positive cross-product values surpass negative cross-product values, the index will be close to zero. The numerator is determined by the variation in order to minimize index values from −1.0 to +1.0.

The ISA mechanism can be the extent to which high (clustered) or low (dispersed) spatial correlations and whether they have been significant or not detected by a peak suggested by the index. This can be both measures of distance are based on the feature centers, and the default start distance (500 m) is the smallest distance (each feature has at least one nearby area). The clustering strength is determined by the returned z-point. Typically, the z-score, indicating an increasing clustering, increases as the gap increases. The z-score usually peaks at a certain point (Jossart et al. 2020). However, when there is more than one statistically significant peak, the clustering at each of these distances is pronounced. Select the maximum distance that best fits the size of the study you want; it is also the first statistically relevant summit that has been identified.

12.3.2.2 Optimized Cluster Analysis

The mapping tools perform cluster analysis in order to determine the location of hotspots, coldspots, spatial outliers, and similar features or areas of statistically significant importanceusing the *Anselin Local Moran's I* statistic (Anselin 1995). The tool is particularly useful for intervention dependent on the location of one or more clusters. This method distinguishes statistically important spatial clusters with high (hot) and low (cold) values. The system aggregates heavy metallic soil data automatically, determines the appropriate analysis scale, and corrects multiple tests as well as spatial dependency.

Since the *Optimized Outlier Analysis* tool uses the nearest average and median next-door calculations for aggregation and also for an adequate scale of analysis, a component for initial data assessment will also identify locations in each soil characteristics at geographical locations. This method measures the average closest distance of each element and compares all the distances spread.

12.3.2.3 Estimation of Spatial Interpolation of Soil Parameters

Spatial patterns, values in unmeasured areas, and the uncertainty associated with a predicted value in unmeasured locations may be determined by Kernal Smoothing (KS). KS is used to model and measure the spatial variability in the sampled places of each of the influential parameters (Gribov and Krivoruchko 2004). The Z vector p() theory is based on both randomly and spatially autocorrelated. The predictions are model-based on:

$$Z(p_i) = \mu + \varepsilon'(p_i)$$

where μ is the constant stationary function (global mean) and $\varepsilon'(p_i)$ is the spatially correlated stochastic part of the variation. The forecasts are collected with:

$$\hat{Z}_{ok}(p_0) = \sum_{i=1}^{n} w_i(p_0) \cdot Z(p_i) = \lambda_0^T \cdot a$$

where λ_0 is the vector of kriging weights (w_i), a is the vector of n observations at primary locations.

The semivariogram is a convenient tool for analyzing spatial dependence structures in geostatistics. It is focused on the basic difference and is defined by:

$$\gamma(h) = \frac{1}{2} Var(Z(p_i) - Z(p_i + h))^2$$

where $Z(p_i)$ is the value of a random variable at some sampled location and $Z(p_i+h)$ is the value of the location at a distance $(p_i + h)$.

The variogram for each parameter was drawn from a Polynomial, Quartic, Exponential, and Gaussian model, based on the shortest distance between points and determining the best variable model feature. To all kernel functions, r is a radius centered at point s, and h is bandwidth for all formulas (Yan, 2009):

$$\text{Polynomial} = \left(1 - \left(\tfrac{r}{h}\right)^3 \left(10 - \left(\tfrac{r}{h}\right)\left(15 - 6\left(\tfrac{r}{h}\right)\right)\right)\right), \ for \ \tfrac{r}{h} < 1.$$

$$\text{Quartic} = \left(1 - \left(\tfrac{r}{h}\right)^2\right)^2, \ for \ \tfrac{r}{h} < 1.$$

$$\text{Exponential} = e^{-3\left(\frac{r}{h}\right)}.$$

$$\text{Gaussian} = e^{-3\left(\frac{r}{h}\right)^2}.$$

12.3.2.4 Cross-Validation

For the evaluation and comparison of model performance, a cross-validation technique was adopted. In the model accuracy assessment, the, root mean squared error (RMSE) and average standard error were identified (Zhang et al. 2018).

12.4 Results

12.4.1 Exploratory Analysis of Soil Variables

The physico chemical characteristics of soil that eventually affect the root growth and mobility of the contaminant can greatly influence the assimilation of heavy metals. Descriptive statistics of all soil variables are presented in Table 12.1. The soil pH of the research area is ranges between 7.5 and 8.3, with a mean value of 7.9 ± 0.19. The mean value of Organic Carbon (OC), Sand, Silt, clay, Phosphorus (P), Potassium (K), Iron (Fe), Manganese (Mn), Zink (Zn), Copper (Cu) is calculated as 0.73%, 40.29%, 36.99%, 22.72%, 19.45 mg/kg, 261.05 mg/kg, 2.31 mg/kg, 7.18 mg/kg,

Table 12.1 Descriptive characteristics of concentration of heavy metals in soils samples

	Mean	Standard error	Median	Standard deviation	Kurtosis	Skewness	Confidence level (95.0%)
pH	7.90	0.02	7.90	0.19	−0.59	−0.01	0.05
OC (%)	0.73	0.04	0.68	0.32	1.03	1.07	0.08
Sand (%)	40.29	1.18	40.30	9.72	0.34	0.57	2.35
Silt (%)	36.99	0.78	36.90	6.41	−0.53	−0.13	1.55
Clay (%)	22.72	0.73	23.00	5.99	−0.61	−0.26	1.45
Phosphorus (mg/kg)	19.45	1.97	11.20	16.25	0.13	1.09	3.93
Potassium (mg/kg)	261.05	16.16	249.71	133.27	3.10	1.42	32.26
Fe (mg/kg)	2.31	0.08	2.28	0.68	0.20	0.66	0.16
Mn (mg/kg)	7.18	0.49	6.04	4.06	2.53	1.61	0.98
Zn (mg/kg)	2.84	0.46	1.09	3.78	1.58	1.68	0.92
Cu (mg/kg)	1.16	0.04	1.13	0.29	−0.34	0.47	0.07

2.84 mg/kg, 1.16 mg/kg, respectively. The highest standard deviation is calculated for potassium (±133.27), followed by sand (±9.72), silt (±6.41), and clay (±5.99). The negative kurtosis and skewness are calculated for pH, Silt, Clay, and Cu. The maximum kurtosis is estimated for K (3.10), followed by Mn (2.53) and Zn (1.58). The maximum skewness is calculated for Zn (1.68), followed by Mn (1.61) and K (1.42).

12.4.2 Pearson's Correlation

The correlation coefficient for different soil properties results is presented in Fig. 12.2. The coefficient of correlation between the pH and zinc ($r = 0.41$) has been found to be positive ($P < 0.05$). However non-significant ($P > 0.05$) relationship was found with OC ($r = 0.23$) and Silt ($r = 0.27$). Results also showed strong negative correlation between sand, silt ($r = -0.80$) and clay ($r = -0.77$). There is moderate positive correlation was calculated between sand and zinc; whereas, a negative relationship was found between clay and zinc. Similarly, the meager positive correlation was observed between OC, K, and Mn with available P. A meager negative correlation is observed between available K and zinc, and a positive correlation is calculated with OC. Fe shows a moderate negative correlation with the clay and available K; and a

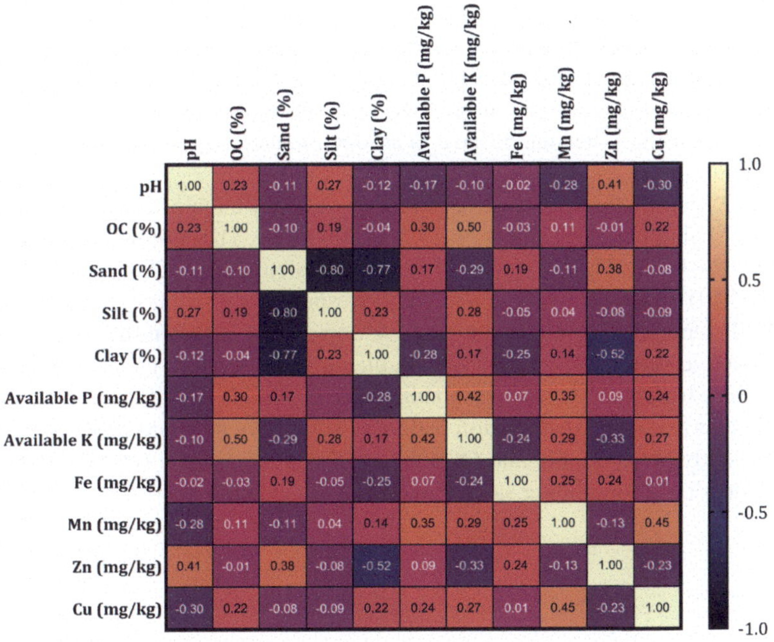

Fig. 12.2 Cross-correlation matrix of heavy metals in soils samples ($n = 68$)

Table 12.2 Spatial autocorrelation of heavy metals in soils samples

Parameters	Moran's Index	Expected Index	Variance	z-score	P-value
pH	0.749	−0.015	0.041	3.789	0.000
OC (%)	0.392	−0.015	0.040	2.045	0.041
Sand (%)	0.589	−0.015	0.040	3.015	0.003
Silt (%)	0.196	−0.015	0.041	1.047	0.295
Clay (%)	0.820	−0.015	0.041	4.142	0.000
Phosphorus (mg/kg)	0.138	−0.015	0.040	0.762	0.446
Potassium (mg/kg)	−0.041	−0.015	0.038	−0.133	0.894
Fe (mg/kg)	0.360	−0.015	0.040	1.870	0.061
Mn (mg/kg)	0.118	−0.015	0.039	0.677	0.498
Zn (mg/kg)	0.317	−0.015	0.039	1.671	0.095
Cu (mg/kg)	0.298	−0.015	0.040	1.554	0.120

positive correlation with the Mn and Zn. Mn shows a moderate positive correlation with the available P, K, and Fe. Zn shows a moderate positive correlation with the pH, sand, and Fe; whereas, a negative correlation is calculated for clay, available K, and Cu. Statistical outcomes exhibited that Cu, influencedby pH and $CaCO_3$ levels, increased with the moving soil fractions bonding.

The soil pH and the available proportion Cu had a negative association, as per the observations. Based on the bioavailability and chemical processes of heavy metals, binding in various fractions vary considerably. As a result of the apparent competition between dissolved metals, the adsorption of heavy metals has been shown to decrease. Heavy metals on the negative surfaces formed on organic colloidal materials and on minerals, on the other hand, have been documented to adsorb electrostatically (Sungur et al. 2014).

Table 12.2 shows the spatial autocorrelation of heavy metals of soil samples in the part of Neyshabur plain of Khorasan-e-Razavi Province. The significance of Moran's I is tested ($P < 0.05$). The maximum *Moran's I* is calculated for clay (8.20), followed by pH (0.749) and sand (0.589), and the corresponding *P-values* are calculated as <0.0001. It represents a significant positive correlation between the sample values and is clustered pattern. Moreover, the calculated value of Moran's *I* of silt, P, Mn, K are very close to zero, and the corresponding Z-score and *P-values* are not significant. This indicates a uniform distribution pattern of soil heavy metal contents (silt, P, Mn, K) in the study area. Zn, Fe, and OC have moderate significant ($P < 0.05$) positive spatial autocorrelation among sample values in the study area.

12.4.3 Spatial Autocorrelation

The statistical existence of data sets based on the distance between high autocorrelation in the spatial area is taken into consideration. It typically can be accomplished through an iterative and data-led process that determines how spatial autocorrelation occurs differs at various distances. For each increase in distance, space autocorrelation measures the associated *Moran's I*, Expected Index, Variance, z-score, and p-values for a number of distance increments and reports. High Z-scores value suggested statistical significance ($P < 0.05$). This means that heavy metal concentrations are higher in Z-score based upon allocations of spatial variability and metal heterogeneity at different soil depths (Ren et al. 2016). The threshold value of the beginning distance is considered as 500 m, and the incremental threshold distance specified as 1030, 1560, 2091, 2621, 3152, 3682, 4213, 4743, and 5274 m (Table 12.3). The high value of *Moran's I* indicate the distance at which the clustering of the data is more affirmed. The highest *Moran's I* value at 500 m is calculated for pH (Z-score—3.065; *P*-value <0.002) and clay (Z-score—3.496; *P*-value <0.000), followed by Sand (Z-score—2.218; *P*-value <0.033) and OC (Z-score—2.316; *P*-value <0.020). At 1030 m distance, the maximum *ISA* is calculated for clay (*Moran's I*—0.664; Z-score—6.891; *P*-value <0.000), followed by Zn (*Moran's I*—0.424; Z-score—4.547; *P*-value <0.000) and F (*Moran's I*—0.403; Z-score—4.277; *P*-value <0.000). At a distance of 1560 m, maximum ISA is calculated for clay (*Moran's I*—0.597; Z-score—9.492; *P*-value <0.000), followed by Zn (*Moran's I*—0.475; Z-score—7.717; *P*-value <0.000) and pH (*Moran's I*—0.372; Z-score—6.002; *P*-value <0.000). However, the derived output of ISA value for clay and Zn is maximum at each threshold distance. Moreover, the minimum estimated ISA value is calculated for P and Mn at each distance band at which the sample locations are uniformly distributed.

12.4.4 Cluster Analysis

The GiZ-Score map is generated by the optimal cluster analysis (OCA) tool, which shows the hot and cold locations in the study area. It also provides point features in the research region that signify hot and cold locations (Fig. 12.3). GiZScore is a tool that creates a z-score value for each sampling location, which serves to identify the statistical significance of feature clusters and, ultimately, hot and cold locations. Heavy metallic parameters of soil characteristics with a high positive z-score are designated as hotspots (red), while heavy metallic elements of soil features with a low z-score are designated as cold spots (blue). The z-score is used to determine whether the sampling location exhibit a random pattern or statistically significant clustering or dispersion, indicating a spatial process at work. As a result, the greater the value for a statistically significant positive z-score, the more intense the clustering of the hotspot.

Table 12.3 Incremental spatial autocorrelation analysis of heavy metals in soils samples

Distance (m)	500			1030			1560			2091			2621		
Soil parameter	Moran's index	z-score	p-value	Moran's index	z-score	p-value	Moran's index	z-score	p-value	Moran's index	z-score	p-value	Moran's index	z-score	p-value
pH	0.607	3.065	0.002	0.393	4.139	0.000	0.372	6.002	0.000	0.264	5.824	0.000	0.246	6.474	0.000
OC (%)	0.401	2.128	0.033	0.012	0.286	0.775	0.054	1.090	0.276	0.019	0.722	0.470	0.013	0.691	0.489
Sand (%)	0.446	2.316	0.020	0.307	3.292	0.001	0.289	4.747	0.000	0.205	4.625	0.000	0.176	4.785	0.000
Silt (%)	0.218	1.190	0.233	−0.064	−0.496	0.620	−0.012	0.044	0.965	−0.015	0.008	0.993	−0.016	−0.021	0.983
Clay (%)	0.696	3.496	0.000	0.664	6.891	0.000	0.597	9.492	0.000	0.526	11.290	0.000	0.469	12.035	0.000
Phosphorus (mg/kg)	0.227	1.288	0.197	0.275	2.973	0.003	0.054	1.087	0.277	0.000	0.313	0.754	−0.038	−0.583	0.559
Potassium (mg/kg)	−0.015	0.065	0.948	0.038	0.563	0.573	0.092	1.711	0.087	0.051	1.422	0.155	0.056	1.798	0.072
Fe (mg/kg)	0.400	2.092	0.036	0.403	4.277	0.000	0.352	5.725	0.000	0.235	5.256	0.000	0.248	6.581	0.000
Mn (mg/kg)	0.184	1.094	0.274	0.163	1.854	0.063	0.156	2.707	0.006	0.142	3.347	0.000	0.130	3.691	0.000
Zn (mg/kg)	0.266	1.451	0.146	0.424	4.547	0.000	0.475	7.717	0.000	0.440	9.641	0.000	0.429	11.225	0.000
Cu (mg/kg)	0.312	1.676	0.094	0.234	2.538	0.011	0.167	2.828	0.004	0.172	3.924	0.000	0.160	4.361	0.000
Distance (m)	3152			3682			4213			4743			5274		
Soil parameter	Moran's index	z-score	p-value	Moran's index	z-score	p-value	Moran's index	z-score	p-value	Moran's index	z-score	p-value	Moran's index	z-score	p-value
pH	0.216	6.664	0.000	0.178	6.302	0.000	0.154	6.123	0.000	0.143	6.203	0.000	0.132	6.289	0.000
OC (%)	0.005	0.582	0.558	0.031	1.533	0.125	0.032	1.706	0.088	0.016	1.264	0.205	−0.021	−0.249	0.802

(continued)

Table 12.3 (continued)

Distance (m)	3152			3682			4213			4743			5274		
Soil parameter	Moran's index	z-score	p-value	Moran's index	z-score	p-value	Moran's index	z-score	p-value	Moran's index	z-score	p-value	Moran's index	z-score	p-value
Sand (%)	0.158	5.036	0.000	0.137	5.002	0.000	0.132	5.349	0.000	0.082	3.833	0.000	0.038	2.298	0.022
Silt (%)	−0.015	−0.021	0.982	−0.016	−0.022	0.983	0.008	0.821	0.411	−0.017	−0.086	0.931	−0.025	−0.437	0.662
Clay (%)	0.420	12.570	0.000	0.362	12.302	0.000	0.326	12.310	0.000	0.268	11.152	0.000	0.204	9.360	0.000
Phosphorus (mg/kg)	−0.037	−0.652	0.514	−0.031	−0.543	0.587	−0.035	−0.741	0.459	−0.042	−1.059	0.289	−0.028	−0.577	0.564
Potassium (mg/kg)	0.090	3.123	0.002	0.102	3.920	0.000	0.094	4.058	0.000	0.076	3.717	0.000	0.059	3.273	0.001
Fe (mg/kg)	0.185	5.817	0.000	0.135	4.908	0.000	0.095	3.999	0.000	0.070	3.390	0.000	0.043	2.489	0.012
Mn (mg/kg)	0.101	3.429	0.000	0.099	3.809	0.000	0.073	3.281	0.001	0.072	3.536	0.000	0.061	3.350	0.000
Zn (mg/kg)	0.404	12.331	0.000	0.350	12.101	0.000	0.308	11.861	0.000	0.259	10.950	0.000	0.206	9.595	0.000
Cu (mg/kg)	0.119	3.903	0.000	0.082	3.139	0.001	0.047	1.733	0.082	0.029	1.733	0.082	0.016	1.343	0.017

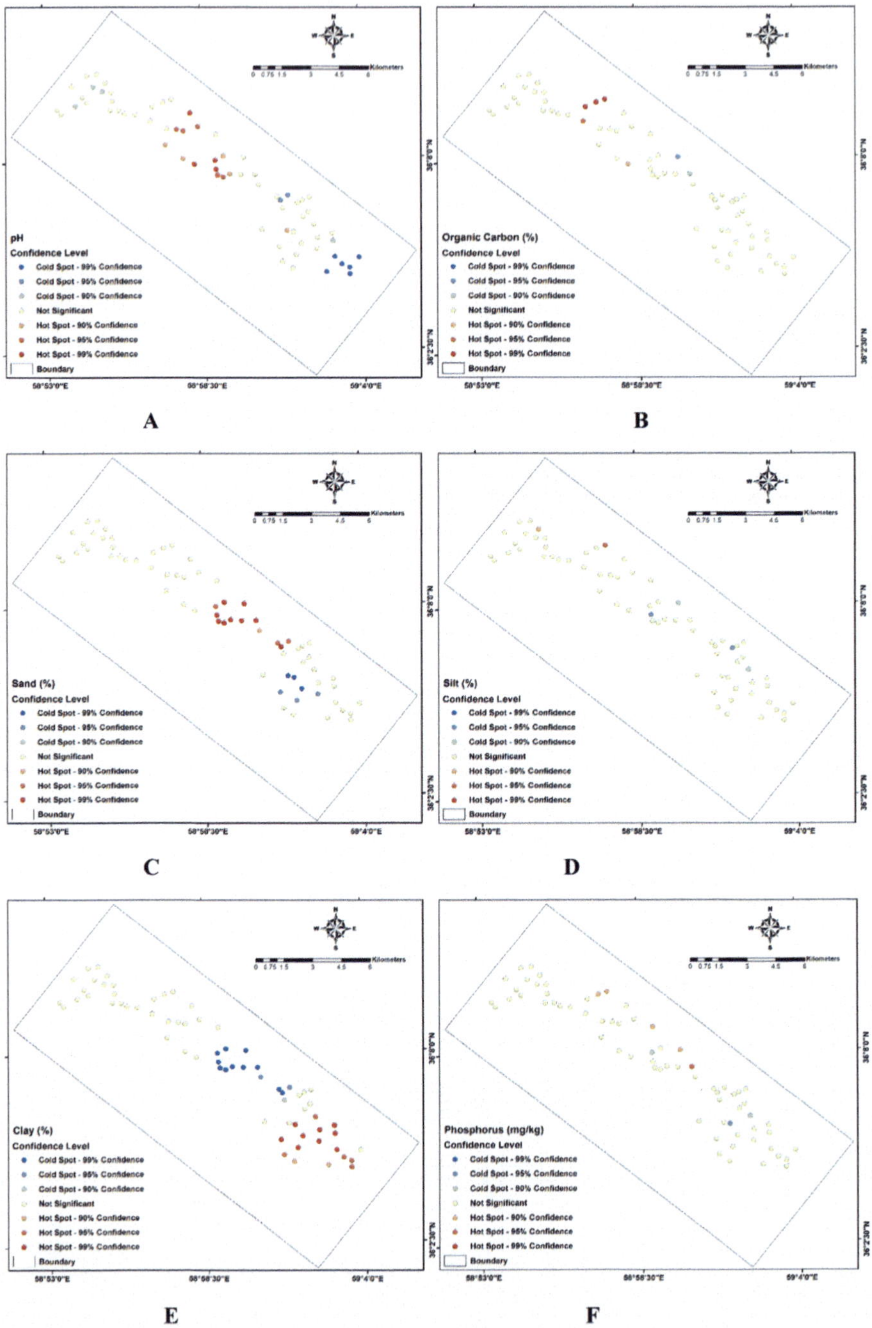

Fig. 12.3 Optimized cluster analysis of soil samples using Getis-Ord Gi* statistics

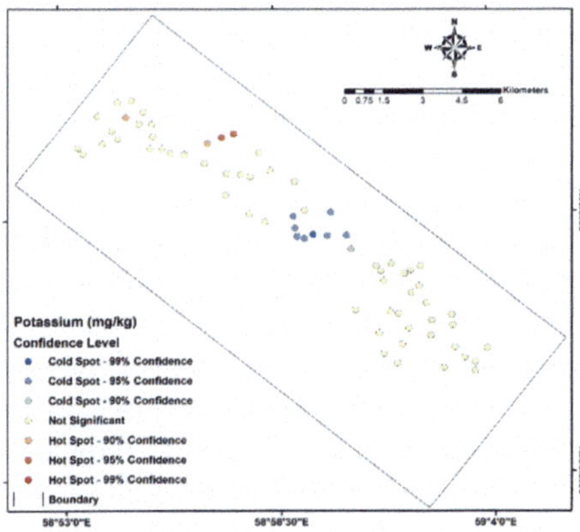

G

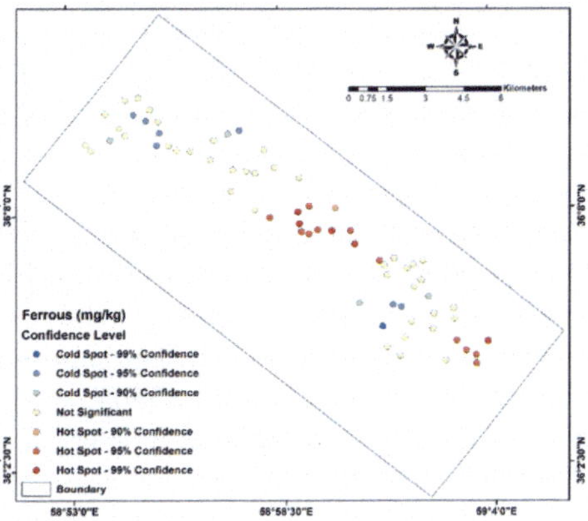

H

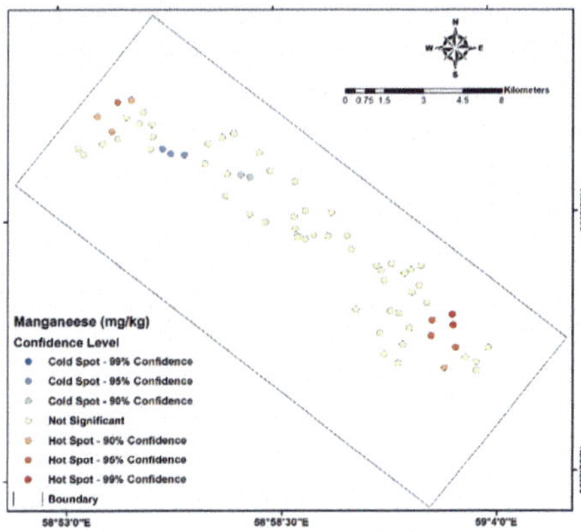

I

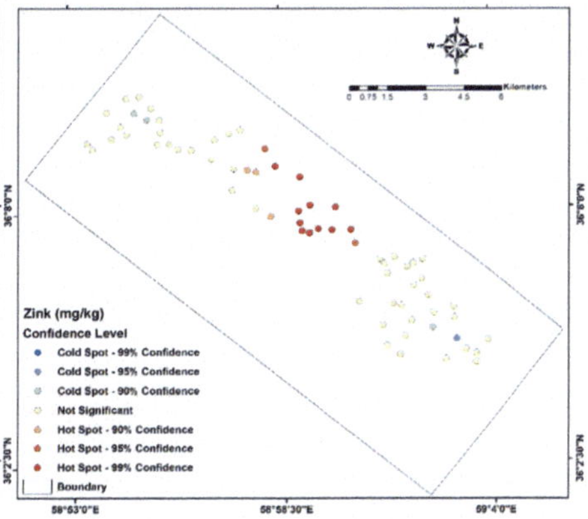

J

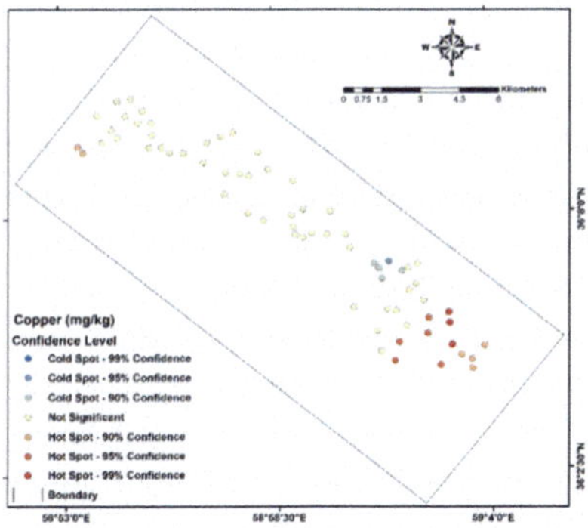

K

Fig. 12.3 (continued)

In Fig. 12.3a, it can be seen that hotspots and coldspots are presented by pH where the values were 2.946422 > z > 1.767853 (red) and −3.17487 < z < −1.785019 (blue) standard deviations, respectively. The hotspots and coldspots of OC are presented as 3.101711 > z > 1.952999 (red) and −2.117656 < z < −1.714938 (blue) standard deviations, respectively (Fig. 12.3b). The significance of hotspots and coldspots of sand are presented as 4.249433 > z > 2.203002 (red) and −3.046544 < z < −2.02786 (blue) standard deviations, respectively (Fig. 12.3c). The standard deviations of the hotspots and coldspots of Silt are 2.1069 > z > 1.793934 (red) and −2.514968 < z < −1.886327 (blue) (Fig. 12.3d). The significance of hotspots and coldspots of clay are presented as 3.736432 > z > 1.858045 (red) and −4.870407 < z < −2.229149 (blue) standard deviations, respectively (Fig. 12.3e). Hotspot and cold spot as indicated by P, with standard deviations of 2.263199 > z > 1.767853 (red) to −2.233934 < z < −1.848018 (blue) (Fig. 12.3f). K showing 2.266854 > z > 1.797812 and −2.688332 < z < −1.710025 (blue) standard deviations are the hotspots and coldspots (Fig. 12.3g). The hotspots and coldspots are presented by Fe, where the values were 2.959069 > z > 1.846312 (red) and −2.60816 < z < −1.876887 (blue) standard deviations, respectively (Fig. 12.3h). Mn shows the hotspots and cold spots with standard deviations of 3.227574 > z > 1.757108 (red) and −2.232847 < z < −1.700522 (blue) (Fig. 12.3i). Zn displays hotspots and coldspots with 4.940435 > z > 1.759745 (red), and −1.968598 < z < −1.817021 (blue), respectively, standard deviations (Fig. 12.3j). Coldspots and hotspots are presented with the values of Cu with standard deviations of 2.977647 > z > 1.917672 (red) and −2.112617 < z < −1.763338 (blue) (Fig. 12.3k).

12.4.5 Spatial Distribution

The aim is to test the performance in heavy metal parameters, combination with estimation and simulation, of four different semivariogram models to explain their uncertainty and spatial heterogeneity. In Table 12.4, along with their respective best-fit results, the simulated semivariogram for the Polynomial, Quartic, Exponential, and Gaussian models is evaluated in order to illustrate the spatial dependence of heavy metal accumulation in soil. The exponential and gaussian models looked similar except for the slight difference in soil heavy metals distribution patches.

In order to map the metal content and delineate the polluted areas, a spatial correlation between the data available with the kernel smoothing technique was used. The results showed that the exponential model was well-matched with the soil heavy metal data (Fig. 12.4). The highest value of pH is observed in the central part and a small pocket north of the study area. The low pH value is observed in the south and east of the region. The pH value of 7.95–8.1 is also observed in the middle of the study area. The maximum concentration of OC is found in the central and northwest of the study area. The medium concentration of OC is portrayed in the central and eastern parts of the study area. The minimum concentration of OC is found in the west and south of the region. The maximum concentration of sand is observed in the

Table 12.4 Best-fitted models used for soil characteristics

Soil samples	Model	Mean	RMSE	Avg. SE
pH	Polynomial5	−0.011	0.159	0.425
	Quartic	−0.011	0.161	0.423
	Gaussian	−0.014	0.160	0.433
	Exponential	−0.013	0.158	0.427
Organic Carbon (%)	Polynomial5	0.111	0.365	0.315
	Quartic	0.112	0.363	0.313
	Gaussian	0.112	0.363	0.307
	Exponential	0.109	0.358	0.312
Sand (%)	Polynomial	1.888	9.312	55.05
	Quartic	1.918	9.27	55.03
	Gaussian	1.972	9.23	54.9
	Exponential	1.888	9.118	54.11
Silt (%)	Polynomial5	0.804	7.076	39.031
	Quartic	0.782	6.985	38.755
	Gaussian	0.780	6.851	38.426
	Exponential	0.745	6.896	38.269
Clay (%)	Polynomial5	0.301	4.192	18.064
	Quartic	0.286	4.185	18.049
	Gaussian	0.260	4.192	18.072
	Exponential	0.243	4.088	17.71
Phosphorus (mg/kg)	Polynomial5	8.863	18.597	59.975
	Quartic	8.991	18.44	60.105
	Gaussian	9.986	18.685	61.16
	Exponential	8.518	18.339	61.201
Potassium (mg/kg)	Polynomial5	55.869	174.706	2361.018
	Quartic	57.395	172.825	2352.581
	Gaussian	60.023	166.594	2339.885
	Exponential	55.860	174.14	2308.157
Fe (mg/kg)	Polynomial5	0.068	0.553	0.819
	Quartic	0.068	0.552	0.825
	Gaussian	0.072	0.544	0.841
	Exponential	0.067	0.555	0.828
Mn (mg/kg)	Polynomial5	1.503	4.346	12.266
	Quartic	1.554	4.496	11.764
	Gaussian	1.869	4.579	12.593
	Exponential	1.365	4.433	12.214

(continued)

Table 12.4 (continued)

Soil samples	Model	Mean	RMSE	Avg. SE
Zn (mg/kg)	Polynomial5	1.452	3.519	5.384
	Quartic	1.462	3.521	5.456
	Gaussian	1.566	3.529	5.589
	Exponential	1.421	3.516	5.421
Cu (mg/kg)	Polynomial5	0.053	0.277	0.292
	Quartic	0.053	0.277	0.292
	Gaussian	0.060	0.281	0.295
	Exponential	0.049	0.274	0.291

northcenter of the study area. However, the minimum sand concentration is observed south and northwest of the study area. The spatial distribution of silt is varied in the study area, whereas the maximum sand distribution is found in the south-center and northwest and its gradually decreases from the central region. There is a heterogeneous distribution of Zn concentration in the study area. Fe is heterogeneously distributed in the study area. The maximum Mn is portrayed in the north and east of the region, andtheminimum Mn is observed in the central and west of the region. The highest Fe is found in the central and east of the study site, and the south and west part is recorded as low Fe concentration. The maximum concentration of Zn is found in the east and west of the study area whereas, the central part of the region is recorded as low concentration of Zn. The highest concentration of Cu is observed in the east and central north of the region and south of the study area with low Cu concentration.

12.5 Discussion

In several regions of the world, particularly in developing countries, metal soil pollution has become a major and pervasive challenge. Farming may be a source of heavy metals in the soil (Huang and Jin 2008), urbanization, industrial development, andmining (Zhong et al. 2012). Heavy metal pollution is mainly due to urban and industrial aerosols, burning of fuel, liquid and solids, mining waste, industrial and farm chemicals, etc. The soils are primarily drained by different soil areas where either the inorganic or organic colloids are preserved very strongly. The weathering of the parents' materials means that heavy metals are existing in all uncontaminated soils. Chemical waste inorganic contaminants cause serious waste disposal issues. Superphosphate, phosphoric acid, aluminum, steel, and ceramics industry fluorides can be found in the atmosphere. 55.89% of the total samples haveapH value >7.90. Most trace elements' solubility decreases when soil pH rises, resulting in low quantities in soil solution (Kabata-Pendias 2011). Any change in the pH of the soil has

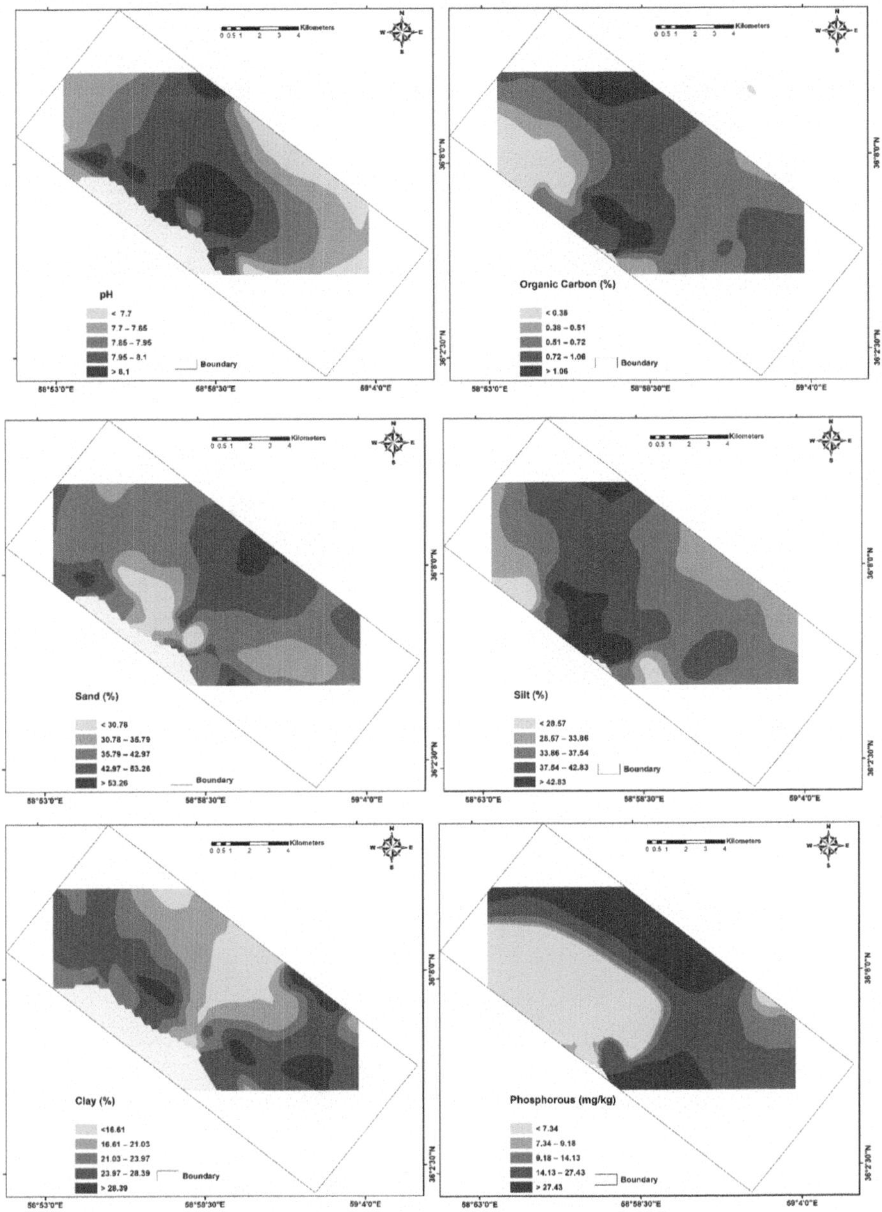

Fig. 12.4 Spatial variability maps of physico-chemical parameters and heavy metals

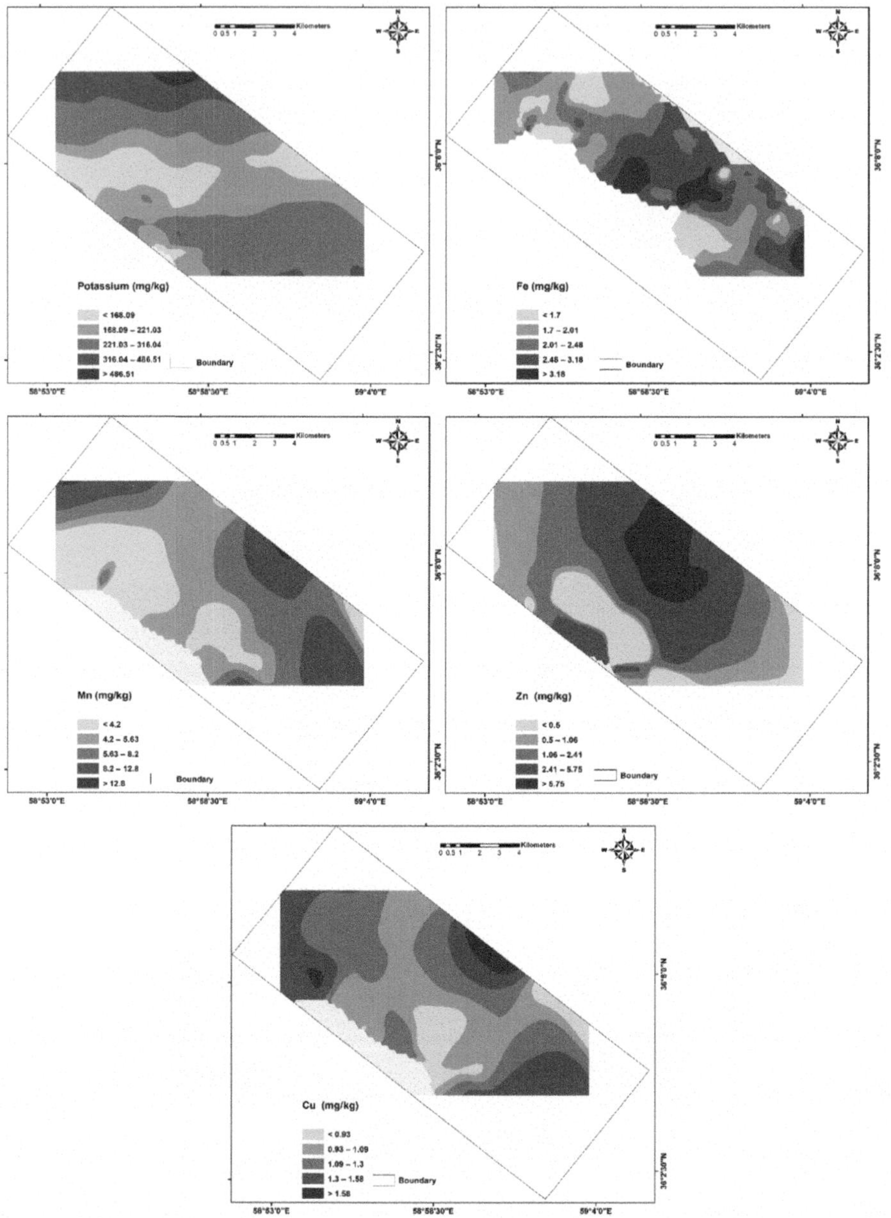

Fig. 12.4 (continued)

an impact on the solubility of metals. This is likely to be dependent on the metals' ionic species and the pH change's direction. 42.65% P concentration is significantly increased (>16 mg/kg) from the topsoil to the subsoil, indicating that the subsoil aced as sink or source for P leaching. As a result, subsurface products may be evaluated for the implementation of mitigation measures to prevent P leaching in soil horizons based on the P content and the soil saturation threshold (Andersson et al. 2015). Soils can be very acidic by sulfur dioxide emitted by plants and thermal plants. These metals damage the leaf and destroy plant life (Richardson et al. 2006). Water irrigation with sewage is responsible for deep irrigation vicissitudes in the soils. Numerous variations in the soil as an offset of sewage irrigation comprise physical changes such as leaching, humus deviations, porosity, etc., and chemical changes such as soil reactions, soil base exchanges, salinity, nutrient quantities, and nutrient accessibility nitrogen, potash, phosphorus and so forth. This can result in plant phytotoxicity.

Metal ions reach the water of the soil at various such concentrations by which they may either stay in the water or reach drainages to be consumed by plants increasing on the soil or be preserved in sparingly soluble or insoluble soil types (Urzelai et al. 2000). This soil's organic matter is very similar to heavy metal cations, which form stable complexes and thus reduce its nutrient content. However, one or two of the elements in agricultural soils can be concentrated in several ways, such as chemicals, sewage sludge, farm slurries, etc. Increased doses of fertilizer, pesticides, or agricultural chemicals are added over a period of time to contaminate soils using heavy metals. There are also cadmium residuesin some phosphatic fertilizers that can be found in these soil areas. Soil micronutrient accessibility measures parent materials, the effects of soil redox potentials, pH, soil microbial activity, interactions with coexisting ions, soil mineral reactions, and organic matter, as well as the effects of soil edaphic and biological factors activity in the study area. Pearson's correlation coefficient analyzed the relationship between different physico-chemical properties and heavy metal values. A bivariant approach is used for defining the interaction between two different parameters. Agricultural practices, including soil water control application and soil modifications, will make soil micronutrients available. Crop residues are a major source of many micronutrients. Roughly 50–80% of the rice and wheat crops used in Zn, Cu, and Mn may be recovered by incorporating residues (Dhaliwal et al. 2019).

In order to transform test point data on sampling sites into thematic maps showing the geographical variation, traditional interpolator processes like the Generalized Kriging method, a polynomial method, and the Inverse Distance Weighting (IDW) approaches were widely used (Rodrigo-Comino et al. 2019). In comparison, the variogram model is used by the Kriging interpolation method to illustrate the structure of the geographical change of assessed values and the spatial autocorrelation in the modeling of the surface (Wang et al. 2017). The kriging technique includes a collection of methods of stochastic-based interpolation, such as ordinary kriging, cokriging, universal kriging, simple kriging, residual methods, and regression methods (Li and Heap 2014).

Soils of low Zn may have low total Zn (some acidic leached soils in the tropical world) or may have a relatively high total Zn content; however, because of the soil chemistry, a plant-accessible fraction that favors the synthesis of poorly soluble Zn complexes (Rengel 2002). Intensive cultivation of high yield rice and wheat threatened the continued high levels of food production with a Zn and Fe shortfall in rice and Mn weight of wheat. It is widely known to be one of the most operative procedures of growing OC levels and improving soil quality in the application of organic materials. A significant cultivation activity in terms of crop yield and efficiency, climate conservation, and soil regeneration is the use of appropriate quantities of fertilizer (Oenema et al. 2009; Atafar et al. 2010). As a guideline, 15 kg of both phosphorus and sulfur are available to plants for every ton of carbon in OC as organic matter is demolished (Hoyle et al. 2013). As a result, locations with a lot of heavy metal would pollute the soil and put people's and other living organisms' health at risk. The soil concentrations of these metals were higher than critical concentrations in the majority of the field, but their presence in the soil reduces solubility and bioavailability (Krami et al. 2013). The low metal contamination levels in other parts of the region indicate that those areas are safe regions.

12.6 Conclusion

In the present study, geostatistical techniques, correlation matrix, spatial autocorrelation, and spatial modeling were used to analyze the geographical distribution pattern and concentration of heavy metals in Neyshabur plain region in Iran. The outcomes of the geostatistical techniques have confirmed the gravity of pollution and their anthropogenic impact based on spatial changes in contamination levels. The genesis of the pollution process was influenced by natural factors (e.g., the high soil shale, the sandstone, the calcareous and the metamorphic parents and the background values) as well as by anthropogenic factors (e.g., waste disposal, extraction from mines of special mineral ores and high, unmanaged uses of fertilizer). Although nearly all the monitoring classes of land use suffered from contamination by heavy metals, farmland was the most polluted. This information will help land use planners and environmental risk administrators.

Disclosure Statement No potential conflict of interest was reported by the author(s).

References

Andersson H, Bergström L, Ulén B, Djodjic F, Kirchmann H (2015) The role of subsoil as a source or sink forphosphorus leaching. J Environ Qual 44:535–544

Anselin L (1995) Local indicators of spatial association-LISA. Geogr Anal 27(2):93–115

Atafar Z, Mesdaghinia AR, Nouri J, Homaee M, Yunesian M, Ahmadimoghaddam M, Mahvi AH (2010) Effect of fertilizer application on soil heavy metal concentration. Environ Monit Assess 160(1–4):83–89

Bagherzadeh A, Ghadiri E, Darban ARS, Gholizadeh A (2016) Land suitability modeling by parametric-based neural networks and fuzzy methods for soybean production in a semi-arid region. Model Earth Syst Environ 2(2):104

Bhattacharya A, Routh J, Jacks G, Bhattacharya P, Mörth M (2006) Environmental assessment of abandoned mine tailings in Adak,Västerbotten district (northern Sweden). Appl Geochem 21:1760–1780

Dhaliwal SS, Naresh RK, Mandal A, Singh R, Dhaliwal MK (2019) Dynamics and transformations of micronutrients in agricultural soils as influenced by organic matter build-up: a review. Environ Sustain Indicators 1–2:100007

Doabi SA, Karami M, Afyuni M, Yeganeh M (2018) Pollution and health risk assessment of heavy metals in agricultural soil, atmospheric dust and major food crops in Kermanshah province, Iran. Ecotoxicol Environ Saf 163:153–164

Doabi S, Karami M, Afyuni M (2019) Heavy metal pollution assessment in agricultural soils of Kermanshah province, Iran. Environ Earth Sci 78:70

Gee GW, Bauder JW (1986) Methods of soil analysis: part 1. Agronomy handbook 9. In: Klute A (ed) Particle size analysis. American Society of Agronomy and Soil Science Society of America, Madison (WI), pp 383–411

Gribov A, Krivoruchko K (2004) Geostatistical mapping with continuous moving neighbourhood. Math Geol 36(2)

Heidari A, Kumar V, Keshavarzi A (2019) Appraisal of metallic pollution and ecological risks in agricultural soils of Alborz province, Iran, employing contamination indices and multivariate statistical analyses. Int J Environ Health Res 1–19

Hou D, O'Connor D, Nathanail P, Tian L, Ma Y (2017) Integrated GIS and multivariate statistical analysis forregional scale assessment of heavy metal soil contamination: a critical review. Environ Pollut 231:1188–1200

Hoyle FC, Antuono MD, Overheu T, Murphy DV (2013) Capacity for increasing soil organic carbon stocks in dryland agricultural systems. Soil Res 51:657–667

Huang S-W, Jin J-Y (2008) Status of heavy metals in agricultural soils as affected by different patterns of land use. Environ Monit Assess 139(1–3):317–327. https://doi.org/10.1007/s10661-007-9838-4

Jossart J, Theuerkauf SJ, Wickliffe LC, Morris Jr. JA (2020) Applications of spatial autocorrelation analyses for marine aquaculture siting. Front Mar Sci. https://doi.org/10.3389/fmars.2019.00806

Kabata-Pendias A (2011) Trace elements in soils and plants. CRC Press, Boca Raton, FL, USA

Khamesi A, Khademi H, Zeraatpisheh M (2020) Biomagnetic monitoring of atmospheric heavy metal pollution using pine needles: the case study of Isfahan, Iran. Environ Sci Pollut Res 27:31555–31566. https://doi.org/10.1007/s11356-020-09247-5

Krami LK, Amiri F, Sefiyanian A, Shariff AR, Tabatabaie T, Pradhan B (2013) Spatial patterns of heavy metals in soil under different geological structures and land uses for assessing metal enrichments. Environ Monit Assess 185(12):9871–9888. https://doi.org/10.1007/s10661-013-3298-9

Li J, Heap AD (2014) Spatial interpolation methods applied in the environmental sciences: a review. Environ Model Softw 53:173–189

Lu X, Wang L, Li LY, Lei K, Huang L, Kang D (2010) Multivariate statistical analysis of heavy metals instreet dust of Baoji, NW China. J Hazard Mater 173:744–749

Moghtaderi T, Mahmoudi S, Shakeri A, Masihabadi MH (2018) Heavy metals contamination and human health risk assessment in soils of an industrial area, Bandar Abbas-South Central Iran. Hum Ecol Risk Assess 24:1058–1073

Moghtaderi T, Mahmodi S, Shakeri A, Masihabadi MH (2019) Contamination evaluation, health and ecological risk index assessment of potential toxic elements in the surface soils (case study: Central Part of Bandar Abbas County). J Soil Water Conserv 8:51–65

Oenema O, Witzke HP, Klimont Z, Lesschen JP, Velthof GL (2009) Integrated assessment of promising measures to decrease nitrogen losses from agriculture in EU-27. Agric Ecosyst Environ 133:280–288

Okrent D (1999) On intergenerational equity and its clash with intragenerational equity and on the need for policies to guide the regulation of disposal of wastes and other activities posing very long time risks. Risk Anal 19:877–901

Olsen SR, Cole CV, Watanabe FS, Dean LA (1954) Estimation of available phosphorus in soils by extraction with sodium bicarbonate. Government Printing Office, Washington. D.C

Parizanganeh A, Hajisoltani P, Zamani A (2010) Assessment of heavy metal pollution in surficial soils surrounding zinc industrial complex in Zanjan-Iran. International society for environmental information sciences 2010 annual conference (ISEIS). Procedia Environ Sci 2:162–166

Ravankhah N, Mirzaei R, Masoum S (2016) Spatial eco-risk assessment of heavy metals in the surface soils ofindustrial city of Aran-o-Bidgol, Iran. Bull Environ Contam Toxicol 96:516–523

Reghunath R, Murthy TRS, Raghavan BR (2002) The utility of multivariate statistical techniques in hydrogeochemical studies: an example from Karnataka, India. Water Res 36:2437–2442

Ren B, Chen Y, Zhu G, Wang Z, Zheng X (2016) Spatial variability and distribution of the metals in surfacerunoff in a nonferrous metal mine. J Anal Methods Chem 2016:4515673

Rengel Z (2002) Agronomic approaches to increasing zinc concentration in staple food crops. In: Cakmak I, Welch RM (eds) Impacts of agriculture on human health and nutrition. UNESCO, EOLSS Publishers, Oxford, UK

Richardson GM, Bright DA, Dodd M (2006) Do current standards of practice in Canada measure what is relevant to human exposure at contaminated sites? II: oral bio-accessibility of contaminants in soil. Hum Ecol Risk Assess 12:606–618

Rodrigo-Comino J, Keshavarzi A, Zeraatpisheh M, Gyasi-Agyei Y, Cerdà A (2019) Determining the best ISUM (Improved stock unearthing Method) sampling point number to model long-term soil transport and micro-topographical changes in vineyards. Comput Electron Agric 159:147–156. https://doi.org/10.1016/j.compag.2019.03.007

Santos-Francés F, Martínez-Graña A, Zarza CÁ, Sánchez AG, Rojo PA (2017) Spatial distribution of heavymetals and the environmental quality of soil in the Northern Plateau of Spain by geostatistical methods. Int J Environ Res Public Health 14:568

Simeonov V, Simeonova P, Tzimou-Tsitouridou R (2004) Chemometric quelity assessment of surface waters: two case studies. Chem Eng Ecol 11:449–469

Soffianian A, Bakir HB, Khodakarami L (2015) Evaluation of heavy metals concentration in soil using GIS, RS and geostatistics. IOSR J Environ Sci Toxicol Food Technol 9(12):61–72

Soffianian A, Madani ES, Arabi M (2014) Risk assessment of heavy metal soil pollution through principal components analysis and false color composition in Hamadan Province, Iran. Environ Syst Res 3:3. https://doi.org/10.1186/2193-2697-3-3

Sungur A, Soylak M, Ozcan H (2014) Investigation of heavymetal mobility and availability by the BCR sequential extraction procedure: relationship between soilproperties and heavy metals availability. Chem Speciation Bioavail 26(4):219–230. https://doi.org/10.3184/095422914X14147781158674

Thomas GW (1996) Methods of soil analysis: part 2. Agronomy handbook 9. In: Page AL (ed) Soil pH and soil acidity. American Society of Agronomy and Soil Science Society of America, Madison (WI), pp 475–490

Urzelai A, Vega M, Angulo E (2000) Deriving ecological risk-based soil quality values in the Basque Country. Sci Total Environ 247:279–284

Walkley A, Black IA (1934) An examination of the Degtjareff method for determining soil organic matter and a proposed modification of the chromic acid titration method. Soil Sci 37:29–38

Wang G, Zhang S, Xiao L, Zhong Q, Li L, Xu G, Deng O, Pu Y (2017) Heavy metals in soils from a typicalindustrial area in Sichuan, China: Spatial distribution, source identification, and ecological risk assessment. Environ Sci Pollut Res 24:16618–16630

Xiao R, Bai J, Huang L, Zhang H, Cui B, Liu X (2013) Distribution and pollution, toxicity and risk assessment of heavy metals in sediments from urban and rural rivers of the Pearl River delta in southern China. Ecotoxicology 22(10):1564–1575

Yan X (2009) Linear regression analysis: theory and computing. Published by World Scientific Publishing Co. Pte. Ltd. 5 Toh Tuck Link, Singapore 596224

Zhang LM, Liu YL, Li XD, Huang LB, Yu DS, Shi XZ et al (2018) Effects of soil map scales on simulating soil organic carbon changes of upland soils in eastern China. Geoderma 312:159–169

Zhong L, Liu L, Yang J (2012) Characterization of heavy metal pollution in the paddy soils of Xiangyin County, Dongting lake drainage basin, central south China. Environ Earth Sci. https://doi.org/10.1007/s12665-012-1671-6

Chapter 13
Soil Quality Assessment in Hilly and Mountainous Landscape

Anu David Raj and Suresh Kumar

Abstract Mountains and hills play a crucial function in controlling the hydrological cycle and water availability to meet requirement of all biological components of the ecosystem. Unsustainable land management practices, steep slopes and climate change deteriorate the hilly and mountainous ecosystems. Soil erosion in the hilly and mountainous region is a significant threat to soil quality. Soil quality index is the most widely used method to measures soil quality based on soil quality indicators. Site-specific external factors are necessary for soil quality indices while determining the potential indicators for soil quality assessment. Soil hydrological properties are identified as the most potential indicators which can be used for soil quality assessment in the hilly and mountainous region. Soil erosion models can also provide comprehensive information on soil quality degradation. The geo-spatial technologies can provide a more explicit spatial distribution of soil quality. The broadened attributes acquired from geo-spatial technologies can improve the soil quality assessment of inaccessible regions in the hilly and mountainous terrain with the help of spectral indices, geo-statistics, multispectral remote sensing. The GIS interfaced soil erosion models can provide clear visualisation of soil degradation. The holistic, multifaceted and novel methods can make soil quality assessment much more manageable. Climate change and rising global food requirements demand governments and policymakers to be more vigilant in enhancing soil quality in hilly and mountainous regions.

Keywords Soil quality · Soil erosion modelling · Geo-spatial methods · Soil degradation · Hilly and mountainous region

A. David Raj · S. Kumar (✉)
Agriculture & Soils Department, Indian Institute of Remote Sensing, Indian Space Research Organization (ISRO), 4-Kalidas Road, Dehradun 248001, India
e-mail: suresh_kumar@iirs.gov.in

A. David Raj
e-mail: anudraj@iirs.gov.in

P. K. Shit et al. (eds.), *Soil Health and Environmental Sustainability*,
Environmental Science and Engineering,
https://doi.org/10.1007/978-3-031-09270-1_13

13.1 Introduction

Mountains cover one by fourth (25%) of the entire land surface of the Earth. Around 12% of the global inhabitants are reliant on mountain assets. A considerably more significant proportion of the population relies on other mountain resources, most conspicuously water. Moreover, mountains are rich in biological diversity and endangered species (Agenda 21 UN 1992). Mountains are considered the world's water towers, thus contributing 50% of the total population with freshwater. In addition, it contributes a significant percentage of the runoff from the river basins. Water is the essential source of energy production in the mountain region. Thus, mountains are inevitable to the global ecosystem's sustainability. They are highly susceptible to ecological changes triggered by climate change and anthropogenic disruptions. The deterioration of highland ecosystems will exacerbate natural resource degradation and increase the frequency of calamities like landslides and floods. Tarolli and Straffelini (2020) state that when an undulating topography is combined with unsustainable agricultural techniques and climate change, soil erosion grow to be a severe crisis in the hilly region which must be addressed with extreme sensitivity. Soil erosion in the hilly and mountainous region is a significant threat to soil quality. To grasp the gravity of the situation, it is vital to point out that worldwide potential soil loss grew by 2.5% from 2001 to 2012 (Borrelli et al. 2017). Although, sustainable land use planning and conservation measures can reduce the adverse effects of soil erosion.

Globally, limited resources and a growing population emphasised sustainability in all domains. Sustainable resource management aids in the reduction of natural resource degradation while also conserving resources for future utilisation. Sustainable land management practices improve soil quality, providing higher productivity. The primary way of ensuring ecosystem sustainability is by enhancing soil quality. However, due to their highly dynamic and vulnerable nature, the soils in hilly and mountainous regions require additional consideration. According to the UN's 17 sustainable development goals, the second goal is mentioning the 'Zero Hunger' of the population through sustainable land (Agriculture) and ecosystem management. At the same time, the thirteenth goal is also crucial for achieving sustainable agriculture management because the influence of climate is a prime factor for agriculture. The better policies with implementing institutions will lead to a sustainable hilly and mountainous ecosystem towards the UN's 'Zero Hunger' goal (Fig. 13.1). (Doran and Zeiss 2000) stated that soil health must be assessed using various soil quality

Fig. 13.1 An approach to the sustainable improvement of agro-ecosystem and natural ecosystems in the hilly and mountainous regions

indicators to reveal the extended time scale variability of agricultural lands. Thus, the monitoring and assessment of soil quality are essential for improving soil health.

Healthy soil can supply pure air and water by filtering, buffering, groundwater recharge, nutrient cycling and provide diverse habitats for species. It has a significant role in conserving land quality through performing ecosystem functions. Land quality and soil quality are interrelated, whereas it is a component of land quality. Land quality comprises weather, topography, hydrology, geology, land cover and ultimately soil quality. The soil quality is associated to all of the characteristics/components of soil. The soil properties perform the ecosystem function and which will directly and indirectly increases the environmental quality. There are numerous methods for identifying the quality and deterioration of the soil. The researchers employ various techniques based on the accessibility of data, the cost of analysis, the ease of analysis, and the study region's suitability. However, the complexity and interdependence of the processes impacting soil quality and functions make it challenging to determine it directly.

13.1.1 Soil Quality

Soil quality is the "capacity of a specific kind of soil to function within natural or managed ecosystem boundaries to sustain plant and animal productivity, maintain or enhance water and air quality and support human health and habitation" (Karlen et al. 1997). Some others stated as "soil's capacity or fitness to support crop growth without resulting in soil degradation or otherwise harming the environment or more simply, fit for purpose" (Oliver et al. 2013). The soil quality represents how well soil performs its functions to an ecosystem (Greiner et al. 2017). Soils have enormous potential to carry and perform vast functions in the ecosystem. It has the capability to sustain biological diversity, activities and productivities. Drobniket al. (2018) comprehensively described the soil functions and corresponding soil quality indicators, and they established a soil quality index depend on soil quality functions. Muñoz-Rojas et al. (2017) stated that soil quality indicators could act as a tool for ecosystem restoration.

Soil quality is a characteristic of the land that influences the capability of land at different levels, such as storage of water and nutrients, plant growth, and nutrient cycling. It has greater weight on soil biodiversity and ecological functions, which makes soils live. Soil quality is composed of the inherent and dynamic type of quality. The inherent soil quality is that, as the name indicates, it is the natural ability or quality of a specific soil, such as soil texture and drainage properties. In contrast, the dynamic soil quality is reliant on management practices. It can easily be changed according to which type of management is adopted in the field. The quantity of organic matter, soil structure and soil depth are the crucial dynamic soil qualities. Improvement of soil quality is a fundamental approach to attain agricultural and economic sustainability at different levels. Therefore, the foremost objective of soil quality assessment is to identify the management practices which improves the soil quality.

Agricultural production is affected by soil degradation due to a decreased soil quality and ecosystem services (Lal 2015). Climate change may directly or indirectly affect carbon and nitrogen mineralisation via the variation in temperature, soil moisture and mineralisation rates through the changes in the soil quality (Keller et al. 2004). Thus, continuous monitoring is necessary for improving soil quality. There is always a need for a site-specific method to enhance soil quality by reinstating the physicochemical, biological and hydrological components (Gregory et al. 2012). Soil quality assessment/valuation is the process of measuring/quantifying the dynamic/management derived changes in the soil. Larson and Pierce (1994) state that soil quality is measured in terms of specific soil functions. However, the qualitative and complex nature, the quality of the soil couldn't be determined straightway. Thus, we assess the soil quality based on the soil's physical, chemical, hydrological, and biological properties and processes; hence, we called it indicators. These indicators are quantifiable soil or terrain properties, which provides an overview of soil health/soil quality. These indicators must be easy to quantify, detect vicissitudes in soil functions related to ecosystem processes, and need sensitivity to the climate and soil management.

13.2 Soil Quality Indicators

We depend on soil to perform many functions. Healthy soil can provide enormous amount benefits to humans and other environmental systems. The critical functions of soil are comprehended as nutrient cycling, buffering and filtering, water relations and physical stability. The primary soil quality indicators and their functions or processes involved are described in Table 13.1.

As already stated in the preceding section, the soil quality indicators are divided into physical, chemical and biological indicators. Each soil quality indicator has its own or combined functions in the ecosystem. The physical properties of soil quality indicators involve bulk density, soil structure, porosity, hydraulic conductivity, depth of soil, infiltration, water holding capacity, aggregate stability, soil texture and water storage. It primarily affects the root movement, seedling emergence and movement of water. The chemical properties include soil organic carbon, pH, electrical conductivity, cation exchange capacity, available nitrogen, phosphorus, potassium, and other macro and micronutrients. The chemical condition affects the soil and plant relationship and mobility of nutrients. Among the different kinds of soil quality indicators, the biological indicators are rarely used than others and comprise microbial biomass carbon and nitrogen, soil respiration, number of earthworms and nitrogen mineralisation. The respiration and the decomposition are primarily affected due to biological indicators. All these indicators perform together and perform ecosystem functions properly.

Table 13.1 Soil quality indicators and their functions suggested by Doran and Parkin (1994)

	Indicators	Functions/processes involved
Physical	Soil texture	Holding and transport of water and chemicals
	Rooting depth	Productive environment for microbes
	Bulk density and water infiltration	Water movement and productivity
	Water holding capacity	Retention and workability
	Aggregate stability	Aeration and erosion resistivity
Chemical	pH	Chemical and biological activities thresholds and nutrient availability
	Electrical conductivity	Optimum condition required for plant and microbial processes
	Organic carbon	Soil fertility, stability and erosion resistance
	Extractable N P K	Plant available nutrients for vital functions and productivity
	Minor elements	Micronutrient availability
Biological	Soil respiration	Microbial biomass activity
	Mineralizable N	Soil productivity and N supplying
	Microbial biomass C and N	Soil productivity, microbial catalytic and N supplying potential

13.2.1 Soil Quality Indicators Relevant to the Hilly and Mountainous Region

The mountain soils are considered as slowly forming due to the lower temperature and lower biological activity. They are usually shallow, less developed and often relatively low in fertility than the other soils. The soils in hilly and mountainous regionsvary with slope and elevation. According to Harden (2001), enhanced soil erosion is frequent and environmentally damaging, particularly in mountainous areas. Soil erosion and decreased soil moisture content are severe problems in the hilly and mountainous regions (Gupta et al. 2019). However, the soils in the mountain have a significant potential to increase the ecosystem functions needed to support the Earth. Several mountains are covered with natural vegetation (forest), which helps to improve soil quality and reduce soil erosion in those regions.

The moisture content of the soils in the hilly region (especially agricultural land) is found as the limiting factor for the soil quality. Apart from other soil physical and chemical properties, most studies included the hydrological parameters for assessing soil quality in total data sets (TDS) and minimum data set (MDS). The majority of studies used minimum data set (MDS) to select the capable indicators which affect the soil quality. The soil moisture (Bo-Jie et al. 2004), permanent wilting point (Govaerts et al. 2006), available water content (Erkossa et al. 2007; Nosrati and

Collins 2019), soil porosity (Tesfahunegn 2014; Nabiollahi et al. 2018), saturated hydraulic conductivity (Sofi et al. 2016), effective porosity (Dai et al. 2018) and infiltration rate (Mandal et al. 2011) are the significant hydrological indicators identified by researchers. As erosion is the primary threat to the landscape in the hilly and mountainous region, the majority of studies included soil organic carbon, available P, K and total N. As per the Pausas et al. (2007) and Pham et al. (2018), soil organic carbon changes be influenced by the topographic characteristics, such as aspects and slope. Because of the influence of other soil parameters and crop production differences, it is identified that, slope gradient is one among the most critical topography feature element determining soil quality (Jakšić et al. 2021). As previously stated, the vast majority of studies addressed soil quality assessment in hilly and mountainous regions, but the majority of studies omitted to explore topographical parameters (slope, aspect, elevation) and their relationship to soil quality (Brejda et al. 2000; Govaerts et al. 2006; Pal et al. 2012; Singh et al. 2014; Tesfahunegn 2014). At the same time, Ghosh et al. (2014), Nabiollahi et al. (2018) and Nosrati and Collins (2019) evaluated the soil quality based on topography and soil erosion vulnerability. The selection of appropriate indicators and soil quality assessment based on different topographical parameters and erosion processes are necessary for reliable soil quality estimates in the hilly and mountainous region.

13.2.2 Soil Organic Carbon as an Indicator of Soil Quality

Carbon deposited in the form of soil organic matter is referred to as total organic carbon. The decomposition of plant and animal dead wastes, as well as soil microorganisms, stores carbon in the soil. It is considered the critical supply of energy for soil microorganisms. It can influence plant development as an energy source and nutrient availability via mineralisation; thus, soil organic carbon (SOC) is considered one among the utmost significant soil quality elements. Humus helps to maintain aggregate stability and nutrient and water storage capacity. Bastida et al. (2008), Lal (2016) and Hueso-González et al. (2018) designated organic matter/carbon as a star indicator in soil quality assessments over agro-ecosystems, and they have also stated that organic carbon is essential for the growth of crops and plants in natural vegetation. According to Egli and Poulenard (2016), organic matter is the extremely dynamic and crucial constituent of soils due to its participation in all kind of processes. Even in younger mountain environments, extensive carbon stocks and sequestration rates can be explored. Soil organic carbon is essential for climate change adaptation and mitigation methods because it is associated with a diverse of soil processes and also serves as a primary carbon sink in terrestrial ecosystems (Lozano-García et al. 2017; Muñoz-Rojas et al. 2017).

According to Jones et al. (2014), in large-scale surveys, dissolved organic carbon quality rather than quantity gives a more meaningful soil quality index. Van-Camp et al. (2004) state that SOC promotes soil nutrient availability encouraging plant production; it enhances water holding capacity, reduces runoff and soil erosion.

According to Rajan et al. (2010), soil organic carbon is the most discriminating soil quality measure in the eroded terrain. As a result, SOC is a considerably more dependable soil quality indicator, and its continuous examination reflects whether the land is improving, worsening, or remaining static. Total organic carbon in the soil has been demonstrated to be meticulously associated with the amount of organic matter contributed to the soils in agricultural residues, manure, or other sources. Soil type, climate, region, land use type, and land management all impact SOC stock. Therefore, SOC is a vital variable in soil management (Bot and Benites 2005). Thus, total carbon/organic matter is one among the commonly utilised indicators (Bünemann et al. 2018).

13.3 Methods of Assessing Soil Quality

The soil's physicochemical and biological characteristics perform as an indicator of soil quality. There are various approaches available to evaluate the soil quality illustrated in Fig. 13.2. The most straightforward method of assessment is through the use of soil quality indices. Karlen et al. (1994) assessed soil quality using a multi-parametric index. Previously, simple quotient-type soil quality indices were commonly utilised (Insam and Domsch 1988). Bastida et al. (2008) provided a detailed description of the vastly utilised soil quality indices. The multi-criteria decision-making techniques are often used to assign weight to soil quality indicators. Furthermore, soil quality can be examined using modelling techniques. In this

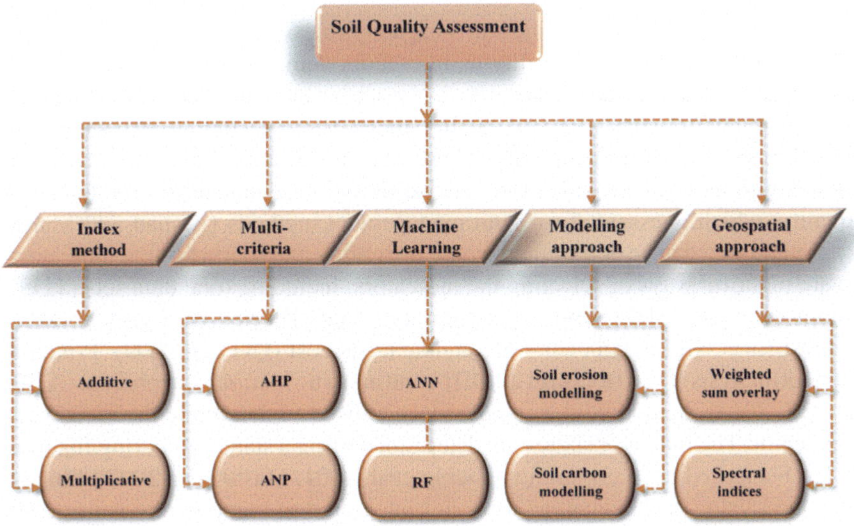

Fig. 13.2 Flowchart showing overall general approaches for soil quality assessment

setting, scientists develop models that accurately represent real-world processes. This approach incorporates the soil erosion models and soil carbon dynamics models. Soil erosion is the primary land degradation process that occurs in the hilly and mountainous region. For example, the Erosion Productivity Impact Calculator (EPIC) model (Liu and Han 2020) can forecast the impact of soil degradation due to soil erosion. CENTURY is a site-specific complex model used to simulate the dynamics of carbon, nitrogen, phosphorus, and sulphur in soil (Smith et al. 2009). There are several geo-spatial techniques available to assess soil quality. Only two methods are illustrated in Fig. 13.2. Apart from this, geostatistical methods are also widely used to find the spatial distribution of soil properties. Geo-spatial technologies can aid in the assessment of soil quality both directly and indirectly. The spectral indices produced by remote sensing techniques are incredibly dependable and can measure soil quality over a vast area. For these types of analyses, high-resolution satellite data and different spectral resolution satellite sensors are required. For example, Kalambukattu et al. (2018b) created a digital soil map of a Himalayan watershed using Landsat 8-OLI 30 m resolution data. The different methods used to assess soil quality in hilly and mountainous regions are described below. While last few years Artificial intelligence (AI), especially machine learning techniques is widely used for the digital soil quality parameter mapping on a larger geographical region. Artificial neural network (ANN), random forest (RF), support vector machine (SVM) etc. are the major algorithm using for the mapping.

13.3.1 Index Method

The term "soil quality" refers to the soil's capacity to execute its essential services. Most researchers examine soil quality in agro-ecosystems rather than other ecosystems since it has a more significant direct impact on humans than others. Among the several methodologies, the soil quality index-based technique is extensively used worldwide due to its simplicity of use and quantitative flexibility (Qi et al. 2009). In addition, it can provide more precise site-specific indications for assessing soil quality, which is essential for land management methods (Arshad and Martin 2002). Soil quality can be determined by the weighting and scoring of the indicators. There are plentiful ways of selecting the indicators, including total data sets (TDS), minimum data sets (MDS) and expert opinion (EO) (Herrick et al. 2002). MDS is the most successful in dependability and analysis cost (Qi et al. 2009). Principal component analysis (PCA) (Vasu et al. 2016) and analytic hierarchical process (AHP) (Lotfi et al. 2013) methodologies were widely utilised to determine the weights of each indicator. The indicator scores are determined using a standard scoring function (SSF) has been used by many studies (Liebig et al. 2001; Masto et al. 2007). Finally, the weighted value and indicator scores are integrated to compute the soil quality index for the region of interest. There are primarily two approaches for integrating soil quality indices, namely additive and multiplicative.

Bo-Jie et al. (2004) evaluated the soil quality regime associated with land cover and slope positions along an incredibly reformed slope landscape. Due to its highly dynamic nature, the slope plays a significant role in distributing soil quality. The principal component analysis produced the soil quality index based on the land cover type by using simple individual indicator scoring and weight assessment. They observed that there is considerable variance in soil quality between the slopes. Similarly, Ghosh et al. (2014) conducted a study to establish the consequence of elevation and slope aspects on India's northwestern Himalayan soil quality. The microclimate, vegetation establishment, water movement, and erosion are entirely influenced by topographic characteristics such as the hill slope's directional orientation and steepness (Birkeland 1984). Soil quality degradation is caused mainly by increased soil erosion rates on the steep slopes of hills and mountains. Thus, the distribution of soil physical, chemical, biological, and hydrological properties vary significantly within a narrow range, particularly in mountainous terrains. Campos et al. (2007) used the principal components to assess the soil quality, which varies due to the elevation change and land use. They found that the changes in land use resulted in a decrease in organic matter content, particularly on the middle slope and lower slope. They identified 62.1% of data variation due to these six variables (C, N, Mg, effective CEC, Total P, bulk density, and microbial quotient). According to the soil quality indices developed, the higher slope class had a tremendous soil loss rate with less soil quality in the hilly and mountainous region of Kurdistan, Iran. On the other hand, the lesser slope class had the highest soil quality grades (Nabiollahi et al. 2018).

Erkossa et al. (2007) aim to study the effect of land managing practices in the central highlands of Ethiopia. The additive method (Andrews and Carroll 2001) was used for the integration of indicators. Indicator scoring is the only step needed for the additive method, and the indicators selected are based on non-linear scoring curves explained by Andrews et al. (2002). This scoring curve employs a site-specific coefficient in nature, considering factors such as region, climate, crop and, most significantly, slope. Due to erosion, the slope is a significant element in hilly and mountainous landscapes, as many soil nutrients are removed from the surface. Therefore, they considered slope and other indirect site-specific factors while measuring soil quality. It contributes to the improvement of a more reliable soil quality index in hilly terrain. Nosrati and Collins (2019) aim to develop a new soil quality index based on principal component and classification analysis (PCCA) and general discriminant analysis (GDA) for the Southern Alborz Mountains in Iran. Correlations and multiple regression models demonstrated that the soil quality index is sensitive to changes in soil surface factors caused by soil erosion in the study area's various land use categories.

Numerous soil quality evaluation studies based on soil quality indices have been undertaken worldwide in hilly or mountainous terrain (Table 13.2). It was identified that encroachment in these areas degrades soil quality via natural vegetation removal or intensive agricultural land management techniques. These are particularly sloping regions; removal of natural vegetation leads to direct rainfall interception, which deteriorates the physical and chemical qualities of the soil. As a result, measuring

Table 13.2 Soil quality indexing methods adopted by different studies conducted in the mountainous region

Country	Objective	Indictor selection	Indicator scoring	Integration	SQI	Assumed beneficiaries*	Source
China	Soil quality regime with respect to slope and land use	TDS	Linear	Multiplicative (PCA)	$QI = \sum_{i=1}^{n} W_i \times Q(x_i)$ *– for land use-based soil quality index*	Farmers/Government	Bo-Jie et al. (2004)
				Relative difference	$PI = \sum_{i=1}^{n} (((x_i - x_i^{'})/x_i^{'}) \times 100)/n$ *– for slope-based soil quality index*		
Ethiopia	Consequence of land preparation method on soil quality	MDS (Literature)	Non-linear	Additive	$SQI = \left(\frac{\sum_{i=1}^{n} S_i}{n}\right) \times 10$	Farmers/Decision-making	Erkossa et al. (2007)
Mexico	Change of land use on soil quality on a slope by comparing PCs	Potential indicator (PCA)	–	–	Comparison of PC (PCA)	Farmers/Decision-making	Campos et al. (2007)
India	Influence of topographic features and INM on crop productivity and soil quality	MDS (PCA)	–	Weighted (PCA)	$SQI = \sum_{i=1}^{n} W_i \times S_i$	Farmers/Decision-making	Ghosh et al. (2014)

(continued)

Table 13.2 (continued)

Country	Objective	Indictor selection	Indicator scoring	Integration	SQI	Assumed beneficiaries*	Source
Ethiopia	Approaches for evaluating soil quality and monitoring soil degradation	MDS (PCA and EO)	Linear and non-linear	Additive and multiplicative (PCA)	$Unscreened SQI = \frac{\sum_{i=1}^{n} S_i}{n}$ $PCA - SQI = \sum_{i=1}^{n} W_i \times S_i$	Farmers/Decision-making	Tesfahunegn (2014)
India	Impact of soil quality on cropping system	MDS (PCA)	Linear	Multiplicative (PCA)	$SQI = \sum_{i=1}^{n} W_i \times S_i$	Farmers/Decision-making	Sofi et al. (2016)
Iran	Effects of slope gradient and land use change on soil quality	TDS and MDS (PCA)	Linear and non-linear	Additive and multiplicative (FA)	$SQI_a = \left(\frac{\sum_{i=1}^{n} N_i}{n}\right) \times 10$ $SQI_w = \sum_{i=1}^{n} W_i \times N_i$ $SQI_n = (\sqrt{Pavg^2 + Pmean^2/2}) \times \frac{n-1}{n}$	Farmers/Decision-making	Nabiollahi et al. (2018)
Tibet	Assessing soil quality for sustainable cropland management	MDS	Fuzzy set techniques	Multiplicative (PCA)	$SQI = \sum_{i=1}^{n} W_i \times \mu(x_i)$	Decision/Policymaking	Dai et al. (2018)
Iran	SQI for land use and soil erosion classes	MDS (expert opinion)	SSF	Multiplicative (PCCA and GDA)	$SQI = 0.91(urease) - 0.42(clay) + 0.30(\beta glucosidase) + 0.21(sand) + 0.03(alkaline\ phosphatase) - 0.02(dehydrogenase)$	Decision/Policymaking	Nosrati and Collins (2019)

(continued)

Table 13.2 (continued)

Country	Objective	Indictor selection	Indicator scoring	Integration	SQI	Assumed beneficiaries*	Source
Brazil	Assessing soil degradation of extensive pastures in hilly landforms region using Soil quality index	MDS	–	Partial least square	–	Decision/Policymaking	Burak et al. (2021)

* The beneficiaries for each literature are assumed by the authors and is not stated by the authors of respective articles. The majority of studies has not stated the beneficiary explicitly

the soil quality in the mountain region is critical, mainly using more straightforward and dependable methods, such as the soil quality index.

13.3.2 Multi-criteria Method

The multi-criteria decision making is used to derive the best soil quality based on several criteria (soil quality indicators/processes) available for the study. There are diverse methods and principles available for multi-criteria decision making. In soil quality analysis, analytical hierarchy processes (AHP) is seen as the most common multi-criteria method. In most cases, It was used as part of the soil quality indexing method. It is mainly used to derive potential soil quality indicators and deduce their weight concerning each other (Lotfi et al. 2013). However, it requires expert knowledge to determine the pairwise comparison of different indicators. To begin, create a comparison matrix based on a set of pairwise comparisons using the AHP preference scale (Saaty 1987). Then, the Delphi approach, also known as the expert judgment system, was used to govern the comparative value of each factor. Using the analytic hierarchy technique, Qi et al. (2009) determined the weight contributions of individual indicators. Furthermore, the AHP ensures that the decision maker's valuations are consistent, decreasing preconceived notion in the decision-making procedure. Xue et al. (2019) developed a method for assessing soil health by combining meta-analysis to recognize MDS, AHP and expert valuing to allocate weight, for identifying the long-term effect of organic farming.

While there are various ways for determining weights, the analytic network process (ANP) is one of the best for dealing with multiple and heterogeneous elements (Saaty 2005). Seyedmohammadi et al. (2019) used the fuzzy-ANP technique to determine land suitability and discovered that the ANP could effectively govern criteria weights with varying relative relevance depending on skilled opinion. Integrating the ANP with a fuzzy set technique resulted in a realistic estimation of yield. Demirel and Tüzün (2011) assessed the problem using a multi-criteria decision-making process. Climate, terrain, soil, land use, and human activities are some of the primary characteristics considered. Because of the problems in a hierarchical structure, this study also offers certain sub-criteria with it. The problem was solved using fuzzy-ANP and soil erosion prevention strategies evaluated. Dai et al. (2018) also assessed soil quality for sustainable farmland management in the Lhasa River Valley, Tibetan plateau, using factor analysis and fuzzy sets. The inherent correlation and variation between soil quality indicators were determined using factor analysis and the primary components determining soil quality. In addition, they found that fuzzy sets can be an effective tool for scoring soil quality indicators.

13.3.3 Modelling Approach

13.3.3.1 Soil Erosion Modelling

Soil erosion will drastically diminish the soil quality in hilly regions; thus, the soil erosion model can systematically assess the soil quality deterioration. These kinds of models are classified into empirical, semi-empirical, and process-based models. These models use soil, topography, land use/land cover, meteorological and hydrological features to simulate the soil erosion processes. In addition, these models simulate soil erosion/nutrient erosion using field observations. Thus, they can quickly analyse the soil quality and optimum management strategies for a specific landscape or a crop without conducting field experiments. Soil quality assessment has become increasingly integrated into land evaluation procedures in recent years. However, modelling can assess the trend of soil quality where significant degradation occurs and, more explicitly, the degradation process. Soil erosion is the primary hazard to soil quality in mountain environments. Soil erosion is indistinguishably linked to decreased crop production and impairs the soil's ability to execute its function. Identifying and measuring areas prone to soil erosion is required to mitigate against soil degradation and implement management practices to enhance soil quality. This section discusses several soil erosion models used to assess soil erosion in hilly and mountainous regions.

As mentioned previously, most soil quality indices based studies conducted in hilly and mountainous regions did not consider topographic factors when measuring soil quality. The soil, water, and atmosphere are all interrelated and function as a unified system. Thus, the main limitation of the soil quality index is that it merely provides an indicator value and does not explain the relationship between the various external elements or components. The universal soil loss equation (USLE) is a pioneering attempt to quantify soil degradation caused by soil erosion. It simulates or calculates soil erosion from a field/watershed using rainfall data, soil parameters, topographical factors, land use/land cover, and conservation practices in an area. While the attributes utilised for the soil quality index and the modelling technique are compared, the soil quality index reflects only the dynamic aspects affecting the soil and ignores intrinsic and external elements that directly or indirectly affect soil quality. Thus, erosion models have more potential than soil quality indices to identify soil degradation status and mitigation measures against it.

Sun et al. (2014) try to determine land use and topography impact on soil erosion in China. They assessed soil erosion in the steep watershed using RUSLE. The findings revealed that under the same land use type, a rising trend in soil erosion was determined as the slope gradient increased. Similarly, Prasannakumar et al. (2012) employed the revised universal soil loss equation (RUSLE) to quantify soil erosion risk in a sub-watershed of Kerala. Shrestha (1997) used MMF to measure the soil erosion in Likhu Khola valley in Nepal. The Morgan-Morgan Finney (MMF) model is a semi-empirical model used to assess annual soil loss from field-sized areas on the hill slope. Gupta and Kumar (2017a) try to identify climate change impact on soil erosion over the mid-Himalayas using the RUSLE. Also, Kalambukattu et al. (2021) studied the soil erosion risk assessment in a Himalayan state, Uttarakhand, India using the RUSLE. Nevertheless, these empirical and semi-empirical models lack the complete process representation, whereas the process-based model can represent the physical processes based on the law of conservation of mass, energy and momentum.

The physically-based models produce more trustworthy output than empirical and semi-empirical models. However, physically-based models require a large quantity of data on rainfall, temperature, evapotranspiration, land use/cover characteristics, soil physical, chemical, biological, hydrological properties, and topographical factors. For example, Hessel et al. (2006) assess the Limburg Soil Erosion Model (LISEM) on a watershed in Kenya and Tanzania. Another benefit of process-based models is that they enable us to identify potential sources of soil erosion in a given watershed. However, assessing land management approaches is frequently time-consuming and costly, while process-based models can readily simulate soil erosion based on inputs. For example, Sooryamol (2020) used the ArcSWAT model to forecast the future climate influence on soil erosion in the mid-Himalayan environment. Similarly, Singh (2012) uses the process-based model APEX to predict surface runoff, soil erosion, and nutrient loss at the watershed scale. Recently, various studies quantified the climate change impact on soil erosion using calibrated erosion models in Lesser and Shiwalik Himalayas (David Raj et al. 2022; Sooryamol et al. 2022). Gupta and Kumar (2017b) studied the climate change impact recently on soil carbon sequestration in a Himalayan landscape using the CENTURY, a process-based model. They have successfully calibrated and validated the carbon sequestration potential at field conditions. Then SDSM derived future climate scenario was used to model the future carbon sequestration scenario and found that soil organic carbon in soils will be decreasing in.

Soil erosion models can be used as a versatile tool in assessing soil quality. The different modelling studies conducted in the hilly region is described in Table 13.3. Primarily it computes erosion, runoff and critical source areas. As we examine the methods for assessing soil quality in greater detail, modelling tools provide a broader perspective on soil quality and related aspects. These techniques improve reliability, are less time-intensive, and are more cost-effective. This technique may replicate both real-world and future events by calibrating and validating the model against site-specific properties. The primary advantage of this approach is that it enables us to simulate the outcome of a larger region in a concise amount of time and cost. However, due to the inaccessible terrain and a lack of specific information on soil erosion variables, assessing soil erosion in hilly and mountainous terrain was complicated for academicians and policymakers. Today, most soil erosion models operate via a spatial interface for data input and output. The GIS interface visualises the geographic distribution of soil and other attributes, providing a complete picture of the area of interest.

Table 13.3 Showing some selected empirical, semi-empirical and process-based models in each group with details

Type	Model	Inputs	Calibration/validation	Objectives	Country	Source
Empirical	RUSLE	R, K, LS, C, P	Nil	Soil erosion risk assessment	India	Kalambukattu et al. (2021)
Semi-empirical	MMF	Rainfall characteristics, slope, soil detachability index, BD and cover factor	Nil	Soil erosion prediction	Nigeria	Ande et al. (2009)
Physically-based	SWAT	Topography, weather, soil properties, land use/land cover and hydrological properties	Yes	Surface runoff and sediment yield modelling	Ethiopia	Asres and Awulachew (2010)
Process-based	APEX	Topography (DEM), weather, soil properties, land use/land cover and hydrological properties	Yes	Surface runoff, soil erosion, water quality and BMP estimation	India	Kumar et al. (2021)
Physically-based	LISEM	Rainfall data, plant characteristics and soil characteristics	Yes	Soil erosion induced by rainfall	Italy	Cuomo et al. (2015)
Process-based	CENTURY	Soil, weather, field management practices	Yes	Simulating climate change influence on soil carbon sequestration	India	Gupta and Kumar (2017b)

13.3.3.2 Soil Carbon Modelling

Soil carbon has a decisive role in crop productivity and improving the environmental quality of a region. The soil carbon dynamics has a decisive role in controlling climate change. The soil can act as a sink for the carbon in the atmosphere by various processes. As we know, most of the hilly and mountainous regions are covered with forest cover and act as a sink for carbon. The plant and animal residue makes the forest soils more productive. Changes to boost vegetative inputs to soils from plant production and steps to prevent soil carbon losses due to decomposition, erosion, or other disturbances are the management strategies for raising soil carbon. Soils are essential constituents of the carbon cycle. Increased soil organic carbon content improves soil quality, diminishes soil erosion and degradation, enriches surface water quality and boosts soil productivity (Lal 2002). There several carbon dynamics model that is used to simulate the soil carbon processes. It can accurately simulate the dynamics of carbon in soil and atmosphere.

Falloon et al. (2000) used the Rothamsted carbon model to govern the importance of the inert organic matter in predictive soil carbon modelling. Easter et al. (2007) used the Global Environment Facility Soil Organic Carbon (GEFSOC) model to conduct regional-scale inventories of soil carbon and analyse the effects of land use variation on soil carbon. Farage et al. (2007) investigated the effects of changing agricultural methods to surge soil carbon stocks using two well-proven soil organic carbon models (CENTURY and RothC). APSIM, RothC, and CENTURY were used to model soil carbon dynamics by Ranatunga et al. (2001), and the models provided a good depiction of the pattern of soil carbon reduction under continuous farming. The CENTURY model was employed to evaluate the effect of climate change and soil carbon sequestration Himalayan croplands in Uttarakhand, India (Gupta and Kumar 2017b). These studies proved that soil carbon dynamic models efficiently simulate soil carbon processes and suggest sustainable soil organic carbon-storing and enhancing methods.

13.3.4 Geo-spatial Applications in Soil Quality Mapping

The geo-spatial approaches of soil quality assessment are current trending methods used widely around the globe. The main advantage of this method is, of course, cost-effective and getting to know about the spatial distribution of soil quality and other properties related to soil. The above-discussed majority of soil erosion models utilised the GIS interface for modelling soil erosion. However, the soil qualities differ in space and time from a small field to a larger area. Thus, capturing the variation of soil quality in space and time, geo-spatial technologies are highly required. This section only analyses literature that primarily focuses on the geo-spatial approach to quantify and map soil quality.

Physical characteristics of soil take longer to assess than chemical characteristics. Furthermore, soil chemical characteristics change dramatically over time and space

(Minasny and Hartemink 2011). Remote sensing techniques have provided several solutions for quick soil assessment over the last few decades (Vasques et al. 2010). It is possible to accomplish that near-infrared spectroscopy (NIRS) can be employed in precision farming, and it is a quick and ecologically acceptable tool for predicting soil quality indices. Yunus et al. (2019) used near Infrared Spectroscopy for the estimation of soil quality indices. The results of this study reveal that NIRS technology can be utilised to monitor and quickly assess the NPK contents of soil samples with high accuracy and robustness. Similarly, Paz-Kagan et al. (2015) used the VIS–NIR-SWIR spectral band to create a soil quality index map for the Agricultural Research Site in the Northwestern Negev Desert, Israel. It was found that by utilising the spectral-spatial data delivered through infrared spectroscopy (IS) technology, soil quality may be monitored efficiently. Thus, the IS-based soil classification system provides the foundation for spatially precise and assessable strategies for examining soil quality and purpose at the local scale.

Geostatistics is another technique mainly used to create semi-variograms to depict spatial patterns and forecast soil property values at unsampled places. Kumar and Singh (2016) used terrain features and geostatistical approaches to discover the spatial distribution of soil nutrients in a Himalayan watershed. The findings revealed that using auxiliary terrain variables can significantly enhance soil property prediction. The necessity to identify crucial areas for sustainable management in the watershed has arisen due to the spatial forecast of soil nutrients. To forecast the geographical variation of SOM, Zhang et al. (2012) used ordinary kriging (OK), multiple linear stepwise regressions (MLSR), and regression kriging (RK). To forecast soil qualities (SOM, CEC, Mg, K, pH), Khanal et al. (2018) employed multispectral aerial photographs, topography data, and machine learning. Ballabio (2009) used support vector regression to test the spatial prediction of soil parameters in temperate mountain environments. Dharumarajan et al. (2017) utilised random forest approaches to forecast the spatial distribution of significant soil parameters such as SOC, pH, and EC. Pande et al. (2021) used multispectral satellite pictures, and wavelet transforms methods to estimate soil chemical characteristics such as SOC, pH, and EC.

Kalambukattu et al. (2018a) aim to identify the geographic heterogeneity of soil quality measures in a Sub-Himalayan landscape. The inverse distance weighted (IDW) interpolation technique was employed to develop spatial distribution maps of SOC content, SOC stratification ratio, and CN ratio. Kalambukattu et al. (2018b) also used an artificial neural network model to develop a digital soil mapping of a Himalayan watershed. The indices developed from the Landsat-8 satellite and topography metrics were developed using CartoDEM (30 m). Tunçay et al. (2021) also used kriging and IDW interpolation to develop the spatial distribution of soil properties to create a soil fertility index. Thus, the geo-spatial methods also need the validation procedure to ensure the reliability of the prediction.

Ennaji et al. (2018) applied multi-criteria approach in a GIS environment for land suitability analysis for sustainable agricultural practices. By superimposing the indicator and sub-indicator weights on the soil quality map in a GIS environment, the weighted sum overlay analysis was utilised to produce the soil quality map. The AHP multi-criteria method has also proven helpful for intensive agriculture as

Table 13.4 Geo-spatial approaches adopted by different studies conducted in the mountainous region

Geo-spatial method used	Objective	Region/Country	Source
Inverse distance weighted (IDW) interpolation approach	Spatial variability analysis of soil quality	India	Kalambukattu et al. (2018a)
Spectral indices and ANN	Digital soil mapping	India	Kalambukattu et al. (2018b)
Near-infrared spectral data to predict soil properties	Rapid prediction of soil quality indices	Indonesia	Yunus et al. (2019)
Spatial thematic layer of slope, elevation, soil, land use/land cover and spectral indices	Mapping soil erosion	Morocco	El Jazouli et al. (2017)
IDW and kriging	Assessing soil fertility index	Turkey	Tunçay et al. (2021)

a soil quality and land suitability evaluation methodology. Likewise, Yalew et al. (2016) used remote sensing, GIS, and AHP approach to investigate agricultural land suitability. On a GIS platform, these criteria were pre-processed as raster layers, and the weights assigned to the raster layers in establishing compatibility were derived using the analytic hierarchy approach. Also, GIS interfaced erosion models provide various capabilities for assessing soil erosion in different scales. El Jazouli et al. (2017) applied the USLE model in combination with GIS and spectral indices to determine the geographic distribution of erosion-prone areas. The steep slopes and inadequate plant cover facilitate soil loss. The spectral index method, which allows for a qualitative water erosion assessment.

Remote sensing and GIS are necessary components of soil quality evaluation in terms of dependability, the time required, spatial perspective, and visualisation. Recently, in the soil quality index and other approaches, numerous researchers have used (Table 13.4) geo-spatial methods to govern the spatial distribution of soil quality indicators and produce soil quality index maps for the study region at various scales. Kumar and Kalambukattu (2021) provide a detailed description of geo-spatial technology and their application in soil quality assessment. The RS and GIS is an emerging technique and will provide the spatial distribution of properties accurately. Hence, it will provide a more comprehensive understanding of degradation in the area of interest.

13.3.5 Novel Approach

Smart farming is a new concept which will help agricultural monitoring and are helpful in soil quality assessment. It is possible to combine sensors (soil), software, telematics and data analytics solutions (Ali et al. 2020). There is a need for a cost-effective and fast analysis technique for soil quality assessment (Bünemann et al. 2018). The electrochemical sensor can detect soil properties in real-time and rapidly and port one place to another (Alahi et al. 2016). It can be used as a long-term soil quality monitoring system, although it has still some limitations (Hashemi et al. 2011). Combining sensors with the internet of things (IoT) can contribute to smart agriculture and soil quality assessment. The farmers can access the data related to their farm/field soil quality parameters using the mobile phone (Rajalakshmi and Mahalakshmi 2016). The soil quality management over the inaccessible hilly and mountainous terrains can help this kind of approach. As a result, wireless with IoT in agriculture can aid in agricultural harvesting and global production. In other words, smart investigative approaches for soil health monitoring and management can be utilised by IoT assisted electrochemical soil sensing. To designate it more accessible to small-scale farmers, genuine public, private and social collaboration efforts are desirable (Ali et al. 2020).

Considering the importance, the electrochemical sensor for soil quality monitoring receives insufficient attention. It may be due to a deficiency of familiarity between farmers and an absence of marketing strategies. Furthermore, this approach calls for significant scientific, administrative, and policy efforts to undertake and supervise future research in innovative electrochemical sensing systems for active and proficient soil quality evaluation required for smart farming. Such developments will be beneficial in lowering farming costs, reducing unnecessary pesticide and fertiliser use, managing crop growth, and ensuring good food for a healthy future (Ali et al. 2020). Furthermore, these electrochemical sensors and other satellite-based monitoring can also be integrated with IoT-based smart farming systems. Therefore, the smart farming-based soil quality assessment can contribute to a great extent to the development of the sustainable hilly and mountainous ecosystem.

Apart from this, machine learning approaches are widely used for spatial distribution of soil quality assessment in various studies. The vegetation, soil spectral indices, terrain indices etc. can be used in machine learning technique to identify the soil quality after proper training of different model with known data points. It can deliver more accurate spatial distribution of soil quality parameter for the area of interest. For example, Paul et al. (2020) employed Random Forest (RF) algorithm to asses the soil quality in a river basin located in eastern India and also found that RF provide acceptable performance of soil quality assessment.

13.4 Case Study: Soil Quality Assessment in a Watershed of Himalayan Region-India

13.4.1 Study Area

A study was carried out in a watershed representing hilly and mountainous landscape comprising the forest and crop land. The watershed lies in the landscape of the Shiwalik range located in the Hamirpur district of Himachal Pradesh, India. Geographically, the watershed extends 31° 43′ 26″ N to 31° 45′ 00″ N latitude and 76° 26′ 25″ E to 76° 28′ 10″ E longitude, covering an area of 500 ha with an elevation ranging from 475 up to 823 m. Sandy loam is the predominant soil texture. The climate belongs to the humid subtropical zone, which is characterised as hot, and the rainy season generally extends from July to September. The average annual rainfall in the watershed is 1342 mm, and more than 75% of the rainfall is gained from July to September. Paddy and maize are the major Kharif (rainy season) crops, and Rabi (winter) crop is wheat. The forest type in the area is dry mixed deciduous. The main species of forest trees are Khair (*Senegalia catechu*), Chir pine (*Pinusroxburghii*), Sheesham (*Dalbergiasissoo*) and Neem (*Azadirachta indica*).

13.4.2 Methodology

CartosatDEM digital elevation model was used for slope analysis and topography delineation in the watershed. The land use/land cover map was derived from the IRS LISS-IV false colour composite. The detailed methodology is illustrated in Fig. 13.3. The soil-landscape/physiography map of the watershed was generated by intersecting the topography map, slope map, and land use map. According to this, the watershed has 11 major landscape/physiography classes (Fig. 13.4) are identified. Each class has unique soil physicochemical properties. Soil samples (52 sites) were collected from the soil-landscape/physiographic unit, namely upper and middle hillslope with upper and lower piedmont, including all land use/land cover of the watershed. Soils were analysed for the physical properties (soil depth, bulk density, sand, silt, clay, and aggregate stability (0.50 mm), chemical properties (pH, EC, SOC, N, P, K) and hydrological properties (infiltration rate and unsaturated hydraulic conductivity).

The minimum data sets (MDS) and total data sets (TDS) are the effective methods of selecting the indicators. Qi et al. (2009) reported that MDS could adequately represent the quality and thus saves time and money, but care should be taken to define which indicators are comprised in the MDS method. Plenty of methods are available to screening the parameters viz., analytical hierarchy process (AHP) (Saaty 1994), and principal component analysis (PCA) (Masto et al. 2007). The PCA was used to screen the most affected parameters of soil quality. Scoring can be done for selected parameters, and the indicators are selected based on their importance for soil functions (Kumar et al. 2019; Vasu et al. 2016). We used the Glover LSF method”

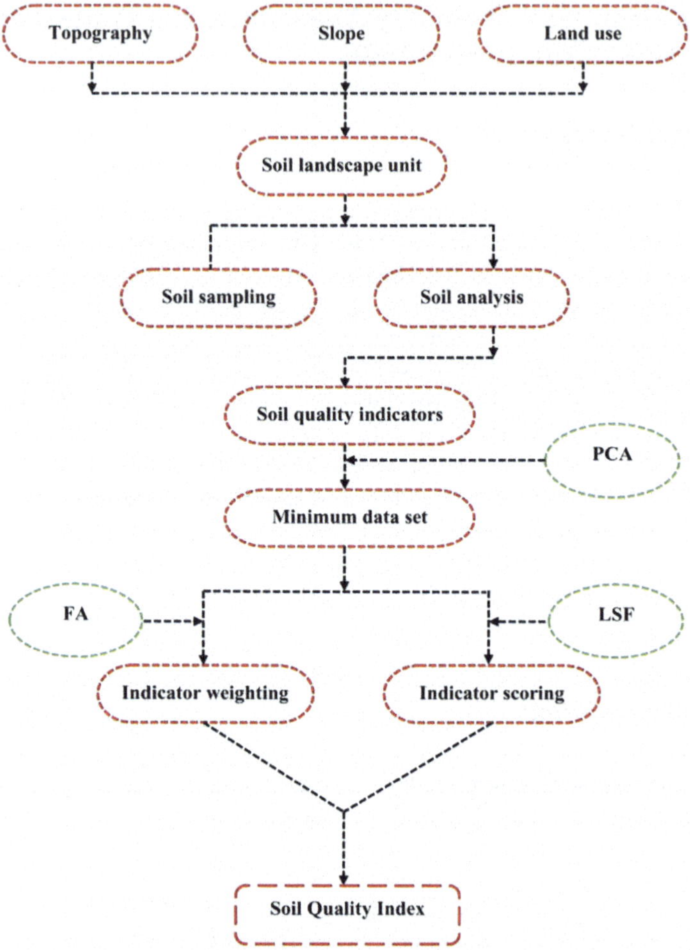

Fig. 13.3 Flowchart of the methodology used in the case study

for the indicator scoring (Glover et al. 2000; Masto et al. 2007). The linear score is Y, the soil attribute value is x, and the lower and upper threshold values are s and t. The score is 0 for numbers below and above the threshold. For "more is better," Eq. (13.1) was utilised, while (13.2) was used for "less is better," and a combination of both was used for "optimum is better."

Linear Scoring function

$$LSF(Y) = \frac{(x - s)}{(t - s)} \tag{13.1}$$

Linear Scoring function

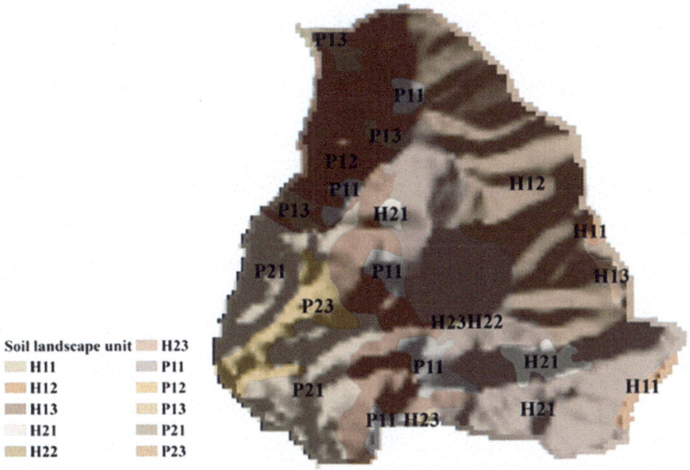

Fig. 13.4 Soil-landscape/physiographic map unit of the watershed

$$LSF(Y) = 1 - \frac{(x - s)}{(t - s)} \tag{13.2}$$

After extracting the minimum data set, the indicator was scored by the linear scoring method and weighted based on the communalities derived from the factor analysis (FA). The soil quality index was calculated based on the integration of weights (W_i) and scores (S_i) (13.3). The spatial distribution of the soil quality map was also generated using the IDW interpolation technique.

$$SQI = \sum_{i=1}^{n} W_i \times S_i \tag{13.3}$$

13.4.3 Results and Discussion

Five principle components with eigenvalues ≥1 were selected for the MDS indicators (Table 13.5). The PC1 comprises 29% of variation followed by 27.89%, 15.93%, 11.35% and 9.1% for PC2, PC3, PC4 and PC5 respectively. The first five PCs can explain approximately 93.5% variation in soil quality. The scree plot describes the principal component and their eigenvalues (Fig. 13.5). Phosphorus and potassium, soil depth, and infiltration rate were all heavily weighted factors in the first principle component (PC1). The most weighted variable, available phosphorus, was preferred for the MDS as these are the utmost characteristic of that set. For PC2, PC3, PC4

Table 13.5 Principal component analysis (PCA) of soil quality indicators from a hilly and mountainous watershed

PCs	PC1	PC2	PC3	PC4	PC5
Eigenvalue	4.676	4.4643	2.549	1.816	1.456
Percent	29.223	27.895	15.934	11.353	9.103
Cumulative percent	29.223	57.118	73.052	84.405	93.507
Eigenvectors					
Sand	0.392	0.823	−0.297	0.189	−0.127
Silt	−0.55	−0.689	0.379	−0.027	−0.165
Clay	0.047	−0.702	0.03	−0.388	**0.578**
BD	−0.347	−0.024	0.23	**0.685**	0.334
AggS	0.512	−0.044	**−0.736**	−0.156	0.316
pH	0.444	−0.068	−0.294	0.658	0.492
EC	0.583	0.759	−0.093	0.25	−0.017
SOC	0.591	0.183	−0.132	**−0.695**	0.125
TN	0.016	−0.034	**0.719**	0.034	0.536
P	**0.906**	−0.049	0.352	−0.039	0.088
AK	0.756	0.2	0.529	−0.049	0.108
Infiltration rate	0.712	−0.493	0.418	0.088	−0.177
Soil depth	0.765	−0.587	−0.035	−0.009	−0.131
Kusat	0.652	−0.386	0.26	0.289	−0.5
Slope	−0.133	**0.848**	0.465	−0.167	0.045
Elevation	−0.138	0.838	0.479	−0.185	0.057

PC-principal component; Underlined eigenvectors (factor loadings) are considered highly weighted. The bold eigenvectors are the indicators included in the MDS. BD-Bulk density; AggS-Aggregate stability; EC-Electrical conductivity; OC-Soil organic carbon; TN-Total nitrogen; P-phosphorus; AK-Available potassium; Kusat-Unsaturated hydraulic conductivity

and PC5, percentage of sand, slope, elevation and clay, total nitrogen, and aggregate stability, BD and soil organic carbon, clay were highly weighted. All PCs have more than one extremely weighted indicator was observed. As the PC3 and PC4 have lesser factor loadings than PC1 and PC2, we have selected two highly weighted eigenvectors (soil indicators) from the PC3 and PC4. In PC2, the slope and elevation were the highly weighted component. It indicates the role of slope and elevation in distributing soil quality over a hilly and mountainous region. The final MDS was thus comprised of available phosphorus, slope, aggregate stability, total nitrogen, SOC, BD and percentage of clay. Thus, soil texture, nutrients and topography are the prime components determining soil quality in the experimental watershed. The communality and corresponding weightage are provided in Table 13.6. The highest weightage was obtained for slope and clay, followed by available phosphorus, aggregate stability, soil organic carbon, total nitrogen and least with bulk density. The

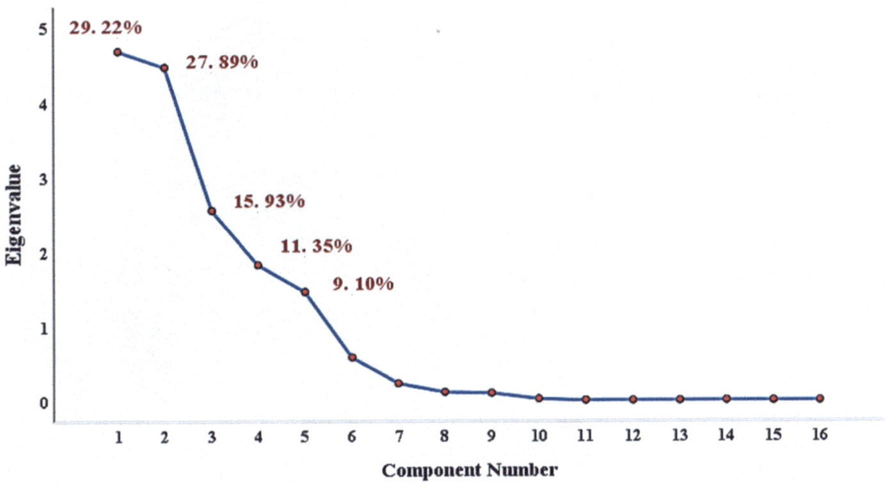

Fig. 13.5 Scree plot result of principal component analysis (PCA)

Table 13.6 The communality and weight value of each soil quality indicator in MDS indicator methods

Indicator	Communality	Weight
P	0.957	0.152
Slope	0.983	0.156
AggS	0.930	0.147
TN	0.807	0.128
BD	0.754	0.119
SOC	0.900	0.143
Clay	0.980	0.155

scores were assigned for each sample based on the soil-landscap/physiographye units.

13.4.3.1 Soil Quality Index

The soil quality index is normalised into a scale of 0 with least quality and 1 with the highest quality. The experimental sites' soil quality index value ranges from 0.26 to 0.57, with an average of 0.41. The higher soil erosion rate is the primary reason for the declined soil quality. The spatial distribution of SOC is depicted in Fig. 13.6, and the highest SOC was observed in the agricultural and forest land, with the least was observed from the scrub land. The spatial distribution of SQI within the watershed was also generated by IDW interpolation (Fig. 13.7), which depicted higher SQI values at the watershed's higher and middle elevation regions than the lower region. The higher and middle elevations are covered with forest, and the human interventions are

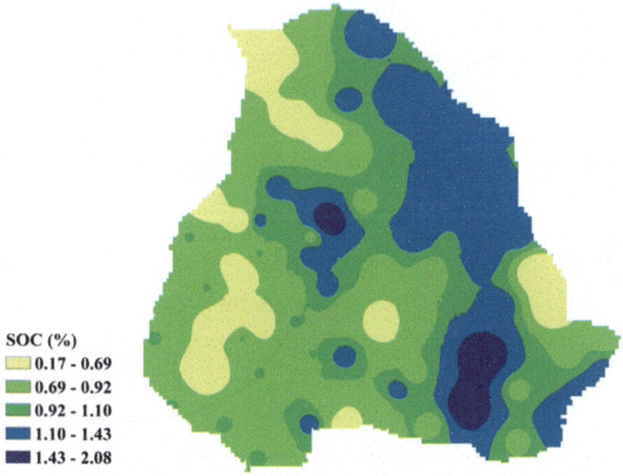

Fig. 13.6 Spatial distribution of soil organic carbon (SOC)

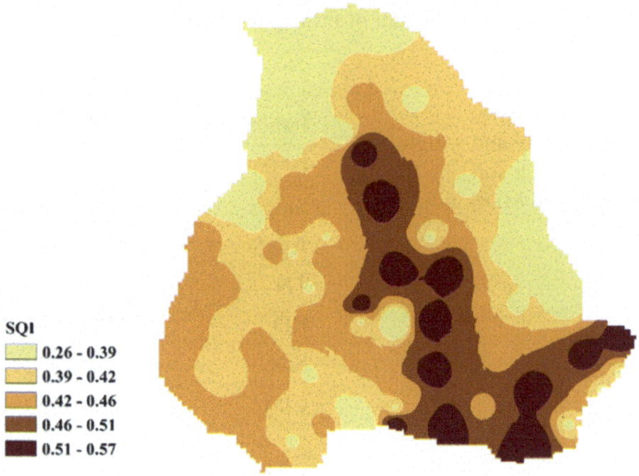

Fig. 13.7 Spatial distribution of soil quality index (SQI)

significantly less. Also, some agricultural patches show a higher soil quality index because of fertiliser application in fields. The Northwest side of the watershed has lower soil quality because of the scrubland. The scrub land has a lesser cover, and being directly exposed to rainfall leads to removal of top soil and resulted in a lower index. The East side of the study area is characterised as the highest elevation and slope, while the dense forest conserves the top soil. Although, some patches resulted in a lower index due to scrub land and agricultural field, causing higher erosion from these areas. The slope, land use type and soil erosion are the primary limiting factors for soil quality. For example, the less covered scrub land at higher hillslope observed lesser index while less covered scrub land at lower piedmont has a higher index than others.

Soil quality index of the experimental site was classified into four soil quality classes (Fig. 13.8). The highest grade in the watershed is represented as grade I

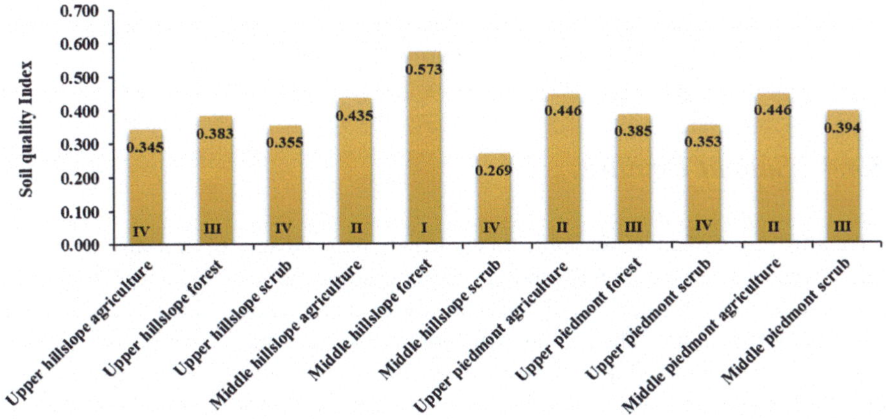

Fig. 13.8 The soil quality index (SQI) and soil quality grade* of various soil-landscape units. *Labelled at the lower portion of the bar

and least with grade IV. The middle hillslope forest received the highest soil quality grade (I) as it has good cover, relatively less slope and the least disturbed ecosystem from humans. The II quality grade was received by middle hillslope agriculture, upper piedmont agriculture and middle piedmont agriculture. The farmyard manure, fertiliser application and conservation practices (terraces) in the agricultural field provided good soil quality. The third soil quality grade covers upper hillslope forest, upper piedmont forest and middle piedmont scrub. The higher slope and grazing (disturbance) in the forest land and the less covered scrubland with soil erosion caused lower grades. Similarly, scrub land in upper and middle hillslope and middle piedmont has the least soil quality grade in the watershed. Soil erosion, less cover, grazing and other human disturbances lead to the reduction of soil quality. Soil quality based on different slope positions exhibited that the middle hillslope and middle piedmont had better soil quality than the upper hillslope and upper piedmont. Middle piedmont had less soil erosion, and the middle hillslope had fewer human disturbances with high vegetation cover. Bo-Jie et al. (2004) reported that disturbance in the forest ecosystem due to humans could reduce the soil quality.

Nearly 49.72% of the study area had a soil quality grade of III, whereas 25.30% of the area had a soil quality grade of I, 20.58% had a soil quality grade of II, and 4.40% had a soil quality grade IV (Table 13.7). Thus, it shows that a large area of the

Table 13.7 Distribution of soil quality grades in the watershed

Soil quality grade	Average SQI	Area (%)	Area (ha)
I	>0.573	25.30	129.06
II	0.442	20.58	104.97
III	0.387	49.72	253.62
IV	<0.330	4.40	22.45

watershed had a soil quality grade of III, which necessitates soil quality improvement through better management practices and proper land use planning.

13.4.4 Salient Findings

Soil quality assessment is inevitable for environmental and agricultural sustainability. It affects water resources, biodiversity, climate system and food security. Therefore, the researchers and farmers should quantify the best soil quality index to understand and transmit to real life. The spatial distribution of soil attributes provides a comprehensive perception of soil quality in an area. This study analyses the soil quality of a Shiwalik Himalayan landscape in India using the most widely accepted soil quality index (SQI). Principle component analysis (PCA) was used to select the minimum data sets (MDS) indicators, and a linear scoring function was used for indicator scoring.

The study revealed that approximately 74% of the watershed is below the average soil quality grade II (SQI- < 0.442). The forest has the highest soil quality and is least contributed by the scrubland. The farming practices adopted by farmers were helping to enhance the soil quality in agricultural fields. At the same time, the disturbance in the upper and lower elevation forests shows a decline in soil quality. Deforestation might lead to declined soil quality and further to an unsustainable ecosystem. The soil quality assessment indicates the need for soil and water conservation measures for the watershed. The unsustainable land management practices and steep slopes with climate change can further decline soil quality. However, this study gives only a preview of the variation in SQI due to soil, slope and land management practices for the Shiwalik Himalayan region.

13.5 Summary and Conclusion

Land abandonment, unsustainable land practices, steep slopes, and climate change hasten the soil degradationin hilly areas. Soil quality monitoring is the better way to ameliorate the degradation problem. Continuous monitoring will help us to assess whether the soil quality degrading or enhancing. It supports continuous soil quality improvement using different management and conservation practices. There are several methods widely available to assess soil quality. We divided the approaches into soil quality indices, multi-criteria decision making, soil erosion modelling, soil carbon dynamic modelling and geo-spatial techniques, although no definite boundary was considered to delineate between these methods. Because presently, the majority of studies using a combination of these methods for more reliable explicit estimation.

Site-specific external factors are necessary for soil quality indices while determining the potential indicators for soil quality assessment. The majority of studies used the MDS method for assessing soil quality, and many studies found it was more efficient than the TDS method and also it can provide site-specific most potential indicators. Apart from this, the weighting method provides more preference to suitable indicators. Several studies did not consider topographic parameters for assessing soil quality. Because the topography, climate, and land use play a significant role in this, the soil properties may change with the influence of these factors. Especially in hilly and mountainous terrain, the variability is higher than in the plain terrain. Thus, the selection and scoring of indicators are very vital for soil quality assessment. The scoring of indicators based on the optimum value of soil properties for the corresponding ecosystem is necessary rather than simple scoring, and it can provide the actual soil quality condition of the field by comparing the benchmark soil properties. Simple scoring may misrepresent the field soil quality conditions. The hydrological properties of soils are found to be the most potential soil quality indicators in hilly and mountainous regions. Several studies identified soil moisture, wilting point, available water content, soil porosity and infiltration rate as the potential soil quality indicators. It may be because the water content in hilly regions is identified as the limiting factor. Apart from this, the star indicator, soil organic carbon, was also a potential indicator in these regions.

The advancement of geo-spatial technologies is beneficial in soil quality assessment. They can capture the soil quality continuously with respect to space and time. The spectral indices-based techniques can provide the characteristics of soil quality with limited field and soil data. The comprehensive attributes developed from geo-spatial technologies can enhance the soil quality assessment of inaccessible regions in the hilly and mountainous terrain with the help of spectral indices, geostatistics, multispectral remote sensing. The geostatistics will help to enhance the point scale soil quality data to the continuous spatial distribution data. It will reduce cost and time while assessing the soil quality. The digital elevation models and terrain models provide explicit information about the terrain morphological characteristics. Apart from this, the weighted sum overlay geo-spatial technology also allows us to integrate different soil properties according to their weighted priority.

Soil erosion is the major threat of soil quality degradation in the hilly and mountainous ecosystem. The modelling approaches account for soil properties, topography, climate, and land use/land cover characteristics; thus, modelling can quickly assess soil deterioration. Process-based models can simulate the processes efficiently with the law of conservation of mass, energy and momentum. The proper calibration and validation of the process-based model provides a more realistic and reliable condition of soil degradation. The majority of research used process-based models that have been calibrated and verified using site-specific soil, climate, land use, and other properties. Currently, the majority of erosion models are developed with the GIS interface. It will allow us to provide the comprehensive, explicit spatial data to the model which is developed from remote sensing and other field data collection methods in the form of raster images. It will enhance the efficiency of output and time required to process the real-world scenarios. It can also provide better visualisation of erosion processes in complex terrains with the help of a digital elevation model (DEM). ArcSWAT, ArcAPEX, GeoWEPP and GEPIC are some examples of efficiently used spatial interfaced soil erosion models.

A case study was carried out in a watershed representing hilly and mountainous landscape comprising the forest and crop land at different topographic positions to assess the soil quality. The MDS was selected based on the PCA analysis. The first principal component contributes the soil nutrients (P and K) and hydrological parameters (soil depth, infiltration rate and hydraulic conductivity). The second principal component is mainly contributed by the slope and elevation factor. It indicates the importance of topographic factors while assessing soil quality in the hilly region. We found that majority of the region is characterising diminished soil quality because of soil erosion. The undisturbed forest ecosystem situated relatively lesser slope has the highest soil quality followed by the agricultural ecosystem. The fertiliser and ploughing activities enhanced the soil quality in the agricultural field. At the same time, less covered scrub land witnesses lower soil quality. The study recommends the proper land use planning, conservation measures, and afforestation/reforestation to enhance the soil quality of the study area.

References

Alahi ME, Xie L, Zia AI, Mukhopadhyay S, Burkitt L (2016) Practical nitrate sensor based on electrochemical impedance measurement. In: 2016 IEEE International instrumentation and measurement technology conference proceedings. IEEE, pp 1–6

Ali MA, Dong L, Dhau J, Khosla A, Kaushik A (2020) Perspective—electrochemical sensors for soil quality assessment. J Electrochem Soc 167(3):037550

Ande OT, Alaga Y, Oluwatosin GA (2009) Soil erosion prediction using MMF model on highly dissected hilly terrain of Ekiti environs in southwestern Nigeria. Int J Phys Sci 4(2):53–57

Andrews SS, Carroll CR (2001) Designing a decision tool for sustainable agro-ecosystem management: soil quality assessment of a poultry litter management case study. Ecol Appl 11(6):1573–1585

Andrews SS, Karlen DL, Mitchell JP (2002) A comparison of soil quality indexing methods for vegetable production systems in Northern California. Agr Ecosyst Environ 90(1):25–45

Arshad MA, Martin S (2002) Identifying critical limits for soil quality indicators in agro-ecosystems. Agr Ecosyst Environ 88(2):153–160

Asres MT, Awulachew SB (2010) SWAT based runoff and sediment yield modelling: a case study of the Gumera watershed in the Blue Nile basin. Ecohydrol Hydrobiol 10(2–4):191–199

Ballabio C (2009) Spatial prediction of soil properties in temperate mountain regions using support vector regression. Geoderma 151(3–4):338–350

Bastida F, Zsolnay A, Hernández T, García C (2008) Past, present and future of soil quality indices: a biological perspective. Geoderma 147(3–4):159–171

Birkeland PW (1984) Soils and geomorphology. Oxford University Press, New York (NY)

Bo-Jie FU, Shi-Liang LIU, Li-Ding CHEN, Yi-He LÜ, Jun QIU (2004) Soil quality regime in relation to land cover and slope position across a highly modified slope landscape. Ecol Res 19(1):111–118

Borrelli P, Robinson DA, Fleischer LR, Lugato E, Ballabio C, Alewell C, Panagos P (2017) An assessment of the global impact of 21st century land use change on soil erosion. Nat Commun 8(1):1–13

Bot A, Benites J (2005) The importance of soil organic matter: key to drought resistant soil and sustained food and production. Food and Agriculture Organization of the United Nations (FAO), Rome

Brejda JJ, Moorman TB, Karlen DL, Dao TH (2000) Identification of regional soil quality factors and indicators I. Central and southern high plains. Soil Sci Soc Am J 64(6):2115–2124
Bünemann EK, Bongiorno G, Bai Z, Creamer RE, De Deyn G, de Goede R, Brussaard L (2018) Soil quality—a critical review. Soil Biol Biochem 120:105–125
Burak DL, Monteiro EDC, Passos RR, Mendonça EDS (2021) Soil quality index for extensive pastures in hilly landforms region of highly weathered soils in an Atlantic forest biome, Brazil. Afr J Range Forage Sci 1–12
Campos A, Oleschko K, Etchevers J, Hidalgo C (2007) Exploring the effect of changes in land use on soil quality on the eastern slope of the Cofre de Perote Volcano (Mexico). For Ecol Manag 248(3):174–182
Cuomo S, Della Sala M, Novità A (2015) Physically based modelling of soil erosion induced by rainfall in small mountain basins. Geomorphology 243:106–115
Dai F, Lv Z, Liu G (2018) Assessing soil quality for sustainable cropland management based on factor analysis and fuzzy sets: a case study in the Lhasa River Valley, Tibetan Plateau. Sustainability 10(10):3477
David Raj A, Kumar S, Sooryamol KR (2022) Modelling climate change impact on soil loss and erosion vulnerability in a watershed of Shiwalik Himalayas. CATENA, 214:106279
Demirel T, Tüzün S (2011) Multi criteria evaluation of the methods for preventing soil erosion using fuzzy ANP: the case of Turkey. In: Proceedings of the world congress on engineering, vol 2, pp 6–8
Dharumarajan S, Hegde R, Singh SK (2017) Spatial prediction of major soil properties using random forest techniques—a case study in semi-arid tropics of South India. Geoderma Reg 10:154–162
Doran JW, Parkin TB (1994) Defining and assessing soil quality. Defin Soil Qual Sustain Environ 35:1–21
Doran JW, Zeiss MR (2000) Soil health and sustainability: managing the biotic component of soil quality. Appl Soil Ecol 15(1):3–11
Drobnik T, Greiner L, Keller A, Grêt-Regamey A (2018) Soil quality indicators—from soil functions to ecosystem services. Ecol Ind 94:151–169
Easter M, Paustian K, Killian K, Williams S, Feng T, Al-Adamat R, Batjes NH, Bernoux M, Bhattacharyya T, Cerri CC, Cerri CEP (2007) The GEFSOC soil carbon modelling system: a tool for conducting regional-scale soil carbon inventories and assessing the impacts of land use change on soil carbon. Agr Ecosyst Environ 122(1):13–25
Egli M, Poulenard J (2016) Soils of mountainous landscapes. In: International encyclopedia of geography: people, the earth, environment and technology, pp 1–10
El Jazouli A, Barakat A, Ghafiri A, El Moutaki S, Ettaqy A, Khellouk R (2017) Soil erosion modeled with USLE, GIS, and remote sensing: a case study of Ikkour watershed in Middle Atlas (Morocco). Geosci Lett 4(1):1–12
Ennaji W, Barakat A, El Baghdadi M, Oumenskou H, Aadraoui M, Karroum LA, Hilali A (2018) GIS-based multi-criteria land suitability analysis for sustainable agriculture in the northeast area of Tadla plain (Morocco). J Earth Syst Sci 127(6):1–14
Erkossa T, Itanna F, Stahr K (2007) Indexing soil quality: a new paradigm in soil science research. Soil Res 45(2):129–137
Falloon P, Smith P, Coleman K, Marshall S (2000) How important is inert organic matter for predictive soil carbon modelling using the Rothamsted carbon model? Soil Biol Biochem 32(3):433–436
Farage PK, Ardö J, Olsson L, Rienzi EA, Ball AS, Pretty JN (2007) The potential for soil carbon sequestration in three tropical dryland farming systems of Africa and Latin America: a modelling approach. Soil Tillage Res 94(2):457–472
Ghosh BN, Sharma NK, Alam NM, Singh RJ, Juyal GP (2014) Elevation, slope aspect and integrated nutrient management effects on crop productivity and soil quality in North-west Himalayas, India. J Mt Sci 11(5):1208–1217

Glover JD, Reganold JP, Andrews PK (2000) Systematic method for rating soil quality of conventional, organic, and integrated apple orchards in Washington State. Agr Ecosyst Environ 80(1–2):29–45
Govaerts B, Sayre KD, Deckers J (2006) A minimum data set for soil quality assessment of wheat and maise cropping in the highlands of Mexico. Soil Tillage Res 87(2):163–174
Gregory PJ, Hester R, Harrison R (2012) Soils and food security: challenges and opportunities. Soils Food Secur 1–30
Greiner L, Keller A, Grêt-Regamey A, Papritz A (2017) Soil function assessment: review of methods for quantifying the contributions of soils to ecosystem services. Land Use Policy 69:224–237
Gupta S, Kumar S (2017a) Simulating climate change impact on soil erosion using RUSLE model—a case study in a watershed of mid-Himalayan landscape. J Earth Syst Sci 126(3):43
Gupta S, Kumar S (2017b) Simulating climate change impact on soil carbon sequestration in agro-ecosystem of mid-Himalayan landscape using CENTURY model. Environ Earth Sci 76(11):1–15
Gupta AK, Negi M, Nandy S, Alatalo JM, Singh V, Pandey R (2019) Assessing the vulnerability of socio-environmental systems to climate change along an altitude gradient in the Indian Himalayas. Ecol Ind 106:105512
Harden CP (2001) Soil erosion and sustainable mountain development. Mt Res Dev 21(1):77–83
Hashemi P, Walsh PL, Guillot TS, Gras-Najjar J, Takmakov P, Crews FT, Wightman RM (2011) Chronically implanted, nafion-coated Ag/AgCl reference electrodes for neurochemical applications. ACS Chem Neurosci 2(11):658–666
Herrick JE, Brown JR, Tugel AJ, Shaver PL, Havstad KM (2002) Application of soil quality to monitoring and management: paradigms from rangeland ecology. Agron J 94(1):3–11
Hessel R, Van den Bosch R, Vigiak O (2006) Evaluation of the LISEM soil erosion model in two catchments in the East African Highlands. Earth Surf Process Landf J Br Geomorphol Res Group 31(4):469–486
Hueso-González P, Muñoz-Rojas M, Martínez-Murillo JF (2018) The role of organic amendments in drylands restoration. Curr Opin Environ Sci Health 5:1–6
Insam H, Domsch KH (1988) Relationship between soil organic carbon and microbial biomass on chronosequences of reclamation sites. Microb Ecol 15(2):177–188
Jakšić S, Ninkov J, Milić S, Vasin J, Živanov M, Jakšić D, & Komlen V (2021) Influence of slope gradient and aspect on soil organic carbon content in the region of Niš, Serbia. Sustainability 13(15):8332
Jones DL, Simfukwe P, Hill PW, Mills RT, Emmett BA (2014) Evaluation of dissolved organic carbon as a soil quality indicator in national monitoring schemes. PLoS ONE 9(3):e90882
Kalambukattu JG, Kumar S, Ghotekar YS (2018a) Spatial variability analysis of soil quality parameters in a watershed of Sub-Himalayan Landscape—a case study. Eurasian J Soil Sci 7(3):238–250
Kalambukattu JG, Kumar S, Raj RA (2018b) Digital soil mapping in a Himalayan watershed using remote sensing and terrain parameters employing artificial neural network model. Environ Earth Sci 77(5):1–14
Kalambukattu JG, Kumar S, Hole RM (2021) Geo-spatial modelling of soil erosion and risk assessment in Indian Himalayan region—a study of Uttarakhand state. Environ Adv 4:100039
Karlen DL, Cc Wollenhaupt N, Erbach DC, Berry EC, Swan JB, Eash NS, Jordahl JL (1994) Crop residue effects on soil quality following 10-years of no-till corn. Soil Tillage Res 31(2–3):149–167
Karlen DL, Mausbach MJ, Doran JW, Cline RG, Harris RF, Schuman GE (1997) Soil quality: a concept, definition, and framework for evaluation (a guest editorial). Soil Sci Soc Am J 61(1):4–10
Keller JK, White JR, Bridgham SD, Pastor J (2004) Climate change effects on carbon and nitrogen mineralisation in peatlands through changes in soil quality. Glob Change Biol 10(7):1053–1064
Khanal S, Fulton J, Klopfenstein A, Douridas N, Shearer S (2018) Integration of high resolution remotely sensed data and machine learning techniques for spatial prediction of soil properties and corn yield. Comput Electron Agric 153:213–225
Kumar S, Kalambukattu JG (2021) Geospatial modelling for soil quality assessment. In: Geospatial technologies for crops and soils. Springer, Singapore, pp 387–420

Kumar S, Singh RP (2016) Spatial distribution of soil nutrients in a watershed of Himalayan landscape using terrain attributes and geostatistical methods. Environ Earth Sci 75(6):473

Kumar S, Singh RP, Kalambukattu JG (2021) Modelling daily surface runoff, sediment and nutrient loss at watershed scale employing APEX model interfaced with GIS—a case study in Himalayan landscape. https://doi.org/10.21203/rs.3.rs-232906/v1

Kumar U, Kumar N, Mishra VN, Jena RK (2019) Soil quality assessment using analytic hierarchy process (AHP): a case study. In: Interdisciplinary approaches to information systems and software engineering. IGI Global, pp 1–18

Lal R (2015) Restoring soil quality to mitigate soil degradation. Sustainability 7(5):5875–5895

Lal R (2016) Soil health and carbon management. Food Energy Secur 5(4):212–222

Lal R (2002) Soil carbon dynamics in cropland and rangeland. Environ pollut 116(3):353-362

Larson WE, Pierce FJ (1994) The dynamics of soil quality as a measure of sustainable management. Defin Soil Qual Sustain Environ 35:37–51

Liebig MA, Varvel G, Doran J (2001) A simple performance-based index for assessing multiple agro-ecosystem functions. Agron J 93(2):313–318

Liu M, Han G (2020) Assessing soil degradation under land-use change: insight from soil erosion and soil aggregate stability in a small karst catchment in southwest China. PeerJ 8:e8908

Lotfi AZ, Esmali OA, Hashemimajd K, Najafi N (2013) Soil fertility evaluation of Ardabil plain for wheat and potato based on some soil chemical properties by AHP and GIS techniques. J Water Soil 27(1):45–53

Lozano-García B, Muñoz-Rojas M, Parras-Alcántara L (2017) Climate and land use changes effects on soil organic carbon stocks in a Mediterranean semi-natural area. Sci Total Environ 579:1249–1259

Mandal UK, Yadav SK, Sharma KL, Ramesh V, Venkanna K (2011) Estimating permanganate-oxidisable active carbon as quick indicator for assessing soil quality under different land-use system of rainfed Alfisols. Indian J Agric Sci 81(10):927–931

Masto RE, Chhonkar PK, Singh D, Patra AK (2007) Alternative soil quality indices for evaluating the effect of intensive cropping, fertilisation and manuring for 31 years in the semi-arid soils of India. Environ Monit Assess 136(1–3):419–435

Minasny B, Hartemink AE (2011) Predicting soil properties in the tropics. Earth Sci Rev 106(1–2):52–62

Muñoz-Rojas M, Abd-Elmabod SK, Zavala LM, De la Rosa D, Jordán A (2017) Climate change impacts on soil organic carbon stocks of Mediterranean agricultural areas: a case study in Northern Egypt. Agr Ecosyst Environ 238:142–152

Nabiollahi K, Golmohamadi F, Taghizadeh-Mehrjardi R, Kerry R, Davari M (2018) Assessing the effects of slope gradient and land use change on soil quality degradation through digital mapping of soil quality indices and soil loss rate. Geoderma 318:16–28

Nosrati K, Collins AL (2019) A soil quality index for evaluation of degradation under land use and soil erosion categories in a small mountainous catchment, Iran. J Mt Sci 16(11):2577–2590

Oliver DP, Bramley RGV, Riches D, Porter I, Edwards J (2013) soil physical and chemical properties as indicators of soil quality in Australian viticulture. Aust J Grape Wine Res 19(2):129–139

Pal S, Panwar P, Bhardwaj DR (2012) Soil quality under forest compared to other land uses in acid soil of North Western Himalaya, India. Ann For Res 56(1):187–198

Pande CB, Kadam SA, Jayaraman R, Gorantiwars S, Shinde M (2021) Prediction of soil chemical properties using multispectral satellite images and wavelet transforms methods. J Saudi Soc Agric Sci

Paul GC, Saha S, Ghosh KG (2020) Assessing the soil quality of Bansloi river basin, eastern India using soil-quality indices (SQIs) and random forest machine learning technique. Ecol Ind 118:106804

Pausas GJ, Casals P, Camarero L, Huguet C, Sebastia MT, Thompson R, Romanya J (2007) Soil organic carbon storage in mountain grasslands of the Pyrenees: effects of climate and topography. Biogeochemistry 82(3):279–289

Paz-Kagan T, Zaady E, Salbach C, Schmidt A, Lausch A, Zacharias S, Karnieli A (2015) Mapping the spectral soil quality index (SSQI) using airborne imaging spectroscopy. Remote Sens 7(11):15748–15781

Pham TG, Nguyen HT, Kappas M (2018) Assessment of soil quality indicators under different agricultural land uses and topographic aspects in Central Vietnam. Int Soil Water Conserv Res 6(4):280–288

Prasannakumar V, Vijith H, Abinod S, Geetha NJGF (2012) Estimation of soil erosion risk within a small mountainous sub-watershed in Kerala, India, using revised universal soil loss equation (RUSLE) and geo-information technology. Geosci Front 3(2):209–215

Qi Y, Darilek JL, Huang B, Zhao Y, Sun W, Gu Z (2009) Evaluating soil quality indices in an agricultural region of Jiangsu Province, China. Geoderma 149(3–4):325–334

Rajalakshmi P, Mahalakshmi SD (2016) IOT based crop-field monitoring and irrigation automation. In: 2016 10th international conference on intelligent systems and control (ISCO). IEEE, pp 1–6

Rajan K, Natarajan A, Kumar KA, Badrinath MS, Gowda RC (2010) Soil organic carbon—the most reliable indicator for monitoring land degradation by soil erosion. Curr Sci 823–827

Ranatunga K, Hill MJ, Probert ME, Dalal RC (2001) Comparative application of APSIM, RothC and Century to predict soil carbon dynamics. In: Proceedings of the international congress on modelling and simulation (MODSIM'01), pp 733–738

Saaty RW (1987) The analytic hierarchy process—what it is and how it is used. Math Model 9(3–5):161–176

Saaty TL (1994) Fundamentals of decision making and priority theory with the AHP. RWS Publications, Pittsburg

Saaty TL (2005) Theory and applications of the analytic network process: decision making with benefits, opportunities, costs, and risks. RWS Publications

Seyedmohammadi J, Sarmadian F, Jafarzadeh AA, McDowell RW (2019) Integration of ANP and Fuzzy set techniques for land suitability assessment based on remote sensing and GIS for irrigated maise cultivation. Arch Agron Soil Sci 65(8):1063–1079

Shrestha DP (1997) Assessment of soil erosion in the Nepalese Himalaya: a case study in Likhu Khola Valley, Middle Mountain Region. Land Husb 2(1):59–80

Singh RP (2012) Surface runoff, soil erosion and water quality estimation using APEX model integrated with GIS—a case study in Himalayan Watershed. M. Tech. (RS & GIS) thesis, Andhra University, Visakhapatnam, 98 pp

Singh AK, Bordoloi LJ, Kumar M, Hazarika S, Parmar B (2014) Land use impact on soil quality in eastern Himalayan region of India. Environ Monit Assess 186(4):2013–2024

Smith WN, Grant BB, Desjardins RL, Qian B, Hutchinson J, Gameda S (2009) Potential impact of climate change on carbon in agricultural soils in Canada 2000–2099. Clim Change 93(3):319–333

Sofi JA, Bhat AG, Kirmai NA, Wani JA, Lone AH, Ganie MA, Dar GIH (2016) Soil quality index as affected by different cropping systems in northwestern Himalayas. Environ Monit Assess 188(3):1–13

Sooryamol KR (2020) Potential impact of climate change impact on surface runoff and sediment yield in a watershed of lesser Himalayas. MSc (climate change adaptation) thesis, Kerala Agricultural University, Thrissur, 97 pp

Sooryamol KR, Kumar S, Regina M et al (2022) Modelling climate change impact on soil erosion in a watershed of north-western Lesser Himalayan region. J Sediment Environ 7:125–146. https://doi.org/10.1007/s43217-022-00089-4

Sun W, Shao Q, Liu J, Zhai J (2014) Assessing the effects of land use and topography on soil erosion on the Loess Plateau in China. CATENA 121:151–163

Tarolli P, Straffelini E (2020) Agriculture in hilly and mountainous landscapes: threats, monitoring and sustainable management. Geogr Sustain 1(1):70–76

Tesfahunegn GB (2014) Soil quality assessment strategies for evaluating soil degradation in Northern Ethiopia. Appl Environ Soil Sci

Tunçay T, Kılıç Ş, Dedeoğlu M, Dengiz O, Başkan O, Bayramin İ (2021) Assessing soil fertility index based on remote sensing and gis techniques with field validation in a semiarid agricultural ecosystem. J Arid Environ 190:104525

UN, Agenda 21 (1992) Chapter 13: Managing fragile ecosystems: sustainable mountain development

Van-Camp L, Bujarrabal B, Gentile AR, Jones RJ, Montanarella L, Olazabal C, Selvaradjou SK (2004) Technical working groups established under the thematic strategy for soil protection

Vasques GM, Grunwald S, Harris WG (2010) Spectroscopic models of soil organic carbon in Florida, USA. J Environ Qual 39(3):923–934

Vasu D, Singh SK, Ray SK, Duraisami VP, Tiwary P, Chandran P, Anantwar SG (2016) Soil quality index (SQI) as a tool to evaluate crop productivity in semi-arid Deccan plateau, India. Geoderma 282:70–79

Xue R, Wang C, Liu M, Zhang D, Li K, Li N (2019) A new method for soil health assessment based on analytic hierarchy process and meta-analysis. Sci Total Environ 650:2771–2777

Yalew SG, Van Griensven A, Mul ML, van der Zaag P (2016) Land suitability analysis for agriculture in the Abbay basin using remote sensing, GIS and AHP techniques. Model Earth Syst Environ 2(2):1–14

Yunus Y, Satriyo P, Munawar AA (2019) Rapid prediction of soil quality indices using near infrared spectroscopy. In: IOP conference series: earth and environmental science, vol 365, no 1. IOP Publishing, p 012043

Zhang S, Huang Y, Shen C, Ye H, Du Y (2012) Spatial prediction of soil organic matter using terrain indices and categorical variables as auxiliary information. Geoderma 171:35–43

Chapter 14
Soil Pollution Due to Sewage Sludge and Industrial Effluents

Ayush Bahuguna, S. K. Singh, Sachin Sharma, Arvind, Astha Pandey, Basant Kumar Dadarwal, Bharti Yadav, Akshita Barthwal, and Raghu Nandan Singh Khatana

Abstract In recent times the issue of soil pollution is becoming important as all the nutrient are taken up by plant from the soil. The wastewater treatment produces sewage sludge as an end product, but its primary source is from domestic effluents as well as industrial effluents. Water makes up more than 90% of the sewage water produced in India, 40–50% organics, 30–40% inert compounds, 10–15% bio-resistant organics, and 5–8% miscellaneous components make up the solid fraction. Rapid urbanisation, industrialisation and increasing population generate a huge amount of waste and the waste from industry and the sewage released into the soil as a source of irrigation without being treated properly causes accumulation of toxic heavy metals, persistent organic pollutant, microplastics and high salt that leads to the reduction in the quality of the soil. Similarly, the effluents released by the industries such as distillery, tanning, textile also contains different sources of heavy metals, sodium and other salts and organic chemicals. This chapter providesa detailed aspect of soil pollution from the industry as well as sewage sludge.

Keywords Microbial pollution · Microplastic · Organic pollutant · Wastewater

14.1 Introduction

Soil pollution in recent times is becoming analarming issue as the build-up of contaminant either from industry disposal or from the sewage leads to the accumulation of potentially toxic element, organic pollutant and microplastics, which in long term affect the soil quality and health (Chowdhary et al. 2018). In developing countries, sewage treatment plants have not received much attention, and all industrial effluents

A. Bahuguna (✉) · S. K. Singh · S. Sharma · Arvind · A. Pandey · B. K. Dadarwal · B. Yadav
Department of Soil Science of Agricultural Chemistry, Institute of Agricultural Sciences, Banaras Hindu University, Varanasi, U.P 221 005, India
e-mail: ayushbahuguna1995@gmail.com

A. Barthwal · R. N. S. Khatana
Department of Soil Science of Agricultural Chemistry, Naini Agriculture Institute Sam Higginbottom University of Agriculture, Technology and Sciences, Prayagraj, U.P 211 007, India

P. K. Shit et al. (eds.), *Soil Health and Environmental Sustainability*, Environmental Science and Engineering,
https://doi.org/10.1007/978-3-031-09270-1_14

are routinely dumped into the sewage system. Sewage and industrial effluents contain a wide range of useful and toxic substances (Tripathi et al. 2011). Because some of these effluents are high in plant nutrients, the soil is the appropriate place for them to be disposed of, but some untreated and contaminated sewage or industrial effluent may contain a huge quantity of concentration of heavy metal such as Cd (cadmium), Cr (chromium), Ni (Nickel) and Pb (lead) (Arora et al. 1985). Anti et al. (2004) reported that regular disposal of sewage and industrial effluent lead to the sickness of soil and accumulation of toxic heavy metal in soil. Depending on the source of the sewage, sludge may also contain organic pollutants such as polychlorinated biphenyls (PCBs), dioxins, and furanes (Hembrock and König 1990). Irrigation using various industrial effluents has the potential to alter the physical, chemical, and biological aspects of soil. These soil characteristics have a critical role in the transformation of nutrients present in irrigation effluent (Chowdhary and Bharagava 2019). Irrigation with this type of water could change the structure of soil and affect the diversity and function of the soil microbial population (Mani et al. 2019). Similarly, the effluents released by the industries such as distillery, tanning, textile also contains different sources of heavy metals, high biochemical oxygen demand and chemical oxygen demand, sodium and other salts and organic chemicals. Different properties of soil are also affected either biological, chemical and physical, when the untreated waste released directly into the soil and on the long-term application their build up in the soil caused decline in the soil health. This chapter goes through in-depth about how several industry effluents, such as distillery, textile, pulp and paper, tanning, and sewage sludge, pollute the soil.

14.2 Soil Pollution Due to Industry Effluents

Industrial effluents are a major source of direct and continuous pollutant inputs into the soil and water systems. The main source of soil pollution is the improper disposal of industrial waste. Numerous industries are located and run within cities in many sections of the country. Industries such as tanneries, textile mills, paper mills, breweries, and pharmaceutical enterprises inadvertently endanger human health by introducing pollutants such as metals in excess of safe levels into industrial effluent, causing detrimental implications for the ecosystem. These effluents have variable chemical properties and toxic metal contaminants that may prove detrimental effect to the soil environment. Soil pollution is caused by the untreated release of industrial wastes into the soil, which include high quantities of dangerous contaminants. Industrial wastes contain a variety of hazardous compounds and chemicals that, when deposited in soil, change the strength of the topsoil soil layer, reducing soil fertility and biological activity. Industrial pollutants are mainly released from diverse origins including pulp and paper mills, chemicals, oil refineries, sugar mills, textiles, steel, distilleries, fertilizers, pesticides, mining industries, coal and mining, drugs, glass and cement industries, etc. (Fig. 14.1).

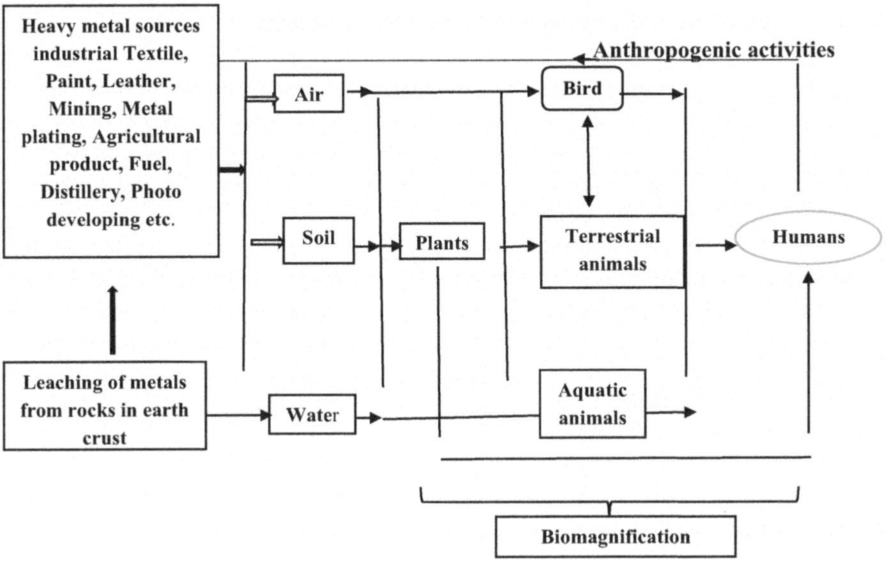

Fig. 14.1 Ecological flow diagram of industrial effluents and heavy metal impacting the environment and living system. *Sources* Vincent (2014)

14.3 Sources of Industry Causing Soil Pollution

14.3.1 Tannery Industry

The tanning/leather processing industry is a significant polluter of the environment and a long-term source of soil pollution. Tannery effluents are ranked as the highest pollutant amongst all industrial wastes. The effluents from tanning contain many dissolve salts, sulphide and higher suspended solids, organic and inorganic compound and toxic metallic substances. During the leather processing, different toxic pollutant is used such as NaCl, H_2SO_4, $Cr_2(SO_4)_3$, dyes etc., which resulted in effluent being rich in chromium and salts (SO_4 and Cl of Na) (Srinivasa et al. 2010). The intense colour (reddish dull-brown), high BOD, high pH, and high total dissolved solids differentiate tannery waste.

Chromium in the effluent is largely in the less hazardous trivalent form (Cr (III)), but when discharged into the soil, Cr (III) is oxidized to the toxic hexavalent form, which stays as Cr (VI) due to variable environmental factors (Srinivasa et al. 2010). Zayed and Terry (2003) reported that in India approximately 2000 to 30,000 tons of elementary chromium enter the environment through the uncontrolled release of contaminated sludge from the tannery industry.

14.3.1.1 Effect on Soil Physical and Chemical Property

The tannery industry is the prime source of toxic metal contamination in surrounding areas of agricultural lands and regular utilisation of wastewater for irrigation can cause heavy metal accumulation inthe soil profile. The application of tannery effluents has several changes in soil properties, such as total porosity and hydraulic conductivity decreased and bulk density of soil increased. Because of the reduction in the soil's heavy metal retention capacity, this heavy metal is released into groundwater and is available in soil solution for plant uptake Buckland et al. (2002) found that the effluent from tannery results in an increase of salinity and sodicity in soil which leads to reduced aggregate stability and infiltration properties. When wastewater rich in tannery effluent applied to the soil the concentration of chromium increases in the soil (Mohd 2008), which results in long term toxic effect on the soil ecosystem (Harshita et al. 2015).

14.3.1.2 Effect on Soil Microorganism

According to Shi et al. (2002), excessive levels of chromium alter the microbial population and have a deleterious impact on cell metabolism. Chromium addition also hinders soil respiration, soil microbial biomass, nitrification and nitrogen mineralization (Hassan et al. 2016). Also, the negative effect of chromium on soil microbial behaviour was reported by Liu et al. (2014) (Table 14.1).

Table 14.1 Characteristic of combined tannery industry effluents (all value is in mg L^{-1} except pH and COD sulphate ratio)

S. no	Characteristics	India	International
1	Sulphides	25–220	88–200
2	TKN (total kjeldal nitrogen)	250–400	267–400
3	NH_4-N	100–300	89–100
4	Chromium (III)	60–75	67–100
5	Chloride	6000–9500	3044–5700
7	TDS (total dissolved salts)	10,000–21,000	8000–13,899
8	pH	7–8.5	7–7.5
9	Suspended Solid	2000–3000	1844–3311
10	COD (chemical oxygen demand)	3000–6000	3222–5133
11	BOD (biological oxygen demand)	1200–2700	1111–1911
12	COD/sulphate ratio	1000–3000/2–3	1156–2444/2.1–2.8

Sources Sabumon (2016)

14.3.2 Paper Mills Industry

This industry discharges a huge quantity of solid waste, brownish coloured effluents containing toxic organic compounds like chlorinated hydrocarbons, lignin and polymerized tannins. Papermills consume a large amount of water, ranging from 300 to 450 m^3 t^{-1}, with 220–350 m^3 of water discharged as wastewater per tonne of paper produced (Singh et al. 2013). The effluents are influencingthe soil and aquatic ecosystem by imparting the large biological oxygen demand (BOD), toxicity and colour. The toxicity is mostly due to tannins, wood resins and chlorinated phenols. The pulp & paper industry yearly produces over a thousand tonnes of dry solid waste. Disposal of large volumes of wastewater is one of the mostimportant problem which the industry confronting. Sundari and Kanakarni (2001) reported that the elements found in pulp mill wastewater such as magnesium (Mg), sodium chloride (NaCl) and sulphur (S) can cause increase soil salinity, nutrient imbalances in crops and deteriorate soil structure.

Dissolved particles such as chlorides and sulphates of sodium and calcium, as well as varying levels of suspended organic compounds, are abundant in this wastewater (Singh et al. 2019). The reaction of effluents is alkaline having high biological oxygen demand (BOD) and chemical oxygen demand (COD). Papermill wastewater contains a significant amount of carbonate and bicarbonate alkalinity and tends to precipitate calcium in the soil as $CaCO_3$, increasing the soil solution's sodium to calcium, magnesium, and sodium adsorption ratio (Sharma et al. 2014). Singh et al. (2019) reported the effect of paper and pulp mill effluents disturbing the soil qualityparameters like:

- Raising soil pH.
- Altering soil physical properties like colour, texture etc.
- Inadequate macro-and micronutrient balance in the soil.
- Negative impact on soil microbial activity and disruption of all-natural cycles.
- A rise in organic load.
- Reduction of supply of oxygen in the soil.

The presence of lignin and its by-products, which are difficult to degrade due to the presence of strong intra-molecular C–C linkages. Due to their non-biodegradable nature, lignin and its derivatives are transferred to the neighbouring soil over layers affecting soil quality (Phukan and Bhattacharyya 2003). The pulp mills using chlorine bleaching have been known to adversely affect the environment through discharge of effluent containing toxic polychlorinated dioxins, furans, and polychlorinated biphenyl (PCB). The mill disposes of a large number of solid wastes as landfill in the roadside low-lying areas, which are contiguous with vast tracts of agricultural land used for paddy rice cultivation. When tannery wastewater was applied to the soil there was a decrease in moisture content and water holding capacity of the soil (Miller and Turk 2002). Kumar and Chopra (2015) also found higher concentrations of heavy metals such as Fe, Zn, Cd, Cu, and Pb in irrigated soil from paper mill effluent (Table 14.2 and Fig. 14.2).

Table 14.2 Major characteristics of paper and pulp industry effluents

S. no	Parameter	Value
1	pH	7.65
2	Biological oxygen demand (mg L^{-1}))	176
3	Chemical oxygen demand (mg L^{-1})	534
4	BOD/COD ratio	0.32
5	TOC (total organic carbon) (mg L^{-1})	209
6	Color (CPU)	1154
7	TDS (total dissolved salts) (mg L^{-1})	1858
8	Conductivity (ms)	3.32

Sources Kumar and Sharma (2019)

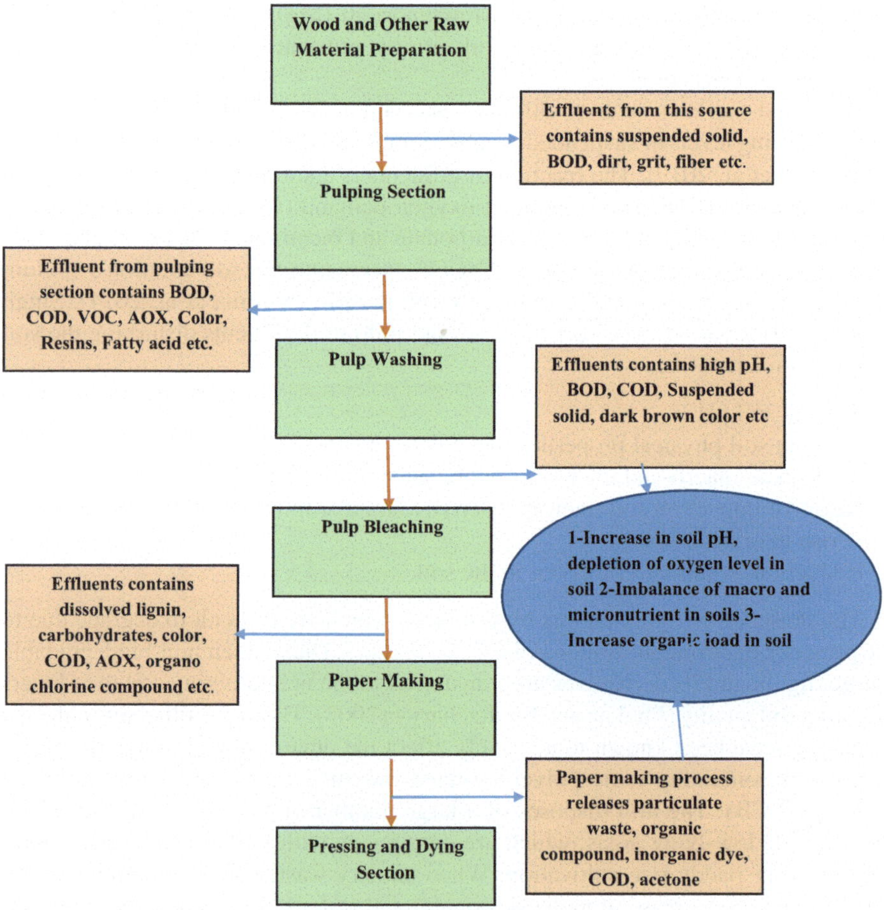

Fig. 14.2 Paper and pulp industry wastewater treatment and its effect on soil. *Sources* Ram et al. (2020)

14.3.3 Textile Industry

The effluents discharged from the textile industry contain a huge amount of residual dye, toxic metals, alkalies, starches cellulose, soluble salts, oil and other contaminations. Textile industries use a lot of water and generate a lot of wastewater, which is usually discharged into an industrial area's common effluent drain Toxic colours, organic pollutants, and high concentrations of heavy metals are found in the composite effluents from India's textile factories, which may have an impact on the region's surface water, soil, groundwater, and plant tissues. The effluents released from textile industries createsoil, surface and groundwater pollution, besides causing several adverse effects on agricultural products, animals and the health of people living in that area (Priya et al. 2017). The application of effluents of the textile industry increases the number of water-soluble salts, organic matter and concentration of sodium and calcium. After that, the occurrence of salinization and alkalinization of groundwater and irrigation of textile wastewater soil becomes saline and alkaline.

Malik and Bharti (2013) reported that the toxic pollutants may percolate down through the soil profile and reach groundwater, posing health risks to humans and livestock after consumption as part of daily drinking requirements. Contaminated groundwater has deteriorated the soil systems, drinking waterand crop productivity (Table 14.3).

Table 14.3 The standard for discharge of effluents from the textile industry

S. mo	Parameter	Standard
	Treated effluents	Maximum concentration values in mg L^{-1} except for pH, colour, and SAR
1	pH	6.5–8.5
2	Suspended solid	100
3	Colour, PCU (platinum cobalt unit)	15
4	BOD	30
5	COD	250
6	Oil and grease	10
7	Total chromium (Cr)	2.0
8	Sulphides	2.0
9	Phenolic compound	1.0
10	Total dissolved salts (TDS)	2100
11	Sodium adsorption ratio (SAR)	26
12	Ammonical nitrogen	50

Source Ministry of Environment Forest and Climate Change (2016)

14.3.3.1 Effect on Soil Properties

Kaur and Sharma (2014) reported that thesoil irrigated with textile wastewaterincreased the pH of the soil and the pH ranges between 8.0 to 8.9. Similarly, the increasing concentration of chemical salts emitted by the textile industries increased the EC values of the soil (Shellina and Joshi 2019). Soil bulk density, pH, total porosity, and hydraulic conductivity have all been observed to decrease as a result of irrigation with textile effluent, resulting in reduced soil productivity. Field clogging with effluent turns fertile land into barren land, and when wastewater is dumped into agricultural fields, heavy metals in the wastewater build up in the soil (Mobaret al. 2015). Nawaz et al. (2006) reported that the. long-term usage of effluent containing even trace amounts of lead (Pb) can cause bioaccumulation of the metal in soil and living organisms, disrupting ecosystem processes and lowering agro-ecological production.

14.3.4 Distillery Effluents

There are about 319 distilleries in India, which produces 3.25 billion litres of alcohol and generates about 40 billion litres of wastewater yearly (Chandra et al. 2012). The distillery unit isone of the major industries, these generate a huge amount of organic and inorganic pollutant such as polysaccharides, reduced sugar, proteins, melanoidin, waxes, N, K, PO_4^{3-}, Ca, SO_4^{2-} etc. during the alcohol production process and disposal of industrial waste is a major problem responsible for soil pollution. The distilleries using molasses, a by-product of the sugar industry, for the production of alcohol by fermentation and distillation process, generate highly organic and coloured wastewater. Because of the presence of water-soluble recalcitrant colouring organic and inorganic compounds in distillery wastewater, it primarily affects the soil (Chowdhary et al. 2018).

The industrial effluent from distilleries is non-toxic, biodegradable, and totally of plant origin, including high amounts of soluble organic matter and plant nutrients that the plants can use for growth and yield. The only issue with distillery effluent is that it has a high biological oxygen demand (BOD), chemical oxygen demand (COD), and salt content, all of which are considered environmentally unfavourable. The proportion of wastewater which is generally the spent wash is approximately 15 times than the total alcohol production and 100 times more concentrated than domestic sewage (Haniffa and Sundaravadhanam 1977). This effluent if without any treatment released in the soil affects the soil to a greater extent. The raw spent wash is acidic with dark brown colour and unpleasant odour, high BOD and COD (45,000 mg L^{-1} and 1,00,000 mgL^{-1}), high dissolved and suspended solids. The general composition of distillery spent wash is shown in Table 14.4.

Table 14.4 General composition of distillery spent wash

Chemical parameter	Mg L^{-1}
pH	3.9–4.3
Electrical conductivity	28,700
Total solids	35,340
Total dissolved solids	27,240
Total suspended solid	9980
Settleable solids	9860
Chemical oxygen demand (COD)	30,520
Biochemical oxygen demand (BOD)	15,300
Carbonate	Negligible
Bicarbonate	12,200
Total phosphorus	28.36
Total potassium	6500
Calcium	920
Magnesium	753.25
Sulphate	5100
Sodium	420
Chloride	5626
Iron	6.3
Manganese	1429
Copper	0.265
Cadmium	0.036
Lead	0.19
Chromium	0.067
Nickel	0.145

Source Rath et al. (2010)

14.3.4.1 Distillery Effluent Causing Soil Pollution

The spent wash generated in the distillery when released in the soil system without any treatment affects the soil physical and chemical status such as infiltration rate, hydraulic conductivity, pH, electrical conductivity, water retention capacity, available nutrient status. Spent wash has a high ability to extract out manganese from the soil, which in long term used resulted in the deficiency of manganese in the soil (Sophie et al. 2017). It has an adverse effect on the microbial population as well as microbial biomass in the soil, which in turn affect the fertility status of the soil. Soil porosity and permeability are essential parameters to be considered for liquid waste disposal to the soil. In soils, a high dose of organic carbon compound from the disposal of distillery waste can lead to a high demand for oxygen from bacterial activity under anaerobic conditions, resulting in a decrease in the infiltration rate and a decline in

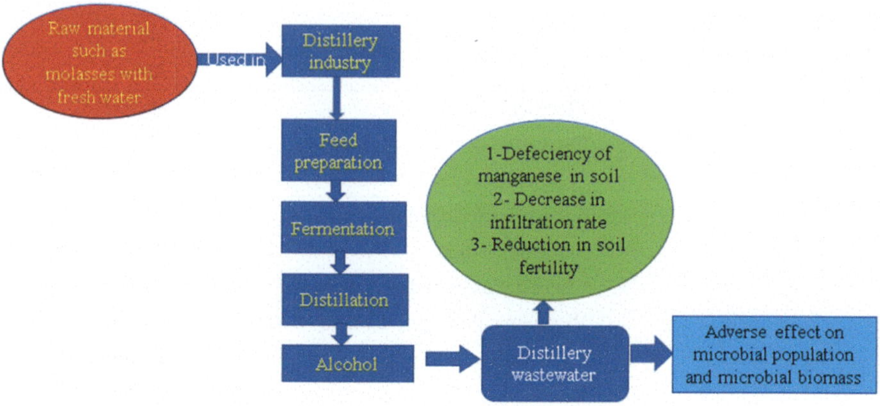

Fig. 14.3 Effect of distillery wastewater on the soilgiven by Chowdhary et al. (2018)

soil fertility. Below is a flowchart representing the effect of distillery wastewater on the soil (Fig. 14.3).

14.4 Scenario of Sewage Sludge Generation in World and India

The world's population is growing and concentrated in cities. This trend is particularly evident in emerging countries, where an additional 2.1 billion people are expected to reside in cities by 2030 (Xu et al. 2014; Mateo-Sagasta et al. 2015). When suspended solids are removed from the wastewater and soluble organic compounds are converted to bacterial biomass, (STPs) sewage treatment plants produce sludge which is a semi-solid substance referred to as sewage sludge (Mateo-Sagasta et al. 2015). In India, due to rapid urbanisation the number of wastewater treatment plant increasing which leads to more production of sewage sludge. Kaur et al. (2012) mention that the 38.354 million litres day^{-1} (MLD) sewage generated in India. Only about 11.786 million litres per day (MLD) of this is treated at sewage treatment plants (STPs), with the rest discharged without treatment. India produces 12 Mt of municipal solid waste each year (MSW). In India, sewage treatment plants in 893 Class I towns (population 1–10 lakes) and Class II towns (population 0.5–1 lake) are estimated to produce 29,129 MLD of sewage (CPCB 2005). Similarly, sewage sludge produced during sewage effluent treatment is dried in sludge beds and used to fertilize agricultural crops (CPHEEO 2012). Available potassium (44–60 mg Kg^{-1}), available phosphorus (44–60 mg Kg^{-1}), available nitrogen (4600–6300 mg Kg^{-1}) and totalnitrogen (15,400–1920 mg Kg^{-1}) were all present in sewage sludge, characterize in India. Dubey et al. (2006) also found increased levels of heavy metals such as arsenic (8–23mg Kg^{-1}), Ni (12–596 mg Kg^{-1}), Pb (26–154 mg Kg^{-1}) Cr (66–1098 mg Kg^{-1}), Hg

Table 14.5 Composition of sewage sludge of some selected countries

Properties	Thailand	Spain	India
Exchangeable Ca (mg/kg)	8332	–	–
Calcium (%)	–	–	1.62
Exchangeable K (mg/kg)	870	–	–
Copper (mg/kg)	801	174	700
Potassium (%)	–	0.2	0.42
Manganese (mg/kg)	2621	–	400
Zinc (mg/kg)	1326	445	1900
pH	6.82	8.6	7.1
Cadmium (mg/kg)	1.22	1.00	1.00
Total phosphorus (%)	–	1.06	1.34
Organic matter (%)	19.82	43.4	23.2
Total nitrogen (%)	3.43	2.5	2.6

(7–32 mg Kg^{-1} and Cd (2–9 mg Kg^{-1}). In municipal sewage sludge, Kumar and Chopra (2015) and Kumar et al. (2016) reported high levels of nutrients and heavy metals. Microplastics are also present in the sewage sludge. Table 14.5 shows the composition of sewage sludge, while Table 14.6 shows the elemental makeup of sewage sludge.

14.5 Source and Types of Sewage

14.5.1 Domestic Source

It comes from toilets, sinks, showers, washing machines and industrial processes and was historically called sewage. Wastewater produced due to human activities in households is called domestic wastewater i.e., wastewater from the kitchen, shower, washbasin, toilet and laundry. In this domestic or household activities are the source of wastewater such as if it is collected from flush toilets, human excreta (faces, urine, blood, and other biological fluids) are frequently mixed with used toilet paper or wet wipes; this is termed as blackwater. Water used for washing such as clothes, floor, car, dishes etc., known as greywater or sullage.

14.5.2 Commercial Source

Urban wastewater also includes commercial waste, which is liquid-transported waste from retailers and service providers serving the immediate communities. Industrial

Table 14.6 Total elemental composition of sewage sludge

Compound	Min. conc. (μg/g)	Max. conc. (μg/g)	Median conc. (μg/g)
Inorganic C	0.3	553	1.4
Organic C	6.5	48	30.4
NH_4-N	<0.1	6.7	1.0
Total N	<0.1	17.6	3.3
Total P	<0.1	14.3	2.3
NO_3-N	<0.1	0.5	<0.1
Total S	0.6	1.5	1.1
Inorganic P	<	2.4	1.6
Mg	0.03	2.0	0.4
Ca	0.10	25.0	3.9
Na	0.01	–	0.2
K	0.02	–	0.3
Al	0.10	–	0.4
Fe	<0.10	15.3	1.1
Cu	84	10,400	850
Zn	101	27,800	1,740
Mn	18	7,100	260
Cd	3	3,410	16
Cr	10	93,000	890
Ni	2	3,515	82
Co	1	18	4
Ba	21	8980	162
Hg	<1	10,600	5
As	6	230	10
Mo	5	39	30
Pd	13	19,730	500

Source Singh and Agarwal (2008)

wastewater, on the other hand, is classed as industrial wastewater rather than wastewater, and it is normally collected separately and filtered or processed. Industrial wastewater is produced by different of processes.

14.5.3 Urban Source

In mixed sewers, sewage and surface runoff can mix unless rainwater is collected separately. Storm surge or overland flow are other terms for surface runoff. It refers to the amount of precipitation that falls swiftly to a specific channel on the earth's surface. Precipitation absorbs gases and particles from the atmosphere, dissolves and leaches materials from vegetation and soil, suspends items on land, and washes away pollution and garbage from metropolitan areas. It is primarily a mixture of municipal and industrial garbage in cities.

14.6 Sewage Sludge as a Source of Soil Pollutant

Because sewage sludge is composed of organic matter and is high in macro-and micronutrients, its disposal in agriculture and forestry is highly recommended. Soils treated with sewage sludge differ from their equivalent unsludged control soils in that they tend to have higher amounts of organic matter with varying decomposition rates and higher concentrations of macronutrients; higher concentrations of micronutrients and non-essential trace elements; pH value can be increased/decreased; and soil microorganism activity can be different. Organic and inorganic pollutants and pathogens (Escherichia coli or Salmonella typhimurium); potential hazardous elements (e.g., zinc, copper, cadmium, lead, silver, etc.); polycyclic aromatic hydrocarbons (PAH); polychlorinated and phytopharmaceuticals (PCP); plastic and microplastic. As sewage sludge contains these contaminants its use in agriculture and forestry should be done cautiouslyHowever, the application of sewage sludge to agricultural land can lead to the spread of a variety of undesirable elements on soils that can be utilised in the food productions. Pollutants (potentially poisonous elements (PTE) such metals, trace organic chemicals (TrOC), and pathogenic microorganisms) can be damaging to the environment (Andreoli et al. 2017). If not treated effectively, toxic pollutants in sewage sludge can exacerbate pre-existing Secondary pollution and poisoning are affected by environmental issues.

14.6.1 Sewage Sludge as a Source of Chemical Contaminant

Initial concentrations (both in soils and in sludge), pace of application (cumulative effects), management tactics, and losses all impact the environmental risk of sludge contaminants and their concentrations in the soil following land application. Volatile and quickly degradable pollutants may indeed be an environmental problem in the case of high starting concentrations and repeated applications (Harisson et al. 2006).

Two environmental and public health hazards arise from the usage or disposal of wastewater treatment facility biosolids:

- Potentially toxic elements (PTEs)
- Organic pollutant.

14.6.1.1 Potential Toxic Element

Sewage sludge adds heavy metal such as chromium, copper, zinc, lead, nickel and cadmium. The application of heavy metal to the soil is the primary source of heavy metal in the soil. Metallic qualities such as ductility, malleability, conductivity, and ligand affinity are considered potentially hazardous in elements with a high density and relative atomic weight (Omprakash 2018). Increased extractable amounts of Zn and Cd have been observed in sewage sludge–amended soils that last for a long time even after sludge applications are discontinued (McGrath et al. 2000). Effect of potential toxic element onsoil properties discussed below.

14.6.1.2 Soil Microbial Activity

When there is an increasing level of heavy metal contamination the viability of microbes decreases. The effect of sewage sludge on the agricultural soil for a period of 20 years, decreased the microbial biomass (Brookes and Mcgrath 1984). Also, there is a decrease in the colony-forming unit of fungi and bacteria in the forest soil mixed with Pb-Cu sludge. According to Giller et al. (1989), Because of earlier sewage sludge applications, which resulted in just one strain of rhizobium surviving in the metal contaminated soil, the long-term consequence of metal toxicity on the soil microbial community is a loss of variety in rhizobium. Heavy metals influence the microbial community that synthesises enzymes, which has an indirect effect on soil enzymatic activity. The potential toxic element reduced the richness of bacterial species while increasing the relative abundance of soil actinomycetes or the heavy metal reduced both thebiomass and diversity of the bacterial community. The activity of enzymes is more affected by the type of metal ion species such as cadmium is more toxic compared to lead as it has a little affinity towards soil colloid and greater mobility. Cr (VI) is a highly toxic oxidising agent, whereas Cr (III) is a micronutrient and a non-hazardous species that is ten to hundred times less toxic than Cr (VI). At high concentrations, Cr (VI) has been shown to cause changes in the composition of soil microbial populations and to have a negative impact on microbial cell metabolism (Singh and Kalamdhad 2011).

14.6.1.3 Effect on Enzymatic Activity

The heavy metal such as Cu suppresses the action of the enzyme-glucosidase, whereas lead considerably reduces the activity of urease, catalase, invertase, and acid phosphatase. Cadmium contamination adversely affect the activities of urease, protease, alkaline phosphatase and arylsulphatase. The order of inhibition of urease activity normally decreases in the order of Cr > Cd > Zn > Mn > Pb (Singh and Kalamdhad 2011).

14.6.1.4 Effect on Soil Microbial Composition

These heavy metals affect the quantity and quality of soil fungus, soil bacteria, soil actinomycetes, and other populations of microbes when they accumulate in the soil over their acceptable limits. They also change the chemical and biological properties of the soil and cause distinct microbial community patterns. When metal-enriched sewage sludge is added to soils, microbial biomass drops and microbial community structure alters (Omprakash 2018). The effectiveness of microbial communities in organic mineralization is inversely related to the organic carbon content of the soil, which is an indicator of the consequence of potentially hazardous element contamination (Omprakash 2018). Ultimately it is these potential toxic elements present in sludge restricts its use for the agriculture purpose. That is why its field application should be carefully managed to avoid its effecton soil quality and soil health.

14.6.2 Persistent Organic Pollutant

Sewage sludge contains a significant amount of organic pollutant (Guo et al. 2009). Discharges from domestic and industrial facilities, as well as atmospheric deposition and urban runoff, are the sources of these pollutants in wastewater. So, they are concentrated in the sewage sludge during the treatment of wastewater. Among these polycyclic aromatic hydrocarbonsis the important type of organic pollutant. Many persistent organic pollutants (POPs) are found in sewage, and they can resist treatment methods such as anaerobic digestion before accumulating in soils where sewage sludge is applied. The persistent chemicals are generally hydrophobic and linked to soil organic matter, but their hydrophobicity and volatility varies widely. Organochlorine pesticides (OCPs), polychlorinated biphenyls (PCBs), polycyclic aromatic hydrocarbons (PAHs), hexachlorocyclohexane (HCHs) isomers, and dichlorodip are all organic pollutants that may be deposited in soil (Roccaro and Vagliasindi 2014). Sewage sludge is regarded as a material capable of increasing persistent organic pollutants (POP) input into the soil (Elskens et al. 2013).

The half-life in sewage sludge of polychlorinated dibenzo-p-dioxins/furans and polychlorinated biphenyls varies between 9 and 12 years, according to Zennegg et al. (2013). As a result, repeated applications of sewage sludge to the soil can increase abundance of these compounds in the soil. In addition to these pollutants, newly discovered pollutants such PBDEs (polybrominated diphenyl ethers) should be considered. Lee et al. (2014) published a study on the distribution of PBDE (polybrominated diphenyl ethers) congeners in sewage sludge, which revealed that BDE-209 (brominated diphenyl ether) was the most abundant in all sludge samples.

14.6.2.1 Sewage Sludge as a Sink for the Organic Pollutant in Soil

A long-term field experiment started in 1996 at the Czech Republic showed that there is an increase in the polychlorinated biphenyls content in the soil treated with sewage sludge as well as the increase was related to the application of sewage sludge in soil. In the plot where sewage sludge was applied three times, the concentration of polychlorinated biphenyls was found to be 1.8 times higher than the general sewage sludge rate. The polychlorinated biphenyls in sludge (Jana et al. 2019) had 150 times the amount of PCBs found in the soil, showing that sewage sludge could be a substantial source of polychlorinated biphenyls (PCBs) in agricultural land. When compared to mineral fertiliser, Umlauf et al. (2011) sewage sludge-treated soils were found to be greater value of dioxin-like polychlorinated biphenyls (DL-PCBs).

Wang et al. (2011) found that the higher values of PBDEs i.e. 17 $\mu g\ kg^{-1}$ dw in a rice field. Similarly, initial brominated flame retardant (BFRs) in sludge/soil mixture (1:2) remained constant over a 3-year monitoring period (Venkatesh and Halden 2014). As a result, if sewage sludge is used regularly, it may be a source of additional chemical build-up in the soil. Long-term sewage sludge application may cause these chemicals to build up in the soil, leading the soil to pollute to a greater level over time. Figure 14.4 is a flowchart representing the effect of the potential toxic element as well as a persistent organic pollutant on soil Fig. 14.4.

14.7 Microplastics in Sewage Sludge

Direct use of sewage sludge for agricultural fertilisation has been scientifically proven to be a source of microplastic in the soil (Weithmannet et al. 2018). All over the world, microplastics have been observed in sewage sludge. Approximately 90% of the microplastic in the wastewater is retained during the treatment process and accumulates in the sewage sludge (Carr et al. 2016; Haernvall et al. 2018; Tagg and Labrentz 2018). Synthetic polymers are also applied on a regular basis during the sewage sludge drainage and treatment process (Stubenrauch and Ekardt 2020). Every year, between 0.2 and 8 μg of microplastics per hectare per person are projected to be released into agricultural soils across Europe (Nizzeto et al. 2016), depending on the frequency of sewage sludge fertilization. Sludge, also known as biosolids, is either disposed of in a landfill, incinerated, or put to the land once it has been processed, resulting in three paths for MPs (microplastics) to reach the environment. MPs have been found in soil that received biosolids from WWTPs (wastewater treatment plants), highlighting the function of WWTPs as contributors to the release of environmental MPs in some situations. Because of their chemical makeup and hydrophobicity, microplastics can leak into the environment and pose a chemical and physical threat (Rolsky et al. 2020). Figure 14.5 representing environment risk associated with the sludge amended to the soil.

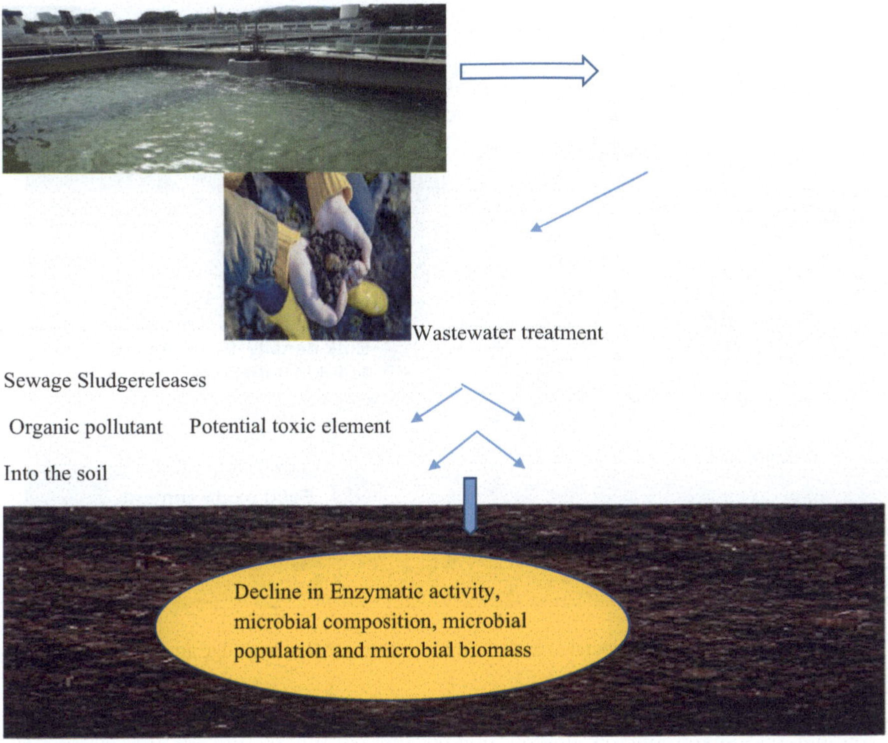

Fig. 14.4 Effect of potential toxic element and persistent organic pollutant on soil present in sewage sludge

14.7.1 Effect of Microplastics on Soil Physical and Chemical Property

Microplastics have been detected in sludge-affected soil, and they have been demonstrated to detract from the benefits of biosolids by reducing soil water holding capacity, microbial activity, and bulk density. Because of their ability to withstand microbial assimilation. They destroy the soil structure, impact water holding capacity, reduce infiltration rate. Microplastics have a negative impact on soil organic carbon, nitrogen, nutrient transfer and soil microbial activity. They have no ion exchange capacity, water holding capacity as well not act as a carbon source for microbial growth (Baile et al. 2019). The polyacrylic fibres largely affect the water-stable aggregates and resulted in the formation of soil clumps that leads to erosion. The fragmentation of (previously larger) pieces into microplastic increases the likelihood of entering deeper soil layers. Wind and water transport may be reduced further, while biological transport (in the digestive tracts of soil organisms) may become an important factor. Longer dwell times of microplastic particles mean longer exposure times for the various soil organisms affected. The alteration in physical properties greatly

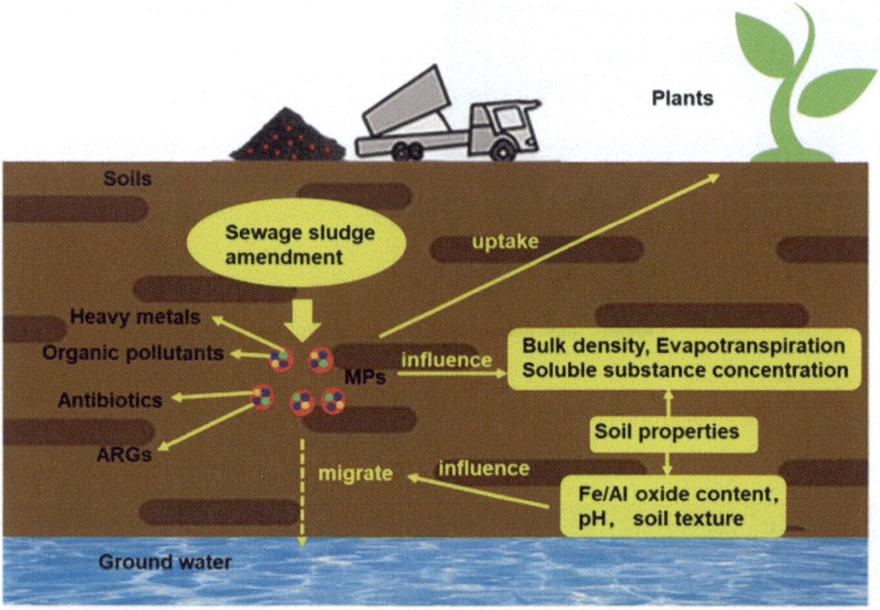

Fig. 14.5 Environment risk associated with the microplastics (MPs) in sludge amended soil (*Source* Gao et al. 2020)

affects the soil organisms by decreasing water holding capacity and increase evaporation, which reduces the water availability to plant results in reduced plant growth and ultimately less production of root exudates. This leads to reduced microorganism growth (Guo et al. 2020).

14.7.1.1 Effect on Soilbiota

The ingestion of microplastics by earthworms (*L. terristris*) only leads to mortality after long-term exposure to microplastic concentrations >27% total litter. The growth of *Eisenia foetida* is also inhibited and mortality is increased at a higher microplastic concentration (>1%) in the soil. Metal transfer analysis from microplastics to earthworms has also revealed that MPs may have higher zinc desorption than soils (Welden and Lusher 2020). Veresoglou et al. (2015) figure out that the alteration in pore space caused by microplastics resulted in microhabitat loss and extinction of indigenous microorganism in soil. Microplastics impacted the microbial community structure and lowered substrate-induced respiration (SIR) rates, indicating that the microplastics influenced soil microbial activity (Guo et al. 2020).

14.7.1.2 Effect on the Sorption Capacity of Soil

While microplastics contain additives like diethylhexyl phthalate (DEHP), a common organic pollutant during plastic production, Heavy metals like zinc, copper, and lead, as well as hazardous pollutants like polybrominated diphenyl ether (PBDE) and perfluorochemicals (PFOS), are adsorbable (Guo et al. 2020). It can easily spread in the soil thus hampering the soil quality and health. Huffer et al. (2019) found that adding 10% polyethene to soil reduced the sorption capacity of the soil overall. PE pollution may increase the mobility of organic pollutants in soil, diminishing the soil's natural retention capacity, as evidenced by the reported decrease in sorption. Figure 14.6 depicts a schematic overview of the environmental dangers and processes associated with microplastics in soil.

More importantly, one recent study found that sludge-based MPs (microplastics) had higher heavy metal adsorption capacity than virgin MPs (microplastics). This implies that MPs (microplastics) in sewage sludge after processing may have a greater influence on the transport behaviour of toxic metals in the environment, resulting in more negative effects than MPs (microplastics) from other sources. Despite this, little is known about the fate and properties of MPs in sewage sludge as a result of various treatments.

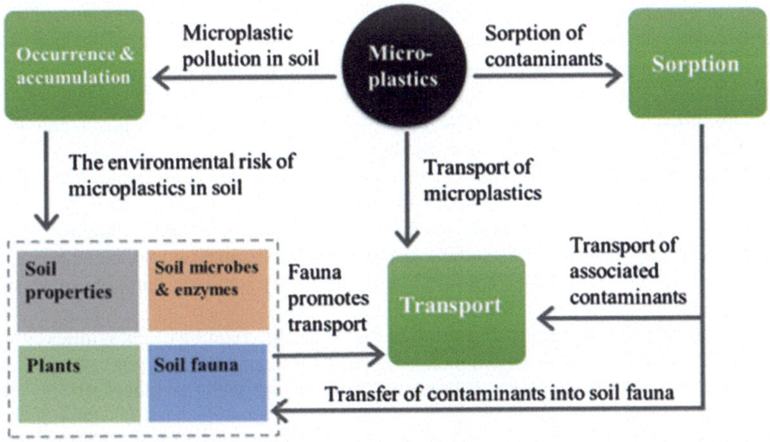

Fig. 14.6 Environmental dangers and processes associated with microplastics in soil are depicted in this diagram. Microplastics' presence and build-up can have negative consequences for soil systems, including soil characteristics, bacteria and enzymes, plants, and animals (*Source* Baile et al. 2019)

14.8 Conclusion

In the current scenario, soil pollution from industries, as well as sewage sludge, has become a major problem as they are not treated properly and released directly or indirectly in the soil system. The waste released in the soil system contains heavy metal, organic pollutant and microplastics which results in the decline in microbial population, bulk density, enzymatic activity and fertility of soil. The long-term application of sewage sludge as well as industrial effluents in the soil affect the soil quality and soil health. So, to overcome this problem their treatment should be done or the permissible limit of each heavy metal to be determined before using it for soil application. So that their concentration not build up in the soil.

References

Andreoli CV, Von Sperling M, Fernandes F, Ronteltap M (2017) Sludge treatment and disposal. IWA Publishing

Antil S, Kumar V, Kathpal TS, Narwal RP, Sharma SK, Mittal SB, Kuhad MS (2004) Extent of land degradation in different agro-climatic zones of Haryana. Fertiliser News 49(3):47–60

Arora BR, AzadA S, Singh B, SekhonGS (1985) Pollution potential of municipal waste waters of Ludhiana, Punjab. Indian J Ecol 12(1):1–7

Bansal OP (2018) The influence of potentially toxic elements on soil biological and chemical properties, metals in soil—contamination and remediation. IntechOpen, Rijeka

Baskar M, Saravanan A, Chitra L, Dhevagi P, Lenin RD, Pandiyarajan P, Ambast SK (2013) Ecofriendly utilization of distillery waste water in agriculture. AICRP on management of salt affected soils and use of saline water in agriculture 1–50

Bharti PK, Kumar P, Singh V (2013) Impact of industrial effluents on ground water and soil quality in the vicinity of industrial area of Panipat city, India. J Appl Nat Sci 5(1):132–136

Buckland GD, Rodney B, Mikalson D, Jong DE, Chang C (2002) Soil salinization and sodication from alternate irrigations with saline-sodic water and simulated rain. Can J Soil Sci 82(3):297–309

Carr SA, Liu J, Tesoro AG (2016) Transport and fate of microplastic particles in wastewater treatment plants. Water Res 91:174–182

Chandra R, Bharagava RN, Kapley A, Purohit HJ (2012) Characterization of Phragmites cummunis rhizosphere bacterial communities and metabolic products during the two-stage sequential treatment of post methanated distillery effluent by bacteria and wetland plants. Biores Technol 103(1):78–86

Chowdhary M, Mostafa MG, Biswas TK, Mandal A, Saha AK (2015) Characterization of the effluents from leather processing industries. Environ Process 2(1):173–187

Chowdhary P, Raj A, Bharagava RN (2018) Environmental pollution and health hazards from distillery wastewater and treatment approaches to combat the environmental threats: a review. Chemosphere 194:229–246

Chowdhary P, Bharagava RN (2019) Toxicity, beneficial aspects and treatment of alcohol industry wastewater. In: Emerging and eco-friendly approaches for waste management. Springer, Singapore, pp 83–97

CPCB (2005) Parivesh sewage pollution—newsletter. Central Pollution Control Board, Ministry of Environment and Forests, Govt. of India

CPCB (2009) Status of water supply, wastewater generation and treatment in Class I cities and Class II towns of India. Series: CUPS/70/2009–10. Central Pollution Control Board, India

CPHEEO (2012) Manual on sewerage and sewage treatment, part a: engineering final draft, central public health and environmental engineering organisation, ministry of urban development, New Delhi

Elskens M, Pussemier L, Dumortier P, Van Langenhove K, Scholl G, Goeyens L, Focant JF (2013) Dioxin levels in fertilizers from Belgium: Determination and evaluation of the potential impact on soil contamination. Sci Total Environ 454:366–372

Frostegård Å, Tunlid A, Bååth E (1993) Phospholipid fatty acid composition, biomass, and activity of microbial communities from two soil types experimentally exposed to different heavy metals. Appl Environ Microbiol 59(11):3605–3617

Gao D, Li XY, Liu HT (2020) Source, occurrence, migration and potential environmental risk of microplastics in sewage sludge and during sludge amendment to soil. Sci Total Environ 140355

Guo JJ, Huang XP, Xiang L, Wang YZ, Li YW, Li H, Wong MH (2020) Source, migration and toxicology of microplastics in soil. Environ Int 137:105–263

Haernvall K, Zitzenbacher S, Biundo A, Yamamoto M, Schick MB, Ribitsch D, Guebitz GM (2018) Enzymes as enhancers for the biodegradation of synthetic polymers in wastewater. Chem Bio Chem 19(4):317–325

Harrison EZ, Oakes SR, Hysell M, Hay A (2006) Organic chemicals in sewage sludges. Sci Total Environ 367(2–3):481–497

Hassan Z, Ali S, Farid M, Khan MM, Abid M, Omar AM, Maheshwari RK (2016) Effect of chromium (Cr) on the microbial and enzymatic activities in the soil: a review. Biodiversity conservation in changing climate. Delhi, India Lenin Media Private Limited, pp 305–323

Hembrock HA, König W (1990) Vorkommen und Transfer von polycyclischen aromatischen Kohlenwasserstoffen in Böden und Planzen. VDI Berichte 837:815–830

Hüffer T, Metzelder F, Sigmund G, Slawek S, Schmidt TC, Hofmann T (2019) Polyethylene microplastics influence the transport of organic contaminants in soil. Sci Total Environ 657:242–247

Kaur R, Wani S P, Singh A K, Lal K (2012) Wastewater production, treatment and use in India. In: National report presented at the 2nd regional workshop on safe use of wastewater in agriculture. Available at: http://www.ais.unwater.org/ais/pluginfile.php/356/mod_page/content/111/CountryReport_India

Kaur V, Sharma G (2014) Effects of industrial effluent on soil characteristics: a review. Int J Adv Eng Sci Technol 3:201–207

Khan S, Joshi N (2019) Effect of textile wastewater irrigation on soil properties in Western Rajasthan. J Exp Biol Agric Sci 7(4):489–493

Kumar V, Chopra AK (2015) Fertigation with agro-residue-based paper mill effluent on a high-yield spinach variety. Int J Veg Sci 21(1):69–97

Kumar D, Sharma C (2019) Remediation of pulp and paper industry effluent using electrocoagulation process. J Water Resour Prot 11(03):296

Kumar V, Chopra AK, Srivastava S (2016) Assessment of heavy metals in spinach (*Spinacia oleracea* L.) grown in sewage sludge–amended soil. Commun Soil Sci Plant Anal 47(2): 221–236

Liu J, Zhang Y Q, Zhang L M, Zhou XB, Shi XJ (2014) Impact of Cr $^{3+}$ pollution on microbial characteristics in purple paddy soil. Pakistan J Pharmaceutical Sci 27

Mani S, Chowdhary P, Hare V (2019) Industrial effluents: impact on agricultural soils and microbial diversity. Plant Biotic Interact 43–60

Mateob SJ, Raschid SL, Thebo A (2015) Global wastewater and sludge production, treatment and use. Wastewater. Springer, Dordrecht, pp 15–38

McGrath SP, Zhao FJ, Dunham SJ, Crosland AR, Coleman K (2000) Long-term changes in the extractability and bioavailability of zinc and cadmium after sludge application. Am Soc Agron 29(3):875–883

Millar CE, Turk LM (2002) Fundamentals of soil science. Daya Books

Ministry of Environment (2016) Forest and Climate Change

Mobar S, Kaushik P, Bhatnagar P (2015) Physiochemical comparison of textile effluent impacted and un-impacted agricultural soil of Jaipur city, India. Int J Recent Sci Res 6(3):3090–3093

Nawaz A, Khurshid K, Arif MS, Ranjha AM (2006) Accumulation of heavy metals in soil and rice plant (*Oryza sativa* L.) irrigated with industrial effluents. Int J Agric Biol 8(3):391–393
Nigam H, Das M, Chauhan S, Pandey P, Swati P, Yadav M, Tiwari A (2015) Effect of chromium generated by solid waste of tannery and microbial degradation of chromium to reduce its toxicity: a review. Adv Appl Sci Res 6(3):129–136
Nizzetto L, Futter M, Langaas S (2016) Are agricultural soils dumps for microplastics of urban origin?
Phukan S, Bhattacharyya KG (2003) Modification of soil quality near a pulp and paper mill. Water Air Soil Pollut 146(1):319–333
Pokhriya P, Punetha D, Arunachalam K, Arunachalam A (2017) Can we use textile effluent as a source of irrigation: a case from Bhagwanpur, Uttarakhand (India). Int J Appl Environ Sci 12(3):527–540
Pulkrabová J, Černý J, Száková J, Švarcová A, Gramblička T, Hajšlová J, Tlustoš P (2019) Is the long-term application of sewage sludge turning soil into a sink for organic pollutants? evidence from field studies in the Czech Republic. J Soils Sediments 19(5):2445–2458
Ram C, Rani P, Gebru KA, Abrha MGM (2020) Pulp and paper industry wastewater treatment: use of microbes and their enzymes. Phys Sci Rev 5(10)
Rath P, Pradhan G, Mishra M K (2010) Effect of sugar factory distillery spent wash (DSW) on the growth pattern of sugarcane (*Saccharum officinarum*) crop. J Phytol 2(5)
Roccaro P, Vagliasindi FG (2014) Risk assessment of the use of biosolids containing emerging organic contaminants in agriculture. Chem Eng 37
Rolsky C, Kelkar V, Driver E, Halden RU (2020) Municipal sewage sludge as a source of microplastics in the environment. Curr Opin Environ Sci Health 14:16–22
Sabumon PC (2016) Perspectives on biological treatment of tannery effluent. Adv Recycl Waste Manage 1:3–10
Shellina K, Joshi N (2019) Effect of textile wastewater irrigation on soil properties in Western Rajasthan. J Exp Biol Agric Sci 7(4):489–493
Shi W, Bischoff M, Turco R, Konopka A (2002) Long-term effects of chromium and lead upon the activity of soil microbial communities. Appl Soil Ecol 21(2):169–177
Singh J, Kalamdhad AS (2011) Effects of heavy metals on soil, plants, human health and aquatic life. Int J Res Chem Environ 1(2):15–21
Singh PK, Ladwani K, Deshbhratar PB, Ramteke DS (2013) Impact of paper mill wastewater on soil properties and crop yield through lysimeter studies. Environ Technol 34(5):599–606
Singh P, Srivastava N, Singh P, Geetha S, Usharani N, Jagadish RS, Upadhyay A (2019) Effect of toxic pollutants from pulp and paper mill on water and soil quality and its remediation. Int J Lakes Rivers 12:1–20
Singh RP, Agrawal M (2008) Potential benefits and risks of land application of sewage sludge. Waste Manage 28(2):347–358
Sophie PL, Kokila C, Meena, K, Gomathi A (2017) Impact of distillery effluent on soil quality
Stubenrauch J, Ekardt F (2020) Plastic pollution in soils: governance approaches to foster soil health and closed nutrient cycles. Environments 7(5):38
Sundari S, Kanakarani P (2001) The effect of pulp unit effluent on agriculture. J Ind Pollut Control 17(1):83–97
Tadesse GL, Guya TK, Walabu M (2017) Impacts of tannery effluent on environments and human health: a review article. Adv Life Sci Technol 54:10
Tagg AS, Labrenz M (2018) Closing microplastic pathways before they open: a model approach
Tripathi DM, Tripathi S, Tripathi BD (2011) Implications of secondary treated distillery effluent irrigation on soil cellulase and urease activities. J Environ Prot 2(05):655
Umlauf G, Christoph EH, Lanzini L, Savolainen R, Skejo H, Bidoglio G, Scherer H (2011) PCDD/F and dioxin-like PCB profiles in soils amended with sewage sludge, compost, farmyard manure, and mineral fertilizer since 1962. Environ Sci Pollut Res 18(3):461–470

Venkatesan AK, Halden RU (2014) Brominated flame retardants in US biosolids from the EPA national sewage sludge survey and chemical persistence in outdoor soil mesocosms. Water Res 55:133–142

Veresoglou SD, Halley JM, Rillig MC (2015) Extinction risk of soil biota. Nat Commun 6(1):1–10

Vijay S, Garg UK, Deepak A (2014) Impact of pulp and paper mill effluent on physico-chemical properties of soil. Arch Appl Sci Res 6(2):12–17

Vincent S (2014). Int J Waste Resour

Weithmann N, Möller JN, Löder MG, Piehl S, Laforsch C, Freitag R (2018) Organic fertilizer as a vehicle for the entry of microplastic into the environment. Sci Adv 4(4):806

Welden NA, Lusher A (2020) Microplastics: from origin to impacts. In: Plastic waste and recycling, pp 223–249

Xu GH, Wang YK, Sheng GP, Mu Y, Yu HQ (2014) An MFC-based online monitoring and alert system for activated sludge process. Sci Rep 4(1):1–7

Xu B, Liu F, Cryder Z, Huang D, Lu Z, He Y, Xu J (2020) Microplastics in the soil environment: occurrence, risks, interactions and fate–a review. Crit Rev Environ Sci Technol 50(21):2175–2222

Zayed AM, Terry N (2003) Chromium in the environment: factors affecting biological remediation. Plant Soil 249(1):139–156

Chapter 15
Spatial Distribution and Radiological Risk Quantification of Natural Radioisotopes in the St. Martin's Island, Bangladesh

Rahat Khan, Md. Abu Haydar, Sudipta Saha, Md. Masud Karim, Md. Ahosan Habib, Md. Bazlar Rashid, Abubakr M. Idris, and Debasish Paul

Abstract The radioactivity concentrations of ^{226}Ra, ^{232}Th and ^{40}K were measured by HPGe gamma-ray spectroscopy in beach sand and water samples collected from and around the only coral reefed Island (St. Martin's), in the Bay of Bengal. Average activity concentrations of ^{226}Ra, ^{232}Th and ^{40}K are 15.53, 15.42 and 372.32 Bq kg^{-1} for beach sand samples, and 4.96, 4.67 and 22.78 Bq kg^{-1} for water samples, respectively. No artificial radionuclides (e.g., ^{134}Cs, ^{137}Cs) were detected in any of the analyzed samples. Lower activity concentrations of sand samples compared to those of other coastal areas of the Bay of Bengal may be due to the thick coral reef of this island. The estimated radiation hazard parameters including radium equivalent activity, radiation hazard index, annual effective dose rate, absorbed dose rate and excess lifetime cancer risk are lower than the permissible limits. In terms of radiological parameters, this island is quite safe for tourism.

Keywords Coral reefed island · Naturally occurring radionuclides · Beach sand and water · Radiological hazard indices · St. Martin's Island · Bangladesh

R. Khan (✉) · Md. A. Haydar · S. Saha · Md. M. Karim · D. Paul
Institute of Nuclear Science & Technology, Bangladesh Atomic Energy Commission, Savar, Dhaka 1349, Bangladesh
e-mail: rahatkhan.baec@gmail.com

Md. A. Habib · Md. B. Rashid
Geological Survey of Bangladesh, Segunbaghicha, Dhaka 1000, Bangladesh

A. M. Idris
Department of Chemistry, College of Science, King Khalid University, Abha 62529, Saudi Arabia
e-mail: abubakridris@hotmail.com

Research Center for Advanced Materials Science (RCAMS), King Khalid University, Abha 62529, Saudi Arabia

P. K. Shit et al. (eds.), *Soil Health and Environmental Sustainability*, Environmental Science and Engineering,
https://doi.org/10.1007/978-3-031-09270-1_15

15.1 Introduction

Beach sands are mostly composed of those minerals which are resistant to wave abrasion, (e.g., quartz, feldspar). A combination of weathering, degradation and fragmentation processes supply those wave resistant minerals to the coastal areas (Papadopoulos et al. 2016 and the reference therein). A number of coastal areas in and around the Bay of Bengal have already been reported for higher level of natural radioisotopes (e.g., ^{226}Ra, ^{232}Th and ^{40}K) owing to the presence of wave resistant placer minerals such as ilmenite, zircon, monazite, garnet, rutile etc. (Khan et al. 2019a, 2021; Rao et al. 2009; Mohanty et al. 2004; Kannan et al. 2002; Alam et al. 1999). These naturally occurring radionuclides (NORMs) could cause numerous radiological health risks including different types of cancers, kidney malfunction, bone deformities etc. (e.g., Habib et al. 2022; Khan et al. 2019b and the references therein) owing to the ionizing radiation emission and radon inhalation. Thus considering the highest contribution of external dose to the human by the natural radiation, the assessments of radiological distribution along with their associated potential health risks are of huge importance (Habib and Khan 2021; UNSCEAR 2000).

Unlike the other islands of the Bay of Bengal, St. Martin's Island (Bangladesh) is one of the few islands in the world which is surrounded by thick coral reefs. St. Martin's Islandis of great ecological importance as it is the only fossiliferous marine island in the Bay of Bengal which possesses huge areas of sandy beach and mangrove formations (Tomascik 1997). This is one of the most beautiful domestic and foreign tourist destinations in Bangladesh since it possesses attractive natural sceneries, clear sea-site, and natural beauties of coral colonies. Coral reefs as 'rainforests of the sea' engage only <0.1% of the world's total marine area, but they supply accommodations, breeding environments and food to the more than 25% of all aquatic botanical and zoological species (Islam et al. 2019 and the reference therein). Considering the marine ecological significance, millions of peoples dependence (as tourists and/or inhabitants) and scientific importance of coral reefs, several studies on heavy metal accumulations in corals, marine sediments and sea water along with their associated health and ecological risk assessment have been performed (Joy et al. 2019; Jafarabadi et al. 2017a, 2018a, b; Prouty et al. 2013; Mokhtar et al. 2012). Along with the heavy metal distribution, *n*-alkanes, polycyclic aromatic hydrocarbons and persistent organic pollutants distribution in coral-associated environmental compartments have also been reported by Jafarabadi et al. (2017b, 2018c, d). Furthermore, geochemistry and precise elemental ratios in coral skeletons and associated environmental compartments have long been used to explain the climate change, to reconstruct the temporal pollution history and to assess the sea water quality (Saha et al. 2016, 2018, 2019; Lewis et al. 2018; Prouty et al. 2010). However, the studies on NORMs distribution in the coral reefed Island and their associated radiological health risks (Lin et al. 2019) coral reefs in the South China Sea) are very much scarce. Islam et al. (2019) reported radioactivity concentrations of coral skeleton and the marine sediments around the St. Martin's Island, leaving the beach sand (or, soil) and the seawater in and around the Island. Thus, geological variation and variability of environmental compartments (e.g., marine sediments, coral skeletons, seawater, beach sands or soils) invoke similar radiological studies in various environmental compartments in and around the coral reefs of the world to provide a comprehensive scenario to the scientific community as well as to the tourists and local inhabitants.

Study of radiological health hazards owing to the NORMs is very much important for the safety of the tourists as well as the local inhabitants. It is also essential to measure the baseline level of NORMs in different environmental sections (e.g., beach sand, water, marine sediment and coral backbone) before any pollution events (e.g., nuclear weapon test, nuclear reactor accidents etc.) commence nearby. However, a comprehensive study on the distribution of NORMsin the beach-sand and water samples across and around this island has not hitherto been done. Thus, the present study aims at the determination of NORMs (^{226}Ra, ^{232}Th and ^{40}K) as well as search for the anthropogenic radionuclides (e.g., ^{134}Cs, ^{137}Cs) in beach sand and water samples taken from the St. Martin's Island'ssurroundings to evaluate the baseline distribution of radionuclides and to estimate the potential radiological risks.

15.2 Materials and Methods

15.2.1 Area of Interest

St. Martin's Island (area: ~8 km^2) resides in the north-eastern side of the Bay of Bengal and southernmost part of Bangladesh which is about 9 km south of the Cox's Bazar-Teknaf peninsula (Fig. 15.1). Length of this Island (south-north side) is approximately 5.6 km and the width (east–west side) varies from 200 to 700 m. The island is almost flat and is about 3.6 m above the average sea level (Akhtar 1992). St. Martin's Island represents the westernmost extent of the Arakan Yoma uplift and it is ringed by a boulder field in the intertidal zone along the southern and western shore of the Island (Khan 1964).

15.2.2 Sampling and Sample Processing

Samples (sand and water) were collected along the coast line of St. Martin's Island by January 2017. Beach sand samples were taken from 15 different sampling stations (separated each other by $\geq$500 m) covering both the east and west side of the sandy beaches as shown in Fig. 15.1. From each sampling points, approximately 1.5 kg of superficial sand samples were taken (sampling depth: ~10 cm). Collected samples were then instantly preserved in airtight clean and properly marked zip-lock polyethylene bags and moved to the sample preparation laboratory for subsequent analysis. After eliminating the extraneous matters including stones, gravels, pebbles, roots and botanical debris, the samples were homogenously powdered, weight and dried

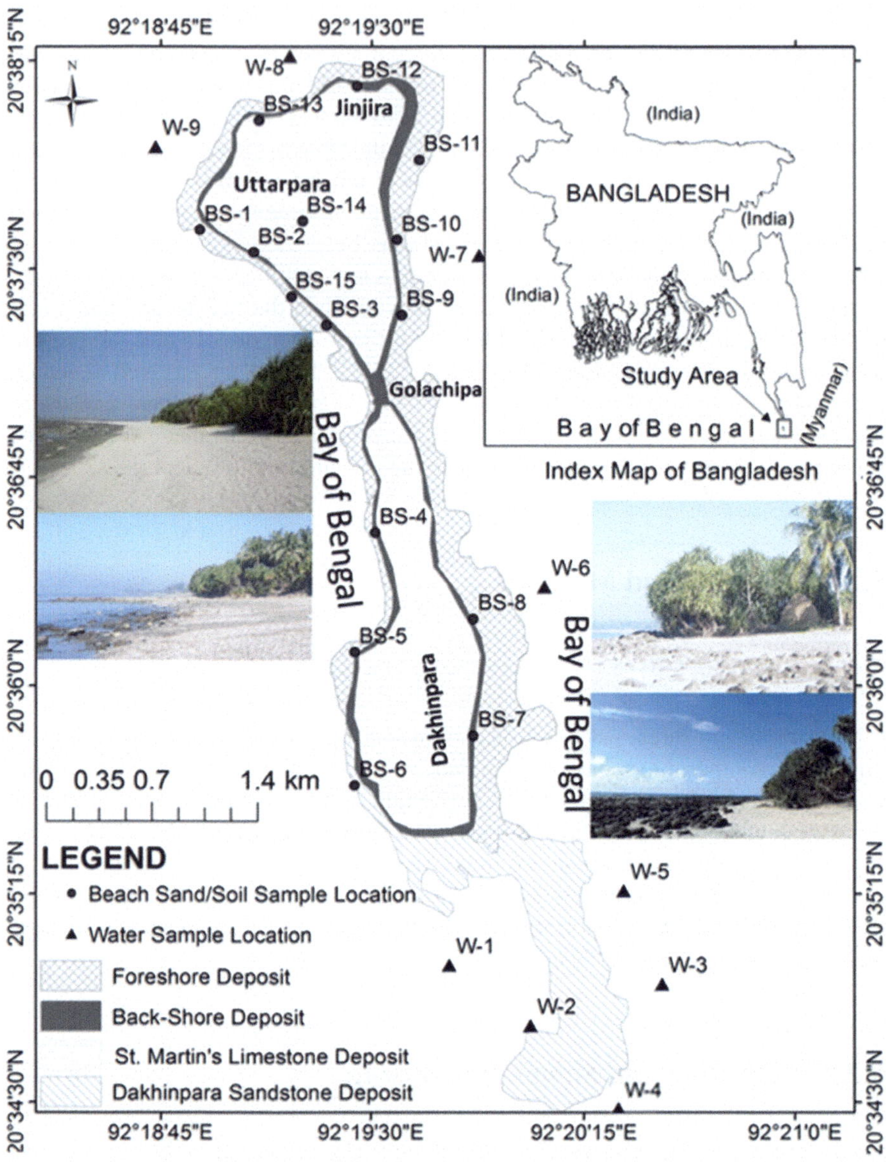

Fig. 15.1 Map shows the sampling points at St. Martin's Island, Bangladesh

(at 105 °C) until attaining the constant weight. Thereafter homogenous dried powder samples (~250 g) were hermetically packed in a cylindrical plastic pot (dimensions of the pot are identical to that of Khan et al. 2019b) and then sealed hermetically to avoid the loss of NORMs (as radon) and subsequently stored for at least 28 days at room temperature to attain the secular equilibrium among ^{238}U and ^{232}Th decay series along with the respective daughter products (Habib et al. 2018, 2019) prior to being measured. The details of the sampling preparation procedures were previously reported elsewhere (Begum et al. 2022; Habib et al. 2018, 2019; Khan et al. 2022, 2019b).

Sea surface water samples were collected from 9 different spots (Fig. 15.1) around the Island. From each spot, about 1.5 L of water was taken in clean, acid (diluted HNO_3) rinsed and dried plastic container. The collected water samples in the plastic container were immediately acidified (pH ~ 1) with nitric acid to prevent adsorption of NORMs onto the walls of containers (Agbalagba and Onoja 2011) and transferred to the sample processing laboratory. The collected samples were then poured into cylindrical plastic containers of equal size and shape (of volume 260 ml). Sealing and storing of the water sample containers prior to the gamma-ray counting are same as those of sand samples.

15.2.3 *Analytical Procedure*

15.2.3.1 Radioactivity Measurement by Gamma Spectrometry

Analytical procedures of measuring the NORMs (^{226}Ra, ^{232}Th and ^{40}K) and searching the potential artificial radionuclides (^{134}Cs and ^{137}Cs) were identical to those of our previous studies (Khan et al. 2019b; Majlis et al. 2022; Habib et al. 2018, 2019). Briefly, coaxial p-type high purity Ge gamma detector with 40% relative efficiency was used in this study. In direct measurement of ^{226}Ra and^{232}Th (measuring from the activities of their progenies) and direct measurement of ^{40}K were performed by following Khan et al. (2019b). Other than the naturally occurring radionuclides (^{226}Ra, ^{232}Th and ^{40}K), artificial radionuclides such as ^{134}Cs (604.5 and 795.8 keV) and ^{137}Cs (661.6 keV) were also searched in each analyzed samples. Blank correction, energy and efficiency calibration were similar to that of Khan et al. (2019b).

Radioactivity concentrations of NORMs in the beach sand and the seawater samples were measured from the net count rate, counting efficiency and emission probability of specific radionuclides and mass (for beach sand) or volume (for water sample) of the sample by the Eqs. (15.1) and (15.2):

$$\text{A (Bq)} = \frac{\text{cps}_{\text{sample}} - \text{cps}_{\text{BG}}}{\varepsilon(\text{E}) \times \text{P}_{\gamma}} \tag{15.1}$$

$$\text{AC}\left(\text{Bq kg}^{-1}\right) = \frac{\text{A}}{\text{m}} \tag{15.2}$$

where, A is the radioactivity (in Bq); AC, radioactivity concentration (in Bq kg^{-1}); cps_{sample}, counts per second for the sample (in s^{-1}); cps_{BG}, Counts per second for the background (in s^{-1}); ε (E_{γ}), counting efficiency of the HPGe gamma-ray detector; P_{γ}, the emission probability; m, sample mass (kg). In this study, the minimum detectable activity (MDA) for the gamma-ray measuring system was computed by the Eq. (15.3) (Asaduzzaman et al. 2015; Khandaker et al. 2012, 2016, 2017):

$$\mathrm{MDA}\left(\mathrm{Bq\ kg^{-1}}\right) = \frac{\mathrm{C_F} \times \sqrt{\mathrm{B}}}{\varepsilon\left(\mathrm{E_\gamma}\right) \times \mathrm{P_\gamma} \times \mathrm{t} \times \mathrm{m}} \tag{15.3}$$

where, C_F is the statistical coverage factor (=1.64) for 95%confidence level; B is the background counts over the region of interest for each radionuclide; ε (E_γ), the absolute efficiency of the HPGe detector; P_γ, the emission probability; t, the measuring time in seconds and m, the mass of the sample (in kg).

15.2.4 Data Presenting Processes

Basic descriptive analysis was done for measured NORMs (variables) in our sample suits. Pearson's correlation analysis was performed to define the degree of association and interdependency existing among the determined and estimated variables parameters both radionuclides and corresponding radiological hazard indices using SPSS (version 20) software. The Inverse Distance Weighting (IDW) method was applied to interpolate the measured parameters at unmeasured locations from the observations of its values at nearby points by using ArcGIS 10.3software. It is commonly used for displaying the spatial distribution of interested parameters in the determined beach sand samples of the mapped area (Habib et al. 2018).

15.3 Results and Discussion

15.3.1 Distribution of Radionuclides

In beach sand samples, the radioactivity level (Table 15.1) of ^{226}Ra, ^{232}Th and ^{40}K were found to be 8.79 ± 2.45 to 29.12 ± 2.66 Bq kg^{-1}, 8.68 ± 3.41 to 24.72 ± 8.70 Bq kg^{-1}and 166.17 ± 68.01 to 472.53 ± 74.44 Bq kg^{-1}, respectively. The obtained average values for these nuclides (Table 15.2) along with their standard deviations were 15.53 ± 5.09, 15.42 ± 5.61 and 372.32 ± 78.87 Bq kg^{-1}, respectively. The activities of ^{226}Ra, ^{232}Th and ^{40}K across the sampling points did not vary widely (Fig. 15.2, Table 15.2). Spatial distributions of ^{226}Ra, ^{232}Th and ^{40}K in our studied area are showed by inverse distance weighting (IDW) map (Fig. 15.3). Figure 15.3a, b, c and d represent the spatial distribution patterns of activity contents of ^{226}Ra, ^{232}Th,^{40}K and absorbed dose rate, respectively in the mapped area. In the Island, radioactivity distribution maps show almost uniform partitioning. Touristic areas (northern part of the Island: Jinjira and Uttarpara) possess relatively lower activity of ^{226}Ra (Fig. 15.3a). This work pointed some trivial hot spots in some sites (eastern side

Table 15.1 Radioactivity concentrations in beach sand (BS) and soil (SL) samples collected from St. Martin Island, Bangladesh along with their associated radium equivalent activity (Ra_{eq} in Bq kg^{-1}), external hazard index (H_{ex}), absorbed dose rate (D in ηGy h^{-1}), annual effective dose rate (E_{ff} in ηGy h^{-1}) and excess lifetime cancer risk (ELCR)

Sample ID	Activity concentrations (Bq kg^{-1})						Radiological indices				
	^{226}Ra	±	^{232}Th	±	^{40}K	±	Ra_{eq}	H_{ex}	D	E_{ff}	ELCR
BS-1	9.45	2.37	16.27	4.39	264.38	67.20	53.07	0.143	25.44	0.031	1.10 × 10^{-4}
BS-2	14.17	2.47	15.18	1.40	361.87	70.86	63.74	0.172	30.97	0.038	1.33 × 10^{-4}
BS-3	14.18	2.29	12.11	1.40	293.59	64.91	54.10	0.146	26.24	0.032	1.13 × 10^{-4}
BS-4	13.81	2.35	8.68	3.41	416.90	69.76	58.32	0.158	29.03	0.036	1.25 × 10^{-4}
BS-5	16.66	2.42	14.23	1.48	389.11	68.41	66.97	0.181	32.65	0.040	1.41 × 10^{-4}
BS-6	12.52	2.59	10.67	1.42	472.53	74.44	64.16	0.173	31.98	0.039	1.38 × 10^{-4}
BS-7	16.27	2.45	24.69	4.58	437.28	69.81	85.25	0.230	40.99	0.050	1.76 × 10^{-4}
BS-8	29.12	2.66	20.42	1.61	355.70	71.14	85.71	0.231	40.87	0.050	1.76 × 10^{-4}
BS-9	18.25	2.55	10.99	1.37	409.07	71.92	65.46	0.177	32.20	0.040	1.39 × 10^{-4}
BS-10	13.86	2.48	12.61	1.49	355.77	69.66	59.29	0.160	28.97	0.036	1.25 × 10^{-4}
BS-11	8.79	2.43	24.72	8.70	406.81	71.53	75.46	0.204	36.29	0.045	1.56 × 10^{-4}
BS-12	18.48	2.49	24.63	4.93	398.48	70.06	84.38	0.228	40.36	0.050	1.74 × 10^{-4}
BS-13	14.8	2.39	14.8	1.48	429.92	68.64	69.07	0.187	33.84	0.042	1.46 × 10^{-4}
BS-14	21.62	3.15	12.01	1.85	427.25	88.72	71.69	0.194	35.14	0.043	1.51 × 10^{-4}
BS-15	10.93	2.38	9.30	1.31	166.17	68.01	37.02	0.100	17.71	0.022	0.76 × 10^{-4}

Associated uncertainties are due to the counting statistics

of Dakhinpara) of the island showing relatively higher activity of measured radionuclides, except for ^{40}K (which distributed mainly in the western side of Dakhinpara). However, trivial hot spots and consequential minute inhomogeneous distribution of NORMs do not essentially invoke any radiological risk.

The descriptive statistics of the measured values of our study are compared to those of previously published works (Rudnick and Gao 2014; Huang et al. 2015; Ghosal et al. 2017; Alam et al. 1999; Kannan et al. 2002; Khandaker et al.

Table 15.2 Comparison of radioactivity concentrations and associated radiological hazard indices of this study to those of literature data

Sample ID	Activity concentrations (Bq kg^{-1})			Radiological indices				
	^{226}Ra	^{232}Th	^{40}K	Ra_{eq}	H_{ex}	D	E_{ff}	ELCR
This study								
Mean (n = 15)	15.53	15.42	372.32	66.25	0.179	32.18	0.040	1.39×10^{-4}
SD (1σ)	5.09	5.61	78.87	13.33	0.036	6.33	0.008	0.27×10^{-4}
RSD (%)	32.8	36.4	21.2	20.1	20.1	19.7	19.7	19.7
Median	14.18	14.23	398.48	65.46	0.177	32.20	0.040	1.39×10^{-4}
Min.	8.79	8.68	166.17	37.02	0.100	17.71	0.022	0.76×10^{-4}
Max.	29.12	24.72	472.53	85.71	0.231	40.99	0.050	1.76×10^{-4}
Literature data								
World avg.[a]	35	30	400	370	1	55	0.06	2.90×10^{-4}
UCC[b]	33	43	720	149.93	0.40	72.0	0.09	3.09×10^{-4}
Xiamen Island, China[c]	14.6 (7.9 – 25.7)	10.9 (6.7 – 41.4)	396.4 (197.4 – 487.6)	60.71	0.164	29.9	0.04	1.29×10^{-4}
Coastal Odisha, India[d]	273.93 (24.6 – 532.7)	2489.23 (394.5 – 4520.8)	683.23 (26.6 – 1295.4)	3886.14	10.493	1705.4	2.10	73.4×10^{-4}
Beach soil, Cox's Bazar, Bangladesh[e]	18.9 (10.8 – 27.3)	36.7 (27.4 – 49.4)	458.2 (117 – 688)	106.66	0.288	50.5	0.06	2.18×10^{-4}
Black sand, Langkawi Island, Malaysia[f]	1478.2 (451 – 2411)	718.2 (232 – 1272)	102.8 (61 – 136)	2513.14	6.789	1133.1	1.39	48.8×10^{-4}
White sand, Langkawi Island, Malaysia[f]	9.78 (8.3 – 13.7)	5.87 (4.5 – 9.4)	102 (85 – 133)	26.03	0.070	12.4	0.02	0.53×10^{-4}
Giresun sea beach, Turkey[g]	21 (2 – 44)	14 (2 – 27)	531 (23 – 1306)	81.91	0.221	40.4	0.05	1.74×10^{-4}
Penang Island, Malaysia[h]	23 (8 – 43)	19 (9 – 43)	243 (68 – 478)	68.88	0.186	32.5	0.04	1.40×10^{-4}

(continued)

Table 15.2 (continued)

Sample ID	Activity concentrations (Bq kg^{-1})			Radiological indices				
	^{226}Ra	^{232}Th	^{40}K	Ra_{eq}	H_{ex}	D	E_{ff}	ELCR
Miami Bay, Penang, Malaysia[h]	1023 (24 – 2641)	2086 (37 – 5622)	381 (296 – 495)	4035.32	10.898	1787.0	2.20	7.69×10^{-4}
Preta beach, Brazil[i]	121 (54 – 180)	239 (128 – 349)	110 (47 – 283)	471.24	1.273	209.2	0.26	9.01×10^{-4}
Dois Rios beach, Brazil[i]	40 (6 – 78)	48 (12 – 87)	412 (269 – 527)	140.36	0.379	65.4	0.08	2.82×10^{-4}
Kalpakkam beach, India[j]	16 (5 – 71)	119 (15 – 776)	406 (200 – 854)	217.43	0.587	98.3	0.12	4.23×10^{-4}
West coast, SriLanka[k]	299 (BDL – 1243)	1032 (14 – 6257)	335 (BDL – 647)	1800.56	4.862	794.6	0.98	34.2×10^{-4}
Black sea shore, Romania[l]	6.7 (2.9 – 14)	3.7 (1.2 – 8.5)	69 (9 – 282)	17.30	0.047	8.3	0.01	0.36×10^{-4}
Marine sediment, St. Martin's Island[m]	30.7 (24.5 – 39.5)	36.8 (28.3 – 47.4)	388 (358 – 422)	113.20	0.306	53.1	0.07	2.29×10^{-4}
Coral skeletons, St. Martin's Island[m]	16.5 (11.7 – 21.9)	28.7 (23.3 – 32.8)	334 (322 – 346)	83.26	0.225	39.3	0.05	1.69×10^{-4}

[a] UNSCEAR (2000)
[b] Upper Continental Crust (UCC-calculated data): Rudnick and Gao (2014)
[c] Huang et al. (2015)
[d] Ghosal et al. (2017), Alam et al. (1999)
[f] Khandaker et al. (2018)
[g] Kucukomeroglu et al. (2016)
[h] Shuaibu et al. (2017)
[i] Freitas and Alencar (2004)
[j] Kannan et al. (2002)
[k] Mahawatte and Fernando (2013)
[l] Margineanu et al. (2013)
[m] Islam et al. (2019)

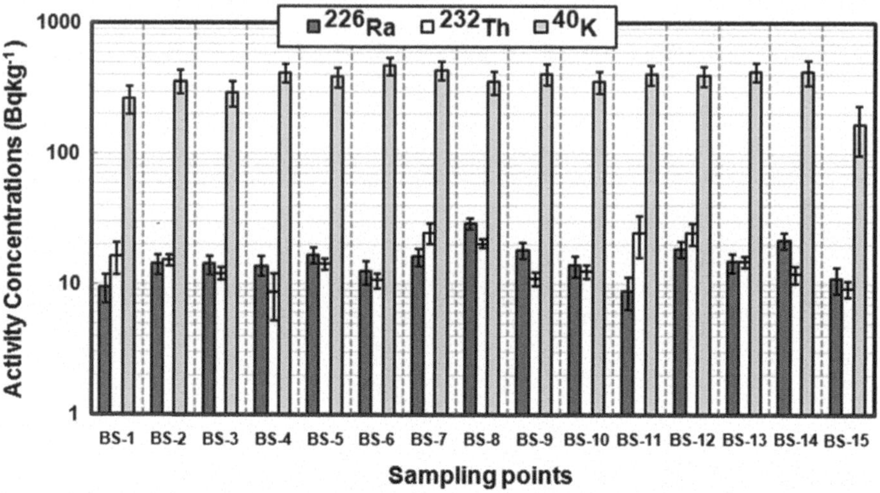

Fig. 15.2 Variation of activity concentrations of beach sand and soil samples of St. Martin's Island, Bangladesh at different sampling locations

2018; Kucukomeroglu et al. 2016; Shuaibu et al. 2017; Freitas and Alencar 2004; Mahawatte and Fernando 2013; Margineanu et al. 2013; Islam et al. 2019) in Table 15.2. The radionuclides concentrations of St. Martin's Island of this work are significantly lower than the other coastal areas known for higher background radiation (e.g., Khandaker et al. 2018; Shuaibu et al. 2017; Freitas and Alencar 2004). Specific activities of ^{226}Ra and ^{40}K of this study are comparable to those of beach sand of Cox's Bazar, Bangladesh (Alam et al. 1999) whereas activity concentration of ^{232}Th shows opposite trend which implies that Th-rich monazite is less abundant in St. Martin's Island compared to the mainland (Cox's Bazar) of Bangladesh. Disregarding the variation of activity concentrations, Kalpakkam beach, India (Kannan et al. 2002) shows similar trend as those of Cox's Bazar beach, Bangladesh with our study. However, the activity concentrations in our study are significantly lower than the other coastal areas of the Bay of Bengal, including Coastal Odisha, India (Ghosal et al. 2017) and West coast, Sri Lanka (Mahawatte and Fernando 2013). A reasonable assumption for such lower activity concentrations of our study compared to those of other coastal areas of the Bay of Bengal (Alam et al. 1999; Kannan et al. 2002; Ghosal et al. 2017; Khan et al. 2017, 2018) can be explained in terms of coral abundances and the ocean current dynamics around the St. Martin's Island. Activity concentrations of ^{226}Ra and ^{232}Th in the beach sands of the present study are significantly lower (Table 15.2) than the marine sediments (Islam et al. 2019) around the St. Martin's Island, while the activity concentration of ^{40}K are almost comparable among them. ^{232}Th-radioactivity is ~1.8 times lower in the beach sand than that of coral skeleton, whereas the specific activities of ^{226}Ra and ^{40}K in beach sand are comparable to those of coral skeleton. The comparable radioactivity concentration of ^{40}K in beach sands, marine sediments and coral skeleton may be explained in

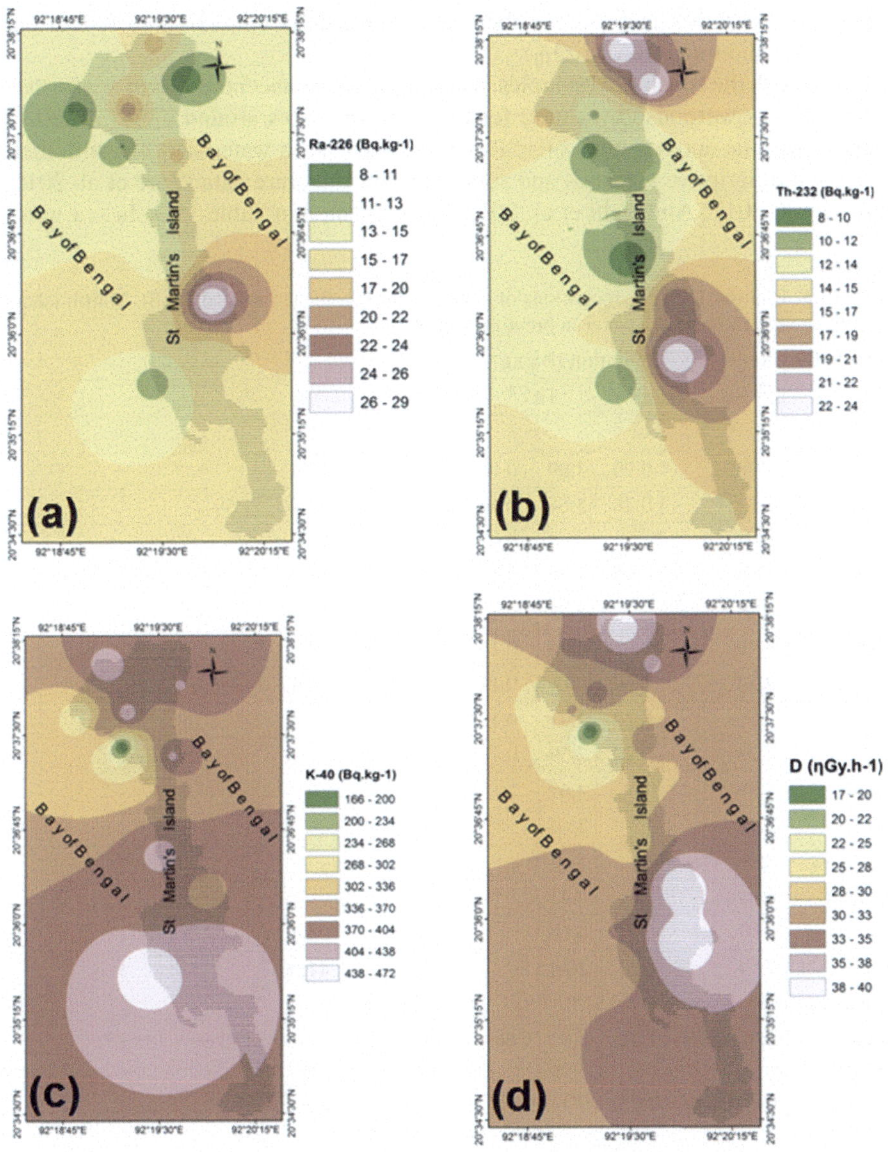

Fig. 15.3 Inverse distance weighting (IDW) map for the spatial distribution of activity concentrations of **a** ^{226}Ra, **b** ^{232}Th, **c** ^{40}K and **d** estimated dose distribution in the St. Martin's Island, Bangladesh

terms of K-solubility and the distribution of terrigenous minerals like K-feldspar ($KAlSi_3O_8$) and mica ($KAlSi_4O_{10}$).

Along with the beach sand samples, the radioactivity concentrations of sea surface water samples were also measured for the 9 marine spots around the St. Martin's Islands. Specific activity level of ^{226}Ra, ^{232}Th and ^{40}K in water samples with their associated descriptive statistics and some relevant literature data (Zare et al. 2015; Baltas et al. 2017; Almayahi et al. 2012) are presented in Table 15.3. In sea-water

Table 15.3 Radioactivity concentrations of water samples collected around the St. Martin Island are compared to those of sea water in previous works

Sample ID	Activity concentration (Bq kg^{-1})					
	^{226}Ra	±	^{232}Th	±	^{40}K	±
This study						
W-1	7.78	0.29	3.89	0.06	BDL	
W-2	1.32	0.28	5.68	0.14	41.00	1.59
W-3	BDL		0.68	0.07	BDL	
W-4	1.36	0.22	4.23	0.13	BDL	
W-5	BDL		5.48	0.15	BDL	
W-6	6.39	0.22	4.64	0.16	BDL	
W-7	9.54	0.30	10.41	0.17	BDL	
W-8	0.56	0.22	4.41	0.12	BDL	
W-9	7.78	0.29	2.59	0.10	BDL	
Mean	**4.96**		**4.67**			
SD (1σ)	3.75		2.64			
RSD (%)	75.6		56.5			
Median	6.39		4.41			
Min.	BDL		0.68		BDL	
Max.	9.54		10.41		41.00	
Literature data						
Oman Sea[a]	2.50 (2.19 – 2.82)		1.90 (1.66 – 2.17)		141.48 (132.60 – 148.87)	
Black Sea in Rize, Turkey[b]	0.26 (0.16 – 0.63)		0.11 (0.07 – 0.17)		5.42 (3.44 – 6.20)	
Northern Peninsula, Malyasia[c]	3.46 (2.33 – 7.03)		3.63 (1.58 – 8.64)		190.2 (150 – 220)	

BDL: Bellow detection limit
[a] Zare et al. (2015)
[b] Baltas et al. (2017)
[c] Almayahi et al. (2012)

samples the specific activities of ^{226}Ra, ^{232}Th and ^{40}K were found to be below detection limit (BDL) to 9.54 ± 0.30 Bq kg^{-1}, 0.68 ± 0.07 to 10.41 ± 0.17 Bq kg^{-1} and BDL to 41.00 ± 1.59 Bq kg^{-1}, respectively. None of the analyzed samples (both sand and water) contains detectable amount of artificial radionuclides (here, ^{134}Cs and^{137}Cs). The average values of minimum detectable activities (MDAs) of ^{226}Ra, ^{232}Th, ^{40}K, ^{134}Cs and ^{137}Cs in the determined sand samples are 0.48, 0.35, 18.8, 0.50 and 0.41 Bq kg^{-1}, respectively whereas for water samples MDAs for those radionuclides are 0.52, 0.38, 20.2, 0.53 and 0.45 Bq L^{-1}, respectively.

15.3.2 Radiological Risk Assessment

To assess the potential radiological risks owing to the natural radioisotope in beach sand of this highly touristic area (St. Martin's Island, Bangladesh), radium equivalent activity (Ra_{eq}), radiation hazard index (H_{ex}), absorbed dose rate (D), annual effective dose rate (E_{ff}) and excess lifetime cancer risk (ELCR) were calculated in the current research.

15.3.2.1 Radium Equivalent Activity (Ra_{eq})

To assess the combined radiological threat to the population, Ra_{eq} has widely been used which is attained from the activity concentrations of ^{226}Ra, ^{232}Th and ^{40}K (Khan et al. 2019b and the references therein). Most of the radiation dose received by human being is due to the emission of gamma radiation from natural radiation sources (Tufail 2012), including ^{40}K and the progeny of the ^{238}U and ^{232}Th decay series. Owing to the potential disequilibrium among^{226}Ra and its progenies, radionuclides (^{226}Ra, ^{232}Th) may not be evenly distributed in the environmental geochemical samples (e.g., sediment, soil, sand, etc.) (Ahmed et al. 2014). Thus, for homogeneous exposure calculation, the radioactivity concentrations are expressed as Ra_{eq} (in Bq kg^{-1}) which can be calculated by the following expression:

$$Ra_{eq}\left(Bq\ kg^{-1}\right) = A_{226_{Ra}} + 1.43A_{232_{Th}} + 0.077A_{40_K} \leq 370 \tag{15.4}$$

where, $A_{226_{Ra}}$, $A_{232_{Th}}$ and A_{40_K} are activity concentrations of ^{226}Ra, ^{232}Th and ^{40}K (in Bq kg^{-1}), respectively. In our studied area Ra_{eq} ranges from 37.02 to 85.71 Bq kg^{-1} (Table 15.1) with a mean value of 66.25 ± 13.33 (SD) Bq kg^{-1} (Table 15.3), which are significantly lower than the prescribed value of 370 Bq kg^{-1} (UNSCEAR 2000).

15.3.2.2 Radiation Hazard Index (H_{ex})

External hazard index (H_{ex}) has commonly been employed (Agbalagba et al. 2012; Iqbal et al. 2000) to estimate the external exposure, which is defined as follows:

$$H_{ex} = \frac{A_{226Ra}}{370} + \frac{A_{232Th}}{259} + \frac{A_{40K}}{4810} \leq 1 \tag{15.5}$$

where, H_{ex} is a dimensionless quantity, since the unit of the denominator of Eq. (15.5) is also Bq kg^{-1}, (Farai and Ademola 2005). Corresponding to the upper allowable value of Ra_{eq}, the highest permissible value of H_{ex} can be ≤ 1 (Merdanoğlu and Altinsoy 2006). Tables 15.1 and 15.2 provide the H_{ex} for the beach sand and soil samples of our study, varying from 0.100 to 0.231 with an average value of 0.179 ± 0.036. All the calculated values of H_{ex} are less than 1, which implies that this island is radiologically safe for the local inhabitants as well for the tourists.

15.3.2.3 Absorbed Dose Rate (D)

The geographical characteristics govern the distribution of radiation exposure in a given place. Following UNSCEAR (2000) guideline, NORMs are supposed to be homogeneously distributed and the absorbed dose rates (D)owing to the terrestrial gamma radiations (from^{226}Ra, ^{232}Th and ^{40}K) at 1 m above the ground level for public can be estimated using the following Eq. (15.6):

$$D(\eta Gyh^{-1}) = 0.462A_{226Ra} + 0.621A_{232Th} + 0.0417A_{40K} \tag{15.6}$$

where, 0.462, 0.621 and 0.0417 are the respective dose conversion factors transforming the radioactivities of NORMsinto dose rates (in nGy h^{-1}). The D-value in air owing to the NORMs in the beach sand samples of our studied area range from 17.71 to 40.99 nGy h^{-1} with a mean value of 32.18 nGy h^{-1} (Table 15.2), which are significantly lower than the permissible value of 55 nGy $h^{-1}$for the public (Table 15.2) as prescribed in the UNSCEAR (2000). To pictorially represent the spatial distribution of cumulative contribution of NORMs, in Fig. 15.3d, absorbed dose rates are shown by IDW map.

15.3.2.4 *Annual Effective Dose Rate (E_{ff})*

Two aspects should be considered while calculating E_{ff} in outdoor (UNSCEAR 2000)-(a) the conversion factor from absorbed dose in air to the effective dose (0.7 Sv Gy^{-1}) and (b) the indoor occupancy factor (0.2). Therefore, the E_{ff} (in mSv y^{-1}) can be estimated by using the succeeding relation (15.7):

$$E_{ff}(mSv\,y^{-1}) = D(\eta Gy\,h^{-1}) \times 8760\,h\,yr^{-1} \times 0.7 \times \left(10^3 \frac{mSv}{10^9}\right)\eta Gy^{-1} \times 0.2 \tag{15.7}$$

Considering 8760 h as the total number of hours per year, the estimated E_{ff} from the beach sand samples vary from 0.022 to 0.050 mSv y^{-1} with an average value of

0.040 ± 0.008 mSv y^{-1} which are significantly lower than the quoted world mean value of 0.07 mSv y^{-1} (Table 15.2).

15.3.2.5 Excess Lifetime Cancer Risk (ELCR)

The ELCR can be computed by the following Eq. (15.8) (ICRP 1990):

$$\text{ELCR} = E_{\text{ff}} \times \text{ALT} \times \text{RF} \tag{15.8}$$

where, average life time (ALT) is assumed to be 70 years and the risk factor (RF) is 5.0×10^{-5} for the public exposure (ICRP 1990). In this study, the computed values of ELCR for exposure to beach sand samples were found to be varied from 0.76×10^{-4} to 1.76×10^{-4} with an average value of $1.39 \times 10^{-4} \pm 0.27 \times 10^{-4}$ (Table 15.3), which are considerably lower than the average for world value 2.90×10^{-4} (Table 15.2). All the estimated radiological hazard indices are within the permissible limits. In terms of radiological safety, it can be concluded that the samples from St. Martin tourist area do not endanger human health and threat to the ambient environment.

15.3.3 Correlation Matrix Analysis

To identify the source of radionuclides and their relationship with the radiological hazard indices, the calculated correlation coefficients are appeared in Table 15.4. In the current work, a significant positive relationship was found among the measured radionuclides and risk indices which suggested that the emission of gamma radiation is from all radionuclides. While the determined radionuclides show a weak degree of association or insignificant correlation among them. It indicated that ^{226}Ra and ^{232}Th decay series exist in different mineral suites/rock types in the beach sand samples and differences in geochemical behaviors of these radionuclides were assumed.

Table 15.4 Mutual correlation matrix of radionuclides and sand properties of the St. Martin's Island, Bangladesh

	^{226}Ra	^{232}Th	^{40}K	Ra_{eq}	Dose
^{226}Ra	1				
^{232}Th	0.16	1			
^{40}K	0.259	0.223	1		
Ra_{eq}	0.596*	0.764**	0.688**	1	
Dose	0.593*	0.726**	0.735**	0.998**	1

* Correlation is significant at the 0.05 level (2-tailed)
** Correlation is significant at the 0.01 level (2-tailed)

15.4 Conclusion

This study reveals the distribution of ionizing radiation emitting NORMs (^{226}Ra, ^{232}Th and ^{40}K) in the beach sand and water of and around the only coral reefed island in the Bay of Bengal for the first time. Artificial radionuclides (^{134}Cs, ^{137}Cs) have not been detected in the present study. Activity concentrations of the radionuclides are almost homogenously distributed across the island. The results of this study are assessed to check the compatibility of international and national values. Unlike the high background coastal areas of the world and the Bay of Bengal, activity concentrations of coastal areas of the St. Martin's island is significantly low and are within the limit of UNSCEAR (2000). Coral reefs of this island are assumed to obstacle the gathering of wave resistant heavy minerals (which are enriched with the NORMs). The estimated radiation hazard parameters including radium equivalent activity (Ra_{eq}), radiation hazard index (H_{ex}), absorbed dose rate (D), annual effective dose rate (E_{ff}) and excess lifetime cancer risk (ELCR) are lower than the admissible recommended limits. Results of this study will form reference data for the only coral reefed Island in the Bay of Bengal and will be considered as the baseline data for the future works.

Acknowledgements We are thankful to the technical personnel associated with this study, especially to the persons who helped us during sampling. Financial support from the Atomic Energy Research Establishments (AERE) of Bangladesh Atomic Energy Commission is also thankfully acknowledged.

Conflicts of Interest The authors declare that they have no conflict of interest.

Funding The authors express their gratitude to Research Center of Advanced Materials, King Khalid University, Saudi Arabia, for support (award number KKU/RCAMS/22).

References

Agbalagba EO, Onoja RA (2011) Evaluation of natural radioactivity in soil, sediment and water samples in Niger Delta (Biseni) flood plain lakes, Nigeria. J Environ Radioact 102(7):667–671

Agbalagba EO, Avwiri GO, Chad-Umoreh YE (2012) γ-Spectroscopy measurement of natural radioactivity and assessment of radiation hazard indices in soil samples from oil fields environment of Delta State, Nigeria. J Environ Radioact 109:64–70

Ahmed MM, Das SK, Haydar MA, Bhuiyan MMH, Ali MI, Paul D (2014) Study of natural radioactivity and radiological hazard of sand, sediment, and soil samples from Inani Beach, Cox's Bazar, Bangladesh. J Nucl Part Phys 4(2):69–78

Akhtar A (1992) Palynology of Girujan Clay, St. Martin's Island, Cox's Bazar District, Bangladesh. Geological survey of Bangladesh, Dhaka, vol 7(2), p 24

Alam MN, Chowdhury MI, Kamal M, Ghose S, Isam MN, Mustafa MN, Miah MMH, Ansary MM (1999) The ^{226}Ra, ^{232}Th and ^{40}K activities in beach sand minerals and beach soils of Cox's Bazar, Bangladesh. J Environ Radioact 70:2652–2660

Almayahi BA, Tajuddin AA, Jaafar MS (2012) Radiation hazard indices of soil and water samples in Northern Malaysian Peninsula. Appl Radiat Isot 57:109–119

Asaduzzaman Kh, Mannan F, Khandaker MU, Farook MS, Elkezza A, Amin YM, Sharma S, Kassim HA (2015) Assessment of natural radioactivity levels and potential radiological risks of common building materials used in Bangladeshi dwellings. PLoS ONE 10(10):1–16

Baltas H, Kiris E, Sirin M (2017) Determination of radioactivity levels and heavy metal concentrations in seawater, sediment and anchovy (*Engraulis encrasicolus*) from the Black Sea in Rize, Turkey. Mar Pollut Bull 116(1–2):528–533

Begum M, Khan R, Hossain SM, Al Mamun SMMA (2022) Redistributions of NORMs in and around a gas-field (Shabazpur Bangladesh): radiological risks assessment. J Radioanal Nucl Chem 331(1):317–330. https://doi.org/10.1007/s10967-021-08107-x

Farai IP, Ademola JA (2005) Radium equivalent activity concentrations in concrete building blocks in eight cities in Southwestern Nigeria. J Environ Radioact 79:119–125

Freitas AC, Alencar AS (2004) Gamma dose rates and distribution of natural radionuclides in sand beaches—Ilha Grande, Southeastern Brazil. J Environ Radioact 75:211–223

Ghosal S, Agrahari S, Guin R, Sengupta D (2017) Implications of modeled radioactivity measurements along coastal Odisha, Eastern India for heavy mineral resources. Estuar Coast Shelf Sci 184:83–89

Habib MA, Khan R (2021) Environmental impacts of coal-mining and coal-fired power-plant activities in a developing country with global context. Spatial modeling and assessment of environmental contaminants (Chapter 24), environmental challenges and solutions, Springer Nature Switzerland AG. https://doi.org/10.1007/978-3-030-63422-3_24

Habib MA, Khan R, Phoungthong K (2022) Evaluation of environmental radioactivity in soils around a coal burning power plant and a coal mining area in Barapukuria Bangladesh: radiological risks assessment. Chem Geol 600:120865. https://doi.org/10.1016/j.chemgeo.2022.120865

Habib MA, Basuki T, Miyashita S, Bekelesi W, Nakashima S, Phoungthong K, Khan R, Rashid MB, Islam ARMT, Techato K (2018) Distribution of naturally occurring radionuclides in soil around a coal-based power plant and their potential radiological risk assessment. Radiochim Acta (accepted). https://doi.org/10.1515/ract-2018-3044

Habib MA, Basuki T, Miyashita S, Bekelesi W, Nakashima S, Techato K, Khan R, Majlis ABK, Phoungthong K (2019) Assessment of natural radioactivity in coals and coal combustion residues from a coal-based thermoelectric plant in Bangladesh: implications for radiological health hazards. Environ Monit Assess (accepted). https://doi.org/10.1007/s10661-018-7160-y

Huang Y, Lu X, Ding X, Feng T (2015) Natural radioactivity level in beach sand along the coast of Xiamen Island, China. Mar Pollut Bull 91:357–361

ICRP (1990) Recommendations of the international commission on radiological protection. 21(1–3), publication 60

Iqbal M, Tufail M, Mirza SM (2000) Measurement of natural radioactivity in marble found in Pakistan using a NaI (Tl) gamma-ray spectrometer. J Environ Radioact 51:255–265

Islam AAMS, Khandaker MU, Miah MH, Hossain S (2019) Radioactivity in coral skeletons and marine sediments collected from the St. Martin's Island of Bangladesh. J Radioanal Nuclear Chem (accepted). https://doi.org/10.1007/s10967-019-06582-x

Jafarabadi AR, Bakhtiyari AR, Toosi AS, Jadot C (2017a) Spatial distribution, ecological and health risk assessment of heavy metals in marine surface sediments and coastal seawaters of fringing coral reefs of the Persian Gulf, Iran. Chemosphere 185:1090–1111

Jafarabadi AR, Bakhtiari AR, Aliabadian M, Toosi AS (2017b) Spatial distribution and composition of aliphatic hydrocarbons, polycyclic aromatic hydrocarbons and hopanes in superficial sediments of the coral reefs of the Persian Gulf, Iran. Environ Pollut 224:195–223

Jafarabadi AR, Bakhtiari AR, Maisano M, Pereira P, Cappello T (2018a) First record of bioaccumulation and bioconcentration of metals in Scleractinian corals and their algal symbionts from Kharg and Lark coral reefs (Persian Gulf, Iran). Sci Total Environ 640–641:1500–1511

Jafarabadi AR, Bakhtiari AR, Spanò N, Cappello T (2018b) First report of geochemical fractionation distribution, bioavailability and risk assessment of potentially toxic inorganic elements in sediments of coral reef Island of the Persian Gulf, Iran. Mar Pollut Bull 137:185–197

Jafarabadi AR, Bakhtiari AR, Hedouin L, Toosi AS, Cappello T (2018c) Spatio-temporal variability, distribution and sources of *n*-alkanes and polycyclic aromatic hydrocarbons in reef surface sediments of Kharg and Lark coral reefs, Persian Gulf, Iran. Ecotocxicol Environ Saf 163:307–322

Jafarabadi AR, Bakhtiari AR, Aliabadian M, Laetitia H, Toosi AS, Yap CK (2018d) First report of bioaccumulation and bioconcentration of aliphatic hydrocarbons (Ahs) and persistent organic pollutants (PAHs, PCBs and PCNs) and their effects on alcyonacea and scleractinian corals and their endosymbiotic algae from the Parsian Gulf, Iran: inter and intra-species differences. Sci Total Environ 627:141–157

Joy A, Anoop PP, Rajesh R, Mathew J, Mathew A, Gopinath A (2019) Spatial variation of trace element concentration and contamination assessment in the coral reef sediments of Lakshsdweep Archipelago, Indian Ocean. Mar Pollut Bull 146:106–116

Kannan V, Rajan MP, Iyengar MAR, Ramesh R (2002) Distribution of natural and anthropogenic radionuclides in soil and beach sand samples of Kalpakkam (India) using hyper pure germanium (HPGe) gamma ray spectrometry. Appl Radiat Isot 57:109–119

Khan FH (1964) Geology of St. Martin's Island: the geological survey of Pakistan records, vol X, pt 2-B, p 12

Khan R, Islam HMT, Islam ARMT (2021) Mechanism of elevated radioactivity in Teesta river basin from Bangladesh: radiochemical characterization provenance and associated hazards. Chemosphere 264:128459. https://doi.org/10.1016/j.chemosphere.2020.128459

Khan R, Rouf MA, Das S, Tamim U, Naher K, Podder J, Hossain SM, (2017) Spatial and multilayered assessment of heavy metals in the sand of Cox's-Bazar beach of Bangladesh. Reg Stud Mar Sci 16:171–180. https://doi.org/10.1016/j.rsma.2017.09.003

Khan R, Ghosal S, Sengupta D, Tamim U, Hossain SM, Agrahari S, (2018) Studies on heavy mineral placers from eastern coast of Odisha, India by instrumental neutron activation analysis. J Radioanal Nucl Chem (accepted). https://doi.org/10.1007/s10967-018-6250-1

Khan R, Das S, Kabir S, Habib MA, Naher K, Islam MA, Tamim U, Rahman AKMR, Deb AK, Hossain SM (2019a) Evaluation of the elemental distribution in soil samples collected from shipbreaking areas and an adjacent island. J Environ Chem Eng (accepted). https://doi.org/10.1016/j.jece.2019a.103189

Khan R, Islam HMT, Apon MAS, Islam ARMT, Habib MA, Phoungthong K, Idris AM, Techato K (2022) Environmental geochemistry of higher radioactivity in a transboundary Himalayan river sediment (Brahmaputra Bangladesh): potential radiation exposure and health risks. Environ Sci Pollut Res. https://doi.org/10.1007/s11356-022-19735-5

Khan R, Parvez MS, Jolly YN, Haydar MA, Alam MF, Khatun MA, Sarker MMR, Habib MA, Tamim U, Das S, Sultana S, Islam MA, Naher K, Paul D, Akter S, Khan MHR, Nahid F, Huque R, Rajib M, Hossain SM (2019b) Elemental abundances, natural radioactivity and physicochemical records of a southern part of Bangladesh: implication for assessing the environmental geochemistry. Environ Nanotechnol Monit Manag (accepted). https://doi.org/10.1016/j.enmm.2019b.100225

Khandaker MU, Jojo PJ, Kassim HA, Amin YM (2012) Radiometric analysis of construction materials using HPGe gamma-ray spectrometry. Radiat Prot Dosim 152:33–37

Khandaker MU, Mohd Nasir NL, Asaduzzaman K, Olatunji MA, Amin YM, Kassim HA, Alrefae T (2016) Evaluation of radionuclides transfer from soil-toedibleflora and estimation of radiological dose to the Malaysian populace. Chemosphere 154:528–536

Khandaker MU, Mohd Nasir NL, Zakirin NS, Kassim HA, Asaduzzaman K, Bradley DA, Zulkifly MY, Hayyan A (2017) Radiation dose to the Malaysian populace via the consumption of bottled mineral water. Radiat Phys Chem 140:173–179

Khandaker MU, Asaduzzaman K, Sulaiman AFB, Bradley DA, Isinkaye MO (2018) Elevated concentrations of naturally occurring radionuclides in heavy mineral-rich beach sands of Langkawi Island, Malaysia. Mar Pollut Bull 127:654–663

Kucukomeroglu B, Karadeniz A, Damla N, Yesilkanat CM, Cevik U (2016) Radiological maps in beach sands along some coastal region of Turkey. Mar Pollut Bull 112:255–264

Lewis SE, Lough JM, Cantin NE, Matson EG, Kinsley L, Bainbridge ZT, Brodie JE (2018) A critical evaluation of coral Ba/Ca, Mn/Ca and Y/Ca ratios as indicators of terrestrial input: New data from the Great Barrier Reef, Australia. Geochim Cosmochim Acta 237:131–154

Lin W, Yu K, Wang Y, Liu X, Ning Q, Huang X (2019) Radioactive level of coral reefs in the South China Sea. Mar Pollut bull 142:43–53. https://doi.org/10.1016/j.marpolbul.2019.03.030

Mahawatte P, Fernando KNR (2013) Radioactivity levels in beach sand from the West Coast of Sri Lanka. J Natn Sci Found Sri Lanka 41(4):279–285

Majlis ABK, Habib MA, Khan R, Phoungthong K, Techato K, Islam MA, Nakashima S, Islam ARMT, Hood MA, Hower JC (2022) Intrinsic characteristics of coal combustion residues and their environmental impacts: a case study for Bangladesh. Fuel 324:124711. https://doi.org/10.1016/j.fuel.2022.124711

Margineanu RM, Duliu OG, Blebea-Apostu AM, Gomoiu C, Bercea S (2013) Environmental dose-rate distribution along the Romanian Black Sea shore. J Radioanal Nucl Chem 298:1191–1196

Merdanoğlu B, Altinsoy N (2006) Radioactivity concentrations and dose assessment for soil samples from Kestanbol granite area, Turkey. Radiatprot Dosim 121(4):399–405

Mohanty AK, Sengupta D, Das SK, Saha SK, Van KV (2004) Natural radioactivity and radiation exposure in the high background area at Chhatrapur beach placer deposit of Orissa, India. J Environ Radioact 75:15–33

Mokhtar MB, Praveena SM, Aris AZ, Yong OC, Lim AP (2012) Trace metal (Cd, Cu, Fe, Mn, Ni and Zn) accumulation in Scleractinian corals: a record for Sabah, Borneo. Mar Pollut Bull 64:2556–2563

Papadopoulos A, Koroneos A, Christofides G, Papadopoulou L, Tzifas I (2016) Assessment of gamma radiation exposure of beach sands in highly touristic areas associated with plutonic rocks of the Atticocycladic zone (Greece). J Environ Radioact 162–163:235–243

Prouty NG, Field ME, Stock JD, Jupiter SD, McCulloch M (2010) Coral Ba/Ca records of sediment input to the fringing reef of the southshore of Moloka'i, Hawai'i over the last several decades. Mar Pollut Bull 60:1822–1835

Prouty NG, Goodkin NF, Jones R, Lamborg CH, Storlazzi CD, Hughen KA (2013) Environmental assessment of metal exposure of corals living in Castle Harbour, Bermuda. Mar Chem 154:55–66

Rao NS, Sengupta D, Guin R, Saha SK (2009) Natural radioactivity measurements in beach sand along southern coast of Orissa, eastern India. Environ Earth Sci 59:593–601

Rudnick RL, Gao S (2014) Composition of the continental crust. Treatise on geochemistry, 2nd edn, Chapter 4, pp 1–64

Saha N, Webb GE, Zhao J (2016) Coral skeletal geochemistry as a monitor of inshore water quality. Sci Total Enviorn 566–567:652–684

Saha N, Rodriguez-Ramirez A, Nguyen AD, Clark TR, Zhao J, Webb GE (2018) Seasonal to decadal scale influence of environmental drivers on Ba/Ca and Y/Ca in coral aragonite from the southern Great Barrier Reef. Sci Total Environ 639:1099–1109

Saha N, Webb GE, Zhao J, Nguyen AD, Lewis SE, Lough JM (2019) Coral-based high resolution rare earth element proxy for terrestrial sediment discharge affecting coastal seawater quality, Great Barrier Reef. Geochim Cosmochim Acta 254:173–191

Shuaibu HK, Khandaker MU, Alrefae T, Bradley DA (2017) Assessment of natural radioactivity and gamma-ray dose in monazite rich black sand beach of Penang Island, Malaysia. Mar Pollut Bull 119(1):423–428

Tomascik T (1997) Management plan for coral resources of Narikel Jinjira (St. Martin's Island). Final Report, National Conservation Strategy Implementation Project-1, Ministry of Environment and Forest, Government of Bangladesh, p 125

Tufail M (2012) Radium equivalent activity in the light of UNSCEAR report. Environ Monit Assess 184:5663–5667

UNSCEAR (2000) Sources and effects of ionizing radiation; United Nations; Report to the General Assembly, with Scientific Annexes. United Nations (A/55/46), New York

Zare MR, Kamali M, Omidi Z, Khorambagheri M, Mortazavi MS, Ebrahimi M, Akbarzadeh G (2015) Evaluation of natural radioactivity content in high-volume surface water samples along the northern coast of Oman Sea using portable high-resolution gamma-ray spectrometry. J Environ Radioact 144:134–139

Chapter 16
Risk Assessment of Heavy Metal Contaminations in Soil and Water Ecosystem

Akansha Mishra, Jiban Kumar Behera, Pabitra Mishra, Manojit Bhattacharya, Bhaskar Behera, and Niladri Bhusan Kar

Abstract Devastating outcome of environmental contamination through soil, water, air and food has been still the most important subjects in recent years holistically. With the alarming rise in human population, faster industrial development and urbanization have resulted elevation of various heavy metal in our environment. Although a very low concentration of several heavy metals are present in our diet, it needs for healthiness. But relatively high concentration of these metals are present in food that can arise acute or chronic poisoning in human body. Accumulation of high concentrations heavy metal reduces the fertility of soil as well as it poisons the crop or plant. In the plant tissue such metals create hazards, when it used as food for animal consumption. Continuous release of the chemicals, especially heavy metals and pesticides, damage water quality, become unsuitable and toxic for aquatic organisms, simultaneously possess a threat of human life. Taking these considerations, a valid effort is made to deal with different aspects of soil and water pollution by heavy metals which directly or indirectly affect the survival of human beings in natural habitat. Here we are mainly focused on level of contaminations, exposure sources and its effect on environment. This study will be definitely helpful especially for the developing countries, which during their economic progress, often close the eyes to the consequences for environmental quality.

Keywords Contamination · Heavy metals · Accumulation · Exposure · Consequences · Toxic

---

A. Mishra · J. K. Behera · P. Mishra · M. Bhattacharya (✉) · N. B. Kar
Department of Zoology, Fakir Mohan University, Vyasa Vihar, Balasore 756020, Odisha, India
e-mail: mbhattacharya09@gmail.com

B. Behera
Department of Biosciences and Biotechnology, Fakir Mohan University, Vyasa Vihar, Balasore 756020, Odisha, India

P. K. Shit et al. (eds.), *Soil Health and Environmental Sustainability*,
Environmental Science and Engineering,
https://doi.org/10.1007/978-3-031-09270-1_16

16.1 Introduction

Disorganised growth in population demands enlargement of industries and urbanization to enhance a standard life for living. As a consequence the level of pollutants has progressively increased with increasing of polluted society. Ultimately it creates a series of environmental problems and health risks which has become a major issue in recent time. Especially developing countries, during their economic progress, often pay no attention to environmental issues which can affect the welfare of its people (Chi 1994). Normally all the unwanted elements or particles arise from our machine made activities and which have negative impact on our surroundings are considered as pollutants.

Among all the pollutants, heavy metals are one of the major contaminating agents of our environment especially in food supply (Khairy 2009; Gholizadeh et al. 2009). Traditional discharge of industrial and municipal effluents as well as solid wastes into the nearby aquatic sources and grasslands without any treatment, which is accumulate in soils and water (Chen et al. 2005; Zhuang et al. 2009; Zhang et al. 2008). Tropic transfer of heavy metals through food chain creates serious hazard in various plants and animals (Fig. 16.1). When these toxic elements are release into the aquatic body and even into irrigation channels cause significant contamination of soil and water. It results less harvest or production of toxic plants, seeds and other plant products (Khan et al. 2009; Sawut et al. 2018).

Again the poor impact of wastewater is not limited in aquatic animals and their environment but it also directly affects those animals, who intake water and fishes from these polluted aquatic sources (Chagas et al. 2019; Paital and Rivera-Ingraham 2016; Manzano et al. 2015). Humans are exposed to a range of metals as well as metalloids such as cadmium (Cd), aluminium (Al), nickel (Ni), tin (Sn), uranium (U), mercury (Hg), arsenic (As), silicon (Si) etc. associated with environmental, occupational, and dietary exposure (Fig. 16.2) (Dorne et al. 2015). Unnecessary

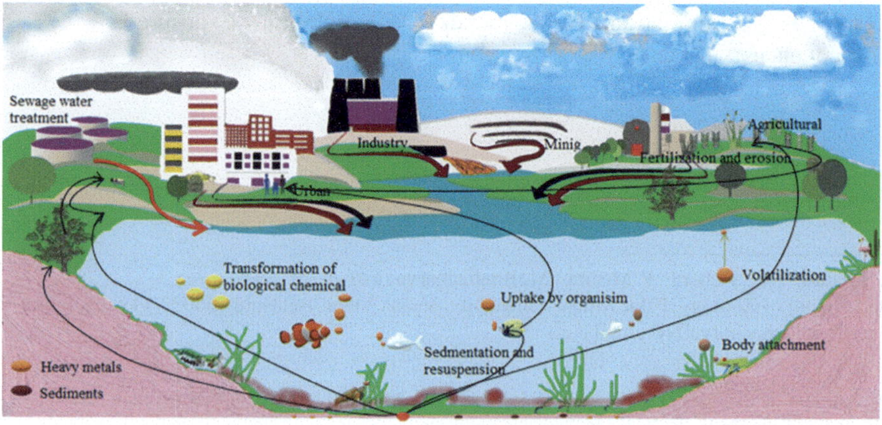

Fig. 16.1 Trophic transfer of heavy metals through fish and plants to human beings

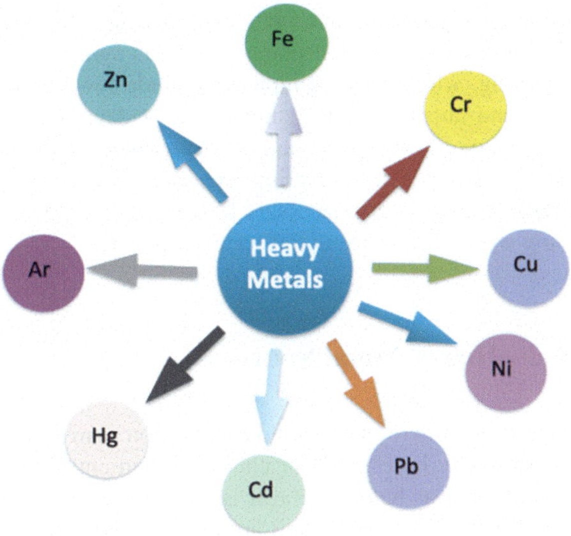

Fig. 16.2 The most common type of heavy metals from natural and anthropogenic sources

accumulation of these persistent toxicants in human bodies creates severe problems like low energy levels, damage to brain and affect central nervous system function, reproductive issues, and damage to blood cells, kidney, lungs, liver and other vital organs (Järup 2003; Pichhode and Gaherwal 2020).

Whereas a very less amount of heavy metals such as iron (Fe), zinc (Zn), copper (Cu) etc. are essential for usually body development and functions of living beings (Muhammad et al. 2011). The long term exposure of Cadmium (Cd) can cause severe health issues, which is sometimes lead to death of the individual (Järup et al. 2000). It causes Itai-Itai (Ouch-Ouch) diseases associated with skeletal and kidney damage (Nordberg et al. 2002). In addition to copper (Cu) and iron (Fe) can cause vomiting, diarrhoea, loss of strength, dysfunction of liver and cardiovascular system when it exposed at very high concentration (Chiarugi et al. 2002; El-Hioui et al. 2008; Georgopoulos et al. 2001). Another element is mercury (Hg) which is lethal at very small extent. Exposure to mercury, including its compounds can cause a severe neurological disease called Minamata disease (Ekino et al. 2007; Eto 2000).

Accumulation of lead (Pb) and chromium (Cr) can cause damage to tissues of kidney, liver, lungs and other vital organ failure (Saleh et al. 2019; Shah et al. 2020; Goodale et al. 2008). Lead poisoning mainly affects children under 5 years which harmful consequences include respiratory disorders, neurological problems, mental retardation and cancer. Acute as well as chronic arsenic (As) exposure can cause skin and respiratory infections, cardiovascular problems, neurological and haematological disorders, renal failure, developmental delays, reproductive difficulties, gastrointestinal issues along with many diseases related to mutagenesis (Patel et al. 2005; Kapaj et al. 2006). Beside these, accumulation of many other toxic metals such as aluminium (Al), zinc (Zn), tin (Sn), cobalt (Co) etc. create many long term diseases which are sometimes fatal in the body tissues of humans and other living organisms (Bull and Cox 1994). In plants chlorosis symptom is the main outcome of metal poisoning, where other symptoms include low biomass accumulation, imbalanced water and nutrients supply, inhibition of development and photosynthesis process etc. Intake of toxic plants and plant products by the herbivores create severe health issues like hyper pigmentation, anaemia, kidney and liver dysfunctioning in them. Thus it becomes very necessary to study about detrimental effect of heavy metal on living world through soil and water impurities.

16.2 Heavy Metal Sources

Even though there is no specific definition conveying metals and metalloids, but in generally any base metal or chemical element with relatively high solidity which is mentioned as heavy metal.It is toxic or fatal at very small concentration (Oves et al. 2012). These persistent are the contaminants of the earth crust which cannot be destroyed or degraded (Ernst 1998; Bradl 2005). Heavy metals are come to our environment through both natural and anthropogenic processes and accumulate in different environmental compartments for example air, water, soil etc.

16.3 Natural Sources of Heavy Metals

Natural outrush of heavy metals from their endemic sphere to our background compartments take place under certain and different environmental circumstances. Such actions include volcanic eruptions, forest fires, weathering of rocks, Sea sprays near the coastline, soil erosion etc. (Fig. 16.3) (Mohammed et al. 2011; He et al. 2013). Through these processes the most common heavy metals for instance lead (Pb), nickel (Ni), arsenic (Ar), zinc (Zn), coper (Cu) etc. can be found in traces, they still affect living beings (Agarwal 2009; Wang et al. 2015; Duruibe et al. 2007).

16.4 Anthropogenic Sources of Heavy Metals

Manmade activities have been found to contribute more heavy metal contamination due to the daily manufacturing of goods to meet the requirement of the large population. Excess utilization of mines and smelters, the combustion of fossil fuel, application of metal containing fertilizers and pesticides, metal based paints, metal bearing sewage water in agricultural use, metallurgical industries, military training processes, domestic sewages, manufacture, use, and disposal of electronic things as well as clinical or hospital products etc. (Fig. 16.3) are the important anthropogenic

Fig. 16.3 Natural and anthropogenic sources of heavy metals production

emergence which radically add to the heavy metal contaminations in the environment (Tsakona et al. 2007; Alloway 1995; Sitaramaiah and Kumari 2014).

16.5 Heavy Metals in Soil

Mainly contamination of soil occurs through anthropogenic sources as compare to contaminations through natural sources (Dixit et al. 2015). Application of phosphate fertilizers, waste water use in irrigation, pesticide and insecticide treatment etc. in intensive as well as commercial farming process increase lead (Pb), chromium (Cr), arsenic (Ar), and cadmium (Cd) like heavy metals in cultivation land soils (Liu et al. 2013; Nicholson et al. 2003; Mortvedt 1996). Anthropogenic activities such as mining and smelting are responsible for increase in cadmium (Cd) like heavy metals in soil of China (Chen et al. 2015).

The studies on soil samples collected from various urban sites throughout the world showed considerably elevated of varieties of metals concentration such as cadmium (Cd), copper (Cu), Aluminium (Al), lead (Pb), zinc (Zn), arsenic (Ar), mercury (Hg) etc. (Fig. 16.2) mostly due to mining and smelting activities (Karim et al. 2015; Hutchinson and Whitby 1974; Kapusta and Sobczyk 2015). Combustion of fossil fuel release nickel (Ni), vanadium (V), mercury (Hg), selenium (Se) and tin (Sn) (Mohammed et al. 2011; Guan et al. 2014; Kelepertzis 2014). Nickel (Ni), lead (Pb) and zinc (Zn) mainly comes from traffic, while battery manufacturing and oil or paint factories release lead (Pb) and cadmium (Cd) (Authman et al. 2015). Again animal manures, sewage sludge and compost are also responsible for heavy metal contamination in soil. From the metal bearing rocks heavy metals like copper (Cu), nickel (Ni), cadmium (Cd), cobalt (Co) disintegrate into small or fine particles and enter the soil background by several natural processes such as volcanic eruption, terrestrial erosion, leaching, surface winds, meteorites etc. (Muradoglu et al. 2015).

16.6 Heavy Metals in Water

As every single life requires water to survive, aquatic system contaminated by heavy metalwhich is truly affects all the living organisms on our planet. Although source of water contamination through heavy metals are plentiful, industrialization and urbanisation are two of the main culprits. Various industrial untreated water effluents contain heavy metals like lead (Pb), copper (Cu), Zinc (Zn), cadmium (Cd), mercury (Hg) etc. (Fig. 16.2) mix with nearby pond, river, lake etc. (Modoi et al. 2014; Bagul et al. 2015).

Again urban sewages release into the sea water body has contain household waste water, construction debris, clinical and hospital waste products etc. which are rich in various toxic metal like iron (Fe), copper (Cu), aluminium (Al) and lead (Pb) (Hadi et al. 2019). Metal containing fertilizers, pesticides and insecticides flow through agricultural runoffs to the nearby aquatic sources. Surface water of several rivers are nearer to urban areas in India has contain metallic chemicals for instance Cr, Mn, Cu, Al, Fe, Pb, Ni and Zn, as effluents from industries, domestic sewages and agricultural return flows are generally discharged into the nearby aquatic sources (Hejabi et al. 2011).

16.7 Impact of Soil Pollution Through Heavy Metals on Human Being and Other Living Organisms

Soils are one of the major medium for heavy metals released into the environment mainly by anthropogenic activities. Over the past few decades metal pollution in soil has become a global threat for the safety of agricultural goods (Hu et al. 2017). Plants are sessile life forms. With respect to their living and reproduction plants adapt to different composition of the soils. But unnecessary use of essential along with additional elements in soil may cause harm to them (Vardhan et al. 2019). Micronutrients such as potassium, nitrogen, sulphur, phosphorus, magnesium, calcium, silicon etc.

and macronutrients like sodium, nickel, manganese, zinc, iron, molybdenum, chlorine, copper and boron are very necessary for normal growth and development of the plants (Asdeo and Loonker 2011). But presence of these elements are in high concentrations in soil not only affected growth and development of the plants but also accumulate in various tissues of plants (Flores-Magdaleno et al. 2011). Later it creates various hazards when enters in to the bodies of human beings and other animals as they take these toxic heavy metal containing vegetables, fruits and other plant products as food (Table 16.1). Beside plants and humans, number of creatures living in soil suffers a lot directly by the existence of abnormal quantity of metals in soil.

Soil is considered as the most complex ecosystem containing a great variety of terrestrial lives and earthworms are the most common and beneficial organisms of this ecosystems. They play a very important role in husbandry by maintaining fertility quality of the soil. Exposed high concentration of zinc (Zn), lead (Pb), copper (Cu), mercury (Hg) and cadmium (Cd) like heavy metals in soil increase the mortality rate of these organisms (Uwizeyimana et al. 2017; Sivakumar 2015). Because of deadly effect of metal pollution in soil a range of burrowing animals like rabbits, rodents, reptiles and insects are being lost. Metal contamination in urban area and agricultural soils are progressively getting worse globally with the overpopulation, intensive cultivation, industrial and mining activities. These heavy metals in urban

Table 16.1 Several heavy metals have harmful effects on the human body

Heavy metals	Harmful effects
Pb	Anorexia, high blood pressure, skin allergies, reduced fertility, renal failure, neurological problem, lung fibrosis, chronic nephropathy, hyperactivity, hair loss
Cd	Bone demineralization, kidney and lungs damage, severe headache, hypertension, testicular atrophy, lung and prostate cancer, emphysema, coughing, Itai-Itai disease, lymphocytosis, microcytic hypochromic anaemia
Hg	Hearing, vision and memory loss, attention deficit, ataxia, gingivitis, kidney diseases, nausea, gastrointestinal irritation, diarrhoea, blood pressure issues, vomiting, skin rashes, Minamata disease
As	Brain damage, cardiovascular disorder, breathing problem, conjunctivitis, dermatitis, skin cancer
Ni	Chest pain, cardiovascular diseases, dry cough and shortness of breath, renal failure, lungs and nasal cancer, nausea, dermatitis, headache, dizziness
Cu	Arthritis, autism, delayed growth and development, diarrhoea, liver damage, fatigue, abdominal pain, headache, hypertension, panic attack, fears, nausea, anaemia, vomiting
Cr	Damage to blood cells, liver, kidney and heart, bronchopneumonia, chronic bronchitis, vomiting, skin irritation, circulatory and nerve problems, emphysema
Zn	Loose motion, gastrointestinal infections, jaundice, metal fume fever, impotence, posterior and stomach cancer, lethargy, muscular degeneration, seizures, muscle pain, indigestion

soils may go into the human body directly through skin absorption and inhalation of dusts which ultimately damage, especially children's health (Su 2014).

16.8 Impact of Water Pollution Through Heavy Metals on Human Being and Other Living Organisms

The natural resources of aquatic ecosystem for example ponds, rivers, lakes and seas are contaminated more rapidly than other ecosystems. Release of both solid and liquid wastes containing large amount of heavy metals enters into the water bodies that cause an enormous damage to that ecological community (Briffa et al. 2020; Alkarkhi et al. 2008). Existence of toxic chemical elements in rivers and other water reservoirs disturb the regular lives of native people. They are always depending on these water sources for consumption of water (Rai 2008). Thus supply of pure drinking water to the community has a big challenge almost in every country around the world (Izah et al. 2016; Chaturvedi and Dave 2012).

Heavy metals in water bodies pose a serious treats to lives by accumulating in tissues of various aquatic flora and fauna. Fish, prawn and crabs etc. are the direct victims of the noxious effect of these pollutants in water. Growth and population of a variety of water birds like ducks, swans, cranes, kingfishers and crocodiles, snakes like reptiles are severely affected by the presence of unnecessary quantity of metals in water. When they entering into the food chain, cause mutations and diseases in the entire food chain (Erchull and Fisher 2016). Again waste waters containing heavy metals to croplands results toxic food productions (Singh et al. 2004; Flores-Magdaleno et al. 2011; Sharma et al. 2008). Through these poisonous crops, vegetables and fruits etc. heavy metals come into the human beings and create many fatal diseases. Direct consumption of water containing high level of different heavy metals can reduce or damage normal function of various vital organs in human beings and other animals (Table 16.1) (Lone et al. 2008; Lu et al. 2015; Halder and Islam 2015). Domestic water supply from ponds and rivers containing industrial effluents results severe skin diseases not only in human beings but also in domestic animals (Cheung et al. 1990). As every single life depends on water for survival, it becomes very necessary to manage and reduce water pollution for a healthy earth.

16.9 Human Health Risk Assessment

Heavy metal deposition in soil and aquatic mediums are considerably higher than the atmospheric deposition, which is the mainly contributed by anthropogenic activities. Recent this is the most serious exertion that has raised doubt for the safe existence of humans along with other living creatures on earth (Maldonado et al. 2008; Marshall et al. 2007; Chary et al. 2008). Thus that situation demands development and applied methodology of novel and efficient techniques to detect and remove metal contaminants from our surroundings. So that the risk of heavy metal poisoning could not be partially affect but at least to some degree minimized (Muchuweti et al. 2006; Eriyamremu et al. 2005).

Risk has been defined as a function of hazard and exposure (Dorne et al. 2015). Method of evaluation to chance of impairment caused to residents outcome of exposure to different contaminants at a location is addressed as human health risk assessment (Sexton et al. 1995; Koki et al. 2015). This area employs advice to employees of different disciplines such as science, engineering, and statistics. By working together they can find out and measure probable paths of exposure and ultimately utilize those details to estimate a numerical value to characterize the possible hazard (Sauvé et al. 1998; Lushenko 2010). Identification and physical, chemical as well as biological analysis of the concentration of the pollutants arising from sources like air, water, vegetation, sediments are may be remarkable examples for health risk assessment of humans. Again we can take various human bio monitors such as plasma, hair, nails, human milk and adipose tissues for examination purpose (Vaajasaari et al. 2002; Paustenbach et al. 1997).

Hazard identification, dose response assessment, exposure assessment, toxicity assessment and risk characterization are the steps respectively for human health risk assessment (Oves et al. 2012). In the 1st step, risk factor which may be a thing or situation that has potential to harm the health of inhabitants at the site is identified. Further scientific investigations are done in case of requirements. In 2nd step the relationship between the amount of dose and its negative impact on the community health is critically analyse here to evaluate the toxic efficiency of a particular substance. The 3rd step includes the process of estimating the magnitude, frequency and duration of exposure to an agent, along with the population number and characteristics. Again the 4th step provides the estimation result of negative health effects arising from the exposure to different heavy metal. The last step is determination of the nature and magnitude of the risks obtained which help us to adopt appropriate protection measure against the individual pollutant (Sexton et al. 1995).

Health risk assessment helps to classify chemical elements as carcinogenic elements having no effective threshold or there is a risk of cancer and non-carcinogenic elements (Lushenko 2010).

16.10 Environmental Legislation

Nowadays humans have tough control on more or less every major aquatic and soil ecosystem. But the regular activities of human population have dramatically altered the quality of waters as well as cultivating soils worldwide (Nriagu 1996; Panda and Panda 2002). Several heavy metals concentrations and range vary from level to level in the surface of soil and water system (Fig. 16.4) (Lenart and Wolny-Koładka 2013; Ogbonna et al. 2011). When heavy metals reach into the aquatic environment, deteriorate the life sustaining quality of water and cause great damages to both flora

Fig. 16.4 Different levels of concentration of heavy metals in soil and water system

and fauna (Sharma et al. 2016). Presence of heavy metals in soil decreases its fertility quality and also contaminates the vegetables and crops grown in it. Moreover directly or indirectly heavy metals contamination in soil and water has a negative impact on human health.

Thus artificial use of these elements has been restricted and the maximum permissible quantities of heavy metals in water and soil are decided on the basis of human safety (Table 16.2) (Chiroma et al. 2014). Due to differences in risk communication, risk issues in ecology, and acceptable risk levels in different countries, a large variation in regulations and their execution were observed throughout the world. Hence most current legislations are even based on every concentration of contaminants in water and soil (Tianlik et al. 2016). Use of mercury and lead are banned in almost every country, except specific circumstances and their emissions are also monitored to reduce metal pollution. On the basis of concentrations five heavy metals are in

Table 16.2 Permissible limits of heavy metals in soil and water

Heavy metal	Maximum permissible level in water (μg/ml)	Maximum permissible level in soil (μg/g)
Pb	0.065	100
Cd	0.01	3
Fe	0.50	50,000
As	0.10	20
Ni	1.40	50
Cu	0.017	100
Cr	0.55	100
Zn	0.20	300

the order of Pb > As > Cr > Cd > Hg and the contamination state stands in the order of Cd > As > Hg > Pb > Cr according to the GB15618-1995, the standard of China (Cheng et al. 2007). Not only soils but also aquatic systems are drastically polluted by Pb, Ar, Hg, and Cd near the factory belt (Solgi et al. 2012; Buccolieri et al. 2006).

As they pose severe carcinogenic risks to public health, they are known as the priority control heavy metals(Yang et al. 2018; Xianjin et al. 2010; Egashira et al. 2012). The metal concentrations in the soils of Shenyang (China) is compared with Chennai (India), Thrace (Turkey), Mortagne du Nord (France), Ibadan (Nigeria), Zagreb (Croatia), and Creswick-Ballarat (Australia) report suggests that the concentration of Cu, Cd, Pb, Zn are higher in Shenyang (China) than in almost all those given regions of the soil sample of Chennai (India) contains much more concentration of Zn and Cu as compared to the soil sample of Shenyang (China) (Sun et al. 2010). The concentration of 8 heavy metals i.e. Cr, Mo, Cu, Cd, Pb, Ni, Ba, Zn is higher than the minimal concentration in drinking water samples collected from various aquatic sources of Khorramabad, Iran (Mohammadi et al. 2019). Range of heavy metal contamination of soil and water are found all over the world, many developed countries have made their laws and regulations on artificial use of heavy metals. Several other countries have also taken vital steps of limited use of heavy metals an industrial purposes (Tianlik et al. 2016).

16.11 Management of Heavy Metal Pollution in Soil and Water

In the present situation, several methods and strategies are following by many countries to resolve the problems of soil and water contamination by various heavy metals (Lu et al. 2015). The two major technology as in-situ and ex-situ remediation is helping in the management of heavy metal pollution in our environment. In-situ strategies are used for pollutants treatment without removing contaminated soil or groundwater from their place. But the ex-situ remediation for polluted soil or water removing from its original place and taken to another location for treatment (Kpan et al. 2014). These remediation processes of soil and water are accomplishing by physical, chemical, and biological methods. Though proper way the soil or water remediation methods are selected after analysing of several basic things such as the characteristics of the selected location, type of contaminant to be separate, the concentration of the pollutant, and the last use of the contaminated medium.. The World Health Organization guideline applied for water and soil quality where the heavy metal concentration increased (Gyamfi et al. 2019).

16.12 Conclusion

Although there is a range of risk factors associated with pollution, exposure to heavy metal pollution remains a vital source of health risk throughout the world, especially in developing countries where poverty is more and investment in modern technologies is less. Hence it becomes necessary for more depth study on heavy metal pollution presently. Soil and water are higher contamination through heavy metals than air which eventually warns the safe and sound existence of lives on earth. Our review focused on the level of heavy metal contamination, exposure sources, and its adverse impact on the environment, it will be help humanity. The majority of heavy metals are toxic at low concentrations. Because of anthropogenic activities, they accumulate in the ecosystem in concentrations much higher than their background values. Their destructive nature not only harms lives on earth but also is the cause of unbalanced biodiversity. Throughout the chapter, we have found out the effects of various heavy metal pollutions in aquatic and soil ecosystem. Heavy metals are one of the major causes of the extinction of key species from earth. By reviewing several pieces of literature, it found that the major causes of heavy metals entering into the human body, by the consumption of drinking water, irrigation water, cultivation of land soil, and sediments. To prevent or control the metal poisoning strong legislation should be passed mainly in developing countries. Our review will provide overall knowledge about the detrimental effect of heavy metals on the environment and help to create awareness against pollution, and also helpful for future research perspectives.

References

Agarwal SK (2009) Heavy metal pollution. APH Publishing Corporation, ISBN:9788131304846, 8131304841:1–270

Alkarkhi FA, Ismail N et al (2008) Assessment of arsenic and heavy metal contents in cockles (Anadara granosa) using multivariate statistical techniques. J Hazard Mater 150(3):783–789

Alloway B (1995) Soil processes and the behaviour of heavy metals in heavy metals in soils by. Blackie and Sons Ltd., London, pp 18–24

Asdeo A, Loonker S (2011) A comparative analysis of trace metals in vegetables. Res J Environ Toxicol 5(2):125–132

Authman MM, Zaki MS et al (2015) Use of fish as bio-indicator of the effects of heavy metals pollution. J Aquacult Res Dev 6(4):1–13

Bagul V, Shinde D et al (2015) New perspective on heavy metal pollution of water. J Chem Pharm Res 7(12):700–705

Bradl H (2005) Sources and origins of heavy metals. In: Interface science and technology, vol 6. Elsevier, pp 1–27

Briffa J, Sinagra E et al (2020) Heavy metal pollution in the environment and their toxicological effects on humans. Heliyon 6(9):e04691

Buccolieri A, Buccolieri G et al (2006) Heavy metals in marine sediments of Taranto Gulf (Ionian Sea, southern Italy). Mar Chem 99(1–4):227–235

Bull PC, Cox DW (1994) Wilson disease and Menkes disease: new handles on heavy-metal transport. Trends Genet 10(7):246–252

Chagas TQ, da Silva Alvarez TG et al (2019) Behavioral toxicity of tannery effluent in zebrafish (Danio rerio) used as model system. Sci Total Environ 685:923–933
Chary NS, Kamala C et al (2008) Assessing risk of heavy metals from consuming food grown on sewage irrigated soils and food chain transfer. Ecotoxicol Environ Safety 69(3):513–524
Chaturvedi S, Dave PN (2012) Removal of iron for safe drinking water. Desalination 303:1–11
Chen Y, Wang C et al (2005) Residues and source identification of persistent organic pollutants in farmland soils irrigated by effluents from biological treatment plants. Environ Int 31(6):778–783
Chen Z, Zhao Y et al (2015) Cadmium (Cd) localization in tissues of cotton (Gossypium hirsutum L.), and its phytoremediation potential for Cd-contaminated soils. Bull Environ Contam Toxicol 95(6):784–789
Cheng J-L, Zhou S et al (2007) Assessment and mapping of environmental quality in agricultural soils of Zhejiang Province, China. J Environ Sci 19(1):50–54
Cheung W, Chang K et al (1990) Health effects of beach water pollution in Hong Kong. Epidemiol Infect 105(1):139–162
Chi C-C (1994) Growth with pollution: unsustainable development in Taiwan and its consequences. Stud Comp Int Dev 29(2):23–47
Chiarugi A, Pitari GM et al (2002) Effect of prolonged incubation with copper on endothelium-dependent relaxation in rat isolated aorta. Brit J Pharmacol 136(8):1185–1193
Chiroma T, Ebewele R et al (2014) Comparative assessment of heavy metal levels in soil, vegetables and urban grey waste water used for irrigation in Yola and Kano. Int Refereed J Eng Sci 3(2):01–09
Dixit R, Malaviya D et al (2015) Bioremediation of heavy metals from soil and aquatic environment: an overview of principles and criteria of fundamental processes. Sustainability 7(2):2189–2212
Dorne J-LC, Kass GE et al (2015) 2 human risk assessment of heavy metals: principles and applications. In: Metal ions in toxicology: effects, interactions, interdependencies, De Gruyter, pp 27–60
Duruibe JO, Ogwuegbu M et al (2007) Heavy metal pollution and human biotoxic effects. Int J Phys Sci 2(5):112–118
Egashira R, Tanabe S et al (2012) Adsorption of heavy metals in mine wastewater by Mongolian natural zeolite. Procedia Eng 42:49–57
Ekino S, Susa M et al (2007) Minamata disease revisited: an update on the acute and chronic manifestations of methyl mercury poisoning. J Neurol Sci 262(1–2):131–144
El-Hioui M, Ahami AOT et al (2008) Iron deficiency and anaemia in rural school children in a coastal area of Morocco. Pakistan J Nutr 7:400–403
Erchull C, Fisher L (2016) Remedying and regulating the unintended consequences of subtherapeutic dosing of livestock with antibiotics: can the EPA's implementation of the clean water act reign in the problem. W New Eng l Rev 38:397
Eriyamremu G, Asagba S et al (2005) Evaluation of lead and cadmium levels in some commonly consumed vegetables in the Niger-Delta oil area of Nigeria. Bull Environ Contam Toxicol 75(2):278–283
Ernst W (1998) The origin and ecology of contaminated, stabilized and non-pristine soils. Metal-Contam Soils In Situ Inactivat Phytorestoration 17–29
Eto K (2000) Minamata disease. Neuropathology 20:14–19
Flores-Magdaleno H, Mancilla-Villa OR et al (2011) Heavy metals in agricultural soils and irrigation wastewater of Mixquiahuala, Hidalgo, Mexico. Afr J Agric Res 6(24):5505–5511
Georgopoulos G, Roy A, Yonone-Lioy MJ, Opiekun RE, Lioy PPJ (2001) Environmental copper: its dynamics and human exposure issues. J Toxicol Environ Health Part B Crit Rev 4(4):341–394
Gholizadeh A, Ardalan M et al (2009) Solubility test in some phosphate rocks and their potential for direct application in soil. World Appl Sci J 6(2):182–190
Goodale BC, Walter R et al (2008) The cytotoxicity and genotoxicity of hexavalent chromium in medaka (Oryzias latipes) cells. Aquat Toxicol 87(1):60–67
Guan Y, Shao C et al (2014) Heavy metal contamination assessment and partition for industrial and mining gathering areas. Int J Environ Res Public Health 11(7):7286–7303

Gyamfi E, Appiah-Adjei EK et al (2019) Potential heavy metal pollution of soil and water resources from artisanal mining in Kokoteasua, Ghana. Groundw Sustain Dev 8:450–456

Hadi M, Fadlillah L et al (2019) Heavy metal pollution and water quality assessment in Belik river Yogyakarta. In: IOP conference series: earth and environmental science, IOP Publishing

Halder JN, Islam MN (2015) Water pollution and its impact on the human health. J Environ Human 2(1):36–46

He B, Yun Z et al (2013) Research progress of heavy metal pollution in China: sources, analytical methods, status, and toxicity. Chin Sci Bull 58(2):134–140

Hejabi AT, Basavarajappa H et al (2011) Heavy metal pollution in water and sediments in the Kabini River, Karnataka, India. Environ Monit Assess 182(1):1–13

Hu B, Chen S et al (2017) Application of portable XRF and VNIR sensors for rapid assessment of soil heavy metal pollution. PLoS ONE 12(2):e0172438

Hutchinson T, Whitby L (1974) Heavy-metal pollution in the Sudbury mining and smelting region of Canada, I. Soil and vegetation contamination by nickel, copper, and other metals. Environ Conserv 1(2):123–132

Izah SC, Chakrabarty N et al (2016) A review on heavy metal concentration in potable water sources in Nigeria: human health effects and mitigating measures. Exposure Health 8(2):285–304

Järup L (2003) Hazards of heavy metal contamination. Brit Med Bull 68(1):167–182

Järup L, Hellström L et al (2000) Low level exposure to cadmium and early kidney damage: the OSCAR study. Occupat Environ Med 57(10):668–672

Kapaj S, Peterson H et al (2006) Human health effects from chronic arsenic poisoning–a review. J Environ Sci Health Part A 41(10):2399–2428

Kapusta P, Sobczyk Ł (2015) Effects of heavy metal pollution from mining and smelting on enchytraeid communities under different land management and soil conditions. Sci Total Environ 536:517–526

Karim Z, Qureshi BA et al (2015) Geochemical baseline determination and pollution assessment of heavy metals in urban soils of Karachi, Pakistan. Ecol Ind 48:358–364

Kelepertzis E (2014) Accumulation of heavy metals in agricultural soils of Mediterranean: insights from Argolida basin, Peloponnese, Greece. Geoderma 221:82–90

Khairy H (2009) Toxicity and accumulation of copper in Nannochloropsis oculata (Eustigmatophyceae, Heterokonta). World Appl Sci J 6(3):378–384

Khan S, Farooq R et al (2009) Health risk assessment of heavy metals for population via consumption of vegetables. World Appl Sci J6(12):1602–1606

Koki IB, Bayero AS et al (2015) Health risk assessment of heavy metals in water, air, soil and fish. Afr J Pure Appl Chem 9(11):204–210

Kpan JD, Opoku BK et al (2014) Heavy metal pollution in soil and water in some selected towns in Dunkwa-on-Offin District in the Central Region of Ghana as a result of small scale gold mining. J Agric Chem Environ 3(02):40

Lenart A, Wolny-Koładka K (2013) The effect of heavy metal concentration and soil pH on the abundance of selected microbial groups within ArcelorMittal Poland steelworks in Cracow. Bull Environ Contam Toxicol 90(1):85–90

Liu X, Song Q et al (2013) Human health risk assessment of heavy metals in soil–vegetable system: a multi-medium analysis. Sci Total Environ 463:530–540

Lone MI, He Z-L et al (2008) Phytoremediation of heavy metal polluted soils and water: progresses and perspectives. J Zhejiang Univ Sci B9(3):210–220

Lu Y, Song S et al (2015) Impacts of soil and water pollution on food safety and health risks in China. Environ Int 77:5–15

Lushenko MA (2010) A risk assessment for ingestion of toxic chemicals in fish from Imperial beach, California, San Diego State University

Maldonado V, Arias HR et al (2008) Heavy metal content in soils under different wastewater irrigation patterns in Chihuahua, Mexico. Int J Environ Res Public Health 5(5):441–449

Manzano BC, Roberto MM et al (2015) Evaluation of the genotoxicity of waters impacted by domestic and industrial effluents of a highly industrialized region of São Paulo State, Brazil, by the comet assay in HTC cells. Environ Sci Pollut Res 22(2):1399–1407
Marshall F, Holden J et al (2007) Contaminated irrigation water and food safety for the urban and peri-urban poor: appropriate measures for monitoring and control from field research in India and Zambia. Inception Rep DFID Enkar 8160(3)
Modoi O-C, Roba C et al (2014) Environmental risks due to heavy metal pollution of water resulted from mining wastes in NW Romania. Environ Eng Manage J (EEMJ) 13(9)
Mohammadi AA, Zarei A et al (2019) Carcinogenic and non-carcinogenic health risk assessment of heavy metals in drinking water of Khorramabad, Iran. Methods X6:1642–1651
Mohammed AS, Kapri A et al (2011) Heavy metal pollution: source, impact, and remedies. In: Biomanagement of metal-contaminated soils. Springer, pp 1–28
Mortvedt J (1996) Heavy metal contaminants in inorganic and organic fertilizers. In: Fertilizers and environment. Springer, pp 5–11
Muchuweti M, Birkett J et al (2006) Heavy metal content of vegetables irrigated with mixtures of wastewater and sewage sludge in Zimbabwe: implications for human health. Agr Ecosyst Environ 112(1):41–48
Muhammad S, Shah MT et al (2011) Health risk assessment of heavy metals and their source apportionment in drinking water of Kohistan region, northern Pakistan. Microchem J 98(2):334–343
Muradoglu F, Gundogdu M et al (2015) Cadmium toxicity affects chlorophyll a and b content, antioxidant enzyme activities and mineral nutrient accumulation in strawberry. Biol Res 48:1–7
Nicholson FA, Smith SR et al (2003) An inventory of heavy metals inputs to agricultural soils in England and Wales. Sci Total Environ 311(1–3):205–219
Nordberg G, Jin T et al (2002) Low bone density and renal dysfunction following environmental cadmium exposure in China. AMBIO J Human Environ 31(6):478–481
Nriagu JO (1996) A history of global metal pollution. Science 272(5259):223
Ogbonna PC, Emea R et al (2011) Heavy metal concentration in soil and woody plants in a quarry. Toxicol Environ Chem 93(5):895–903
Oves M, Khan MS et al (2012) Soil contamination, nutritive value, and human health risk assessment of heavy metals: an overview. Toxicity Heavy Metals Legumes Biorem 1–27
Paital B, Rivera-Ingraham GA (2016) High speed urbanization and its effects on aquatic food chain especially on fish in Bata river of Odisha, India. J Fisheries sciences. Com 10(4):1
Panda BB, Panda KK (2002) Genotoxicity and mutagenicity of metals in plants. In: Physiology and biochemistry of metal toxicity and tolerance in plants. Springer, pp 395–414
Patel K, Shrivas K et al (2005) Arsenic contamination in water, soil, sediment and rice of central India. Environ Geochem Health 27(2):131–145
Paustenbach DJ, Panko JM et al (1997) Urinary chromium as a biological marker of environmental exposure: what are the limitations? Regul Toxicol Pharmacol 26(1):S23–S34
Pichhode M, Gaherwal S (2020) Effect of heavy metal toxicity, arsenic trioxide on the biochemical parameter of fresh water fish, Clarias batrachus. Poll Res 39:123–125
Rai PK (2008) Heavy metal pollution in aquatic ecosystems and its phytoremediation using wetland plants: an ecosustainable approach. Int J Phytorem 10(2):133–160
Saleh HN, Panahande M et al (2019) Carcinogenic and non-carcinogenic risk assessment of heavy metals in groundwater wells in Neyshabur Plain, Iran. Biol Trace Element Res 190(1):251–261
Sauvé S, Dumestre A et al (1998) Derivation of soil quality criteria using predicted chemical speciation of Pb2+ and Cu2+. Environ Toxicol Chem Int J 17(8):1481–1489
Sawut R, Kasim N et al (2018) Pollution characteristics and health risk assessment of heavy metals in the vegetable bases of northwest China. Sci Total Environ 642:864–878
Sexton K, Beck BD et al (1995) Chemical mixtures from a public health perspective: the importance of research for informed decision making. Toxicology 105(2–3):429–441

Shah N, Khisroon M et al (2020) Assessment of copper, chromium, and lead toxicity in fish (Ctenopharyngodon idella Valenciennes, 1844) through hematological biomarkers. Environ Sci Pollut Res 27(26):33259–33269

Sharma A, Kaur M et al (2016) Heavy metal pollution: a global pollutant of rising concern. In: Toxicity and waste management using bioremediation. IGI Global, pp 1–26

Sharma RK, Agrawal M et al (2008) Heavy metal (Cu, Zn, Cd and Pb) contamination of vegetables in urban India: a case study in Varanasi. Environ Pollut 154(2):254–263

Singh KP, Mohan D et al (2004) Impact assessment of treated/untreated wastewater toxicants discharged by sewage treatment plants on health, agricultural, and environmental quality in the wastewater disposal area. Chemosphere 55(2):227–255

Sitaramaiah Y, Kumari MK (2014) Impact of electronic waste leading to environmental pollution. J Chem Pharm Sci ISSN 974:2115

Sivakumar S (2015) Effects of metals on earthworm life cycles: a review. Environ Monit Assess 187(8):1–16

Solgi E, Esmaili-Sari A et al (2012) Soil contamination of metals in the three industrial estates, Arak, Iran. Bull Environ Contam Toxicol 88(4):634–638

Su C (2014) A review on heavy metal contamination in the soil worldwide: situation, impact and remediation techniques. Environ Skept Crit 3(2):24

Sun Y, Zhou Q et al (2010) Spatial, sources and risk assessment of heavy metal contamination of urban soils in typical regions of Shenyang, China. J Hazard Mater 174(1–3):455–462

Tianlik T, Norulaini NARN et al (2016) Risk assessment of metal contamination in soil and groundwater in Asia: a review of recent trends as well as existing environmental laws and regulations. Pedosphere 26(4):431–450

Tsakona M, Anagnostopoulou E et al (2007) Hospital waste management and toxicity evaluation: a case study. Waste Manage 27(7):912–920

Uwizeyimana H, Wang M et al (2017) The eco-toxic effects of pesticide and heavy metal mixtures towards earthworms in soil. Environ Toxicol Pharmacol 55:20–29

Vaajasaari K, Joutti A et al (2002) Comparisons of terrestrial and aquatic bioassays for oil-contaminated soil toxicity. J Soils Sediments 2(4):194–202

Vardhan KH, Kumar PS et al (2019) A review on heavy metal pollution, toxicity and remedial measures: current trends and future perspectives. J Mol Liq 290:111197

Wang Q, Xie Z et al (2015) Using ensemble models to identify and apportion heavy metal pollution sources in agricultural soils on a local scale. Environ Pollut 206:227–235

Xianjin T, Chaofeng S et al (2010) Inorganic and organic pollution in agricultural soil from an emerging e-waste recycling town in Taizhou area, China. J Soils Sediments 10

Yang Q, Li Z et al (2018) A review of soil heavy metal pollution from industrial and agricultural regions in China: pollution and risk assessment. Sci Total Environ 642:690–700

Zhang Y, Dai J et al (2008) Effects of long-term sewage irrigation on agricultural soil microbial structural and functional characterizations in Shandong, China. Eur J Soil Biol 44(1):84–91

Zhuang P, Zou B et al (2009) Heavy metal contamination in soils and food crops around Dabaoshan mine in Guangdong, China: implication for human health. Environ Geochem Health 31(6):707–715

Chapter 17
Microplastics, Their Toxic Effects on Living Organisms in Soil Biota and Their Fate: An Appraisal

Sourav Bhattacharyya, Sanjib Gorain, Monoj Patra, Anup Kumar Rajwar, Dinesh Gope, Santosh Kumar Giri, Jayeeta Pal, Madhumita Mahato, Shuli Barik, and Surjyo Jyoti Biswas

Abstract Microplastics are miniature plastic fragments that originate as a result of the advancement of commercial products as well as the breakdown of bigger plastics. Microplastics have been identified as a serious global environmental issue due to its poor waste management. This review covers the impact of microplastics on the soil ecosystem, their transit behaviour, and their impact on numerous organisms. The impact of microplastics on soil animals and plants, is influenced by the size, shape, and concentration of microplastics in the soil. Microplastic has been found in a variety of soil types, including agricultural, industrial, and coastal soils. Plastic particles in soil have increased, posing a major threat to soil ecosystem functioning, including the soil microbial population, nitrogen cycle, and higher organisms. The current review highlights and assimilates the findings of other scientists so that it can serve as a resource for readers and scientists dealing with microplastics, including toxicity, risk assessment in the environment, and microplastic treatment options.

Keywords Polyethylene · Polypropylene · Terephthalate · Contamination · Vertebrates · In-vertebrates

17.1 Introduction

Plastics have found their way into our daily lives because they are light, flexible, non-corrosive, and long-lasting. Overuse of plastics resulted in a ubiquity of plastics in the oceans and soils (ECHA 2019; Wang et al. 2019). As a result, when plastic particles enter the environment, they may pose a threat to ecosystems (both flora and fauna) as well as human health (Simon et al. 2018; Stubenrauch and Ekardt 2020). Micro and nano-plastic particles have been observed to invade the human

S. Bhattacharyya · S. Gorain · M. Patra · A. K. Rajwar · D. Gope · S. K. Giri · J. Pal · M. Mahato · S. Barik · S. J. Biswas (✉)
Genetics and Cell Biology Laboratory, Department of Zoology, Sidho-Kanho-Birsha University, Ranchi Road, Purulia 723104, West Bengal, India
e-mail: surjyo@rediffmail.com; surjyo-jyoti-biswas@skbu.ac.in

P. K. Shit et al. (eds.), *Soil Health and Environmental Sustainability*, Environmental Science and Engineering,
https://doi.org/10.1007/978-3-031-09270-1_17

food chain through various trophic levels, and plants have also been reported to absorb these hazardous compounds. The term microplastic was introduced to the world by a marine ecologist Professor Richard Thompson in 2004 who was working in University of Plymouth, UK. "Microplastics" are small plastic particle which are smaller than 5 mm in diameter (Thompson et al. 2004).

Rillig (2012) was the first to address microplastic contamination, and since then, an increasing number of research have focused on microplastic pollution of soil and the environment (Chae and An 2018). It is estimated that plastics which are released in terrestrial environment are 4-23 fold greater than marine environment. As a result, the United Nations Environment Program (UNEP) has advocated for additional investigation into the consequences of microplastic pollution on soil ecosystems (UNEP 2018; Horton et al. 2017a, b).

Microplastic can enter through many routes including sewage sludge, irrigation, littering, atmospheric deposition, flooding, plastic mulching (Jiang et al. 2017; Blasing and Amelung 2018). Compost and sewage sludge have been used as fertilisers in Europe and North America, and they are a substantial source of soil plastic pollution in these locations (EPA 2015). China, Japan, and South Korea accounted for 80% of all plastic mulching (Espi et al. 2006).

Plastic mulching remains in the soil and is destroyed by ultraviolet (UV) light and other physical abrasion processes, resulting in microplastic buildup (Blasing and Amelung 2018). When microplastic is introduced to the soil ecosystem, it has a detrimental effect on soil ecology. Microplastic can alter soil fertility, nutrient cycling process, disrupt microbial community (Awet et al. 2018; Mai et al. 2018). Earthworm and some other soil animals were found to ingest microplastics. It spreads across many terrestrial food systems, posing a risk to soil animals and humans. (Lwanga et al. 2017). Earthworm movement can provide a path for microplastic to enter deeply into soil and disturb groundwater system (Yu et al. 2019). At the same time, toxic chemicals such as DDT, polycyclic aromatic hydrocarbons (PAHs), pharmaceuticals and personal care products (PPCPs), polychlorinated biphenyls (PCBs), HCHs, pesticides, perfluoroalkyl substances (PFASs), and heavy metals can be absorbed by microplastic (Engler 2012; Velzeboer et al. 2014; Wu et al. 2016).

17.2 Microplastic Occurrence, Source, and Properties

Microplastic pollution are comparatively a new topic in soil science. Based on previous reports plastic litter were categories into three main types—microplastic (<5 mm), mesoplastic (5–25 mm), macroplastic (>25 m). Microplastic pollution has been identified in a wide variety of proportions in agricultural lands, greenhouses, home gardens, coastal, industrial, and floodplain soils around the planet. Deeper layers of soil have been found to have lower concentrations of microplastics (Liu et al. 2018), while sewage sludge and wastewater result in a higher number of microplastic pollution concentrations (Corradini et al. 2019; Xu et al. 2020).

The degradation of minor bits of plastic debris is another noteworthy source of microplastic entering the soil. It has been postulated by scientist that a garden with waste mismanagement contains more concentrations of microplastic than a farmland soil (Lwanga et al. 2017). Microplastic pollution in coastal soils may have resulted from the fragmentation or decomposition of big plastic waste (Zhou et al. 2016, 2018). Recently it has been reported by other investigators that organic fertilisers are major source for microplastic that contaminates the terrestrial environment (Weithmann et al. 2018). Records revealed that polyethylene (PE), polypropylene (PP) are predominant microplastics in the soil ecosystem (Andrady and Neal 2009). Microplastics have been detected in agricultural and floodplain soils in several assessments (Liu et al. 2018; Piehl et al. 2018; Scheurer and Bigalke 2018; Xu et al. 2018; Sarker et al 2020). Microplastic and mesoplastic may have similar polymer composition but, it is not difficult to conclude that microplastic evolved from mesoplastics. But more information like, polyethylene (PE) polymer is most abundant in microplastic and mesoplastic. This information provides that mesoplastic could be a possible source of microplastic.

Polyethylene, polypropylene, and polystyrene particles are microplastics that are commonly found in cosmetic and medicinal items. (Horton et al. 2017b). Due to their adverse effect on soil and human and animal population, these microplastic containing products have been barred in several countries (Ballent et al. 2016).

Physical, chemical, and biological progressions produce secondary microplastics, which are then fragmented into plastic waste (Thompson 2006; Ryan et al. 2009). Photooxidation of plastics in open places takes place due to Ultraviolet (UV) radiation, hence they become brittle and fragments into microplastics. Plastic disintegrate relatively slowly into plastic particles in the aquatic environment due to cold, anoxic conditions and sediments. Some scientist suggested that heat and daylight are perfect settings for producing microplastics by the mechanism of iterative fragmentation processes (Harshvardhan and Jha 2013; Zhang 2017). Microplastics are generally available in variety of morphologies, which includes pellets, threads, and pieces, depending on their sources of origin (Thompson and Law 2014; Klein et al. 2015).

The leading contaminants of the aquatic environment and soil are household sewage or spilled plastic resin powders; sewage sludge containing synthetic fibres or sedimented microplastic from personal care or household items also contribute to soil pollution (Gregory 1996; Horton et al. 2017b). It was reported by others that fibres from synthetic clothes and hosiery reaches the soil due to continuous abrasion and release from effluents from washing machine effluents (Browne et al. 2011; Duis and Coors 2016; Napper and Thompson 2016; Horton et al. 2017b). So, it can be said that secondary sources of microplastics contributes to larger amount of microplastic pollution. Secondary microplastic contaminate soil due to anthropogenic activities such as disposal of municipal solid trash (Horton et al. 2017b). Agriculture is one of the main anthropogenic causes of microplastic pollution in soil due to sewage sludge, tiny plantations using plastic containers, and overuse of agricultural plastic mulches, according to recent studies (Nizzetto et al. 2016). Tyres and road colorations also cause microplastic pollution, because heavy downpour often causes runoff and they

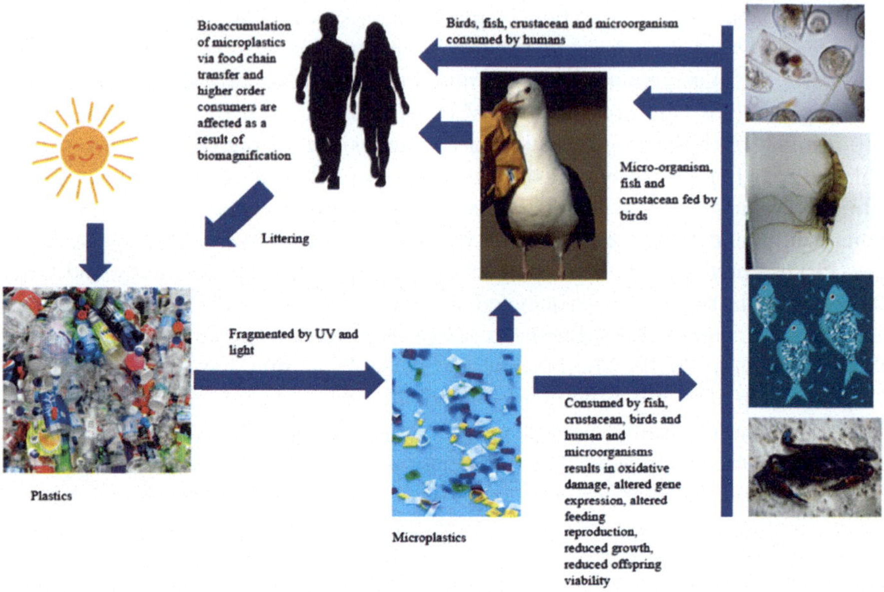

Fig. 17.1 Showing degradation of microplastic and trophic transfer in various organisms

contribute to transport of the weaned-out particles of tyres and road paints (Horton et al. 2017a). Investigation suggests that in highly urbanized areas huge quantity of plastic fibres are being transported, via atmospheric fallout (Dris et al. 2017). Figure 17.1 represents degradation of microplastics and trophic transfer.

17.3 Methods for Detection and Quantification of Microplastics

Two key techniques which are usually used in laboratories for detection of microplastics are Infrared Spectroscopy and Raman Spectroscopy techniques whose principle relies on sample illumination to promote molecular vibrations and then the spectrum so produced was read. The spectrum comprises a wide range of peaks that generate a fingerprint that may be compared to spectral libraries to determine the precise polymer (Toussaint et al. 2019). Documentation and quantification of plastic fragments were also conducted routinely and these are thermo analytical techniques, where microplastics' physical and chemical properties are linked to their thermal stability, which was assessed as a function of time or temperature in the presence of an inert gas (Dumichen et al. 2015). Microplastics can be analysed using GC–MS (Silva et al. 2010). However, there remain certain drawbacks of using these techniques, such as the fact that they are destructive and do not provide data on the size and shape of the plastic fragments, but gives mass concentration result (Kappler et al. 2016). Differential Scanning Calorimetry (DSC) was also favoured for documentation of primary microplastics by some investigators along with pyrolysis gas chromatography mass spectrometry (Shim et al. 2016). Further, it was reported that hyperspectral imaging technology and chemometrics a swift technique to screen microplastics in soil (Zhao et al. 2018). Some laboratories use pyrolysis gas chromatography mass spectrometry which is coupled with FTIR (Fourier Transform Infrared Spectrophotometer) or mass spectrometry but the major disadvantages of employing these techniques is they are expensive and takes a long time to run the samples (He et al. 2018; Toussaint et al. 2019).

17.4 Advance Methods

By bombarding the surface of the specimen with a finely focused pulsed ion beam, the ToF–SIMS method provides information concerning the chemical structure of the surface areas (5 nm depth) of solid materials (Toussaint et al. 2019). This technology is used to deal with and capture particles in the low micrometre range as well as nano-particles that are difficult to identify.

17.5 Effect on Physical and Chemical Properties of Soil

Microplastic in the soil poses a threat to the terrestrial ecology, as well as causing harm to soil fauna and plants (de souza Machado et al. 2018). Several research yielded mixed results when it came to the impact of soil physical features. An existing report stated, the impact of microplastic on soil assets is dependent on the type of microplastic (Liu et al. 2016, 2019). Microplastics with a shape and size similar to soil particles have a smaller impact on soil structure and the water cycle (Liu et al. 2019). Polyester fibres reduce soil density and enhance water holding capacity and evapotranspiration, whereas PE (polyethylene) fragments and PA (polyamide) beads have similar effects but to a lesser degree (de Souza Machado et al. 2018). Further, it was also noted that soil texture is a key factor in determining the impact of microplastics on soil parameters (Liu et al. 2019). Others have observed that polyester fibres influence the pore structure and size of a clay loamy soil and that PE film increases the rate of water evaporation from clayey soil (Zhang and Liu 2018; Zhang et al. 2019; Wan et al. 2019). Similarly, plastic film residues can alter soil water distribution, bulk density, total porosity, and soil water content (Liu et al. 2017). Plastics dispersion in sediments is a serious problem all around the world (Rillig et al. 2017). The typical microplastic concentration in beach deposit through the Belgian coast was detected to be 92.8 particles kg1 of dry sediment, principally made up of plastic fibres (Claessens et al. 2011). Due to the presence of a high ratio of microscopic fragments, the beaches of Guanabara Bay (Brazilian coast) have been designated as one of the most contaminated beaches with microplastics. Microplastic fragments accounted for 56% of total plastic debris measured, followed by styrofoam (26.7%), pellets (9.9%), and fibres (9.9%) (7.2%). On some beaches, microplastic concentrations ranged from 12 to 1300 particles per square metre (Carvalho and Neto 2016). Further, it was reported by others that microplastics assist subtle variations in the organic properties of soil by increasing the concentration of dissolved organic carbon, inorganic nitrogen and total phosphorus in soil. Therefore, from the above studies it can be stated that presence of microplastics could pose a potential threat to soil ecosystem (Liu et al. 2019; Ng et al. 2018; Qi et al. 2020).

17.6 Effect on Soil Fauna

Soil biota plays a vital role in the soil ecosystem, delivering a diverse range of ecosystem services such as organic matter decomposition, nutrient cycling, and the suppression of soil-borne infections and infestations (Brussaard 1997). Microplastics have a wide range of consequences on biota, according to certain review publications (Horton et al. 2017a). The consequences of microplastics in the soil environment are depicted in Fig. 17.2. However, our understanding of the consequences of microplastics on soil is insufficient at this time. Nematode (*C. elegans*), Annelida (earthworm), Molluscs (snail), and Arthropods (mites, collembolans) are examples of soil organisms (Chae and An 2018). Herein emphasis was given to the effects of microplastics pollution on each of the individual group of soil microorganism.

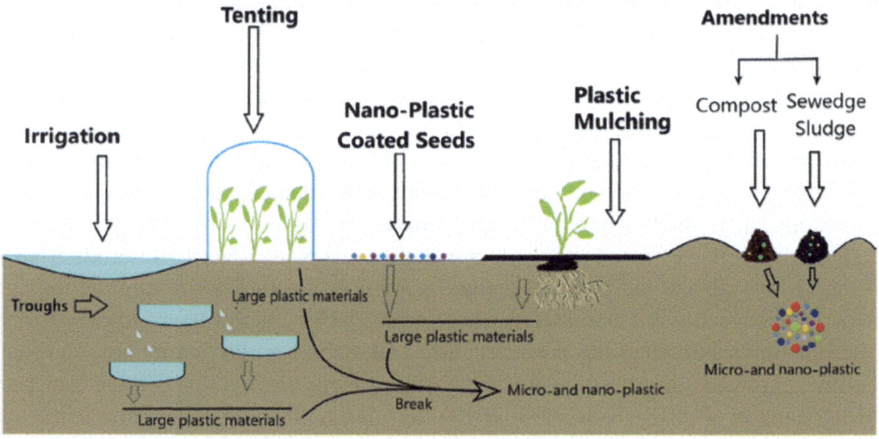

Fig. 17.2 Various causes of microplastic contamination in soil ecosystem (*Source* Iqbal et al. 2020)

17.6.1 Nematode

Recent investigations revealed that microplastics are ingested by Nematodes *C. elegans* (Lei et al. 2018). Impact of microplastics on nematodes is size dependent. On *C. elegans*, it promotes intestinal damage, oxidative stress, and a reduction in intestinal calcium levels, along with elevated expression of the oxidative stress gene gst-4. Further, microplastics causes a decline in longevity, decreases body length and lowers reproduction rate in these soil nematodes. The same research group also discovered that both polystyrene nano-plastics and microplastics cause size-dependent excitatory toxicity during locomotion, as well as downregulation of the unc-17 and unc-47 genes, resulting in cholinergic and GABA neuron damage (Lei et al. 2018).

17.6.2 Arthropod

Microplastics of two different sizes of < 100 μm and 100-200 μm were ingested by two collembolan species *Folsomia candida* and *Proisotom minuta* (Maa et al. 2017). When *Folsomia candida* were exposed to PVC they found appreciable changes in gut microbiota content, body weight and in reproductive capability (Zhu et al. 2018a, b).

17.6.3 Mites

According to Zhu et al. (2018a), mobility of soil microplastics by a soil creature can alter microplastic exposure to other species in the soil as well as change the biophysical characteristics of the soil. In a recent study on soil mite *Hypoasis aculeifer* the investigators found that transport of microplastics improved when predator and prey co-exist at a place rather than predator or prey existing separately in soils (Zhu et al. 2018b).

17.6.4 Annelida

Polyurethane and polybrominated diphenyl ether (PBDE) were also studied for their effects on the earthworm *Eisenia fetida*. The chemical release from microplastics gets accumulated in earthworms, according to the results of their experiment (Gaylor et al. 2013). Scientists administered zinc and high-density polyethylene (HDPE) to *Lumbricus terrestris* in another study, and discovered that microplastics serve as a pathway for bioavailable metals in the soil environment (Hodson et al. 2017).

They did not observe any change in body weight or survivability of earthworm. It was also observed by others that when *Lumbricus terrestris* were exposed to low density polyethylene microplastics (LDPE) for two months they exhibited increased morality rate and formation of tunnel. Earthworm consume microplastic selectively, primarily depending on their size and transport it to other types of soil organisms in ecosystem (Lwanga et al. 2017). Another study discovered that the earthworm L. terrestris carried microplastics from the soil to their burrows, affecting groundwater accessibility and ultimately the terrestrial food web (Lwanga et al. 2017). There were no obvious differences in survival, number of juveniles, or ultimate weight of adult earthworms after exposure to varied concentrations of microplastics for 28 days in earthworms *Eisenia andrei.* Scientist have also reported ingestion of microplastics by earthworm retarded the reproduction, altered the histopathology and at the same time decreased their immune system functioning (Rodriguez-Seijo et al. 2017).

17.6.5 Isopods

For a few days, scientists exposed isopods *Procellio scaber* to polyethylene microplastics and found no influence on rate of food intake, faeces, food assimilation, body mass change, or survival rates (Jemec Kokalj et al. 2018). The aquatic organisms, (both vertebrates and some invertebrates), are harmed by land next to water that has been contaminated by microplastics. Several investigations have also been undertaken earlier, that showed presence of microplastics in the guts of fish (Biginagwa et al. 2016).

17.6.6 Effect on Flora

Several experiments employed sorghum, grains, mung bean, lettuce, rice, faba bean, and spring onion to see how microplastics and nanoplastics affect plant growth. Microplastics have been shown to interfere with seed germination (Sforzini et al. 2016), reduce root length (Bosker et al. 2019), substantially reduce biomass, reduce leaf surface area, decreases leaf number (Qi et al. 2018) and fresh weight, and cause genotoxicity (Qi et al. 2018; Jiang et al. 2019). As demonstrated in Table 17.1, the concentration and size of plastic particles have an impact on a number of the aforementioned aspects. A 72-h study on both *Sorghum saccharaum* and *Lepidium sativum* showed that incubation with microplastics (bioplastic) reduces the seed germination rate (Sforzini et al. 2016; Bosker et al. 2019). The growth of *Vicia faba* bean was also suppressed at concentrations of 10, 50, and 100 mg L-1, while nanoplastics had comparable effects but at a level of 100 mg/L. (Jiang et al. 2019).

Shape of microplastics also modifies the effect of root length and seed germination respectively. A study with 0.5 and 4.8 μm sized plastics where the concentration and

Table 17.1 Various effects on plants used for ecotoxicology tests with microplastics

Species	Microplastic composition	Shape	Size (μm)	Concentration of significant effect	Exposure condition	Endpoint	References
Sorghum saccharatum	Bioplastic	Film	–	(12.5 g kg^{-1})	Artificial soil, for 72 h	Decrease Seed germination rate	Sforzini et al. (2016)
Lepidium sativum	Bioplastic	Film	–	(12.5 g kg^{-1})	Artificial soil, for 72 h	Decrease Seed germination rate	Sforzini et al. (2016)
	PS	Particle	0.5	10^7 particles ml^{-1}	Petri dishes, for 72 h	Effect on Root length	Bosker et al. (2019)
	PS	Particle	4.8	10^7 particles ml^{-1}	Petri dishes, for 72 h	Effect on seed germination	Bosker et al. (2019)
Vicia faba	PS	Sphere	5	50 and 100 mg L^{-1}	Beaker, incubation, 48 h	Fresh weight and length of root	Jiang et al. (2019)
	PS	Sphere	0.1	100 mg L^{-1}	Beaker, incubation, 48 h	Toxicity and fresh weight	Jiang et al. (2019)
	PS	Sphere	0.1 and 5	10 mg L^{-1}	Beaker, incubation, 48 h	Oxidative stress	Jiang et al. (2019)

Source Liu et al. (2019)

incubation time was kept fixed showed that it altered root length and seed germination in *Lepidium sativum* (Bosker et al. 2019).

17.6.7 Effect on Microbes

It has been stated by de Souza Machado et al. (2018) that the size, shape of plastics and soil type affects microbial community. The same group stated that polyacrylic and polyester reduced the activity of microbes that thrive in the agricultural land while polyethylene induces no such effect. It was also observed by them that polyester and polypropylene in agricultural soil increases root colonization of microbes while polyethylene terephthalate depletes colonization of microbes. The metabolic activity of microbes declines in presence of polyethylene terephthalate while it augmented in the presence of polyacrylic, polyester and high-density polyethylene (de Souza Machado et al. 2019). Table 17.2 represents the effect of micro and nano plastics on different soil organism.

Table 17.2 Table summarizing the effect of micro and nano plastic on soil organism and their activities

Polymers	Size	Media	Effects	References
Polyacrylic (PA)	PA: 18 µm	Agricultural soil	Microbes activity has been decreased by PA and PS but PE had no effect	de Souza Machado et al. (2018)
Polyester (PS) and polyethylene (PE)	PS: 8 µm and PE: 643 µm			
Polyethylene terephthalate (PET), Polyester (PES) and Polypropylene (PP)	PET: 222–258 µm PES: 5000 µm PP: 647–754 µm	Agricultural soil	Root colonization decreased by PET while increased by PES and PP	de Souza Machado et al. (2019)
Polyamide (PA), Polyethylene terephthalate (PET), Polyester (PES), Polypropylene (PP), Polystyrene (PS) and high density polyethylene (PEHD)	PA: 15–20 µm PET: 222–258 µm PES: 5000 µm PP: 647–754 µm PS: 547–555 µm HDPE: >800 µm	Agricultural soil	Metabolic activity decreased by PET while increased by PA, PEHD and PES	de Souza Machado et al. (2019)

Source Iqbal et al. (2020)

17.6.8 Impact on Nitrogen Cycle

Nitrogen is one of the most essential nutrients for ecosystem productivity. Polystyrene and polyethylene tend to decrease the action of enzymes that regulate nitrogen cycle (Awet et al. 2018). But polypropylene tends to accumulate in soils where the concentrations of C, N, P is high in soil (Liu et al. 2017). Further, it was reported that application of polylactic and biodegradable plastic decreased soil ammonium concentrations, increase in NO_3 and NO_2 concentration (Chen et al. 2020). As reported by Rillig (2012), microplastic had severe consequences on genes, it effects the genes which are primarily involved in nitrogen cycle like- nifH (in N_2 fixation), nosZ (in N_2O reduction), while nirS (a denitrification gene) expression increases while the expression of nirK decreases.

17.6.9 Other Impact of Plastics on Vertebrates

Animals misapprehend plastics for food and it enters the body of animal accidentally in the course of consuming left overs such as dogs, crows, racoons, bear, cow or buffalo. It affects wild and domesticated animals in the process of grazing. Plastics hinders agility of animals and hence face difficulty in getting proximity to food and water resources, wild animals on the other hand are an easy prey to the predator. Plastics also limit movement and hinders their ability to fly and often they get entangled with them. It also harms young birds because it was reported by others that bird often prepare nest with these plastics. It hampers their foraging behaviour, diet, breeding and distribution. Plastics often gets stuck in their throat, accumulate in stomach in case of ruminants, causes incision and infection. Presence of microplastics in soil affects the incubation microenvironment in loggerhead turtles because microplastics hold large amount of heat and the turtles lay their eggs in sand. Therefore, it modifies the sex ratio of turtles. Microplastic contamination in domesticated duck was also reported by Susanti et al. (2021) where they observed 25 duck samples and found that microplastic debris was found in most of them. Reynolds and Ryan (2018) reported that West African duck species contained microplastic fibres. Microplastic fibres were identified in roughly 5% of faecal samples ($n = 283$) and 10% of feather brushings ($n = 408$) after an analysis of duck faeces and feather brushings. Microplastic concentrations are directly proportional to human density, activity, industrialization, tourism, and environmental pressure (Wardle 2013; Gündodu and Cevik 2017). Duncan et al. (2019) reported that Olive Ridley turtles were contaminated with microplastics since they feed on various types of jelly fish, bivalves, sea urchins and tunicates thus their reproductive ability was vulnerable. The crabs and other crustacean which lives in soil contaminated with microplastics are eaten by various avian species and hence immediate assessment of birds is a necessity. Several species of bivalves are known to have loads of microplastic content and people feeding on these bivalves gets contaminated accidentally and are biomagnified through trophic transfer. Sand crabs are considered indicator species because their health reflects the health status of the ecosystem. It was reported that microplastics contamination increased mortality and decreased reproductive performance in sand crabs and these sand crabs are preferred food for shorebirds and mammals. Consumption of sand crabs might have potential deleterious effects on other species higher up on the food chain and which also needs further investigation.

17.7 Future Research Prospects

The present review summarizes how soil ecosystem is being contaminated by microplastic and its biomagnification. In a state of inadequate mitigation success, the challenge is to recuperate, augment, develop and enhance a justifiable and economic possibility that the public can accept, install and maintain themselves to reduce soil microplastic pollution however there are few challenges which are given below:

1. Limited information is available about source of microplastics, its distribution and transport to different medium (be it soil or water) and how these microplastics degrade in soil. How microplastics are transported horizontally and vertically by wind and water in soil biota. It still remains to be elucidated as how natural processes affect microplastic fate in soil ecosystems.
2. It is important to investigate properly the soil biota because soil organisms also face potential toxicity and it can be transferred to higher trophic levels and thereby hampering food safety.
3. Government and citizens should work together to reduce plastic use and create awareness among masses so as not to use or dispose plastics in soil.
4. Government should ban single use plastics and impose extended producer guideline, simultaneously enforce a law which would make producer/manufacturer accountable for handling plastic waste after a consumer has used their merchandise.
5. Development of simple methods that takes little time for detection, quantification and degradation of microplastics is desired.

17.8 Conclusion

Soil microplastic pollution is global environmental problem and, in this review, emphasis was given to terrestrial pollution by microplastics and we encountered that microplastics obstruct sustainable crop production and thus affects food safety. In such a scenario, research on microplastic should be high on priority with special emphasis on their distribution, toxicity and distribution/transport pathways so that we can overcome this perilous problem. The review discussed several facets of plastic pollution and is designed to support as a resource for academics and readers engaged in pollution research, as well as for the betterment of society and biodiversity conservation.

Acknowledgements Authors are thankful to Department of Science and Technology, New Delhi for infrastructural support in the form of DST-FIST (SR/FST/LS-I/2018/173), and to DST-Govt. of West Bengal for financial support [261(Sanc./ST/P/ S&T/1G-59/2017].

References

Andrady AL, Neal MA (2009) Applications and societal benefits of plastics. Philos Trans R Soc B Biol Sci 364(1526):1977–1984

Awet TT, Kohl Y, Meie F, Straskraba S, Grün AL, Ruf T, Jost C, Drexel R, Tunc E, Emmerling C (2018) Effects of polystyrene nanoparticles on the microbiota and functional diversity of enzymes in soil. Environ Sci Eur 30(1):11. https://doi.org/10.1186/s12302-018-0140-6

Ballent A, Corcoran PL, Madden O, Helm PA, Longstaffe FJ (2016) Sources and sinks of microplastics in Canadian Lake Ontario nearshore, tributary and beach sediments. Mar Pollut Bull 110(1):383–395

Biginagwa FJ, Mayoma BS, Shashoua Y, Syberg K, Khan FR (2016) First evidence of microplastics in the African Great Lakes: recovery from Lake Victoria Nile perch and Nile tilapia. J Great Lakes Res 42(1):146–149

Bläsing M, Amelung W (2018) Plastics in soil: analytical methods and possible sources. Sci Total Environ 612:422–435

Bosker T, Bouwman LJ, Brun NR, Behrens P, Vijver MG (2019) Microplastics accumulate on pores in seed capsule and delay germination and root growth of the terrestrial vascular plant Lepidium sativum. Chemosphere 226:774–781

Browne MA, Crump P, Niven SJ, Teuten E, Tonkin A, Galloway T, Thompson R (2011) Accumulation of microplastic on shorelines woldwide: sources and sinks. Environ Sci Technol 45(21):9175–9179

Brussaard L (1997) Biodiversity and ecosystem functioning in soil. Ambio 563–570.

Chae Y, An YJ (2018) Current research trends on plastic pollution and ecological impacts on the soil ecosystem: a review. Environ Pollut 240:387–395

Chen H, Wang Y, Sun X, Peng Y, Xiao L (2020) Mixing effect of polylactic acid microplastic and straw residue on soil property and ecological function. Chemosphere 243:125271

Claessens M, De Meester S, Van Landuyt L, De Clerck K, Janssen CR (2011) Occurrence and distribution of microplastics in marine sediments along the Belgian coast. Mar Pollut Bull 62(10):2199–2204. https://doi.org/10.1016/j.marpolbul.2011.06.030

Corradini F, Meza P, Eguiluz R, Casado F, Huerta-Lwanga E, Geissen V (2019) Evidence of microplastic accumulation in agricultural soils from sewage sludge disposal. Sci Total Environ 671:411–420

de Souza Machado AA, Kloas W, Zarfl C, Hempel S, Rillig MC (2018) Microplastics as an emerging threat to terrestrial ecosystems. Glob Change Biol 24(4):1405–1416

Dris R, Gasperi J, Mirande C, Mandin C, Guerrouache M, Langlois V, Tassin B (2017) A first overview of textile fibers, including microplastics, in indoor and outdoor environments. Environ Pollut 221:453–458. https://doi.org/10.1016/j.envpol.2016.12.013

de Souza Machado AA, Lau CW, Kloas W, Bergmann J, Bachelier JB, Faltin E, Becker R, Görlich AS, Rillig MC (2019) Microplastics can change soil properties and affect plant performance. Environ Sci Technol 53(10):6044–6052. https://doi.org/10.1021/acs.est.9b01339

Duncan EM, Broderick AC, Fuller WJ, Galloway TS, Godfrey MH, Hamann M, Limpus CJ, Lindeque PK, Mayes AG, Omeyer LCM, Santillo D, Snape RTE, Godley BJ (2019) Microplastic ingestion ubiquitous in marine turtles. Glob Chang Biol 25(2):744–752. https://doi.org/10.1111/gcb.14519

Duis K, Coors A (2016) Microplastics in the aquatic and terrestrial environment: sources (with a specific focus on personal care products), fate and effects. Environ Sci Eur 28(1):1–25

Dümichen E, Barthel AK, Braun U, Bannick CG, Brand K, Jekel M, Senz R (2015) Analysis of polyethylene microplastics in environmental samples, using a thermal decomposition method. Water Res 85:451–457

ECHA (2019) proposes to restrict intentionally added microplastics ECHA/PR/19/03. Accessed 30 June 2021

Engler RE (2012) The complex interaction between marine debris and toxic chemicals in the ocean. Environ Sci Technol 46(22):12302–12315

Environmental Protection Agency (2015) Urban waste water treatment in 2014: a report for the year 2014. Accessed 30 June 2021

Espí E, Salmerón A, Fontecha A, García-Alonso Y, Real AI (2006) Journal of Plastic Film and. J Plastic Film Sheet 22:59

Gaylor MO, Harvey E, Hale RC (2013) Polybrominated diphenyl ether (PBDE) accumulation by earthworms (Eisenia fetida) exposed to biosolids-, polyurethane foam microparticle-, and penta-BDE-amended soils. Environ Sci Technol 47(23):13831–13839

Gregory MR (1996) Plastic 'scrubbers' in hand cleansers: a further (and minor) source for marine pollution identified. Mar Pollut Bull 32(12):867–871

Harshvardhan K, Jha B (2013) Biodegradation of low-density polyethylene by marine bacteria from pelagic waters, Arabian Sea, India. Mar Pollut Bull 77(1–2):100–106

He D, Luo Y, Lu S, Liu M, Song Y, Lei L (2018) Microplastics in soils: analytical methods, pollution characteristics and ecological risks. TrAC, Trends Anal Chem 109:163–172

Hodson ME, Duffus-Hodson CA, Clark A, Prendergast-Miller MT, Thorpe KL (2017) Plastic bag derived-microplastics as a vector for metal exposure in terrestrial invertebrates. Environ Sci Technol 51(8):4714–4721

Horton AA, Svendsen C, Williams RJ, Spurgeon DJ, Lahive E (2017a) Large microplastic particles in sediments of tributaries of the River Thames, UK–abundance, sources and methods for effective quantification. Mar Pollut Bull 114(1):218–226

Horton AA, Walton A, Spurgeon DJ, Lahive E, Svendsen C (2017b) Microplastics in freshwater and terrestrial environments: evaluating the current understanding to identify the knowledge gaps and future research priorities. Sci Total Environ 586:127–141

Iqbal S, Xu J, Allen SD, Khan S, Nadir S, Arif MS, Yasmeen T (2020) Unravelling consequences of soil micro-and nano-plastic pollution for soil-plant system with implications for nitrogen (N) cycling and soil microbial activity. Chemosphere 127578

Jemec Kokalj A, Horvat P, Skalar T, Kržan A (2018) Plastic bag and facial cleanser derived microplastic do not affect feeding behaviour and energy reserves of terrestrial isopods. Sci Total Environ 615:761–766. https://doi.org/10.1016/j.scitotenv.2017

Jiang XJ, Liu W, Wang E, Zhou T, Xin P (2017) Residual plastic mulch fragments effects on soil physical properties and water flow behavior in the Minqin Oasis, north western China. Soil and Tillage Research 166:100–107

Jiang X, Chen H, Liao Y, Ye Z, Li M, Klobučar G (2019) Ecotoxicity and genotoxicity of polystyrene microplastics on higher plant *Vicia faba*. Environ Pollut 250:831–838

Käppler A, Fischer D, Oberbeckmann S, Schernewski G, Labrenz M, Eichhorn KJ, Voit B (2016) Analysis of environmental microplastics by vibrational microspectroscopy: FTIR, Raman or both? Anal Bioanal Chem 408(29):8377–8391

Klein S, Worch E, Knepper TP (2015) Microplastics in the Rhine-Main area in Germany: occurrence, spatial distribution and sorption of organic contaminants. Environ Sci Technol 49:2–3

Lei L, Liu M, Song Y, Lu S, Hu J, Cao C, Xie B, Shi H, He D (2018) Polystyrene (nano) microplastics cause size-dependent neurotoxicity, oxidative damage and other adverse effects in *Caenorhabditis elegans*. Environ Sci Nano 5(8):2009–2020

Liu H, Yang X, Liu G, Liang C, Xue S, Chen H, Ritsema CJ, Geissen V (2017) Response of soil dissolved organic matter to microplastic addition in Chinese loess soil. Chemosphere 185:907–917. https://doi.org/10.1016/j.chemosphere

Liu K, Wang X, Fang T, Xu P, Zhu L, Li D (2019) Source and potential risk assessment of suspended atmospheric microplastics in Shanghai. Sci Total Environ 675:462–471. https://doi.org/10.1016/j.scitotenv.2019.04.110

Liu L, Fokkink R, Koelmans AA (2016) Sorption of polycyclic aromatic hydrocarbons to polystyrene nanoplastic. Environ Toxicol Chem 35(7):1650–1655

Liu M, Lu S, Song Y, Lei L, Hu J, Lv W, Zhou W, Cao C, Shi H, Yang X, He D (2018) Microplastic and mesoplastic pollution in farmland soils in suburbs of Shanghai, China. Environ Pollut 242(Pt A):855–862. https://doi.org/10.1016/j.envpol.2018.07.051

Lwanga H, Mendoza Vega EJ, Ku Quej V, Chi J, Sanchez Del Cid L, Chi C, Escalona Segura G, Gertsen H, Salánki T, van der Ploeg M, Koelmans AA, Geissen V (2017) Field evidence for transfer of plastic debris along a terrestrial food chain. Sci Rep 7(1):14071. https://doi.org/10.1038/s41598-017-14588-2

Maa BS, Daphi D, Lehmann A, Rillig MC (2017) Transport of microplastics by two collembolan species. Environ Pollut 225:456–459

Mai L, Bao LJ, Wong CS, Zeng EY (2018) Microplastics in the terrestrial environment. In: Microplastic contamination in aquatic environments. Elsevier, pp 365–378

Napper IE, Thompson RC (2016) Release of synthetic microplastic plastic fibres from domestic washing machines: effects of fabric type and washing conditions. Mar Pollut Bull 112(1–2):39–45
Ng EL, Lwanga EH, Eldridge SM, Johnston P, Hu HW, Geissen V, Chen D (2018) An overview of microplastic and nanoplastic pollution in agroecosystems. Sci Total Environ 627:1377–1388
Nizzetto L, Futter M, Langaas S (2016) Are agricultural soils dumps for microplastics of urban origin?
Piehl S, Leibner A, Löder MG, Dris R, Bogner C, Laforsch C (2018) Identification and quantification of macro-and microplastics on an agricultural farmland. Sci Rep 8(1):1–9
Qi Y, Yang X, Pelaez AM, Lwanga EH, Beriot N, Gertsen H, Garbeva P, Geissen V (2018) Macro-and micro-plastics in soil-plant system: effects of plastic mulch film residues on wheat (Triticum aestivum) growth. Sci Total Environ 645:1048–1056
Qi Y, Beriot N, Gort G, Lwanga EH, Gooren H, Yang X, Geissen V (2020) Impact of plastic mulch film debris on soil physicochemical and hydrological properties. Environ Pollut 266:115097
Reynolds C, Ryan PG (2018) Micro-plastic ingestion by waterbirds from contaminated wetlands in South Africa. Mar Pollut Bull 126:330–333
Rillig MC (2012) Microplastic in terrestrial ecosystems and the soil?
Rillig MC, Ingraffia R, de Souza Machado AA (2017) Microplastic incorporation into soil in agroecosystems. Front Plant Sci 8:1805
Rodriguez-Seijo A, Lourenço J, Rocha-Santos TAP, Da Costa J, Duarte AC, Vala H, Pereira R (2017) Histopathological and molecular effects of microplastics in Eisenia andrei Bouché. Environ Pollut 220:495–503
Ryan PG, Moore CJ, Van Franeker JA, Moloney CL (2009) Monitoring the abundance of plastic debris in the marine environment. Philoso Trans R Soc B Biol Sci 364(1526):1999–2012
Sarker A, Deepo DM, Nandi R, Rana J, Islam S, Rahman S, Hossain MN, Islam MS, Baroi A, Kim JE (2020) A review of microplastics pollution in the soil and terrestrial ecosystems: a global and Bangladesh perspective. Sci Total Environ 733:139296. https://doi.org/10.1016/j.scitotenv.2020.139296
Simon M, van Alst N, Vollertsen J (2018) Quantification of microplastic mass and removal rates at wastewater treatment plants applying focal plane array (FPA)–based Fourier transform infrared (FT–IR) imaging. Water Res 142:1–9. https://doi.org/10.1016/j.watres.2018.05.019
Sforzini S, Oliveri L, Chinaglia S, Viarengo A (2016) Application of biotests for the determination of soil ecotoxicity after exposure to biodegradable plastics. Front Environ Sci 4:68
Shim WJ, Song YK, Hong SH, Jang M (2016) Identification and quantification of microplastics using Nile Red staining. Mar Pollut Bull 113(1–2):469–476
Silva MF, Doménech-Carbó MT, Fuster-López L, Mecklenburg MF, Martin-Rey S (2010) Identification of additives in poly (vinylacetate) artist's paints using PY-GC-MS. Anal Bioanal Chem 397(1):357–367
Susanti R, Yuniastuti A, Fibriana F (2021) The Evidence of microplastic contamination in Central Javanese local ducks from intensive animal husbandry. Water Air Soil Pollut 232:178. https://doi.org/10.1007/s11270-021-05142-y
Thompson RC (2006) Plastic debris in the marine environment: consequences and solutions. Mar Nat Conserv Eur 193:107–115
Thompson RC, Law KL (2014) Microplastics in the seas. Science 345(6193):144–145
Thompson RC, Olsen Y, Mitchell RP, Davis A, Rowland SJ, John AW, McGonigle D, Russell AE (2004) Lost at sea: where is all the plastic? Science 304(5672):838. https://doi.org/10.1126/science.1094559
Toussaint B, Raffael B, Angers-Loustau A, Gilliland D, Kestens V, Petrillo M, Rio-Echevarria IM, Van den Eede G (2019) Review of micro-and nanoplastic contamination in the food chain. Food Addit Contam Part A 36(5):639–673
UNEP (2018) Plastic planet: how tiny plastic particles are polluting our soil. Accessed 29 June 2021
Velzeboer I, Kwadijk CJAF, Koelmans AA (2014) Strong sorption of PCBs to nanoplastics, microplastics, carbon nanotubes, and fullerenes. Environ Sci Technol 48(9):4869–4876

Wan Y, Wu C, Xue Q, Hui X (2019) Effects of plastic contamination on water evaporation and desiccation cracking in soil. Sci Total Environ 654:576–582
Wang J, Liu X, Li Y, Powell T, Wang X, Wang G, Zhang P (2019) Microplastics as contaminants in the soil environment: a mini-review. Sci Total Environ 15(691):848–857. https://doi.org/10.1016/j.scitotenv.2019.07.209
Wardle DA (2013) Communities and ecosystems: linking the aboveground and belowground components (MPB-34). Princeton University Press
Weithmann N, Möller JN, Löder MG, Piehl S, Laforsch C, Freitag R (2018) Organic fertilizer as a vehicle for the entry of microplastic into the environment. Sci Adv 4(4):eaap8060
Wu C, Zhang K, Huang X, Liu J (2016) Sorption of pharmaceuticals and personal care products to polyethylene debris. Environ Sci Pollut Res 23(9):8819–8826
Xu B, Liu F, Brookes PC, Xu J (2018) The sorption kinetics and isotherms of sulfamethoxazole with polyethylene microplastics. Mar Pollut Bull 131:191–201
Xu B, Liu F, Cryder Z, Huang D, Lu Z, He Y … Xu J (2020) Microplastics in the soil environment: occurrence, risks, interactions and fate–a review. Crit Rev Environ Sci Technol 50(21):2175–2222
Yu M, Van Der Ploeg M, Lwanga EH, Yang X, Zhang S, Ma X, Ritsema CJ, Geissen V (2019) Leaching of microplastics by preferential flow in earthworm (*Lumbricus terrestris*) burrows. Environ Chem 16(1):31–40
Zhang GS, Liu YF (2018) The distribution of microplastics in soil aggregate fractions in southwestern China. Sci Total Environ 642:12–20
Zhang GS, Zhang FX, Li XT (2019) Effects of polyester microfibers on soil physical properties: perception from a field and a pot experiment. Sci Total Environ 670:1–7
Zhao J, Liu L, Zhang Y, Wang X, Wu F (2018) A novel way to rapidly monitor microplastics in soil by hyperspectral imaging technology and chemometrics. Environ Pollut 238:121–129
Zhou Q, Zhang H, Fu C, Zhou Y, Dai Z, Li Y, Luo Y (2018) The distribution and morphology of microplastics in coastal soils adjacent to the Bohai sea and the Yellow sea. Geoderma 322:2101–2208
Zhu BK, Fang YM, Zhu D, Christie P, Ke X, Zhu YG (2018a) Exposure to nanoplastics disturbs the gut microbiome in the soil oligochaete *Enchytraeus crypticus*. Environ Pollut 239:408–415
Zhu D, Bi QF, Xiang Q, Chen QL, Christie P, Ke X, Wu LH, Zhu YG (2018b) Trophic predator-prey relationships promote transport of microplastics compared with the single *Hypoaspis aculeifer* and *Folsomia candida*. Environ Pollut 235:150–154
Zhou Q, Zhang H, Zhou Y, Yuan L, Xue Y, Fu C, Chen T, Luo Y (2016) Separation of microplastics from a coastal soil and their surface microscopic features. Chinese Sci Bull 61:1604–1611
Zhang H (2017) Transport of microplastics in coastal seas. Estuar Coast Shelf Sci 199:74–86

Part III
Soil Health and Sustainable Management

Chapter 18
Sustainable Land Use, Landscape Management and Governance

Roshan M. Bajracharya

Abstract The soil and other land-based resources such as forests, shrub and grass lands have nurtured and supported humans from the beginning of civilization. However, following the industrial revolution in the mid-1800s, extensive deforestation, over-grazing of pastures and agricultural intensification have caused soils and the land resource base to become degraded. The increasing demands of a-growing population, leading to over-exploitation of natural resources, as well as, human-induced climate change imply major consequences for sustaining human health and habitation on Earth. In many part of the world, much of the cultivable land has been fully exploited, but the requirement of food and fiber production for meeting human nutrition and livelihoods continues to increases. Meeting these insatiable demands requires that farmers must produce more food, fodder and fiber from the same parcel of land. Hence, approaches and technologies for intensive cropping at scale without damaging the soil and land resource base is ever more vital for the survival of humans in the future. Conventional farming and livestock rearing practices resulted in decline of soil fertility, productivity and the general land quality while contributing to climate change due to net emission of greenhouse gases from soil and land. Therefore, a paradigm shift in sustainable soil as well as land management is undoubtedly required without delay. Regenerative agriculture involving holistic and integrated farming practices offer a promising solution. This approach includes agricultural practices that incorporate agroforestry, agri-slivopastoral systems, along with diversified cropping, as well as, soil and water conservation practices. Such restorative practices offer the possibility of numerous simultaneous benefits while preserving and enhancing the very resources that human societies depend on.

Keywords Agro-forestry · Permaculture · Diversified cropping · Holistic land management · Integrated farming · Regenerative agriculture

R. M. Bajracharya (✉)
Kathmandu University, Dhulikhel, Nepal
e-mail: rmbaj@ku.edu.np

Tribhuvan University, Kathmandu, Nepal

P. K. Shit et al. (eds.), *Soil Health and Environmental Sustainability*,
Environmental Science and Engineering,
https://doi.org/10.1007/978-3-031-09270-1_18

18.1 Introduction

During pre-historic times, before the advent of settled agriculture, primitive communities were comprised of small, scattered populations that relied primarily upon hunting and gathering for their survival. The activities of such human communities exerted minimal impact on the physical environment and left it relatively unaltered (Braidwood 1960; Piggot 1961). As a result, they lived in relative harmony with their surroundings and nature. However, following the establishment of settled agriculture, about 10,000 years ago, people came to depend upon productive soils to support their growing communities. Early humans gradually learned to cultivate and manage their lands to produce food and fiber for ever-expansive settlements.

Soil and land have formed the backbone of human civilizations, which flourished with access to fertile soils and inevitably declined as a result of degraded agricultural lands, lack of sufficient water resources and destruction of productive soils (Hillel 1992, 2007). As settled agriculture spread across the globe, humans began to inhabit ever-larger communities, ultimately resulting in the establishment of large towns and cities. Moreover, increasing technological advancement and inputs, such as motorized vehicles, tractors and tillage implements, meant that people could shape their physical environment to their preferences, and thus, had ever-increasing impacts on their surroundings. In modern times, human activities have had far-reaching, global-scale consequences leading to significant alteration of our planet (Thomas et al. 1956; Darlington 1969). The industrial revolution of the 1800s, agricultural revolution of the 1900s, and the population explosion, have led to large-scale monoculture and intensified cropping and animal husbandry (Boserup 1965; Carswell 1997; Dahal et al. 2009). Intensification of agriculture involves planting multiple crops (three or more) on the same parcel of land, simultaneously or in rotation, often accompanied by extra inputs, such as, fertilizers, pesticides and labor (Dahal et al. 2008; Bajracharya and Dahal 2012).

The modernization of agricultural operations in most advanced and emerging nations involve large-scale mono-culture cropping (planting a single crop) over thousands of hectares using heavy machinery and widespread application of chemical fertilizers and pesticides. New and 'improved' hybrid varieties of crops, as well as, genetically modified organisms (GMOs) mean that production methods have increasingly shifted away from natural or organic approaches. Excessive tillage and exposure of soil has led to the erosion of soil and degradation of vast tracts of land in many parts of the world (Lal 2011). The above factors, along with the clearing of more and more forests for settlements and livestock rearing, especially in the tropics, has led to ecological imbalance and destruction of natural food chains and food webs. As land becomes degraded through over-exploitation and misuse, and soils lose their fertility and productive capacity, the entire production system falls into a vicious cycle of degradation that reinforces itself through a positive feedback loop (Bajracharya 2021).

In many parts of the world declining productivity of the land has become confounded by climate change impacts. Climate change is leading to evermore unpredictable and increased frequency of severe weather events often causing prolonged droughts in some regions of the globe, such as Africa and the Middle East, while simultaneously leading to floods in other parts, such as South and Southeast Asia. Furthermore, the ever-growing waste generation, leading to the pollution of air, water and land, from sprawling urban concentrations of populations has created unnatural environments making living in mega-cities in emerging economies hazardous to human health. Hence, all of the above activities and adverse impacts have grave implications for the very survival of the human race. Evidently, humankind has begun to influence and alter natural systems in ways that are in fact "changing the face of the Earth" (Revelle and Suess 1957; Thomas et al. 1956). It has, therefore, become evident that production strategies must go beyond mere sustainability toward a regenerative approach (RAI/TCU 2017), which will require a paradigm shift in agriculture, animal husbandry, and, indeed even diets and lifestyles. Choosing to take the path of regenerative land management could reverse the cycle of degradation and lead to sustainable agricultural production as depicted in Fig. 18.1. If unchecked this cycle could spiral out of control in a self-perpetuating positive feedback loop.

18.2 Paradigm Shift in Managing Land-Based Resources

There can be no doubt that a paradigm shift towards alternative approaches to managing our soil and land resources is urgently required. As terrestrial beings that are heavily dependent upon the land for sustenance, humans must prioritize the proper management of the land and soils for which there is no viable substitute, especially for food production. Indeed, soils have many vital roles and provide ecosystem services that help sustain life on Earth. These include nutrient and organic carbon cycling, climate regulation via fluxes in carbon and nitrogen, water absorption and flow regulation, pest and disease control for humans, animals and plant; and buffering and decontamination of the environment (Brevik et al. 2015; Melecis 2010). As documented by Ohlson (2014) who quotes Prof. Lal of the Ohio State University, "nothing in nature repeatedly and regularly turns over the soil to the specified depth of 15–20 cm. Therefore, neither plants nor soil organisms have evolved or adapted to this drastic perturbation." It is well known that plowing damages the soil structure and exposes soil organic matter to decomposition (releasing carbon to the atmosphere). Bare land causes the soil microorganisms to become starved, whereas, an abundance of plants (living and dead) and roots (exudates) nurture the microbes. Furthermore, numerous research findings have confirmed that long-term no-till farming leads to loose, porous, organic matter rich and deep soils that absorb and retain water (like a sponge) rather than generating runoff or causing puddles on the soil surface (Lal 2009; Grogg 2013). Soils rich in carbon can help to buffer against both droughts and floods. The process of photosynthesis removes vast amounts of CO_2 from the atmosphere,

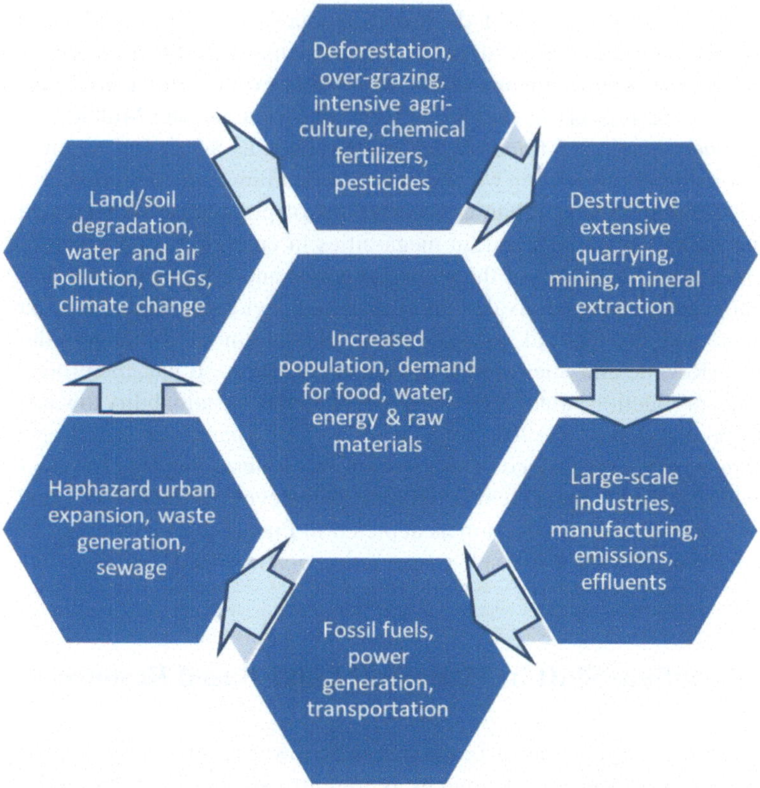

Fig. 18.1 The cycle of degradation set in motion by population pressures and unsustainable exploitation of land-based resources

accumulating carbon in the soil; which is Mother Nature's approach for sequestering carbon and balancing the atmospheric carbon budget. Important approaches and practices for managing various resource categories that need to be implemented simultaneously in order to achieve sustainable soil and land management are shown in Fig. 18.2.

Modern-day farming practices, extensive water use, energy use, urbanization, waste generation; and, indeed our very way of life, which emphasizes reliance on technological solutions to problems, are all causing an imbalance in natural ecosystems with the cumulative effect having global implications. Moreover, conventional, large-scale livestock rearing also upsets the carbon balance. As pointed out by Savory (1999) "under natural conditions in the past, large herds of ruminant herbivores roamed the prairies and savannas in close groups, constantly moving for protection against predators; they ate only the tops of the grass/plants encouraging rapid regeneration. But by domestication, herds of cattle left to graze securely in a large area could eat plants right to the ground or even pull the roots out." Range management expert Allen Savory of the Africa Center for Holistic Management near Victoria Falls,

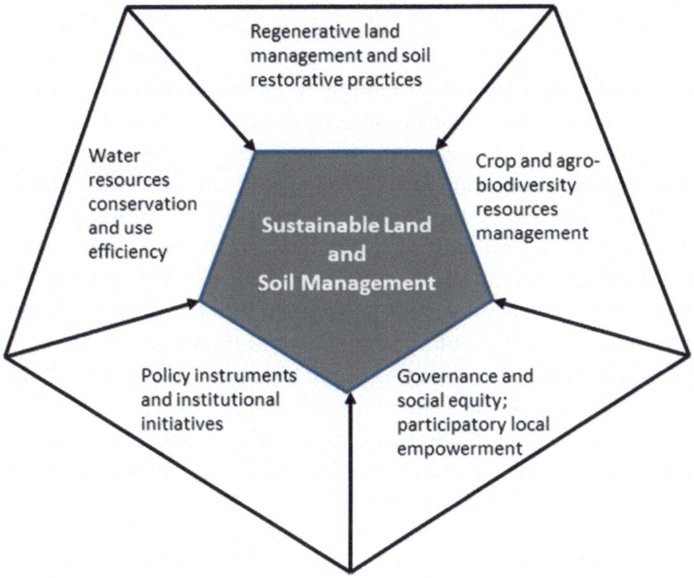

Fig. 18.2 Practices and approaches concomitantly required to achieve sustainable land management

S. Africa, demonstrated that large herds moving in tightly packed groups trample the ground and incorporate plant and animal litter into the soil. Over time such an approach can lead to significant improvements in the pasture land soil quality, species composition, biodiversity and ultimately ecosystem health (Savory 1999).

18.3 Alternatives and Approaches for Achieving Sustainability

Ohlson (2014) further reports that after decades of government programs and commercial interests to promote chemical fertilizers, pesticides and 'improved' seed, individuals and interest groups in the US and other parts of the world have sought alternative approaches. These include 'regenerative agriculture' and 'holistic management'. The pioneers in the field refer to themselves as: 'carbon farmers or ranchers' and 'microbe farmers', referring to the "vast underground kingdom" of trillions of soil microorganisms that are the key to soil quality. Plants need microbes (bacteria, fungi, protozoa, rotifers and actinomycetes) to thrive, just as microorganisms depend on plants for a steady supply of nutrients. Together they lead to healthy productive soils that sustain life on Earth. This is the ultimate synergistic relationship which forms a "Biological partnership"—coevolution and mutualism.

The 'Future Farmers of America' advocate establishing "Carbon Ranches", as opposed to conventional ranching. Here the goal is to establish 'cover crop cocktails' of dozens and even up to 140 different grass and legume species instead of just a few preferred species. Also, on farms, instead of mechanical clearing of fields, they recommend sending in cows to eat the crop residues and stalks resulting in in-situ manuring and incorporation of residues into the soil. It is now well established that chemical fertilizers are highly inefficient, for example, 32 billion pounds of agrochemicals are used in the mid-west of which most runoff into the River Mississippi and has caused a dead zone of 6000 square miles in the Gulf of Mexico (eutrophication). Therefore, there is an urgent need to eliminate chemicals and allow the soil biology to function as it should. Gabe Brown of the Future Farmers of America claims that this is the only way the system can become sustainable (Ohlsen 2014).

Jeff Moyer of the Rodale Institute emphasizes that "Good organic farming isn't just the absence of synthetic chemicals; rather it's an approach that works with biological processes and regards the soil as a complex system of living organisms". This alludes to maintaining the overall 'soil health'. Research done by Matt Leibmann, an ecologist at Iowa State University shows that, 3- and 4-crop rotations, with the incorporation of legumes and diverse crops (compared to the convention 2 crop rotations) can reduce chemical fertilizer need by 90%. The use of natural enemies for pest control offers a viable alternative to chemical pesticides. If farms are left in weedy fallow (without fall tillage), field mice would consume as much as 70% of the weed seeds reducing the need for herbicides. Hence, there are highly complex ecological interactions and food chains in nature and mono-culture destroys these, thereby, creating more problems than it solves! The global food crisis is, then, not really a problem of production, but of distribution (Millennium Institute); data indicates that we already produce enough food to feed a world population of 9 billion people (Ohlsen 2014).

Apart from the technical and ecological measures required to achieve sustainable land use and management, simultaneous initiatives at the policy and governance levels are also needed to ensure the adoption and implantation of beneficial practices and approaches. Table 18.1 provides a brief overview of some of the governance and policy issues that must be addressed. A decentralized mode of governance should be promoted so that the local agencies and organizations, which are best placed to implement locally suitable and site-specific measures, have the requisite authority to do so. Moreover, through local involvement in decision-making, communities can feel empowered and have a sense of ownership of the various regulations and policies that are promulgated. Also, through a participatory approach, local user groups and management committees can oversee the activities, ensuring equitable resource use, as well as, as gender and social inclusion.

Policies that could be effective in promoting and encouraging sustainable and sound practices and approaches include incentives and sanctions, along with institutional strengthening and market access and support mechanisms. Incentives may take the form of tax breaks, subsidies, carbon credits, technical support for activities requiring specialized skills, such as, pest control, conservation structures, animal breeding, crop breeding and varietal selection, etc. These could be provided through

Table 18.1 Governance and policy initiatives needed to promote sustainable and restorative land management

Initiative	Approach	Activities/description
Governance	Decentralized modality	Hand over authority to local governments, agencies and organizations
	Local involvement and initiatives	Community-led decision making; local empowerment and ownership
	Participatory planning and management	Local user groups and management committees; equitable resource access and distribution; gender equity and social inclusion
Policies	Incentivizing sound practices	Tax breaks, carbon credits, technical support, agricultural and forestry extension services
	Sanctions for adverse impacts	Disincentives, fines, pollution taxes, carbon tax
	Institutional strengthening	Capacity enhancement, training of staff, investment in research, public–private partnerships with businesses and universities
	Market access and support mechanisms	Rural infrastructure, transport and cold storage facilities, cooperatives and credit access

agricultural and forestry extension (outreach) services. Sanctions or disincentives could involve fines, pollution taxes, etc. Institutional strengthening typically entails the enhancement of local capacity to deal with issues such as sustainable intensification of production systems, climate change adaptation and mitigation, and other specific technical training. Here public–private partnerships could be very effective in the implementation of training programs, as well as, for investment in research on different aspects of the local issues. Finally, market access and support mechanisms are critical to the success of any interventions and are required to enable local farmers and businesses to procure returns from their products. These may include upgrading of rural infrastructure, like roads and bridges, or expediting transport of goods or cold storage facilities for perishable items. Apart from physical infrastructure, capital access through credit and cooperative organizations may also play a crucial role in aiding local production and business ventures.

18.4 Regenerative Land Management—A Restorative Approach

Regenerative management of land and associated resources involves a set of approaches that contribute to enhancing soil fertility and overall quality; increases retention and percolation of water; and increases biodiversity, ecosystem health and resilience. It contributes to reversing climate change through soil organic matter accumulation and restoring degraded soil biodiversity, which leads to carbon sequestration and improving the water cycle (Savory 1999; RAI/TCU 2017). Regenerative Agriculture has been defined by RAI/TCU (2017) as "a holistic land management practice that leverages the power of photosynthesis in plants to close the carbon cycle, and build soil health, crop resilience and nutrient density". It includes various practices that improve soil biodiversity and ecosystem integrity, such as, minimum or zero tillage, application of organic residues, biochar, compost and animal manure, crop rotations and cover cropping, agroforestry and permaculture systems, and well-managed grazing practices like agri-silvopastoral systems (Roberts 2017; Regeneration International 2018; Novak et al. 2015; Penn State 2018). The beneficial effects of various regenerative land management practices are shown in Table 18.2.

Diversified cropping with mixed and relayed crop rotations, as well as, cover crop plantation involves careful selection and planting of different types of crops during a cropping cycle (typically annual). This practice offers the opportunity to incorporate leguminous crops (able to fix atmospheric nitrogen making it available to plants) into the rotation which is highly beneficial in managing plant nutrients and reduces the need for chemical fertilizer additions. Moreover, changing the type of crops planted breaks the cycle of continuous cropping and, thereby, offers protection and resilience against crop pests, such as, insects, diseases and weeds. Cover crops, on the other hand, provide protection for the soil and prevent exposure of bare soil in between major cropping periods. Therefore, cover crops are not harvested, rather they cultivated into the soil prior to planting of the next crop and serve as a source of organic manure that eventually decomposes and releases nutrients for the subsequent crop (Bajracharya et al. 2014; Sherchan and Karki 2006).

The use of organic rather than chemical forms of fertilizers for meeting plant nutrient requirements has numerous benefits. Organic fertilizers include crop residues, animal manure, biogas slurry, and compost made from farmyard wastes like weeds and leaf-litter. Nutrients released from the decomposition of organic materials become gradually available to plants throughout the course of the cropping season. Thus, leaching losses in storm water runoff is minimized and any excess or residual nutrients remain available for subsequent crops. The use of organic fertilizers also lead to an increase in the organic matter (humus) content of soils rendering them more stable, porous and fertile. Moreover, high organic matter and readily available organic residue in the soil enhances and supports microbial and biological activity, which promotes nutrient cycling (RAI/TCU 2017; Lal 2009).

Apart from the application of organic manures or farmyard manure, the use of biochar as a soil amendment has gained considerable accolade in recent years as

Table 18.2 The beneficial impacts of regenerative land management practices relative to conventional agriculture

Land/water/crop Mgmt. practice	Increased SOM and C-sequestration	Enhanced microbial activity	Improved aggregation and soil structure	Increased water use efficiency and conserv	Improved nutrient availability	Increased agro-biodiversity	Enhanced fertility and productivity
Conventional farming	O	O	O	O	O	O	O
Crop rotation, cover crops	**	**	**	**	*	*	*
Mixed/relay cropping	**	**	**	*	*	**	**
Compost, manure appl	***	**	***	*	***	O	**
Biochar appl	**	**	**	**	***	O	**
Reduced; zero tillage	**	**	**	**	*	O	**
Agroforestry; permaculture	***	***	**	**	**	***	***
Agri-silvo-pasture	***	***	**	**	**	***	***
Holistic range management	***	***	**	**	**	*	**
Aquaculture; aquaponics	O	O	O	***	**	*	**

Note O indicates that the practice does not have beneficial effects; * indicates slightly beneficial; ** indicates moderate beneficial effects; and *** indicates highly beneficial a promising method of enhancing the carbon accumulation and long-term retention in soils, and concomitantly improving the fertility and productive capacity of soils (Regeneration International 2018; Novak et al. 2015; Bajracharya et al. 2014). A pyrolysis product of biomass, biochar has been applied to soils by ancient civilizations in the Amazon, North West Europe and the Andes (Downie et al. 2011; Sandor and Eash 1995). In recent decades it has received scientific recognition as a simple yet potentially powerful technique for climate change mitigation which also simultaneously contributes to sustainable agricultural production (Downie et al. 2011; IBI 2012). The benefits of biochar has been reported to derive from its high stability, porosity and resistance to microbial breakdown, in soils, it provides sites for enhanced microbial activity along with increased water and nutrient retention (Novak et al. 2015; Sohi 2012). Thus, biochar serves the function of a catalyst for biochemical reactions in the soil which improves plant nutrient availability.

Reduced or zero tillage is the method of producing grain crops on large tracts of land with a minimum of disturbance to the soil. Seeds are planted into the ground by no-till seed drill that only pierces a hole in the soil and drops the seed into it. If done manually, the seed are simply scattered over the soil surface by hand-tossing or dropped into a hole in the ground made using a pointed stick or rod. By avoiding over-turning of the soil or other major manipulation, soil biological communities and microorganisms are preserved while disturbance to the physical structure of the soil is reduced. The method of crop cultivation reduces the soil respiration and emission of carbon dioxide to the atmosphere, as well as, maintains good soil aggregation and porosity which enhances water percolation (RAI/TCU 2017; Bajracharya et al. 2014; Lal 2009).

Agroforestry and permaculture systems have good potential to serve as both a climate adaptive strategy, as well as, a climate change mitigation option, particularly in mountainous regions (Gautam et al. 2017; Nair 2011). These systems incorporate perennial trees or shrubs into farm plots along with other annual crops, thereby, diversifying the production. In the event of severe weather such as droughts or floods, such diversification leads to the possibility that all the crops will not fail, hence a total loss can be prevented. Furthermore, tree and shrub crops can lead to sequestration of carbon, both in the biomass as well as the soil, thereby contributing to climate change mitigation. The tree or shrub crops may be of high value species, like fruits or medicinal plants, or be used as fodder for livestock, which augment the farm income enabling improved livelihood (Penn State 2018; Rhodes 2012). The interactions among agricultural crops, forest and livestock are shown in (Fig. 18.3).

Pasture cropping or agri-silvopastoral systems offer considerable promise as an integrated and holistic land management approach that combines livestock grazing with crop production. After harvesting of the annual crop, the livestock are released into the field and allowed to graze freely on the plant stubble and residues. This practice leads to the control of weeds and utilization of crop residues while simultaneously fertilizing the field through the urine and manure deposited by the animals. Therefore, dual benefits of improved crop yields along with gains in animal products can be obtained through such integrated practices (Roberts 2017; White 2012).

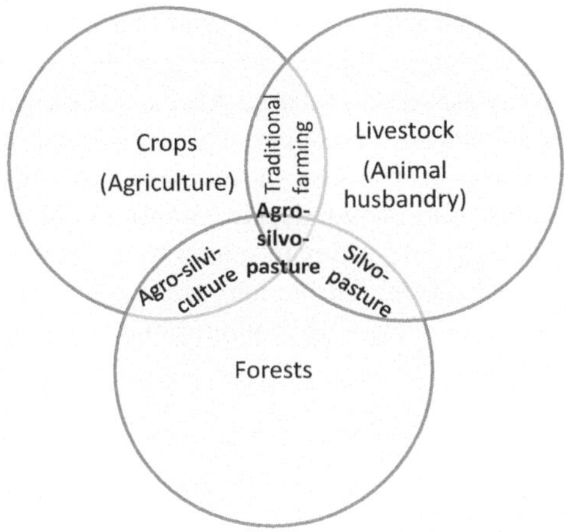

Fig. 18.3 Spheres of interaction among various agroforestry and land production systems (modified from Tejwani 1994)

As land for agriculture becomes increasingly scarce, new approaches for food production will undoubtedly be required. These will likely include, vertical farming, hydroponics, aquaculture, and aquaponics. Vertical farming involves growing crops in a vertically stacked manner with or without soil (Birkby 2016; Skar et al. 2019). Hydroponics is the cultivation of crops, especially leafy vegetables, in water with the provision of dissolved nutrients, thereby, eliminating the need for soil (USDA 2020). In addition, aquaculture, which involves keeping fish in artificial ponds or in indoor aquaria, is also a promising production system which can be done with minimal land (EPA 2020). Moreover, these types of fisheries can be combined with crop production that utilizes a common water resource, and is termed aquaponics (Birkby 2016; North 2016). These innovative emerging alternative food production systems can pave the way to circular economies which offer sustainability through efficient water and nutrient use, low cost, space saving, reuse and recycling of resources, and climate change adaptation.

18.5 Conclusions

Recognition regarding the need for sound management of land, soil, and water resources is not new, but can be traced back to the early 20th Century. More recently research and field experience over the past few decades has clearly shown that only practices sustaining the diverse microbial populations in the soil and retaining high organic matter status will lead to long-term productivity and soil health (Lal 2009; Stockwell and Bitan 2011; Grogg 2013). In order to ensure that humans will continue to be able to feed the growing world population well into the future, it is imperative that we adopt intensive but integrated and holistic soil and land management practices that increase soil organic carbon, enhance soil biological and microbial activity and plant diversity (The Royal Society 2009; Chen et al. 2011; Tilman et al.

2011; Bajracharya 2021). Soils have the capacity to mitigate climate change by sequestering soil organic carbon in the degraded soils across many parts of the world such as Asia, sub-Saharan Africa, and Central America through appropriate soil and land restorative practices. This, along with simultaneous reduction of fossil fuel use and other greenhouse gas emission, minimizing agro-chemicals, and protecting soil humus from the wind, rain, runoff and other unwise development as well as disturbances, would ultimately lead to a reversal of global warming. It is only through the realization that humans are not separate from Nature and that it is in our best interest to work with Nature, rather than against it, that our long-term survival can be assured.

References

Bajracharya RM (2021) Regenerative approach for sustainability and climate resilience of mountain agroecosystems. In: Bajracharya RM, Sitaula BK, Raut N, Gurung S (eds) Sustainable natural resource management in the Himalayan Region: livelihood and climate change. Nova Science Publisher Inc., New York, pp 42–56

Bajracharya RM, Dahal BM (2012) Agricultural intensification. The Berkshire Encyclopedia of sustainability: ecosystem management and sustainability. Berkshire Publishing Group, New York, NY, USA, pp 7–10

Bajracharya RM, Sherchan DP, Dahal BM, Raut N (2014) Soil management for sustainable agricultural intensification in the Himalayan region. In: Lal R, Stewart BA (eds) Soil management of smallholder agriculture. Taylor & Francis Publisher, Boca Raton, FL, USA, pp 143–164

Birkby J (2016) Vertical farming. AATRA Sustainable Agriculture. www.attra.ncert.org. Accessed 30 Mar 2020

Boserup E (1965) The conditions of agricultural growth: the economics of agrarian change under population pressure. George Allen & Unwin Ltd., London, p 108p

Braidwood RJ (1960) The agricultural revolution. Sci Am 203(3):131–148

Brevik EC, Cerda A, Mataiz-Solera J, Pereg L, Quinton JN, Six J, Van Oost K (2015) The interdisciplinary nature of SOIL. Soil 1:117–129. Copernicus Publisher, European Geosciences Union

Carswell G (1997) Agricultural intensification and rural sustainable livelihoods: a "think piece". IDS Working Paper No. 64

Chen X-P, Cui Z-L, Vitousek PM, et al (2011) Integrated soil-crop system management for food security. PNAS 108(16):6399–6404. www.pnas.org/cgi/doi/10.1073/pnas.1101419108

Dahal BM, Sitaula BK, Bajracharya RM (2008) Sustainable agricultural intensification for livelihood and food security in Nepal. Asian J Water Environ Pollut 5(2):1–12

Dahal BM, Nyborg I, Sitaula BK, Bajracharya RM (2009) Agricultural intensification: food insecurity to income security in a mid-hill watershed of Nepal. Int J Agric Sustain 7(4):249–260

Darlington CD (1969) The evolution of man and society. Simon and Schuster Publ, New York, pp 69–70

Downie AE, Van Zwieten L, Smernik RJ, Morris S, Munroe RR (2011) Terra Preta Australis: reassessing the carbon storage capacity of temperate soils. Agric Ecosyst Environ 140:137–147

EPA (2020) Urban Agriculture. Environmental Protection Agency. https://www.epa.gov/agriculture/agricultural-crops#UrbanAgriculture. Accessed 30 Mar 2020

Grogg P (2013) No-till holds the key to food security. Inter Press Service News Agency

Gautam DK, Bajracharya RM, Sitaula BK (2017) Boichar amendment of soil and its effect on crop production of small holder farms in Rasuwa District of Nepal. Int'l J Ag Environ Biotech 2(2):200–215

Hillel D (1992) Out of the earth: civilization and the life of the soil. University of California Press, Berkeley, CA, USA, p 321
Hillel D (2007) Soil in the environment: crucible of terrestrial life. Academic Press, MA, USA, p 320
IBI [International Biochar Initiative] (2012) Biochar. http://www.biochar-international.org/
Lal R (2009) Ten tenets of sustainable soil management. J Soil Water Conserv 64(1):20A-21A
Lal R (2011) Soil degradation and food security in South Asia. In: Lal R, Sivakumar MVK, Faiz SMA, Mustafizur Rahman AHM, Islam KR (eds) Climate change and food security in South Asia. Springer, pp 137–152
Melecis V (2010) Ecosystem services. In: Klavins M, Filho WL, Zaloskins J (eds) Environment and sustainable development. Academic Press of University of Latvia, Riga, pp 29–47
Nair PKR (2011) Agroforestry system and environment quality: introduction. J Environ Qual 40:784–790
Novak J, Ro K, Ok YS, Sigua G, Spokas K, Uchimiya S, BolanN (2015) Biochar's multifunctional role as a novel technology in the agricultural, environmental, and industrial sectors.Chemosphere 142.https://doi.org/10.1016/j.chemosphere.2015.06.066
North D (2016) What is aquaponics and how does it work? https://www.permaculturenews.org/2016/05/30/what-is-aquaponics-and-how-does-it-work/. Accessed 30 Mar 2020
Ohlson K (2014) The soil will save us—how scientists, farmers, and foodies are healing the soil to save the planet. Rodale Inc., NY, USA, p 242p
Penn State (2018) Agroforestry systems may play vital role in mitigating climate change. ScienceDaily. www.sciencedaily.com/releases/2018/02/180201115554.htm. Accessed 1 Feb 2018
Piggot S (ed) (1961) The dawn of civilisation. Thames and Hudson Publisher, London, UK
RAI/TCU (2017) What is regenerative agriculture? Regenerative Agriculture Initiative, California State University, Chico. http://www.csuchico.edu/sustainablefuture/aginitiative/; The Carbon Underground https://thecarbonunderground.org/. Accessed 16 Feb 2017
Regeneration International (2018) What is biochar? http://www.regenerationinternational.org/what-is-biochar/. Accessed 16 May 2018
Revelle R, Suess HE (1957) Carbon dioxide exchange between the atmosphere and ocean and the question of an increase in atmospheric CO2 during the past decade. Tellus 9(Issue I)
Rhodes CJ (2012) Feeding and healing the world: through regenerative agriculture and permaculture. Sci Prog 95(2):101–201. www.scienceprogress.co.uk
Roberts T (2017) Silvopasture: a sustainable way to raise large livestock. Permaculture Research Institute. http://www.regenerationinternational.org/2017/08/13/silvopasture-sustainable-way-raise-large-livestock/. Accessed 13 Aug 2017
Sandor JA, Eash NS (1995) Ancient agricultural soils in the Andes of Southern Peru. Soil Sci Soc Am J 59:170–179
Savory A (1999) Holistic management. A framework for decision making. Island Press, Washington D.C.
Sherchan DP, Karki BK (2006) Plant nutrient management for improving crop productivity in Nepal. Improving plant nutrient management for better farmer livelihoods, food security and environmental sustainability. Procs Regional Workshop Beijing 12–16 December 2005. FAO RAP Publ 2006/27 pp 41–57
Skar SLG, Pinieda-Martos R, Timpe A, et al (2019) Urban agriculture as a keystone contribution towards securing sustainable and healthy development for cities in the future. Blue Green Syst 2(1):1–26. https://iwaponline.com/bgs/. Accessed 30 Mar 2020
Sohi SP (2012) Carbon storage with benefits. Science 338:1034–1035. Publ. on-line by AAAS. http://www.sciencemag.org/cgi/collection/ecology. Accessed 10 Jan 2013
Stockwell R, Bitan E (2011) Future friendly farming: seven agricultural practices to sustain people and the environment. National Wildlife Federation, Reston, VA
Tejwani KG (1994) Agroforestry in India. Oxford and IBH Publishing Co. Pvt. Ltd., New Delhi, India, p 233p

Tilman D, Balze C, Hill J, Befort BL (2011) Global food demand and the sustainable intensification of agriculture. PNAS 108(50):20260–20264
The Royal Society (2009) Reaping the benefits: science and the sustainable intensification of global agriculture. Royal Society, London, UK, p 72
Thomas WL, Sauer CO, Bates M, Mumford L (eds) (1956) Man's role in changing the face of the earth. Chicago University Press, IL, USA
USDA (2020) Hydroponics, United States Department of Agriculture. https://www.nal.usda.gov/afsic/hydroponics. Accessed 30 March 2020
White C (2012) Pasture cropping: a regenerative solution from down under. Acres USA 42(7):5p

Chapter 19
Characterization and Mapping of Soils for Sustainable Management Using Geospatial Techniques: A Case Study of Northeastern Bihar, India

S. K. Reza, S. Mukhopadhyay, D. C. Nayak, T. Chattopadhyay, S. K. Singh, and B. S. Dwivedi

Abstract Poor knowledge on location specific data, mostly on soils, and of situation-specific recommendations has been the causes of failure for most of the agricultural related development schemes that operated in the country in the past. The land resource inventory (LRI) may be filled these gaps by generating data on location specific soil and other land resources. LRI involves systematic surveys of soils on 1:10,000 scale for land use planning scientifically in the GIS platform. The present work was undertaken in Kadwa block, Katihar district in northeastern Bihar, India. Four major landforms like old alluvial plains (9.12%), young alluvial plains (24.46%), meander plains (39.48%) and flood plains (4.61%) were identified after visual interpretation of Indian Remote Sensing Satellite (IRS) R2-LISS-IV data in conjunction with cadastral map. The detailed soil survey was carried out and eight soil series viz. Chauni, Sitalpur, Kumaripur, Asiani, Kaliganj, Sikarpur, Dangi and Mahinagar were identified in different landforms and mapped into 14 soil mapping units (phases of series). Soils developed on meander plains are very deep, moderately well drained, brown to gray, loam to silt loam texture with reddish brown mottles and

---

S. K. Reza (✉) · S. Mukhopadhyay · D. C. Nayak · T. Chattopadhyay
ICAR-National Bureau of Soil Survey and Land Use Planning, DK-Block, Sector-II, Salt Lake, Kolkata, West Bengal, India
e-mail: S.Reza@icar.gov.in

S. Mukhopadhyay
e-mail: subratajee@hotmail.com; Subrata.Mukhopadhyay@icar.gov.in

D. C. Nayak
e-mail: dulalnayak@yahoo.co.in

T. Chattopadhyay
e-mail: taritchattopadhyay@gmail.com; Tarit.Chatopadhyay@icar.gov.in

S. K. Singh · B. S. Dwivedi
ICAR-National Bureau of Soil Survey and Land Use Planning, Maharashtra Nagpur, India
e-mail: skcssri@gmail.com; SK.Singh1@icar.gov.in

B. S. Dwivedi
e-mail: brahma.dwivedi@icar.gov.in

P. K. Shit et al. (eds.), *Soil Health and Environmental Sustainability*,
Environmental Science and Engineering,
https://doi.org/10.1007/978-3-031-09270-1_19

classified as Coarse-loamy, mixed hyperthermic Fluventic Endoaquepts (Kumaripur series), Coarse-loamy, mixed, hyperthermic Aquic Haplustepts (Asiani series) and Fine-loamy, mixed, hyperthermic Aeric Endoaquepts (Kaliganj series). Young alluvial plains soils are very deep, well to somewhat poorly drained, yellowish brown to dark gray, silt loam in texture with brown mottles and classified as Coarse-loamy, mixed, hyperthermic Typic Ustifluvents (Sikarpur series) and Fine-loamy, mixed, hyperthermic Typic Haplustepts (Dangi series). Soils developed on old alluvial plains are very deep, moderately well drained, light brownish gray to dark gray, silt loam to clay loam in texture dark brown mottles and classified as Coarse-loamy, mixed, hyperthermic Typic Haplustepts (Chauni series) and Fine-loamy, mixed, hyperthermic Typic Endoaquepts (Sitalpur series). Flood plains soil are very deep, well drained, light yellowish brown to brown, silt loam surface texture, severe erosion and very frequent flooding and classified as mixed, hyperthermic Typic Ustipsamments (Mahinagar series). Surface soils of the block were grouped into eight soil reaction classes. It was observed that very strongly acid to moderately acidic soils are occupying 50.84% and neutral soils 7.53% of total geographical area (TGA). Organic carbon status (medium to high) occupied 64.10% and available phosphorus is low in 45.82% of TGA. Based on interpretation of soil survey data, the study area is divided into three land capability classes viz. II, III and IV. The results on suitability indicates that crops grown in the study area are moderate to marginally suitable due to coarse texture, fertility and ponding of water for long period limitations. Considering the major problems and potentials four land management units (LMUs) were identified and suggested alternate land use for each LMU of the study area.

Keywords Land resource inventory · Remote sensing · GIS · Mapping · Land evaluation · Crop suitability · Sustainable land use options

19.1 Introduction

Land is a delineable area of the earth's surface and the basic unit of all material production. The limited and inexpansible land resource has to be used very judiciously to meet the expectations of the people and competing demands. Though, India represents only 2.4% of the geographical area but it supports 17.5% of the total world's population (Mythili and Goedecke 2016; Jangir et al. 2020). Globally, present-day global crisis on food, fuel and energy, increasing food prices in the international market, conversion of arable lands to several non-agricultural uses, demand of good quality agricultural land for industry and urbanisation etc., the growing population need to be fed with shrinking and deteriorating land and water resources. Therefore, a systematic survey of the land resources and their mapping are essential for managing the resources in a sustainable way (Sarkar 2011; Supriya et al. 2019).

Soil mapping is basically an inference process. Soil is described as a function of climate, organisms, relief, parent material and time, referred to as CLORPT (Jenny 1941) and interactions among these soil-forming factors is potentially important because it is a possible source to understand soil pattern (McBratney et al. 2003) and thereby assist in mapping the distribution of various soils. In small areas where climate, parent material, and time are almost similar, the major factors influencing the soil properties can be attributed to variation in relief and flora and fauna (Dobos et al. 2000; Srivastava and Saxena 2004). Depending upon the requirement of the users, soil mapping can be done at various scales, such 1:250,000, 1:1,000,000 or smaller scale, medium scale like 1:100,000, 1:50,000, and large scale like 1:25,000, 1:10,000 or larger scale (Srivastava and Saxena 2004; Sharma et al. 2019). Soil survey has been carried out by different sources and at various scales for the Bihar state. However, to increase the, productivity of crops and other farm produce at block level, detailed information on soil landscape features, soils, land use, etc. are essential for the overall development of the region.

The satellite remote-sensing data products are widely accepted for small (country level) and medium scale (district level) soil mapping (Soil Survey Division Staff 2000). But, their utility is limited for large scale soil mapping due to the large resolution of satellite data. Previously large scale soil mapping was mostly done with conventional methods. These were time consuming, expensive with low repetitive value especially in hilly and mountainous regions, wetlands and other problematic areas (Adam et al. 2010). In the course of time, with advent of high spatial, spectral and radiometric resolution satellite data/remote sensing data along with stereo capabilities and digital elevation model, new studies have been undertaken to characterize soils at large scale through the physiography-land use-soil relationship. The technique of large scale soil mapping (1:12,500 scale) was discussed by Srivastava and Saxena (2004) in a basaltic terrain with a physiography-land use (PLU) approach and differentiated soil types using topographic information available in the Survey of India toposheet and land use/land cover information from IRS-1C PAN merged data. In a basaltic terrain with a PLU approach using landform, slope, and land use/land cover, large scale (1:5000 scale) soil map was also prepared by Nagaraju et al. (2014).

The entire state of Bihar has been mapped at 1:250,000 scale with soil series association as soil mapping units (Haldar et al. 1996). It provided information on physiographic units and soil information at smaller scale. Land resource inventory (LRI) on 1:10,000 scales provides adequate information on characteristics and spatial distribution soils and properties of soils that support land management in sustainable manner, that includes possibilities of irrigation, control of soil erosion, management of soil fertility and choice of crops (van de Wauw et al. 2008; Seid et al. 2013; Singh et al. 2016). After characterization from soil resources the land evaluation is essential to know the suitability of a particular crop or a group of crops. For evaluation of capability and suitability of the soils for particular land use, the detailed studies on specific soil-related constraints like soil fertility, available water content, degradation hazards and soil erosion are necessary (AbdelRahman et al. 2016; Fekadu et al. 2020; Mandal et al. 2020).

Further, this information's are pre-requisite for developing a land use plan for a block. Land use planning involves right land use and right technology in site-specific mode may be one of the options that may help in meeting out the demand of food as well as in preserving the quality of land for future. LRI on 1:10,000 scales is helped in developing such site-specific information, which paves the way for applying right land use, right technology at the right place. Hence, the present study is proposed as an attempt to supplement the information gap in the Kadwa block, Katihar district, Bihar especially in characterization and mapping soil resources at 1:10,000 scale, modelling soil physiographic relation, finding crop suitability, land-use options and conservation of natural resources.

19.2 Study Area

The area selected for investigation belongs to the Kadwa block, Katihar district, northeastern Bihar, India extended to 25° 30′–25° 47′ N latitude and 87° 35′–87° 55′ E longitude covering an area of 340.47 km^2 (Fig. 19.1). It is bounded by Baisi and Dagarua blocks of Purnia district in the north, south by Azamnagar block of Katihar district, east by Balrampur and Barsoi blocks of Katihar district, and west Hasanganj and Dandkhora blocks of Katihar district. The topography of the study area is more or less flat topography (1–3% slope) with the slope gradient towards south. In other words areas towards north are at higher elevation than those at south. The regional slope takes a slight tilt from west to east. The entire Kadwa block is underlain by

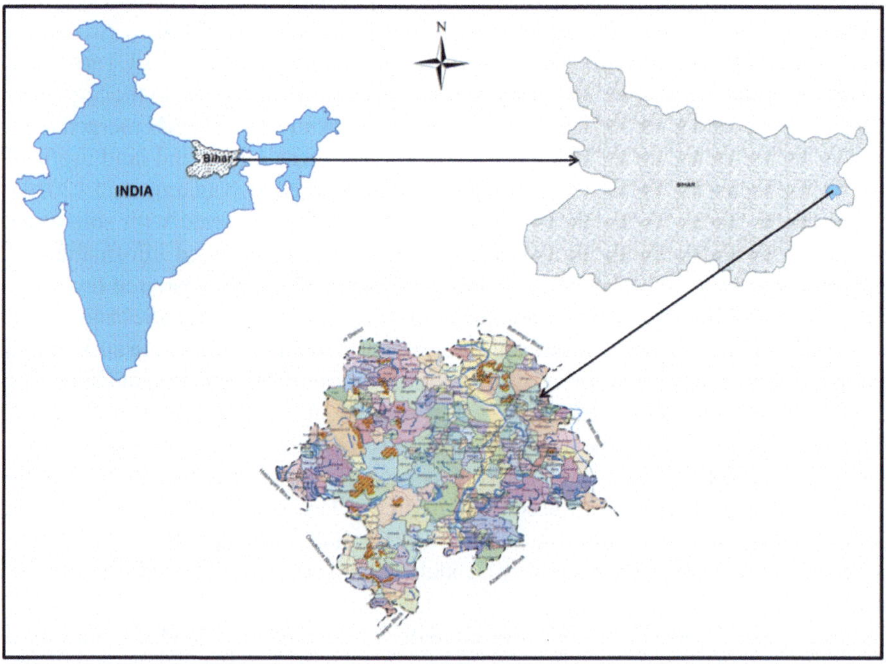

Fig. 19.1 Location map of the study area

thick unconsolidated sediment of Quaternary period (GSI 1998). Climate is moderate during the winter and hot in summer, the maximum mean temperature is 43 ºC and minimum mean temperature is 8 ºC. The mean annual rainfall is 2194 mm and the majority (about 85%) of rainfall is from south-west monsoon during the months of July to September (Reza et al. 2021). The mean summer and mean winter soil temperature difference in the block is more than 5 °C; hence, the soil temperature class is "hyperthermic". The soil moisture regime is "aquic" and "ustic" (Soil Survey Staff 2003). The area belongs to agroecological sub-region (AESR) 13.1, North Bihar and Avadh plains, hot dry to moist subhumid transitional ecological sub-region with deep, loamy alluvium-derived soils. Natural vegetation of the block consists of trees, shrubs, grasses and weeds. The major tree species are Mango (*Mangifera indica*), Jamun (*Syzigium cumini*), Arjun (*Terminalia arjuna*), Date palm (*Phoenix sylvestris*), Neem (*Azadirachta indica*), *Babul* (*Acacia nilotica*), Aswatha (*Ficus religiosa*), Ber (*Zizyphus mauritiana*), Sajina (*Moringa oleifera*), Bamboo (*Bambusa sp.*) etc.

19.3 Materials and Methods

19.3.1 Preparation of Base Maps

Toposheets of Survey of India on 1:50,000 scale, IRS-R2 LISS-IV data (5.8 m resolution) of 9th November 2013 and 13th February 2014 (Fig. 19.2) were georeferenced using WGS 84 datum, Universal Transverse Mercator (UTM) projection and ground control points (GCPs) (Nagaraju et al. 2014). The village map of the block was scanned and co-registered using orthorectified LISS-IV data as a reference. The rasterized village map was digitized on-screen after the geo-referencing. Land use/land cover, landform analysis was carried out by onscreen visual interpretation using IRS-P6 LISS-IV data in ArcGIS software. A LEU layer was prepared by integrating the landform, slope, and land-use/land cover layers in a GIS environment (ArcGIS ver. 10.5) and LEU units are relatively homogeneous in terms of the main

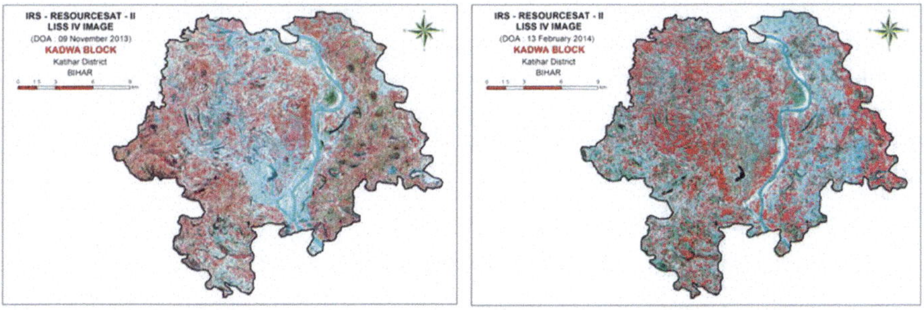

Fig. 19.2 IRS-P6 LISS-IV satellite data (9th November 2013 and 13th February 2014) of Kadwa block

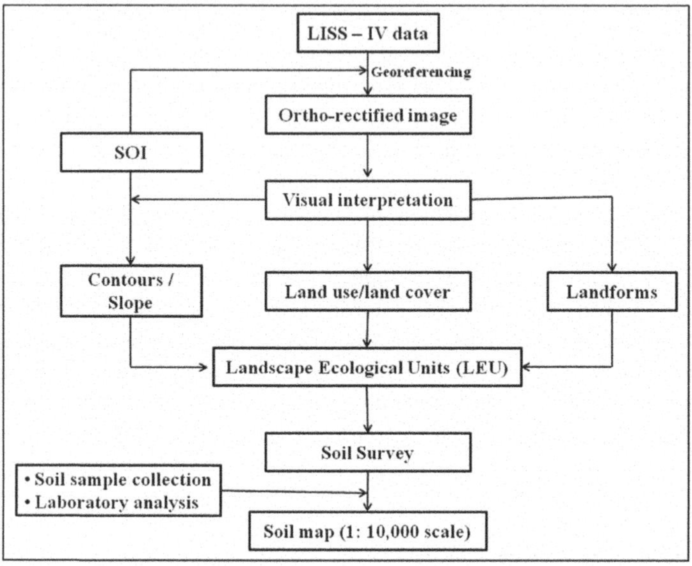

Fig. 19.3 Flowchart of the methodology

factors of soil formation and typical predictors of soil characteristics and used as a base map for ground truth verification and soil mapping. The flow chart showing the methodology of detailed soil mapping using LISS-IV data derived products was presented in Fig. 19.3.

19.3.2 Ground-Truth Verification

The identified different landform units, slope and present land use/land cover of the study area was traversed and correlated with image interpretation units. The originally derived boundaries of base maps were verified and corrected wherever necessary. Representative sites on each physiographic units using handheld Global Positioning System (GPS) were selected to understand the soil variability in the study area and profiles observations have taken as per variation in phases and were described for site and soil characteristics such as depth, colour (matrix and mottle), boundary, structure, texture, cutans, etc. following the guidelines for field soil descriptions (Soil Survey Staff 1995).

19.3.3 Soil Sampling and Analysis

The soil samples collected during the soil survey fieldwork were air dried at room temperature in the laboratory. The dried soil samples were grounded using a wooden pestle and mortar, sieved through a 2 mm sieve. After properly labeled the samples were stored in polythene bags for laboratory analysis. Standard procedures were used for analysis of soil physical and chemical parameters in the laboratory. International pipette method was used for particle size analysis. A combined glass-calomel electrode was used to determine the pH of aqueous suspensions (1:2.5 soil/solution ratio). Organic carbon (OC) was determined using the wet digestion method of Walkley and Black (1934) whereas, available nitrogen (N) was measured by the alkaline permanganate method as described by Subbiah and Asija (1956). Bray II method (Bray and Kurtz 1945) was used for estimation of available phosphorus (P). Cation exchange capacity (CEC) of soil was determined as per the standard procedure outlined by Jackson (1976). For determination of exchangeable cations [calcium (Ca), potassium (K), and magnesium (Mg)] soil samples were extracted with 1 M ammonium acetate (NH_4OAc) (pH 7.0). K content was estimated by flame photometry (Rich 1965), while Ca and Mg were determined in ethylene diamine tetra acetic acid (EDTA) titration. 1 N potassium chloride (KCl) solution was used for extracted of exchangeable Al and titrated the aliquot with 0.1 N sodium hydroxide (NaOH) solution. Soils of the study area were classified as per guidelines outlined by Keys to Soil Taxonomy (Soil Survey Staff 2014).

19.3.4 Soil Classification

The following criteria (Soil Survey Staff 2014) are used to classify the soils of the study area.

Order: Inceptisols-presence cambic (Bw) horizon and structural development in subsurface horizon and Entisols—soils that do not show any profile development and no diagnostic horizons, and most are mostly unconsolidated sediments with little or no alteration from their parent materials.

Suborders: Aquepts—Inceptisols that have a aquic soil moisture regime, Ustepts—Inceptisols that have a ustic soil moisture regime, Aquents—Entisols that have a aquic moistire regime, Psamments—Other Entisols that doesn't fit in any suborder and within the particle-size control section have a texture class of loamy fine sand or coarser in all layers (sandy loam lamellae are permitted), Fluvents—Entisols that show decrease in organic carbon content irregular within a depth of 25 cm and either a depth of 125 cm below the mineral soil surface.

Great groups: Haplustepts—Ustepts, which doesn't meet the requirement of the great group of Ustepts suborder; Endoaquepts—Aquepts with endo-saturation; Endoaquents—Aquents with endo-saturation; Ustifluvents—other Fluvents that having an ustic soil moisture regime and Ustipsamments—Psamments, which doesn't meet the requirement of the great group of Psamments suborder.

Subgroups: Typic Haplustepts—Haplustepts, which doesn't meet the requirement of the subgroup of Haplustepts greatgroup; Typic *Endoaquepts—Endoaquepts*, which doesn't meet the requirement of the subgroup of *Endoaquepts* great group; Fluventic Endoaquepts—Endoaquepts that show decrease in organic carbon content irregular within a depth of 25 cm and either a depth of 125 cm below the mineral soil surface.; Aquic Haplustepts—Haplustepts great groups that have redoxmorphic characteristics with chroma of 2 or less in one or more horizons within 75 cm of the mineral soil surface and also some time in normal years having aquic conditions; Aeric Endoaquepts—Endoaquepts that have Chroma value 2 or more in one or more than one horizons between the A or Ap horizon and below the mineral soil surface of a depth of 75 cm; Typic Ustifluvents- Ustifluvents, which doesn't meet the requirement of the subgroup of Ustifluvents great group and Typic Ustipsamments—Ustipsamments, which doesn't meet the requirement of the subgroup of Ustipsamments great group.

19.3.5 Development of Soil Mapping Legend

Phases of the soil series were considered as mapping units in the present study. The soil series may be defined is a group of soils or polypedons that have similar arrangement and in differentiating characteristics in horizons and sets of properties with narrow range (Soil Survey Division Staff 2000). Soil depths, surface texture, slope, erosion and flooding criteria were considered for defining the soil phases within a soil series (Nagaraju et al. 2014). The pedons were studied during the soil survey work and correlated for identification of soil series in each major landform. The extension of soil series were verified using the diagnostic soil characteristics from soil profile and augur observations. A soil map were prepared at 1:10,000 scale showing soil series and their phases and the soil legend code developed indicates the name of the series followed by surface texture, slope, erosion and flooding (Singh et al. 2016).

19.3.6 Land Evaluation

Land capability classification (LCC) was used to find out the general capability of soil resources of an area which was suitable for agricultural, forestry and other uses. Based on their limitations to field crops and the way they respond to management the mapping units were grouped into various capability units. The capability classes were identified based on their inherent soil characteristics, external land features and environmental factors that limit the use to of land (AIS&LUS 1970). The characteristics used to group the land resources identified in the study area are: texture, slope, erosion and drainage. In the capability system, mapping units are generally grouped at three levels- capability class, sub-class and unit. The broadest groups of capability classes were designed by Roman numerals I to VIII and increasing the numerals indicate progressively greater limitations and narrow choice for practical use. The eight classes Class I–VIII were used in the system whereas, Class I–IV indicates these were suitable for cultivation with increasing limitations. They are capable of producing commonly cultivated crops of the region under good management. Classes V to VII are suited to adopted native plans, pasture or forestry. Class VIII was not suitable for agriculture as well as for silviculture.

Capability sub-classes were described based on the limitations observed within the capability classes. There are designed by adding a lower case letter like e, s, w or c to the class numeral. For example in sub class IVe, the letter e shows that the main hazard in class IV land is the risk of erosion. Similarly, the symbol 'w' indicates drainage or wetness as a limitation for plant growth; the symbol 's' indicate root zone limitations and 'c' indicates climate or rainfall with short growing period.

Some important soil characteristics namely soil texture, depth, available water retention capacity of soils, salinity, infiltration and permeability were used for soil irrigability classification. In addition to soil irrigability class, land irrigability classification was made by taking the consideration the quantity of water as well as quality, requirement of drainage, topography and economic condition. Criteria for classes are qualitatively defined in such a way that a soil can qualify for only one class (AIS&LUS 1970). The most limiting property is determined for classification. For example, the soil may have all the properties of the most desirable class except one, but due to one undesirable property it is assigned to a lower class. Irrigability classes are further divided in su-bclasses to indicate the dominant limitations such as 's' for soil limitation, 'd' for drainage limitation and 't' for topography limitation.

Evaluation of soil site suitability has been done by maximum likelihood method (Sys et al. 1993; Naidu et al. 2006). The suitability criteria included climatic attributes (c) viz. rainfall and temperature; wetness aspect (w) viz. drainage and flooding; physical condition (s) viz. surface texture, rooting condition as soil depth (d) and soil fertility factor (f) viz. pH, OC, apparent CEC, base saturation and sum of cations. Soils have been evaluated as highly suitable (S1), moderately suitable (S2), marginally suitable (S3), temporarily not suitable (N1) and permanently not suitable (N2) classes.

19.4 Results and Discussion

19.4.1 Land Use/Land Cover

IRS-P6 LISS-IV data (09th November 2013 and 13th February 2014) were interpreted and four land use/land cover classes were identified. The land use data (Fig. 19.4) indicates that about 77.36% of total geographical area (TGA) of the

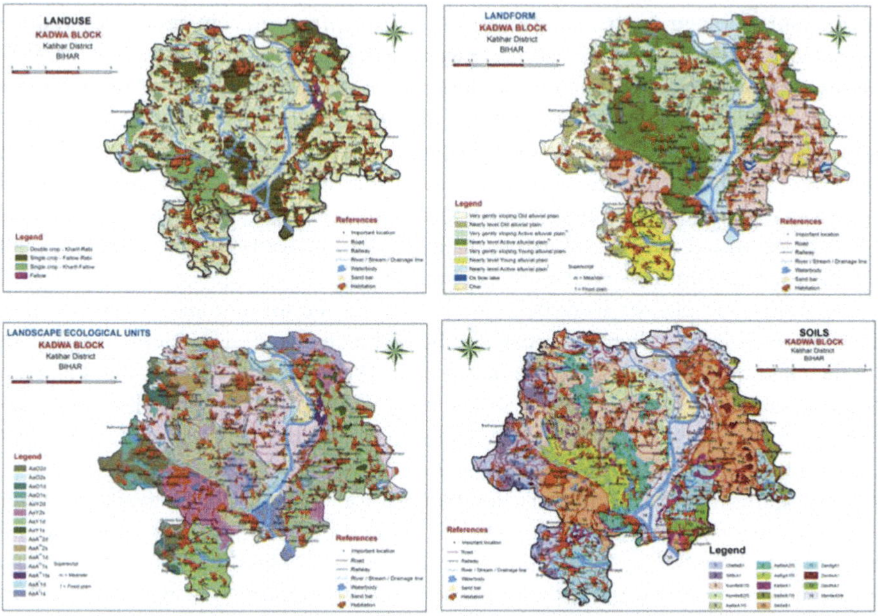

Fig. 19.4 Land use/land cover, landform, landscape ecological units and soil maps of Kadwa block

study area is under agriculture, 0.31% is under fallow and 22.33% TGA under miscellaneous viz. habitation, ox bow lake, sand bar and river system.

19.4.2 Landform and Landscape Ecological Units (LEUs)

Visual interpretation of LISS-IV data indicated that the block was characterized into old alluvial plain, active alluvial plain, young alluvial plain, ox bow lake and char. The major landforms were further subdivided based on elevation, land-uses and other local features. The old alluvial plain were sub-divided into very gently sloping and nearly level old alluvial plain, young alluvial plain into very gently sloping and nearly level young alluvial plain and active alluvial plain into very gently sloping and nearly level active alluvial plain (meander plain) and nearly level active alluvial plain (flood plain) (Fig. 19.4). The landform, slope and land-use/land-cover maps were integrated in ArcGIS and LEU map was prepared. Based on integration, 16 LEU units were delineated in the study area and the characteristics of each LEU unit was described (Table 19.1) and mapped (Fig. 19.4). On the alluvial plain, eight LEU units were identified based on two slope classes (0–1 and 1–3%) and two land use/land cover classes (single crop and double crop). Five LEU units were identified on the meander plain with two slope classes (0–1 and 1–3%) and three land use/land cover classes (single crop, double crop and fallow). Three LEU units were identified

Table 19.1 Landscape ecological units (LEU) of Kadwa block

Landscape ecological unit (LEU)	Area (ha)	TGA (%)
Very gently sloping old alluvial plain under double crop (AaO2d)	1824	5.36
Very gently sloping old alluvial plain under single crop (AaO2s)	48	0.14
Nearly level old alluvial plain under double crop (AaO1d)	548	1.61
Nearly level old alluvial plain under single crop (AaO1s)	684	2.01
Very gently sloping young alluvial plain under double crop (AaY2d)	4440	13.04
Very gently sloping young alluvial plain under single crop (AaY2s)	1916	5.63
Nearly level young alluvial plain under double crop (AaY1d)	1679	4.93
Nearly level young alluvial plain under single crop (AaY1s)	295	0.86
Very gently sloping active alluvial plain under double crop (AaA^m2d)	4757	13.97
Very gently sloping active alluvial plain under single crop (AaA^m2s)	962	2.83
Nearly level active alluvial plain under double crop ((AaA^m1d))	5299	15.56
Nearly level active alluvial plain under single crop ((AaA^m1s))	2317	6.81
Nearly level active alluvial plain under fallow (AaA^m1fa)	105	0.31
Nearly level active alluvial plain plain under double crop (AaA^f1d)	734	2.15
Nearly level active alluvial plain under single crop (AaA^f1s)	838	2.46
Ox bow lake (AaOx)	127	0.38
Total cultivated area	26,573	78.05
Miscellaneous	7474	21.95
Total area	34,047	100

with two land use/land cover classes (single crop and double crop) and single slope class (0–1%) and one LEU unit in ox bow lake.

19.4.3 Soil-Landform Relationship

Soils of old alluvial plains and meander plains are very deep and having two genetic horizons A–B with clear smooth and gradual smooth boundary in surface and subsurface horizons, respectively (Table 19.2). The pedons showed difference in surface and subsurface matrix colour. The surface horizon colours are dark yellowish brown (10YR 4/4) and dark grey (2.5Y 4/0) in old alluvial plains (Chauni and Sitalpur series), and light brownish grey (10YR 6/2) and greyish brown (10YR 5/2) in meander plains (Kumaripur, Asiani and Kaliganj series). Whereas, the subsurface colour for old alluvial plains and meander plains varied from brown (10YR 5/3) to dark grey (2.5Y 4/0) and brown (10YR 5/3) to dark grey (10YR 4/1), respectively. Soils of young alluvial and flood plains deep to very deep and having A horizon except Dangi series of young alluvial plains (Table 19.2) with clear smooth and gradual smooth boundary in surface and subsurface horizons, respectively indicating the

Table 19.2 Morphological properties of soils

Depth (cm)	Horizon	Boundary	Matrix colour (moist)	Texture	Structure	Mottle colour	Roots
Old alluvial plains							
Chauni series: *Coarse-loamy, mixed, hyperthermic Typic Haplustepts*							
0–13	Ap	cs	10YR 4/4	Silt loam	m2sbk	–	cf
13–31	Bw1	gs	10YR 4/3	Silt loam	m2sbk	7.5YR 4/6	fvf
31–55	Bw2	gs	10YR 4/3	Silt loam	m2sbk	7.5YR 4/4	fvf
55–85	Bw3	gs	10YR 4/3	Silt loam	m2sbk	7.5YR 4/4	–
85–115	BC1	gs	10YR 4/4	Loam	f1sbk	–	–
115–155	BC2	–	10YR 5/3	Sandy loam	f1sbk	–	–
Sitalpur series: *Fine-loamy, mixed, hyperthermic Typic Endoaquepts*							
0–13	Ap	cs	2.5Y 4/0	Silty clay loam	m2sbk	–	fm
13–45	Bw1	gs	2.5Y 4/1	Clay loam	c2sbk	7.5YR 5/8	ff
45–68	Bw2	gs	2.5Y 4/2	Loam	f1sbk	7.5YR 5/8	fvf
68–110	Bw3	–	2.5Y 4/2	Sandy loam	f1sbk	7.5YR 5/8	–
Meander plains							
Kumaripur series: *Coarse-loamy, mixed hyperthermic Fluventic Endoaquepts*							
0–26	Ap	cs	10YR 6/2	Silt loam	m2sbk	–	cm
26–54	Bw1	gs	10YR 5/2	Silt loam	c2sbk	5YR 4/4	ff
54–79	Bw2	gs	10YR 4/2	Silt loam	c2sbk	5YR 3/4	ff
79–115	Bw3	–	10YR 4/1	Silt loam	c2sbk	5YR 3/4	ff
Asiani series: *Coarse-loamy, mixed, hyperthermic Aquic Haplustepts*							
0–17	Ap	cs	10YR 5/2	Silt loam	m2sbk	–	cf
17–46	Bw1	gs	10YR 5/1	Silt loam	m2sbk	7.5YR 5/8	fvf

(continued)

Table 19.2 (continued)

Depth (cm)	Horizon	Boundary	Matrix colour (moist)	Texture	Structure	Mottle colour	Roots
46–77	Bw2	gs	10YR 5/3	Silt loam	f1sbk	7.5YR 4/4	fvf
77–112	BC	–	10YR 5/2	Loam	f1sbk	7.5YR 4/4	
Kaliganj series: *Fine-loamy, mixed, hyperthermic Aeric Endoaquepts*							
0–18	Ap	cs	10YR 6/2	Silty clay	m2sbk	–	cm
18–40	Bw1	gs	10YR 5/1	Silty clay loam	m2sbk	7.5YR 4/6	fvf
40–70	Bw2	gs	10YR 4/3	Silty clay	m2sbk	7.5YR 4/4	fvf
70–110	Bw3	–	10YR 4/3	Silt loam	m2sbk	7.5YR 4/4	–
Young alluvial plains							
Sikarpur series: *Coarse-loamy, mixed, hyperthermic Typic Ustifluvents*							
0–16	Ap	cs	10YR 4/1	Silt loam	m2sbk	–	mf
16–38	C1	gs	10YR 4/4	Sandy loam	massive	–	ff
38–80	C2	gs	10YR 4/4	Silt loam	m2sbk	–	ff
80–121	2C3	gs	10YR 4/3	Silt loam	m2sbk	5YR 2.5/2	–
121–176	3C4	–	10YR 4/4	Silt loam	m2sbk	–	–
Dangi series: *Fine-loamy, mixed, hyperthermic Typic Haplustepts*							
0–15	Ap	cs	10YR 5/1	Clay loam	m2sbk	-	cm
15–45	Bw1	gs	10YR 5/3	Silty clay loam	m2sbk	7.5YR 4/4	fvf
45–72	Bw2	gs	10YR 5/3	Silty clay loam	m2sbk	7.5YR 3/4	fvf
72–109	Bw3	gs	10YR 4/4	Silt loam	f1sbk	7.5YR 4/4	-
109–151	BC	-	10YR 4/4	Silt loam	f1sbk	7.5YR 3/4	-
Flood plains							
Mahinagar series: *mixed, hyperthermic Typic Ustipsamments*							

(continued)

Table 19.2 (continued)

Depth (cm)	Horizon	Boundary	Matrix colour (moist)	Texture	Structure	Mottle colour	Roots
0–20	Ap	gs	10YR 4/4	Silt loam	f1sbk	–	–
20–54	C1	gs	10YR 6/1	Loamy sand	sg	–	–
54–90	C2	gs	10YR 6/1	Loamy sand	sg	–	–
90–150	C3	-	10YR 5/2	Loamy sand	sg	–	–

these soils were developed under fluvial process. The soils of young alluvial plains showed difference in surface and subsurface matrix colour. The surface horizons are dark grey (10YR 4/1) (Sikarpur series) and grey (10YR 5/1) (Dangi series) with subsurface colour dark yellowish brown (10YR 4/4) to brown (10YR 5/3). The soils of flood plains (Mahinagar series) are light yellowish brown (surface) and brown (subsurface) colour with brown mottles.

Sitalpur series of old alluvial plains, and Kumaripur series and Kaliganj series of mender plains showed grey matrix colour with chroma 0–2, which indicates that these soils were under prolonged submergence and subsequently developed under reducing conditions during flooding. The low chroma of soils judged the severity of gleying due to poor drainage conditions high groundwater table in lower topographical position (Stoop and Eswaran 1985). When these soils become dry, the reduced iron (Fe^{3+}) is oxidized and precipitates by releasing of H^{+} ions to acidify and disintegrate the clay. Under saturated condition for a long time these soils developed distinctive gley horizons resulting from oxidation and reduction process and has iron and manganese mottles or streaks in B horizons due to slow diffusion process (Ponnamperuma 1972, 1985). These soils are also known as hydromorphic soils and gleization as the major pedogenic process operating for their developments (Khan et al. 2012).

The variation of soil properties with depths indicates the dominant soil processes operating over the course of profile development. In initial stage, the OM input and mineral weathering occurs in weakly developed soils (Entisols and Inceptisols). Soils of the study area varied to a great extent with depths due to different pedogenic processes. Kumaripur series of meander plains (coarse-loamy, mixed hyperthermic Fluventic Endoaquepts), Sikarpur series of young alluvial plains (coarse-loamy, mixed, hyperthermic Typic Ustifluvents) and Mahinagar series of flood plains (mixed, hyperthermic Typic Ustipsamments) showed irregular distribution of OC, clay content and CEC (Table 19.3). Such irregular distribution could be attributed to the pedogenic processes namely, mass movement, periodic flooding and deposition of alluvium brought down by water during different fluvial cycles (Huggett et al. 1975, 1976). However, soils of Chauni series (coarse-loamy, mixed, hyperthermic Typic Haplustepts) and Sitalpur series (fine-loamy, mixed, hyperthermic

Table 19.3 Physical and chemical properties of soils

Depth (cm)	Horizon	Sand (%)	Silt (%)	Clay (%)	pH H_2O (1:2.5)	OC (%)	Exchangeable cations				CEC
							Ca	Mg	Na	K	
							cmol(p+) kg^{-1}				
Old alluvial plains											
Chauni series: *Coarse-loamy, mixed, hyperthermic Typic Haplustepts*											
0–13	Ap	7.0	78.2	14.8	5.5	0.52	1.2	0.6	0.1	0.2	3.5
13–31	Bw1	3.4	81.6	15.0	5.9	0.42	3.1	0.9	0.1	0.1	6.8
31–55	Bw2	4.0	76.0	20.0	5.8	0.34	3.8	2.4	0.1	0.1	9.0
55–85	Bw3	4.1	79.9	16.0	6.3	0.25	0.7	0.4	0.1	0.1	3.8
85–115	BC1	51.3	36.6	12.1	7.4	0.07	0.8	0.7	0.1	0.1	2.7
115–155	BC2	54.6	37.2	8.2	7.1	0.03	0.7	0.5	0.1	0.1	2.4
Sitalpur series: *Fine-loamy, mixed, hyperthermic Typic Endoaquepts*											
0–13	Ap	19.6	49.3	31.1	5.4	0.95	1.7	1.2	0.2	0.4	6.6
13–45	Bw1	21.2	46.1	32.7	6.7	0.30	3.5	1.6	0.1	0.3	6.7
45–68	Bw2	30.0	47.5	22.5	7.3	0.18	2.2	1.0	0.2	0.2	4.7
68–110	Bw3	42.0	40.5	17.5	7.5	0.12	2.0	1.5	0.1	0.2	4.0
Meander plains											
Kumaripur series: *Coarse-loamy, mixed hyperthermic Fluventic Endoaquepts*											
0–26	Ap	14.2	69.6	16.2	6.0	0.67	1.7	1.2	0.2	0.4	7.4
26–54	Bw1	23.4	63.8	12.8	6.6	0.31	1.5	1.6	0.1	0.3	4.9
54–79	Bw2	21.7	61.1	17.2	6.9	0.18	2.2	1.0	0.2	0.2	6.0
79–115	Bw3	29.7	53.5	16.8	6.7	0.32	2.0	1.5	0.1	0.2	5.8

(continued)

Table 19.3 (continued)

Depth (cm)	Horizon	Sand (%)	Silt (%)	Clay (%)	pH H_2O (1:2.5)	OC (%)	Exchangeable cations				CEC
							Ca	Mg	Na	K	
							cmol(p+) kg^{-1}				
Asiani series: *Coarse-loamy, mixed, hyperthermic Aquic Haplustepts*											
0–17	Ap	18.1	59.2	22.7	5.9	0.68	3.4	1.0	0.1	0.2	6.9
17–46	Bw1	19.5	62.9	17.6	7.2	0.23	2.9	0.8	0.1	0.1	5.3
46–77	Bw2	13.3	71.7	15.0	7.3	0.18	2.4	1.1	0.1	0.1	4.9
77–112	BC	37.2	49.1	13.7	7.1	0.10	2.0	1.0	0.1	0.1	4.3
Kaliganj series: *Fine-loamy, mixed, hyperthermic Aeric Endoaquepts*											
0–18	Ap	1.5	55.6	42.9	5.5	0.52	1.2	0.6	0.1	0.2	3.5
18–40	Bw1	3.0	60.4	36.6	5.9	0.32	3.1	0.9	0.1	0.1	6.8
40–70	Bw2	8.8	50.3	40.9	5.8	0.31	3.8	2.4	0.1	0.1	9.0
70–110	Bw3	1.4	66.5	32.1	6.3	0.25	0.7	0.4	0.1	0.1	3.8
Young alluvial plains											
Sikarpur series: *Coarse-loamy, mixed, hyperthermic Typic Ustifluvents*											
0–16	Ap	21.8	68.8	9.4	4.4	0.43	1.6	0.7	0.4	0.4	4.7
16–38	C1	47.5	46.6	5.9	6.2	0.13	1.7	0.8	0.3	0.1	4.0
38–80	C2	26.7	66.4	6.9	6.3	0.12	2.0	0.9	0.2	0.2	4.5
80–121	2C3	9.5	76.3	14.2	6.3	0.18	2.4	1.4	0.2	0.3	6.0
121–176	3C4	26.2	69.4	4.4	6.8	0.07	1.3	1.0	0.5	0.2	3.9

(continued)

Table 19.3 (continued)

Depth (cm)	Horizon	Sand (%)	Silt (%)	Clay (%)	pH H_2O (1:2.5)	OC (%)	Exchangeable cations				CEC
							Ca	Mg	Na	K	
							cmol(p+) kg^{-1}				
Dangi series: *Fine-loamy, mixed, hyperthermic Typic Haplustepts*											
0–15	Ap	22.0	49.0	29.0	6.0	0.67	4.9	1.0	0.3	0.3	8.5
15–45	Bw1	16.1	55.2	28.7	7.0	0.30	4.7	1.1	0.2	0.3	7.4
45–72	Bw2	10.5	58.9	30.6	7.1	0.28	4.6	1.5	0.1	0.2	7.4
72–109	Bw3	5.5	73.7	20.8	6.8	0.18	4.0	1.4	0.2	0.1	7.1
109–151	BC	9.3	71.5	19.2	6.8	0.16	3.0	1.3	0.2	0.2	6.0
Flood plains											
Mahinagar series: *mixed, hyperthermic Typic Ustipsamments*											
0–20	Ap	31.2	54.2	14.6	5.5	0.17	0.16	0.63	0.24	0.10	6.9
20–54	C1	80.0	17.5	2.5	5.4	0.26	0.32	0.32	0.18	0.05	3.2
54–90	C2	83.3	12.7	4.0	4.7	0.30	0.48	0.79	0.14	0.04	3.0
90–150	C3	78.3	20.7	4.0	5.6	0.13	0.32	0.63	0.20	0.03	3.8

Typic Endoaquepts) in old alluvial plains, Asiani series (coarse-loamy, mixed, hyperthermic Aquic Haplustepts) and Kaliganj series (fine-loamy, mixed, hyperthermic Aeric Endoaquepts) in meander plains and Dangi series (fine-loamy, mixed, hyperthermic Typic Haplustepts) in young alluvial plains shows systematic variation with depth may be due to presence of uniform parent materials from where soil profiles developed and reflect pedogenesis (Table 19.3).

19.4.4 Soil Mapping

Soil is an open system and its properties are related to the functions operating in the system (Jenny 1941). With the changing in the system the soil properties change and directly depend to soil formation factors of which was expressed as follows:

$$S = (cl, o, r, p, t, \ldots)$$

where, S denotes soil property; cl, climate (rainfall and temperature); o, organisms (flora and fauna); r, relief; p, parent material; and t, time or age.

The present study area is almost similar in climate, parent material and time or age, the soil properties varies mainly depends on variation in relief or topography (r) and flora and fauna biosphere organisms (o). Hence, the LEU concept was used in the study area for mapping soils. The morphological characteristics observed during soil survey and analyzed soil properties, the soils were classified up to family level as per the Keys to Soil Taxonomy (Soil Survey Staff 2014). Eight soil series have been identified and mapped on 1:10,000 scale with 14 soil mapping units (phases of series) (Fig. 19.4). The brief description of the soil series identified along with their taxonomic classification is given in the mapping legend (Table 19.4).

19.4.5 Soil Survey Interpretation

Soil maps and other its interpretation maps are the ultimate products of soil survey. They provide valuable information on various aspects like physiography/landform, geology, vegetation, soils, drainage, etc. and are useful to the planners, administrators and other user agencies. Land use/agricultural planning of any particular area are largely based on soil resource interpretations (site characteristics and soil properties).

Following the criteria outlined in the Field Manual (Sehgal et al. 1987) and Hand Book of Agriculture (Takkar 2009), various thematic maps such as surface texture, slope, drainage, soil reaction (pH), OC, available N, P and K have been prepared. The site characteristics and the soil properties of the surface soils of each soil phases have been considered for the preparation of different thematic maps.

Table 19.4 Soil series and phases of Kadwa block

Landform	LEU map unit	Soil series	Soil map unit	Mapping legend	Brief description of soil series	Area (ha)	TGA (%)
Very gently sloping old alluvial plain	AaO2d	Chauni	1	Cha6eB1	Very deep, moderately well drained, yellowish brown to dark gray, medium texture soils on very gently sloping old alluvial plain with silt loam surface texture and slight erosion (*Coarse-loamy, mixed, hyperthermic Typic Haplustepts*)	1872	5.50
Nearly level old alluvial plain	AaO1d	Sitalpur	2	Sit6kA1	Very deep, poorly drained, dark gray to dark grayish brown, fine texture soils on nearly level old alluvial plain with silty clay surface texture and slight erosion (*Fine-loamy, mixed, hyperthermic Typic Endoaquepts*)	1232	3.62

(continued)

19.4.5.1 Surface Texture

The particle-size distribution like sand, silt and clay relatively expressed as soil texture and one of the most important soil physical variable that governing nearly all properties of soils (Zhai et al. 2006; Adhikari et al. 2009). Soil texture affects water availability and retention in soil, and transform (Katerji and Mastrorilli 2009; Reza et al. 2016), leaching and erosion potential (Reza et al. 2011, 2018), plant nutrient storage (Kettler et al. 2001), and organic matter dynamics (Kong et al. 2009), it plays

Table 19.4 (continued)

Landform	LEU map unit	Soil series	Soil map unit	Mapping legend	Brief description of soil series	Area (ha)	TGA (%)
Very gently sloping active alluvial plain (meander pain)	AaA^{m}2d	Kumaripur	3	Kum6eB1f3	Very deep, well drained, brown to dark yellowish brown, medium texture soils on very gently sloping meander plain with silt loam surface texture, slight erosion and frequent flooding (*Coarse-loamy, mixed hyperthermic Fluventic Endoaquepts*)	3361	9.87
	AaA^{m}2s		4	Kum6eB2f3	Same as Kumaripur series with moderate erosion	1225	3.60
Nearly level active alluvial plain (meander pain)	AaA^{m}1d	Asiani	5	Asi6eA1f3	Very deep, moderately well drained, brown to gray, medium texture soils on nearly level meander plain with silt loam surface texture, slight erosion and frequent flooding (*Coarse-loamy, mixed, hyperthermic Aquic Haplustepts*)	1493	4.38
	AaA^{m}1d		6	Asi6eA2f3	Same as Asiani series with moderate erosion	1787	5.25

(continued)

Table 19.4 (continued)

Landform	LEU map unit	Soil series	Soil map unit	Mapping legend	Brief description of soil series	Area (ha)	TGA (%)
	$AaA^{m}1d$		7	Asi6gA1f3	Same as Asiani series with silty clay loam surface texture	1685	4.95
Nearly level active alluvial plain (meander scars)	$AaA^{m}1s$	Kaliganj	8	Kal6eA1	Very deep, poorly drained, brown to gray, fine texture soils on nearly level meander scar with silt loam surface texture and slight erosion (*Fine-loamy, mixed, hyperthermic Aeric Endoaquepts*)	871	2.56
Very gently sloping young alluvial plain	AaY2d	Sikarpur	9	Sik6eB1f3	Very deep, well drained, brown to dark yellowish brown, medium texture soils on very gently sloping young alluvial plain with silt loam surface texture, slight erosion and frequent flooding (*coarse-loamy, mixed, hyperthermic Typic Ustifluvents*)	627	1.84
	AaY2d		10	Sik6eB1	Same as Sikarpur series with no flooding	4365	12.82

(continued)

Table 19.4 (continued)

Landform	LEU map unit	Soil series	Soil map unit	Mapping legend	Brief description of soil series	Area (ha)	TGA (%)
Nearly level young alluvial plain	AaY1d	Dangi	11	Dan6gA1	Very deep, moderately well drained, brown to dark yellowish brown, fine texture soils on nearly level young alluvial plain with silty clay loam surface texture and slight erosion (*Fine-loamy, mixed, hyperthermic Typic Haplustepts*)	2912	8.55
	AaY1s		12	Dan6eA1	Same as Dangi series silt loam surface texture	387	1.14
	AaY1d		13	Dan6kA1	Same as Dangi series silty clay surface texture	534	1.57
Nearly level flood plain	AaA^f1d	Mahinagar	14	Mah6eA3f4	Very deep, well drained, light yellowish brown to brown, coarse texture soils on nearly level flood plain with silt loam surface texture, severe erosion and very frequent flooding (*mixed, hyperthermic Typic Ustipsamments*)	4095	12.03
Total cultivated area						26,446	77.67
Miscellaneous						7601	22.33
Total area						34,047	100

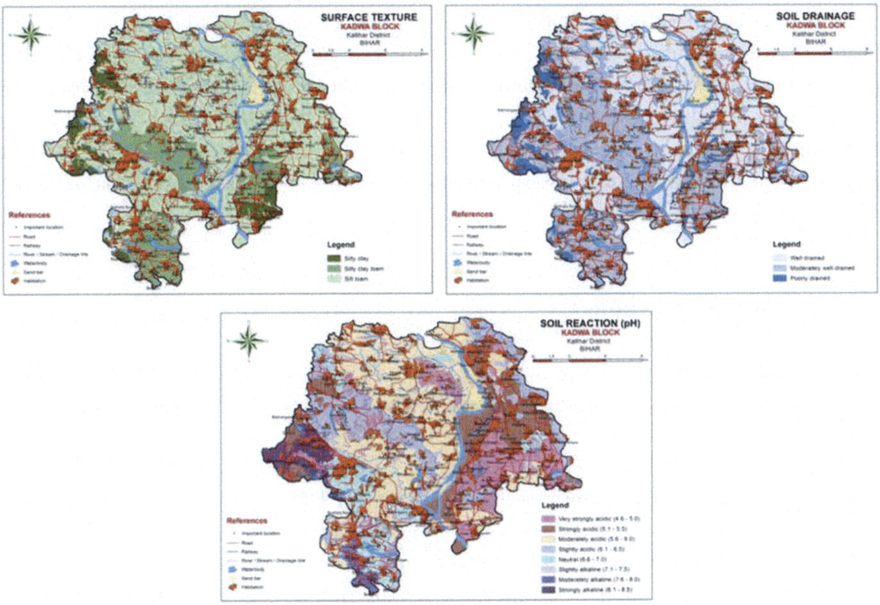

Fig. 19.5 Surface texture, soil drainage and soil reaction (pH) maps of Kadwa block

a key role in total behaviour of soil. Based on soil texture the study area grouped into three classes (Fig. 19.5). Soils are dominantly silt loam (58.99% TGA) followed by silty clay loam (13.50% TGA) and silty clay (5.19% TGA).

19.4.5.2 Drainage

Soil texture, landscape position and ground water depth directly influence the internal drainage of the soils (Reza et al. 2014a). Interpretation of data showed that three soil drainage classes were dominant in the study area (Fig. 19.5). Well drained soils occupied 40.15% of TGA followed by moderately well drained (31.34% TGA) and poorly drained (6.18% TGA), respectively.

19.4.5.3 Soil Reaction (pH)

The intensity of soil acidity or alkalinity is a measured soil reaction (pH). It acts as an indicator to assess the availability of different plant nutrients and also the percentage base saturation (Black 1968). The pH value also helps to determine the amount of various amendments to be added to the soils for acidity or alkalinity. Soils of the study area were grouped into eight soil reaction classes (Fig. 19.5). It is observed that very strongly acid to moderately acid soils occupy 50.58% of TGA, neutral soils

8.22% of TGA and soils in alkaline range occupy only 4.88% of TGA. The large extents of acid soils in the block is due to Mahananda river originated near Chimli, east of Kurseong in Darjeeling district from the Himalaya range at an elevation of 2100 m and sediments carrying by the rivers and their tributaries are acidic in nature and deposited in the study area (Kumari 2014; Reza et al. 2017a) as well as application of high dose of N fertilizer in rice–wheat cropping system (Yadav et al. 1998; Reza et al. 2017a).

19.4.5.4 Organic Carbon (OC)

Organic matter serves as a reservoir of soil nutrients that are essential for plant growth and is therefore, considered as the vital and essential soil attribute controlling productivity (Reza et al. 2019a, 2020a). OC in the study area were grouped into three organic carbon classes (Fig. 19.6). OC status in soils of study area was low to high. Data indicated that high to medium level of OC occupied 64.10% of TGA and only 13.57% of TGA was low (<0.5%). Maximum area in the S-E quadrant of the study area was high in OC due to balanced application of NPK and NP in rice–wheat cropping sequence can increase production of root biomass and stubbles (Subramaniam and Kumarswamy 1989), which may have increased OC.

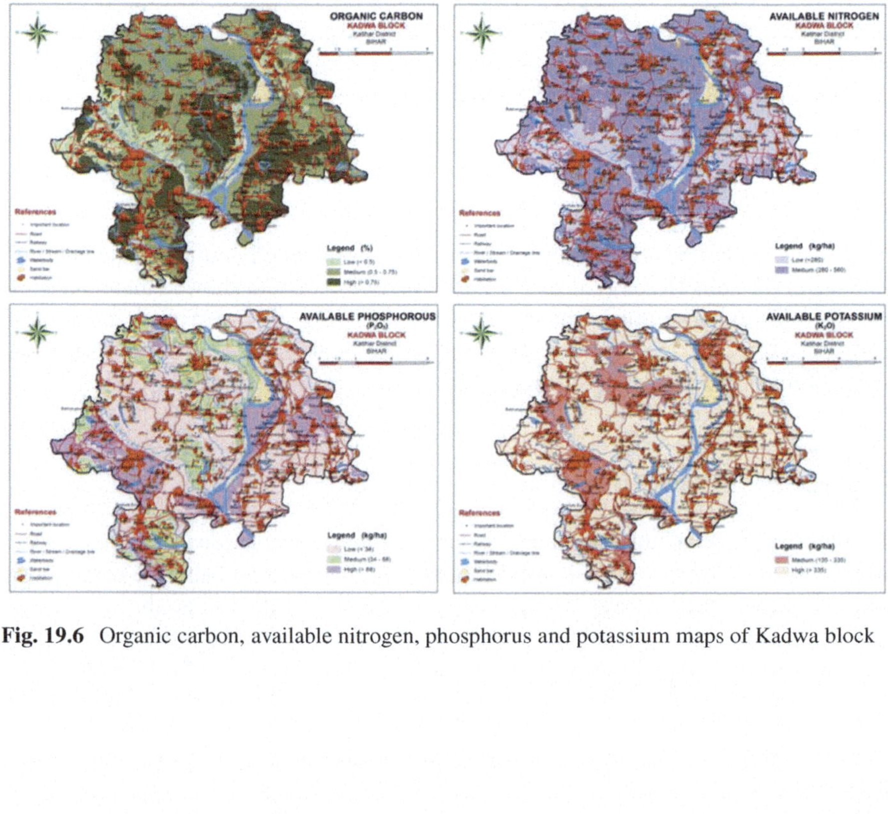

Fig. 19.6 Organic carbon, available nitrogen, phosphorus and potassium maps of Kadwa block

19.4.5.5 Available Nitrogen (N)

Soil nitrogen (N) is important macronutrients and play important role for crop growth and development. The presence of N in soil also governed the yield of crop and their yield attributes. However, it's adverse affects in crop production and productivity may also observe due to imbalance use in soil (Reza et al. 2019b, c). In the study area soils were grouped into two classes (Fig. 19.6). It was observed that 51.90% of TGA were medium in N, whereas 25.77% of TGA was low in category.

19.4.5.6 Available Phosphorous (P)

Among three major nutrients, phosphorus (P) plays an important role to complete the life cycle of a plant. Its functions start right from the stimulation of root growth to proper seed filling and seed setting, in addition to it is an indispensable constituent of genetic material (Khasawneh et al. 1986). It also plays a vital role in photosynthesis, carbohydrate breakdown and transfer of energy in the form of ATP and ADP compounds in various metabolic processes. P content of surface soils of the study area were grouped into three classes (Fig. 19.6) viz. low, medium and high. It was observed that 45.82% of TGA were low in P, whereas 15.87% of TGA and 15.98% of TGA comes under medium and high categories, respectively.

19.4.5.7 Available Potassium (K)

The importance potassium (K) is well recognized in agriculture (Krauss and Johnson 2002) and it is an essential nutrient for plant growth. Exchangeable K i.e. available K is widely used to evaluate the soil K status and to predict the crop K requirements (Askegaard and Jørgen 2002; Reza et al. 2014b, c). K content of soils was grouped into medium and high and is depicted in Fig. 19.6. It was observed that about 67.35 of TGA of the study area were high in K, whereas 10.32% of TGA comes under medium category. Subba Rao et al. (2011) were also reported similar observation for alluvial soils of India.

19.4.6 Land Capability Classification

Land capability classification is an interpretative grouping made primarily for broad agricultural and non-agricultural uses. The United States Department of Agriculture (USDA) was placed the arable lands into I-IV classes according to their limitations, grazing and forestry into class V-VII and class VIII lands for recreation having maximum limitations, wild life and quarrying. The capability classes were further sub-divided into sub-classes based on dominant limiting factors, such as erosion (e), soil (s), climate (c) and wetness (w). It was observed that the soils of the study area

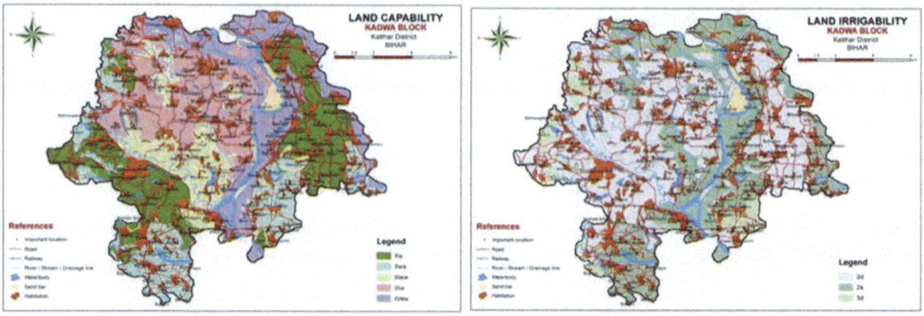

Fig. 19.7 Land capability and land irrigability maps of Kadwa block

divided into three land capability classes viz. II, III and IV. The major limiting factors are erosion and drainage. Five land capability sub-classes were recognized viz. IIw (19.46% TGA), IIws (13.74% TGA), IIIw (17.85% TGA), IIIew (12.76% TGA) and IVew (13.87% TGA) (Fig. 19.7).

19.4.7 Land Irrigability Classification

The study showed that soils were grouped into two irrigability classes which further sub-divided into three sub-classes based on the limitation of soils and site characteristics. The data revealed that about 44.82% of TGA was under 2d sub-classes followed by 2 s (27.67% TGA) and 3d (5.19% TGA) (Fig. 19.7).

19.4.8 Soil Suitability for Crops

Soil and climatic conditions play a vital role for optimal crop growth. The physico-chemical characteristics and micro-environments of soils were largely influenced by water and plant nutrients availability. As such, soil depth, subsoil texture, fertility and drainage conditions etc. are taken into account for soil site evaluation, so that soil maps can be interpreted in terms of suitability for agricultural crops for better socio-economic upliftment. The results showed that soils of the study area was moderately suitable for paddy (53.6% TGA), jute (54.8% TGA), maize (69.9% TGA), wheat (55.0% TGA), mustard (69.9% TGA) and potato (68.8% TGA) due to coarse texture, fertility and wetness (flooding) limitations whereas, 24.1, 22.9, 7.8, 22.7, 7.8 and 8.9% area of TGA were marginally suitable for paddy, jute, maize, wheat, mustard and potato, respectively due to coarse texture and wetness limitations (poor drainage and reduced matrix colour) (Fig. 19.8).

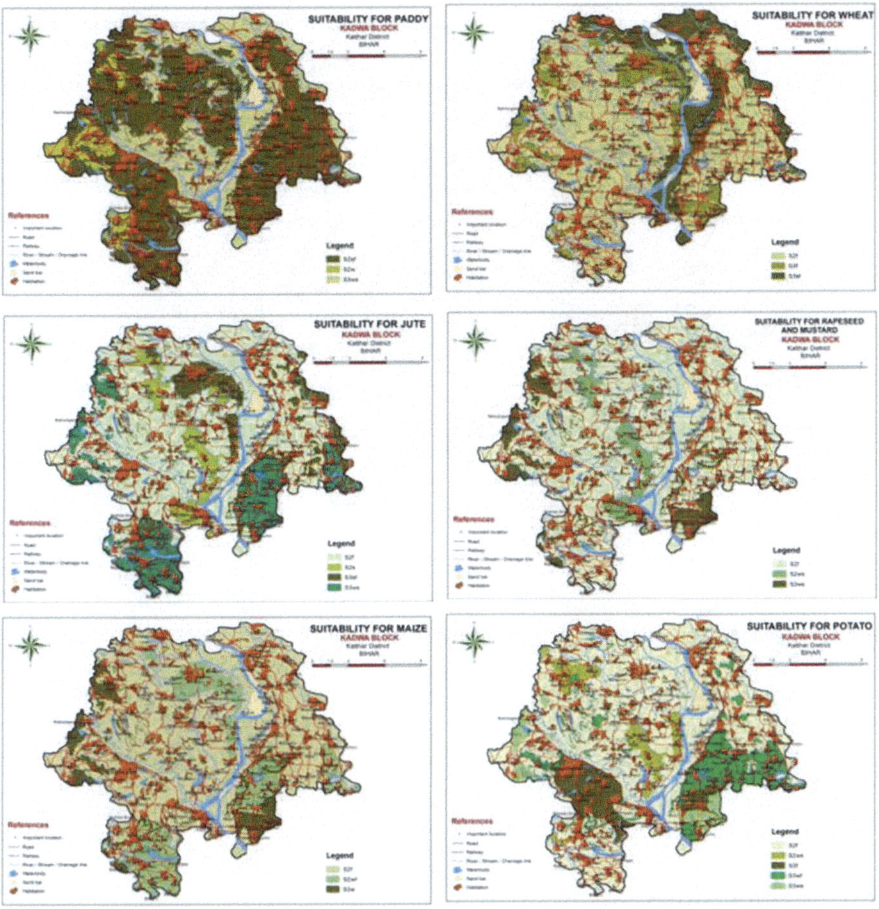

Fig. 19.8 Crop suitability (paddy, jute, maize, wheat, rapeseed and mustard, and potato) maps of Kadwa block

19.4.9 Identification of Alternate Land Use Options Based on Problems and Potentials of Soils

In the study area the low productivity of cultivated agricultural crops is due to the combined effect of the soil and water (Reza et al. 2017b). Erosion, soil acidity, light texture, low fertility status (Reza et al. 2016, 2017a) is the major dominant soil problems. Based on above mentioned characteristics four land management units (LMUs) were identified and mapped after merging 8 soil series (Fig. 19.9) in the study area. Hence, after carefully merging of soil series with similar range of soil characteristics like soil texture, soil pH, internal soil drainage conditions and status of fertility and erosion the LMUs were mapped (Ghosh et al. 2018; Reza et al. 2020b). In each LMU the present and alternate land use option was proposed in Table 19.5

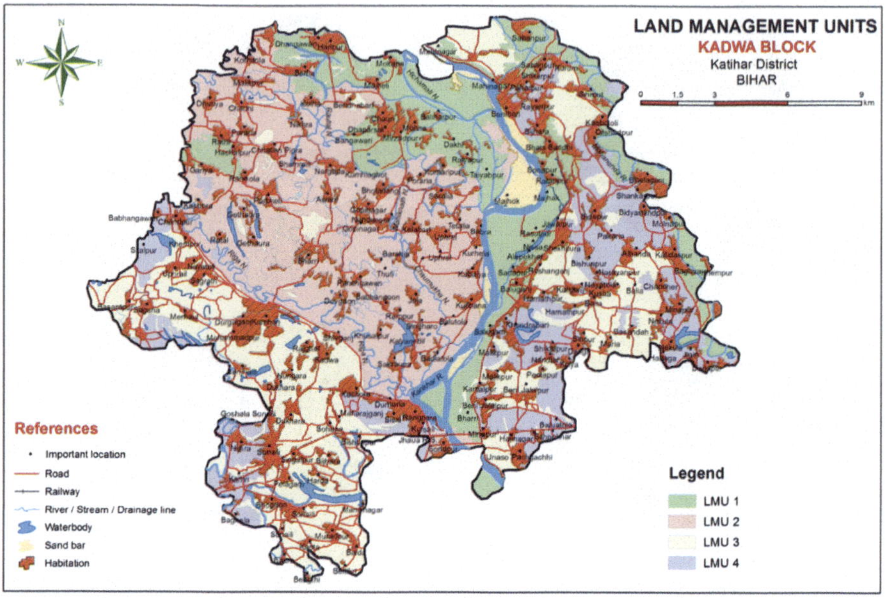

Fig. 19.9 Land management units (LMUs) map of Kadwa block

for the study area. The adaptation of LRI based land use plan will help the farmers to increase the productivity and profitability as compared to traditional based land use system.

19.5 Conclusions

In this study a detailed land resource inventory was carried out at 1:10,000 scales. Geomorphologically, the study area represents 1–3% slope means flat topography with regional slope decreased from north to south. The entire study area is underlain by thick unconsolidated sediment of Quaternary period and belongs to agroecological sub-region (AESR) 13.1, North Bihar and Avadh plains, hot dry to moist subhumid transitional ecological sub-region. Three types of land use/land cover were observed viz. (i) single crop paddy (*kharif*), (ii) double crop (paddy followed by *rabi* crops) and fallow. Visual interpretation of LISS-IV data indicated that the block was characterized into flood plain, meander plain young alluvial plain and old alluvial plain. In ArcGIS landform, slope and land use/land cover maps were integrated and LEU map was prepared with 15 LEU units. At 1:10,000 scale, eight soil series were identified and mapped into 14 soil mapping units (phases of series). Fertility status of the soils indicate that soils of the study area was wide range in soil reaction (very strongly acidic to strongly alkaline), low to high in organic carbon, low to medium in available nitrogen and available phosphorus. Soil survey interpretation showed that

Table 19.5 Present and suggested land-use of Kadwa block

LMUs	Present land use	Suggested land use options
1	Only potato/vegetable/maize cultivation in *rabi* season	• After short duration maize, summer vegetables like bottle guard, snake guard, cucumber and watermelon can be grown Management: It is recommended that some preventive measures are necessary in this unit to maintain the pH of the surface soils in near neutral range which will help to increase the efficiency of phosphatic fertilizers
2	Kharif paddy/fallow—mustard/maize—boro paddy/fallow	• If no flooding then kharif paddy—lathyrus/bengal gram as paira crop—maize—boro paddy • If heavy flooding occurs then early vegetable/mustard/potato—wheat/maize—boro paddy Management: early rice varieties like Prabhat, Dhanlaxmi, Richharia and Turanta; Wheat varieties like HD-2733, PBW-343 and PBW-502
3	Kharif paddy—maize/wheat/potato	• kharif paddy—lathyrus/bengal gram as paira crop—maize/wheat—boro paddy Management: early rice varieties like Prabhat, Dhanlaxmi, Richharia and Turanta; Wheat varieties like HD-2733, PBW-343 and PBW-502
4	Kharif paddy—maize/wheat/potato	• kharif paddy—lathyrus/bengal gram as paira crop—maize / wheat—boro paddy Management: It is recommended that some preventive measures are necessary in this unit to maintain the pH of the surface soils in near neutral range which will helped the phosphatic fertilizers to increase its efficiency; reduced the use of nitrogenous fertilizer; the drainage may be improved by installing surface and sub-surface drainage channels

the study area was divided into three land capability classes viz. II, III and IV and two irrigability classes which further sub-divided into three sub-classes based on the limitation of soils and site characteristics. The suitability for different crops grown in the study area showed marginally suitable due to coarse texture, fertility and wetness (flooding) limitations. Finally, based on major problems and potential the study area were divided into four soil management units and suggested the alternative land-use options LMU wise for the study area.

References

AbdelRahman MAE, Natarajan A, Hegde R (2016) Assessment of land suitability and capability by integrating remote sensing and GIS for agriculture in Chamarajanagar district, Karnataka, India. Egypt J Remote Sens Space Sci 19(1):125–141

Adam E, Mutanga O, Rugege D (2010) Multispectral and hyperspectral remote sensing for identification and mapping of wetland vegetation: a review. Wetlands Ecol Manage 18:281–296

Adhikari K, Guadagnini A, Toth G, Hermann T (2009) Geostatistical analysis of surface soil texture from Zala County in western Hungary. In: Proceedings of international symposium on environment, energy and water in Nepal: recent researches and direction for future. 31March–1April 2009, Kathmandu, pp 219–224

AIS&LUS (All India Soil and Land Use Survey Organization) (1970) Soil survey manual. IARI, New Delhi

Askegaard M, Jørgen E (2002) Exchangeable potassium in soil as indicator of potassium status in an organic crop rotation on loamy sand. Soil Use Manage 18(2):84–90

Black AA (1968) Soil-plant relationships. Wiley, New York

Bray HR, Kurtz LT (1945) Determination of total organic and available forms of phosphorus in soil. Soil Sci 59(1):39–46

Dobos E, Micheli E, Baumgardner MF, Biehl L, Helt T (2000) Use of combined digital elevation model and satellite radiometric data for regional soil mapping. Geoderma 97(3–4):367–391

Fekadu E, Negese A, Yildiz F (2020) GIS assisted suitability analysis for wheat and barley crops through AHP approach at Yikalo sub-watershed, Ethiopia. Cogent Food Agric 6(1)

Ghosh BN, Das K, Bandyopadhyay S, Mukhopadhyay S, Nayak DC, Singh SK (2018) Impact assessment of GIS based land resource inventory towards optimizing agricultural land use plan in Dandakaranya and Easternghats physiographic confluence of India. J Ind Soc Remote Sens 46(4):641–654

GSI (Geological Survey of India) (1998) 1:2 million scale geological map of North-East. Geol Surv Ind, Kolkata, India

Haldar AK, Srivastava R, Thampi CJ, Sarkar D, Singh DS, Sehgal J, Velayutham M (1996) Soils of Bihar for optimising land use. NBSS Publ. No, ICAR-National Bureau of Soil Survey and Land Use Planning, Nagpur, p 50b

Huggett RJ (1975) Soil landscape systems: a model of soil genesis. Geoderma 13(1):1–22

Huggett RJ (1976) Lateral translocation of soil plasma through a small valley basin in the Northway Great Wood Hertfordshire. Earth Surf Process 1(2):99–109

Jackson ML (1976) Soil chemical analysis. Prentice Hall, Englewood Cliffs, N.J.

Jangir A, Tiwari G, Sharma G, Dash B, Paul R, Vasu D, Malav LC, Tiwary P, Chandran P (2020) Characterization, classification and evaluation of soils of Kamrej Taluka in Surat district, Gujarat for sustainable land use planning. J Soil Water Conserv 17(1):15–24

Jenny H (1941) Factors of soil formation. A system of quantitative pedology. McGraw-Hill, New York

Katerji N, Mastrorilli M (2009) The effect of soil texture on the water use efficiency of irrigated crops: results of multi-year experiment carried out in the Mediterranean region. Eur J Agron 30(2):95–100

Kettler TA, Doran JW, Gilbert TL (2001) Simplified method for soil particle-size determination to accompany soil-quality analyses. Soil Sci Soc Am J 65(3):849–852

Khan ZM, Hussain MS, Otiner F (2012) Morphogenesis of three surface-water gley soils from the Meghna floodplain of Bangladesh. J Biol Sci 21(1):17–27

Khasawneh FE, Sample EC, Kamprath EJ (1986) The role of phosphorus in agriculture. In: Proceeding of symposium of phosphorus. NEDC, Muscle Shoals, Albana, USA, pp 48–56

Kong X, Dao TH, Qin J, Qin H, Li C, Zhang F (2009) Effects of soil texture and land use interactions on organic carbon in soils in North China cities urban fringe. Geoderma 154(1–2):86–92

Krauss A, Johnson AE (2002) Assessing soil potassium; can we do better? In: 9th international congress of soil science. March 18–20, Faisalabad

Kumari A (2014) Encyclopaedia of Bihar. Prabhat Books Publisher, New Delhi
Mandal VP, Rehman S, Ahmed R, Masroor M, Kumar P, Sajjad H (2020) Land suitability assessment for optimal cropping sequences in Katihar district of Bihar, India using GIS and AHP. Spat Inf Res 28:589–599
McBratney AB, Santos MLM, Minasny B (2003) On digital soil mapping. Geoderma 117(1–2):3–52
Mythili G, Goedecke J (2016) Economics of land degradation in India. In: Nkonya E, Mirzabaev A, von Braun J (eds) Economics of land degradation and improvement—a global assessment for sustainable development. Springer, Cham, pp 431–469
Nagaraju MSS, Kumar N, Srivastava R, Das SN (2014) Cadastral-level soil mapping in basaltic terrain using Cartosat-1-derived products. Int J Remote Sens 35(10):3764–3781
Naidu LGK, Ramamurthy V, Challa O, Hegde R, Krishnan P (2006) Soil site suitability criteria for major crops. ICAR-National Bureau of Soil Survey and Land Use Planning, Nagpur. NBSS Publ. No.129
Ponnamperuma FN (1972) The chemistry of submerged soils. Adv Agron 24:29–96
Ponnamperuma FN (1985) Chemical kinetics of wetland rice soil relative to fertility. In: Wetland soils-characterization, classification and utilization. International Rice Research Institute (IRRI), Philippines, pp 71–90
Reza SK, Bandyopadhyay S, Ray P, Ramachandran S, Mukhopadhyay S, Ray SK (2020b) Delineation of land management units for alternate land use options: a case study of Bishalgarh block in Sepahijala district of Tripura. Agric Observ 1(3):118–123
Reza SK, Baruah U, Chattopadhyay T, Sarkar D (2014c) Distribution of forms of potassium in relation to different agroecological regions of North-Eastern India. Arch Agron Soil Sci 60(4):507–517
Reza SK, Baruah U, Dutta D, Sarkar D, Dutta DP (2014b) Distribution of forms of potassium in Lesser Himalayas of Sikkim. India. Agropedology 24(1):106–110
Reza SK, Baruah U, Nath DJ, Sarkar D, Gogoi D (2014a) Microbial biomass and enzyme activity in relation to shifting cultivation and horticultural practices in humid subtropical North-Eastern India. Range Mgmt Agrofor 35(1):78–84
Reza SK, Baruah U, Nayak DC, Dutta D, Singh SK (2018) Effects of land-use on soil physical, chemical and microbial properties in humid subtropical northeastern India. Natl Acad Sci Lett 41(3):141–145
Reza SK, Baruah U, Sarkar D, Dutta DP (2011) Influence of slope positions on soil fertility index, soil evaluation factor and microbial indices in acid soil of Humid Subtropical India. Indian J Soil Conserv 39(1):44–49
Reza SK, Nayak DC, Chattopadhyay T, Mukhopadhyay S, Singh SK, Srinivasan R (2016) Spatial distribution of soil physical properties of alluvial soils: a geostatistical approach. Arch Agron Soil Sci 62(7):972–981
Reza SK, Nayak DC, Mukhopadhaya S, Singh SK (2021) Soil-site suitability of major crops for suitable crop planning in Kadwa block of Katihar district in Bihar. Agric Observ 2(2):7–11
Reza SK, Nayak DC, Mukhopadhyay S, Chattopadhyay T, Singh SK (2017b) Land resources inventory for alternative land use options in Kadwa block, Katihar district. Bihar. Popular Kheti 5(1):132–137
Reza SK, Nayak DC, Mukhopadhyay S, Chattopadhyay T, Singh SK (2017a) Characterizing spatial variability of soil properties in alluvial soils of India using geostatistics and geographical information system. Arch Agron Soil Sci 63(11):1489–1498
Reza SK, Dutta D, Bandyopadhyay S, Singh SK (2019c) Spatial variability analysis of soil properties of Tinsukia district, Assam, India, using geostatistics. Agric Res 8(2):231–238
Reza SK, Ray P, Ramachandran S, Bandyopadhyay S, Mukhopadhyay S, Sah KD, Nayak DC, Singh SK, Ray SK (2019a) Profile distribution of soil organic carbon in major land use systems in Bishalgarh block. Tripura. J Ind Soci Soil Sci 67(2):236–239
Reza SK, Ray P, Ramachandran S, Bandyopadhyay S, Mukhopadhyay S, Sah KD, Nayak DC, Singh SK, Ray SK (2019b) Spatial distribution of soil nitrogen, phosphorus and potassium contents and stocks in humid subtropical North-eastern India. J Ind Soc Soil Sci 67(1):12–20

Reza SK, Ray P, Ramachandran S, Jena RK, Mukhopadhyay S, Ray SK (2020a) Soil organic carbon fractions in major land use systems in Charilam block of Tripura. J Ind Soc Soil Sci 68(4):458–461

Rich CI (1965) Elemental analysis by flame photometry. In: Black CA (eds) Methods of soil analysis, part 2: chemical and microbiological properties. Am Soc Agron Madison,Wisc, 849–864

Sarkar D (2011) Geo-informatics for appraisal and management of land resources towards optimizing agricultural production in the country-issues and strategies. J Ind Soc Soil Sci 59(supplement):35–48

Sehgal JL, Saxena RK, Vadivelu S (1987) Field manual. Soil resource mapping of different states of India. ICAR-National Bureau of Soil Survey and Land Use Planning, Nagpur. NBSS Publ No13

Seid NM, Yitaferu B, Kibret K, Ziadat F (2013) Soil-landscape modeling and remote sensing to provide spatial representation of soil attributes for an Ethiopian watershed. Appl Environ Soil Sci 1–11

Sharma RP, Singh RS, Naitam RK, Singh RS (2019) Technique of large scale soil mapping using remote sensing satellite data in basaltic terrain of peninsular region in the north-west Gujarat, India. J Ind Soc Soil Sci 67(2):151–159

Singh SK, Chatterji S, Chattaraj S, Butte PS (2016) Land Resource Inventory on 1:10000 scale, why and how? ICAR-national bureau of soil survey and land use planning. Nagpur. NBSS Publ. No. 172:1–110

Soil Survey Division Staff (2000) Soil survey manual. Soil conservation service. USDA Handbook 18, Revised. Scientific Publishers, Jodhpur

Soil Survey Staff (1995) Soil survey manual, USDA, agricultural handbook No. 18, New Revised Edition, Scientific Publishers, Jodhpur

Soil Survey Staff (2003) Keys to soil taxonomy, 9th edn., USDA-natural resources conservation service, Washington, DC

Soil Survey Staff (2014) Keys to soil taxonomy, 12th edn., USDA-Nat Res Conserv Serv, Washington, DC

Srivastava R, Saxena RK (2004) Technique of large-scale soil mapping in Basaltic terrain using satellite remote sensing data. Int J Remote Sens 25(4):679–688

Stoops G, Eswaran H (1985) Morphological characteristics of wet soils. In: Wetland soils-characterization, classification and utilization. International Rice Research Institute (IRRI), Philippines, pp 177–189

Subba Rao A, Srinivasarao Ch, Srivastava S (2011) Potassium status and crop response to potassium on the soils of Agroecological regions of India. IPI Research Topics No. 20. 2nd, revised. International Potash Institute, Horgen

Subbiah BV, Asija GL (1956) A rapid procedure for estimation of available nitrogen in soils. Curr Sci 25:259–260

Subramaniam KS, Kumarswamy K (1989) Effect of continuous cropping and fertilization on chemical properties of soils. J Indian Soc Soil Sci 39:171–173

Supriya K, Naidu MVS, Kavitha P, Reddy MS (2019) Characterization, classification and evaluation of soils in semi-arid region of Mahanandi mandal in Kurnool district of Andhra Pradesh. J Ind Soc Soil Sci 67(2):125–136

Sys C, Van Ranst E, Debaveye J, Beernaert F (1993) Land evaluation, Part-III. Crop requirements. Agricultural Publications No.7, GADC, Brussels, Belgium

Takkar PN (2009) Soil fertility, fertilizer and integrated nutrient use. In: Rai et al (eds) Hand book of agriculture, six (revised) editions. Indian Council of Agriculture, New Delhi pp 516

Van de Wauw J, Baert G, Moeyersons J, Nyssen J, De Geyndt K, Taha N, Zenebe A, Poesen J, Deckers J (2008) Soil-landscape relationships in the Basalt-dominated highlands of Tigray, Ethiopia. Catena 75(1):117–127

Walkley A, Black IA (1934) An examination of the Degtjareff method for determining soil organic matter and a proposed modification of the chromic acid titration method. Soil Sci 37(1):29–38
Yadav RL, Yadav DS, Singh RM, Kumar A (1998) Long term effects of inorganic fertilizer inputs on crop productivity in a rice-wheat cropping system. Nutr Cycl Agroecosyst 51:193–200
Zhai Y, Thomasson JA, Boggess JE, Sui R (2006) Soil texture classification with artificial neural networks operating on remote sensing data. Comput Electron Agric 54(2):53–68

Chapter 20
Soil Pollution by Industrial Effluents, Solid Wastes and Reclamation Strategies by Microorganisms

Sourav Singha and Sabyasachi Chatterjee

Abstract Soil receives enormous pollutants from industrial effluents, agricultural & municipal wastes at a higher rate every day and cause accumulation of toxic heavy metals (Cr, Hg, Cd, Pb, & As etc.), radioactive nuclei, halogenated compounds, aromatic hydrocarbons and phenolic compounds etc. At elevated concentration, these pollutants are proved to be having an adverse effect on soil health, resulting in to unnatural changes in soil physiology affecting all forms of life directly or indirectly. Therefore, it is imperative to mitigate soil pollution aiming to restore soil ecosystem. Several study suggested, bioremediation have been extensively explored to reclaim soil & showed favourable outcome. Especially, microbial based techniques used to remove, reduce or transform noxious pollutants & are considered as most efficient, reliable & eco-friendly approach. The decontamination of soil is confined to bioavailability of pollutants. However, it is induced by type, chemical characteristics & concentration of pollutants, considering soil physical conditions. Microbes, especially bacteria, fungi & algae adopted different of bioremediation strategies. This study provides a comprehensive insight on occurrence of organic & inorganic soil pollutants, their characteristics & impact on soil health. We also discuss about the in situ & ex situ remediation methods and their applications with special emphasis on advance techniques. Moreover, this review will give a definite idea of microbial processes that would aid in selection of a competent approach (s) combating soil contamination effectively.

Keywords Soil pollution · Bioremediation · Organic pollutant · Industrial effluent · Soil reclamation · Heavy metals

S. Singha
Department of Microbiology, Bankura Sammilani College, Bankura, W.B, India

S. Chatterjee (✉)
Department of Botany (PG), Ramananda College, Bishnupur, Bankura, W.B, India
e-mail: schatterjeebiotech@gmail.com

P. K. Shit et al. (eds.), *Soil Health and Environmental Sustainability*,
Environmental Science and Engineering,
https://doi.org/10.1007/978-3-031-09270-1_20

20.1 Introduction

Soil is a natural habitat of living organisms that contributes to basic needs like food and water. Soil accounts for sustaining the ecosystem & managing biodiversity to a great extent. It also acts as a vital resource that essentially contributes to the perseverance of life on Earth. Especially agricultural practices like food grain cultivation, horticulture & vegetation etc. solely dependent on the physicochemical properties of soil (Mishra et al. 2016). The inherent soil property has a direct influence on soil behaviour and nature; hence the comprehensive knowledge of soil properties, nature & behaviour becomes imperative managing environment in sustainable way (Sonwane et al. 2010).

Biotic & abiotic components of soil also have a role to play in soil health. Living components including plants, flora & fauna equally contribute to soil functioning. Soil acts as a major harbouring site for interactions where processes like decomposition, humification, solubilization & mineralization are taken place. These processes impact soil fertility by the reciprocal action of soil biota with humus materials, minerals and maintaining soil structure (Xue et al. 2021). Soil physicochemical characteristics such as pH, water content, availability of nutrients including the amount of carbon (C), nitrogen (N) and Potassium (K), etc. are very essential parameters. A slight imbalance causes a notable change in soil which directly or indirectly hinders its habitants. Therefore, soil quality necessarily has to maintain to confer its native functioning (Vincent et al. 2018).

In the last decade or so degradation of soil quality become a global issue of concern, where the soil is exceedingly contaminated by industrial effluents and solid wastes. Unplanned urbanization with booming industrialization, improper waste disposal, and anthropogenic activities had caused unsettling of soil composition & ended up with soil pollution. Industries without proper waste management systems are the biggest contributors to soil pollution (Lavanya et al. 2019; Kumar and Agrawal 2020). Industries like textiles, metallurgy, tannery, battery manufacturing industries, glass factories, microelectronics, paper processing plants, iron & steel plants, coal burning thermal plants, nuclear power stations, petroleum industries & plastics manufacturing etc. producing more pollutants which directly or indirectly released into the soil. The by-products of these industries are disposed of inappropriate manner as a form of effluent contains several organic and inorganic pollutants including toxic heavy metals and other non-biodegradable substances (Chhonkar et al. 2010; Zhan et al. 2015). The bioaccumulation of organic & inorganic waste materials & heavy metals in the environment exert toxicity & causing several health issues to the living world (Tchounwou et al. 2012; Jaishankar et al. 2014; Engwa et al. 2019; Zwolak et al. 2019). Especially heavy metals, pesticides & other xenobiotic compounds present in industrial effluents, are not biodegradable and have the tendency to persist in the environment, and their concentrations can be magnified significantly with time. These pollutants are not water-soluble thus they primarily accumulate on top layer of soil (Mishra et al. 2016). An elevated concentration of these could cause severe damage to the living cells by showing extreme toxicity due to inhibition of metabolic reactions

(Vongdala et al. 2018). Plants' lifecycles are shortened when they are exposed to such high contamination due to the inability to adapt that abrupt change in soil chemistry. Even the plant-associated microorganisms found in soil (Fungi and bacteria) begin to decline; their natural interactions disrupted which creates additional problems to the soil. It slowly hampers fertility and converts land unsuitable for agriculture and any vegetation to survive.

Urban & rural household waste materials also cause problems as they are been discharged in the environment in an uncontrolled manner. Sewages & garbages from domestic as well as commercial waste sources primarily consists of plastics, papers, discarded food, clothes, metallic cans, sludge, glasses, fibers, bottles, rubbers, etc. Among these, a few are biodegradable and are recycled by composting, while non-biodegradable materials are disposed of in landfills. Landfills are common in practice and economical but uncontrolled disposal of solid wastes gives rise to major consequences related to soil sustainability. It creates nuisance and has considerable environmental impacts by unsettling the soil ecosystem. These kinds of open landfills produce sanitary problems and act as a harbour of insect vectors & major sources of vector-borne pathogens. These waste dumps also produce several organic acids that percolate into the soil and cause underground water contamination (Chadar and Chadar 2017). Several reports suggested that the production of acids resulting in an acidic environment may inhibit biodegradation of waste materials by inherent microbial communities. In due course those soil ecosystems destroyed fully and are converted to the barren and unfertile land, unable to support any life on it.

Recent studies revealed that the presence of radioactive nuclei impacted soil degradation greatly, which is one of the pivotal factors of soil pollution generates both naturally and in a technogenic way. Emission of radioactive elements like ^{3}C, ^{60}Co, ^{90}Sr, ^{137}Cs, ^{226}Ra, ^{232}Th, ^{238}U and ^{239}Pu, etc. from nuclear power plants contaminate soil and accumulated in the vegetables and crops grown on that contaminated soil (Aleksakhin 2009; Ali et al. 2019).

Several bodies are formed in many countries in order to regulate and minimize the pollution level. Environmental Protection Agency (EPA) is one such organization working on the restoration of the environment by making a perfect balance of sustainability, economy & Society (report.epa.gov 2016). Published literature suggested that potential biological remediation strategies can be employed to retrieve soil native nature. Biological operations like microbial remediation or phytoremediation are effectively used for the removal of soil contaminants to a great extent. Especially microbial cells exert various processes including oxidative reduction, precipitation, mineralization, biosorption, complexation and enzymatic transformation by which hazardous pollutants are removed from soil efficiently (Ojuederie and Babalola 2017; Igiri et al. 2018).

The prime focus of this study is to recount the profuse sources & nature of soil contamination through industrial effluent and solid wastes & plausible soil restoration strategies (Fig. 20.1).

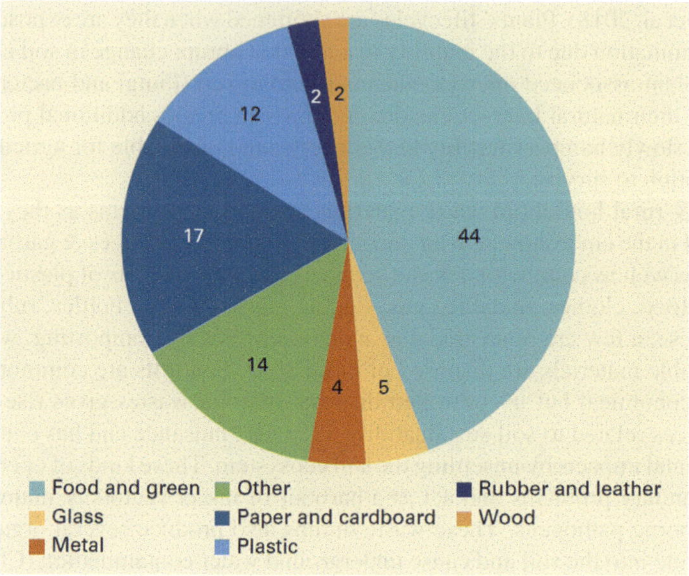

Fig. 20.1 Composition of Solid waste as per USEPA 2016

20.2 Nature, Composition & Characteristics of Industrial Effluent

Industries generally discharge wastewater in untreated form into the environment. Including India, worldwide wastewater generation from industries and production plants is common in practice. It has been reported by several researchers that, due to the shortage of requisite space or lack of proper disposal management system, a huge amount of toxic liquid is produced and enters into the open environment. Most industries disposed of their raw effluents in nearby water channels, drains or open soil (Ahmed et al. 2016). According to the published data of CPCB in the year 2010, 13,500 million litre industrial wastewater produced per day in India. These effluents typically consist of organic & inorganic materials which exert high toxicity (Table 20.1). Organic pollutants mainly includes phenolic compounds, hydrocarbons, pesticides, azo dyes, esters, etc. (Bhargava & Saxena 2020). Heavy metals are major constituent of inorganic pollutants. Commonly found heavy metals in industrial effluents are arsenic (As), nickel (Ni), chromium (Cr), lead (Pb), mercury (Hg), and cadmium (Cd), etc. Certain free living electrolytes (K^+, Ag^{2+}, Na^{2+}, Mg^{2+}, Ca^{2+}, Cl^-, CO^{3-}, HCO^{3-}, Cl^-) are also likely to be found in the form of inorganic pollutants (Subramani et al. 2014; Ahmed et al. 2016; Tejaswi et al. 2017) (Table 20.2).

The composition & chemical nature of the effluents varies according to the industries it released. Generally, industries like paper mills & Zn smelter release acidic (pH 3 to 5) drain water while the textile wastewater is alkaline in nature. On the other

Table 20.1 Different types of industrial effluent and their characteristics

Different sources of industrial waste water	Composition & chemical properties
Petrochemical industry waste water	Primarily hydrocarbon compounds, BTEX (benzene, toluene, ethylbenzene, xylenes) & dioxins, presence of phenolic compounds with metals. High in sulphides High COD with alkaline pH
Paper & pulp industry wastewater	Ligno-cellulose components, phytosterols phenol and chlorophenols compounds with resin acids & fatty acids. Highly toxic chlorinated compounds (dioxins & furan). Elevated BOD & COD with alkaline pH
Steel Industry waste water	Iron (Fe) & chromium (Cr) are the main components. bag filter dust (BFD)
Textile and fabric industry wastewater	Principal pollutants are harmful residual chlorinated dyes (Azo, diazo & anthraquinone), several carcinogenic metal complexes (As, Hg, Ni, and Cr etc.) High TDS & BOD
Tannery or leather industry wastewater	High concentration of toxic metals (Cr, Cu, Pb and As) & organic compounds (phosphate, bicarbonate and chloride) with salts (sodium, chloride, and sulphide), high in BOD, COD TDS and TSS, presence of hazardous phenolic compounds & phthalates (cause neuro toxicity)
Distillation process waste water (stilage)	Dark brown coloured effluent contains several organic compounds in higher concentration like polysaccharides, protein residues, polyphenols, waxes & melanoidins (condensation of sugar & amino acids). high in BOD, COD & TDS
Wine industry wastewater	Acidic effluent contains high amount of sugars, ethanol polyphenols, tannin and lignin & short chain fatty acids, presence of Na, K & phenolic compounds elevates COD and TSS level. Presence of several toxic heavy metals including Co, Pb, Cd, Ni & Cr etc.
Pharmaceutical Industry wastewater	Containing hazardous organic solvents like petroleum ether, ethanol, benzene, chloroform, with several organic compounds such as steroids, antibiotics, analgesics, drug residues and pharma-metabolites along with significant amount of metalloids like mercury, chromium, copper etc.
Paint manufacturing industry wastewater	Presence of high amount of dyes, colorant, adhesives, trace of oils (hydrocarbons) & grease, organic solvents (toluene and methyl ethyl ketone).toxic heavy metals (chromium and lead) and dissolved solids
Abattoir (slaughterhouse) wastewater	High concentration of suspended organic materials contributed to high COD & BOD, excessive nutrients promote the growth of pathogenic and non-pathogenic microorganisms, presence of toxic heavy metals and other materials like fats, oil, and grease (FOG)
Landfill drainage water	Ample amount of organic compounds comprising proteins, carbohydrate, aromatic hydrocarbons, contains toxic metals, low molecular acids & gases (CO_2 & H_2), Volatile fatty acids (VFA) etc. high in total suspended solids leads to increased COD & BOD

(continued)

Table 20.1 (continued)

Different sources of industrial waste water	Composition & chemical properties
Mining industry drainage (acid mine)	Dark colour, acidic (pH below 2) effluent contains high concentrations of toxic metals such as Iron, Cadmium, Lead, Nickel & Copper, Cobalt etc. Presence of hydrated sulphates (SO_4^{2-}) is characteristics of the effluent

Adapted from Ahmed et al. (2016), Bhargava & Saxena (2020)

Table 20.2 Different organic contaminants in industrial waste water with their sources & functions

Type of organic pollutants	Sources and characteristics
Phenol & Chlorinated phenols	Major sources are pesticides, pharmaceuticals wastes, petroleum refineries, distilleries, pulp and paper mills, wood preservation plants & coal excavation sites It causes various skin related problems like dryness and burn even hamper central nervous system Chlorinated phenols are potent carcinogenic and mutagenic agents
Nitro-aromatic (Azo dyes)	Released from industries which uses different colorant materials frequently such as textile & fabric, pharmaceutical, cosmetics, paint, plastics, leather & paper industries etc., Non-biodegradable aromatic amines have severe health hazards in humans and animals such as neurotoxicity, digestive tract discomfort, nausea, vomiting etc., indiscriminate exposure of these may lead to liver and kidney dysfunction in human

hand, oil refineries, paper sugar mills, distillery and effluents possess much higher organic carbon. These effluents also contain xenobiotic compounds like aromatic hydrocarbons, metalloproteins and phenol compounds (Ahmed et al. 2016).

BOD and COD are the crucial parameters used to determine the wastewater characteristics. Several reports suggested that the abnormalities in BOD & COD values (Chhonkar et al. 2010) of untreated industrial effluents are very high contributed by various organic acids (Table 20.3).

20.3 Sources, Composition & Nature of Solid Wastes

Generation of waste material is an unavoidable phenomenon where a huge amount of waste is produced through industrial processes, from manufacturing units, or in the form of municipal and urban garbage. But the problem arises when these toxic & hazardous solid wastes are disposed of in an open environment without any proper treatments (Agarwal 2016; Kumar and Agrawal (2020)). These untreated solid wastes cause several complications. Generally, developing countries do have problems with

Table 20.3 Types of inorganic contaminants in industrial effluent with their sources & functions

Type of inorganic pollutants	Environmental pollution and toxicity profile
Cadmium (Cd)	Emission of Cd is greatly contributed by fuel combustion & waste incinerations along with steel industries, phosphate fertilizer manufacturing unit & paint sludge. Accumulation of Cd can cause severe problems like muscle cramp, stomach pain with vomit tendency, psychological disorders & damage of neuro system etc.
Chromium (Cr)	Major sources are glass factories, wood preservation plants, paint manufacturing units, tanneries, steel & alloy industries & mining. Inhalation of Cr can have lethal effect like respiratory distress, perforation in lungs, significant dysfunction of several organs e.g. renal failure, cardiovascular damage etc.
Arsenic (As)	Majority of As contamination occurs from fuel combustion, coal burning power plants, mining & metal extraction processes It cause systematic disruption of internal body parts like lungs, liver, spleen etc. High exposure of As leads to anaemia, cardiovascular malfunction, disruption in neuro transmission, gastrointestinal lesions and even death of individual
Lead (Pb)	Pb mostly released from battery wastes, ceramic industry, pesticides industry, fuel combustion, smelting operations, thermal power plants etc. Hypertensions, renal dysfunctions, abdominal discomfort, encephalopathy, hearing loss, reduced consciousness, CNS dysfunction & difficulties with concentration etc. are the major Pb associated problems arose upon exposure
Mercury (Hg)	Contamination Hg of rises from several industries like chemical processing, pharmaceuticals, coal based power stations, chemical metal extraction processes, electronic wastes, agricultural wastes, & hospital waste etc. Mercury has drastic impact on human heath like development of odd metal taste, frequent vomiting, breathing problems including neurological disorder which leads to loss of vision & hearing with speech slurring

Adapted from Tchounwou et al. (2012), Ahmed et al. (2016), Tarekegn et al. (2020)

waste management where solid waste materials are dumped in a specific site or they can be used as landfill materials. Lack of space near-source stations is a major reason for that (Lavanya et al. 2019). Preferably waste materials are transported to outskirts areas of cities where landfills or dumpsites are located. According to Shankar and Shikha, in India, it is only about 40% of total municipal solid wastes are collected and dumped in specified sites in daily basis. Insufficient infrastructure adding up more problems and ended up with Piling up of hazardous materials (abdel-Shafy and Mona Mansour 2018; Ferronato and Torretta 2019) (Table 20.4).

A massive amount of waste materials emancipate openly from industries can be categorized as hazardous and non-hazardous. Waste materials like papers, plastics, wood, cardboard, packaging materials are relatively less harmful and can be utilized further or recycled. However trashes of heavy industries like coal ash from thermal power plants, steel melting slag, scrap metal & blast furnace slag from the steel manufacturing unit, lime from pulp and paper industries, gypsum from allied industries, red mud and tailings other than Iron (e.g. aluminium, zinc and copper) from

Table 20.4 Type of solid wastes and their characteristics

Solid waste material	Characteristics
Steel and Blast furnace slug	Scrap materials produced during Iron & steel making, rich in minerals (mainly silicate, iron, aluminium, calcium & magnesium)
Brine sludge mud	Organic rich semi solid waste from Chlorine-alkali & Caustic soda industry, major component soda ash
Copper slag	By product of various metallurgical processes present As, Fe, Cu etc.
Fly ash	By product of Coal-combustion contains ample Pb & Al
Lime sludge waste	Prime component is Calcium carbonate ($CaCO_3$) released generally from different industries like pesticide, sugar, soda ash & paper
Mica mine scrape	Mining of Mica produce silica, aluminium, oxides of potassium and Iron
Phospho-gypsum	Known as calcium sulphate, produced during processing of phosphate & phosphoric acid, contains radioactive elements
Bauxite mining waste (Red mud)	Metallic waste (contains-iron, silica, titanium & alumina) produced from bauxite ore
Coal dust	Mining of coal produce fine particles
Iron ore tailing	Solid by product of iron ore processing, typically posses alumina & iron oxide

Adapted from Basu et al. Brifa et al. (2020) & Tarekegn et al. (2020)

metal industries are really creates environmental problems (Agarwal 2016; Lavanya et al. 2019).

20.4 Impact of Industrial Effluent & Solid Waste on Soil Health

The solid & liquid industrial waste are rich in chemicals, which are non-biodegradable and exert toxicity. At elevated concentrations of these ingredients of wastes exert an adverse impact on soil health. The components present in effluents tend to change the chemical makeup of the soil. Overabundance may influence soil stability by altering composition and physical factors like pH, salinity, etc. Deposition of organic and inorganic materials into the soil also amend the microenvironment of soil which indeed very essential for crop production. Several instances proved that the precipitation of fly ash on topsoil nearby industrial belts result in the loss of fertility. The immediate consequence of that is the production of barren lands (Bhat 2015).

It is evident that bioaccumulation of heavy metals have shown phytotoxicity (Hiroki 1992; Ahmed et al. 2016). Many researchers have highlighted the lethal effect of heavy metals on biological systems. The physiology of cell interior (organelles) markedly affected by these toxic ions (Jayashankar et al. 2014; Brifa et al. 2020; Tarekegn et al. 2020). In general, metals are indispensable for plant growth. Physiological & biological processes are highly dependent on metal concentration with in the cell. Depending upon the dose and exposure these chemicals started affecting plant health & disintegrate soil natural microbiota functioning. The presence of contaminants like inorganic metals affect adversely & causes various plant diseases such as high concentration of Cd result chlorosis, excess Cu produces oxidative stress etc. (Ahmed et al. 2016).

Man made organic chemicals such as halogenated organic pollutants (HOPs), polychlorinated biphenyls (PCBs), polycyclic aromatic hydrocarbons (PAHs), BTEX (benzene, toluene, ethylbenzene, and xylenes), nitro-aromatic compounds and organophosphorus compounds are found in soil in large quantity. Their high molecular weight and poor water solubility makes biologically unavailable and therefore, tend to persist in the environment. These organic chemicals are potentially mutagenic and carcinogenic, often accumulates in vegetables & fruits and cause a major threat to humans (Perelo 2010; Ali et al. 2019).

20.5 Reclamation of Soil by Microbial Remediation of Industrial Effluent & Solid Waste Contaminants

Microbial remediation is considered as effective techniques for removing soil pollutants. One of the key attribute of microorganism is the capacity to transform soil pollutants into harmless entity by exploring their wide metabolic range. Especially, fungi and bacteria able to produce variety of extracellular enzymes and low molecular weight organic acids that can somehow modify organic pollutants. (Rajendran et al. 2003). Therefore, *in-situ* & *ex-situ* treatment of pollutants proven as a cost-effective, eco-friendly & sustainable approach (Megharaj et al. 2011).

20.5.1 Microbial Remediation of Heavy Metals

Heavy metal pollutants can be partially or completely removed from soil by utilizing the metabolic activity of microbes. It is an entirely sustainable process i.e. no harm to the environment compare to other physical or chemical processes. Microbes are employed to remove, reduce, transform or completely remove the heavy metals from the soil. Several genes either present in the genome or plasmid are responsible for these physiochemical activities (Rajendran et al. 2003).

Efficient microorganisms including *Bacillus* sp., *Arthrobactor* sp. *Pseudomonas* sp., *Staphylococcus* sp. *Streptomyces* sp., *Aspergillus* sp., *Rhizopus* sp., *Sacharomyces* sp. *Penicillium* sp. etc. are widely distributed in soil and effectively remediate soil under natural conditions (Table 20.5).

Table 20.5 Microorganisms & respective metals they remediate

Microorganisms	Remediating heavy metals
1. Bacteria	
Pseudomonas aeruginosa	Hg(II)
Pseudomonas sp.	Pb(II)
Bacillus sp.	Pb(II)
Arthrobacter viscosus	Cr(VI)
Staphylococcus epidermidis	Cr(VI)
Eichhornia sp.	Cu(II)
Brevibacterium sp.	Zn(II)
Rhodobacter capsulatus	Zn(II)
Pseudomonas aeruginosa	Cd(II)
Bacillus cereus	Cd(II)
Ochrobactrum sp.	Cd(II)
Sporosarcina ginsengisoli	As (III)
Bacillus cereus	Cr (VI)
Kocuria flava	Cu(II)
Pseudomonas veronii	Cd (II), Zn, Cu
Actinomycetes sp	Cd (II)
Stenotrophomonas maltophilia	Pb (II)
Enterobacter cloacae	Cr (VI)
Rhodopseudomonas sp	Co
Bacillus subtilis	Cr (VI)
Cupriavidus metallidurans	Se (VI)
Bacillus megaterium	Cr (VI)
Pseudomonas aeruginosa	Cr (VI)
2. Fungi	
Aspergillus versicolor	Ni, Cu
Aspergillus niger	Cr (VI)
Aspergillus foetidus	Cr (VI), Pb (II)
Aspergillus fumigatus	Pb
Drechslera rostrata	Cr (VI)
Gloeophyllum sepiarium	Cr (VI)
Rhizopus oryzae	Cr (VI)
Penicillium canescens	Hg
Sacharomyces cerevisiae	Pb, Cd
Rhizopus stolonifer	Cd, Pb, Zn
Rhizopus arrhizus	Hg, Pb, Cd

(continued)

Table 20.5 (continued)

Microorganisms	Remediating heavy metals
3. Algae	
Spirogyra sp.	Cd, Hg, Pb
Cladophora glomerata	Cu, Pb, Cd
Spirullina sp.	Pb, Cr, Cu, Fe, Zn
Hydrodictylon sp.	As
Rhizoclonium sp	As
Oedogonium rivulare	Cu, Pb, Cd, As

Adapted from: Dwivedi (2012), Snehalata et al. Rodriguez et al. Igiri et al. (2018)

Tabak et al. (2005) described different mechanisms of bioremediation by which soil microbes can minimize the effect of heavy metals including bioaccumulation, bioprecipatation, biosorption, transformation, immobilization & cometabolism etc. Under an intuitive environment, microbes adopt one of these techniques and make toxic metals biologically unavailable.

20.5.1.1 Biosorption

Biosorption or bioabsorption refers to the physical attachment of metals on the cell exterior by extra cellular polymeric substances (Tabak et al. 2005; Tarekegn et al. 2020). Biosorption is strictly dependent on physicochemical properties of the host cell. The ion absorbing efficacy is greatly vary upon composition of cell wall, temperature & pH of the surroundings, surface area for contact and metal gradient, exposure time, ionic strength as well as the chemical nature of the metal ions, etc. (Shamim 2018). Biosorption is a very common technique employed by many fungal, algal or bacterial species to defend themselves against cadmium, silver, lead, or nickel etc. (Tabak et al. 2005; Tarekegn et al. 2020). According to Shamim (2018), the accumulation of metal ions is not ATP dependent process rather the concentration of metals in the exterior, i.e. chemo osmotic pressure greatly influences the uptake capacity. The ionic nature of the membrane along with the gradient created on either side helps in specific and nonspecific metal sorption. Especially the presence of peptide chain linked repeated unit of NAG (N-acetyl glucosamine) & NAM (1,4-N-acetylmuramic acid) make bacteria more negative charge which attracts positively charged metallic ions (Shamim 2018) (Fig. 20.2).

20.5.1.2 Bioaccumulation

Microorganisms can retain toxic heavy metals within their biomass in a physical manner. It is evident that microbial cells are able to uptake metals through the cell

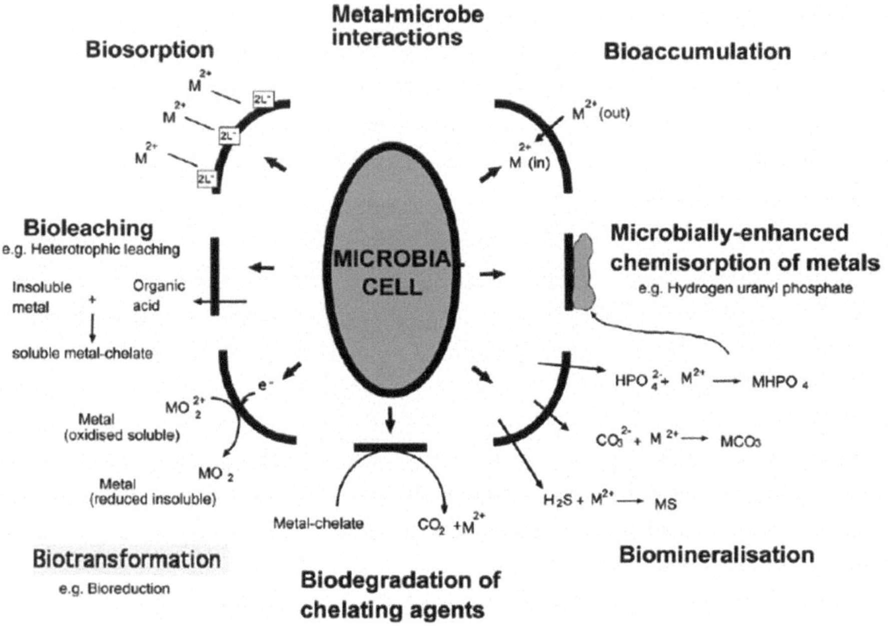

Fig. 20.2 Interactions of metals and microbes affecting Bioremediation. *Source* Tabak et al. (2005)

membrane due to several compounds released by the cell (Tabak et al. 2005; Banerjee et al. 2015). Several indigenous soil bacterial genera accumulates toxic metals such as *Escherichia hermannii* and *Enterobacter cloacae* showed resistance against Cd and Ni, *Bacillus cereus & Citrobacter* sp. uptake Pb and Cd, *Thiobacillus ferrooxidans* & *Bacillus subtilis* absorb Ag & Cr respectively. Similarly, *Pseudomonas aeruginosa* (U) & *Micrococcus luteus* (Sr) are also reported to show bioaccumulation. Certain fungal species efficiently deal with metals e.g. *Saccharomyces cerevisiae* act on U (Urenium), *Rhizopus arrhizus* act on Hg and *Aspergillus niger* on Th (Thorium) etc. (Juwarkar 2010).

20.5.1.3 Biotransformation

Microbiological transformations deals with the conversion of notorious pollutants (heavy metals) which can participate in the metabolic process. This technique is very useful to detoxify hazardous metals by reducing them enzymatically. Microorganisms takes up metals ions and then undergo various reactions such as oxidation, reduction, alkylation or methylation (Tabak et al. 2005). For example, *Corynebacterium* sp. shows biotransformation & reduce Chromium from its toxic form (Cr^{6+}) to less toxic form(Cr^{3+}) (Zhao et al. 2021). Similarly, *Bacillus licheniformis* cells can reduce of Pb^{2+} to Pb^{0} enzymatically (Jin et al. 2018).

20.5.1.4 Bioprecipitation

Various microbial activity may result in the precipitation or crystallization of metallic compounds which facilitate transformation of noxious metals into comparatively harmless one (Tarekegn et al. 2020). Eltarahony et al. (2020) reported that growing microbial cells secrete carbonate compounds which trap heavy metals causing precipitation. Such depositions of metals are greatly elevated when microorganisms tend to produce secondary metabolites. Previous researchers have shown that bio precipitation of Pb in a compound form ($PbHP_O{}^4$) that precipitates on the cell surface of by *Citrobacter* sp. & *Bacillus* sp (Peens et al. 2018).

20.5.1.5 Bioleaching of Metals

Bioleaching or biomining is the extraction of specific metal from mineral-rich natural compounds (ore) through microbial transformation. Bacteria like A*cidophilus ferrooxidans* & *Thiobacillus* sp. are capable to extract Cu, As, Hg, Pb, Fe, Ni etc. efficiently from mineral ore (Jerez 2017). Biomining widely used as a replacement of conventional chemical mining proved to be cost-effective & hazard-free. Several reports suggested that the microorganisms which are associated with bioleaching tend to have tolerance towards heavy metals. Since, this process produce certain organic acids like citric acid, gluconic acid & oxalic acid etc. which aids the mineralization of insoluble metal sulfides into soluble one.

20.5.1.6 Biomineralization

Biomineralization is the transformation process by which metallic compounds turns into crystalline precipitates. Microbial induced mineralization mainly based on cellular metabolism where metals are subjected to modify chemically and partially precipitates on the cell surface.

20.5.1.7 Cometabolism

Cometabolism is the process where degradation of one compound dependent on another compound (Hazen and Terry 2015). Usually, it is referred to as the simultaneous degradation of two compounds where the first substrate is fortuitously degraded by an enzyme which is the metabolic product of another compound (the secondary substrate). Typically, the microorganisms involved in it having no direct benefit from each other. Such co-metabolism strategies explored to cope with complex pollutants (Daniel et al. 2019; Zhao et al. 2021).

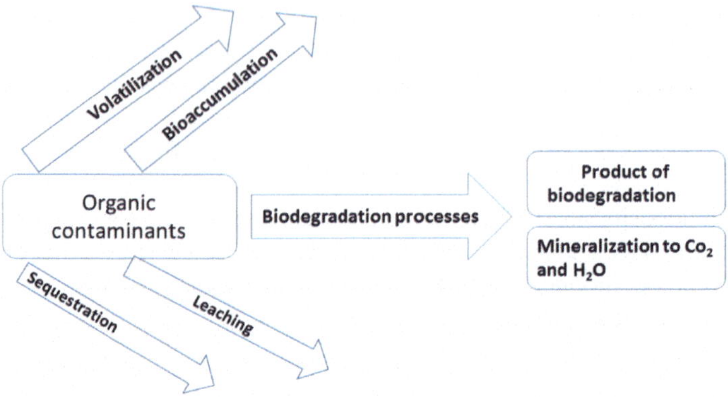

Fig. 20.3 Scheme of Microbial biodegradation of organic pollutants. Adapted from Tabek et al. (2005)

20.5.2 Remediation of Organic Pollutants

The major contaminants like Poly aromatic hydrocarbons (PAHs) and Polychlorinated biphenyls (PCBs) popularly known as Persistent Organic Pollutants (POPs) are found frequently and are considered as recalcitrant due to their high molecular weight & low water solubility (Mir and Gulfishan 2020). However, certain indigenous microorganisms have shown the potential to degrade these materials partially without hampering the native ecosystem and make these carcinogenic biologically unavailable (Perelo 2010; Megharaj et al. 2011; Mir and Gulfishan 2020) (Fig. 20.3).

Microbes mediated biodegradation of organic pollutants primarily occurs by anaerobic or aerobic metabolism and are mainly based on various processes including Monitored natural recovery (MNR), biostimulation & bioaugmentation & addition of compost material etc. (Kang 2014). Under controlled physical conditions microorganisms utilizing catabolic enzymes like oxygenase or dioxygenase to transform pollutants and ultimately the products of the microbial activity incorporated in the metabolic pathway (Perelo 2010). Bacterial species like *Pseudomonas sp., Burkholderia sp., Methococcus sp., Bacillus sp.* etc. were studied for their biodegradation capacity of PAHs, & PCBs (Kang 2014) (Fig. 20.4).

20.5.2.1 Monitored Natural Recovery (MNR)

Monitored natural recovery (MNR) is a sustainable process of remediating polluted sediments (Perelo 2010). A combined approach (biological & chemical) is adopted to treat contaminated site for a time span under close monitoring. MNR often employed indigenous factors which minimize the ecological and human health related risk significantly.

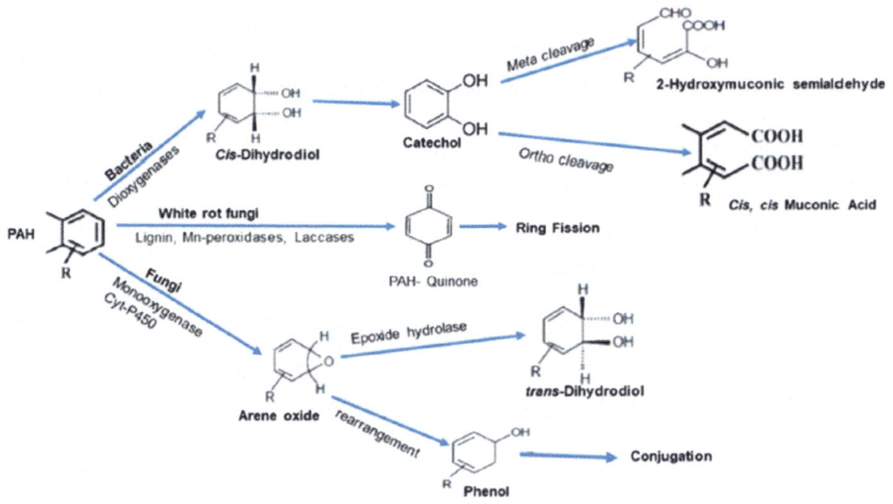

Fig. 20.4 Microbial Biodegradation Pathways of PAHs. Adapted from Sayara and Sanchez

20.5.2.2 Biostimulation

Biostimulation is the moderation of the growth parameters of microorganisms to enhance the rate of the bioremediation process in soil. Various nutrients such as phosphorus, nitrogen, oxygen, or carbon supplemented as stimulants for microorganisms (Ratnakar et al. 2016; Goswami et al. 2018). Preferably under controlled environment addition of the stimulants improves potential growth affecting biomass & accelerates bioremediation. (Igiri et al. 2018).

20.5.2.3 Addition of Compost

Many researchers have reported that the addition of inoculum in compost form in contaminated soil has shown a significant response in terms of bioremediation. (Kästner and Miltner 2016). Compost bioremediation has proven to be effective procedure for minimizing the toxicity of many types of contaminants, especially chlorinated and non-chlorinated hydrocarbons. This process works in a precise manner as it treats specific contaminants at specific sites therefore it is often called to as "tailored" or "designed" compost (Ratnakar et al. 2016).

20.5.2.4 Bioaugmentation

Bioaugmentation is the incorporation of exogenous microorganisms or genetically modified strains to contaminated sites to get rid of pollution. The idea behind this is to speed up the biotransformation of the hazardous elements into less toxic substances under optimized conditions (Kastner and Miltner 2016). These transformed substances can be further utilized by other microbes and be incorporated into metabolism (Smitha et al. 2017). This process is effectively used where other bioremediation processes failed to show satisfactory results due to the lack of sufficient microbial populations or efficacy (Megharaj et al. 2011).

20.6 Conclusion & Future Aspects

Rapid industrial development and unimpeded urbanization in an unplanned manner are producing enormous wastes and continuous uncontrolled dumping of these wastes affects soil physicochemical properties and productivity. There is no doubt about the need for industrialization at this progressive era but conservation of natural resources also indispensable & equally important. Thus, proper management and safe removal of wastes can be ensured to diminish soil pollution-related problems.

Acknowledgements The authors are grateful to the Bankura Sammilani College, Bankura & Ramananda College, Bishnupur for library assistance and necessary support to carry out this work.

References

Abdel-Shafy HI, Mona Mansour SM (2018) Solid waste issue: sources, composition, disposal, recycling, and valorization. Egypt J Pet 27:1275–1290

Acosta-Rodríguez I, Cárdenas-González JF, Rodríguez Pérez AS, Oviedo JT, Martínez-Juárez VM (2018) Bioremoval of different heavy metals by the resistant fungal strain aspergillus niger. Bioinorganic Chem Appl 1–7. https://doi.org/10.1155/2018/3457196

Agarwal V (2016) Industrial solid waste: emerging problems, challenges and its solution. Int J Manag Appl Science 2(7):68–72

Ahmad HR, Aziz T, Zia-ur-Rehman M, Sabir M, Khalid H (2016) Sources and composition of waste water: threats to plants and soil health. In: Soil science: agricultural and environmental prospectives, pp 349–370. https://doi.org/10.1007/978-3-319-34451-5_16

Aleksakhin RM (2009) Radioactive contamination as a type of soil degradation. Eurasian Soil Sci 42(12):1386–1396

Ali H, Khan E, Ilahi I (2019) Environmental chemistry and ecotoxicology of hazardous heavy metals: environmental persistence, toxicity, and bioaccumulation. J Chem. https://doi.org/10.1155/2019/6730305

Banerjee S, Gothalwal R, Sahu PK, Sao S (2015) Microbial observation in bioaccumulation of heavy metals from the Ash Dyke of thermal power plants of Chhattisgarh, India. Adv Biosci Biotechnol 6:131–138. https://doi.org/10.4236/abb.2015.62013

Bharagava RN, axena G (2020) Progresses in bioremediation technologies for industrial waste treatment and management: challenges and future prospects. Bioremediat Ind Waste Environ Safety. https://doi.org/10.1007/978-981-13-3426-9_21

Bhat RR (2015) Industrial effluents: a major threat to India. Int J Res 2(3):115–119

Briffa J, Sinagra E, Blundell R (2020) Heavy metal pollution in the environment and their toxicological effects on humans. Heliyon 6(9):e04691 https://doi.org/10.1016/j.heliyon.2020.e04691

Chadar SN, Chadar K (2017) Solid waste pollution: a hazard to environment. Recent Adv Petrochem Sci 2(3):41–43

Chhonkar PK, Datta S, Joshi HC, and Pathak S (2010) Impact of industrial effluents on soil health and agriculture—Indian experience: part II-tannery and textile industrial effluents. J Sci Ind Res 59(6):446–454

Daniel D, Jegathambal P, Bevers B (2019) In situ bioremediation of textile dye effluent-contaminated soils using mixed microbial culture. Int J Civ Eng 17:1527–1536. https://doi.org/10.1007/s40999-019-00414-5

Dwivedi S (2012) Bioremediation of heavy metal by algae: current and future perspective. J Adv Lab Res Biol 3(3):195–199

Eltarahony M, Zaki S, Abd-El-Haleem D (2020) Aerobic and anaerobic removal of lead and mercury via calcium carbonate precipitation mediated by statistically optimized nitrate reductases. Sci Rep 10:4029. https://doi.org/10.1038/s41598-020-60951-1

Engwa GA, Ferdinand PU, Nwalo FN and. Unachukwu MN (2019) mechanism and health effects of heavy metal toxicity in humans https://doi.org/10.5772/intechopen.82511

Ferronato N, Torretta V (2019) Waste mismanagement in developing countries: a review of global issues. Int J Environ Res Public Health. https://doi.org/10.3390/ijerph16061060

Fu X, Qiao Y, Xue J, Cheng D, Chen C, Bai Y, Jiang Q (2021) Analyses of community structure and role of immobilized bacteria system in the bioremediation process of diesel pollution seawater. Sci Total Environ 799:149439. https://doi.org/10.1016/j.scitotenv.2021.149439

Goswami M, Chakraborty P, Mukherjee K, Mitra G, Bhattacharyya P, Dey S, Tribedi P (2018) Bioaugmentation and biostimulation: a potential strategy for environmental remediation. J Microbiol Experimentation 6(5):223–231

Hiroki M (1992) Effects of heavy metal contamination on soil microbial population. Soil Plant Nutr 8(1):141–147

Igiri BE, Okoduwa SIR, Idoko GO, Akabuogu EP, Adeyi AO, Ejiogu IK (2018) Toxicity and bioremediation of heavy metals contaminated ecosystem from tannery wastewater: a review. J Toxicol 2018:1–16

Jaishankar M, Tseten T, Anbalagan N, Mathew BB, Beeregowda KN (2014) Toxicity, mechanism and health effects of some heavy metals. Inter Discipline Toxicol 7(2):60–72

Jerez CA (2017) Biomining of metals: how to access and exploit natural resource sustainably. Microb Biotechnol 10(5):1191–1193. https://doi.org/10.1111/1751-7915.12792

Jin Y, Luan Y, Ning Y, Wang L (2018) Effects and mechanisms of microbial remediation of heavy metals in soil: a critical review. Appl Sci 8(8):1336. https://doi.org/10.3390/app8081336

Juwarkar AA, Yadav SK (2010) Bioaccumulation and biotransformation of heavy metals. Bioremediat Technol 266–284. https://doi.org/10.1007/978-90-481-3678-0_9

Kang JW (2014) Removing environmental organic pollutants with bioremediation and phytoremediation. Biotech Lett 36(6):1129–1139

Kastner M, Miltner A (2016) Application of compost for effective bioremediation of organic contaminants and pollutants in soil. Appl Microbiol Biotechnol 100(8):3433–3449

Kumar A, Agrawal A (2020) Recent trends in solid waste management status, challenges, and potential for the future Indian cities—a review. Curr Res Environ Sustain. https://doi.org/10.1016/j.crsust.2020.100011

Lavanya T, Purushothaman R, Nathiya S, Gandhi RS, Sivaraja M (2019) Impact of industrial solid waste on soil and subsurface water nearby TNPL. Int J Mach Const Engg 6(1):2394–3025

Megharaj M, Ramakrishnan B, Venkateswarlu K, Sethunathan N, Naidu R (2011) Bioremediation approaches for organic pollutants: a critical perspective. Environ Int 37(8):1362–1375

Mishra S, Pradhan N, Panda S, Akcil A (2016) Biodegradation of Dibenzothiophene (DBT) and its application in the production of clean coal. Fuel Process Technol 152:325–342. https://doi.org/10.1016/j.fuproc.2016.06.025

Mir RA, Gulfishan M (2020) The biodegradation of organic pollutants. Eur J Mol & Clin Med 7(10):3552–3562

Ojuederie OB and Babalola OO (2017) Microbial and plant-assisted bioremediation of heavy metal polluted environments: a review. Int J Environ Res Public Health 14(12). https://doi.org/10.3390/ijerph14121504

Peens J, Wu Y, Brink H (2018) Microbial Pb (II) precipitation: the influence of elevated Pb (II) concentrations. Chem Eng Trans 64:583–588

Perelo LW (2010) Review: In situ and bioremediation of organic pollutants in aquatic sediments. J Hazard Mater 177(1–3):81–89

Rajendran P, Muthukrishnan J, Gunasekaran P (2003) Microbes in heavy metal remediation. Indian J Exp Biol 41:935–944

Ratnakar A, Shankar S, Shikha (2016) An overview of biodegradation of organic pollutants. Int J Sci Innov Res 2016(1):73–91

Shamim S (2018) Biosorption of heavy metals. Biosorption. https://doi.org/10.5772/intechopen.72099

Smitha MS, Singh S, Singh R (2017) Microbial biotransformation: a process for chemical alterations. J Bacteriol Mycol Open Access 4(2):47–51

SnehLata KHP, Mishra T (2019) Cadmium bioremediation: a review. Int J Pharm Sci Res 10(9):4120–4128

Sonawane DV, Lawande SP, Gaikwad VB, and Kuchekar SR (2010) Impact of industrial waste water on soil quality and organic matter around Kurkumbh industrial area Daund, Pune District (MS). Int J Chem Sci 8(1): 97–102

Subramani T, Mangaiyarkarasi M, Kathirvel C (2014) Impact of sewage and industrial effluent on soil plant health act on environment. J Eng Res Appl 4(2): 270–273

Tabak H, Lens P, Hullebusch EV, Dejonghe W (2005) Developments in bioremediation of soils and sediments polluted with metals and radionuclides-1. Microbial processes and mechanisms affecting bioremediation of metal contamination and influencing metal toxicity and transport. Rev Environ Sci Bio/Technol 4:115–156

Tarekegn MM, Salilih FZ, Ishetu AI (2020) Microbes used as a tool for bioremediation of heavy metal from the environment. Cogent Food Agric. https://doi.org/10.1080/23311932.2020.1783174

Tchounwou PB, Yedjou CG, Patlolla AK, Sutton DJ (2012) Heavy metal toxicity and the environment. Exp Suppl 101:133–164

Tejaswi PB, Sharma KL. Mandal UK, Solomon Raju1 AJ, Venkanna K., Masanne R, Samuel J, Karthikeyan K (2017) Soil quality and heavy metal contamination of soils in mindi industrial area, Visakhapatnam, Andhra Pradesh, India. Indian J Ecol 44(4):774

Vincent Q, Auclerc A, Beguiristain T, Leyval C (2018) Assessment of derelict soil quality: abiotic, biotic and functional approaches. Sci Total Environ 613–614:990–1002. https://doi.org/10.1016/j.scitotenv.2017.09.118

Vongdala N, Tran HD, Xuan T, Teschke R, Khanh T (2018) Heavy metal accumulation in water, soil, and plants of municipal solid waste landfill in Vientiane, Laos. Int J Environ Res Public Health 16(1):22. https://doi.org/10.3390/ijerph16010022

Zwolak A, Sarzyńska M, Szpyrka E, Stawarczyk K (2019) Sources of soil pollution by heavy metals and their accumulation in vegetables: a review. Water, Air, Soil Pollut 230(7). https://doi.org/10.1007/s11270-019-4221-y

Zhao MM, Kou J, Chen Y, Xue L, Fan TT, Wang S (2021) Bioremediation of wastewater containing mercury using three newly isolated bacterial strains. J Cleaner Prod 299:126869. https://doi.org/10.1016/j.jclepro.2021.126869

Zhan Q, Qian C, Wang M (2015) In situ bioremediation of heavy metals in contaminated soil using microbial agents and planting experiments. Polish J Environ Stud 24(3):1395–1400. https://doi.org/10.15244/pjoes/36987

Chapter 21
Pollutants Bioremediation Using Biosurfactants: A Novel Approach for Improving Soil Health

Varun Dhiman, Anand Giri, and Deepak Pant

Abstract Heavy metals and hydrocarbons are considered significant soil pollutants. Their continued interaction with soil biota represents a genuine risk to the soil ecosystem. Subsequently, the removal of these toxins is a must to improve soil health and henceforth utilization of biosurfactants gives an option yet amazing novel methodology for remediation of contaminated soil. Attributable to exceptionally different applications, biosurfactants end up being profoundly viable in managing the persisting challenge. Biosurfactants are a group of surface-active substances produced by microorganisms that have a wide range of structural characteristics and amphiphilic structure in nature. Because of their potential benefits, biosurfactants are widely employed in a variety of industries, including agriculture, food production, chemistry, cosmetics, and pharmaceutics. Different examples determine the efficiency of biosurfactants in soil bioremediation. Recent studies found that bacterial biosurfactants remove aliphatic and aromatic compounds from oil-contaminated soil with the removal efficiency of 81.6% and 43.8% respectively. It has been further observed that the precipitation method using biosurfactant solution can remove heavy metals (zinc, arsenic, and cadmium) from the affected soil with 84.5–100% efficacy. The current chapter provides a comprehensive overview of biosurfactant production by microbes and their role in heavy metal and hydrocarbon remediation from polluted soil.

Keywords Biosurfactants · Heavy metals · Hydrocarbons · Microbes · Bioremediation

---

V. Dhiman · A. Giri (✉) · D. Pant
Department of Environmental Sciences, Central University of Himachal Pradesh, Dharamshala, India
e-mail: anandgiri40855@gmail.com

P. K. Shit et al. (eds.), *Soil Health and Environmental Sustainability*, Environmental Science and Engineering,
https://doi.org/10.1007/978-3-031-09270-1_21

21.1 Introduction

Biosurfactants are versatile biological surface-active molecules synthesized by a variety of bacterial, fungal, and yeast species (Karlapudi et al. 2018; Ghasemi et al. 2019). Their amphipathic nature permits them to interact in both polar and non-polar media (Santos et al. 2016a). They can work with air–water, oil–water, air–solid, and liquid–solid interfaces. This is because the polar and non-polar ends reduce the repulsive forces between different interfaces and accordingly enhance their interactions and mixing (Soberón-Chávez and Maier 2011). The higher surface tension reduction potential of biosurfactants identifies with its higher proficiency in various working conditions (Desai and Banat 1997). Already, synthetic surfactants were in well-known use. These were essentially oil inferred surfactants with their environmental concerns. Due to their non-biodegradable nature, persistence, and higher toxicity, scientists develop environment-friendly biosurfactants of microbial origin. These biosurfactants are known for their structural diversity (Fig. 21.1), low toxicity, higher biodegradability, lower CMC, pH tolerance, potent selectivity, renewable byproducts, and thermostability. All these parameters draw the attention of numerous industries subsequently extended their assorted use in agriculture, pharmaceuticals, cosmetic formulations, food processing, and environmental bioremediation. The soil remediation using a sustainable management process reduces the risk imposed on human health and the environment. The bio-based surfactants provide a potential solution for soil treatment. The waste material employed for the production of biological surfactants provides a cheap and effective bioremediation method for the metal–organic contaminants removal and soil washing (Mulligan 1993). Their use increases the soil affinity towards numerous metals present in the soil environment which in turn enhances the process of soil remediation.

The use of biosurfactants is proved to be more effective when they are used and selected based on pollutants characteristics and properties, efficiency, time constraints, time, and regulatory requirements. Further, appropriate biosurfactant selection can be done through the understanding of the interaction mechanism between the pollutants and biosurfactants in varied environmental conditions (Mulligan 2021). Recently, researchers enhanced the biosurfactant yield using the *Pseudomonas sp.* CQ2. The obtained biosurfactant was found to exhibit higher removal and bioleaching efficiency of heavy metals from the soil (Sun et al. 2021). The biosurfactants are also used as an additive to enhance pollutants solubility. Recently, Rhamnolipid biosurfactants were used as additives with the biochar and nitrogen as an integrative tool to remediate crude oil-contaminated coastal marshy soil (Wei et al. 2020). Following current research on biosurfactants-assisted bioremediation summaries in Table 21.1.

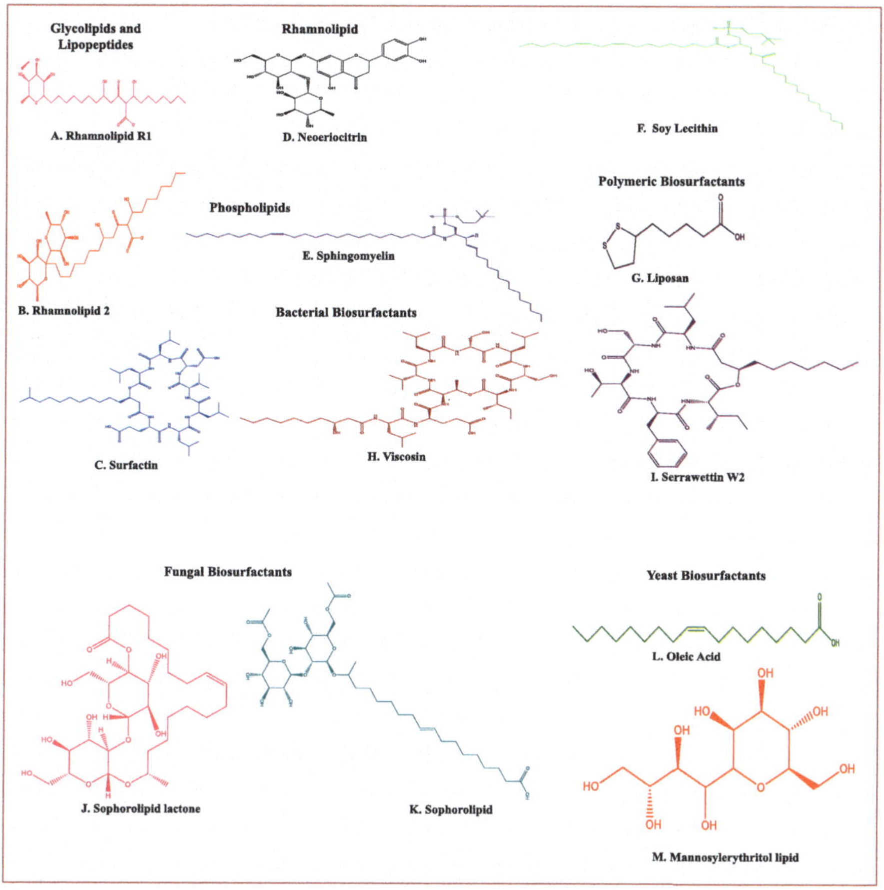

Fig. 21.1 Structural diversity of biosurfactants representing different bacterial, fungal, and yeast biosurfactants (*Source* Perfumo et al. 2010; Gudiña et al. 2013; Singh et al. 2018)

21.1.1 Physiochemical Properties

The practical implications require the basic understanding of the physiochemical aspects of biosurfactants which develops the knowledge of their molecular characteristics and interfacial behavior. Some of the important physicochemical properties have been discussed below (Fig. 21.2):

A. Reduced Surface and interfacial tension

As previously discussed, biosurfactants have both polar and non-polar ends. They have the potential of reducing surface and interfacial tension between different interfaces. They form kinetically stabilized emulsions by performing adsorption at different interfaces. Also, the hydrophobic moiety and the size of the head group

Table 21.1 Recent studies on biosurfactants-assisted soil bioremediation

Pollutant	Bioremediator	Maximum removal	References
Creosote PAHs	*Bacillus cereus* SPL-4	79% of 5- and 6-ring PAH	Bezza and Chirwa (2017)
Dichlorodiphenyltrichloroethane (DDT)	*Pseudomonas* sp. SB	> 60%	Wang et al. (2017)
Heavy metals (Ni, Cu, Cd)	Rhamnolipid biosurfactant	–	Lee and Kim (2019)
Oily sludge	*Pseudomonas stutzeri* Z12	71.9%	Pourfadakari et al. (2020)
PHC	Rhamnolipids	92.3%	Li et al. (2018)
Petroleum hydrocarbons	*Pseudomonas Aeruginosa* SR17	86.1%	Patowary et al. (2018)
Pyrene	Rhamnolipid type biosurfactant, *Pseudomonas aeruginosa strain* R_4	82%	Ahmadi et al. (2021)
Dichlorophenol	*Bacillus axarquiensis*	80–85%	Christopher et al. (2021)
Pyrene	*Pseudomonas* sp. ISTPY2	97%	Swati et al. (2020)
Lead	*Bacillussubtilis* SASCBT01	65%	Selvam et al. (2021)
TPH	Surfactin and rhamnolipid	86%	Fanaei et al. (2020)
Hg, Pb	Rhamnolipid	50.2%, 62.5%	Chen et al. (2021)
Pb	*Bacillus cereus* NWUAB01	69%	Ayangbenro and Babalola (2020)
4-Chloroaniline	*Bacillus* sp + lipopeptide surfactant	100%	Femina Carolin et al. (2021)

determine the biosurfactant's behavior for reducing the interfacial tension. In the case of microbial bioremediation of hydrophobic contaminants, biosurfactants modify the surface energy of microbial cells (De et al. 2015).

B. Micellization

The process of Micellization occurred at the bulk concentration of solvent and above the CMC value (Perinelli et al. 2020). The Micellization is spontaneous in these conditions (Santos et al. 2016b). The process develops a thermodynamically

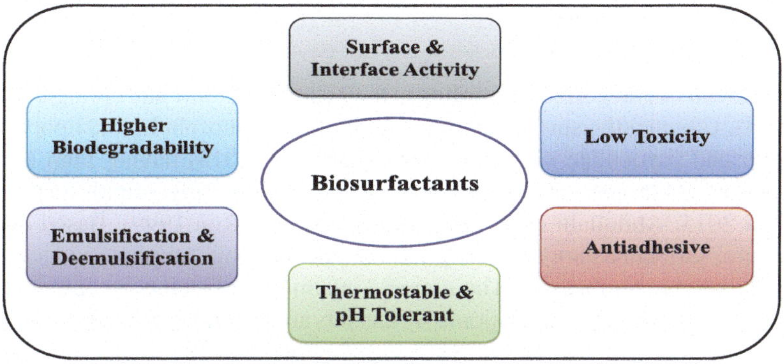

Fig. 21.2 Physiochemical properties of biosurfactants (*Source* Soberón-Chávez and Maier 2011; Morais et al. 2017; Jahan et al. 2020)

stable nanostructure due to equilibrium (Hanafy et al. 2018). The self-assembly has been observed in the biosurfactants which spontaneously give rise to hydrophobic, hydrogen bonding, and Vander Waals interactions resulting in the formation of micelles having a typical radius of a few nanometers (Lombardo et al. 2015). As the surfactant concentration goes on the rise, an increase in the number of micelles has also been observed (Jahan et al. 2020).

C. Solubility

The hydrophobic compounds can be solubilized by self-assembled micelles. The relative concentration, pH, and salinity are the major factors to determine the solubilization behavior of hydrophobic compounds (Abouseoud et al. 2010). Different studies show the role of increasing pH. The increase in pH ranges from 4 to 11 lowers the surface tension and enhances the emulsion stability (Cameotra et al. 2010; Benderrag et al. 2016).

D. Emulsification and Demulsification

Biosurfactants show effective emulsification and demulsification behavior (Rocha e Silva et al. 2017). The formed emulsions are kinetically stable emulsions and their chemical structures and physicochemical properties are regulated by different emulsion components and their varied composition. Different factors such as temperature and pressure, input energy, used equipment, and mixing time will also influence the required conditions for the process of emulsification (Amodu et al. 2014) while different mechanisms such as flocculation, coagulation, creaming, coalescence, and Ostwald ripening are essential in causing demulsification (Liang 2015). The bacterial strain *Bacillus subtilis* LAM005produces Surfactin biosurfactant shows a higher emulsification index ($E_{24} > 50\%$) on kerosene and soybean oil (Hsieh et al. 2004; Campos et al. 2014) while other strains *P. aeruginosa* LBI produce Rhamnolipids which are highly efficient in the emulsification of numerous hydrocarbons at saline conditions with a pH > 5 (Benincasa et al. 2004).

21.1.2 Classification

Surfactants can be categorized according to their chemical, ionic and biological origin (Fig. 21.3). Chemically, surfactants include lipolipids, phospholipids, lipoproteins, polymeric and particulate surfactants (Shekhar et al. 2015) having high and low molecular weight molecules (e.g. polymeric and particulate, glycolipids respectively Matsuoka 2015; Abdullahi et al. 2020; Rosenberg and Ron 1999). Based on their ionic nature, surfactants are categorized as anionic, cationic, amphoteric, non-ionic, and combinational surfactants. The amine-containing surfactants are cationic (e.g. alkyl pyridinium) in nature and anionic surfactants are generally synthetic, traditional soaps, detergents, sulfates, and sulphonate compounds. Compounds like ethoxylates and alkyl ethoxy sulfates are examples of non-ionic and combinational surfactants respectively. The quantity and nature of the biosurfactants produced is mainly depending upon the source and microbial interactions. Different energy substrates of C, N, P, Mg, Fe, and Mn influence biosurfactants production.

21.2 Synthesis of Precursor Molecule for Biosurfactants Production

Synthesis of polar moiety and cellular metabolic reactions are regulated by microbes using hydrophilic substrates while the hydrocarbon part of surfactant is formed with the help of hydrophobic substrates (Weber et al. 1992; Desai and Banat 1997). The nature of the carbon source defines the nature of metabolic pathways (Giri et al.

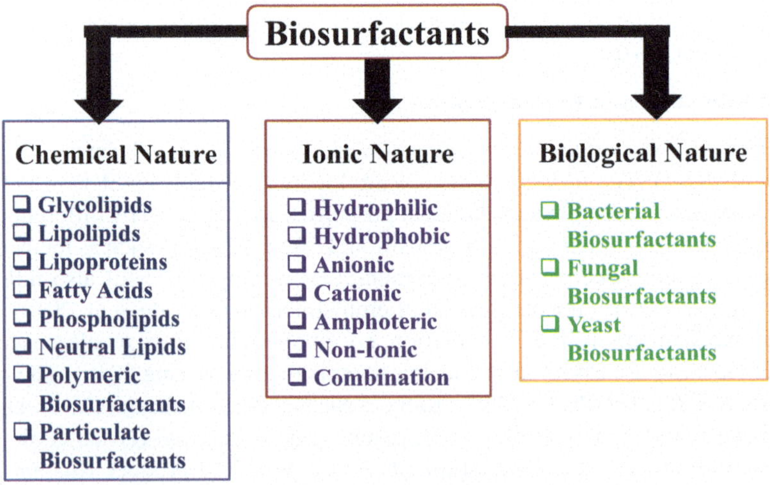

Fig. 21.3 Diagrammatic representation of biosurfactant classification (*Source* Shekhar et al. 2015)

2018; Giri and Pant 2019). For example, glycolipids production depends upon the carbohydrate as a carbon source.

Glycerol and glucose are hydrophilic substrates that formed several intermediates subsequently such as glucose and fructose 6-phosphate during the glycolytic pathway. In this process, the glucose is transformed to pyruvate and then to acetyl-CoA through glycolysis followed by malonyl-CoA formation. The malonyl-CoA is further associated with OAA to form fatty acid which acts as an important precursor molecule for lipid synthesis (Hommel and Huse 1993). Figure 21.4 illustrates the lipid formation using the glucose molecules.

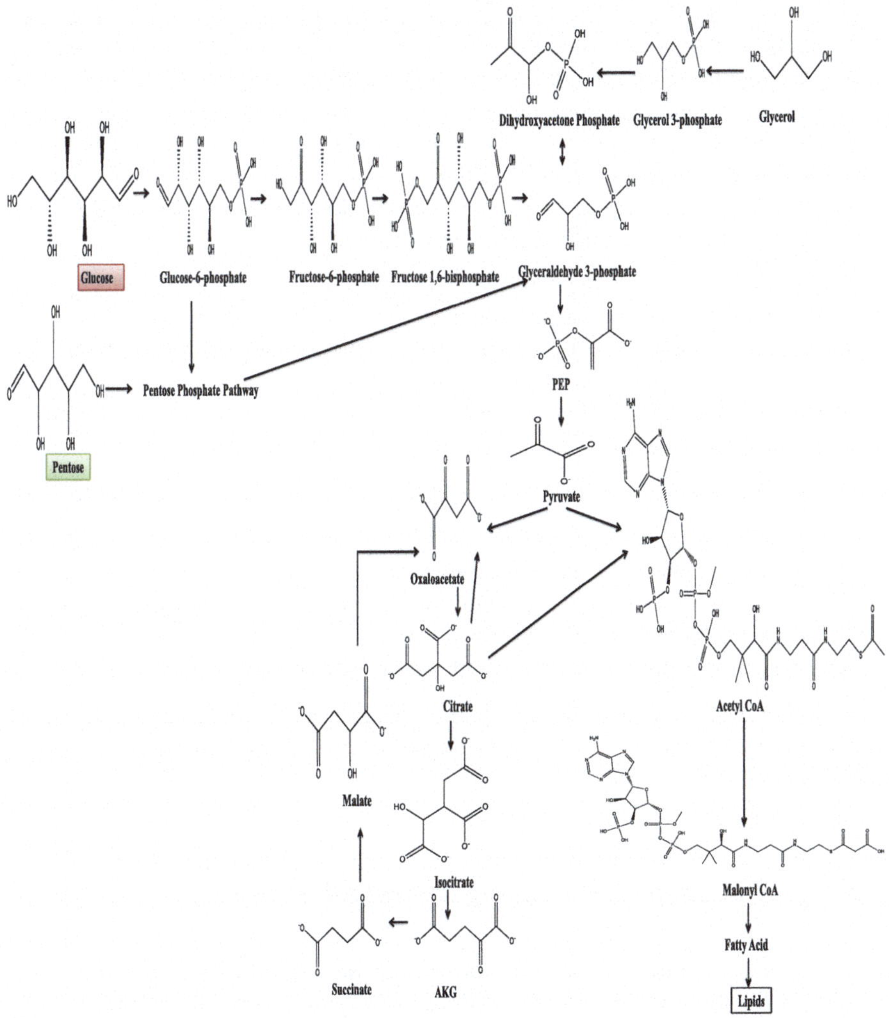

Fig. 21.4 Synthesis of precursor molecule for biosurfactants production (*Source* Haritash and Kaushik 2009; Santos et al. 2016a, b)

21.2.1 Microbial Synthesis

Biosurfactants are amphipathic compounds produced by microbes such as bacteria. fungi, yeasts, etc., and consist of both hydrophobic and hydrophilic domains. The hydrophobic domain of amphipathic molecules (long-chain hydroxy fatty acids) and hydrophilic domain (carboxylic, carbohydrate, alcohol, and phosphate group) are responsible for the synthesis of various biosurfactant moieties and their applications (Marcelino et al. 2020). Carbohydrates and lipids are the main carbon sources for biosurfactant production, however many remediation techniques also used hydrocarbon. Agro-industrial by-products as a sugar and lipid sources can also be used for biosurfactant production (Nitschke et al. 2004) for cell development and biosurfactant production. A previous study explained that microbial surfactants can induce by using hydrophobic substrate and different polar carbon compounds (Sharma et al. 2020). Thus, hydrocarbons, polar carbon sources, and oils can be used strategically in the synthesis of these byproducts. Microbial synthesis of surfactants is a complex process and products from catabolic pathways can act as precursors of biosurfactant synthesis. The surfactants producing microorganisms can be divided into the following three different groups as per alkane utilization.

1. Some microorganisms like *Arthrobacter sp.*, *corynebacterium sp.*, and *Nocardia sp.* produce surfactants during growth on alkane.
2. *Pseudumonas aeruginosa* produces surfactants during growth on alkane and water-soluble compounds (Sharma et al. 2021).
3. Some microbes like *Bacillus subtilis* produce surfactants during growth on water-soluble compounds.

There are four main possibilities for the production of such amphiphilic molecules:

1. De novo synthesis of hydrophilic and hydrophobic moieties through independent pathways and induced by the substrate.
2. The hydrophilic moiety is dependent on the presence of a substrate, whereas the hydrophobic moiety is synthesis de novo
3. The carbon substrate affects the synthesis of both residues.

The following mechanism is important for large-scale production and design of growth media and conditions as well as for the insertion of precursor molecules to induce metabolic pathways.

21.2.2 Biosurfactant Producing Microbial Strains

Table 21.2 shows some of the most promising and well-studied biosurfactants. Glycolipids are the most common microbial surfactants, while lipopeptide biosurfactants are more structurally diverse.

Table 21.2 Some of the most promising and well-studied biosurfactants

Sr. no	Microorganism	Biosurfactanttype	References
1	*Candida lipolytica*	Glycolipid	Santos et al. (2016a, b)
2	*Candida sphaerica*UCP 0995	Glycolipid	Santos et al. (2014)
3	*Starmerellabombicola* ATCC 22,214	Glycolipid	Liu et al. (2018)
4	*Ustilago sp.*	Cellobioselipids	Sineriz et al. (2001)
5	*Arthrobacter sp.*	Corynomycolates	
6	*Candida sp.*	Mannosylerythriol Lipids	
7	*Pseudomonas sp.*	Rhamnoselipids	
8	*Torulopsis sp.*	Sophoroselipids	
9	*Bacillus licheniformis*	Lipopeptides	
10	*Bacillus subtilis*	Surfactatin	

21.3 Environmental Bioremediation and Soil Health

The presence of heavy metals, hydrocarbons, and other organic pollutants in different environmental spheres has caused serious environmental concerns (Goswami et al. 2019; Giri and Pant 2018) These pollutants adversely affect the terrestrial environment and degrade the soil's health by binding with terrestrial mineral particles, soil organic matter, and its local flora and fauna. They interact with these components and cause toxic impacts to them. Different studies proved the beneficiary role of numerous microbial species in the bioremediation of different organic pollutants persists in different environmental spheres. For example, bacterial strains proved fruitful in remediating the existed hydrocarbons and their derivatives from the soil and improving its health. The emulsifying action of different microbial biosurfactants makes hydrocarbon more bioavailable for their degradation purpose. Hydrocarbon degradation, pseudo-solubilization, and their adhesion help to form small oil droplets which are easily removed from the contaminated soil (Paleček et al. 2015). The chemical composition is the major factor that defines surfactant bioactivity. They act intracellularly or extracellularly (Antoniou et al. 2015). The organic pollutant phenanthrene was degraded with the help of non-ionic bacterial surfactants. The use of 0.5% of Tween 80 in combination with the fungi *Polyporus* species, S133 increases the phenanthrene biodegradation by accelerating fungal growth from 56 to 88% (ref 1). Also, the aromatic hydrocarbons bioremediate from the polluted soil using the trehalose-5, 5'-dicorynomycolates surfactant (Itrich et al. 2015). In another study, bacterial surfactants show their higher potential to remediate PAHs from contaminated soil sites by producing surface-active glycolipids (Chakrabarti 2012). Bacterial biosurfactants develop biofilms with the pollutant surface and alter their wettability and surface properties. The microbial strains such as *Acinetobacter hemolysis*, *Pseudomonas ML2*, and *Pseudomonas aeruginosa* seem to degrade numerous hydrocarbons (Karlapudi et al. 2018). Similarly, *Pseudomonas* ML2 biosurfactant reported a reduction of 39–71% of hydrocarbons from the contaminated soil (ref 2). Addition of rhamnolipid biosurfactant to the hydrocarbons containing silt loam and sandy loam soil results in the hydrocarbon recovery of 25–70% and 40–80% respectively (ref 3). In another study, 82% of tetradecane was observed in the soil when treated with *Pseudomonas aeruginosa* UG2 biosurfactant during the 2 months incubation period (ref 4). Further, JBR-425, an anionic biosurfactant was found to be efficient in reducing heavy metals such as zinc, copper, lead, and cadmium in the affected soil by 39, 56, 68, and 43%, respectively (ref 5).

21.4 Bioremediation Mechanism

It has been shown that biosurfactants exhibit higher biodegradation efficiency of pollutants removal by solubilizing or emulsifying numerous pollutants present in the soil environment. The solubilization and emulsification can be done with the help of enhancing soil enzyme activity or interfacial uptake of pollutants by microbial species. Figure 21.5 shows the mechanism of multilevel interactions between biosurfactants, pollutants, microbial species, and soil environment in the bioremediation process.

The soil bioremediation potential of the biosurfactants has received considerable interest and attention among the scientific community which has been reported in many studies and scientific literature (Table 21.1). The researchers proposed the use of biosurfactants to enhance the pollutants solubility which was likely to be responsible for their efficient biodegradation and hence, removal from the contaminated soil.

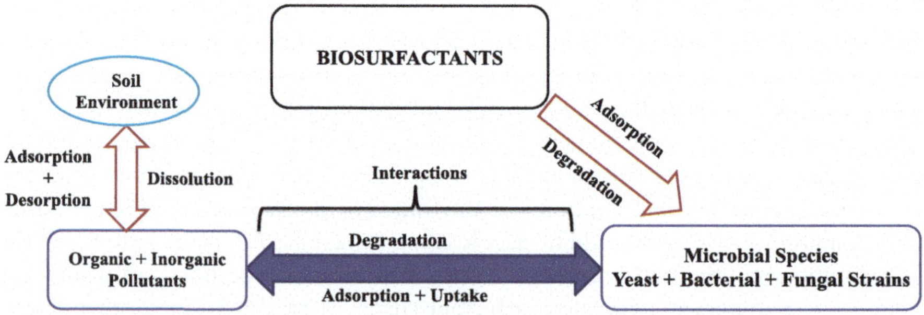

Fig. 21.5 Mechanism of multilevel interactions between biosurfactants, pollutants, microbial species, and soil environment in the bioremediation process. Adapted from Liu et al. 2018

21.5 Conclusion and Future Directions

Now's days, the development of biosurfactants attracts more attention towards sustainable and environmentally friendly development. Biosurfactants offer the possibility of replacing chemical surfactants and higher potential to remediate PAHs from contaminated soil. From the above, it is concluded that biosurfactants are less damaging to the environment, and due to their wide applicability, they have been used to improve soil health as demonstrated in their potential of hydrocarbon biodegradation and heavy metal removal through their surface-active compounds that enhance the bioavailability, bacterial growth, and bioremediation rates. From the future perspective, much work is needed to enhance the biosurfactants process sustainability. It is important to develop alternative substrates for their sustainable and economical production. In-situ biosurfactant addition to the contaminated soil may further enhance the bioremediation efficiency and also reduce the contamination risk. It is further needed to deeply understand the environmental fate of soil pollutants and their interactions with the biosurfactants that will likely open new ways for the development of soil bioremediation strategies.

References

Abdullahi W, Crossman M, Griffiths PC (2020) Surfactant-Modulation of the Cationic-Polymer-Induced Aggregation of Anionic Particulate Dispersions. Polym. 12

Abouseoud M, Yataghene A, Amrane A, Maachi R (2010) Effect of pH and salinity on the emulsifying capacity and naphthalene solubility of a biosurfactant produced by Pseudomonas fluorescens. J Hazard Mater 180:131–136. https://doi.org/10.1016/j.jhazmat.2010.04.003

Ahmadi M, Niazi F, Jaafarzadeh N et al (2021) Characterization of the biosurfactant produced by Pesudomonas areuginosa strain R4 and its application for remediation pyrene-contaminated soils. J Environ Heal Sci Eng 19:445–456. https://doi.org/10.1007/s40201-021-00617-w

Amodu OS, Ntwampe SK, Ojumu TV (2014) Emulsification of hydrocarbons by biosurfactant: Exclusive use of agro-waste. BioResources 9:3508–3525

Antoniou E, Fodelianakis S, Korkakaki E, Kalogerakis N (2015) Biosurfactant production from marine hydrocarbon-degrading consortia and pure bacterial strains using crude oil as carbon source. Front Microbiol 6:274

Ayangbenro AS, Babalola OO (2020) Genomic analysis of Bacillus cereus NWUAB01 and its heavy metal removal from polluted soil. Sci Rep 10:19660. https://doi.org/10.1038/s41598-020-75170-x

Benderrag A, Daaou M, Bounaceur B, Haddou B (2016) Influence of pH and cationic surfactant on stability and interfacial properties of algerian bitumen emulsion. Chem Pap 70:1196–1203. https://doi.org/10.1515/chempap-2016-0061

Benincasa M, Abalos A, Oliveira I, Manresa A (2004) Chemical structure, surface properties and biological activities of the biosurfactant produced by Pseudomonas aeruginosa LBI from soapstock. Antonie Van Leeuwenhoek 85:1–8. https://doi.org/10.1023/B:ANTO.0000020148.45523.41

Bezza FA, Chirwa EMN (2017) The role of lipopeptide biosurfactant on microbial remediation of aged polycyclic aromatic hydrocarbons (PAHs)-contaminated soil. Chem Eng J 309:563–576. https://doi.org/10.1016/j.cej.2016.10.055

Cameotra SS, Makkar RS, Kaur J, Mehta SK (2010) Synthesis of Biosurfactants and Their Advantages to Microorganisms and Mankind BT - Biosurfactants. In: Sen R (ed) Springer. New York, NY, New York, pp 261–280

Campos JM, Stamford TLM, Sarubbo LA (2014) Production of a Bioemulsifier with Potential Application in the Food Industry. Appl Biochem Biotechnol 172:3234–3252. https://doi.org/10.1007/s12010-014-0761-1

Chakrabarti S (2012) Bacterial biosurfactant: Characterization, antimicrobial and metal remediation properties

Chen Q, Li Y, Liu M et al (2021) Removal of Pb and Hg from marine intertidal sediment by using rhamnolipid biosurfactant produced by a Pseudomonas aeruginosa strain. Environ Technol Innov 22:101456. https://doi.org/10.1016/j.eti.2021.101456

Christopher JM, Mohan M, Sridharan R et al (2021) Biosurfactant matrix for the environmental clean-up of dichlorophenol from aqueous medium and soil. Environ Sci Pollut Res. https://doi.org/10.1007/s11356-021-15265-8

De S, Malik S, Ghosh A et al (2015) A review on natural surfactants. RSC Adv 5:65757–65767. https://doi.org/10.1039/C5RA11101C

Desai JD, Banat IM (1997) Microbial production of surfactants and their commercial potential. Microbiol Mol Biol Rev 61:47 LP – 64

Fanaei F, Moussavi G, Shekoohiyan S (2020) Enhanced treatment of the oil-contaminated soil using biosurfactant-assisted washing operation combined with H2O2-stimulated biotreatment of the effluent. J Environ Manage 271:110941. https://doi.org/10.1016/j.jenvman.2020.110941

Femina Carolin C, Senthil Kumar P, Chitra B, et al (2021) Stimulation of Bacillus sp. by lipopeptide biosurfactant for the degradation of aromatic amine 4-Chloroaniline. J Hazard Mater 415:125716. https://doi.org/10.1016/j.jhazmat.2021.125716

Ghasemi A, Moosavi-Nasab M, Setoodeh P et al (2019) Biosurfactant Production by Lactic Acid Bacterium Pediococcus dextrinicus SHU1593 Grown on Different Carbon Sources: Strain Screening Followed by Product Characterization. Sci Rep 9:5287. https://doi.org/10.1038/s41598-019-41589-0

Giri A, Banerjee UC, Kumar M, Pant D (2018) Intracellular carbonic anhydrase from Citrobacter freundii and its role in bio-sequestration. Biores Technol 267:789–792

Giri A, Pant D (2018) Inhalation dose due to Rn-222, Rn-220 and their progeny in indoor environments. Appl Radiat Isot 132:116–121

Giri A, Pant D (2019) CO2 management using carbonic anhydrase producing microbes from western Indian Himalaya. Bioresource Technology Reports 8:100320

Goswami M, Pant G, Mansotra DK, Sharma S, Joshi PC (2021) Biochar: a carbon negative technology for combating climate change. In: Advances in carbon capture and utilization. Springer, Singapore, pp 251–272

Gudiña EJ, Rangarajan V, Sen R, Rodrigues LR (2013) Potential therapeutic applications of biosurfactants. Trends Pharmacol Sci 34:667–675. https://doi.org/10.1016/j.tips.2013.10.002

Hanafy NAN, El-Kemary M, Leporatti S (2018) Micelles Structure Development as a Strategy to Improve Smart Cancer Therapy. Cancers (basel) 10:238. https://doi.org/10.3390/cancers10070238

Haritash AK, Kaushik CP (2009) Biodegradation aspects of Polycyclic Aromatic Hydrocarbons (PAHs): A review. J Hazard Mater 169:1–15. https://doi.org/10.1016/j.jhazmat.2009.03.137

Hommel RK, Huse K (1993) Regulation of sophorose lipid production by Candida (Torulopsis) apicola. Biotechnol Lett 15:853–858. https://doi.org/10.1007/BF00180154

Hsieh F-C, Li M-C, Lin T-C, Kao S-S (2004) Rapid Detection and Characterization of Surfactin-Producing Bacillus subtilis and Closely Related Species Based on PCR. Curr Microbiol 49:186–191. https://doi.org/10.1007/s00284-004-4314-7

Itrich NR, McDonough KM, van Ginkel CG et al (2015) Widespread Microbial Adaptation to l-Glutamate-N, N-diacetate (L-GLDA) Following Its Market Introduction in a Consumer Cleaning Product. Environ Sci Technol 49:13314–13321. https://doi.org/10.1021/acs.est.5b03649

Jahan R, Bodratti AM, Tsianou M, Alexandridis P (2020) Biosurfactants, natural alternatives to synthetic surfactants: Physicochemical properties and applications. Adv Colloid Interface Sci 275:102061. https://doi.org/10.1016/j.cis.2019.102061

Karlapudi AP, Venkateswarulu TC, Tammineedi J et al (2018) Role of biosurfactants in bioremediation of oil pollution-a review. Petroleum 4:241–249. https://doi.org/10.1016/j.petlm.2018.03.007

Lee A, Kim K (2019) Removal of Heavy Metals Using Rhamnolipid Biosurfactant on Manganese Nodules. Water, Air, Soil Pollut 230:258. https://doi.org/10.1007/s11270-019-4319-2

Li X, Fan F, Zhang B et al (2018) Biosurfactant enhanced soil bioremediation of petroleum hydrocarbons: Design of experiments (DOE) based system optimization and phospholipid fatty acid (PLFA) based microbial community analysis. Int Biodeterior Biodegradation 132:216–225. https://doi.org/10.1016/j.ibiod.2018.04.009

Liang C (2015) Cationic and Anionic Carbon Dioxide Responsive Switchable Surfactants

Liu G, Zhong H, Yang X et al (2018) Advances in applications of rhamnolipids biosurfactant in environmental remediation: A review. Biotechnol Bioeng 115:796–814. https://doi.org/10.1002/bit.26517

Lombardo D, Kiselev MA, Magazù S, Calandra P (2015) Amphiphiles Self-Assembly: Basic Concepts and Future Perspectives of Supramolecular Approaches. Adv Condens Matter Phys 2015:151683. https://doi.org/10.1155/2015/151683

Marcelino PRF, Gonçalves F, Jimenez IM, Carneiro BC, Santos BB, da Silva SS (2020) Sustainable production of biosurfactants and their applications. Lignocellulosic Biorefin Technol 159–183

Matsuoka H (2015) Polymer Surfactant BT - Encyclopedia of Polymeric Nanomaterials. In: Kobayashi S, Müllen K (eds) Springer. Berlin Heidelberg, Berlin, Heidelberg, pp 1917–1922

Morais IMC, Cordeiro AL, Teixeira GS et al (2017) Biological and physicochemical properties of biosurfactants produced by Lactobacillus jensenii P6A and Lactobacillus gasseri P65. Microb Cell Fact 16:155. https://doi.org/10.1186/s12934-017-0769-7

Mulligan CN (1993) Factors influencing the economics of biosurfactants. Biosurfactants 339–345

Mulligan CN (2021) Sustainable Remediation of Contaminated Soil Using Biosurfactants. Front. Bioeng. Biotechnol. 9:195

Nitschke M, Pastore GM (2004) Biosurfactant production by Bacillus subtilis using cassava-processing effluent. Appl Biochem Biotechnol 112(3):163–172

Paleček E, Tkáč J, Bartošík M et al (2015) Electrochemistry of Nonconjugated Proteins and Glycoproteins. Toward Sensors for Biomedicine and Glycomics. Chem Rev 115:2045–2108. https://doi.org/10.1021/cr500279h

Patowary R, Patowary K, Kalita MC, Deka S (2018) Application of biosurfactant for enhancement of bioremediation process of crude oil contaminated soil. Int Biodeterior Biodegradation 129:50–60. https://doi.org/10.1016/j.ibiod.2018.01.004

Perfumo A, Smyth TJP, Marchant R, Banat IM (2010) Production and Roles of Biosurfactants and Bioemulsifiers in Accessing Hydrophobic Substrates BT - Handbook of Hydrocarbon and Lipid Microbiology. In: Timmis KN (ed) Springer. Berlin Heidelberg, Berlin, Heidelberg, pp 1501–1512

Perinelli DR, Cespi M, Lorusso N et al (2020) Surfactant Self-Assembling and Critical Micelle Concentration: One Approach Fits All? Langmuir 36:5745–5753. https://doi.org/10.1021/acs.langmuir.0c00420

Pourfadakari S, Jorfi S, Ghafari S (2020) An efficient biosurfactant by Pseudomonas stutzeri Z12 isolated from an extreme environment for remediation of soil contaminated with hydrocarbons. Chem Biochem Eng Q 34:35–48. https://doi.org/10.15255/CABEQ.2019.1718

Rocha e Silva FCP, Roque BAC, Rocha e Silva NMP, et al (2017) Yeasts and bacterial biosurfactants as demulsifiers for petroleum derivative in seawater emulsions. AMB Express 7:202. https://doi.org/10.1186/s13568-017-0499-6

Rosenberg E, Ron EZ (1999) High- and low-molecular-mass microbial surfactants. Appl Microbiol Biotechnol 52:154–162. https://doi.org/10.1007/s002530051502

Santos DK, Brandão YB, Rufino RD, Luna JM, Salgueiro AA, Santos VA, Sarubbo LA (2014) Optimization of cultural conditions for biosurfactant production from Candida lipolytica. Biocatal Agr Biotechno 3(3):48–57

Santos DKF, Rufino RD, Luna JM et al (2016a) Biosurfactants: Multifunctional Biomolecules of the 21st Century. Int J Mol Sci 17:401. https://doi.org/10.3390/ijms17030401

Santos MS, Tavares FW, Biscaia EC Jr (2016b) MOLECULAR THERMODYNAMICS OF MICELLIZATION: MICELLE SIZE DISTRIBUTIONS AND GEOMETRY TRANSITIONS. Brazilian J. Chem. Eng. 33:515–523

Selvam K, Senthilkumar B, Selvankumar T (2021) Optimization of low-cost biosurfactant produced by Bacillus subtilis SASCBT01 and their environmental remediation potential. Lett Appl Microbiol 72:74–81. https://doi.org/10.1111/lam.13394

Sharma T, Sharma A, Sharma S, Giri A, Kumar A, Pant D (2020) Recent developments in CO_2-capture and conversion technologies. Chemo-biological Syst for CO_2 Utilization 1–14

Sharma T, Bhardwaj R, Bhardwaj R, Giri A, Pant D, Nadda AK (2021) Progresses in bioenergy generation from CO_2: mitigating the climate change. In: Advances in carbon capture and utilization. Springer, Singapore, pp 297–312

Shekhar S, Sundaramanickam A, Balasubramanian T (2015) Biosurfactant Producing Microbes and their Potential Applications: A Review. Crit Rev Environ Sci Technol 45:1522–1554. https://doi.org/10.1080/10643389.2014.955631

Sineriz F, Hommel RK, Kleber HP (2001) Production of biosurfactants. Encycl Life Support Sys 5:386–392

Singh R, Glick BR, Rathore D (2018) Biosurfactants as a Biological Tool to Increase Micronutrient Availability in Soil: A Review. Pedosphere 28:170–189. https://doi.org/10.1016/S1002-0160(18)60018-9

Soberón-Chávez G, Maier RM (2011) Biosurfactants: A General Overview. In: Soberón-Chávez G (ed) Springer. Berlin Heidelberg, Berlin, Heidelberg, pp 1–11

Sun W, Zhu B, Yang F, et al (2021) Optimization of biosurfactant production from Pseudomonas sp. CQ2 and its application for remediation of heavy metal contaminated soil. Chemosphere 265:129090. https://doi.org/10.1016/j.chemosphere.2020.129090

Swati, Kumari M, Ghosh P, Thakur IS (2020) Evaluation of a biosurfactant producing bacterial strain Pseudomonas sp. ISTPY2 for efficient pyrene degradation and landfill soil bioremediation through soil microcosm and proteomic studies. Bioresour Technol Reports 12:100607. https://doi.org/10.1016/j.biteb.2020.100607

Wang B, Wang Q, Liu W et al (2017) Biosurfactant-producing microorganism Pseudomonas sp. SB assists the phytoremediation of DDT-contaminated soil by two grass species. Chemosphere 182:137–142. https://doi.org/10.1016/j.chemosphere.2017.04.123

Weber L, Döge C, Gün H et al (1992) Oxygenation of Hexadecane in the Biosynthesis of Cyclic Glycolipids in Torulopsis Apicola. Biocatalysis 5:267–272. https://doi.org/10.3109/10242429209014872

Wei Z, Wang JJ, Gaston LA et al (2020) Remediation of crude oil-contaminated coastal marsh soil: Integrated effect of biochar, rhamnolipid biosurfactant and nitrogen application. J Hazard Mater 396:122595. https://doi.org/10.1016/j.jhazmat.2020.122595

Chapter 22
Agroforestry Systems for Carbon Sequestration and Food Security: Implications for Climate Change Mitigation

Gyanaranjan Sahoo, Singam Laxmana Swamy, Afaq Majid Wani, and Alka Mishra

Abstract Increased quantities of greenhouse gases (GHGs) in the atmosphere, primarily carbon dioxide, methane, nitrous oxide, and ozone, as well as their repercussions on global warming and climate change, are one of the key topics of worldwide concern today. Agriculture, forestry, and land use (AFLOU) sectors account for 13% of CO_2 emanations in the atmosphere, contributing considerably to global warming and climate change. Climate change's negative effects may be seen in the falling ecosystem production, both biological and man-made all around the world. Poor farming communities are the most susceptible to climate change, with an estimated 820 million people already suffering from chronic hunger and malnutrition, and a reduction in food production due to climate change might exacerbate food insecurity. Small farmers and destitute people demand long-term, adaptive livelihood arrangements, which always entail reliance on a variety of products. The carbon stored by agroforestry may balance the present US emission rate of 1,600 Tg C/year from fossil fuel combustion (Combustion, petroleum, and natural gas) by 34%. If large nature reserves or plantations are maintained for long-term carbon sequestration and storage, local residents may revert back to other commodities like fibre and sustenance. From different agroforestry systems, taungya agroforestry system accounted for 174 MgC/ha carbon accumulation in agroforestry networks. As a result, carbon offset regulations must account for important local environmental and socioeconomic

G. Sahoo (✉)
Krishi Vigyan Kendra, Odisha University of Agriculture and Technology, Angul, Odisha, India
e-mail: gyanaranjan.sahoo3@gmail.com

S. L. Swamy
Thakur Chedilal Barrister College of Agriculture and Research, Indira Gandhi Agricultural University, Bilaspur, Chhattisgarh, India

A. M. Wani
Department of Forest Biology and Tree Improvement, College of Forestry, Sam Higginbottom University of Agriculture and Sciences, Uttar Pradesh, Prayagraj, India

A. Mishra
Department of Rural Technology, Guru Ghasidas University, Bilaspur, Chhattisgarh, India

P. K. Shit et al. (eds.), *Soil Health and Environmental Sustainability*,
Environmental Science and Engineering,
https://doi.org/10.1007/978-3-031-09270-1_22

concerns, as well as local participation and veto power. Total above—and below-ground tree C stock accounted for 69 and 64% of total system C in the silvopasture and plantation, respectively. Developing appropriate policies to stimulate the use of carbon-sequestering agroforestry methods faces numerous obstacles, including demonstrating additionality, managing the risk of sequestered carbon losses, and involving smallholders and pastoralists with uncertain land tenure. Understanding how climate change affects poverty and livelihoods involves a deep dive into the complexities of poverty and the lives of poor and non-poor people, and even the diverse and bridge linkages of poverty and fortunes with climate change.

Keywords Agroforestry · C sequestration · Greenhouse gases (GHGs) · Food security · Livelihoods · Sustainability

22.1 Introduction

GHG emissions in the atmosphere have grown dramatically from 270 parts per million in the pre modern age to 420 parts per million at current levels, resulting in increased global warming and climate change (IPCC 2019). Apart from fossil fuel burning, the cement and thermal power sectors, as well as agriculture and forestry land use, are all responsible for increasing GHG concentrations in the atmosphere, causing climate change. Agriculture, Forestry, and Land Use (AFLOU) contributed GHG discharges account for 24% of the total, conferring to the Intergovernmental Panel on Climate Change (IPCC) in its sixth valuation statement (IPCC 2021). If rising levels of GHG emissions are not curbed, the earth's temperature has already risen to 1.5 °C and is anticipated to increase to 2–5 °C by the mid-twenty-first century. We are also seeing the environmental change's adverse impacts, such as periodic disasters, glacier melting, irregular rainfall, hailstorms, and cyclones, which are harming not only varied ecosystems but also society's well-being (Dumont et al. 2017). On a global basis, climate change will have an influence on food sustainability, as it will reduce food obtainability and approachability (Sahoo and Wani 2020). Food price fluctuation, when combined with poverty, makes food inaccessible, which is the leading cause of malnutrition (Inder et al. 2018; FAO 2019; Kumar 2010). Mitigation and adaptation goals should be sought simultaneously: Increased food production would exacerbate climate change if no mitigation measures are implemented in the agricultural sector, while yields in sensitive areas are expected to plummet unless adaptation measures are introduced. As the environmental problems become more extreme, adaptation is becoming severely challenged (Ajit et al. 2013).

Most people in tropical nations relies on agriculture and related activities for their livelihoods and household economy, making them particularly sensitive to climate change. Climate change may have an influence on food production, leading to an increase in hunger, poverty, malnutrition, food insecurity, and access to food (IPCC 2019). Coastal subsistence farming is particularly susceptible to heat and water stress, and as a result, growing seasons will be shortened (Ajit et al. 2013; FAO 2019). As a consequence of global warming, rising temperatures are becoming more common and intense, destabilising food prices and hindering the region's growth and development (Inder et al. 2018). Rapid land usage, ineffective land management, overexploitation, and increased fossil fuel burning for industrial and residential purposes are all key causes of climate change in tropical countries.

Habitat destruction and erosion operations in tropical forests are rampant and continuous in order to expand agriculture, plantations, habitats, and industries in order to feed and economically support a growing population. Annually, 13–17 million hectares of land are expected to be lost, with 450 million hectares of forest land destroyed globally in the last two decades (FAO 2020). The health and functioning of native forest ecosystems are being impacted by unprecedented rates of deterioration. Although tropical forests are acknowledged as major carbon sinks, anthropogenic disturbances have resulted in a vast number of forest areas becoming C sources rather than sinks. The C sink potential is rapidly deteriorating, while unscientific tropical agriculture practices are also leading the increased level of C emissions (Guillemot et al. 2018).

The extension of agriculture for food creation basically paddy development under lowered conditions expanding tremendous degrees of methane in south East Asia, where rice is the staple nourishment for countless populace. Additionally, the utilization of agrochemicals is rise causing the emanations of Nitrous oxide, while the methane has multiple times a lot and NO_2 multiple times higher potential for a dangerous atmospheric deviation than Carbon dioxide (IPCC 2020). The energy utilization in agriculture has hugely expanded because of automation of farm activities, while the unmanaged number of ruminant animals additionally contributing for expanding in the degrees of methane emanations. The significant concern is to diminish the discharges from agriculture area, while keeping up with the agrarian creation from the viewpoint of agrobiodiversity and food security is incredible concern (Marone et al. 2017; Middendorp et al. 2018). The substitute techniques for creation and balancing out the GHGs are critical to resolve the consuming issue of environmental change. There is developing need to handle the expanding levels of GHGs through minimal expense feasible advancements which are naturally, financially, socially reasonable promptly adequate by networks. Quick industrialization and absence of eco well-disposed innovations for creation of commodities are likewise causing genuine worry on intensifying levels of GHGs in developing countries.

Ever since Industrialization, anthropogenic activities like as consumption of petroleum derivatives and the amount of CO_2 levels in the surroundings has risen dramatically as a result of changes in land use, environmental methane (EM), carbon monoxide (CO), and other ozone-depleting substances (GHGs) in the atmosphere (Kajembe et al. 2016; Garrity 2004). In the course of recent several years, numerous researchers have contended that this human-prompted variation in barometrical GHGs is generally liable for worldwide environmental change and fluctuation (IPCC 2014). Authorities, natural groups, and the general public are all reacting to these issues (Dhyani et al. 2009). While CO_2 barometric convergences can be reduced by reducing emissions or sequestering it in the environment, the majority of GHG reductions to date have been achieved through energy-related initiatives such as energy efficiency upgrades and interest in sustainable power developments (Singh 2003). However, there is growing interest in studying alternative methods for reducing GHG emissions, with a particular focus on forests as carbon sinks that retain climatic CO_2 through photosynthesis (Albrecht and Kandji 2003; Kumar 2010). The major solutions recommended to counterbalance rising CO_2 emissions have been afforestation and reforestation (Roshetko et al. 2007; Wright and Hons 2005).

CO_2 emissions reductions resulting from the COVID-19 outbreak appear to have peaked in early April, when they were 17% lower than a year ago. Nonetheless, in several countries, daily CO_2 emissions have already recovered to or surpassed levels recorded in 2019 (International Energy Agency 2020). In the first quarter of 2020, China experienced considerable CO_2 emission reductions, but by the middle of March, the country had reverted to business as usual, and by May 2020, it had surpassed 2019 levels. The United States, India, the European Union, and the rest of the world, on the other hand, witnessed the most substantial declines in the second quarter of 2020. Different nations have had a more varied recovery in CO_2 emissions, depending on variables such as national lockdown methods, the course of the epidemic, the underlying emissions profile, and the economic ramifications of COVID-19 (IPCC 2021). In the third quarter of 2020, daily CO_2 emissions in most countries were lower than the previous year, but were higher than before the first lockdown. The course and recovery of the pandemic remain a subject of substantial discussion. Some predictions show emissions increasing at a reduced rate, while others indicate rises that are higher than previous rates, or even an overshoot of previous projections up to 2030. COVID-19 recovery efforts have so far been disproportionately concentrated on emissions-intensive and environmentally detrimental businesses, according to evidence. This might result in a faster rebound in emissions. By the end of 2020, CO_2 emissions are predicted to be lower in all G20 members than they were in 2019, and by roughly 7.5% across the board. Emission reductions are predicted to range from −2.7% in China to −12.3% in Mexico, depending on the country (Fig. 22.1). To discover the exact reasons for disparities, more research and improved assurance on final reductions would be necessary. However, there may be some links to longer-term processes at work, in addition to the pandemic's consequences and reactions (IPCC 2021).

In the adaptation strategies and long-term development, carbon sequestration programmes that include land use, land-use change, and forestry (LULUCF) activities could be a win–win situation. Projects that are well-designed can help to conserve and/or expand carbon stores while also improving rural livelihoods. The design of a project is quite important (FAO 2016). This includes establishing a baseline for carbon stocks, monitoring embellished and permeability, and utilising methodologies to analyse the greater environmental and socioeconomic impacts. This enables the maintenance or expansion of carbon stocks to be measured while simultaneously enhancing low-income rural communities' participation in sustainable forestry, agroforestry, and other rural development (Kaur et al. 2002).

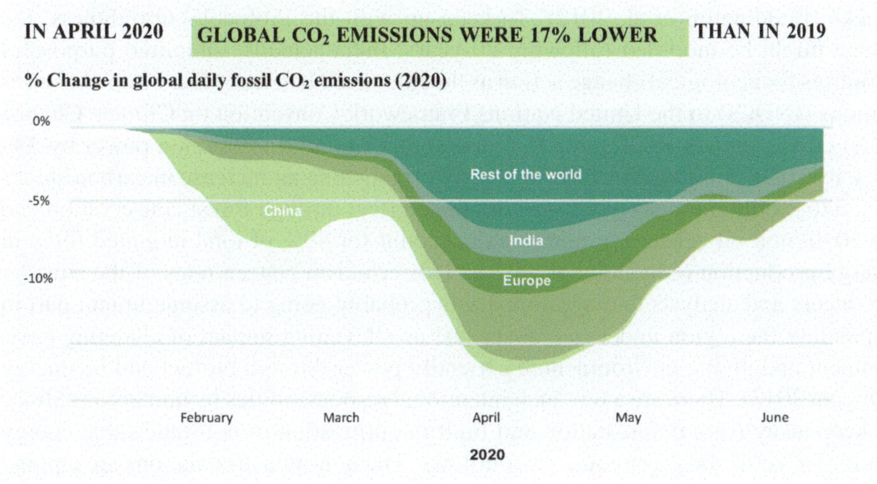

Fig. 22.1 % Change in global daily fossil CO_2 emissions

Similar tasks have been developed and carried out in a variety of countries with a variety of ecosystems and group contexts. They don't actually comply with the Kyoto Protocol's CDM, Bonn, or Paris Agreement's present legally regulating carbon market (de Coninck and Puig 2015). This will ultimately work on the comprehension of the connections between expanding carbon sinks and supportable vocations in local area based regular asset the board. Moreover, it is opportune to investigate vital approaches to move toward future obligatory just as intentional business sectors (Morillas 2020).

Deforestation avoidance and forest protection techniques are not qualified under CDM, at least not during the fundamental obligation period. In actuality, tropical deforestation is the single largest source of CO_2 emissions not yet addressed by the Kyoto Protocol (Bustamante et al. 2014). Woods carbon initiatives with security objectives and significant business benefits are both achievable and enticing in varied tropical settings. They may mitigate natural change by sequestering carbon in the atmosphere, similar to how present terrestrial carbon stores are conserved (Santoro et al. 2020). The projects could be built to meet tight barometrical, ecological, and social criteria, resulting in credits that can be sold across different economic sectors while not fully conforming to the Kyoto regulations. Brazil, the world's largest forested country, proposed a remunerated deforestation reduction approach at COP9 (Schuman et al. 2002). The structure in this suggestion is taken from satellite symbolisms of average annual deforestation over the 1980s. Tropical nations who choose to reduce their public outflows from deforestation below the standard and demonstrate performance during a responsible period are permitted to issue a "carbon testimony." Several countries agree to avoid (or at least reduce) deforestation throughout their respective responsibility epochs. It was also advised that the IPCC establish standard and proportionality criteria for deforestation and carbon stocks (Beddington et al. 2012). To keep up with the Protocol's uprightness, the gauge might be modified following 20 years. India actually submitted purposeful promises for ecological change action as the Intended Nationally Determined Contributions (INDCs) to the United Nations Framework Convention on Climate Change (UNFCCC). India's INDCs aim to reduce India's GDP transmission power by 33–35% by 2030, compared to 2005 levels, by establishing an increase in carbon stocks of 2.5 to 3 billion tonnes of CO_2 as the same effect through forest conservation and by 2030, non-oil subsidiary power will account for 40% of total mounted force in energy production (Dhyani et al. 2020). This criticism concentrates of the strategy producers and analysts that agroforestry is probably going to assume urgent part in expanding the region under tree cover just as alleviating impact of changing environment and giving environmentally friendly power through biofuel and bioenergy (Dhyani 2012). There are a few such intentional responsibilities by numerous nations to keep away from deforestation and limit the utilization of non-renewable energy sources to settle the greenhouse gas outflows. The non-industrial nations are tapping the capability of existing area use frameworks and making approaches and focuses to change over them in to C pools (Josh and Bardhan 2012).

Land management solutions that could provide chances to offset emissions without harming food production and livelihoods should be explored. Agroforestry appears to be a low-cost practical solution for reducing emissions, as trees can absorb huge amounts of CO_2 from the sky and store it as organic C in vegetative biomass and soil, thereby regulating GHG atmospheric concentrations (Nair et al. 2009; Shepherd and Montagnini 2001). Agroforestry, above all natural techniques, offers a win–win situation for achieving carbon sequestration and climate change moderation and edition goals. Regardless of whether agroforestry systems (AFS) are principally built for many products and services rather than carbon sequestration, a growing body of evidence suggests that agroforestry systems can contribute significantly to carbon sequestration in both aboveground and belowground biomass (Murthy et al. 2013).

Agroforestry is another name for an old practice where farmers spatially and transiently coordinate trees, bushes, and animals with crop production on a similar unit of land to guarantee an all-year supply of food and additionally pay, forestall soil debasement, keep up with soil ripeness, enhance pay sources, and work on the effective utilization of soil supplements, water, and radiation and give customary business (Dhyani et al. 2016). Agriculture and forestry synergy is usually viewed as a way to increase food security while also providing a variety of environmental advantages to society. It has the ability to sequester carbon while also providing a variety of financial, ecological, and societal profits (Newaj and Dhyani 2008). The woody perennials are equipped for retaining a lot of climatic CO_2 through photosynthesis and supply C in seemingly perpetual and fleeting biomass parts as well as enhancing the dirt usefulness. For example, trees under agroforestry framework further develop soil richness through enhancing soil natural matter and working on physical and organic properties, through N obsession, removal of supplements from profound soil skylines, and advancement of more shut supplement cycling (Montagnini and Nair 2004). Agroforestry systems that are well-designed and operated, combined tree, agricultural, and pastoral yields could be higher than in solo systems due to improved and effective resource usage. Agroforestry systems increase perseverance of climate change by extending the production base and reducing the dangers of monocropping. Agroforestry systems with microclimate effort expectancy, stop soil deterioration, restore soil fertility, and diversify opportunities to earn money were overwhelmingly proved (Meragiaw 2017; Marone et al. 2017; Middendorp et al. 2018). Agroforestry can limit weakness; fabricate strength of cultivating frameworks and cushion families against environment related dangers (Dhyani 2014). Agroforestry gives a remarkable chance to join the twin destinations of environmental change transformation and relief. It can upgrade the flexibility of the framework for adapting to the unfriendly effects of environmental change. Agroforestry frameworks offer significant freedoms for making collaborations between both transformation and relief activities. The modern agroforestry estates in India set up during the most recent five years have been assessed to sequester 3.65 Mg of C and address the natural concerns (IEA 2020). The job of agroforestry in balancing out the CO_2 levels and expanding the C sink probable has drawn in extensive consideration, particularly after the Kyoto Protocol to the United Nations Framework Convention on Climate Change (UNFCCC). Agroforestry is frequently viewed as a budget-friendly plan for environmental change alleviation. Agroforestry systems collect biomass in the soil and forest residues, as well as reducing ozone-depleting chemical outputs from soils. Larger part of the agroforestry frameworks can possibly sequester carbon, which has the potential to change as per tree species (Prasad et al. 2012) and the board rehearses embraced (Newaj and Dhyani 2008). The potentials of some promising agroforestry practices for C sequestration across the world are presented in Table 22.1.

In agroforestry frameworks, despite the fact that trees impound supplementary carbon, so far crops likewise fix and store carbon in impressive sums. Harvesting occurs with organic substances in soil, which accounts for a significant portion of the earth's Carbon source. An augmentation in carbon pool in soil can be accomplished through selection of suitable yield pivots, coordinated soil fertility management (Lal

Table 22.1 C storage under Agroforestry Systems in different Continents

Continents	Eco-region	System	Mg C/ha
Africa	Humid tropical high	Agrosilvicultural	29–53
South America	Humid tropical low dry lands	Agrosilvicultural	39–102
			39–195
Southeast Asia	Humid tropical dry low lands	Agrosilvicultural	12–128
			68–81
Australia	Humid tropical low	Silvopastoral	28–51
North America	Humid tropical high	Silvopastoral	133–154
	Humid tropical low dry lands	Silvopastoral	104–198
Northern Asia	Humid tropical low	Silvopastoral	15–18

Source Albrecht and Kandji (2003)

2004), exact utilization of manures and organic amendments (Schuman et al. 2002), and selection of preservation agribusiness (Lal 2009). Assortment of editing frameworks is one of the primary highlights of Indian farming as the quantity of soil and climatic boundaries, which decide generally agro-biological conditions (Westholm and Ostwald 2020). Carbon sequestration in various agroforestry frameworks happens both subterranean, as upgrade of soil carbon in addition to root biomass and over-the-ground as carbon put away in standing biomass. Probably the most punctual investigations of probable carbon stockpiling in agroforestry frameworks and elective land use frameworks for India had assessed a retention capability of 68–228 MgC/ha (FAO 2020), 25 tC/ha more than 96 Mha of land (Dhyani et al. 2020). Yet, this worth changes in various areas relying upon the biomass creation (Marone et al. 2017). Investigations carried out by Brown et al. (2018) presented that agroforestry could supply almost 83.6 Mg/ha up to 30 cm soil profundity, 26% more carbon contrasted with development in Haryana fields. In any case, the extent of carbon sequestration from ranger service exercises would rely upon the size of activity and the last utilization of wood. The potentials of trees for C sequestration under agroforestry systems are given in Table 22.2.

The process of depositing Carbon dioxide from the air into the earth in a condition that is not promptly re-emitted by crop wastes is known as carbon sequestration in soil, tree roots, and other organic things. Soil carbon sequestration helps to lower CO_2 levels in the atmosphere while simultaneously improving land performance and promote long-term yield (Gusli et al. 2020). Increasing soil biomass, generating soil disturbances to a threshold, preserving soil and water, enhancing morphology of the soil, and increasing the vitality of microbial diversity are all ways to increase soil carbon sequestration. A rise in the overall trend soil carbon sequestration (SCS) in agroforestry related to other land-use choices was discovered in studies on carbon sequestration in soils in India and the remains of the continent (except where woodlands exist). In general, land-use systems were categorized according to their SOC content: forests > agroforests > tree plantations > arable crops (Nair et al. 2010). Additionally, the projected SCS values in AFS differed widely, reflecting a variety of factors such as biological, physical, and socioeconomic aspects of system components, as well as a lack of homogeneity in study techniques. Prospects for C sequestration in and out of the soil is greater in agroforestry systems with better management. The key pathway for initiatives to mitigate climate change and address local and global degradation issues is to promote climate resilience in food, water, and energy. Table 22.3 shows the carbon stocks and sequestration in soil under plants in agroforestry systems.

Crop rotation and soil conservation are the focus of the bulk of studies on SOC sequestration. Erosion management strategies such as hedgerow intercropping, on the other hand, can boost SOC storage and absorption rates (Lenka et al. 2012b) and so operate as a carbon sink, but this has not been thoroughly explored. Permanent grazing and forestry, as well as multipurpose trees (Ramesh et al. 2015), can improve SOC stock, On the slope cultivable areas, meanwhile, the demographic conditions of the people living in the Eastern Ghats heights of Odisha prevent this.

Table 22.2 Carbon sequestration potential (CSP) of multipurpose trees under Agroforestry

Region	Agroforestry system	Tree species	Density (trees/ha)	Age (years)	CSP (Mg C/ha/year)	References
NE Himalaya, India	Block plantation	*Eucalyptus tereticornis*	2500	3.5	4.40	Dhyani et al. (1996)
Indo-Gangetic	Agrisilviculture	*Leucaena leucocephala*	10,666	6	10.48	Mittal and Singh (1989)
		Populus deltoides	400	7	1.98	
	Block plantation	*Acacia nilotica*	1250	7	2.81	Kaur et al. (2002)
		Dalbergia sissoo	1250	7	5.37	
Humid and Sub-humid	Agrisilviculture	*Gmelina arborea*	592	5	3.23	Swamy and Puri (2005)
	Block plantation	*Gmelina arborea*		6	4.01–5.01	Swamy et al. (2003)
Arid semi-arid	Block plantation	*Albizia procera*	312	10	1.79	Pragason and Karthik (2013)
		Eucalyptus tereticornis	320	2	13.86	
	Agrisilviculture	*Leucaena leucocephala*	4444	4	14.42	Prasad et al. (2012)
		Casuarina equisetifolia	833	4	1.57	Viswanath et al. (2004)
	Silvipasture	*Acacia nilotica* + Natural pasture	312	5	1.9–5.4	Rai et al. (2000)
		Dalbergia sissoo + Natural pasture	312	5	2.5	
Tropical	Home garden	Mixed tree species	667	71	1.60	Saha et al. (2009)
	Block plantation	*Eucalyptus* spp.		7–10	3.71	Ajit et al. (2014)
		Acacia mangium	5000	6.5	12.59	

Source Newaj et al. (2017)

Table 22.3 Soil organic carbon (SOC) stock reported in various agroforestry systems

Agroforestry system	Location	Age (Years)	Soil depth (cm)	Soil C (Mg/ha)	References
Mixed stands, *Eucalyptus* + *Casuarina* (C), C + *Leucaena* (L), *Eucalyptus* + L	Puerto Rico	4	0–40	61.9, 56.6, and 61.7	Parrotta (1999)
Agrisilviculture (*Gmelina arborea* Roxb. + eight field crops)	Chhattisgarh, Central India	5	0–60	27.4	Swamy and Puri (2005)
Tree-based intercropping: hybrid poplar + barley	Ontario, Canada	13	0–20	78.5	Peichl (2006)
Alley cropping *Leucaena* – 4-m wide rows	Western Nigeria	5	0–10	13.6	Lal (2005)
Shaded coffee, *Coffea robusta* L. Linden + *Albizia* spp.	South Western Togo	13	0–40	97.27	Takimoto et al. (2008)
Home gardens	Kerala, India	35	0–100	101–126	Saha et al. (2009)
Shaded cacao systems	Bahia, Brazil	30	0–100	302	Gama-Rodrigues et al. (2010)
Silvopasture: slash pine (*Pinus elliottii* Engelm.) + bahia grass (*Paspalum notatum* Flügge)	Florida, USA	8–40	0–125	6.9–24.2	Haile et al. (2008)
Faidherbia albida (Delile) A. Chev. parkland	Ségou, Mali	35	0–100	33.3	Takimoto et al. (2008)

Agroforestry methods have the ability to enhance SOC concentration faster than conventional agriculture. A 15% rise in SOC concentration was detected in a 12-year hedgerow experiment with *Gliricidia sepium* and *Leucaena leucocephala* on a Nigerian Alfisol (Kang et al. 1999). After two years of planting in a poplar-based agroforestry system, the SOC content rose (Arevalo et al. 2011). Lenka et al. (2012a) showed an 89% increase in SOC in a shifting farmed degraded area after 6 years of research under a horti-silvipastoral system. They also discovered that the hedgerows of Indigofera and *Gliricidia* function as a barrier, retaining more SOC than the entire plot.

22.2 Carbon Inputs and Outputs Are Used to Calculate Carbon Stocks

Most habitats receive CO_2 and mineral nutrients from the atmosphere and convert them to items made from natural ingredients, including wooded ecosystems, agro-ecosystems, and grassland ecosystems. By changing species composition and growing conditions in grasslands, carbon uptake is steered toward the production of fibre and feed. The three principal biogenic ozone-harming chemicals (GHGs)—CO_2, nitrous oxide (N_2O), and methane—are a substantial source and sink for biological systems (CH_4). Even in developed, old growth forest ecosystems, carbon take-up from photosynthesis exceeds misfortunes through breath, resulting in a positive carbon balance in undisturbed biological systems (Prusty et al. 2020). Unsettling influence, like fire, dry season, illness or over the top search utilization by brushing, can prompt generous misfortunes of carbon from the two soils and flora. Severity is a characteristic of all biological systems that continues to impact carbon take-up and release that define the environment's covalently linked balance (Pragason and Karthik 2013).

In terms of their influence on environmental biogeochemical cycles, habitat operations are similar to natural activities. When forest vegetation is destroyed, CO_2 is created, and soil carbon stocks are rapidly depleted after soil disturbances. Land-use change, like environmental hazards like burning and famine, has an impact on plant and soil dynamics, resulting in higher carbon outputs and lower carbon absorption (Murthy et al. 2013). Loss of natural prairies as a consequence of climate change, and migration to farming have resulted in biomass and soil carbon losses of 450–800 Gt/CO_2, or 30–40% of total fossil fuel emissions. Outflows from the conversion of forests to agriculture or other land uses have surpassed carbon emissions from ecological systems are on the rise, despite the fact that considerable amounts of carbon have been lost from biomass and farm soils (Singh 2003). The photosynthetic take-up and absorption of CO_2 into natural mixtures, as well as the arrival of vaporous carbon through breath, are the essential cycles that govern the carbon equilibrium of grasslands, just as they are in other habitats (basically CO_2 yet in addition CH_4). Table 22.4 shows the carbon sequestration capacity of several agroforestry frameworks.

Despite the fact that little research has been done on the efficacy of agroforestry systems to retain carbon in the soil, silvicultural farming and other agroforestry techniques are commonly thought to help restore degraded areas' SOC resource. Fallow agroforestry methods have been observed to improve top soil C storage by

Table 22.4 Carbon sequestration potential of different agroforestry system

Agroforestry system	Carbon accumulation
Taungya agroforestry system	174 MgC/ha
Mixed multistory/multistery system	162 MgC/ha
Falcata-coffee multistorey system	92 MgC/ha

up to 1.6 Mg C/ha/year in degraded sub-humid tropics soils as relative to continual maize cropping (Mutuo et al. 2005).

Biomass in field frameworks, being overwhelmingly herbaceous (for example non-woody), is a little, a temporary chemical reservoir (contrasted with backwoods) and thus soils establish the predominant carbon stock. Field frameworks can be useful biological systems, yet limited developing season length, dry spell periods and eating incited shifts in species synthesis or creation can diminish carbon take-up comparative with that in different environments. Soil natural carbon stocks in grasslands have been depleted less than in crops (Lal 2009), and biomass has increased in some locations due to the concealment of disturbing effect and the consequent woody infringement. Differences in management techniques that boost carbon inflows, optimise carbon inside the framework, or minimise carbon wastes with maintaining fibre and silage yields can aid in the recovery of a substantial quantity of carbon from agricultural land soil and biological reserves (Josh and Bardhan 2012).

22.3 Approaches for Carbon Sequestration

22.3.1 Prospects for the Entire World

The dynamic retention of CO_2 from the air by means of photosynthesis and resulting stockpiling in the biomass of developing trees is the focal point of most carbon sequestration drives (and plants). Subsequently, carbon sequestration has would in general be related with tree planting in both regular woods and ranch circumstances. Notwithstanding the way that there are a lot more choices past afforestation or reforestation, evaluations of the worldwide probable for carbon sequestration have begun with the space of land accessible for afforestation (Sahoo and Wani 2019). Then again, these ought to be taken care of with alert. Land that is 'actually appropriate' for supporting backwoods frameworks since it is edaphically and climatically reasonable, and land that is' socio-politically accessible' in light of the fact that it is actually appropriate and accessible for the foundation of timberland and agroforestry frameworks given current social monetary and extremist circumstances, as per Prasad et al. (2012). Aspects including accrued growth in agricultural land, the viability of soils and ecosystems for ranger service, strategic concerns, and biological consequences will very probably limit the practical and financially feasible capability. In the forests, really practicable land is predicted to be between 865 and 3,125 million ha, while social economic viable land is anticipated to be between 300 and 462 million hectares in the forests and 570 million ha globally (Prusty et al. 2020).

Forestry and other land use methods' potential to store carbon varies substantially. While fast-growing genera assimilate carbon at a higher pace, biotic systems' potential to offset released carbon is ultimately determined by their long-term carbon storage capacity. When analysing long-term carbon storage, consider rotation time, biomass ratios between branch, canopy, and rhizomes, lumber porosity, harvesting frequency and intensity, and the lifespan of wood products (Nair et al. 2017).

22.3.2 For Timber Production and Carbon Sequestration, Plantations Are Used

Carbon-based incentives are likely to be considered by investing corporations and organisations as a supplement to the primary source of revenue in plantation-based carbon sequestration projects. The Face Foundation is supporting the Profafor initiative, which aims to grow trees for both extraction of trees and chemical retention. Afforestation with Australian *Eucalyptus, Pinus radiata*, and *Pinus patula* is targeted in the Andes at altitudes of 2,400 to 3,500 m, much above the range for successful agricultural and animal agriculture. Nonetheless, there are generous regions that have been deforested, which are currently powerless to disintegration and resulting avalanches (Sahoo et al. 2020). Whereas this project is now using interesting species that are proven to produce high-quality lumber, Face and the Ecuadorian Forest Service hope to change these with select local species during the next pivot (20–30 years) (Besar et al. 2020). Farmers who complete the work will receive a planting prize of roughly \$100 per ha, as well as planting materials. 'Local gatherings' are encouraged for small farms. Individual landowners with vast holdings are not eligible for the programme.

Farmers receive a cash subsidy for carbon, are willing to trade the timber generated by the plantations, and benefit from some soil erosion protection as a result of such programmes. In this scenario, the carbon component income would be used to fund the public investment needed to build what would ideally get to be a financially sustainable and efficient land use system.

22.3.3 Forestland Rehabilitation

In a foster forest area, most of the biomass is addressed by not many colossal trees. Woods biomass is thusly immovably dependent upon the degree of exacerbation. A timberland can seem unblemished on a satellite picture while camouflaging the way that in numerous spaces an enormous extent of possibly useful woods is halfway or seriously debased (Sahoo and Wani 2019). Since tainted forests in like manner have a lower financial worth, the capacity to trap carbon and build the heap of standard capital looks to be an inland empire upgrading capability. The majority of reclamation initiatives appear to have the opportunity to deliver both financial and carbon gains, especially if the re-established forests are directly valuable to local networks. However, there may be some tension between partners who want to utilise the property for other purposes, such as horticulture, and those who want to restore the forest. Another issue to examine is how to find a balance between long-term carbon sequestration and wood-based product extraction (Nair et al. 2017). Woodlands managed at a monetary pivot length contain less carbon on average than those left undisturbed or exposed to long rotations.

22.4 As a Prospective Abatement Approach, Agroforestry

Excessive carbon retention in on and around the dry matter, as well as in soil natural carbon, agroforestry practises can reduce or eliminate significant metrics of GHGs (IPCC 2019). By incorporating agroforestry with editing and domesticated animal husbandry systems, significant amounts of carbon can be sequestered. Agroforestry practises that sequester CO_2 include home cultivation, limited planting, natural product plantations, riverine, hedgerows, woodlots, and kindling portions. More carbon is stored in agroforestry than in fields and farms with periodic rotations, but it is not quite the same as wooded areas (Verchot et al. 2007).

Similarly significant, agroforestry can further develop jobs in smallholder cultivating frameworks through expanded pay and money crop frameworks (for example cocoa, espresso, nuts), expanded food security and further developed admittance to nutritious food. Trees on homesteads can likewise assist the ranchers with lessening the monetary recuperation time after catastrophic events (Simelton et al. 2015). Environment variation is especially significant for female ranchers as they frequently have less admittance to assets contrasted with their male partners (Paudel et al. 2017). Feminine smallholders produce a significant piece of the food in many creating areas, yet for the most part don't have similar freedoms to work on primary sources of revenue (Agroforestry Network 2018). It is likewise entirely expected that ladies are left responsible for the smallholder ranch when their life partner is relocating for work, and accordingly need greater ability to deal with the expanded responsibility (Leder et al. 2016).

The 'awfulness of the house' is an example of environmental change, whereas alleviation is a public good. If each expert (person, organisation, or nation) acts unilaterally in their own narrow interests, environmental change relief will be impossible, emphasising the necessity of taking action as a group. Carbon dioxide is the most often produced ozone damaging substance. Carbon sequestration is a technique for removing and storing CO_2 from the atmosphere. It's one way to reduce CO_2 content in the air and hence limit rising temperatures (García de Jalón et al. 2017).

22.5 Carbon Sequestration Through Agroforestry System

Deforestation and forest degradation emit more carbon than other sources (Lal 2009). This, however, may be dealt with by ensuring that land and forest resources are managed sustainably. One of the key approaches for lowering greenhouse gas emissions in the environment is to increase forest C stocks through agroforestry. For example, using fossil fuels in the United States contributes around 25% of world CO_2 emissions (Lasco et al. 2014). Perhaps the most important agroforestry's influence to greenhouse gas emissions is its ability to reduce CO_2 emissions by sequestering carbon from the atmosphere in a productive manner.As indicated by Seserman et al. (2019), carbon sequestration in agroforestry is characterized as the admission of environmental CO_2 during photosynthesis and the exchange of fixed C into vegetation, waste, and soil pools for "secure" (for example long term) stockpiling. Agroforestry is an ozone depleting substance relief approach that involves sequestering carbon (C) in biomass and soils while additionally bringing down GHG discharges on horticultural grounds, generally through stayed away from outflows from energy and fuel investment funds (Leder et al. 2016). Agroforestry rehearses have a higher potential for advancing carbon sequestration in farming scenes than monocrop agribusiness.

Carbon was also stored to varying degrees in the following pools: soil (77–92%) > trees (7–22%) > herbaceous vegetation and litter (1%). According to a study conducted in India (IPCC 2014), the efficacy of agroforestry frameworks in storing carbon is dependent on both natural and financial aspects. Agroforestry frameworks can potentially sequester more than 70 Mg/ha in the best 20 cm of the soil in sticky jungles, according to a meta-examination of 427 soil C stock information sets assembled into four basic AF frameworks-rear entryway editing, windbreaks, silvopastures, and home gardens-and assessed changes in AF and nearby control cropland or field. The average soil C stocks in AF (1 m depth) were 126 Mg C/ha, which is 19% higher than in cropland or field. Subtropical home nurseries, AF with more youthful trees, and soils had the highest C stocks in soil (0–20 cm). With the exception of rear entryway editing, expanded soil C stocks in AF were lower than over-the-ground C stocks in most AF frameworks (Blaser et al. 2017). Home nurseries set aside the most raised C in both over-the-ground and underground, particularly in the earth (20–100 cm). AF could store 5.3 × 109 Mg additional C in soil on 944 Mha globally, with most in the wildernesses and subtropics. Rahman et al. (2016) uncovered consistent evaluation, a 13-year-old back road managing structure in Ontario, Canada, was proven to have 11–41% more C, dependent upon tree species, stood out from sole-altering plots. AF structures could altogether add to overall soil C sequestration at whatever point used in greater locales.

22.6 Food Availability and Reducing Carbon Emissions Have a Synergistic Impact

Moderation of climate change has never been a great catalyst of agricultural activity, and it is unlikely to be in the future. Rural households are plainly uninterested in carbon sequestration on farms for climate change mitigation, especially if mitigation approaches do not result in immediate financial or welfare improvements

(Newaj et al. 2017). Smallholder farmers might be reluctant to surrender any of their frequently pitiful homestead benefits to sequester carbon. Carbon-sequestering land use procedures should either be financed to the degree that they are comparable to inevitable benefits from elective land uses, or they should be beneficial by their own doing—with no pay—if such ranchers are to add to relief regardless (IPCC 2014). With biocarbon projects actually attempting to defeat monetary, institutional, and administration hindrances, the most obvious opportunity with regards to sequestering carbon for an enormous scope on Africa's ranches is through advancements that further develop food security while likewise giving supportability administrations (for example expanded parkland tree cover, diverse cultivating, intercropping, land sharing practices, and so on) (Saha and Jha 2012). Agroforestry is one of the few land-use practises that may aid in both energy security and environmental issues adaptation. It's also less likely to harm non-carbon ecosystem services such as water cycle management and species restoration, which are both important components of "climate-smart agriculture," than alternative options.

22.7 What Are the Potential Consequences for Rural Livelihoods?

Agroforestry frameworks are vital given the region presently under farming, the quantity of individuals who rely upon land for their jobs, and the requirement for incorporating food creation with natural administrations. The provincial poor and landless require versatile, reasonable vocation frameworks that are adaptable for the time being—this perpetually implies reliance on numerous items (ICRAF 2021). Nonetheless, if enormous secured regions or manors are overseen for long term carbon sequestration and capacity, neighborhood individuals may lose admittance to different items like fiber and food. Carbon counterbalance strategy should accordingly work in sufficient arrangements concerning neighborhood natural and social components, with applicable nearby support and powers of blackball (Hillbrand 2021). A large part of the gaining from participatory ranger service and ensured region experience is significant and could be joined into carbon counterbalance strategy. Other than C advantages, large numbers of the agroforestry frameworks could build food creation if biophysical associations are appropriately abused (Agroforestry Network 2018). The unmanaged frameworks could diminish the yields as tree concealing, root contest; allelopathic impacts and holding onto the bug irritations are injurious to the efficiency. The advantages of agroforestry frameworks in upgrading crop yields are summed up in Table 22.5.

Table 22.5 Productivity of certain crops under various agroforestry systems

Agroforestry system	Crop	Crop yield	References
Agrisilviculture	*Triticum aestivum*	25.60 q/ha	Singh et al. (2004)
	Oryza sativa	37.07 q/ha	Thaware et al. (2004), Tomar and Bhatt (2004)
Hortisilviculture	*Solanum tuberosum*	131.1 q/ha	Thaware et al. (2004)
Agrihorticulture	*Citrus limon (L.) Burm.F*	27.61 q/ha	Tomar and Bhatt (2004)
	Psidium guajava L	58.11 q/ha	Tomar and Bhatt (2004)
	Zizyphu smauritiana	140.55 q/ha	Zhang et al. (2013)
Home gardens/plantation crop based systems	*Curcuma longa L*	115.22 q/ha	Vanlalhluna and Sahoo (2009)
	Theobroma cacao L	37.28 kg/tree	Isaac et al. (2007)

22.8 Sustainable Livelihoods

Human settlements are uncommon in many regions of protected areas. They are frequently viewed as a threat to the places' long-term viability. Large-scale plantation initiatives may face a similar scenario. The question is how Agroforestry programmes could aid forest-dependent populations. The social, economic, and human components of a project could be identified at the outset and introduced (Sahoo et al. 2020a, b). Various government departments, along with the local community and existing institutions, could be involved in the project design. The case study of a large-scale pulpwood plantation in Indonesia's Riau Province highlights the negative consequences of the industry. This implies that integrating villagers in a climate-change-related plantation project is incompatible with their current practises (for example, shifting cultivation). For locals, the project's value is extremely poor (less than \$0.5 per ha each month). However, the project has harmed biodiversity and reduced the utility of nearby crops (Sahoo and Wani 2020). Further, it brought about farmers land confiscation. So, affecting nearby individuals in such kind of task resembles a neediness trap. Two differentiating experience is found in Mexico when limited scope family-drove was contrasted and publicly drove reforestation. It was exhibited that lessening neediness doesn't involve expanding level of pay. Individual's support, authenticity, and information are central points of interest. For this situation task's authenticity is more challenged in family-drove local area since helpless families are not all around addressed and can't partake in proper neighbourhood organizations (Feliciano et al. 2018). To get provincial livelihoods, recognize an undertaking's normal yields beginning with the plausibility considers. Rustic associations can assume a successful part in building authenticity yet they are not generally comprehensive of every single nearby individual. Nearby political and asset elements should be painstakingly noticed. Upgrading correspondence and exchange is pivotal. It is additionally basic to foster viable courses of action with public organizations and supplement carbon exercises with different ventures since carbon demand is low (Dhyani et al. 2020).

Opportunities for a superior life Climate-related calamity like flooding, dry season, and fire, just as neediness, address an extreme threat to environmental change endeavours that overlook business concerns. Environmental change projects with a solitary reason for bringing down ozone depleting substance emanations from sources and sequestration from sinks aren't by and large remembered for agricultural nation strategy structures (Guillemot et al. 2018). As a result, environment related drives ought to be planned so that they are pertinent to individuals' livelihoods, with a scope of choices relying upon neighbourhood needs.

22.9 Is It Possible for Rural People to Offer Carbon Credits Through Their Agricultural and Forestry Systems?

The drive to develop carbon-free energy stockpiling at the most minimal cost may will in general support enormous, touching regions under straightforward administration and clear residency, that can ensure a solitary, promptly obvious item—carbon stockpiling—for a long time. Clear instances of such conditions remember set-asides of regular woods for ensured regions, or enormous scope estates. Taking everything into account, governments and organizations, instead of little ranchers, are best positioned to profit with such plans.

22.10 Can Carbon Offsets Help Residents in Rural Areas?

Country people may be able to profit directly from balancing measures in remote regions where people need to grow or care for trees for various reasons and where there is balanced governance to guarantee that land use changes do not degrade value (Rosenstock et al. 2013). However, given the numerous complex necessities of carbon balance mediations, the rural poor and landless will have less access to the carbon counterbalance market due to their typically feeble association (or costly exchange charges of further growing association). Different barriers to provincial inclusion are based on their broad limited reach and complex land use practises, as well as the lack of unambiguous residence frameworks (Sahoo et al. 2021). Furthermore, while carbon balancing income may provide some security, cash remuneration is insufficient for the rural poor and landless, who will continue to require access to normal finance in order to survive, versatile vocations.

22.11 Livelihood Impacts of Carbon Projects

Under the CDM, industrialized nations can put resources into the carbon sequestering exercises in non-industrial nations as a trade-off for carbon balances that mean something negative for outflow decrease targets determined by the Kyoto Protocol (Marone et al. 2017). Interests as carbon sequestration projects hence address significant monetary inflows for non-industrial nations. Experience additionally proposes that, whenever attempted with little land holders, carbon sequestration activities can assist with mitigating country destitution and work on neighbourhood livelihoods in agricultural nations. Carbon sequestration ventures may in this manner give a mutually beneficial arrangement between ecological preservation and expanded freedoms for financial advancement in helpless nations (FAO 2020). There are not many investigations on Africa on job effect of carbon projects on neighbourhood networks. Peichl (2006) discovered that the carbon Project end up being valuable to the neighbourhood local area by giving standard type of revenue as carbon payments, raise the efficiency through agroforestry and create substitute methods for jobs, for example, selling of non-wood ranger service items. In any case, he measures the undertaking may likewise present monetary variations among the families, which could make disdain and nearby agitation.

Other examination in Nair et al. (2009) the Plan Vivo project was discovered to be available to helpless limited scope landholders, and that boundaries to passage would just influence a little extent of likely members. Notwithstanding the instalments for carbon sequestration, the venture was found to have different advantages which it brings to members, which add to food and fuel security at the family level and the task gives social and human limit building. Humbo carbon project is giving the immediate and aberrant monetary and social advantages to neighbourhood networks (Lal 2009). This investigation plans to look the effect of advantages on vocation of neighbourhood networks.

22.11.1 Significance of the Study

As of late, carbon sequestration as ranger service projects has advanced into a suitable choice to handle a dangerous atmospheric deviation and environmental change. As per third assessment report of the Intergovernmental Panel on Climate Change, timberlands, farming fields, and other earthly biological systems have enormous carbon abatement potential (Shepherd and Montagnini 2001). The report expresses that notwithstanding decrease in barometrical carbon dioxide, such activities may likewise give other social, financial and ecological advantages like supportable land the executives and country work. Be that as it may, whenever carried out improperly, they may present danger of unfavourable effect like local area disturbance (Montagnini and Nair 2004). Further, such undertakings could possibly become practical if the financial drivers for deforestation and different misfortunes of carbon pools are tended to. Hence, a comprehension of financial cycles, especially the expected advantages and effects of carbon sequestration projects, is fundamental before they are suggested for more extensive replication.

22.11.2 Short-Term Livelihood Impacts on Community Activities and Income

The module has taught the communities the option of expanding their existing exotic planting area or diversifying their on-farm activities. The project subsidies to build the plantation gave local employment and increased income to several of the surveyed community members who were suffering lower revenue from livestock and agricultural activities.

22.11.3 Long-Term Livelihood Impacts on Communities

All people group expected that the estate would produce expanded pay for local area individuals later on. Local area projections of the significance of ranger service exercises, especially as far as commitment to pay, were fluctuated. In a couple of cases, if existing requirements to domesticated animals and rural exercises proceeded, ranger service exercises were probably going to supplant horticultural exercises (Dhyani et al. 2020). A few networks were building up estates to broaden their pay base while others were anticipating that timber revenues should turn into the significant pay hotspot for the local area. Given diminishing gets back from other on-ranch exercises, the undertaking contracts for 15–20 years address a possibly beneficial venture for the overviewed networks, especially to those with manors of fascinating species. Fire, unforgiving climatic conditions, irritations and infections, and admittance to business sectors were noted as the significant dangers to the benefit of the estates. Nonetheless, under the new 99-year contracts, just networks keen on both monetary and ecological advantages are probably going to acquire. As a simply monetary speculation the 99-year contracts are probably going to be unrewarding, particularly if the chance expense of the land expansions later on. Local area individuals would be in an ideal situation building up estates under different plans, where agreement conditions are more adaptable.

22.11.4 Adaptation

Although the net loss of carbon sinks may have an impact on climate, the changing climate has certain sensitive consequences for agro-ecosystems. To reduce the dangers and harmful impacts, adaptation strategies are required. Human cultures' response to climate change will likewise put more pressure on forests. Traditionally, the effects of food shortages and water stress on communities have been alleviated at the expense of trees. Despite the fact that climate change may speed these processes, governments, civil society, and resource managers should consider implementing adequate and cost-effective adaptation methods. It's past time to bring adaptive capacity together approaches, especially in the most vulnerable agricultural ecosystems, where agroforestry practises could assist alleviate the negative consequences.

22.11.5 Recommendations

- More investment in tree-based land use strategies is required to maintain agricultural, forestry, and livestock contributions to net income in the context of global warming.
- Increased investment in rural education and training to boost labour force capacity and equip young people with the information and skills they need to secure good jobs.
- Finally, in the planning, promotion, and implementation of agroforestry and planting activities aimed at increasing rural household participation in the project and ensuring overall benefits to impoverished families in need.

22.12 Conclusions

The functioning of many nutrient cycles, including the Carbon and Nitrogen cycles, has been interrupted due to global carbon emissions from a variety of anthropogenic sources. Greenhouse gases have a critical function in regulating the temperature of the atmosphere and the rest of the earth. Carbon uptake in living biomass and soil carbon is higher in agroforestry systems, indicating that they have the ability to provide carbon sequestration as an environmental benefit. By reducing the amount of forest-based fuelwood burned and conserving soil, agroforestry systems can assist to reduce Greenhouse gasses. In addition to their potential to absorb and store carbon, agroforestry systems might evolve as a scientific opportunity for lowering tropical degradation levels while simultaneously delivering a wide range of facilities to agricultural households. Agroforestry has been highlighted in several studies around the world as having significant promise for climate change mitigation, biodiversity protection, and the extension of numerous ecosystem services required for environmental well-being. Agroforestry systems provide opportunity for underprivileged people to better their livelihoods by providing economic and environmental security. Because of its high carbon absorption capabilities (both above and below ground) and versatility, for its practitioners' profit and long-term viability, agroforestry has emerged as a feasible alternative for mitigating the consequences of global warming. The target of carbon sequestration through agroforestry can only be met by selecting, identifying, and promoting appropriate agroforestry systems, establishing plantings through breeding/biotechnological methods for high carbon capture and storage prospects, easing rules and laws via agroforestry guidelines, and offering opportunities, credit, and insurance cover for agroforestry.

Farmers and environmentally friendly proactive policies will be developed to encourage agroforestry, and forest and agroforestry policies should be reconsidered. The many agroforestry players are aware of the complicated issues, but the synergies and trade-offs must be thoroughly researched. Financial incentives for tree planting, as well as guaranteed payments for environmental services, credit, insurance, and subsidies, will encourage and prepare the way for agroforestry to be revitalised for long-term development. Integrating appropriate agroforestry systems will not only help to mitigate climate change by increasing carbon stocks, but it will also help to diversify the production base and increase resilience, which will help to address food security and rural livelihoods, both of which are critical for meeting the ambitious goals of sustainable development (SDGs). Agroforestry programs have the capability to reduce agricultural systems' vulnerability to climate variability and climate change consequences, as well as their ability to absorb and sequester carbon. The ultimate comparison of achieving prevention and resilience to climate change interaction is agroforestry.

References

Ajit DSK, Newaj R, Handa AK, Prasad R, Alam B, Rizvi RH, Gupta G, Pandey KK, Jain A, Uma (2013) Modelling analysis of potential carbon sequestration under existing agroforestry systems in three districts of Indo-Gangetic plains in India. Agrofor Syst 87(5):1129–1146

Ajit, Dhyani SK, Handa AK, Sridhar KB, Jain AK, Uma, Sasindran P, Kaza M, Sah R, Prasad SMR, Sriram K (2014) Carbon sequestration assessment of block plantations at JSW Steels Limited. In: Compendium of abstracts, 3rd world agroforestry congress, organized by ICAR, WAC and ISAF at Delhi, 10–13 Feb 2014, pp 354–355

Albrecht A, Kandji ST (2003) Carbon sequestration in tropical agroforestry systems. Agric Ecosyst Environ 99:15–27

Arevalo CBM, Bhatti JS, Chang SX, Sidders D (2011) Land use change effects on ecosystem carbon balance: from agricultural to hybrid poplar plantation. Agric Ecosyst Environ 141:342–349

Beddington J, Asaduzzaman M, Clark M, Fernández A, Guillou M et al (2012) Achieving food security in the face of climate change: final report from the commission on sustainable agriculture and climate change. CGIAR Research Program on Climate Change, Agriculture and Food Security (CCAFS). CCAFS, 64 p

Besar NA, Suardi H, Phua MH, James D, Mokhtar MB, Ahmed MF (2020) Carbon stock and sequestration potential of an agroforestry system in Sabah. Forests, Malaysia, p 11

Blaser WJ, Oppong J, Yeboah E, Six J (2017) Shade trees have limited benefits for soil fertility in cocoa agroforests. Agric Ecosyst Environ 243:83–91. https://doi.org/10.1016/j.agee.2017.04.007

Brown SE, Miller DC, Ordonez PJ, Baylis K (2018) Evidence for the impacts of agroforestry on agricultural productivity, ecosystem services, and human well-being in high-income countries: a systematic map protocol. Environ Evid 7:24
Bustamante M, Robledo-Abad C, Harper R, Mbow C, Ravindranat NH, Sperling F, Haberl H, de Pinto AS, Smith P (2014) Co-benefits, trade-offs, barriers and policies for greenhouse gas mitigation in the agriculture, forestry and other land use (AFOLU) sector. Glob Chang Biol 20:3270–3290
deConinck H, Puig D (2015) Assessing climate change mitigation technology interventions by international institutions. Clim Chang 131:417–433
Dhyani SK (2012) Agroforestry interventions in India: focus on environmental services and livelihood security. Indian J Agrofor 13(2):1–9
Dhyani SK, Newaj R, Sharma AR (2009) Agroforestry: its relation with agronomy, challenges and opportunities. Indian J Agrofor 54(3):249–266
Dhyani SK, Ram A, Dev I (2016) Potential of agroforestry systems in carbon sequestration in India. Indian J AgricSci 86(9):1103–1112
Dhyani S, Bartlett D, Kadaverugu R, Dasgupta R, Pujari P, Verma P (2020) Integrated climate sensitive restoration framework for transformative changes to sustainable land restoration. Restor Ecol 28:1026–1031
Dhyani SK (2014) National agroforestry policy 2014 and the need for area estimation under agroforestry. Curr Sci 107(1): 9–10
Dhyani SK, Puri DN, Narain P (1996) Biomass production and rooting behaviour of *Eucalyptus tereticornis* Sm. on deep soils and riverbed bouldery lands of Doon Valley, India. Indian For 122(2):128–136
Dumont ES, Bonhomme S, Pagella TF, Sinclair FL (2017) Structured stakeholder engagement leads to developmentof more diverse and inclusive agroforestry options. Exp Agric. https://doi.org/10.1017/s0014479716000788
FAO (2019) The state of the world's biodiversity for food and agriculture. In: Bélanger J, Pilling D (eds) FAO commissionon genetic resources for food and agriculture assessments, Rome, 572 pp. http://www.fao.org/3/CA3129EN/CA3129EN.pdf
FAO (2020) The state of food and agriculture. Overcoming water challenges in agriculture, Rome, pp 111–120
FAO. State of the World's Forests (2016) Forests and agriculture: land-use challenges and opportunities. Food Agriculture Organization of the United Nations, Rome, Italy
Feliciano D, Ledo A, Hillier J, Nayak D (2018) Which agroforestry options give the greatest soil and above ground carbon benefits in different world regions. Agric Ecosyst Environ 254:117–129
Gama-Rodrigues EF, Nair PKR, Nair VD, Gama-Rodrigues AC, Baligar V, Machado RCR (2010) Carbon storage in soil size fractions under two cacao agroforestry systems in Bahia, Brazil. Environ Manage 45:274–283
García de Jalón S, Graves A, Palma JHN, Williams A, Upson M, Burgess PJ (2017) Modelling and valuing the environmental impacts of arable, forestry and agroforestry systems: a case study. Agrofor Syst 1059–1073
Garrity DP (2004) Agroforestry and the achievement of the millennium development goals. Agrofor Syst 61:5–17
Guillemot J, le Maire G, Munishamappa M et al (2018) Native coffee agroforestry in the Western Ghats of India maintainshigher carbon storage and tree diversity compared to exotic agroforestry. Agric Ecosyst Environ 265:461–469
Gusli S, Sumeni S, Sabodin R, Muqfi IH, Nur M, Hairiah K, Useng D, van Noordwijk M (2020) Soil organic matter, mitigation of and adaptation to climate change in cocoa-based agroforestry systems. Land 9:323
Haile SG, Nair PKR, Nair VD (2008) Carbon storage of different soil-size fractions in Florida silvopastoral systems. J Environ Qual 37(5):1789–1797
Hillbrand A (2021) What is the potential of agroforestry to restore degraded land in Guatemala? FAO, Rome, Italy

ICRAF (2021) Restoring land with agroforestry: new guide published. ICRAF, Bogor, Indonesia
IEA (2020) Global CO_2 emissions in 2019. https://www.iea.org/articles/global-co2-emissions-in-2019
Inder D, Ram A, Bhaskar S, Chaturvedi OP (2018) Role of Agroforestry in current scenario. In Agroforestry for Climate Resilience and Rural Livelihood; Scientific Publishers, Jodhpur, India, pp 1–10
Intergovernmental Panel on Climate Change (IPCC) (2014) Climate change synthesis report-2014
Intergovernmental Panel on Climate Change (IPCC) (2020) Centre for climate and energy solutions. 5th assessment report Kerstin Stendahl, Deputy Secretary. IPCC, Geneva, Switzerland
Intergovernmental Panel on Climate Change (IPCC) (2021) Climate change. Impacts, adaptation, and vulnerability. Mitigation of climate change. 6th assessment report Kerstin Stendahl, Deputy Secretary. IPCC, Geneva, Switzerland
IPCC (2019) Summary for policymakers. In: Climate change and land: an IPCC special report on climate change, desertification, land degradation, sustainable land management, food security, and greenhouse gas fluxes in terrestrial ecosystems. In: Shukla PR, Skea J, Calvo Buendia E, Masson-Delmotte V, Pörtner H-O, Roberts DC, Zhai P, Slade R, Connors S, van Diemen R et al (eds) IPCC press office, Geneva, Switzerland, p 36
Isaac ME, Timmer VR, Quashie-Sam SJ (2007) Shade tree effects in an 8-year-old cocoa agroforestry system: biomass and nutrient diagnosis of *Theobroma cacao* by vector analysis. Nutr Cycling Agroecosyst 78:155–165
Jose S, Bardhan S (2012) Agroforestry for biomass production and carbon sequestration: an overview. Agrofor Syst 86:105–111
Kajembe J, Lupala I, Kajembe G et al (2016) The role of selected agroforestry trees in temperature adaptation on *Coffea arabica*: a case study of the Moshi district, Tanzania. Climate change and multidimensional sustainability in African agriculture. Springer, Cham, pp 553–566
Kang BT, Caveness FE, Tian G, Kolawole GO (1999) Long term alley cropping with four species on an Alfisol in southwest Nigeria—effect on crop performance soil chemical properties and nematode population. Nutr Cycl Agroecosyst 54:145–155
Kaur B, Gupta SR, Singh G (2002) Carbon storage and nitrogen cycling in silvipastoral system on sodic soil North western India. Agrofor Syst 54:21–29
Kumar AK (2010) Carbon sequestration: under explored environmental benefits of Tarai agroforestry. Indian J Soil Conser 38:125–131
Lal R (2005) Soil carbon sequestration impacts on global climate change and food security. Science 304(5677):1623–1627
Lal R (2009) Soil carbon sequestration for climate change mitigation and food security. In: Souvenir, platinum jubilee symposium on soil science in meeting the challenges to food security and environmental quality, Indian society of soil science, New Delhi, pp 39–46
Lasco RD, Delfino RJP, Espaldon MLO (2014) Agroforestry systems: Helping smallholders adapt to climate risks while mitigating climate change. Wires Clim Chang 5:825–833
Leder S, Das D, Reckers A, Karki E (2016) Participatory gender training for community groups. a manual for critical discussions on gender norms, roles and relations. Report from CGIAR research program on Water, Land and Ecosystems
Lenka NK, Choudhury PR, Sudhishri S, Dass A, Patnaik US (2012a) Soil aggregation carbon build up and root zone soil moisture in degraded sloping lands under selected agroforestry-based rehabilitation systems in eastern India. Agric Ecosyst Environ 150:54–62
Lenka NK, Dass A, Sudhishri S, Patnaik US (2012b) Soil carbon sequestration and erosion control potential of hedgerows and grass filter strips in sloping agricultural lands of eastern India. Agric Ecosyst Environ 158:31–40
Marone D, Poirier V, Coyea M et al (2017) Carbon storage in agroforestry systems in the semi-arid zone of Niayes, Senegal. Agrofor Syst 91:941–954. https://doi.org/10.1007/s10457-016-9969-0
Meragiaw M (2017) Role of agroforestry and plantation on climate change mitigation and carbon sequestration in Ethiopia. J Tree Sci 36(1):1–15

Middendorp RS, Vanacker V, Lambin EF (2018) Impacts of shaded agroforestry management on carbon sequestration, biodiversity and farmers income in cocoa production landscapes. Landsc Ecol 33:1953–1974

Mittal SP, Singh P (1989) Intercropping field crops between rows of *Leucaena leucocephala* under rainfed conditions in northern India. Agrofor Syst 8(2):165–172

Montagnini F, Nair P (2004) Carbon sequestration: An underexploited environmental benefit of agroforestry systems. Agrofor Syst 61:281–295

Morillas P (2020) Lessons from a global crisis: coronavirus, the international order and the future of the EU. https://www.g20-insights.org/policy_briefs/lessons-from-a-global-crisiscoronavirus-the-international-order-and-the-future-of-the-eu/

Murthy IK, Gupta M, Tomar S, Munsi M, Tiwari R, Hegde GT, Ravindranath NH (2013) Carbon sequestration potential of agroforestry systems in India. J Earth Sci Clim Change 4:131

Mutuo PK, Cadisch G, Albrecht A, Palm CA, Verchot L (2005) Potential of agroforestry for carbon sequestration and mitigation of greenhouse gas emissions from soils in the tropics. Nutr Cycl Agroecosyst 71:43–54

Nair PKR, Kumar BM, Nair VD (2009) Agroforestry as a strategy for carbon sequestration. J Plant Nutri Soil Sci 172: 10–23

Nair P, Nair V, Mohan Kumar B, Showalter J (2010) Carbon sequestration in agroforestry systems. Adv Agron 108:237–307

Nair PKR, Viswanath S, Lubina PA (2017) Cinderella agroforestry systems. Agrofor Syst 91:901–917

Agroforestry Network (2018) Scaling up agroforestry: potential, challenges and barriers. A review of environmental, social and economic aspects on the farmer, community and landscape level

Newaj R, Dhyani SK (2008) Agroforestry for carbon sequestration: scope and present status. Indian J Agrofor 10:1–9

Newaj R, Rizvi RH, Chaturvedi OP, Alam B, Prasad R, Kumar D, Handa AK (2017) A country level assessment of area under agroforestry and its carbon sequestration potential. Technical bulletin 2/2017, ICAR- Central Agroforestry Research Institute, Jhansi, pp 1–48

Parrotta JA (1999) Productivity, nutrient cycling, and succession in single- and mixed-species plantations of *Casuarina equisetifolia*, *Eucalyptus robusta*, and *Leucaena leucocephala* in Puerto Rico. For Ecol Manage 124:45–77

Paudel D, Tiwari KR, Bajracharyab RM, Raut N, Sitaulac BK (2017) Agroforestry system: an opportunity for carbon sequestration and climate change adaptation in the mid-hills of Nepal. Oct J Env Res 5(1):022–031

Peichl M (2006) Carbon sequestration potentials in temperate tree-based intercropping systems, southern Ontario, Canada. Agroforsyst 66(3):243–257

Pragason A, Karthik A (2013) Carbon stock sequestered by tree plantation in University campus at Coimbatore, India. Int J Environ Sci 3(5):1700–1710

Prasad JVNS, Srinivas K, Srinivasarao C, Ramesh C, Venkatravamma K, Venkateswarlu B (2012) Biomass productivity and carbon stocks of farm forestry and agroforestry systems of *Leucaena* and *Eucalyptus* in Andhra Pradesh, India. Curr Sci 103(5):536–540

Prusty M, Ray M, Sahoo GR (2020) Carbon sequestration-a way to mitigate green house effect. Glocal environmental governance, policies and ethics-II, pp 32–43. ISBN 978-93-5419-920-2

Rahman SA, Sunderland T, Kshatriya M, Roshetko JM, Pagella T, Healey JR (2016) Towards productive landscapes: trade-offs in tree-cover and income across a matrix of smallholder agricultural land-use systems. Land Use Policy 58:152–164

Rai P, Solanki KR, Singh UP (2000) Growth and biomass production of multipurpose tree species in natural grassland under semi-arid condition. Indian J Agrofor 2:101–103

Ramesh T, Manjaiah KM, Mohapatra KP, Rajasekar K, Ngachan SV (2015) Assessment of soil organic carbon stocks and fractions under different agroforestry systems in subtropical hill agroecosystems of north-east India. Agrofor Syst 89(4):677–690

Roshetko JM, Lasco RD, Angeles MSD (2007) Small holder agroforestry systems for carbon storage. Mitig Adapt Strateg Glob Chang 12:219–242

Saha R, Jha P (2012) Carbon sequestration potentials of agroforestry systems under climate change scenario—brief review with special emphasis on North-Eastern Hill Regions. J Agric Phys 12(2):100–106
Saha S, Nair PKR, Nair VD, Kumar BM (2009) Soil carbon stocks in relation to plant diversity of home gardens in Kerala, India. Agrofor Syst 76:53–65
Sahoo GR, Wani AM (2019) Multifunctional agroforestry systems in India for livelihoods. Ann Hortic 12(2):139–149
Sahoo GR, Wani AM (2020) Effect of climate change on land degradation. Int J Innov Eng Manage Res SSRN Elsevier 09(12):483–494
Sahoo GR, Wani AM, Satpathy B (2020a) Greening wastelands for environmental security through agroforestry. Int J Adv Res Sci Technol (IJARST) 7:2581–9429
Sahoo GR, Wani AM, Sharma A (2020b) Enhancing food security through agroforestry for sustainability: a review. Int J Curr Microbiol App Sci 11: 2001–2020b
Sahoo GR, Wani AM, Gupta S, Prusty M (2021) Soil and water conservation measures through agroforestry: a review. PLANTA Res 2:251–262. ISBN 978-81-953419-4-8
Santoro A, Venturi M, Bertani R, Agnoletti M (2020) A review of the role of forests and agroforestry systems in the FAO globally important agricultural heritage systems (GIAHS) programme. Forests 11
Schuman GE, Janzen HH, Herrick JE (2002) Soil carbon dynamics and potential carbon sequestration by rangelands. Environ Pollut 116:391–396
Seserman DM, Freese D, Swieter A, Langhof M, Veste M (2019) Trade-O between energy wood and grain production in temperate alley-cropping systems: an empirical and simulation-based derivation of land equivalent ratio. Agriculture 9:147
Shepherd D, Montagnini F (2001) Above-ground carbon sequestration potential in mixed and pure tree plantations in the humid tropics. J Tropi ForSci 13:450–459
Simelton E, Dam BV, Catacutan D (2015) Trees and agroforestry for coping with extreme weather events: experiencesfrom northern and central Vietnam. Agrofor Syst 89:1065–1082
Singh TP (2003) Potential of farm forestry in carbon sequestration. Ind for 129:839–843
Swamy SL, Mishra A, Puri S (2003) Biomass production and root distribution of *Gmelina arborea* under an agri-silviculture system in sub-humid tropics of central India. New for 26:167–186
Swamy SL, Puri S (2005) Biomass production and C sequestration of *Gmelina arborea* in plantation and agroforestry system in India. Agrofor Syst 64:181–195
Takimoto A, Nair PKR, Nair VD (2008) Carbon stock and sequestration potential of traditional and improved agroforestry systems in the West African Sahel. Agric Ecosyst Environ 125:159–166
Thaware BL, Bhagat SB, Khadtar BS, Jadhav BB, Dhonukshe BL, Jambhale ND (2004) Effect of tree species on growth and yield of rice (*Oryza sativa* L.) in Konkan Region. Ind J Agrofor 6(2):15–18
Tomar JMS, Bhatt BP (2004) Studies on horti-agricultural systems in mid-altitude of Meghalaya. Ind J Agrofor 6(2):35–39
Vanlaluna PC, Sahoo UK (2009) Performance of multipurpose trees and associated crops in agroforestry systems of Mizoram. Ind J For 32(2):191–194
Verchot LV, Van Noordwijk M, Kandji S, Tomich T, Ong C, Albrecht A, Mackensen J, Bantilan C, Anupama KV, Palm C (2007) Climate change: linking adaptation and mitigation through agroforestry. Mitig Adapt Strateg Glob Chang 12(5):901–918
Viswanath S, Peddappaiah RS, Subramoniam V, Manivachakam P, George M (2004) Management of *Casuarina equisetifolia* in wide-row intercropping systems for enhanced productivity. Ind J Agrofor 6(2):19–25
Westholm L, Ostwald M (2020) Food production and gender relations in multifunctional landscapes: a literature review. Agrofor Syst 94:359–374
Wright AL, Hons FM (2005) Tillage impacts on soil aggregation and carbon and nitrogen sequestration under wheat cropping sequences. Soil till Res 84:67–75
Zhang W, Ahanbieke P, Wang BJ, Xu WL, Li LH, Christie P, Li L (2013) Root distribution and interactions in jujube tree/wheat agroforestry system. Agrofor Syst 87(4):929–939

Chapter 23
Alley Cropping Agroforestry System for Improvement of Soil Health

H. C. Hombegowda, Partha Pratim Adhikary, Praveen Jakhar, and M. Madhu

Abstract Practice of agroforestry system on tree-less lands provides a unique opportunity to improve soil health/quality while maintaining crop productivity in additional to provision of tree products to the rural farming communities. In nutrient poor tropical soils, introduction of agroforestry systems is proven to be sustainable and economical practice in many parts of the world. Among different systems/practice of agroforestry, alley cropping is one of the recently developed agroforestry techniques (during 1970's) to overcome the management problems of the upland soils, and to incorporate the benefits of much needed tree fallow component. In alley cropping, agriculture crops are grown between hedgerows of planted shrubs and trees, preferably leguminous species, which are regularly/intermittently pruned to prevent light competition and shading to the companion crop. Alley cropping improves the soil physical, chemical and biological properties by improving the nutrient recycling through the addition of pruned leafy biomass, reduction in nutrient loss by erosion control and reducing leaching losses. The improved soil physical properties like soil aeration, aggregate stability and infiltration rate, in alley cropping may lead to regeneration of the degraded topsoil and thus may produce more stable aggregates and provide favourable soil media for the crop cultivation. The practice of *Gliricidia* and *Leucaena* alley has shown the improvement in the soil moisture availability during the dry season of the cropping period. The use of N fixing trees as hedgerow adds huge amount of nitrogen and potassium through its biological nitrogen fixation and pruned leafy biomass incorporation. While phosphorus is made available by the

H. C. Hombegowda
ICAR-Indian Institute of Soil and Water Conservation, Research Centre, Ooty, Tamil Nadu, India

P. P. Adhikary (✉)
ICAR-Indian Institute of Water Management, Bhubaneswar, Odisha, India
e-mail: ppadhikary@gmail.com; Partha.Adhikary@icar.gov.in

P. Jakhar
ICAR-Central Institute of Women in Agriculture, Bhubaneswar, Odisha, India

M. Madhu
ICAR-Indian Institute of Soil and Water Conservation, Dehradun, Uttarakhand, India

P. K. Shit et al. (eds.), *Soil Health and Environmental Sustainability*,
Environmental Science and Engineering,
https://doi.org/10.1007/978-3-031-09270-1_23

organic acids bind to Al and Fe oxides which released during the biomass decomposition, and reduces Al saturation and toxicity. Alley cropping proven to improve the soil biological properties like microbial biomass and enzyme activity through improvement in soil organic matter content through addition of tree leaf litter, fine roots, and crop residues. The improvement of all three soil properties ultimately improves the soil quality and soil health. The improved soil quality and modified microclimate ultimately increase the productivity of the associated crops. In recent years, alley cropping is recognised as potential agroforestry practice for soil carbon sequestration. Overall, alley cropping has the ability to improve soil quality by modifying soil properties. The improvement of soil properties is highly varied with different alley species, soil type, climate and the management practice. In general, alley cropping responds more effectively to intensive management and may not be a suitable system in arid areas. In alley cropping management, selection of suitable hedge row species based on soil and environmental condition is important to obtain the full benefits of this agroforestry system.

Keywords Carbon sequestration · Hedge row intercropping · Soil quality · Water stable aggregates

23.1 Introduction

The global crop production and productivity have increased manifold since the last quarter of the twentieth century. Much of the credit for this will go to the intensification of agriculture through mechanization, use of chemical fertilizers and pesticides, irrigation, and development of high yielding crop varieties (Lal 2007). This unprecedented increase in food production was able to support the world's growing human and animal population. However, this abundance in production also comes with costs. The production of pesticides, fertilizers and mechanical equipment is mainly relied upon fossil fuel energy. In some cases, energy consumption exceeds food energy production by over ten to one. Such heavy reliance on non-renewable resources may be unsustainable over the long run. Over use of pesticides and fertilizers for crop production has created groundwater and surface water pollution. This pollution has both human health and ecosystem health affects (Adhikary et al. 2012). Intensive agriculture leads to decrease of soil organic matter, especially in warm, humid areas. Soil organic matter is important because it increases the nutrient holding capacity and moisture retention capacity of soils. As farming becomes more intensified, fields became larger and crop diversity declines. The decline in biodiversity along with other factors, including pest resistance, has led to increases in insect damage despite increases in pesticide use (Pimentel 1997). Therefore, modern agriculture has large benefits, but it also has high environmental costs.

A high environmental cost of intensive agriculture has led to increased interest in low-input or organic agriculture. It is less damaging to the environment and human health aspect is maintained here. Here soil amendments are needed to sustain crop yields. Adding organic residues increases retention of soil C and N. Adding compost or manure is one option but these amendments are produced in other places and then must be transported to the farmers' fields. Interest in cover crops has revived in recent years as a method to maintain soil fertility, reduce fertilizer use, and reduce erosion. However, cover crops must be replanted each season that they are used, and land is taken out of production if the cover crop is used as a nutrient amendment. Also inputs from herbaceous cover crops may not be sufficient at supplying nitrogen and phosphorus to a continuous cropping system. Another option to restore the soil health and to reduce soil and nutrient erosion is planting of aromatic grasses (Adhikary et al. 2018). But this option has to compete with field crops as farmers may not prefer these grasses in good cultivable lands. Another organic amendment option is alley cropping agroforestry system such as leguminous hedgerow intercropping (Adhikary et al. 2017; Hombegowda et al. 2020). Around the world, practices of agroforestry tend to improve the soil quality and microclimatic condition that favours the annual crop productivity. Traditional agroforestry system i.e., Thang Bun, practiced in north east India is well known for its improvement of soil physiochemical properties and crop productivity in the nutrient poor acid soil (Hombegowda et al. 2021). In agroforestry systems, complementary use of water resources exists and it depends on the type of tree species (Hombegowda et al. 2019). Natural systems were guides for the development of alley cropping systems. Ecosystem services that are found in forests are brought to agro-ecosystems by planting trees within fields.

Alley cropping is successful on some acidic soils after organic matter accumulates for several years. High yields can also be found on acidic low fertile soils, if fertilizer or lime is added. The success in soil management to maintain soil quality depends on an understanding of how soil responds to agricultural practices over time. For this reason, recent interest in evaluating the quality of our soil resources has been stimulated by increasing awareness that soil is a critically important component of the earth's biosphere, functioning not only in the production of food and fibre but also in the maintenance of local, regional and worldwide environmental quality (Doran and Parkin 1994). On the other hand, it is possible that trees, whether intimately mixed with crops or planted in rows will improve the total water supply by reducing evaporation. Roots play a part in nearly most of the processes, particularly inorganic matter input, soil physical conditions, nitrogen fixation, and nutrient retrieval.

In resource-poor environments, hedgerow roots are more concentrated in upper soil layers, increasing competition with crops. Nutrient availability in alleys is limited when nutrients or soil moisture are low and the trees are severely stressed. Low yields of hedgerow prunings also contribute to low crop yields in infertile soils (Tossah et al. 1999). If degraded topsoil is above a relatively fertile sub-soil, alley cropping may lead to regeneration of the degraded topsoil. In arid or semi-arid areas, the trees compete with the crops for water, and crop yields are suppressed (Odhiambo et al. 2001). Competition for water can be reduced. However, Lehmann et al. (1998) found that root density of hedgerows decreased dramatically after pruning. They also found that hedgerows had more deep roots in the subsoil when intercropped with sorghum than when grown alone. In their opinion, alley cropping utilized soil resources more efficiently than a monoculture. In general alley cropping found very little success in arid areas, but some success has been found with parkland agroforestry. Although alley cropping is generally less effective in harsh environments, modifications or species selection may allow alley cropping to be effective in areas where soils are dry, infertile, or acidic.

In the world, alley cropping has greatest potential as a technique on small farms. It is easier to incorporate alley cropping into farming systems that rely on light machinery and manual labour. Alley cropping can be of interest among small-scale farmers who intensively managed their land. Many farmers can pool their land in a watershed and get benefit from alley cropping system along with other farming systems (Madhu et al. 2016). Several factors make alley cropping appropriate for farmers. Many farmers grow multiple crops in their fields. This familiarity with intercropping may make them more accepting of planting hedgerows within their fields. The greatest costs associated with alley cropping are labour costs and loss of productive land due to hedgerows. However, by adapting alley cropping techniques to mechanical methods and introducing alley cropping to areas with low land values, the techniques may be economically viable. Also, farmers are able to absorb limited increases in costs of production through growing organic products which are high-value crops. Leguminous hedgerows provide nutrient additions similar to other on-site organic inputs and help maintain soil fertility. Hedgerows restore some ecosystem functions to fields by increasing nutrient cycling and maintaining levels of soil organic matter (Adhikary et al. 2017).

In this chapter we examined the ability of the alley cropping systems to provide nutrients to grain crops. Here we attempted to determine the effectiveness of hedgerow intercropping at providing nutrients, increasing crop yields, and improving soil quality. N and P are focused on because these nutrients are often limiting to plant growth. One of the advantages of alley cropping is that it increases long-term sustainability by improving soil quality. Both chemical and physical soil properties are examined to determine changes in soil quality. Alley cropping is not effective if there is competition between crops and the hedgerows. Here, the effects of competition are examined by reducing root competition. Finally, a brief economic examination is conducted to determine if costs of alley cropping is higher than other farming systems.

23.2 Alley Cropping: Soil Properties

Soil quality as defined by Karlen et al. (1997) is the capacity of a soil to function within natural or managed ecosystem boundaries to sustain plant and animal productivity, maintain or enhance water and air quality, and support human health and habitation. Soil quality can be monitored by a set of measurable attributes termed as soil quality indicators. These indicators can be broadly classified as physical, chemical and biological indicators and one can assess overall soil quality by measuring changes in these indicators and transforming them into single value known as soil quality index (Dalal and Moloney 2000). Alley cropping improves the soil properties by making a situation where nutrient recycling becomes easy. Though alley cropping is not the only answer for improving soil quality under all situations, but it is an option. Factors such as climate and type of alley determine the success of alley cropping at particular sites for improving soil quality.

Depending on the hedgerow species selected, alley cropping systems are able to maintain soil quality and increase crop yields in areas with sufficient rainfall and with initial moderate soil fertility through improving soil properties (Tossah et al. 1999). Alley cropping helps to maintain soil quality by increasing nutrient cycling and reducing leaching of nutrients (Adhikary et al. 2017; Hombegowda et al. 2020). When nitrogen-fixing trees were pruned, they shed their nodules providing additional subsoil nutrients. During decomposition, fine roots release nutrients faster than decomposing leaves which increases microbial population required for plant growth (Jose et al. 2000). While it is easy to quantify the amount of nutrients provided by the green manure, below ground interactions are much difficult to describe. In a review article, Kass et al. (1997) suggested that below ground inputs are larger than above ground inputs. However, Nygren and Ramírez (1995) found that roots of periodically pruned *Erythrina* contribute only a small amount of nitrogen compared to the contribution from the leaves. In some cases, root die back after pruning may have led to increased leaching of soil nutrients (Peter and Lehmann 2000). Thus, alley cropping has the ability to improve soil quality by modifying soil properties.

23.2.1 Soil Physical Properties

Soil physical property can be modified by alley cropping system, but its magnitude depends on several factors. The improvement of soil physical property due to alley cropping is very slow in extreme condition because in extreme conditions the crop growth use to suffer and low biomass incorporation results (Rao et al. 1998). The improvement of soil physical quality like aggregate stability under alley cropping system is highly pronounced in humid to sub-humid areas than arid to semi-arid areas. This is even more pronounced in more fertile soils than less fertile soils as initial soil fertility compensate competition between crop and alley tress for water and nutrients (Rao et al. 1998). In resource-poor environments, hedgerow roots are more concentrated in upper soil layers, increasing competition with crops. Thereby instead of forming more stable aggregates at the top soil layer there are more unstable aggregates in alley cropped areas. Nitrogen fixation becomes limited when nutrients or soil moisture are low and the trees are severely stressed. Low yields of hedgerow prunings also contribute to low crop yields in infertile soils and thereby indirectly regulate the formation of stable aggregates (Tossah et al. 1999) because of less favourable soil physical condition which reduces nutrient uptake. If degraded nutrient poor topsoil is situated above a relatively fertile sub-soil, alley cropping may lead to regeneration of the degraded topsoil and thus may produce more stable aggregates. *Gliricidia* is an excellent hedgerow which increases the water stable aggregates. The amount of water stable aggregates formed in the soil profile depends on the number

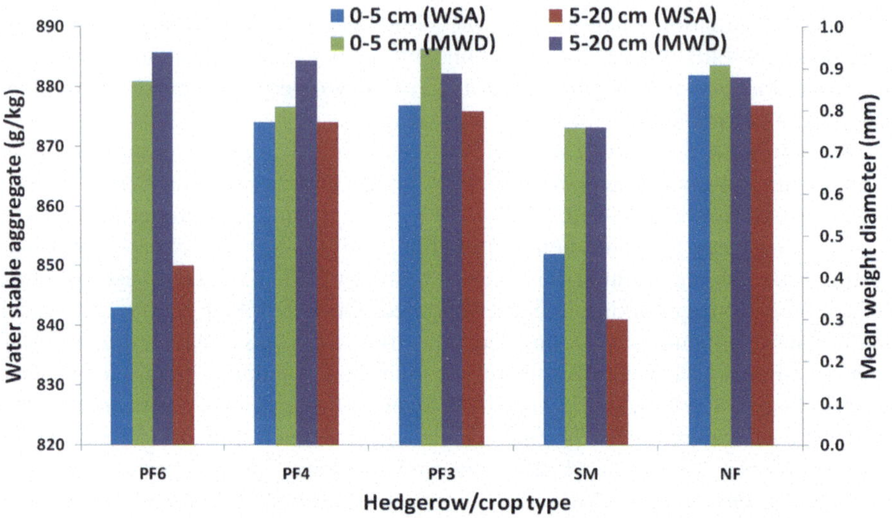

Fig. 23.1 Changes in water stable aggregate (WSA) and mean weighted diameter of soil aggregates (MWD), in the 0–5 cm and 5–20 cm depth, as a function of pruning frequency of *Gliricidia* hedges. PF6, PF4, and PF3 represent 6, 4 and 3 annual prunings, respectively (Modified and adopted from Barreto et al. 2012)

of pruning per year. The water stable aggregates and mean weight diameter of the aggregates are presented in Fig. 23.1.

Alley cropping is helpful to increase soil aeration on some acidic, less aerated soils after organic matter accumulates for several years. Even more in acidic soil the soil aeration rate will grow rapidly, if fertilizer and/or lime are added, because it improves physical quality of the soil. In arid or semi-arid areas the trees compete with the crops for water, and crop yields are suppressed (Odhiambo et al. 2001). Lehmann et al. (1998) found that root density of hedgerows decreased dramatically after pruning thus affecting soil physical property like nodulation. They also found that hedgerows had more deep roots in the subsoil when intercropped with sorghum than when grown alone and thereby increases soil aeration. In their opinion, alley cropping utilized soil resources more efficiently than a monoculture. In general, alley cropping is not a suitable system in arid areas, but some success has also been found where it modifies soil physical quality like nodulation and soil aeration.

Although alley cropping is generally less effective in harsh environments, modifications or species selection may allow alley cropping to be effective in areas where soils are dry, infertile, or physically degraded. Addition of green manure from hedgerows also improves soil physical qualities such as bulk density (Adhikary et al. 2017). The selection of hedgerow species is important to get some effect on soil bulk density. In a study in Oxisol of Nigeria, Hulugalle and Kang (1990) reported that the bulk density decrease was highest for the 0–5 cm of soil depth. Among the different alley cropping species, *Leucaenaleucocephala* has profound effect to reduce soil bulk density. It reduces bulk density of a sandy loam alluvial soil from 1.7 to 1.5 g cm^{-3}(Gangwar et al. 2004).

Soil available water content is also influenced by alley cropping. The focus of the alley cropping is that during dry season the crop should get sufficient soil moisture to survive and grow. Thus, the crop can translocate food to the grain during the dry condition and increases the yield compared to the no alley cropping areas. In an experiment in red Alfisol of India, Adhikary et al. (2017) showed that *Gliricidia* alley have the potential to conserve more soil moisture than Leucaena alley in 5 and 10% land slope condition. The stored soil moisture was sufficient enough to increase upland paddy yield by nearly 15%. In another experiment in Oxisol of Nigeria, Hulugalle and Kang (1990) reported that, the soil moisture content at the surface soil increased due to alley cropping system. In this experiment, *Gliricidia* and *Leucaena* hedgerow species performed equally to improve soil moisture content. Therefore, selection of suitable hedge row species based on soil and environmental condition is important to increase the soil moisture content under alley cropping system.

Many studies have shown that land use patterns are the main factors influencing soil infiltration. Thus, increasing soil infiltration and reducing runoff are crucial for soil and water conservation. In this context agroforestry system like alley cropping can increase the infiltration rate of soil. In an experiment in Loess plateau of China, Wang et al. (2015) reported that walnut wheat alley cropping system has increased the basic infiltration rate significantly than mono cropping system. They also said that with the increase of the age of alley cropping system the infiltration rate also increases. In subsequent time, alley cropping system improves the physical condition of the upper soil layer thus increases the infiltration rate. In different land use patterns and cropping systems, plant root activities are important factors affecting soil infiltration. Therefore, high infiltration rate in alley cropping system may be attributed to the high root activity in this system (Hangen et al. 2002). In alley cropping system the roots of the intercrops rot in the shallow soil and hedge row crop roots were active after the crop harvest. This makes the root channels more in soil and connectivity better, thereby the infiltration rate increased significantly (Wang et al. 2015).Sun et al. (2018) also reported that agro-forest land could have a higher soil infiltration rate compared with wasteland and other land uses. This is consistent with the main activity area of tree roots (10–40 cm) in alley cropping systems (Xu et al. 2013). The alley tree breaks the plough pan which is formed in the normal agricultural plots and increases the infiltration rate (Sun et al., 2018). In a sandy loam soil of Uttar Pradesh, India, alley cropping of *Leucaena leucocephala* hedges increases the infiltration rate by 38.2% (Gangwar et al. 2004).

23.2.2 Soil Chemical Properties

Soil chemical properties are maintained and improved by alley cropping system. Among the soil chemical properties, soil pH is highly influenced by alley cropping. The influence of alley trees in changing soil pH is highly pronounced in alkaline soils where it reduces the soil pH. In an experiment in the alkaline soil of Uttar Pradesh of India Gangwar et al. (2004) reported that the soil pH has been decreased from 8.74 to 8.43 after adding the lopping materials of *Leucaena leucocephala* hedges. The *Leucaena leucocephala* lopping material after decomposing in the soil produces organic acids which reduces the soil pH.

Cation Exchange Capacity or CEC is an important soil property influenced by alley cropping system. Higher root activity in the rhizosphere under alley cropping system increases the cation binding sites in the clay micelle. The organic materials after decomposition produces organic acids also contribute to the higher CEC in the alley cropping system. In an experiment it has been found that the alley cropping system can increase CEC by 34% (Abunyewa et al. 2004). Higher soil organic matter in alley cropping system improves the cation exchange capacity and consequently increases exchangeable bases (i.e., Ca, Mg, and K) in the surface soil until a steady-state is reached (Abunyewa et al. 2004).

In alley cropping system N fixing trees are used as hedgerow which ultimately adds huge amount nitrogen in the soil and increases its level in soil. Two important nitrogen fixing trees, namely, *Leucaena leucocephala* and *Gliricidia sepium* produce high biomass and have been used extensively for soil amelioration and N addition (Casanova-Lugo et al. 2014). In an experiment in the sandy loam alluvial soil of Uttar Pradesh of India Gangwar et al. (2004) reported that addition of *Leucaena leucocephala* lopping material adds an amount of 80 kg ha^{-1} organic nitrogen, which in-tern increase the grain yield of rice and wheat significantly. The *Gliricidia* based alley cropping system can increase maize yield by 54% by increasing the soil nitrogen by 22% (Abunyewa et al. 2004). Incorporation of pruning materials has significant impact on soil chemical properties (Table 23.1).

Soil phosphorus is an important material required by crops is also improved through alley cropping. In acidic soils, large amounts of plant available P in solution become unavailable to plants when P adsorbs to Al or Fe oxides. Adding organic sources of P through alley pruning may reduce the amount of P that becomes fixed and unavailable to plants. As high-quality leaves decompose, organic acids are formed. These organic anions bind to Al and Fe oxides, competing with P for binding sites, reducing P adsorption, and making more P available to crops (Mäder et al. 1999).

Table 23.1 Soil chemical properties change in alley cropping systems as influenced by three different woody species

Treatment	pH	Organic C (%)	Total N (%)	CEC (meq 100 g^{-1})
Leucaena leucocephala	6.1^{a}	0.68^{a}	0.081^{a}	20.32^{ab}
Cajanus cajan	6.0^{a}	0.67^{a}	0.079^{ab}	19.92^{bc}
S. siamea	6.0^{a}	0.67^{a}	0.070^{c}	19.92^{bc}
Control	5.8^{a}	0.58^{c}	0.070^{c}	19.21^{d}
Initial	5.8^{a}	0.65^{b}	0.076^{b}	19.56^{cd}

Source Rahman et al. (2009); Letter denotes difference between alley cropping treatments at the 0.05 level of significance

Additions of organic matter and the binding of organic acids also lead to reductions in Al saturation and toxicity. Organic additions of P may be more available to plants over a season because P is mineralized gradually as the organic matter is decomposed.

Soil potassium has also influenced by alley cropping system. Both *Gliricidia* and *Leucaena* alley cropping system increase the K content in soil. The *Gliricidia* based alley cropping system in a loamy soil can increase the K content by 18.4% (Abunyewa et al. 2004). In a *Leucaena leucocephala* based study Gangwar et al. (2004) reported that the lopping material adds 22.8% more K in the soil than no alley cropping system. In general, hedgerows were most successful at supplying K to the crop. Nutrient balances suggested that nutrients, especially K, were being slowly drained from the soil over several years (Ng et al. 2008).

23.2.3 Soil Biological Properties

Alley cropping system has profound influence on the soil biological property. The differences in litter quality and quantity between the tree and intercropped components can lead to differential enzyme activities and microbial functional diversity in relation to tree rows (Mungai et al. 2005). Plant litter quality influence saprophytic microbes in soil that in turn regulate ecosystem functions such as decomposition and nitrogen (N) mineralization. Alley cropping systems may contribute to soil organic matter (SOM) content through addition of tree leaflitter, fine roots, and crop residues (Mungai et al. 2005). Microbial parameters such as β-glucosidase activities can provide advance evidence of changes in soil organic carbon (SOC) long before it can be accurately measured by routine methodologies. Microclimate differences due to the presence of trees in alley cropping systems can cause variations in soil temperature and water content. Light intensity and soil temperature used to increase with increased distance from the tree's trunk in alley cropping system (Ko and Reich 1993). Soil enzyme activities and parameters used to describe microbial functional diversity were generally higher at the tree row than at the middle of the alley. Microbial functional diversity and activities in soil is influenced by factors such as the availability and quality of organic substances, soil temperature, and soil water content (Mungai et al. 2005). Thus, alley cropping system with its variation on soil quality exerts significant influence on microbial activity or enzyme activity. In an experiment Mungai et al. (2005) observed significant correlation between enzyme activity and alley cropping systems (Table 23.2) under different alley cropping systems in north central Missouri, USA.

Alley cropping with perennial grass has profound influence on the selection of microbes to grow in the soil environment. Grasses have a more continuous supply of organic substrates because of the extensive root system. But as perennial grasses may inhibit nitrifying bacteria leading to low nitrate content (Alexander 1977). Seasonal shift in microbial functional diversity is observed under Pecan alley cropping system because of seasonal cycles of nutrient availability due to temporal patterns in leaf- and root-derived substrates (Myers et al. 2001). Soil temperature and water content

Table 23.2 Correlation coefficients (r) among soil organic carbon and nitrogen, soil enzyme activities, and Shannon diversity index for surfaces oil (0–10 cm) in two alley cropping systems

Site soil property	Pecan		Maple		
	SOC	TKN	SOC	TKN	Yield[a]
β-Glucosidase	0.21	0.49*	0.46*	0.65**	0.59**
Fluorescein DA	−0.04	0.09	0.63**	0.81***	0.71***
Shannon index 48 h[b]	0.52*	0.65**	0.05	−0.17	−0.45*
Shannon index 60 h	0.55*	0.70***	0.11	−0.14	−0.48**
Shannon index 72 h	0.50*	0.69***	0.12	−0.10	−0.47**
Shannon index 84 h	0.51*	0.71***	0.13	−0.04	−0.40*

SOC: Soil organic carbon, TKN: Total Kjeldahl nitrogen, Fluorescein DA: Fluorescein diacetate
[a]Soybean yields, [b]Shannon index at 48, 60, 72 and 84 h of incubation
[*,**,***]Significance at $P < 0.05$, $P < 0.01$, and $P < 0.001$, respectively; *Source* Mungai et al. (2005)

remain high at the middle of the alley, and within the range that would favour optimum microbial growth (Zak et al. 1999). Organic matter decreases with depth and is usually correlated to microbial activities like β-glucosidase activities and biological parameters. Wick et al. (1998) observed reductions in β-glucosidase activities with depth in three agro forestry fields in Nigeria, where, after 4, 10, and 14 years of continuous alley cropping under minimum tillage, no differences were observed among sites in β-glucosidase activities, but activities in the 0–5-cm depth were significantly higher than at the 5–10 cm depth. In a pecan and silver-maple alley cropping experiment, Mungai et al. (2005) observed that soil enzyme activities and microbial functional diversity were higher near alley trees compared to 6 m away of the alley. Microbial activities also varied with time at the alley cropping site. This may have implications for long-term nutrient cycling in alley cropping systems that may require differential nutrient management to maximize productivity.

Microbial biomass carbon (MBC) also varies depending on various agroforestry systems. In alley cropping system the MBC increases a lot. In an experiment on a clay loam soil of USA, Seiter et al. (1999) reported that alley cropping system of *Alnusrubra* and maize has increased the MBC significantly than no alley cropping system. They also pointed out that active fungal and bacterial biomass carbon is higher in between the hedge rows. The MBC content of a new alley cropping system is lower than the same for an older system. Lee and Jose (2003) reported lower soil microbial biomass in a 3-year-old pecan–cotton alley cropping system compared to a similar system that was 47 years old.

Peter and Lehmann (2000) found that pruning *Acacia* hedgerows reduced root development and may have reduced below ground competition with the intercropped plants. The root dynamics of intercropping are still not understood though some studies have found increased yields when hedgerow roots were separated from crop roots. Hedgerows and crops may also compete for light (Friday and Fownes 2001), but light is often a less important interaction than competition for nutrients or water (Mugendi et al. 1999). Competition for light may be more important for ground creeping crops than for tall crops such as maize. In some cases, competition is beneficial. Akobundu et al. (1999) found weeds were better managed in an alley cropping system than in a traditional fallow system. Competition between hedgerows and crops is a problem when resources are severely limited.

23.3 Alley Cropping and Soil Fertility/Nutrient Cycling

In a soil–plant system, plant nutrients are in a state of continuous, dynamic transfer. Plants take up nutrients from the soil and use them for metabolic activities. In-turn, these nutrients are returned back to the soil either naturally as litter fall in unmanaged systems, deliberately as pruning in agroforestry systems or through root senescence in both managed and unmanaged systems. These plant parts are decomposed as a result of microbial activities and release the nutrients held in them into the soil. The nutrient then becomes available for plant uptake once again (Nair et al. 1999). The nutrient cycling in general has been defined as continuous transfer of nutrients that are already present within a soil–plant system (Sanchez and Palm 1996; Buresh and Tian 1998). It involves the continuous transfer of nutrients within and between different components of an ecosystem and include processes such as weathering of minerals, activities of soil biota and other transformation occurring in the biosphere, lithosphere and hydrosphere (Jordan 1985).

The land use systems comprising of trees, crops and pastures play an important role in improving soil fertility and its quality by several ways. Nair (1984) reported that agro-forestry systems have the potential to reduce erosion and runoff, and to maintain soil organic matter, improve soil physical properties and augment nitrogen fixation and promote efficient nutrient cycling. Under different agroforestry research, many workers have emphasized the importance of alley cropping (Kessler and Breman 1991; Jakhar et al. 2017) and agri-horticultural and agro-forestry systems (Das et al. 1993).

23.3.1 Nitrogen

Out of the several benefits accrued from alley cropping under agroforestry systems in terms of soil quality; nutrient cycling is the most predominant one. Natural forest ecosystems of the tropics represent self-sustaining and efficient nutrient cycling systems. These are “closed” nutrient cycling systems with relatively little loss or gain of the actively cycling nutrients and with high rates of nutrients turnover within the system. In contrast, most of the agricultural systems represent ‘open” or “leaky” system with comparatively high nutrient losses and nutrient cycling in agro-forestry systems falls between these “extremes” (Nair et al. 1995).

Under alley system legume hedgerows fixes atmospheric N and supply green manure to the accompanying crop through pruning. The deep-rooted tree legume also serves the function of forest by bringing mineral nutrients such as P, Ca, K, and Mg from the subsoil to the surface layer (Rahaman et al. 2009). These nutrients are then made available to the crop and the hedgerow through decomposition and mineralization of pruning. Such recycling process is more effective in Alfisols than in low base status soils (e.g. Ultisols and Oxisols).

In many strongly acidic soils, a high degree of exchange Al and low exchangeable Ca and Mg content in the sub-soil may prevent normal root growth and penetration. Furthermore, the low C/N ratio of the legume green manure favours rapid decomposition and N loss through leaching hence it is less effective than high C/N materials such as grasses and maize residue for soil organic matter accumulation. In alley cropping system, the rapid decomposition of the legume pruning favours rapid recycling of nitrate through leaching and plant uptake. The maintenance of soil fertility under alley cropping follows a dynamic model; whereas under the bush fallow system, the change in soil fertility status follows an equilibrium model. This is evidenced by the experimental results obtained by the researchers indicating that after several years of alley cropping.

Young (1991) underlined that agroforestry systems have the potential to control both water and wind erosion, which ultimately reduces the loss of soil organic matter and nutrients. Soil organic matter has many roles in maintaining fertility. It is theorised that, under these systems, SOC can be maintained for soil fertility due to the contribution of decomposed residues from the tree component. This contribution may come from above-ground litter and pruning, root residues or indirectly as farmyard manure where pruning are fed to livestock. Organic matter regulates nutrient release pattern by influencing cation exchange capacity. Apart from these, some of the beneficial effects of organic matter have been clearly observed on soil physical properties including water holding capacity and soil microbial activity. The information pertaining to the influence of land use systems on soil fertility and overall soil chemical quality especially in rainfed regions is limited. The tree-based agriculture plays an important role, not only in improving the productivity and overall returns from the system, but also protects the soil from further degradation and improve the quality of the soil across the profile layers (Jakhar et al. 2017).

One of the important aspects of alley systems is contribution towards nitrogen economy through atmospheric nitrogen fixation. Nitrogen, a commonly limiting nutrient in tropical soils, to which growth response is immediately obtained on previously unfertilized soils. Where fertilizers are unavailable to farmers, due to cost or other reasons, improving the nitrogen economy can make a substantial contribution to crop production. Nitrogen fixation by the tree components represents a clear gain to the nutrient economy in agroforestry systems, with substantial economic value (Green 2002). Its effectiveness is proven and research into improvement of rates of fixation, through species selection and inoculation should be continued. Alley systems, on the other hand, can lead to more efficient nutrient cycling, thereby slowing the rate of crop yield decline, or leading to a steady state in low-input systems, or making more effective use of fertilizers in high- input systems. Under low input sustainable agricultural systems, without inorganic fertilizers, crop yields normally

Table 23.3 Nutrient availability under different alley systems

Treatment	Total nitrogen			Total phosphorus			C/N ratio
	N (g ha^{-1})	N (g ha^{-1}) area-based	N (g ha^{-1}) alley-based	P (g ha^{-1})	P (g ha^{-1}) area-based	P (g ha^{-1}) alley-based	
Alley cropping	46[a]	149[a]	198[a]	2.4[a]	7.8	10.4	10.6[a]
Cover cropping	19[b]	84[b]	84[b]	1.2[b]	5.8	5.8	28.3[b]
Inorganic fertilizer	100	45	45	4.36	20	20	–

Source Green (2002); Letter denotes difference between alley cropping and cover cropping treatments at the 0.05 level of significance

decline or to a condition of low-level equilibrium with stable, but unsatisfactory low yields.

Ebeid et al. (2015) in their studies on alley cropping reported the influence of incorporation of *Sesbania* pruning with various rates of nitrogenous fertilizer on growth and yield of lemongrass grown in alley cropping system during 2013/2014 and 2014/2015 seasons. Alley cropping and N fertilizer improved the soil fertility during the two seasons. The growth and yield parameters (plant height, number of tillers/clump, number of leaves/clump as well as herb yield) were significantly enhanced when grown in alleys supplemented with pruning +70 kg N fed during the two seasons. In alley systems, there is a good amount of cycling of basic cations. The cycling of bases in tree litter can assist in (i) ameliorating soil acidity and (ii) reclaiming saline or alkaline soils. Trees have been successfully incorporated in the reclamation of saline and alkaline soils with associated cereal intercropping. Timing of hedgerow pruning to coincide with nutrient demands of the crop led to increased crop yields (Nair et al. 1999). In general, hedgerows were most successful at supplying N to the crop (Table 23.3).

23.3.2 Phosphorus

Adding organic sources of P may reduce the amount of P that becomes fixed and unavailable to plants. As high-quality leaves decompose, organic acids are formed. These organic anions bind to Al and Fe oxides, competing with P for binding sites, reducing P adsorption, and making more P available to crops (Mäder et al. 1999). Additions of organic matter and the binding of organic acids also lead to reductions in Al saturation and toxicity. Organic additions of P may be more available to plants over a season because P is mineralized gradually as the organic matter is decomposed (Green 2002). If nutrient imbalances occur, it may be necessary to occasionally fallow farmland or to add minimal amounts of fertilizer or manure for maintenance of crop yields. In some cases, farmers modify alley cropping systems to include a fallow period (Adesina et al. 2000). Fertilizer additions are also necessary for high crop yields as fertilizer is applied to alley cropping systems, interactions between the leaf litter and the fertilizer should be taken into account. Zaharah and Bah (1997) found that green manure from hedgerows increased solubility of less reactive phosphate rock and had no effect or a slight negative effect on the solubility of reactive phosphate rock.

Nutrient balances suggested that especially P was being slowly drained from the soil over several years (Ng et al. 2008). For alley cropping to be effective, the hedgerows must supply nutrients to the crop without competing for resources. Planting hedgerows with shallow rooted annual crops instead of deep-rooted perennial crops reduces competition. Hedgerow species with greater concentration of roots near the tree and in the sub-soil also reduce competition and increase availability of phosphorus as well as other. Even when roots overlap spatially, there may be limited competition if hedgerow roots die off before crop roots grow (Odhiambo et al. 2001). Under fertile conditions competition is limited though evidence of minor competition has been seen in a few studies. Lupwayi et al. (1999) found that the maize growing closest to hedgerows did not respond to leaf inputs, inorganic fertilizer inputs, or manure additions and presumably the hedgerows were out competing the maize for the nutrient additions. However, the competition effects were easily offset by increased maize yields in the other rows. Trenches or root barriers can be effective at reducing competition for moisture and nutrients (Adhikary et al. 2017).

23.3.3 Potassium

Nutrients released from tree prunings sustain alley cropping system for different nutrients including potassium. Several workers have reported the improvement in K availability under alley system (Kang and Ghuman 1991). Potassium (K) uptake and utilization efficiency were monitored over 16 months in *Gliricidiasepium*, *Leucaenaleucocephala*, and *Albizialebbeck* and found that tree interspecific variation arose in biomass yields and K nutrition. Tissue potassium concentration narrowed within 0.68–1.15% and varied little among tree parts and species over time. Potassium accumulation increased steadily with tree age and significant differences among trees occurred at all ages. *Gliricidiasepium* had a higher yield than the others over the first 8 months, after which the uptake pattern declined drastically to become the lowest at harvest. Differential K partitioning within trees occurred as K in leaves, stems, and roots amounted to 18, 35, and 47% in *Albizia*; 28, 25, and 47% in *Gliricidia*; and 27, 42, and 31% in *Leucaena*, respectively. Higher K allocation into *Leucaena* stems and its low partitioning into *Albizia* leaves were drawbacks for alley cropping. Potassium utilization efficiency decreased inversely to biomass yield and K uptake over time. It differed significantly among trees with Albizia being the most efficient. Low K returns from tree prunings in alley cropping could be due to its uptake potential and partitioning impairment in each species. In Alfisol, significant decreases in exchangeable K and Mg were observed in the surface soils from both fertilized and unfertilized treatments (Yamoah et al. 1986). The low efficiency of N use from pruning, nitrate leaching could be a major factor contributing to the loss of K and Mg from the surface soil under alley cropping.

In the SE periphery of Brazilian Amazonia, low-input agriculture systems on sandy loam soils have very low nutrient use efficiency. The *Acacia* alley system provided better soil coverage throughout the whole corn cycle. Potassium was released faster than nitrogen from the residues. The Leucaena + Acacia treatment was the most effective in increasing post-tasseling N and K assimilation and K use efficiency. This resulted in corn productivity 3.5 times greater (7.3 Mg ha^{-1}) than the control without residue application. The no-till alley cropping of leguminous trees constitutes an important option for low-input farming, its efficiency depends on using a mixture of residues that keeps soil covered and have high rates of both N and K release during the entire crop cycle.

23.4 Alley Cropping: Soil Carbon Stock and Sequestration

Alley cropping has the potential to provide high soil carbon stocks yet agreeable yields while providing numerous environmental benefits at the same time. The soil carbon improvement may be influenced by the species composition, soil type and climate (Hombegowda et al. 2015). Eestablishment of agriculture caused SOC stocks to rebound when planted with trees (Hombegowda et al. 2015). Whereas in degraded soils of sub humid tropics, Mutuo et al. (2005) observed an increase SOC stock up to 1.6 Mg ha^{-1} yr^{-1} under alley cropping with maize intercropping. In hedgerow system due to the barrier effect, most of the fine soil and the associated nutrients that are eroded by runoff water get deposited above the alley row (Adhikary et al. 2017; Hombegowda et al. 2020). The higher stock of SOC near the alley rows over whole plot average was because the higher litter fall and erosion controlling mechanism of hedge row were better attained around the alley (Kanaujia and Bhatia 2001). In addition to that, favourable soil moisture and temperature regime, and higher root biomass turnover were also high near the alley rows. In a hedgerow experiment with *Leucaenaleucocephala* and *Gliricidiasepium* in Nigeria, up to 15% increase in soil carbon concentration was reported during 12-year duration (Kang et al. 1999). Lal (2005) showed an additional increase of 13.6 Mg ha^{-1} of SOC by the practice of *Leucaena* in 5 years duration. Oelbermann et al. (2006) reported that alley cropping of hybrid poplar with wheat intercropping has increased the SOC at a rate of 1.25 Mg C ha^{-1} $year^{-1}$ during the 13-year rotation in Southern Canada. In the same study, similarly *Erythrina poeppigiana* with maize and bean hedgerow intercropping system can improve the SOC stock 1.62 Mg C ha^{-1} $year^{-1}$ in 0–10 cm soil depth. It is also estimated that control of soil erosion in agricultural lands of tropical ecosystem can sequester SOC to the tune of 0.5 Mg C ha^{-1} $year^{-1}$ (Lal 2008).

In alley cropping systems trees grown between the crops are regularly pruned. Pruning productivity in alley cropping systems, and therefore the amount of C returned to the soil, ranges from 0.3 to 4.6MgCha^{-1} (Table 23.4). Climate, soil

Table 23.4 Annual aboveground C inputs (Mg C ha^{-1} year^{-1}) from alley cropping agroforestry tree prunings in tropical alley cropping systems

Location	Tree species	Soil type (FAO)	Age (years)	C input (Mg C ha^{-1} year^{-1})
Costa Rica	*E. poeppigiana*	EutricCambisol	4	1.0
Costa Rica	*E. poeppigiana*	EutricCambisol	10	1.4
Costa Rica	*E. poeppigiana*	EutricCambisol	19	4.0
Costa Rica	*G. sepium*	EutricCambisol	4	0.3
Costa Rica	*G. sepium*	EutricCambisol	10	0.6
Costa Rica	*G. sepium*	EutricCambisol	19	4.6
Nigeria	*G. sepium*	EutricCambisol	7	4.6
Nigeria	*G. sepium*	Planosol	6	2.9[a]
Nigeria	*L. Leucocephala*	Planosol	6	4.2
Central Togo	*G. sepium*	Ferric Acrisol	4	0.8[a]
Peru	*E. poeppigiana*	Planosol	NA	1.6 [a]

Source Oelbermann et al. (2004)

type, tree species variation, and system management result in variable tree productivity (Oelbermann et al. 2004). For example, management factors such as pruning frequency affect the nodulation efficiency in N-fixing species and hence overall tree productivity (Chesney and Nygren 2002). Nygren (1995) showed that C input from prunings of *E. poeppigiana* clones in Costa Rica ranged from 2.3 to 5.2 Mg Cha^{-1} year^{-1}ata tree density of 625 trees ha^{-1}. Oelbermann et al. (2004) determined that C input from *E. Poeppigiana* prunings varied with tree age, ranging from 4.0MgCha^{-1} year^{-1} in 19-year-old trees (555 trees ha^{-1}) to 1.4MgCha^{-1} year^{-1} in 10-year-old trees (833 trees ha^{-1}). The C input from temporal alley cropping system differs from tropical alley cropping system. A 12-year-old hybrid poplar alley crop in southern Canada, on an Albic Luvisol, contributed 0.95 Mg Cha^{-1} year^{-1} within 1 m of the tree row compared to 0.38 Mg Cha^{-1} year^{-1} at a 6.0 m distance (Oelbermann et al. 2004). Zhang (1999) reported litterfall to be 0.63MgCha^{-1} year^{-1} in a 10-year-oldhybrid poplar alley cropping system in southern Canada. Whereas Thevathasan and Gordon (1997) determined that leaves contributed 1.6 Mg Cha^{-1} year^{-1} by collecting all leaves from 7-year-old hybrid poplar at a stand density of 111 trees ha^{-1}.

It is recognized that the magnitude of SOC sequestration was dependent on the quantity of incoming organic matter to soil. The higher SOC stock and build-up rate in alley is due to the higher litter production and green manuring activity. The SOC improvement by *Gliricidia* hedgerow was reported from India by Adhikary et al. (2017) and Hombegowda et al. (2020) with upland paddy and finger millet intercrop, respectively. Under unprotected and higher land slope (10%) condition, the loss of higher quantum of soil reduces the SOC stock rapidly and drastically (Lenka et al. 2012). In addition to SOC, the higher nutrient status principally N in the alley cropping systems is due to its reduced loss through runoff, and improves the SOC sequestration (Lal 2008). Additional impact of erosion control measures primarily develops from the high SOC and the secondary role of SOC in improving soil physical properties, which intern protect soil organic matter and improves the SOC functions (Carter 2002).

23.5 Conclusion

Alley cropping agroforestry as a major viable approach for improvement of soil health is now commonly acknowledged. Many of its positive attributes relate to the management and conservation of marginal soils of the tropics. While evidence exists for the beneficial effects on soils of certain agroforestry technologies (especially on more fertile soils), there is a tendency for over-generalization and extrapolation of soil productivity and sustainability benefits of agroforestry systems to other more marginal soils. The time has come to bring science into the picture and systematically test the effects of agroforestry systems on different soils. Alley cropping system improves the physical condition by increasing infiltration rate and diverse plant components root activities are important factors affecting soil infiltration. Soil chemical properties are highly influenced by alley cropping. Higher soil organic matter in alley cropping system improves the cation exchange capacity and consequently increases exchangeable bases thus maintains soil pH. Nutrient availability increases with dissimilar cycling system of alley cropping vis-à-vis enzymatic activities. Resources are conserved with reduction of erosion and runoff, enhances soil organic matter, improves soil physical properties and augment nitrogen fixation and promote efficient nutrient cycling. In alley cropping systems trees grown between the crops are regularly pruned. Pruning increases the productivity in alley cropping systems and add C to the soil and helps in carbon sequestration. Recognition of what the major soil constraints is in specific areas would improve the design of agroforestry systems. Science-based soil-agroforestry research will provide a realistic site-specific appraisal of whether agroforestry systems improve soil physical properties, maintain soil organic matter or promote nutrient cycling.

References

Abunyewa A, Asiedu EK, Nyamekye AL, Cobbina J (2004) Alley cropping *Gliricidiasepium* with Maize: 1. The effect of hedgerow spasing, pruning height and phosphorus application rate on Maize yield. J Biol Sci 4(2):81–86

Adesina AA, Mbila D, Nkamleu GB, Endamana D (2000) Economic analysis of thedeterminants of adoption of alley farming by farmers in the forest zone ofsouthwest Cameroon. Agr Ecosyst Environ 80:255–265

Adhikary PP, Dash ChJ, Chandrasekharan H, Rajput TBS, Dubey SK (2012) Evaluation of groundwater quality for irrigation and drinking using GIS and geostatistics in a peri-urban area of Delhi, India. Arab J Geosci 5(6):1423–1434

Adhikary PP, Hombegowda HC, Barman D, Jakhar P, Madhu M (2017) Soil erosion control and carbon sequestration in shifting cultivated degraded highlands of eastern India: performance of two contour hedgerow systems. Agrofor Syst 91(4):757–771

Adhikary PP, Hombegowda HC, Barman D, Madhu M (2018) Soil and onsite nutrient conservation potential of aromatic grasses at field scale under a shifting cultivated, degraded catchment in Eastern Ghats, India. Int J Sediment Res. https://doi.org/10.1016/j.ijsrc.2018.01.002

Alexander M (1977) Introduction to soil microbiology, 2nd edn. Wiley, New York

Barreto AC, Chaer GM, Fernandes MF (2012) Hedgerow pruning frequency effects on soil quality and maize productivity in alley cropping with *Gliricidiasepium* in Northeastern Brazil. Soil Tillage Res 120:112–120

Buresh RJ, Tian G (1998) Soil improvement by trees in sub-Saharan Africa. In: Directions in tropical agroforestry research. Springer, Dordrecht, pp 51–76

Carter MR (2002) Soil quality for sustainable land management. Agron J 94:38–47

Casanova-Lugo F, Petit-Aldana J, Solorio-Sanchez FJ, Parsons D, Ramirez-Aviles L (2014) Forage yield and quality of *Leucaena leucocephala* and *Guazumaulmifolia* in mixed and pure fodder banks systems in Yucatan, Mexico. Agroforest Syst 88(1):29–39

Chesney P, Nygren P (2002) Fine root and nodule dynamics of *Erythrinapoeppigiana* in an alley cropping system in Costa Rica. Agrofor Syst 56(3):259–269

Dalal RC, Moloney D (2000) Sustainability indicators of soil health and biodiversity. In: Hale P, Petrie A, Moloney D, Sattler P (eds) Management for sustainable ecosystems. Centre for conservation biology. The University of Queensland, Brisbane, pp 101–108

Das SK, Sharma S, Sharma KL, Saharan N, Nimbole NN, Reddy YVR (1993) Land use options in a semi-arid Alfisols. Am J Altern Agric 8:34–39

Doran JW, Parkin TB (1994) Defining and assessing soil quality. Defining Soil Qual Sustain Environ 35:1–21

Ebeid AFA, Ali EF, Mostafa MM (2015) Impact of alley cropping system amended with Sesbania and/or nitrogenous fertilizer on growth and yield of Cymbopogon citratus (DC) Stapf. J Med Plants Stud 5:07–13

Friday JB, Fownes JH (2001) A simulation model for hedgerow light interception andgrowth. Agric for Meteorol 108:29–43

Gangwar KS, Sharma SA Tomar OK (2004) Alley cropping of Subabul (*Leucaenaleucocephala*) for sustaining higher crop productivity and soil fertility of Rice (*Oryza sativa*) Wheat (*Triticumaestivum*) system in semi- arid conditions. Ind J Agron 49(2):84–88

Green EV (2002) Nutrient addition and crop yield of an alley cropping system in the piedmont of Georgia. M.Sc. thesis, Graduate Faculty of the University of Georgia, Athens, Georgia

Hangen E, Buczko U, Bens O, Brunotte J, Hüttl RF (2002) Infiltration patterns into two soils under conventional and conservation tillage: influence of the spatial distribution of plant root structures and soil animal activity. Soil Tillage Res 63(3–4):181–186

Hombegowda HC, Straaten O, Köhler M, Hölscher D (2015) On the rebound: soil organic carbon stocks can bounce back to near forest levels when agroforests replace agriculture in southern India. Soil Discuss 2:871–902

Hombegowda HC, Köhler M, Röll A, Hölscher D (2019) Tree species and size influence soil water partitioning in coffee agroforestry. Agrofor Syst. https://doi.org/10.1007/s10457-019-00375-7

Hombegowda HC, Adhikary PP, Jakhar P, Madhu M, Barman D (2020) Hedge row intercropping impact on run-off, soil erosion, carbon sequestration and millet yield. Nutr Cycl Agroecosyst 116:103–116. https://doi.org/10.1007/s10705-019-10031-2

Hombegowda HC, Jakhar P, Madhu M, Yearbok M (2021) "Thang Bun": new reporting of indigenous practice of in-situ biochar preparation cum application for improved jhum cultivation in North East India. Curr Sci 120(7):1160–1168. https://doi.org/10.18520/cs/v120/i7/1160-1168

Hulugalle NR, Kang BT (1990) Effect of hedgerow species in alley cropping systems on surface soil physical properties of an OxicPaleustalf in south-western Nigeria. J Agric Sci 114(3):301–307

Jakhar P, Dass A, Adhikary PP, Sudhishri S, Naik BS, Hombegowda HC, Madhu M, Lenka NK et al (2017) Multitier agroforestry system for integrated resource conservation on uplands of Eastern Ghats region in India. Agrofor Syst 97(5):697–712. https://doi.org/10.1007/s10457-016-9976-1

Jordan C (1985) Nutrient cycling in tropical forest ecosystems. Wiley, Chichester

Jose S, Gillespie AR, Seifert MDB, Pope PE (2000) Defining competitionvectors in a temperate alley cropping system in the midwestern USA: 2. Competition for nitrogen and litter decomposition dynamics. Agrofor Syst 48:61–77

Kanaujia VK, Bhatia KS (2001) Effect of alley cropping on splash erosion and growth and yield components in eroded alluvial soils. J Soil Water Conserv 20:109–113

Kang BT, Caveness FE, Tian G, Kolawole GO (1999) Long-term alley cropping with four species on an Alfisol in southwest Nigeria—effect on crop performance soil chemical properties and nematode population. Nutr Cycl Agroecosyst 54:145–155

Kang BT, Ghuman BS (1991) Alley cropping as a sustainable system. Development of conservation farming on hillslopes, pp 172–184

Karlen DL, Mausbach MJ, Doran JW, Cline RG, Harris RF, Schumann GF (1997) Soil quality: a concept, definition and framework for evaluation. Soil Sci Soc Am J 61:4–10

Kass D, Sylvester-Bradley R, Nygren P (1997) The role of nitrogen fixation andnutrient supply in some agroforestry systems of the Americas. Soil Biol Biochem 29(5/6):775–785

Kessler JJ, Breman H (1991) The potential of agroforestry to increase primary production in the Sahelian and Sudanian zones of West Africa. Agrofor Syst 13(1):41–62

Ko LJ, Reich PB (1993) Oak tree effects on soil and herbaceous vegetation in savannas and pastures in Wisconsin. Am Midl Nat 130:31–42

Lal R (2005) Soil erosion and carbon dynamics. Soil Till Res 81:137–142

Lal R (2007) Anthropogenic influences on world soils and implications to global food security. Adv Agron 93:69–93

Lal R (2008) Carbon sequestration. Philos Trans R Soc B 363:815–830

Lee KH, Jose S (2003) Soil respiration, fine root production, and microbial biomass in cottonwood and loblolly pine plantations along a nitrogen fertilization gradient. For Ecol Manage 185(3):263–273

Lehmann J, Peter I, Steglich C, Gebauer G, Huwe B, Zech W (1998) Below-ground interactions in dryland agroforestry. For Ecol Manage 111:157–169

Lenka NK, Dass A, Sudhishri S, Patnaik US (2012) Soil carbon sequestration and erosion control potential of hedgerows and grass filter strips in sloping agricultural lands of eastern India. Agric Ecosyst Environ 158:31–40

Lupwayi N, Haque I, Saka A, Siaw D (1999) Leucaena hedgerow intercropping andcattle manure application in the Ethipoian highlands: II. Maize yields and nutrientuptake. Biol Fertil Soils 28:196–203

Mäder P, Alföldi T, Fließbach PL, Niggli U (1999) Agricultural and ecologicalperformance of cropping systems compared in a long-term field trial. In: Smaling EMA, Oenema O, Fresco LO (eds) Nutrient disequilibria in agroecosystems. EMA Smaling, Cambridge, UK, pp 247–264

Madhu M, Naik BS, Jakhar P, Hombegowda HC, Adhikary PP, Gore KP, Barman D, Naik GB (2016) Comprehensive impact assessment of resource conservation measures in watershed of eastern region of India. J Environ Biol 37:391–398

Mugendi DN, Nair PKR, Mugwe JN, O'Neill MK, Woomer PL (1999) Alleycropping of maize with *calliandra* and *leucaena* in the subhumid highlands ofKenya. Part 1. Soil-fertility changes and maize yield. Agrofor Syst 46:39–50

Mungai NW, Motavalli PP, Kremer RJ, Nelson KA (2005) Spatial variation of soil enzyme activities and microbial functional diversity in temperate alley cropping systems. Biol Fertil Soils 42:129–136

Mutuo PK, Cadisch G, Albrecht A, Palm CA, Verchot L (2005) Potential of agroforestry for carbon sequestration and mitigation of greenhouse gas emissions from soils in the tropics. Nutr Cycl Agroecosyst 71:43–54

Myers RT, Zak DR, White DC, Peacock A (2001) Landscape-level patterns of microbial community composition and substrate use in upland forest ecosystems. Soil Sci Soc Am J 65(2):359–367

Nair PKR, Buresh RJ, Mugendi DN, Latt CR (1999) Nutrient cycling in tropicalagroforestry systems: myths and science. In: Buck LE, Lassoie JP, Fernandes ECM (eds) Agroforestry in sustainable agricultural systems. CRCPress, Boca Raton, FL, pp 1–31

Nair PKR, Kang BT, Kass DBL (1995) Nutrient cycling and soil erosion control in agroforestry system. In: Agriculture and environment: bridging food production in developing countries. ASA special publication no. 60, American Society of Agronomy, Madison, WI, Ch. 7

Nair PKR (1984) Role of trees in soil productivity and conservation. Soil productivity aspects of agro-forestry. The international council for research in agro-forestry. Nairobi, pp 85

Ng SL, Cai QG, Ding SW, Chau KC, Qin EJ (2008) Effects of contour hedgerows on water and soil conservation, crop productivity and nutrient budget for slope farmland in the Three Gorges Region (TGR) of China. Agrofor Syst 74:279–291

Nygren P, Ramírez C (1995) Production and turnover of N2 fixing nodules in relationto foliage development in periodically pruned *Erythrina poeppigiana* (Leguminosae) trees. For Ecol Manage 73:59–73

Odhiambo HO, Ong CK, Deans JD, Wilson J, Khan AAH, Sprent JI (2001) Roots, soil water and crop yield: tree crop interactions in a semi-arid agroforestry systemin Kenya. Plant Soil 235:221–233

Oelbermann M, Voroney RP, Gordon AM (2004) Carbon sequestration in tropical and temperate agroforestry systems: a review with examples from Costa Rica and southern Canada. Agr Ecosyst Environ 104(3):359–377

Oelbermann M, Voroney RP, Thevathasan NV, Gordon AM, Kass DCL, Schlnvoigt AM (2006) Soil carbon dynamics and residue stabilization in a Costa Rican and southern Canadian alley cropping system. Agrofor Syst 68:27–36

Peter I, Lehmann J (2000) Pruning effects on root distribution and nutrient dynamicsin an acacia hedgerow planting in northern Kenya. Agrofor Syst 50:59–75

Pimentel D (1997) Pest management in agriculture. In: Pimentel D (ed) Techniques for reducing pesticide use: economic and environmental benefits. Wiley, New York, pp 1–11

Rahman MA, Miah MG, Yahata H (2009) Maize production and soil properties change in alley cropping system at different nitrogen levels. Agriculturists 7(1):41–49

Rao MR, Nair PKR, Ong CK (1998) Biophysical interactions in tropical agroforestrysystems. Agrofor Syst 38:3–50

Sanchez PA, Palm CA (1996) Nutrient cycling and agroforestry in Africa. UNASYLVA-FAO, pp 24–28

Seiter S, William RD, Hibbs DE (1999) Crop yield and tree-leaf production in three planting patterns of temperate-zone alley cropping in Oregon, USA. Agrofor Syst 40:273–288

Sun D, Yang H, Guan D, Yang M, Wu J, Yuan F, Jin C, Wang A, Zhang Y (2018) The effects of land use change on soil infiltration capacity in China: a meta-analysis. Sci Total Environ 626:1394–1401

Thevathasan NV, Gordon AM (1997) Poplar leaf biomass distribution and nitrogen dynamics in a poplar-barley intercropped system in southern Ontario, Canada. Agrofor Syst 37(1):79–90

Tossah BK, Zamba DK, Vanlauwe B, Sanginga N, Lyasse O, Diels J, Merckx R (1999) Alley cropping in the moist savanna of West-Africa: II. Impact on soilproductivity in a North-to-South transect in Togo. Agrofor Syst 42:229–244

Wang L, Zhong C, Gao P, Xi W, Zhang S (2015) Soil infiltration characteristics in agroforestry systems and their relationships with the temporal distribution of rainfall on the loess Plateau in China. PlosOne 10(4):e0124767

Wick B, Kühne RF, Vlek PL (1998) Soil microbiological parameters as indicators of soil quality under improved fallow management systems in south-western Nigeria. Plant Soil 202(1):97–107

Yamoah CF, Agboola AA, Wilson GF, Mulongoy K (1986) Soil properties as affected by the use of leguminous shrubs for alley cropping with maize. Agr Ecosyst Environ 18(2):167–177

Young A (1991) Soil fertility. In: Avery ME, Cannel MGR, Ong CK (eds) Biophysical research for Asian agroforestry. Winrock International USA and South Asia Books, USA, pp 187–208

Zaharah AR, Bah AR (1997) Effect of green manures on P solubilization and uptakefrom phosphate rocks. Nutr Cycl Agroecosyst 48:247–255

Zak DR, Holmes WE, MacDonald NW, Pregitzer KS (1999) Soil temperature, matric potential, and the kinetics of microbial respiration and nitrogen mineralization. Soil Sci Soc Am J 63(3):575–584

Zhang P (1999) Nutrient inputs from trees via throughfall, stem flow and litterfall in an intercropping system. Doctoral dissertation

Chapter 24
Performance of Rice-Lentil Cropping Under Different Tillage Influencing Soil Suppressiveness: A Short Term Approach

Sk Saruk Islam, Krishnendu Sen, Subrata Dutta, and Sujoy Midya

Abstract Due to causal agents hidden status soil borne pathogens control are difficult. 10–20% annually crop yield losses, because of soil borne pathogen among exists other plant pathogens. Due to melanized sclerotia, higher genetic variability, varied host range and preferable environment these fungal survive a long term in soil. Soil physical property and microbial population influence the suppressiveness and conduciveness of soil towards soil-borne plant pathogens. Spreading rate of soil-borne pathogen depends on different biotic and abiotic factors. A diverse microbial habitat can be control through long term alternation of tillage management. Keeping these points this investigation was undertaken with the objectives of soil microbiological parameters influenced by different tillage management practices associated with *Sclerotium rolfsii* disease suppressiveness. Three types of tillage (Conventional, minimal and zero tillage) were applied in paddy harvested fields where then *S. rolfsii* susceptible crop lentil was transplanted. Collar rot disease incidence, infection foci and sclerotial population was found to be least in zero tillage condition. Temporal changes in microbiological parameters viz., FDA and dehydrogenase activities were found to be higher in zero tillage condition and least activities was noticed under conventional tillage condition. Higher population of total bacteria, *Pseudomonas* and *Actinomyces* were observed under zero tillage condition, whereas, *Bacillus* and total fungi were found to be higher in conventional tillage but *Trichoderma* abundance are variable in different tillage system. The twelve isolates, isolated from experimental field, showed potential antagonistic efficacy in between 27.29 and 71.43% against the tested fungal pathogen, *Sclerotium rolfsii.* Most of the antagonistic bacteria were

S. S. Islam · S. Midya (✉)
Department of Zoology, Raja N.L. Khan Women's College, Midnapore 721102, West Bengal, India
e-mail: sujoy.midya@gmail.com

K. Sen
Department of Microbiology, Vidyasagar University, Midnapore 721102, West Bengal, India

K. Sen · S. Dutta
Department of Plant Pathology, Bidhan Chandra Krishi Viswavidyalaya, Nadia, Mohanpur 741252, West Bengal, India

P. K. Shit et al. (eds.), *Soil Health and Environmental Sustainability*,
Environmental Science and Engineering,
https://doi.org/10.1007/978-3-031-09270-1_24

found to produce secondary metabolites but ZTS27, ZTS9 and MTS39 were potential antagonistic against *S. rolfsii* in terms of theirhighly productivity of secondary metabolites such as HCN, NH_3, siderophore production, SA, IAA, P-sol, protease, cellulose, amylase, pectinase and seedling vigor index by means of plant growth promotion.

Keywords Melanized sclerotia · Microbial enzymatic activity · Minimum tillage · Soil borne pathogen

Abbreviations

TB	Total Bacteria
Temp.	Temperature
GSC	Gmini Spin Column
AUDPC	Area Under Disease Progress Curve
AUSPC	Area Under Suppressivity Progress Curve
DMRT	Duncan's multiple range test
EtBr	Ethidium bromide
MT	Minimal Tillage
CT	Conventional Tillage
ZT	Zero Tillage
DI	Disease Incidence
SA	Salicylic Acid
P-sol	Phosphate solubilization
LMR	Linear Multiple Regression
PGPA	Plant Growth Promoting Activity
CFU	Colony Forming Unit

24.1 Introduction

Sclerotium rolfsii has worldwide extensive host range, *a* fungal pathogen, first reported by Rolfs in Florida in the year of 1892 in tomato plant. It's occurring in the region of tropics and subtropics, also in warm region but rarely occurs in winter region. For that reason, *Sclerotium rolfsii* spread fast in different places of Indian geographical region. Previous study suggests that 500 species of 100 families were affected by these fungal species, but legumes, crucifers, and cucurbits are the most common hosts (Kwon et al. 2009; Choi et al. 2011). *Sclerotium rolfsii* causes Southern Blight disease on different plant species such as on cowpea, peanut, rice, sugar beet, tomato, watermelon and wheat. But till days no others chemical fungicidal kits are established against the fungus *Sclerotium rolfsii* exclusive of biological substance

(Jacobsen et al. 2004). It's well known that chemical substances are more effective rather than biological substance for others micro organisms which are presents in the soil. So, atmospheric condition and antagonist biological populations were playing the major role for treatments of *Sclerotium rolfsii* in present days (Ramarathnam et al. 2011). But the problem is that till day better antagonist population did not found against the fungal pathogen *Sclerotium rolfsii*. So, our work finding is somehow different, it's possible to quire the disease of *Sclerotium rolfsii* infection in to the plant with soil management system? Tillage practices plays a crucial role in agro-eco system. It's having a dynamic effect on physicochemical and biological properties of soil. CT management practices have been testified to violently disturb the soil surface, showed the negative effects on property's soils of agricultural land, including wide-ranging biodiversity of soil reduction and erosion (Kabiri et al. 2016). Tillage soil strongly related with the level compaction of soil. Tillage is an agronomical practice that needs substantial both high-energy inputs and expense to create favorable environments for favorable growth, crop yields and development. Main goals of tillage soil have significantly influenced soil practices and mainly alteration of the physio-chemical, and biological characteristics of soil. Soil microorganisms played significant effects in the agro-ecological practices that have smoothed by soil commotion and contributed to crop quality and growth both directly and indirectly and sustainability of soil efficiency, nutrient cycling (Roger-Estrade et al. 2010). Soil microbial biodiversity are the essential group that contributors to agro-bionetwork function and played a crucial role both in dynamics of soil organic matter and nutrient cycling in agro-ecosystems, inducing microbial aggregation and effectively shifting the environment biochemical parameters of soil (Lammel et al. 2019). Microorganism has unadorned practical redundancy and created durable communications leading to bio-community diversity along with species composition, which was exclusively key elements of agro-ecosystem function. The effect of continuous agricultural soil management practices on the activity and also alignment of soil microbial biodiversity is exclusively important (Liang et al. 2016; Pradhan et al. 2016). The objective of the current delve into was aimed to isolate the bio-agent such as *Bacillus, Pseudomonas*, etc. in Indian agricultural fields and to evaluate its potential in controlling the soil-borne pathogen *Sclerotium rolfsii* and study the tillage management practice associated with *Sclerotium rolfsii* disease suppressiveness.

24.2 Materials and Method

24.2.1 Filled Experiments and Soil Sampling

A field trial was directed to understand the consequence of diverse tillage management practice. We selected twenty-seven same size plot (3 × 10 m) in different part of AB block farm Kalyani, Nadia. In every plot, paddy was the previous crop. After harvesting rice crop, we applied three different (Conventional, Minimum and Zero

Tillage) types of tillage practice. Lentil seedling was transplanted on 23 November 2016 into all the experimental field. Each management treatment carried out in three sites with 3 replications. Different tillage soil samples were collected of a depth 10–15 cm of 3 replicative ways of each 3 subplots of the selected field in sterile container. There after a composite sample was prepared by mixing well of collected soil sample without contamination. To enumeration of microbial population each composite soil sample was subdivided into 3 parts and exactly weighted as 200 gm, also disease index, isolation of antagonist population against *Sclerotium rolfsii* and physico-chemical property of collected soil samples were analysis.

24.2.2 Soil Organic Carbon and Microbial Enzymatic Activity Analysis

24.2.2.1 Estimation of Organic Carbon

Weight exactly 0.5 gm of soil, transfer it in to 500 ml conical flask. Add 10 ml of 1 M $K_2Cr_2O_7$ and mixed properly. After that, add 20 ml concentrate H_2SO_4 and mixed properly to contact the soil with the reagents and stand till 20–30 min. Add 200 ml of distilled water to dilute the solution. Add 10 ml of orthophosphoric acid and 1 ml of diphenylamine indicator. Tritrate the solution by using the 0.5(N) ferrous ammonium sulphate. The titrate reading was record for calculation of soil organic carbon.

24.2.2.2 Fluorescein Di-Acetate (FDA) Hydrolysis Assay of Soil Samples

For measure the enzymatic activity produced by microbes ina soil sample was optimized by Fluorescein di-acetate (FDA) hydrolysis assays. For estimation of FDA, 1 gm of soil sample was dissolved in 25 ml sodium phosphate buffer shake for 5 min. Then 100 μl FDA solutions added with and shake for 2 h at 25 °C. Later 10 ml acetone was added to the suspension and centrifuge at 6000 rpm for 15 min. At first esterases cleave the fluorescein, cells dynamically alter the nonfluorescent FDA to green fluorescent complex "fluorescein". A bright green luminosity was formed during enzymatic action and it was quantified by spectrophotometer JASCO V-630at 490 nm (Bararunyeretse et al. 2017).

24.2.2.3 Dehydrogenase (DHA) Activity Analysis of Collected Soil Samples

Determination of dehydrogenase was carried out by Thalmann method (Lee et al. 2012); according to this method an aqueous 0.2 ml 3% TTC solution recommended in the Thalmann method was added to 1 gm of soil sample. After that 0.5 ml 1% glucose solution was mixed gently and incubated the mixture at 28 °C for 24th hours. After incubation 10 ml methanol was added and placed in shaken condition for 6 h. Pink color supernatant separated and absorbency taken in JASCO V-630 spectrophotometer at 485 nm.

24.2.3 Soil Microbial Community Assay for Culturable Microorganisms

Total bacterial count was carried out with Nutrient agar medium [Hi-media, India] through poured plat technique *Pseudomonas* spp. was counted by King's B. medium [Hi-media, India] through poured plat technique. *Bacillus* spp. was enumerated with Nutrient agar medium [Hi-media, India] through poured plat technique. *Actinomyces* spp. was carried out by MBS agar medium [Hi-media, India] through poured plat technique. Total fungal spp. was counted by Rose Bengal's Media [Hi-media, India] through the poured plate technique. Total *Trichoderma* spp. was enumerated with specific *Trichoderma* agar medium [Hi-media, India] through poured plate technique. After autoclaving, the plate was inoculated with collected soil sample and incubated at 37 °C for 24 h to 96 h. Different types of colonies were appeared on selective medium.

24.2.4 Culture Independent Approach to Obtain the Abundance Label of Pseudomonas, Bacillus and Actinomyces in the Experimental Soil

24.2.4.1 Soil DNA Extraction

Soil microbial community analysis for non-culturable microorganisms was carried out by Soil DNA extraction. For DNA extraction 200 mg of soil sample was taken in a microcentrifuge tube, add 500 μl solution I and vortex vigorously for 1 min. Incubate at 85 °C for 15 min after that add 500 μl solutions II and mixed gently. Then place the tubes at 85 °C again for another 15 min after that add 700 μl buffer solutions III and the tubes immediately. Centrifuge the tubes for 10 min at 1000 rpm, a complete dark brown pellet was formed. Collect the supernatant in a fresh microfuge tube and add 1/5th volume of isopropanol and placed it in room temp. for 5 min. Put on the isopropanol solution to the GSC by petting. Avoid mixing of cell debris with the supernatant as this may clog Gmini Spin Column thus lowering the DNA yield. Wash the GSC with 600 μl Buffer mixture and centrifuge for 2 min and discard the flow through. Place the GSC in a clean 1.5 ml eppendorf and to elute the DNA, added 50 μl nuclease free water into the each GSC, waited for 2 min, later centrifugation for 5 min at 1000 rpm. Finally discard the eluted DNA present in microcentrifuge tube (Rojas-Ruiz et al. 2015).

24.2.4.2 QRT-PCR with Genus Specific Primer

Enumeration of culture independent microorganism's population was carried out by qRT-PCR with genus specific primer. qRT-PCR quantification of soil *Pseudomonas, Bacillus* and *Actinomyces* was performed for each soil DNA sample in 2 replications in a volume of 25 μl containing 10 μl RT-PCR kit Sybr mix, Hi-Media, 0.5 μl (50 ng) of selected different primers, 2 μl of template soil DNA and volume makeup by molecular grade water in an Hi-Media 96 well real-time PCR plate by Applied BioSystem (ABI 7000 Sequence Detection System) RT PCR. The PCR steps were continued in three steps such as, at first denaturation at 95 °C for 5 min, secondly annealing temp. (as per primer, Table 24.1), and lastly extension of reaction at 72 °C each till 1 min. All the analysis was done using threshold cycle (Ct) value. List of genus specific primer with annealing temp. were enclosed in Table 24.1.

24.2.5 In-Vitro Antagonistic Activity of the Native Isolates Against Fungal Pathogen Sclerotim Rolfsii

Antagonistic effects of native isolates against the selected fungal pathogen *Sclerotium rolfsii* was done by dual culture plate technique. Plates were poured with 20 ml PDA

Table 24.1 Description of the primer used in qRT-PCR

Microorganism (Genus specific)	Primer	Sequences (5'-3')	Annealing Temp. (°C)	References
Pseudomonas	PA-GS-F PA-GS-R	GACGGGTGAGTAATGCCTA CACTGGTGTTCCTTCCTATA	54	Spilker et al. (2004)
Bacillus	Bac F R 1378r	GGGAAACCGGGGCTAATACCGGAT CGGTGTGTACAAGGCCCGGGAACG	65	Garbeva et al. (2003)
Actinomyces	F243 R1378r	GCATGAGCCCGCGGCCTA CGGTGTGTACAAGGCCCGGGAACG	63	Heuer et al. (1997)

medium without antifungal and antibiotic. After that, a loopful of native bacterial isolates were streaked linearly on the plate leaving the 1 cm surface margin. Then three days old fungal pathogen culture was placed as 4 mm disc form on the center of petri plate and placed at 28 °C till 3 days for incubation, data were recorded accordingly (Ahmadet al. 2008).

24.2.6 In-Vitro Soil Suppressive Activity on Sclerotium Rolfsii

The soil suppressive activity on fungal pathogen was carried out by natural soil samples which collected from the field. The petri plates are poured with 75 gm of sheave soil samples, without any supplemented ingredients. The fungal pathogen was placed on plate as 4 mm disc since the 3 days old culture at the center of each plate and placed at 28 °C for incubation up to 7 days. The fungal mycelial growth distance was measured from the center to edge of a petri plate (Rodriguez et al. 2011).

24.2.7 Enzymatic Activity of Isolated Antagonist Microorganisms

24.2.7.1 Amylase Activity

Amylase activity test was carried out with Starch agar medium [Hi-Media, India]. The plates were inoculated with the selected bacterial isolates and placed at 28 °C for incubation up to 1 week. For detection of amylase activity, the plate was flooded with iodine solution. If the bacterium hydrolyzed amylase, the growth was surrounded by clear zone, presence of halo zones around the culture growth indicates the amylase activity.

24.2.7.2 Cellulase Activity

Cellulase activity test was carried out with Cellulose agar medium [Hi-Media, India]. The plates were inoculated with the selected bacterial isolates and placed at 28 °C for incubation up to 1 week. After the incubation the culture plate was inundated with the Congo red solution for 15 minto know the Cellulase activity. Their after, plates were de stained by inundating with 1(M) NaCl solution. If the bacterium hydrolyzed cellulose, the growth was surrounded by clear orange zone appearance clear halo zones around the growth indicateCellulase activity (Gerhardt et al. 1994).

24.2.7.3 Pectinase Activity

Pectinase activity test was carried out with pectinase agar medium [Hi-Media, India]. The plates were inoculated with the selected bacterial isolates and placed at 28 °C for incubation up to 2–4 days. For detection of pectinase activity, the plate was then inundated with 1% $C_3H_{10}BrN$ solution till 48 h and poured off. If the bacterium hydrolyzed pectinase, the growth was surrounded by clear zone, appearance of clear zones around the growth indicates pectinase activity (Gerhardt et al. 1994).

24.2.7.4 Lipase Activity

Lipase activity test was carried out with lipase agar medium [Hi-Media, India]. Lipase medium in one flask and Tween-20/80 in another flask was sterilized separately and were mixed together. The plates were inoculated with the test bacterial isolates and placed at 28 °C for incubation up to 2–4 days. For detection of lipase activity opaque zones were appeared around the colonies of selected bacterial culture that produce the enzyme esterase. If the bacterium hydrolysed lipase, an opaque zone develops around colonies of bacteria, appearance of opaque zones around the growth indicates lipase activity (Gerhardt et al. 1994).

24.2.8 Secondary Metabolites Production of Isolated Antagonist Microorganisms

24.2.8.1 Phosphate Solubilization

Phosphate solubilization activity was done with Pikovskaya's medium [Hi-Media, India]. Twenty ml of melted Pikovskaya's agar was poured on Petridishes. After solidification, the selected bacterial isolates were inoculated on the selected medium and placed at 28 °C for incubation up to 2–5 days. The clear zone developed after 2–5 days of incubation and the diameter of the clear zones were measured. Appearance of clear zone indicates the solubilization of insoluble phosphates (Khan et al. 2014).

24.2.8.2 *Production of Ammonia (NH_3)*

Ammonia productiontest was carried out with Peptone water medium [Hi-Media, India]. The cultures were grown in peptone water and incubated at 30 °C up to 4 days. After incubation, 1 ml Nessler's reagent [HiMedia, India: having composition (g/l) of Mercuric chloride, 10.0; Sodium hydroxide, 16.0; Potassium iodide, 7.0; pH (at 25 °C), 13.2 ± 0.05], was added in each tube. Presence of a faint yellow to deep yellow to brownish color indicates maximum ammonia production, appearance of color indicates the ammonia production (Glick et al. 1995).

24.2.8.3 Production of HCN

HCN production test was carried out with picric acid [Hi-Media, India]. solution, King's B broth, amended with 4.4 g/l glycine. The bacterial isolates were inoculated on King's B broth amended with glycine. Picric acid saturated filter paper was placed on top of the conical flask between the glass and cotton and placed it 28 °C till 48 h for incubation. Change of yellow to light brown to brown to reddish-brown color of filter paper recorded as weak, moderate or strong reaction, respectively. Appearance of color changes indicates the production of HCN (Bakker et al. 1987).

24.2.8.4 Production of IAA

A loopful of selected bacterial culture was inoculated in 25 ml of sterilized Nutrient broth [Hi-Media, India] supplemented with L-Tryptophan (0.1 g/l) and then incubated in rotary shaker for 24 h at 28 °C. The bacterial isolates were then centrifuged at 10,000 rpm up to 15 min. 2 ml of supernatant was taken and added 2 drops of O-phosphoric acid.4 ml of the reagent Salkowski was mixed with the aliquot. The resulted sample was incubated for 25 min at room temp. and absorbance was read at 535 nm with the help of spectrophotometer JASCO V-630 (Gordon et al. 1951).

24.2.8.5 Production of Salicylic Acid

Estimation of Salicylic acid was carried out by King's broth medium [Hi-Media, India]. Native isolates were culture in King's-B medium in rotary shaker at 28 ± 2 °C temperature till 48 h, then centrifuged the culture cell at 10,000 rpm for 10 min to collect the supernatant. After that acidified the supernatant by using 1(N) HCl and same amount of chloroform was added to estimate the quantity of salicylic acid. 5 ml 2(M) $FeCl_3$ and 4 ml dH_2O were also added in to the chloroform extract. Absorbance of aqueous solution was taken at 527 nm with the help of spectrophotometer JASCO V-630, and salicylic acid quantity was expressed as mg ml^{-1} (Meyer et al. 1992).

24.2.8.6 Siderophore Production

To estimate the siderophore activity native isolates were grow on Luria broth [Hi-Media, India] for 3 days at 28 °C and centrifuged it at 10,000 rpm till 10 min, after that collect the supernatant for further experiments. Adjusted of pH of collected supernatant at 2.0 by using of 1(N) HCL and adding same amount of ethyl acetate and mixed it properly. Later collected 5 ml of ethyl acetate from the prepared solution and mixes with same amount of Hathway's reagent. Absorbance was measured at 700 nm with the help of spectrophotometer JASCO V-630.The quantity of siderophore activity was expressed as μM benzoic acid/ml.

24.2.9 DNA Extraction and Gel Electrophoresis of the Isolated Antagonist Microorganisms

24.2.9.1 DNA Extraction

1.5 ml of fresh isolated culture (grow in Nutrient medium) was transferred to a 1500 μl eppendorf and centrifuge at 10,000 rpm till 6 min to pellet the cells. Supernatant was discarded and re-suspended the cell pellet in 200 μl proteinage-K solution and was vortex to completely resuspend cell pellet. Then it was incubated at 64 °C for 15 min after that it's transferred to 80 °C for 10 min. After incubation the eppendorf was transferred to cold ice tray for 30 min. Next it was spun at 10,000 rpm for 5 min and carefully taken the DNA supernatant.

24.2.9.2 Gel Electrophoresis

Extracted DNA was checked through agarose gel electrophoresis technique, 0.8% (w/v) agarose gel containing 10X EtBr, in TAE buffer for 1 h 30 min at 3.2 V/cm. DNA fragments were visualized at 312 nm with a UV-transilluminator (DNR Bio imaging system).

24.2.9.3 PCR with Specific Primer of the Isolates

The isolated DNA was amplified with16s-rDNA specific primer pair, forward (27F)5′-AGAGTTTGATCMTGGCTCAG-3′, Reverse (1492R) 5′-TACGGYTACCTTGTTACG-ACT-3′(Lane et al. 1991).The reaction volume contains Template DNA-2μ1 (100 ng), Enzyme: Taq polymerase-0.5 μ1 (3 U/μl) (Genei), 1.5 μ1 of 10 X Taq polymerase buffer (100 mM Tris (pH 9), 500 mMKCl,15 mM $MgCl_2$, 0.1% Gelatin) (Genei), 1.5 μ1 of dNTP mix (10 mM) (Genei), 1 μl each primer (5 pM/μl) (Eurofin).The PCR steps were continued in three steps such as, at first denaturation at 94 °C for 5 min, their after, 94 °C for 45 s., secondly annealing temp. 56 °C for 1 min, and lastly extension of reaction at 72 °C each till 1 min (35 cycles), then 72 °C up to 10 min and final holding at 4 °C. Sequencing of the 16S-rRNA gene amplicons were done by outsourcing (Scigenom, Kochi).

24.2.10 Lentil Seed Germination Assay Supplemented with Isolated Microorganisms

24.2.10.1 Seed Surface Sterilization

Lentil seed surface sterilized by 70% ethyl alcohol. At first lentil seed was dissolved in 70% ethyl alcohol for 2 min. Then the seed wash twice to thrice time with sterile distilled water. Extra surface water of sterile seed soaked by sterilized Whatman filter paper and it's ready for seed germination.

24.2.10.2 Seed Germination

The isolated microorganisms were bio-assayed to know their capability to promote and/or inhibitory action of seedling growth by ESTA method described by Elliot et al. (1985) with few modifications. For seed germination assay growing (Nutrient Broth) young culture was spin at 10,000 rpm for 6 min and discard supernatant take the pellet. Then the pellet was dissolved in sterile distilled water, soak the surface sterilized seed and dissolved pellet for 30 min. Extra water removed from seed and placed on water agar media (0.6% agar). Incubation at room temperature for 5 days and after seed germination measured the plant root length & shoot length also record their germinate percentage and vigor index.

24.2.10.3 Statistical Analysis

AUDPC and AUSPC curve were estimated by a simple midpoint trapezoidal rule (Simko and Piepho et al. 2012).

$$\mathrm{AUDPC} = \sum_{i} -1^{n-1} \frac{Y_i + Y_{i+1}}{2} \times (t_{i+1} - t_i)$$

Duncan, critical difference (CD) and standard error mean (SEm) were determined by DMRT test and univariable analysis using SPSS var21 statistical software (Gomez and Gomez et al. 1984). Pearson correlation and linear regression analysis were also carried out by SPSS var21 software using 2-tailed and stepwise method respectively. Principal component analysis was performed using data reduction and rotation method by varimax and Kaiser Normalization using SPSS (Jolliffe et al. 2002).

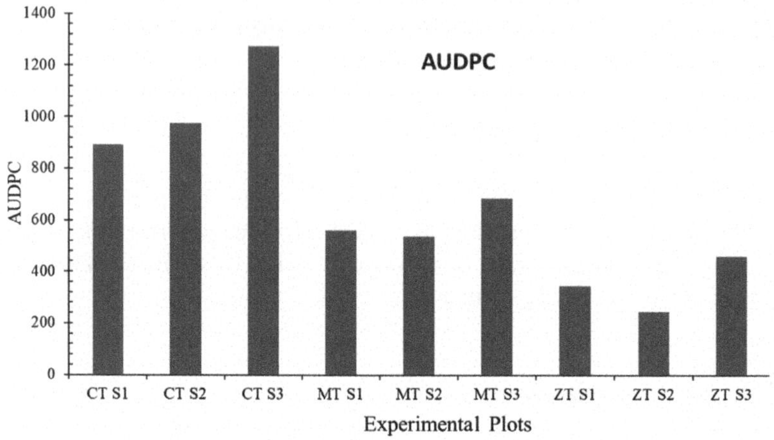

Fig. 24.1 Trapezoidal area under disease progress curve of experimental plot

24.3 Result and Discussion

24.3.1 *Disease Infection and Infectious Loci Affected by Tillage*

Sclerotium rolfsii disease development was highly influenced by applying three different tillage managements. Conventional tillage (CT) had more disease incidence (%) (AUDPC [Trapezoidal area under disease progress curve] = 1047 (±201.87)) than minimal tillage (MT) (AUDPC = 594 (±78.86)) and zero tillage (ZT) (AUDPC = 350 (±106.59)) (Fig. 24.1).

Number of infection loci per field was high for CT plot (Average infection loci = 4.67), moderate for MT plot (Average infection loci = 4) and less for ZT plot (Average infection loci = 3) (Fig. 24.2). Average number of infected plants per loci was high for CT (15.6), moderate for MT (10.5) and slightly less for ZT (9.36) (Fig. 24.2).

Where, X = average value & O = The no. of infected plant per loci.

24.3.2 *Soil Suppressivity Dynamics Following the Tillage Practices*

Soil suppressive index against *Sclerotium rolfsii* was increased by cropping duration upto 60 days after transplanting in case of ZT and MT field. CT field has less influenced soil suppressivity throughout the vegetation period (Fig. 24.3). Average area under suppressivity progress curve (AUSPC) was determined by trapezoidal

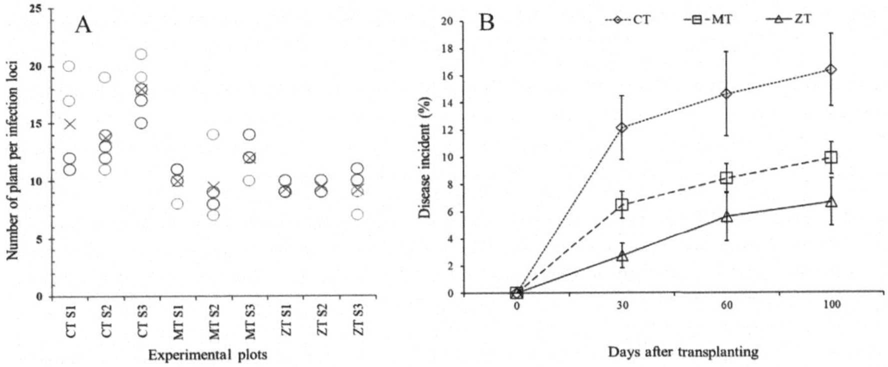

Fig. 24.2 **a** Number of plants per infection loci in experimental plots. **b** Disease incident percentage in experimental plots after transplanting

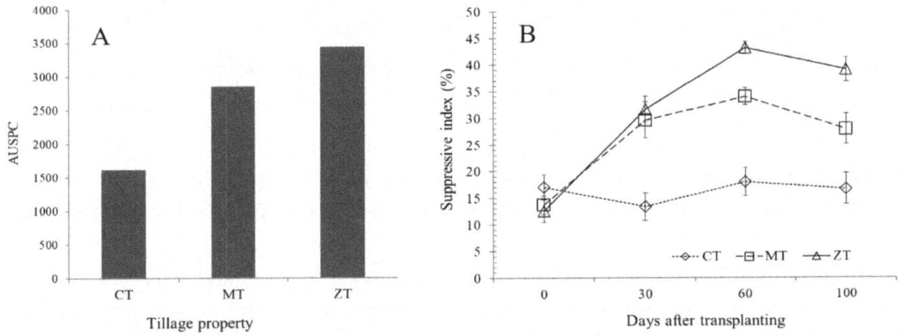

Fig. 24.3 **a** Area Under Suppressivity Progress Curve (AUSPC), and **b** soil suppressive index, against soil born fungal pathogen *Sclerotium rolfsii* in experimental plots

formula. ZT field showed the highest AUSPC than MT and less AUSPC showed by CT field (Fig. 24.3).

24.3.3 Soil Microbiological Parameter Influence by Tillage

Dynamics of soil culturable microbiological parameter was interestingly affected throughout the vegetation period when followed with different tillage management. Initially soil microbial abundance was higher in CT soil than MT and ZT soil. The culturable total bacteria CFU, *Pseudomonas* CFU, *Actinomyces* CFU abundance were remain higher in ZT followed by MT than in CT throughout the lentil vegetation period. Our results reveal that the microbial communities were reduced in CT management system during the lentil cropping period. Previously reported experiment also supports that the microbial population were reduced in CT management system throughout study period (Hassink et al. 1991; Wang et al. 2016). On the other hand, in study, it was found that in ZT management practice the microbial population were higher than other tillage management practice. Increasing of microbial population ZT management practice due to the availability of sufficient organic residues, researcher also supported that in ZT practice the microbial population increase during cropping time due to more organic residues (Johnson et al. 1999). The culturable *Bacillus* CFU, total fungus CFU and *Sclerotium rolfsii* sclerotial population were higher in CT than MT and ZT throughout the cropping period. *Trichoderma* CFU found to be erratic in respect to three tillage managed field. Thirty to sixty days (fruiting stage) of vegetation period seems to be very crucial for influencing the microbiological enrichment. After reaching the harvesting stage microbial abundance declined irrespectively (Fig. 24.4).

24.3.4 Soil Pseudomonas, Bacillus and Actinomyces Abundance Obtain from Culture Independent Quantification by QRT-PCR (on the Basis of Ct Value)

Threshold cycle (Ct) value was obtained from qRT-PCR by using three genus specific primers. Ct value for *Pseudomonas* and *Actinomyces* in CT soil (60 days) was higher than MT and ZT. It indicates that the abundance of *Pseudomonas* and *Actinomyces* (DNA copy number) was high in ZT soil followed by MT and CT. *Bacillus* Ct was higher in MT soil followed by ZT and CT soil. So, in this experiment we got opposite result for *Bacillus* genus abundance than to *Pseudomonas* and *Actinomyces* (Fig. 24.5).

24.3.5 Dynamics of Soil Microbial Enzymatic Activity Affected by Different Tillage Management Practices

FDA hydrolysis and dehydrogenase were two most important enzymatic assay for detection, optimization and abundance of soil microbes present in soil. Tillage operations also affect soil physical parameter, soil biota and microbial abundance and diversity. Both FDA and DHA activities were found to be higher in the soil during the different phases of crop growth under zero and minimal tillage condition as compare to conventional tillage condition (Fig. 24.6). Both FDA & DHA activities raise the highest point at during 30–60 days vegetation period and after when the plant reach to the harvesting stage the both enzymatic activities tend to decline. Tillage arrests the soil interruptions up plant and microbial cells biomass increases at top layer and deep layer, effects on soil temperature regime and hastiness drying of soil (Doran 1982). Tillage soil management systems promotes the releasing and

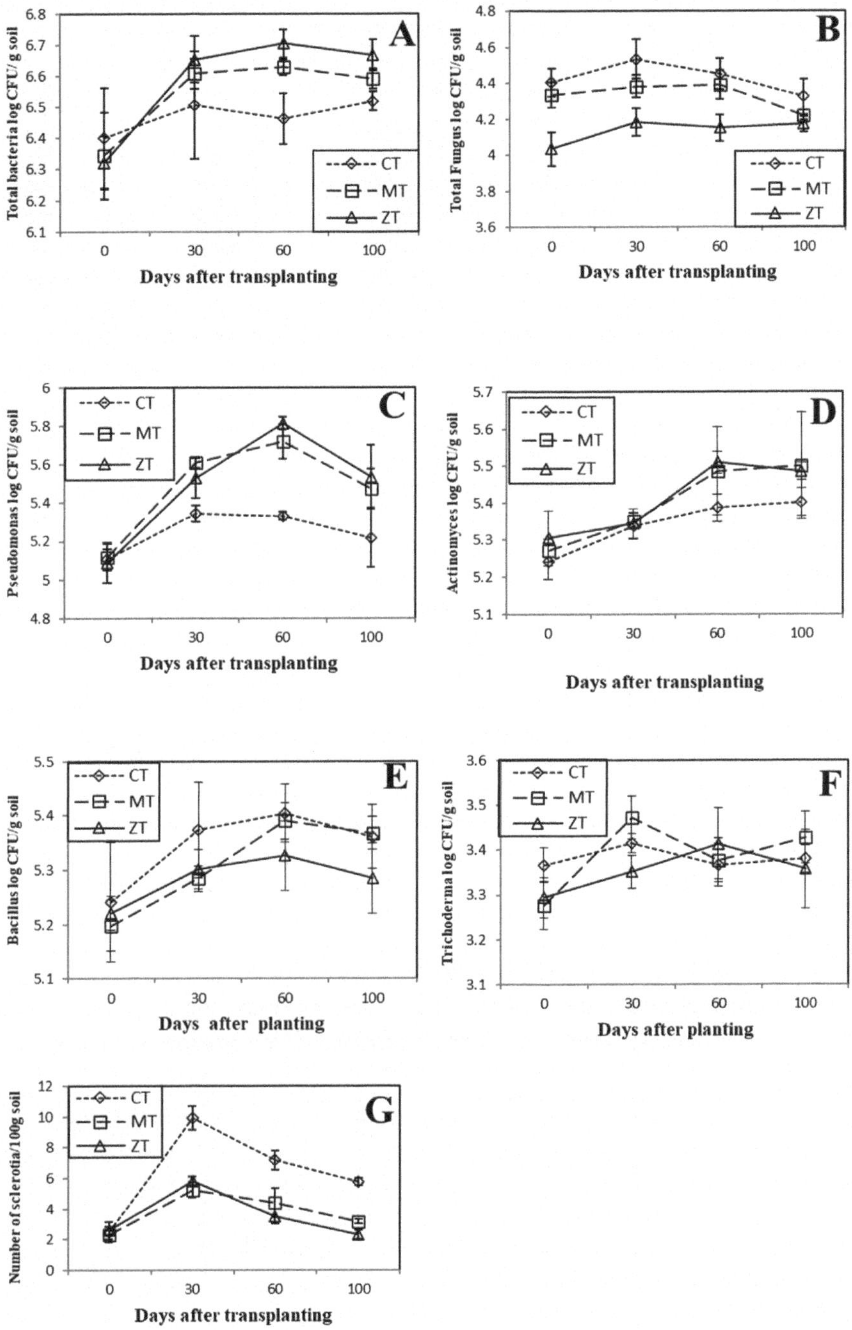

Fig. 24.4 Soil microbiological parameter influence by tillage management system: **a** Total bacteria, **b** Total fungus, **c** *Pseudomonas*, **d** *Actinomyces*, **e** *Bacillus* and **f** Trichoderma population. Where **g**, is number of sclerotia present in experimental plots. Error bar represent the SD value

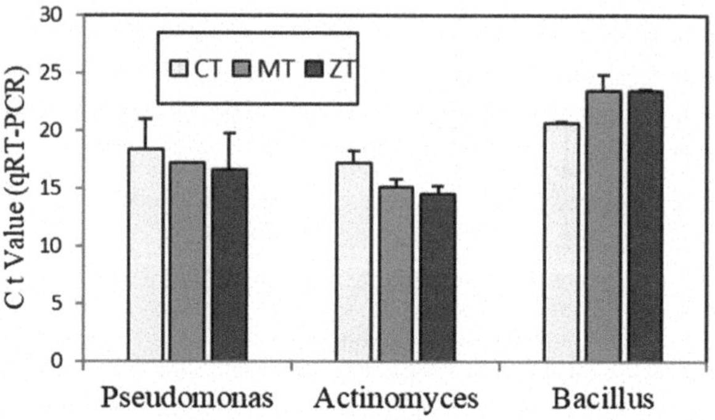

Fig. 24.5 Culture independent quantification by qRT-PCR on the basis of Ct value

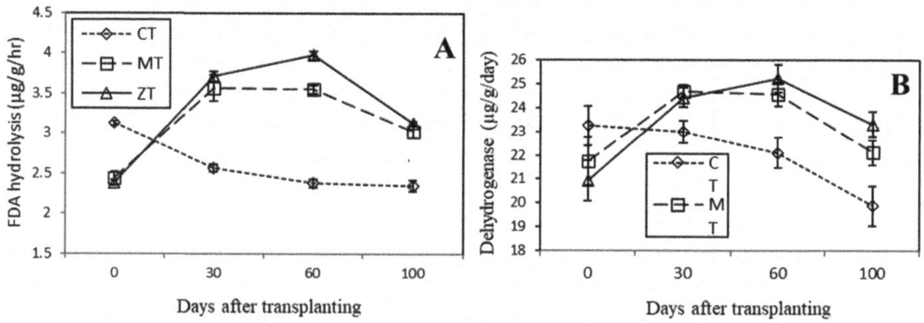

Fig. 24.6 Soil microbial enzymatic activity i.e. **a** FDA hydrolysis, **b** Dehydrogenase activity

degradation of protective organic matters, reduction of microbial biomass (De Luca et al. 1994).

24.3.6 Pearson Correlation Among All the Soil Microbiological, Chemical and Biochemical Parameters with Disease Incidents Percentage

The two tailed Pearson's correlation between soil microbiological, chemical and biochemical parameters and disease incidence caused by *S. rolfsii*. It was observed that that *Actinomyces*, Fluorescent *Pseudomonas*, total bacteria, FDA and dehydrogenase activities were negatively correlated with the DI whereas, infection loci and sclerotial populations were positively correlated with the DI. The PCA analysis also explained the similar observation (Fig. 24.7).

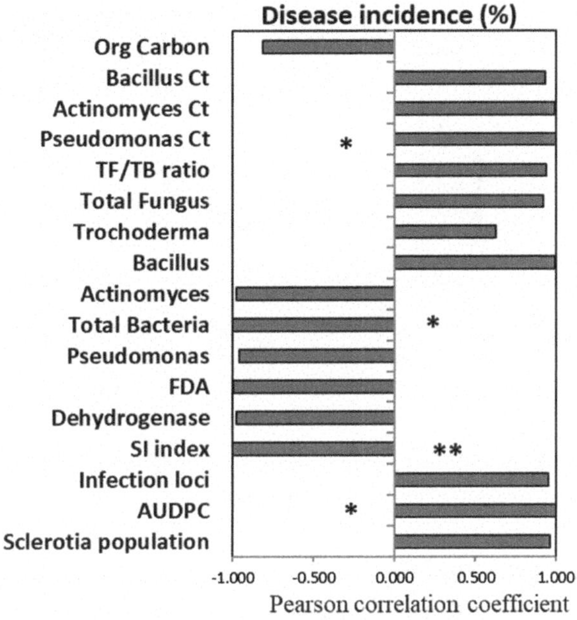

Fig. 24.7 Pearson correlation with disease incidence percentage

24.3.7 Principal Components Analysis of All the Variables (Soil Parameters, Disease Incidence Percentage and Soil Suppressive Index)

Principal component analysis (PCA) is a data reduction method was used in our statistical data analysis considering all the periodic soil microbiological parameter, soil microbial enzymatic activity and suppressive index as variable. The 2-D plot was created according to first two components (component 1 & component 2). The component 1 had eigen value of 15.63 and 91.91% variable coverage whereas component 2 had eigen value of 1.38 and 8.01% variance coverage. According to the rotated component matrix, soil SI Index (−0.821) was closely related to the parameter's lines FDA (−0.890), soil dehydrogenase (−0.924), *Pseudomonas* (−0.973), *Actinomyces* (−0.924), *Bacillus*ct (−0.973) and total bacteria (−0.721) respectively. On the other hand, DI % (0.829), AUDPC (0.817) and Infection Loci (0.621) are closely related with sclerotial population (0.944), *Bacillus* (0.764), *Pseudomonas*ct (0.815), *Actinomyces*ct (0.883), total fungus (0.555) and total fungus and total bacterial ratio (0.588) respectively (Fig. 24.8).

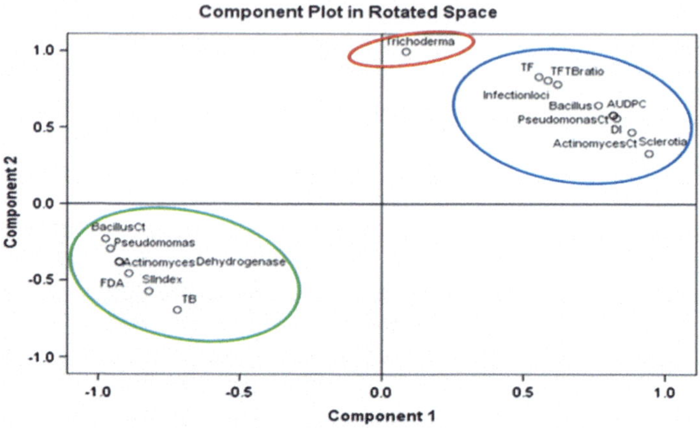

Fig. 24.8 Principal components analysis of all the variables (soil parameters, disease incidence percentage and soil suppressive index)

24.3.8 Biocontrol Potentiality and Biochemical Quantification Observed from the Different Native Biocontrol Bacteria Isolated from the Above Mention Experimental Field

All the soil bacterial isolates in the present studies were tested for their in vitro antagonistic potentiality by dual culture plate assay against *S. rolfsii* (collar rot) pathogen. Mycelial inhibition percentages of the fungal pathogen were calculated. The maximum mycelial inhibition of the *S. rolfsii* was recorded 71.4% and 61.0% respectively challenged with ZTS9 and ZTS43. Experimental data on mycelial inhibition percentage are shown in (Table 24.2). Besides promoting plant growth directly; native soil bacteria can also indirectly promote the plant growth by defensive plants against fungal pathogens. Twelve soil bacterial isolates were screening for the secondary metabolite's trait analysis. They were confirmed for their ability to yield antifungal metabolites and Phyto-hormones—IAA, HCN, Siderophores, SA and P-solactivity as well as to exhibit mucolytic enzyme like lipase, cellulose, amylase and pectinase activity. All the isolates were found to be IAA, siderophore and salicylic acid positive. Whereas, majority of the isolates were cyanogenic except CTS7, MTS39 and ZTS4. All the ioslates are phosphate solubilizers and positive hydrolytic enzymes activity (Fig. 24.9).

Table 24.2 Biocontrol potentiality and biochemical quantification observed from the different native biocontrol bacteria isolated from the above mention experimental field

Isolates name	Inhibition zone	Sidaro-phores	IAA ug/ml	SA ug/ml	NH_3produ-ction	HCN produ-ction	Cellulase zone	Pectinase activity	Amylase zone	Lipase	P-sol
CTS4	27.29	4.84	83.49	21.81	–	+	0.73	2.42	0.43	0.00	0.26
CTS5	43.86	4.95	86.43	23.92	++	++	0.72	2.39	0.23	2.59	0.24
CTS7	51.35	5.66	89.70	22.28	–	–	0.91	2.64	0.64	2.05	0.45
MTS11	50.01	4.93	84.83	22.35	+	++	0.52	0.00	0.00	0.00	0.33
MTS32	57.78	5.22	85.47	21.40	++	+++	0.52	2.12	0.44	0.00	0.43
MTS33	52.38	5.76	90.96	20.62	++	+++	0.71	2.47	0.45	2.29	0.76
MTS39	35.55	5.14	86.77	24.43	–	–	0.51	0.00	0.13	0.00	0.44
ZTS4	52.09	5.69	90.17	24.49	++	–	0.33	0.00	0.00	0.00	0.46
ZTS9	71.43	5.63	88.52	20.78	+++	+++	0.62	2.61	0.34	2.16	0.42
ZTS23	52.72	5.68	90.39	19.53	–	+ + +	0.80	0.00	0.25	1.74	0.35
ZTS27	54.25	5.36	91.45	22.17	++	+ + +	1.22	2.50	0.55	1.85	0.72
ZTS43	61.01	5.32	89.45	21.02	+	+ + +	0.31	1.61	0.14	0.00	0.25

Siderophore activity = uM benzoic acid/ml

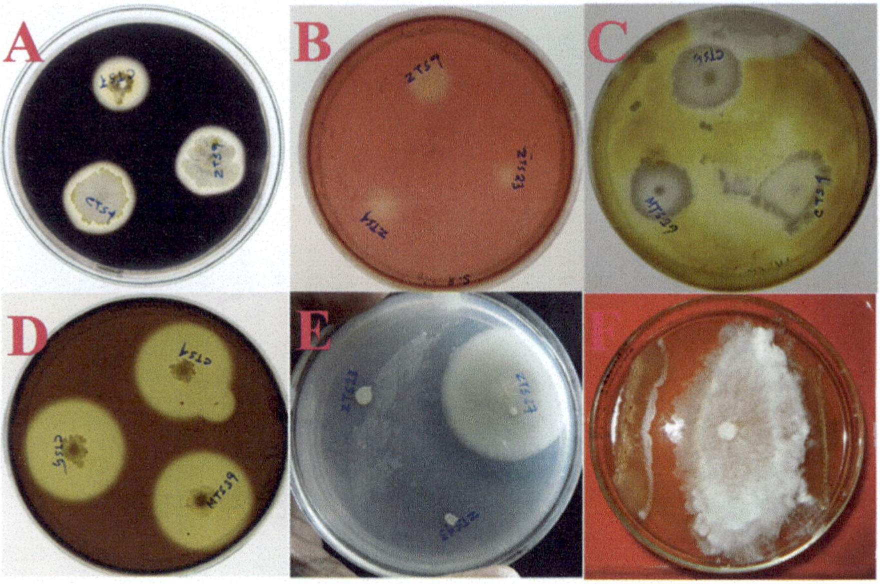

Fig. 24.9 Enzymatic assay i.e. **a** Amylase, **b** Cellulase, **c** Lipase, **d** Pectinase, **e** Phosphate solubilization assay of different biocontrol bacteria. **f** Dual culture plate

24.3.9 Plant Growth Promotion (PGPR) Activity of Different Potential Native Biocontrol Bacteria Isolated from the Above Mention Experimental Field

Lentil seeds were preserved with biotic elicitors and after five days of bacterization of seeds, germination percentage (%), root-shoot length of lentil seedlings was recorded. Observed analysis data revealed that, the performance of lentil seeds on PGPA associated with seed germinability and root-shoot length as well as vigor index in every treatment. From the selected 12 bacterial isolates, isolate ZTS27 from zero tilt soil Kalyani, Nadia was found as the best plant growth promoter bacterial isolates compared to the others soil bacterial isolates. From the observed data it was concluded that, lentil seeds bacterized by using ZTS23 (96.88%) gave maximum per cent of lentil seed germination followed by MTS33, ZTS4, and ZTS27 (Table 24.3). Root and shoot length of lentil seedlings were observed maximum for the native isolate ZTS27 (Fig. 24.10), Where also maximum vigor index was recorded for the native isolates ZTS27 (1366.56) followed by MTS33 (1164.42). Yeole and Dube findings also reported as seed bacterization by using soil bacterial isolates was found to increase germination percentage, root and shoot length of cotton, chilli, groundnut and soybean. There were reports that seed bacterization by using fluorescent *Pseudomonas* and *Bacillus sp.* increases the crop yield and plant growth of field crops like potato, sugar beets (Suslow et al. 1982) and wheat.

Table 24.3 Observed plant growth promotion (PGPR) activity of different potential native biocontrol agent bacteria isolated from the above mention experimental field

Isolates	(%) of Seed germination	Root length (cm)	Shoot length (cm)	Vigor index	Rank
Control	81.25^{e}	1.39^{h}	1.92^{j}	268.83^{j}	12
CTS4	82.29^{e}	0.93^{j}	1.85^{j}	228.09^{j}	13
CTS5	84.38^{e}	1.62^{g}	3.85^{h}	461.48^{g}	10
CTS7	$87.50^{b,c,d,e}$	3.64^{c}	6.24^{d}	865.09^{d}	5
MTS11	$86.46^{c,d,e}$	1.33^{h}	3.16^{i}	387.71^{h}	11
MTS32	$88.54^{b,c,d,e}$	1.42^{h}	5.63^{e}	624.06^{f}	8
MTS33	$94.79^{a,b}$	4.37^{b}	7.91^{a}	1164.42^{b}	2
MTS39	$87.5^{b,c,d,e}$	2.54^{e}	5.36^{a}	690.65^{e}	7
ZTS4	$93.75^{a,b,c}$	3.15^{d}	7.44^{b}	992.7^{c}	4
ZTS9	$88.54^{b,c,d,e}$	2.53^{e}	5.71^{e}	729.28^{e}	6
ZTS23	96.88^{a}	3.15^{d}	7.13^{c}	996.5^{c}	3
ZTS27	$92.71^{a,b,c,d}$	6.89^{a}	7.85^{a}	1366.56^{a}	1
ZTS43	$85.42^{d,e}$	1.84^{f}	5.19^{g}	600.53^{f}	9
CD (<0.05)	4.91	0.07	0.06	40.53	
SeM (±)	2.42	0.04	0.03	19.99	

[a-j] Duncan grouping for all treatments

Fig. 24.10 Germinated plant, after incubation of 5 days, treated with newly isolated biocontrol bacteria from above mention field experiment

24.3.9.1 Identification of All the Potential Biocontrol Isolates by Using 16s-rDNA Sequencing and NCBI Blasting Method

16s-rDNA sequences of the twelve native bacterial isolates have been deposited in NCBI database (Table 24.4). The DNA sequences of the isolates on 16s-rDNA region were searched for homology with Basic Local Alignment Search Tool (BLAST) against the nucleotide data base maintained by National Centre for Biotechnological

Table 24.4 Molecular identification of isolates through 16S- rDNA sequencing

Isolates name	Query cover %	Identy %	Accession	Organism
CTS4	100	99	KY970112.1	*Bacillus mojavensis*
CTS5	100	99	KX588160.1	*Bacillus tequilensis*
CTS7	100	100	KY974386.1	*Bacillus stratosphericus*
MTS11	100	99	KY009934.1	*Bacillus velezensis*
MTS32	100	99	JX027507.1	*Bacillus aerophilus*
MTS33	99	100	JQ659890.1	*Pseudomonas aeruginosa*
MTS39	100	99	KU962126.1	*Pseudomonas aeruginosa*
ZTS4	100	99	KY885171.1	*Pseudomonas aeruginosa*
ZTS9	100	99	CP020704.1	*Pseudomonas aeruginosa*
ZTS23	100	99	KY970137.1	*Bacillus pumilus*
ZTS27	100	99	MF144501.1	*Pseudomonas aeruginosa*
ZTS43	100	99	JX841311.1	*Pseudomonas plecoglossicida*

Information (NCBI), NIH, USA. Native isolates used in the present study exhibited 97–100% sequence similarity to the either *Pseudomonas aeruginosa* or *Pseudomonas sp.* or with different species of *Bacillus* available in NCBI database with lowest E-value and maximum query coverage and maximum identity.

24.3.9.2 *Pearson Correlation Coefficient Among Various Metabolites and Biochemicals of the Biocontrol Bacteria Isolates with Fungal Inhibition and Plant* Vigor *Index*

The two tailed Pearson's correlation between antagonistic activity of soil bacterial isolates and their secondary metabolites and cell wall degrading enzyme production revealed that HCN, NH_3 and siderophores productions of soil bacterial isolates were positively correlated with the antagonistic activity of soil bacterial isolates at 5% level of significance, whereas, IAA productions of soil bacterial isolates were positively correlated with the vigor index at 1% significant level (Table 24.5).

24.3.9.3 *Linear Regression Analysis Among All the Biochemical Parameters When Fungal Inhibition and* Vigor *Index Were the Dependent Variable*

In order to predict the antagonistic activity and Vigor index of soil bacterial isolates with biochemical parameters a LMR models were developed by various biochemical parameters using stepwise method. NH3 production was found to be one of the important predictors of antagonistic activity of soil bacterial isolates. Whereas, IAA productions and phosphate solubilizations were found to be the significant predictors

Table 24.5 Pearson correlation coefficient among various metabolites and biochemicals of the biocontrol bacteria isolates with fungal inhibition and plant vigor index

	Fungus inhibition		Vigor index
NH_3 production	0.633*	Fungus inhibition	0.603*
HCN production	0.578*	Siderophore	0.582*
Siderophore production	0.590*	IAA production	0.795**
		Seed Germination	0.803**
		Root length	0.864**
		Shoot length	0.918**

* 5% level of significance ($p<0.05$)
** 1% level of significant ($P<0.01$)

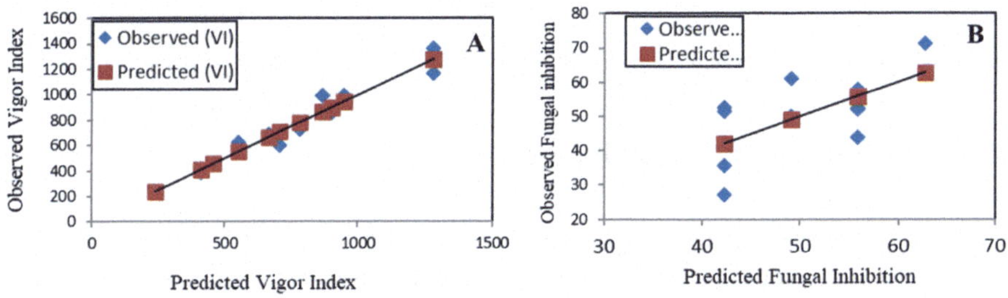

Fig. 24.11 Predicted verses observed curve, **a** Vigor Index **b** Fungal inhibition

and these two parameters possibly will able to explain nearly 95% variation in plant growth promoting potentiality of soil bacterial isolates. The PCA also explained the similar phenomenon.

Linear regression (Fig. 24.11) model equation

Y (Fungal inhibition) = 35.403** + 6.847(NH3)*

where, $R^2 = 0.663$, R^2 Adj = 0.341.

Y (Vigorindex) = −6622.984** + 79.454(IAA)** + 891.423(P-Sol)**

Where, $R^2 = 0.974$, R^2 Adj = 0.949.

24.3.9.4 Principal Component Analysis (PCA) Among All the Biochemical Parameters of the Bacterial Isolates

Principal component analysis among all the biochemical parameters of potential biocontrol bacterial isolates explained that antagonistic activity or fungal inhibition potentiality (0.776) was very much closely related to siderophore (0.869), indole acetic acid (0.879), ammonia (0.527), vigor index (0.836), and phosphate solubilization (0.639) activity. All the analysis was done according to the component 1 which had 4.95 eigen value and 41.26 variance coverage (Fig. 24.12). PCA and two tailed person correlation explained same result in this concern.

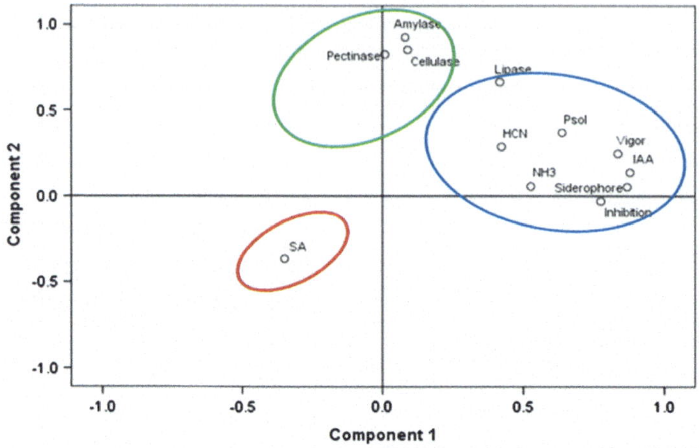

Fig. 24.12 Principal component analysis (PCA) among all the biochemical parameters

24.4 Summary and Conclusion

Collar rot disease incidence and sclerotial population of *S. rolfsii* was found to be higher under conventional tillage system as compared to minimal and zero tillage condition. Collar rot disease incidence, infection foci and sclerotial population was found to be least in zero tillage condition.

Temporal changes in microbiological parameters viz., FDA and dehydrogenase activities were found to be higher in zero tillage condition and least activities was noticed under conventional tillage condition.

Microbial abundance of *Pseudomonas*, *Actinomyces*, *Bacillus*, total bacteria and total fungi were estimated through dilution plating on specific media. Among these higher population of total bacteria, *Pseudomonas* and *Actinomyces* were observed under zero tillage condition, whereas, *Bacillus* and total fungi were found to be higher in conventional tillage but *Trichoderma* abundance are variable in different tillage system.

Total sixty native isolates were screened from lentil soil under three different cropping conditions (ZT, MT & CT) from AB block of Nadia, West Bengal on different specific medium. All the native isolates were screened for in-vitro antagonistic activity against the fungal plant pathogen. From these, twelve isolates were exhibited potential antagonistic activity towards the fungal pathogen, *Sclerotium rolfsii.*

The efficient antagonistic isolates performed antagonistic activity in between the 27.29–71.43% and significantly highest antagonistic potentiality against *S. rolfsii* was observed by ZTS9 isolates (71.43% inhibition potency) in dual culture method.

Among the 12 bacterial isolates, the isolate ZTS27 from lentil soil of Nadia, West Bengal was found to be the best performer in plant growth promoting (PGP) activity allied with seed germination, root-shoot length as well as vigor Index (1366.56) of lentil seedling compared other isolates and control.

Most of the antagonistic bacteria were found to produce secondary metabolites but ZTS27, ZTS9 and MTS39 were actively antagonistic rhizobacterial isolates producing of higher quantity of secondary metabolites such as amylase, cellulase, HCN, IAA, NH_3, protease, pectinase, phosphate solubilization, siderophore and SA.

The two-tailed Pearson's correlation between dependent variable antagonistic activity (percentage of mycelium growth inhibition of *S. rolfsii*) of isolates and the independent variables like secondary metabolites and cell wall degrading enzyme production revealed that NH_3, HCN and siderophore production had positive correlation with the antagonistic activity at 5% level and NH_3 production, an only single variables of soil bacterial isolates depict the variation in antagonistic potentiality of soil bacterial isolates by 66%.

The two tailed Pearson's correlation between dependent variable vigor index of plant and the independent variables like plant growth promotion (PGP) traits of soil bacterial isolates revealed that IAA production, root-shoot length and seed germination were found to be significantly positively correlated with the vigor index even at 1% level and IAA and phosphate solubilization were found two be the two most importance variables whichcould able to explain the variation in vigor index by 97%.

16s-rDNA sequences of native isolates and NCBI database BLASTn search indicated that all the antagonistic isolates belong to different species of *Bacillus* and *Pseudomonas.*

24.5 Future Scope of Research of This Work is that,

- How soil physical parameter including soil porosity, pore network connecting and pore distribution, can affect the soil borne pathogen *S. rolfsii* spreading through soil? is still unknown.
  - So, that can be revealed further to understand the pore geometry and soil structure and its relation towards soil suppressive against *S. rolfsii.*
- How soil pore geometry can influence soil microbiome and differential studies of community level distribution between tilled and non-tilled (lose and compress soil) soil?
- How short-term tillage and non-tillage approaches can influence microbial diversity and community distribution by both culturable and non-culturable approaches is still very unclear?
  - So, it will be very important to understand how the agriculturally beneficial soil microorganism can play the role after treated with various tillage practices.
- To estimate the tenure, require for conversion from tilled to nontilled condition to get enriched with soil beneficial microflora.

Acknowledgements The authors acknowledge Raja N.L. Khan Women's College (Autonomous) and Bidhan Chandra Krishi Viswavidyalaya for supporting with all sorts of library and laboratory facilities.

Funding No financial help from any authorization was received during the course of study as all the authors were self-employed.

Ethics approval and consent to participate Not applicable for this study.

Conflict of Interest Authors have no conflict of interest.

References

Ahmad F, Ahmad I, Khan MS (2008) Screening of free-living rhizospheric bacteria for their multiple plat growth promoting activities. Microbiol Res 163:173–181

Bakker AW, Schippers B (1987) Microbial cyanide production in the rhizosphere in relation to potato yield reduction and *Pseudomonas* spp.-mediated plant growth-stimulation. Soil Biol Biochem 19:451–457

Bararunyeretse P, Yao J, Dai Y, Bigawa S, Guo Z, Zhu M (2017) Toxic effect of two kinds of mineral collectors on soil microbial richness and activity: analysis by microcalorimetry, microbial count, and enzyme activity assay. Environ Sci Pollut Res Inst 24:1565–1577

Choi O, Kwon JH, Min Y, Kim J (2011) First report of stem rot on asiatic dayflower (Commelina-communis L.) caused by *Sclerotiumrolfsii* in Korea. Mycobiology. 39:57–58

De Luca TH, Keeney DR (1994) Soluble carbon and nitrogen pools of prairie and cultivated soils: seasonal variation. Soil Sci Soc Am J 58:835–840

Doran JW (1982) Tilling changes soil. Crops Soils 34:10–12

Elliott GC, Lauchli A (1985) Phosphorus efficiency and phosphate-iron interaction in maize1. Agron J 77(3), 399–403

Garbeva P, Van-Veen J, Van-Elsas J (2003) Predominant bacillus spp in agricultural soil under different management regimes detected via PCR-dgge. Microb Ecol 45:302–316

Gerhardt PE (1994) Methods for general and molecular bacteriology, Rev. Am Soc Microbiol, University of California, Washington

Glick BR (1995) The enhancement of plant growth by free-living bacteria. Can J Microbiol 41:109–117

Gordon SA, Weber RP (1951) Calorimetric estimation of indoleacetic acid. Plant growth-promoting rhizobacteria. Biocontrol Sci Technol 11:557–574

Gomez KA, Gomez AA (1984) Statistical procedure for agricultural research (2nd ed) New York: Wiley

Hassink J, Voshaar JO, Nuhuis EH, Van Veen JA (1991) Dynamics of the microbial populations of a reclaimed-polder soil under a conventional and a reduced-input farming system. Soil Biol Biochem 23:515–524

Heuer H, Krsek M, Baker P, Smalla K, Wellington E. (1997) Analysis of actinomycete communities by specific amplification of genes encoding 16S RNA and gel-electrophoretic separation in denaturing gradients. Appl environ microbial 63(8):3233–41

Jacobsen BJ, Zidack NK, Larson BJ (2004) The role of *Bacillus*-based biological control agents in integrated pest management systems: plant diseases. Phytopathology 94:1272–1275

Johnson AM, Hoyt GD (1999) Changes to the soil environment under conservation tillage. Hort Technol 9:380–393

Jolliffe IT (2002) Principle component analysis, 2nd edn. Springer

Kabiri V, Raiesi F, Ghazavi MA (2016) Tillage effects on soil microbial biomass, SOM mineralization and enzyme activity in a semi-arid Calcixerepts. Agric Ecosyst Environ 232:73–84

Khan MS, Zaidi A, Musarrat J (2014) Phosphate solubilizing Microorganisms: principles and application of microphos technology. Springer

Kwon JH, Chi TT, Park CS (2009) Occurrence of fruit rot of Melon Caused by *Sclerotiumrolfsii* in Korea. Mycobiology. 37:158–159

Lammel DR, Arlt T, Manke I, Rillig MC (2019) Testing contrast agents to improve micro computerized tomography (μct) for spatial location of organic matter and biological material in soil. Front Environ Sci 7:153

Lee S, Kim S, Kim S, Lee I (2012) Effects of soil-plant interactive system on response to exposure to ZnO nanoparticles. J Microbial Biotechnol 22:1264–70

Liang X, Zhang H, He M, Yuan J, Xu L, Tian G (2016) No-tillage effects on grain yield, N use efficiency, and nutrient runoff losses in paddy fields. Env Sci Pollut Res Inst. 23:21451–21459

Meyer JM, Ajelvandre P, Georges C (1992) Iron metabolism in *pseudomonas*: salicylic acid, a siderophore of *pseudomonas*fluorescensCHA0. BioFactors 4:23–27

Pradhan A, Idol T, Roul PK (2016) Conservation agriculture practices in rainfed uplands of India improve maize-based system productivity and profitability. Front Plant Sci 7:1008–1017

Ramarathnam R, Fernando WG, de Kievit T (2011) The role of antibiosis and induced systemic resistance, mediated by strains of *Pseudomonas chlororaphis*, *Bacillus cereus* and *B. amyloliquefaciens*, in controlling blackleg disease of canola. Biocontrol 56:225–235

Rodríguez MA, Cabrera G, Gozzo FC, Eberlin MN, Godeas A (2011) Clonostachysrosea BAFC3874 as a Sclerotiniasclerotiorum antagonist: mechanisms involved and potential as a biocontrol agent. J Appl Microbiol 110:1177–1186

Roger-E J, Christel A, Michel B, Guy R (2010) Tillage and soil ecology: partners for sustainable agriculture. Soil Tillage Res 111(1):33–40

Rojas-Ruiz NE, Sansinenea-Royano E, Cedillo-Ramirez ML, Marsch-Moreno R, Sanchez-Alonso P, Vazquez-Cruz C (2015) Analysis of *Bacillus thuringiensis* population dynamics and its interaction with *Pseudomonas fluorescens* in soil. Jundishapur J Microbiol 8:e27953

Simko I, Piepho HP (2012) The area under the disease progress statistic calculation, advantage and application. Phytopathology 102:381–389

Spilker T, Coenye T, Vandamme P, LiPuma JJ (2004) PCR-based assay for differentiation of pseudomonas aeruginosa from other pseudomonas species recovered from cystic fibrosis patients. J Clin Microbiol 4(5):2074–2079

Suslow TV, Schroth MN, Isaka M (1982) Application of a rapid method for gram differentiation of plant pathogenic and saprophytic bacteria with out staining. Phytopathology 72:917–918

Wang Z, Liu L, Chen Q, Wen X, Liao Y (2016) Conservation tillage increases soil bacterial diversity in the dryland of northern China. Agron Sustain Dev 36:1–9

Chapter 25
Role of Soil Microbes in Soil Health and Stability Improvement

Soumik Chatterjee, Krishna Chandra Mondal, and Sabyasachi Chatterjee

Abstract Soil is a outer most surface on earth which is a home for different microorganisms. Biodiversity of soil is a mixed population of different type biological organisms. It is one of the most biologically diverse upper most part of Earth. Soil structure and health depends on interaction of microbes and soil organic materials. Soil microbes also interact with plants and influence soil health and production of crop. The soil organic mater is a food for soil bacteria and other microorganism. Soil bacteria improve the soil quality by interaction with organic mater as a result increase the entry and storage of soil water, resistance to soil erosion. Different soil bacteria has different ability to react with soil organic mater and control soil health from season to season. Soil microbes play a wide range of essential role with sustainable function on all ecosystems. They are also help to maintain the soil nutrients, nitrogen, phosphorus, soil organic matter, carbon for plants. Some beneficial soil microbes helps to reduce soil-borne disease of plants. By the use of these beneficial microbes inoculums we increase the yield of crop and reduce of plant disease. These advance technology is essential and important resource for the development of sustainable agricultural systems.

Keywords Soil microbes · Soil health · Disease suppression

25.1 Introduction

Soil, outer most layer of earth (pedosphere) is the heterogenous mixture of minerals, organic matter, liquids, gases, and various type of microbes. It is non-renewable resource sustainable ecosystem balance and support life on earth. Rock's weathering and erosion play crucial role in soil formation. Soil function depend on soil structure

---

S. Chatterjee (✉) · S. Chatterjee
PG Department of Botany, Ramananda College, Bishnupur, Bankura 722122, West Bengal, India
e-mail: chatterjee.soumik1996@gmail.com

K. C. Mondal
Department of Microbiology, The University of Burdwan, Burdwan 713104, West Bengal, India

P. K. Shit et al. (eds.), *Soil Health and Environmental Sustainability*,
Environmental Science and Engineering,
https://doi.org/10.1007/978-3-031-09270-1_25

which act as a key factor. Soil particles flocculation, cementation and rearrangement are cause of soil aggregation (Duiker et al. 2003). The expanding population of earth need higher food crop production. This intensification of agricultural system and unscientific use of chemicals gradually derogating soil fertility and soil biodiversity. Therefore, sustainable agro-ecosystem and environmental protection are important. Soil aggregation is facilitated by microflora, organic carbon (SOC) and ionic property of soil etc. (Roselló-Mora and Amann 2001). Dynamics of microorganisms that reside in the soil take prominent role in soil fertility and maintain soil quality. Microbial variety in the soil is more enriched than eukaryotic organisms. Approximately 10 billion of microorganisms present in 1 gm of soil. But among them only 1% are able to visualised under microscope and characterized. Microbes play an important role to maintain agriculture of land ecosystem. In soil rhizosphere different verities of microbes (both harmful and beneficial) are reside in the soil such as endophytes, saprophytes and actinomycetes (Six et al. 2000). Plants microbes interaction in agricultural ecosystem (symbiosis association) are very interesting area of research which attract the many researchers. In natural ecosystem, soil is the heart of verities micobes (beneficial and harmful). Soil rhizosphere bacteria improve the soil health and quality, also increase the yield of crop by control of soil-borne diseases (Antoun and Kloepper 2001). Soil beneficial bacteria control soil-borne disease by the process of biocontrol and increased the crop production. This type of technique are referred as bioagents. Rhizobaceria also play an essential role to maintain soil health by the increasing the production of phosphatase, â-gluconase, dehydrogenase and antibiotics in soil (Antoun and Pre´Vost 2005). This technology is very important to maintain soil health and crop production with out harming environment.

25.2 Soil Structure

Soil structure is the arrangement of soil particles, organic matter and pores within a matrix of solid materials in the solid parts of the soil. Each soil particles bound to each other and form pore in the soil (Bashan and de-Bashan 2005). Many properties like water holding capacity, permeability, root penetration and infiltration depend on quantity, distribution of pores in the soil. Only 50% of solid material present in the soil and another 50% is pore space in the soil. Pore space help the store of water in the soil (Be´langer and Avis 2002) (Fig. 25.1). Living organisms accumulate organic matter and nutrients from the soil. The following diagram (magnified 20 times) shows that how soil particles and pores might arrange in soil. Storage and refuge both are provided by small pores within the aggregates soil. Liquids, gases, roots and organisms are move by the use of large pores in soil aggregates (Haas and Défago 2005). Crop residue, manure of plants and dead animals are added in soil for the source of organic materials. These materials made the surface soil and helps to the development of granular and crumb structure of soil. Soil microorganisms, soil animals help to formation of organic material in soil for subsoil structure development (Hass and Keel 2003). Structure of soil is dynamic, unlike texture of soil (clay, silt

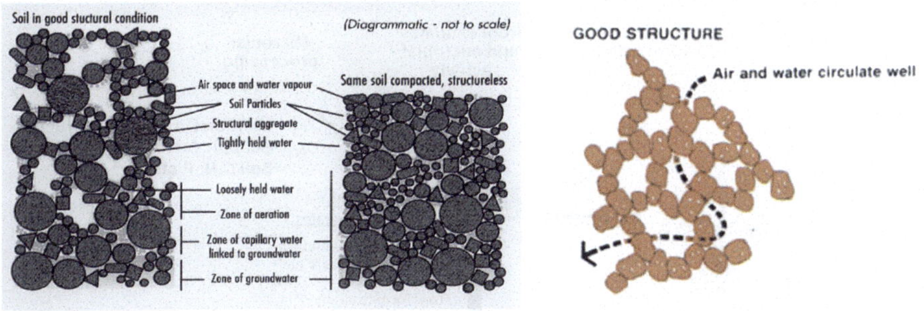

Fig. 25.1 Soil structure (*Source* https://www.slideshare.net/CLAEast/fbu-aylsham-1-johnny-johnston-and-soil-fertility)

and sand) and there structure varies according natural processes such as weather conditions, plants root activity and also for soil-biological activities. Agricultural also influence the soil structure over a time periods. study the soil organisms and agricultural activity on soil structure and soil functions (air permeability, gas diffusion and mechanical properties) is one of the main focuses area of research.

25.3 Soil Microbes Overview

Soil act as a home for many microorganism like a teaspoon of soil (1 g) contains many microbes, bacteria, actinomycetes and fungal filaments. Organic detritus of leaf litter, dung and carrion are used by bacteria as a feeding source. Microbes by secreting digestive enzyme digest their food and absorb. During this enzymatic reaction complex organic molecules breaks down into simpler molecules and then re-absorbs by the microbe for digestion through its cell wall. Microbes have 50 to 60 different types of enzyme for this processes (Fig. 25.2).

25.3.1 Bacteria

In unfavorable environmental condition bacteria form resistant spores for escapes this condition. After unfavorable condition when favorable conditions return the spore of bacteria again germinate and increase population in soil. The main function of bacteria in soil is break organic matter and increase soil fertility. Bacteria act as a main decomposer groups in soil because they produce and secrete huge number of digestive enzymes for break down of organic matter. Some beneficial soil bacteria (Rhizobacteria) can fix N_2 from air in the soil and making it available for plants. Some of these free living soil bacteria e.g. Azotobacter is able to fix nitrogen on the other hand others nitrogen fixing bacteria e.g. Rhizobia living in roots nodules of

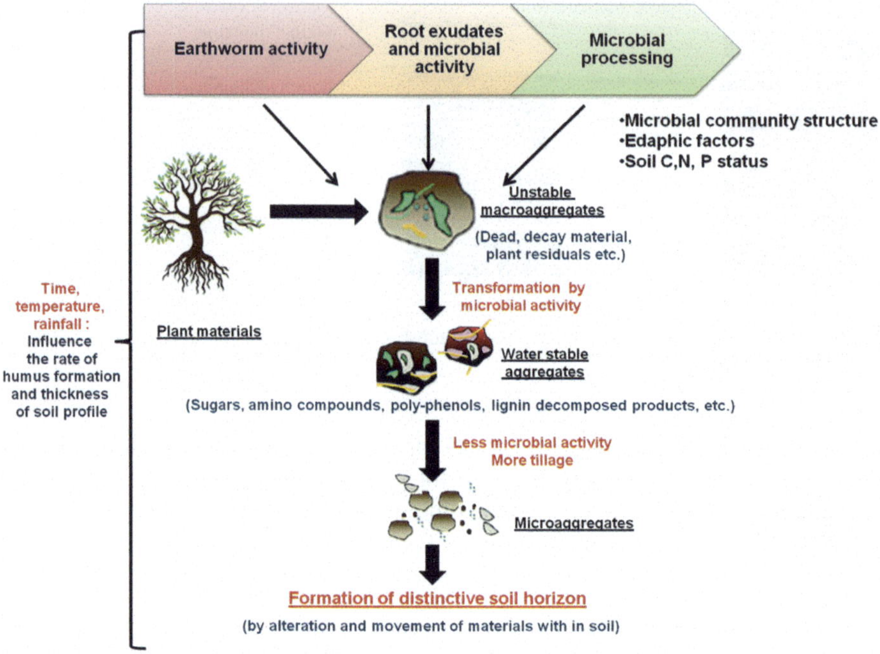

Fig. 25.2 Role of microbes in soil formation (*Source* https://link.springer.com/chapter/10.1007/978-981-13-6480-8_2)

legume plants with symbiotic association (Maksimov et al. 2011). Here bacteria gain advantage by using sugars of plants as a food source and plants take benefits by use fixed nitrogen source which was fixed by the bacteria. This type relationship between bacteria and plants which is called "symbiosis relationship" where both organisms living together and gain a mutual benefit from each other.

25.3.2 Other Microbes

Other then bacteria and fungi many actinomycetes, yeasts and viruses live in soil and help to maintain soil health but not all of these are act as decomposer organisms. Actinomycetes are branched, filamentous, fungus like network which are joined end to end. They have both bacterium and fungus like characters. They are act as good decomposers and are able to break down organic matter such as cellulose and chitin. Soil yeasts is a another class of soil microbes belongs to unicellular group of fungi which is also act as a soil decomposers. Soil viruses are present as soil prey on soil bacteria but has very less information is known about these microorganisms.

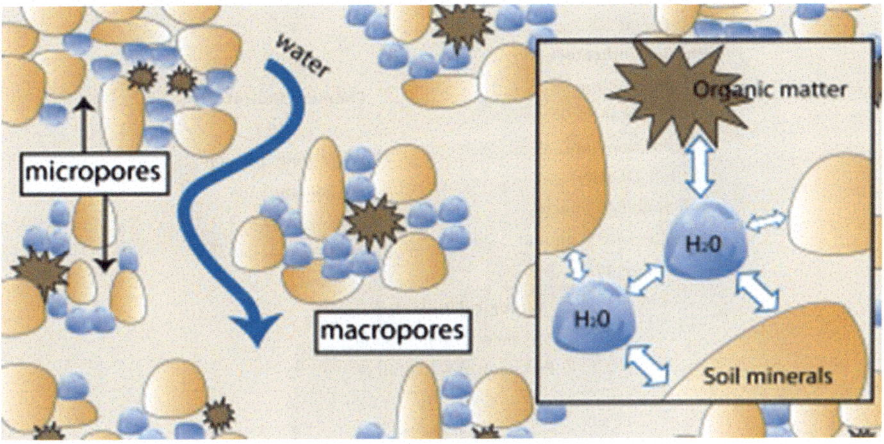

Fig. 25.3 Water move through soil macrospores (*Source* https://www.qld.gov.au/environment/land/management/soil/soil-properties/water)

25.4 Enhancement of Soil Structure

Soil fertility depends on good soil structure. Soil bacteria has major role to improving soil structure and soil health. Soil water holding capacity, water transmission, bulk density and lower potential for soil erosion every thing control by good structure of soil. Soil microorganism make tunnels through the soil by reacting with the organic matter and change the soil structure which is called macrospores (Richardson et al. 2009). Water can easily infiltrate through the macrospores from the surface of soil. This macrospores formation help the soil for water transmission and soil hydrology. Soil erosion is occur when water can easily enter into the soil (Fig. 25.3). Soil microbes mix the each soil layers together and also mixing the organic matter by eating with mineral soil layers. Microbes also maintain the soil stability and fertility by there enzymes. Bacteria help for the formation of water-stable soil aggregates. Bacteria produce and secrete polysaccharide which is mucilaginous substance act as a sticky glue for binding of the soil particles together into aggregates (Rodríguez and Fraga 1999). These structure of soil aggregates are very stable and protect from the action of water for several months and help to prevent dispersion of the soil.

25.5 Indicators of Soil Health

It is clearly understood from many literature that health of soil is depend on balance of the biological (microbial population), chemical and physiological components of soil. So soil health indicators are required for evaluation of soil health and about all components. Since microbial population act as a good monitoring of soil

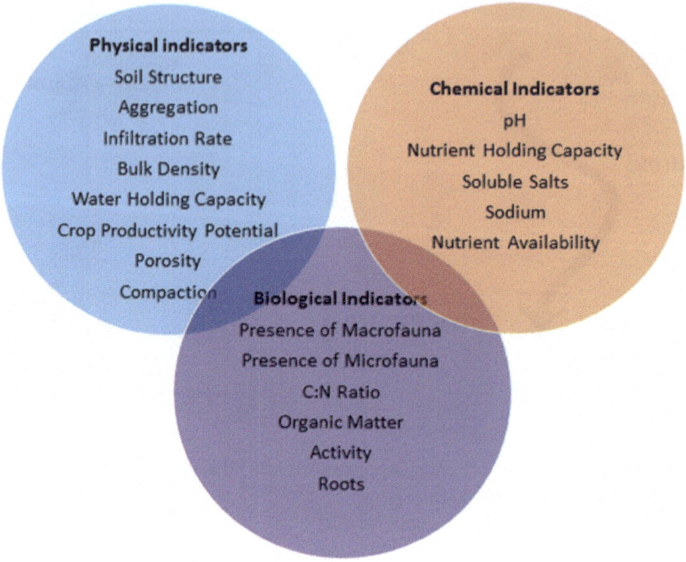

Fig2525..4 Different types of soil health indicators (*Source* https://microbewiki.kenyon.edu/index.php/File:20_indicator_venn.png)

health because they are quickly respond according to changes in the soil ecosystem (Fig. 25.4). If any physical and chemical compounds of soil are change which influence the microbial population and their activity which is providing an early indication for soil health improvement (Siddiqui et al. 2005). These methods were improved on the basis of the repeated identification and documentation of microbial activity in soil.

25.6 Soil Microbes and Their Types

Approximate 70–75% microbial population (bacteria, actinomycetes etc.) constitute living biomass of soil and act as primary decomposers for break down of organic compounds. Two different type of microbe population in soil are:

25.6.1 Disease Causing Microbes

Some microbes are responsible for causing diseases of plants and also damage the quality of soil. Soil-borne pathogenic bacteria can survive and interact with soil for many years. Different types soil pathogenic bacteria are present in soil for this reason it is very difficult to identifies which one is responsible for soil-borne diseases. Some harmful bacterial pathogens like *Erwinia* sp. (soft rot), *Ralstonia solanacearum* (wilt) and *Streptomyces scabies* (potato scab) are responsible for plant diseases and damage crops production (Singh et al. 2015). Lettuce Big Vein Virus (LBVV) and Lettuce Stunt Necrotic Virus (LSNV) are soil-borne virus also cause disease of crops.

25.6.2 Biocontrol Agents Inhabiting in Soil (Resident Biocontrol Agents)

Some microbes (bacteria) play role to controlling plant diseases. Disease control and maintain crop health by the activities of some soil microbes. Some benifical soil microbes helps in production of hydrocyanic acid (HCN), nitrogen assimilation, siderophore, antibiotic and hydrolytic enzyme like lipase, chitinase etc. *Bacillus* spp having PGPR (plant growth-promoting rhizobacteria) and which is excellent biocontrol agents and have plants disease controlling ability (Singh et al. 2013).

25.7 Soil Microbes Can Be Classified as Follows

25.7.1 On the Basis of Microbial Function

25.7.1.1 Decomposers

It is the processes by which microbial population react with organic molecules in aerobic condition and partially oxidized metabolites which secrete bad odour (e.g. ammonia, mercaptans and indole).

25.7.1.2 Synthetic Microbes

Some microbial population has the ability to fixed atmospheric N_2 and CO_2 in soil. Some microorganisms has this biosynthetic ability for utilized metabolic energy. Those soils are termed as synthetic soil which is produced by free living bacteria Azotobacter.

25.7.2 On the Basis of Microbial Activity

25.7.2.1 Disease-Inducing Soil

Disease causing soil contain around 5–20% of pathogenic microbes in total soil microflora. If this types of pathogenic microflora are present is soil they incompletely oxidized fresh organic matter in soil, which are harmful for plants and attacked by harmful pathogens. Such soils may be used for agriculture as diseases suppressive soils by addition of effective microorganisms inoculums (Stein 2005).

25.7.2.2 Disease Suppressive Soils

In this soil some beneficial microbes suppress the activity or growth of pathogenic microbes without secretion of any chemical. Trichoderma, Penicillium and actinomycetes etc. are some antagonistic microbes synthesis sufficient amount of antibiotics in soil which restrict the growth of some pathogens such as Fusarium, Pythium etc. (Tilak et al. 2005). Plants cultivated in this type of soil are rarely infected with diseases.

25.7.2.3 Zymogenic Soils

In this type of soil microbes convert complex substances into simpler ones by the processes of microbial fermentation. Organic compounds like residues of crops, manures of animal and plants, wastes of municipal and green manures are the actual source from which microbes are arise in soil (Van Loon 2007).

25.7.2.4 Synthetic Soils

These soils are enriched with N and C fixing bacteria, they helps to convert complex organic compounds into carbohydrate, proteins and amino acids (Van et al. 2008). Photosynthetic bacteria such as Phycomycetes and Cyanobacteria are important soil microflora.

25.8 Characteristic Features of Soil-Inhabiting PGPR

25.8.1 *Antibiotic Production in Soil*

Rhizobacteria play role in disease suppression and controlling with antibiotic production in soil. Some example of antibiotic such as Phenazines, phloroglucinols, pyrrolnitrin, pyoluteorin, cyclic lipopeptides (diffusible) and HCN are six classes of antibiotic act as biocontrol agents in plants root diseases. *Bacillus subtilis* is a gram-positive soil bacteria involve in maintains of soil structures and is able to synthesize antibiotics against pathogen bacteria (Scher and Baker 1980). *Bacillus* spp. are involved in the synthesis of some useful antibiotics such as polymyxin, circulin and colistin, which were used for killing some Gram positive and Gram-negative pathogenic bacteria and fungi. *Streptomyces hygroscopicus* help to production of geldanamycin antibiotic which control the rhizoctonia root rot disease of Pea plant (Scher et al. 1984).

25.8.2 *Hormone Production*

Plant hormones are very essential for plant survival and developments those are synthesize by PGPR. The most important plant hormone such as auxin which is produced by the help of PGPR. PGPR, which involve in roots growth and development for enabling plants to absorb maximum nutrients from soil for survival (Schippers et al.1987). Two bacterial species such as *Bacillus pumilus* and *Bacillus licheniformis* help in production of another plant hormone gibberellins in the form of gibberellic acid for stem elongation and other function of plants.

25.8.3 *Phosphate Solubilisation*

Inorganic ion phosphorus is one of the most essential element for plant growth and survival. Which is help in cell division of meristematic tissue, photosynthesis for food production, break down of sugar for metabolism also for energy and nutrient transfer for plant growth. Plants absorb and use inorganic phosphate ion in the form of anions which is immobilized and precipitiated with some other cations those are present in soil such as Ca^{2+}, Mg^{2+}, Fe^{3+} and Al^{3+} because phosphate ion is extremely reactive (Schroth and Hancock 1982). Organic compounds are decomposed by the helps of soil rhizobacteria and make phosphorus available for plants. Mineralization of phosphorus is greatly affected by soil microflora. Bacillus and Paenibacillus bacteria are phosphate solubilizing bacteria those are applied to soils specifically for enhancing the phosphorus element in soil and make it available of plants. Some species such as *Pseudomonas* sp., *Bacillus* spp. and *Rhizobium* sp are most beneficial soil bacteria for phosphate solubilizers in cropping system.

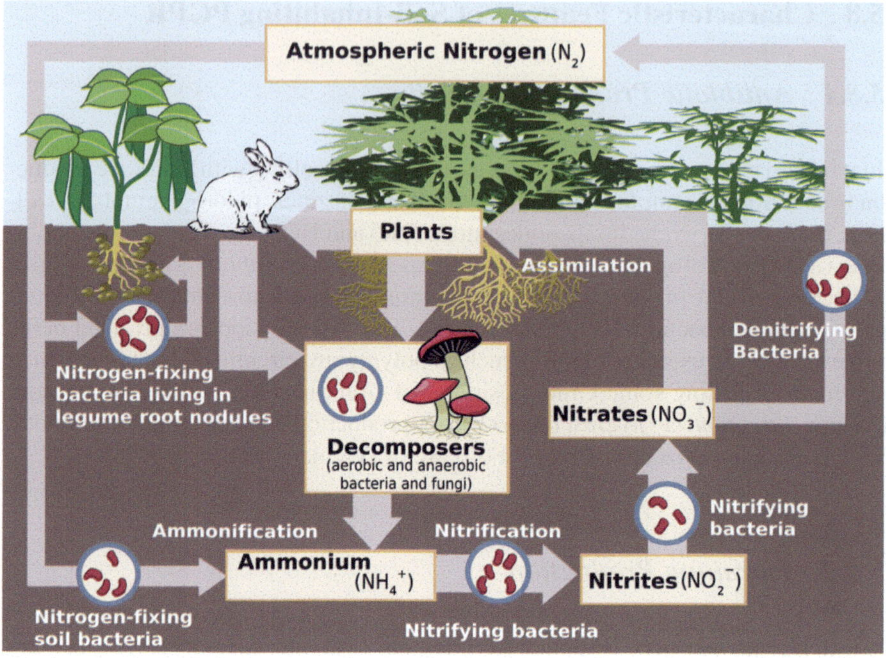

Fig. 25.5 Nitrogen fixation in soil (*Source* https://en.wikipedia.org/wiki/Nitrogen_fixation)

25.8.4 Nitrogen Fixation

Nitrogen amount in soil are very less and plants are not able to use atmospheric diatomic nitrogen gas. Nitrogen is one of the most essential macronutrient for plants growth and development. Some species of bacteria such as Rhizobacteria and Azotobactor are able to fixed nitrogen (bacteria possessing nitrogenase enzyme) in soil and make it available for plant by converting atmospheric dinitrogen (N_2) into ammonia and nitrate (Wilson et al. 2006) (Fig. 25.5).

25.9 Role of Soil Microbes in Disease Control (Suppression).

Beneficial soil microbes influence to plants for absorb nutrients from soil and as a results increase disease suppression crop production (Fig. 25.6). Mainly soil rhizobacteria helps in this processes. *Bacillus subtilis* help in production of more than twenty antibiotics which are use in disease reduction. Antibiotic of *Bacillus* spp. use in crop species like tomato, chilli and brinjal etc. for against harmful pathogens

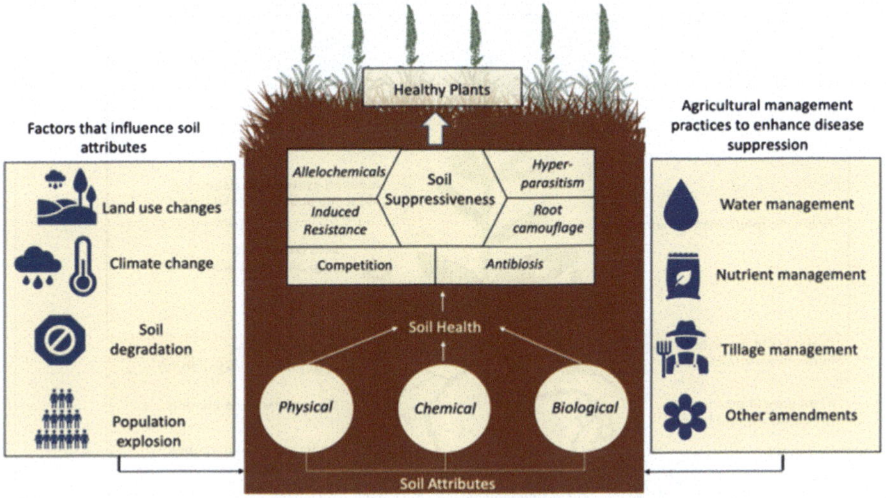

Fig. 25.6 Role of disease-suppressive soils (*Source* author/developed from a number of literature sources)

like *Colletotrichum acutatum*, *C. capsici* and *Pythium aphanidermatum* (Weller et al. 2002).

25.10 Beneficial Effects of Rhizobacteria

Sustainable development of agricultural land is very important and essential in today's world to fulfill our agricultural needs and also for future food security demand. Our traditional agriculture techniques not fulfilling these needs because of many different concerns. We have very urgent need to develop sustainable agricultural land and use effective mechanism for crops production. This type of technique use in special farming for corp production. Sustainable agriculture is beneficial and without affect future generation they use natural resources. Diversity of microbial population (bacteria and actinomycetes) colonizes around the root of plants and interact with plants and help in plants growth. When naturally occurring beneficial soil microbe applied in soil as inoculums to enhance plant growth and development. In addition these microbes in soil to improve soil quality and soil health also influence crop quality. Plants secretes exudates in soil from roots which helps to attract numerous microbes toward roots for crop production (Vidhyasekaran et al. 2001). Rhizobacteria reside around the plant roots because rhizobacteria respond to root exudates and show chemotactic movement toward roots for symbiotic interaction. Some beneficial bacteria presence in close proximity with plants and interact with plants through various methods (Fig. 25.7). They communicate with each other through molecular signaling.

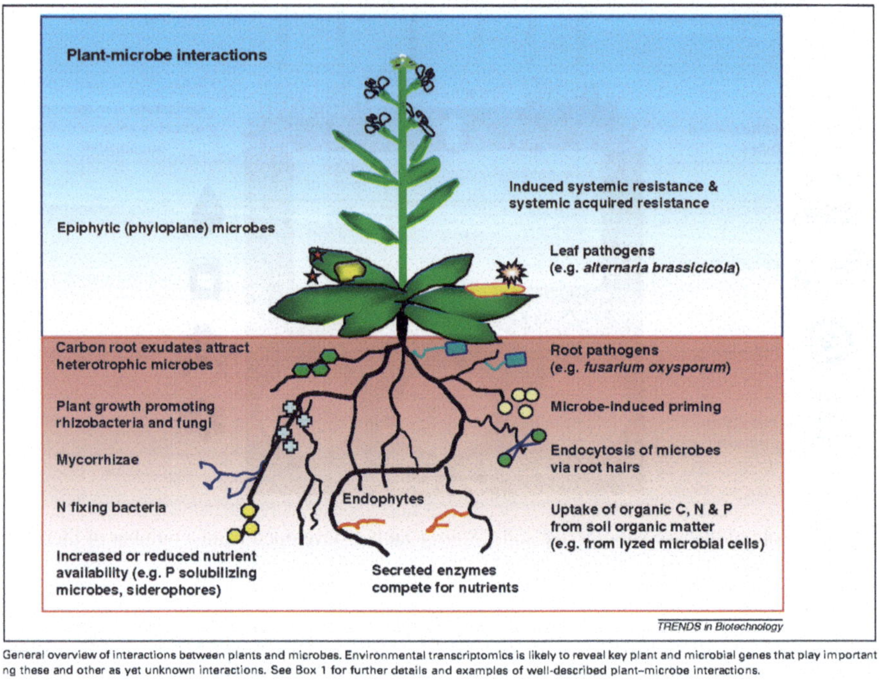

Fig. 25.7 Plant and microbes interaction in soil (*Source* https://doi.org/10.1016/j.tibtech.2011.11.002)

25.11 Future Outlook

Soil bacteria play very important role to control health of soil and also influence plants growth and development by controlling diseases. Many beneficial soil microbes have many side effects on sustainable agriculture development and helps to replacing inorganic fertilizer and pesticides from agricultural lands. Beside this understanding global climatic changes and agricultural practices change the terrestrial ecosystems. Soil health is maintain by the interaction between various soil component with the biotic and abiotic factors of environment. So research in each of these component and interaction between them and on combination of different ecosystem is necessary.

25.12 Conclusion

Soil microbes has very crucial role to maintain and improve soil health in sustainable development of agricultural land. Different beneficial soil bacteria are either present in soil or are exogenously added in soil as a inoculum. Soil microbes are used in soil is a advanced technology for increase yield of crops in field and for protected cultivation. It is not a permanent solution for disease free crop yield through beneficial soil microbes. Some soil microbes reduce the disease of plants. Different climatic conditions like temperature, pH of soil, humidity of air, biotic and abiotic factors influence soil microbial population. Different speciesof microbes are colonize in root area of plant so it is very difficult to say that which rhizobacteria is suitable and which has the most ability to coupled with the effective nutrient availability properties. Use of beneficial bacteria inoculums in soil can increase the quality of soil and health for soil-borne diseases suppressive. To ensure maintaining soil health for long-term and adaptation of microbes by recommended to farmers at commercial level. In this field more research is required. It is essential to increased knowledge in farmer about sustainable agriculture which is microbe-based symbiosis and ensure human and animal food production without effecting environment.

Acknowledgements The authors acknowledged the PG Department of Botany, Ramananda College, Bishnupur, Bankura for providing necessary infrastructure to carry out this survey work.

References

Antoun H, Kloepper JW (2001) Plant growth promoting rhizobacteria (PGPR). In: Brenner S, Miller JM (eds) Encyclopedia of genetics, vol 2, no 1. New York, Academic press, pp 1477–1480

Antoun H, Pre´Vost D (2005) Ecology of plant growth promoting rhizobacteria. In: Siddiqui ZA (ed) PGPR: biocontrol and biofertilization, vol 2, no 1. Springer Dordrecht, pp 1–38

Bashan Y, de-Bashan LE (2005) Bacteria: plant growth-promoting soil. In: Hillel D (ed) Encyclopedia of soils in the environment, vol 1, no 1. Elsevier Oxford UK, pp 103–115

Be´langer RR and Avis TJ (2002) Ecological processes and interactions occurring in leaf surface fungi, vol 1, no 1. In: Lindow SE, Hecht-Poinar EI, Elliot VJ (eds) Phyllosphere microbiology. APS Press St Paul MN, pp 193–207

Duiker SW, Rhoton FE, Torrent J, Smeck NE, Lal R (2003) Iron (hydr)oxide crystallinity effects on soil aggregation. Soil Sci Soc Am J 6(7):606–611

Haas D, Défago G (2005) Biological control of soil-borne pathogens by fluorescent pseudomonads. Nat Rev Microbiol 3(1):307–319

Hass D, Keel C (2003) Regulation of antibiotic production in root colonizing Pseudomonas sp. and relevance for biological control of plant disease. Annu Rev Phytopathol 41(1):117–153

Maksimov IV, Abizgil'dina RR and Pusenkova LI, (2011) Plant growth promoting rhizobacteria as alternative to chemical crop protectors from pathogens (Review). Appl Biochem Microbiol 47(1):333–345

Richardson A, Peter A, Hocking P, Simpson R, George T (2009) Plant mechanisms to optimise access to soil phosphorus. Crop Pasture Sci 60(1):124–143

Rodríguez H, Fraga R (1999) Phosphate solubilizing bacteria and their role in plant growth promotion. Biotechnol Adv 17(1):319–339

Roselló-Mora R, Amann R (2001) The species concept for prokaryotes. FEMS Microbiol Rev 25(1):39–67

Scher FM, Baker R (1980) Mechanism of biological control in a Fusarium-suppressive Soil. Phytopathol 70(5):412–417

Scher FM, Ziegle JS, Kloepper JW (1984) A method for assessing the root-colonizing capacity of bacteria on maize. Can J Microbiol 30(1):151–157

Schippers B, Bakker AW, Bakker PAHM (1987) Interactions of deleterious and beneficial rhizosphere microorganisms and the effect of cropping practices. Ann Rev Phytopathol 25(1):339–358

Schroth MN, Hancock JG (1982) Disease suppressive soil and root-colonizing bacteria. Science 21(6):1376–1381

Siddiqui S, Siddiqui ZA, Iqbal A (2005) Evaluation of fluorescent pseudomonads and Bacillus isolates for the biocontrol of wilt disease complex of pigeon pea. World J Microbiol Biotechnol 2(1):729–732

Singh A, Saha S, Garg R, Pandey AK, Rai AB (2015) Proteins in plant defence system: an overview. In: Roy AK (eds) Emerging technologies of the 21st century. World J Microbiol Biotechnol 1(1):85–102

Singh G, Saha S, Sharma BK, Garg R, Rai AB, Singh RP (2013) Evaluation of selected antagonists against Fusarium wilt disease of tomato. J Interacad 17(2):234–239

Six J, Elliott ET, Paustian K (2000) Soil structure and soil organic matter. II. A normalized stability index and the effect of mineralogy. Soil Sci Soc Am J 6(4):1042–1049

Stein T (2005) *Bacillus subtilis* antibiotics: structures, syntheses and specific functions. Mol Microbiol 56(4):845–857

Tilak KVBR, Ranganayaki N, Pal KK, De R, Saxena AK, Nautiyal CS, Mittal S, Tripathi AK, Johri BN (2005) Diversity of plant growth and soil health supporting bacteria. Curr Sci 8(9):136–150

Van Loon LC (2007) Plant responses to plant growth-promoting rhizobacteria. Eur J Plant Pathol 11(9):243–254

Van Wees SC, M, Van der Ent S and Pieterse CMJ, (2008) Plant immune responses triggered by beneficial microbes. Curr Opin Plant Biol 11(1):443–448

Vidhyasekaran P, Kamala N, Ramanathan A, Rajappan A, Paranidhran V, Velazhahan R (2001) Induction of systemic resistance by Pseudomonas fluorescens Pfl against *Xanthomonas oryzae* pv. oryzae in rice leaves. Phytoparasitica 29(1):155–166

Wilson MK, Abergel RJ, Raymond KN, Arceneaux JEL, Byers BR (2006) Siderophores of *Bacillus anthracis, Bacillus cereus*, and *Bacillus thuringiensis*. Biochem Biophys Res Commun 34(8):320–325

Weller DM, Raaijmakers JM, McSpadden Gardener BB, Thomashow LS (2002) Microbial population responsible for specific soil suppressiveness to plant pathogens. Ann Rev Phytopathol 40(1):309–348

Chapter 26
Rhizospheric Soil–Plant-Microbial Interactions for Abiotic Stress Mitigation and Enhancing Crop Performance

Priyanka Chandra, Arvind Kumar Rai, Parul Sundha, Nirmalendu Basak, and Harshpreet Kaur

Abstract The plant's root zone known as rhizosphere is the area that conjoins and links plants with soil microorganisms which dwell in their root vicinity. The rhizosphere is contemplated as one of the most intricate networks of cross talks between microbes and plants interacting with each other. Plants emanate exudates from roots which act as chemo-attractants for soil microorganisms and leads to the development of their rhizospheric microbiome. The various compounds released as root exudates from plants also alter the physico-chemical characteristics of the soil that also reform the structure of soil microbial structure of the rhizosphere, accordingly. Soil microorganisms inhabiting the rhizosphere are called Plant Growth Promoting Rhizobacteria (PGPR) which possesses the capabilities to enhance crop performance 10–30% as reported by several researchers through various mechanisms and also alleviate the repercussions of abiotic stresses on plants. Hence, this book chapter covers the functions of the rhizosphere which boost to harness plant–microbe interactions and PGPRs for abiotic stress mitigation and enhancing crop performance.

Keywords Abiotic stress · Plant growth promoting rhizobacteria · Rhizosphere · Root exudates

P. Chandra · A. K. Rai · P. Sundha · N. Basak (✉) · H. Kaur
ICAR-Central Soil Salinity Research Institute, Karnal, Haryana 132001, India
e-mail: nirmalendu.basak@icar.gov.in

P. Chandra
e-mail: priyanka.chandra@icar.gov.in

A. K. Rai
e-mail: ak.rai@icar.gov.in

P. Sundha
e-mail: parul.sundha@icar.gov.in

H. Kaur
e-mail: hrshpreet@yahoo.co.in

P. K. Shit et al. (eds.), *Soil Health and Environmental Sustainability*,
Environmental Science and Engineering,
https://doi.org/10.1007/978-3-031-09270-1_26

26.1 Introduction

Soil contains a complex composition of inorganic and organic compounds with varied micro flora and is a reservoir for microorganisms including bacteria, fungi etc. These soil microorganisms play essential roles in the ecosystem with dispense benefit to plants by several mechanisms (Ahkami et al. 2017). The microbes which are habitants of soil usually depend on carbon in the form of sugars, amino acids, and organic acids as rhizodeposits secreted by the plants as a valuable source of nutrition (Nautiyal et al. 2008). However, the rhizodeposition of nutrients not only attracts beneficial microbes but also pathogenic strains. Therefore, a "recognition mechanism" has been evolved by plants to select the compatible microorganisms and discriminate among those causing impairment around their rhizosphere (Nannipieri et al. 2020). This mechanism is felicitated by specific molecules secreted as manifestation of root exudates which helps in shaping the microbial community structure of rhizosphere and creating "rhizosphere microbiome" (Nannipieri et al. 2017; Yin et al. 2020). Secondary metabolites including flavonoids, strigolactones, and terpenoids which are being released as plant root exudates are the components of chemical communication strategies. Various researches on several rhizospheric microbiomes of various plants revealed strong variations in exudates and microbial community which inference that exudates are the form of root derived signals which attracts specific microbes and plays a key role in plant–microbe interactions (Mendes et al. 2013; Pascale et al. 2020; Dessauxet al. 2016).

The vicinity around the roots of plants is known as rhizosphere and microbes dwelling in this area are called Plant Growth Promoting Rhizobacteria (PGPRs) (Meena et al. 2018).They establish a mutualistic symbiosis relationship and can directly or indirectly assist the growth of associated plants through several mechanisms. PGPRs promotes plant growth by solubilizing or mobilizing the unavailable nutrients and facilitating them to the plants, by the process of fixing atmospheric nitrogen called nitrogen fixation, producing cytokinins, auxins and indole-3-acetic acid (IAA) which are plant hormones, and producing extracellular enzymes, 1-aminocyclopropane-1-carboxylate (ACC) deaminase, metabolite such as siderophore (Panda et al. 2016). IAA is a plant hormone known as 'natural auxin' apart of their functioning in cell division, stimulation of cell elongation and differentiation leading to plant growth. They also help in slackening the cell walls of roots which felicitate the increased amount of root exudates for more colonization of PGPR because of availability of surplus nutrients (Gupta et al. 2015; Beneduzi et al. 2012). The enzymes fabricated by PGPR, ACC deaminase aids plants in mitigating stress which helps in cessation of synthesis of ethylene by converting synthesis precursor ACC into α-butanone acid and ammonia; thereby reduce ethylene accumulation which causes stress in plants (Barnawal et al. 2014; Danish et al. 2020). Numbers of PGPR strains have been isolated from the rhizosphere of several crops including wheat (Kumar et al. 2014), maize (Goteti et al. 2013), rice (Aw et al. 2020), sorghum (Mounde et al. 2015), medicinal plants (Swain et al. 2007; Pandey et al. 2018) as well as horticultural (Esitken et al. 2010) and vegetable crops (Walia et al. 2014) and enhance plant growth as well as productivity through several ways. PGPRs enhance plant tolerance against biotic (Kumar et al. 2012; Choudhary and Johri 2009) and abiotic stresses including drought (Chandra et al. 2018; Delshadi et al. 2017), salinity (Tahir et al. 2019; Barnawal et al. 2014), heavy metal (Aw et al. 2020), and nutrient deficits (Panda et al. 2016). Bacterial strains including *Pseudomonas* (Goteti et al. 2014; Karnwal 2021), *Azospirillum* (Dobbeleare et al. 2002), *Azotobacter* (Ahmad et al. 2005), *Enterobacter* (Kumar et al. 2014), *Alcaligenes* (Kakar et al. 2018), *Arthrobacter* (Kumar et al. 2014), *Burkholderia* (Bano et al. 2015), *Bacillus* (Chandra et al. 2018; Delshadi et al. 2017), and *Serratia* (Singh and Jha 2016) are known to have plant growth promoting potential and also been described by several researches to mitigate biotic (Table 26.1) and abiotic stresses (Table 26.2).

26.2 The Rhizosphere

The word "rhizosphere" was coined by the German physiologist Hiltner for the description of plant-root interface. The rhizosphere is used to define the vicinity of plant roots which harbours a unique population of microorganisms whose proliferation has been impacted by the metabolites liberated by plant roots. The clarity of the rhizosphere has become more comprehensible with the course of time and it has been further divided on the basis of their relative distance from the root into three regions (endorhizosphere, rhizoplane and ectorhizosphere). The nearest or internal zone is endorhizosphere which consists of segments of cortex and endodermis including apoplastic space. This "free space" in between the cells is being occupied by beneficial microbes and cations. The medial zone is called rhizoplane which consists of root epidermis and mucilage around the root. The ectorhizosphere is the peripheral area that starts with rhizoplane and ends into the bulk soil.

The 10–40% of the total carbon which has been fixed through the process of photosynthesis carried out by plants is flushed out through roots into the soil (Newman 1985) and released compounds are called rhizodeposits. This photosynthetically fixed carbon contains both organic that is generally low molecular weight and inorganic such as HCO_3 and enacts a crucial role in the ecological processes of the rhizosphere. The liberation of rhizodeposits are influenced by several factors which include environmental conditions, plant type, nutrient availability or deficiency, physicochemical and biological properties of the soil (Barber 1995; Farrar et al. 2003). Root exudates are released from roots and diffused in soil which attracts beneficial microorganisms and serve as "bait" for root pathogens (Hayat et al. 2017). There are two types of root exudates which are bifurcated into high molecular weight organic compounds (HOC) that includes complex molecules which are not easily utilized by soil microbiota and other is low molecular weight organic compounds (LOC) which are effortlessly availed by soil microbiota (Badri and Vivanco 2009; Bertin et al. 2003). HOC improves soil quality through enhancing infiltration rate of water and aeration, also binds with soil particles and forming aggregates. Mucilage, an

Table 26.1 Plant growth promoting rhizobacteria demonstrating resistance against several biotic stresses with following mechanisms

	PGPRs	Biotic stress	Mechanism to control disease	Crops
1	*Stenotrophomonas maltophilia*	*Fusarium graminearum*	Plant growth enhancing potential, oxidative stress tolerance	Wheat
2	Azotobacter and Azospirillum Azotobacter and Azospirillum Azotobacter and Azospirillum *Azotobacter* and *Azospirillium*	*Rhizoctonia solani*	Antibiosis	Wheat
3	*P. aeruginosa, B. subtilis*	*Pythium*	Antifungal compounds included 2,4-DAPG, phenazine, pyochelin, rhamnolipids, pyoverdines, surfactins and AHLs	*Arugula*
4	*Bacillus siamensis* and *Bacillus tequilensis*	*Pestalotiopsis versicolor*	Lipopeptides such as surfactin, iturin, and mycosubtilin chitinase and protease activity	Bayberry
5	*Bacillus amyloliquefaciens*	*Fusarium graminearum*	Lipopeptide antibiotics of iturins	Wheat
6	*Pseudomonas aeruginosa* and *P. fluorescens*	*Meloidogyne javanica*	Salicylic acid (SA)-mediated induction of systemic resistance	Tomato
7	*B. subtilis*	*Fusarium*	Siderophore	Pepper
8	*Bacillus sp.*	*Plasmopara viticola*	Antifungal activity	Grapevine
9	*Bacillus, Brevibacillus, Pseudomonas*	*Ralstonia solanacearum*	Improved expression to defense-related enzymes	Tomato
10	*Pseudomonas*	*Fusarium oxysporum f. sp. Cepa*	Induced systemic resistance	Onion

(continued)

Table 26.1 (continued)

	PGPRs	Biotic stress	Mechanism to control disease	Crops
11	*Bacillus subtilis, B. amyloliquefaciens, Pseudomonas fluorescens, and P. aeruginosa*	*Clavibacter michiganensis subsp. michiganensis*	Siderophores, hydrogen cyanide and indole acetic acid	Tomato
12	*Bacillus subtilis*	*Bipolarissorokiniana*	Antioxidant defense enzymes	Wheat
13	*Streptomyces* spp	*Sclerotiumrolfsii*	Induction of defense regulatory genes	Chickpea
14	*Bacillus* spp,	*Pyriculariaoryzae*	Antioxidant defense enzymes	Rice
15	*Bacillus simplex, B. subtilis*	*Fusarium oxysporum*	Antifungal activity	Pea
16	*Bacillus subtilis, Bacillus amyloliquefaciens*	*Penicillium digitatum Sacc., Penicillium italicum Wehmer, Penicillium crustosum*	Volatile compounds	Citrus fruit

Table 26.2 Bacterial strains with plant growth promoting potential and their mechanism to mitigate abiotic stresses

S.No	Bacterial strains	Mechanism to mitigate abiotic stresses	Abiotic stresses
1	*Pseudomona syringae*	Inhibited influx of sodium due to formation of rhizosheath by exopolysaccharide	Salinity
2	*Pseudomonas aeruginosa*	Bioremoval, Biofilm, EPS production	Cadmium toxiciy
3	*Pseudomonas aurantiaca, Pseudomonas extremorientalis*	Auxin production	Salinity
4	*Pseudomonas chlororaphis*	IAA production	Salinity
5	*Pseudomonas syringae, Pseudomonas fluorescens*	ACC deaminase activity	Salinity
6	*Pseudomonas fluorescens, P. fluorescens, and P. putida*	ACC deaminase activity	Drought
7	*Pseudomonas aeruginosa*	Indole acetic acid, Up regulation of DREB2A, CAT1, DHN, Increased activity of superoxide dismutase, peroxidase, catalase	Drought
8	*Pseudomonas aeruginosa*	Oxidative stress tolerance	High zinc toxicity
9	*Pseudomonas putida*	ACC deaminase	Salt
10	*Pseudomonas pseudoalcaligenes*	Enhanced antioxidant activity	Salt
11	*Pseudomonas fluorescens*	Auxin and phosphate solubilization	Salt
12	*Pseudomonas sp.*	Biosorption of heavy metals	Heavy metal
13	*Pseudomonas putida*	Adsorption	Heavy metal
14	*Azospirillum brasilense*	Abscisic acid accumulation, phytohormones production	Drought
15	*Azosprillum brasilense*	Auxin production and phosphate solubilization	Salt
16	*Azospirillum brasilens*	Improved the antioxidant enzymes and photosynthetic pigments	Water
17	*Azospirillum lipoferum*	Phytohormones production	Salt
18	*A. lipoferum, A. brasilense*	Nitrogen fixation	Salt
19	*Azospirillum sp.*	Production of phytohormones and osmoprotectants	Salt
20	*Azotobacter chroococcum*	Abscisic acid accumulation, positive effect on growth promoting phytohormones	Drought

(continued)

Table 26.2 (continued)

S.No	Bacterial strains	Mechanism to mitigate abiotic stresses	Abiotic stresses
21	*Azotobacter chroococcum*	Increase in polyphenol as well as K^+/Na^+ ratio	Salt
22	*Azotobacter* sp.	IAA, phosphate solubilization and nitrogen fixation	Salt stress
23	*Enterobacter spp.*	ACC deaminase and IAA production	Drought
24	*Enterobacter hormaechei*	ACC deaminase and EPS-producing activity	Drought
25	*Enterobacter sps*	Biosorption of heavy metals	Heavy metal
26	*Bacillus insolitus*	Reduction in ROS activity in plants	Salinity
27	*Bacillus safensis*	Elevation of antioxidant responses	Drought
28	*Bacillus pumulis*	Increase level of proline accumulation and ROS-scavenging enzymes activity	Heavy-metal toxicity
29	*Bacillus subtilis*	Increase K^+/Na^+ ratio	Salt stress
30	*Bacillus megaterium*	Phosphorus solubilization, IAA production, antioxidant enzymes	Drought
31	*Bacillus amylolequifaciens,*	Inhibited influx of sodium due to formation of rhizosheath by exopolysaccharide	Salinity
32	*Bacilluscereus*	genes cAPX, rbcL, and rbcS	Drought
33	*Bacillus thuringiensis*	Biofilm production	Drought
34	*Bacillus spp*	ACC deaminase and IAA production	Drought
35	*Bacillus thuringiensis*	Biosorption	Heavy metal
36	*Bacillus amylolequifaciens*	Inhibited influx of sodium due to formation of rhizosheath by exopolysaccharide	Salinity
37	*Bacillus cereus*	Protease	Heavy metal
38	*Bacillus sp.*	Bioaccumulation and Biosorption	Chromium
39	*Serratia marcescens*	Induced systemic resistance, modulation of antioxidant enzymes, production of ACC deaminase activity, PGP properties	Salinity
40	*Serratia fonticola*	Phytohormones secretion and (ACC deaminase activity, and biofilm formation	Normal

(continued)

Table 26.2 (continued)

S.No	Bacterial strains	Mechanism to mitigate abiotic stresses	Abiotic stresses
41	*Serratia sp.*	genes cAPX, rbcL, and rbcS	Drought

example of HOC is gelatinous and viscous substances released from plant roots and bestow protection to roots from dryness and desiccation, also aids nutrient acquisition to plants. The LOC are more assorted and accomplishes a comprehensive array of promising and potential functions in soil. The LOC includes organic acids, sugar, amino acids, proteins, phenolics and secondary metabolites are being liberated by roots of plant as root exudates (Table 26.3). LOC felicitates the occupancy of the micro (Fe, Zn) and macro (N, P, K) nutrients in plants. LOC functions as an attractant or chemical signal through chemotaxis which fascinate microbes to form symbiotic association/colonization with/on plant roots. These LOC also carried out the process of allelopathy and also caused detrimental effects to the soil pathogens, pests and nematodes (Rohrbacher and St-Arnaud 2016).

26.2.1 Rhizospheric Microbiome Dynamics

The characterization of the rhizosphere microbiome is found to be very specific with plant types which are mainly because of the variation in the root exudates liberated by plants. Root exudates vary among the different genotypes of plants leading to variation in their microbiome. It has been reported by several researches that the roots influence the microbial communities around the rhizospheric zone leading to specific rhizosphere microbiome. Generally greater microbial diversity is being found in rhizospheric zone comparison to the bulk soil without roots. Root exudates consists of HOC (organic compounds) and LOC (organic acids, sugar, amino acids, proteins, phenolics and secondary metabolites) which act as signalling molecules or sometimes inhibitory/antimicrobial agents are accountable for extremely dynamic rhizospheric microbiome (Rohrbacher and St-Arnaud 2016). Different types of PGPRs are being attracted by root exudates which act as chemo-attractant secreted by different plants, indulging them to colonize their roots. Secondary metabolites such as coumarins are the other types of root exudates which are secreted in the rhizosphere from plant roots and influences microbial population, and its repercussions effects shape the rhizosphere microbiome (Igiehon and Babalola 2018). Several crops such as maize and legumes, secretes benzoxazinoids and canavanine in the form of root exudates and exhibit effects on the rhizospheric microbiome (Hayat et al. 2017). It has also been reported that in *Brachypodium* and barley root system architecture defining shape and structure including root type, root length and root hairs has remarkable influence on the conformation of rhizosphere microbial communities. Distribution of microbial biomass around the rhizospheric zone also varies with the differential

Table 26.3 Types of root exudates secreted by plants and their functions

S. No	Compounds	Root exudates	Functions
1	Amino compounds	Asparagine, α- alanine, glutamine, aspartic acid, leucine/isoleucine, methionine, phenylalanine, serine, glycine, cystine/cysteine, tyrosine, threonine, lysine, proline, tryptophan, β-alanine, arginine, homoserine, cystathionine	Significant role in complexing metal ions
2	Fatty acids and sterols	campesterol, stigmasterol, sitosterol linolenic acids, cholesterol, Palmitic, stearic, oleic, linoleic,	Act as signaling molecules
3	Amino acids/compounds	Biotin, thiamine, niacin, pantothenate, choline, inositol, pyridoxine, N-methyl nicotinic acid	Enhance the mobility of plant micronutrients in soils
4	Organic acids	fumaric, glycolic, valeric, malonic propanoic, butyric, succinic, Tartaric, oxalic, citric, malic,	Solubilization of mineral nutrients
5	Carbohydrates/sugars	arabinose, raffinose, galactose, rhamnose, ribose, xylose, Glucose, fructose, sucrose, maltose, oligosaccharide	Supports the growth of soil microorganisms
6	Plant hormones	Auxins	Supports the growth of plant and microorganisms
7	Phenolic compounds	Glycosides, saponin, Caffeic acid, cinnamic acid, coumarin, ferulic acid, salycilic acid, syringic acid, vanillic acid	Act as signaling molecules
8	Flavonoids	Flavonine, flavonols, flavones, flavanols, flavanones, isoflavones, and anthocyanidins	Components of chemical communication strategies
9	Enzymes	Phosphatase, invertase, amylase, protease, polygalacturonase	Enhances nutrient availability

pattern of root exudates secretion (Pascale et al. 2020).The process of chemotaxis carried out by signaling molecules secreted as root exudates attracts microorganisms to the vicinity of surface of the roots which gets impacted by the rate of root elongation (Doornbos et al. 2012). The dynamics of microorganism's adhesion to the root surface and their movement longitudinally along with elongated roots also gets restricted. The bacterial population forms biofilm around the roots which is generally higher in elongation zone and decreases by moving towards matured root zone. Chemotactic movements and dissemination of rhizospheric bacteria governs the composition and shift in the rhizospheric microbiome (Mendes et al. 2013).

Different stages of plant life cycle also affect the root exudation processes and these temporal changes carve the dynamics of the rhizospheric microbiome (Dessaux et al. 2016). Several studies have demonstrated idiosyncratic rhizospheric microbiome at different (early and late) stages of development in different crops including Arabidopsis, rice, barley, strawberry and oats (Berendsen et al. 2012). Variation in conformation of root exudates at different stages sculpts root micro biota consequently (Igiehon and Babalola 2018). Oats also secretes root exudates differentially with respect to their developmental stages, at seedling stage high concentration sucrose in root exudates facilitates the formation of symbiotic relationship with rhizospheric microorganisms while secretion of aromatic compounds and amino acids during the vegetative phase of plant life cycle improves its defence systems (Lu et al. 2020), whereas in Arabidopsis cumulative secretion of amino acids throughout the developmental stages enhances the bacterial root colonization (Korenblum et al. 2020). The defence systems and nutrient acquisition process are correlated with plant different developmental stages along with changes in rhizospheric microbiome. Root associated microorganisms induces systemic resistance in response to root exudation processes. Rhizospheric microorganisms residing around as well as inside the plant roots persuades cavernous variations in the metabolomes, and transcriptomes of roots and shoots of host plants (Farrar et al. 2003).

Secondary metabolites released as root exudates also cause reforms in rhizospheric microbiome. Malic acid as chemo-attractant has been characterized as "signal for help" of the disease infected plants (Doornbos et al. 2012). Root exudates itself owns the defensive capability to control the soil borne pathogens. A caffeic acid derivative called rosmarinic acid released by *Ocimum basilicum* as root exudates has the antimicrobial capabilities to inhibit multiple soil borne pathogens (Bais et al. 2002). In strawberry, rhizosphere microbiome provides resistance against soil-borne pathogens and also mediates nutrient uptake (Lazcano et al. 2021). Plants also have the potential to harbour specific types of microbes as in the wheat rhizosphere proliferation of *Pseudomonas* having ability to produce an antibiotic, 2,4-diacetylphloroglucinol was observed when infected with Take-all fungus disease caused by *Gaeumannomyces graminis* var. *tritici* (Paulin et al. 2017), similarly in the *Arabidopsis* infected with *Pseudomonas syringae* flourishes *Bacillus subtilis* having biocontrol potential in their rhizosphere. Other secondary metabolites such as salicylic acids, jasmonic acids, and chitosans released as root exudates also induces plant defence system and controls the growth of pathogens which includes both fungal and bacterial (Hassan and Mathesius 2012; Mendes et al. 2013). The bacterial quorum sensing process which is mandatory for symbiotic and parasitic interactions with them are also been modulated by the secondary metabolites including rosemarinic acid and naphthoquinones (Trivedi et al. 2020).

26.2.2 Soil Characteristics of Rhizosphere

Apart from biological properties and microbial community, the root exudation process cause changes in soil physical and chemical characteristics also. The root exudates consists of several types of polysaccharides, organic compounds and secondary metabolites alters the soil physico-chemical characteristics in the rhizospheric zone in comparison to the bulk soil (Canarini et al. 2019). Soil properties such as pH and moisture significantly change along with the variation in the process of root exudation (Neina et al. 2019).The most sensitive soil characteristic is the moisture content which mainly gets affected with the root exudates. The mucilage which is a polysaccharide based hexose and pentose sugars and uronic acids mixture which are released by root cells in the form of exudates. It is a "gel like" structure that forms a 'rhizosheath' having water holding capacity and improves moisture content and nutrient movement in the rhizosphere (Nazari et al. 2020). In the rhizospheric zone, moisture content lowers after the absorption of nutrients by plants. Since mucilage binds the soil and root together and the "gel like" structure avoids shrinking of soil and maintains its hydraulic conductivity. Mucilage also promotes soil stability and binding makes soil resistant to erosion. Mucilage aids lubrication of roots mediating availability and absorption of ions, including Fe^{2+}, Fe^{3+}, PO_4^{3-}, and Ca^{2+} in the rhizosphere. Organic acids in the form of root exudates secreted by roots of maize and wheat are being induced by osmotic stress. Furthermore, root exudates significantly affect the soil pH due to release of proton and resulting in acidification and alkalinisation because of which bound nutrients become available in the rhizosphere (Adeleke et al. 2017).

The change in the chemical characteristics of rhizospheric soil takes place analogous with the increase in nutrient availability due to changes in microbial population and biological environment with root exudation process (Chandra et al. 2020; Rai et al. 2020; Phillips et al. 2004). The acidification, changes in the redox potential and chelation of nutrients in the rhizosphere occurs due to the liberation of several chemicals in the form of root exudates. Chemical conditions also change due to dissolution of unavailable/insoluble minerals, desorption of nutrients from clay minerals or organic matter because of root exudates (Pii et al. 2015). Nitrogen availability is generally low in soils due to fixation of ammonium (NH_4^+) in clays and soil organic matter, leaching of soluble nitrate (NO_3^-), and denitrification executed by microbes but rhizospheric environment is different due to root exudation and presence of microorganisms. The presence of ammonium ion in rhizosphere leads to the release of root exudate proton (H^+) which combines with each NH_4^+ for absorption resulting in the reduction in the rhizospheric pH. Correspondingly, the availability of NO_3^- in the rhizosphere leads to the liberation of bicarbonate (HCO_3^-) from root exudates results in escalating rhizospheric pH. The process of changes in pH also enhances the availability of plant nutrients including Zn, Ca, and Mg. Diazotrophic microorganisms possessing dinitrogenase enzyme also changes the soil properties by fixing atmospheric N_2 into ammonia then to nitrite and nitrate in the rhizospheric zone. Root exudates released by plants consist of organic acids and its presence in the rhizosphere helps in overcoming the P deficiency (Ström et al. 2002). Such organic acids which includes citric acid, malic acid, lactic acid etc. effectively reduce pH of the rhizosphere and solubilize P bound in soil minerals and make it available to plants. Piscidic acid, a form of root exudate, chelates iron of $FePO_4$ and makes phosphate available in the rhizosphere (Pantigoso et al. 2020). Enzymes such as acid phosphatase liberated by plant roots also release phosphates from soil organic matter and increase its availability in the rhizosphere (Rai et al. 2021). In Iron deficiency, root exudates such as mugienic acid which is in the form of phytosiderophores are released. These phytosiderophores are complex of amino acids and are non-proteinogenic in nature and possess high affinity for iron. Root exudates also releases protons into rhizosphere hence shooting the reduction power of the rhizodermal cells which makes Fe available in the rhizosphere for plants (Sasse et al. 2018).

26.2.3 Interactions between Plant and Microbes in the Rhizosphere

The interaction among plants and microorganisms in the rhizospheric roots zone is one of the most important ecological functions which occur due to colonization of microbes around/inside the roots of plants (Philippot et al. 2013) (Fig. 26.1). The interaction between plants and microorganisms can be neutral, competition, antagonistic, parasitic and mutualistic (Vishwakarma et al. 2020) (Table 26.4). The relationship between plants and microorganisms is very dynamic as it oscillates between parasitic to symbiotic which are influenced by external milieu (Mhlongo et al. 2018). Several beneficial soil microorganisms dwelling in the plant's rhizospheric zone owns plant growth promotion ability and is known as plant growth-promoting rhizobacteria (PGPR). The beneficial association between plant and soil microorganisms could be symbiotic as well as non-symbiotic (Bhattacharyya and Jha 2012).

The symbiotic interaction involves diazotrophs having the potential to fix atmospheric nitrogen to ammonia and supply it to plants (Terpolilli et al. 2012). In this mechanism, diazotrophic Rhizobium bacteria develops a symbiotic association by entering in the roots of leguminous plants and induces the root nodules formation which is a specialized organ in which the rhizobia resides and converts nitrogen to ammonia through nitrogenase enzymatic system (Franssen et al. 1992). The non-symbiotic interaction includes diazotroph *Azospirillum* which can fix nitrogen in the rhizospheric zone and provides it to the plants without entering into plant tissues (Bashan and Holguin 1997). *Azospirillum* not only promotes growth by fixing nitrogen but also secretes several plant hormones which include auxins, cytokinins

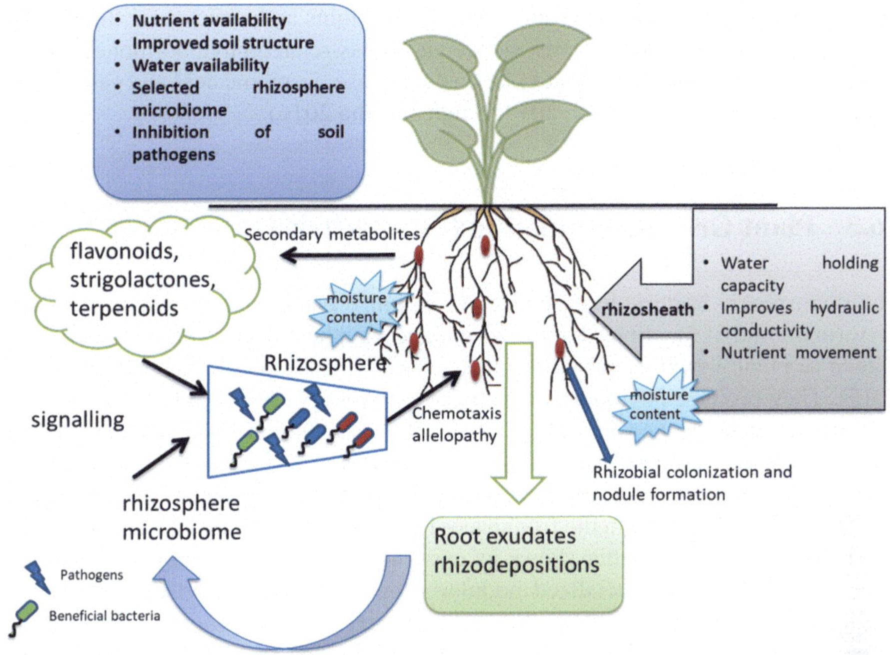

Fig. 26.1 Plant–microbe interactions in the rhizosphere

Table 26.4 Types of plant–microbe interactions in the rhizosphere

S.no	Types of interactions		
1	Neutral	–	Both plant and microbes living together but neither plant nor microbes affecting each other
2	Competition	Negative	Both are adversely affecting to each other for survival, usage of nutrients and resources
3	Antagonistic	Negative	When one is producing such substances which are inhibitory to other then this interaction is Antagonism or Ammensalism
4	Parasitic	Negative	In this interaction one is benefited as it derive nutrition from host (living in their tissues) and causing harm to them
5	Mutualistic	Positive	Both plant and microbe provides benefit to each other. It is an obligatory relationship in which both are metabolically dependent on each other

and gibberellins (Bashan et al. 2004). Auxins are being secreted in highest concentration which stimulates root expansion and growth of the whole plant. PGPR aids plant growth through several mechanisms direct and indirect means; directly they facilitates the acquisition of nutrients (nitrogen, phosphorus, and micronutrients) via biological nitrogen fixation, phosphate solubilization, and siderophore production for iron sequestration (Bhattacharyya and Jha 2012). They also release plant hormones such as auxins, gibberellins, cytokinins and nitric oxide. Indirectly, they enhance induced systemic resistance and stress-related phytohormones like 1-aminocyclopropane-1-carboxylate (ACC) deaminase, jasmonic acid and cadaverine (Ahmad et al. 2005; Tahir et al. 2019; Singh and Jha 2016).

26.3 Plant Growth-Promoting Rhizobacteria

The rhizospheric bacteria has tendency to colonize plant roots or root surface and perform myriad of functions that in turn enhance soil health and crop growth, hence, known as plant growth-promoting rhizobacteria (Chandra et al. 2021; Gupta et al. 2015). They produce phytohormones, make essential plant nutrients available, and inhibit pathogens in the rhizospheric zone. Rhizodeposition is mainly responsible for the survival for these bacteria in the rhizosphere (Ahkami et al. 2017). The PGPRs exists in two forms; extracellular (ePGPR) and intracellular (iPGPR) while former group is dominant in the free spaces between the root cortex cells, rhizoplane and rhizosphere, while, latter shows its existence in the internal parts of root cells which generally form specialized nodular structures (Dakora and Phillips 2002). The genera belongs to ePGPR includes *Agrobacterium, Azotobacte*r, *Azospirillum, Bacillus, Pseudomonas, Burkholderia, Erwinia, Caulobacter, Serratia, Arthrobacter, Micrococcus, Flavobacterium, Chromobacterum*, and *Hyphomycrobium.* Bacteria forming symbiotic relationship with plants fall in the category of iPGPR which includes mainly family *Rhizobiaceae* (*Allorhizobium, Azorhizobium, Bradyrhizobium, Mesorhizobium* and *Rhizobium*) (Bhattacharyya and Jha 2012).These PGPRs improve nutrient acquisition and its assimilation in plants, release secondary metabolites, antibiotics, and signalling molecules simultaneously improve rhizospheric physico-chemical and biological properties (Chandra et al. 2021). These factors lead to enhancement of plant growth and PGPRs not only improve soil and crop productivity, but also succours plants to sustain in biotic and abiotic stresses (Bhattacharyya and Jha 2012) (Fig. 26.2).

26.3.1 Nitrogen Fixing Bacteria

Process of biological nitrogen fixation was discovered by Beijerinck which has been carried out by a particular specialized group of bacteria which are known as nitrogen-fixing bacteria. These have the capabilities of converting atmospheric nitrogen into ammonia which can be assimilated by plants. Two kinds of nitrogen-fixing bacteria have been divided into two categories: first are non-symbiotic, free-living bacteria: which are autotrophic and heterotrophic and do not have any direct association. Free-living N_2 fixers reside in the rhizosphere and rely on organic matter decomposition residues to survive. Another one is mutualistic or symbiotic bacteria which form a collegial relationship with legumes and provide benefits to them (Rai et al. 2020).

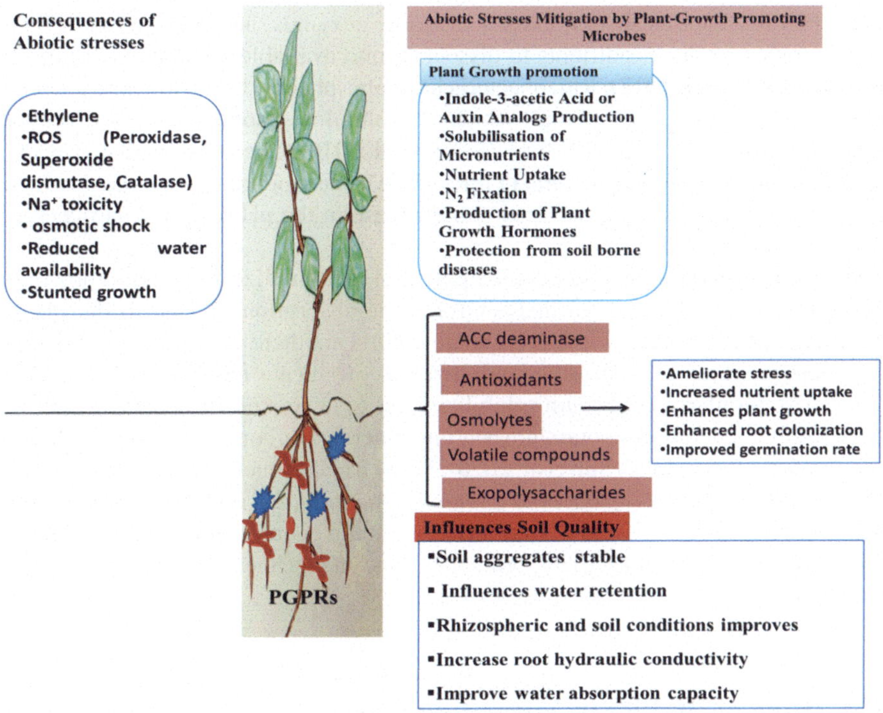

Fig. 26.2 Mechanisms of PGPRs for abiotic stress mitigation

In this process, bacteria infect the root hair cell and enter into the host plants and stimulate the formation of root nodules in which these bacteria reside. Because of this symbiotic relationship, both legume and bacteria bring physical changes in them (Terpolilli et al. 2012). Legume forms a nodule specialized organ which is the home for rhizobia while free-living rod-shaped rhizobia converts itself to bacteroids. This intimate association converts free nitrogen to ammonia under anaerobic environment by nitrogenase enzymes by bacteria within the nodules and ammonia formed is availed by the host for its growth and development while plant provides carbon and nutrition to rhizobia (Franssen et al. 1992).

26.3.2 Nutrient Solubilising Bacteria

Microorganisms that solubilise macro and micro nutrients into accessible form reside in their root zone and interact with the root system in the rhizosphere (Dotaniya and Meena 2015). Therefore, soil microorganisms enhance plant nutrient acquisition by improving its availability in their rhizospheric zone and also felicitate the decomposition of soil organic matter and encourage plant-growth promotion (Ramzan et al.

2015). PGPRs releases several classes of acids, enzymes, and metabolites, which make complex forms of nutrients to dissociate into available form and assimilated by plants (Alori et al. 2017). Rhizobacteria have phosphate solubilizing potential due to the secretion of such compounds which can solubilize insoluble complex form of inorganic phosphate to available form (Panda et al. 2016). Chelation-mediated mechanism is being carried out by several PGPRs while some produces organic acids which have low molecular weight also lead to a decrease in the pH of soil (Sharma et al. 2013; Goteti et al. 2014).

Many rhizobacteria also possess the potential to solubilize zinc and improve its availability in the rhizosphere and reinforce the absorption of zinc to the plants (Ramzan et al. 2015; Karnwal 2021). Some reports are there of silicate-solubilizing rhizobacteria which mediates the solubilization of silicate minerals (Chandrakala et al. 2019). Similarly, potassium solubilizing bacteria are known in the rhizosphere which solubilizes complex potassium bearing minerals and convert it to soluble forms for improved plant uptake (Sun et al. 2020). Many bacteria such as *Bacillus mucilaginosus, Bacillus edaphicus, Bacillus circulans, Acidothiobacillus ferrooxidans*, and *Paenibacillus* are being found to solubilize potassium containing minerals including biotite, feldspar, illite, muscovite, orthoclase, and mica (Sheng and He 2006; Saha et al. 2016).

26.4 Role of PGPRs in Mitigating Stress

26.4.1 Salt Stress

Several microorganisms found in plants growing in high-salt environments have adapted to salinity stress and flourish in these settings (Chandra et al. 2020; Rai et al., 2020). Even at high salt concentrations, halophilic microorganisms maintain protein structure and enzyme activity for numerous metabolic pathways, which is a considerable adaptation. The PGPRs has a pivotal role in preventing salt toxicity in the rhizosphere as they assist in perpetuating osmotic balance, ion homeostasis, and also maintains the turgor pressure of plant cells (Dakora and Phillips 2002). Under salt stressed situations, PGPR augment growth of plants by enhancing the activity of ACC deaminase, producing exopolysaccharides (EPS), biofilms synthesis, releasing osmoprotectants and plant growth regulators, phytohormones, modulating antioxidant enzymes, improving nutrient availability (Singh and Jha 2016).

The bacterial ACC deaminase enzyme lowers the ethylene synthesis, while phytohormones including IAA, zeatin and gibberellins concentration escalate in plant cells under salinity stress in influence of PGPR present in the rhizosphere. PGPRs secrete EPS in the rhizosphere and form biofilm around the roots which forms a bond with excess Ca^{2+}, K^{+}, and Na^{+} present in soil and prevent their uptake to plants. PGPRs present in the rhizosphere modulates the antioxidant enzymes (AO) includes superoxide dismutase (SOD), peroxidase (POD), catalase (CAT), nitrate reductase (NR) and glutathione reductase (GR) under salt stressed condition and neutralizes the pernicious effect of reactive oxygen species (ROS) in plants (Nautiyal et al. 2013).

PGPRs aids osmolytes accumulation in plants in feedback of stresses. These osmolytes includes proline, polyamines, glycine, betaine, quaternary ammonium compounds, and amino acids. These osmolytes helps in maintaining the specific membrane and cell wall structure, and also modifies the intracellular environment by draining toxic ions out of the cell through salt efflux (Tahir et al. 2019). Osmolytes protect the structure of enzymes and proteins so that their proper functioning can be sustained. Osmolytes also binds around the cell's nucleic material and actively destabilize the DNA double helix structure, additionally reducing its temperature of melting (Barnawal et al. 2014).

26.4.2 Drought Stress

Drought affects photosynthesis and cell proliferation, which are two most important activities. Changes in the pool of sugars used for signalling cellular processes or as substrates for biopolymers like cellulose, starch, and proteins are other effects of water scarcity on carbon metabolism. ROS are formed as a result of water scarcity and drought; misdirect the electrons occuring during light reaction of photosynthetic process which causes the creation of ROS (Chandra et al. 2021).

Stressed plants inoculated with nitrogen-fixing bacteria *Azospirillum* and *Azotobacter* showed a considerable increase in AO which includes CAT, SOD and POD (Zakikhani et al. 2012). Another way by which PGPR alleviates drought is by the formation of cytokinins, antioxidants, and the breakdown of the ethylene precursor 1-aminocyclopropane-1-carboxylate by ACC deaminase (Leontidou et al. 2020). PGPR also accumulates abscisic acid (ABA) in leaves, improves cytokinins formation which causes closure of stomata. To survive drought, several microorganisms synthesize exopolysaccharides, stimulate resistance genes, enhance water circulation in plants, and aid in proline production (Tiwari et al. 2016). Moreover, biofilm production under drought conditions also increases the survival rate of plants as it prevents desiccation. Inoculation of beneficial microbes, *Rhizobium* and *Pseudomonas* in *Zea mays* and *Vigna radiate* resulted in greater proline synthesis, which conferred salt tolerance due to the selective uptake of potassium ions and maintained relative water content (Ahmad et al. 2013; Naseem and Bano 2014).

26.4.3 Heavy Metals (HM) Tolerance

Numerous anthropogenic activities out-turned the rise of HM concentrations in soil which adversely affected plant growth and their survival. Heavy metals cannot be eliminated physiologically; instead, can switch to their one oxidation state from another. PGPRs have evolved their physiological machinery to resist HM stress and survive with it (Khanna et al. 2019). There are several mechanisms including sequestration of HM in bacterial cytoplasmic membrane, the polysaccharide layer of PGPRs called capsule entraps the HM in them, biosorption and bioaccumulation of HM in their cell walls, neutralization through microbial methylation process, bio precipitation of HM by the secretion of neutralizing compounds, oxidation–reduction process of HM for their detoxification and transformation of HM (Tak et al. 2013; Nie et al. 2002). Many common PGPRs including *Bacillus*, *Streptomyces*, *Pseudomonas*, and *Methylobacterium* are known to be HM-tolerant microbes and also possess the potential to enhance plant growth and productivity in HM concentric areas (Tiwari and Lata 2018).

26.5 Conclusions

Root exudates which are secreted by plants in the rhizosphere helps in determining microbial community. These nutrient rich root exudates plays an important role in improvising plant and soil health as well as also enhances tolerance in plants against biotic and abiotic s stresses by intensifying water and nutrient availability and soil's buffering capacity. The selective accumulation of abiotic and biotic stresses tolerant and beneficial microbial populations in the rhizosphere improve plant defence and provide tolerance to descendants through legacy effects and soil–plant-microbe interaction. The advantages of the plant–microbe interactions are upcoming as a new system for natural and sustainable agriculture. The literature also recommends that the application of PGPRs can lead to the sustainable and efficient crop production in environmental constraints too. However, it is required extensive studies to understand the mechanism of plant–microbe interactions before application of effective PGPRs in farming practices.

References

Adeleke R, Nwangburuka C, Oboirien B (2017) Origins, roles and fate of organic acids in soils: a review. S Afr J Bot 108:393–406

Ahkami AH, White RA, Handakumbura PP, Jansson C (2017) Rhizosphere engineering: enhancing sustainable plant ecosystem productivity. Rhizosphere 3(2):233–243

Ahmad F, Ahmad I, Khan MS (2005) Indole acetic acid production by the indigenous isolates of *Azotobacter* and fluorescent *Pseudomonas* in the presence and absence of tryptophan. Turk J Biol 29:29–34

Ahmad M, Zahir ZA, Khalid M, Nazli F, Arshad M (2013) Efficacy of Rhizobium and Pseudomonas strains to improve physiology, ionic balance and quality of mung bean under salt-affected conditions on farmer's fields. Plant Physiol Biochem 63:170–176

Alori ET, Glick BR, Babalola OO (2017) Microbial phosphorus solubilization and its potential for use in sustainable agriculture. Front Microbiol 8:971

Badri DV, Vivanco JM (2009) Regulation and function of root exudates. Plant Cell Environ 32:666–681

Bais H, Walker T, Schweizer H, Vivanco J (2002) Root specific elicitation and antimicrobial activity of rosmarinic acid in hairy root cultures of *Ocimum basilicum*. Plant Physiol Biochem 40:983–995

Bano DA, Singh RK, Waza SA, Singh NP (2015) Effect of cowpea *Bradyrhizobium* (RA-5) and *Burkholderia cepacia* (RRE-5) on growth parameters of pigeonpea under salt stress conditions. J Pure Appl Microbio 9:2539–2546

Barber SA (1995) Soil nutrient bioavailability: a mechanistic approach. Wiley, New York, USA

Barnawal D, Bharti N, Maji D, Chanotiya CS, Kalra A (2014) ACC deaminase-containing *Arthrobacter protophormiae* induces NaCl stress tolerance through reduced ACC oxidase activity and ethylene production resulting in improved nodulation and mycorrhization in *Pisum sativum*. J Plant Physiol 171:884–894

Bashan Y, Holguin G (1997) *Azospirillum* – plant relationships: environmental and physiological advances (1990–1996). Can J Microbial 43:103–121

Bashan Y, Holguin G, de-Bashan LE (2004) *Azospirillum*-plant relationships: physiological, molecular, agricultural, and environmental advances (1997–2003). Can J Microbiol 50:521–577

Beneduzi A, Ambrosini A, Passaglia LMP (2012) Plant growth-promoting rhizobacteria (PGPR): their potential as antagonists and biocontrol agents. Genetics Mol Biol 35:1044–1051

Berendsen RL, Pieterse CMJ, Bakker PAHM (2012) The rhizosphere microbiome and plant health. Trends Plant Sci 17:478–486

Bertin C, Yang X, Weston L (2003) The role of root exudates and allelochemicals in the rhizosphere. Plant Soil 256:67–83

Bhattacharyya PN, Jha DK (2012) Plant growth-promoting rhizobacteria (PGPR): emergence in agriculture. World J Microbiol Biotechnol 28:1327–1350

Canarini A, Kaiser C, Merchant A, Richter A, Wanek W (2019) Root exudation of primary metabolites: mechanisms and their roles in plant responses to environmental stimuli. Front Plant Sci 10:157

Chandra P, Tripathi P, Chandra A (2018) Isolation and molecular characterization of plant growth-promoting *Bacillus* spp. and their impact on sugarcane (*Saccharum* spp. hybrids) growth and tolerance towards drought stress. Acta Physiol Plant 40(11):199

Chandra P, Dhuli P, Verma P, Singh A, Choudhary M, Prajapat K, Rai A K, Yadav RK (2020) Culturable microbial diversity in the rhizosphere of different biotypes under variable salinity. Trop Ecol 61:291–300. https://doi.org/10.1007/s42965-020-00089-3

Chandra P, Wunnava A, Verma P, Chandra A, Sharma R K (2021) Strategies to mitigate the adverse effect of drought stress on crop plants—influences of soil bacteria: a review *Pedosphere* 31(3):496–509

Chandra P, Khobra R, Sundha P, Sharma RK, Jasrotia P, Chandra A, Singh DP, Singh GP (2021) Plant growth promoting Bacillus-based bio formulations improve wheat rhizosphere biological activity, nutrient uptake and growth of the plant. Acta Physiol Plant 43:139

Chandrakala C, Voleti SR, Bandeppa S et al (2019) Silicate solubilization and plant growth promoting potential of Rhizobium Sp. isolated from rice rhizosphere. Silicon 11:2895–2906

Choudhary DK, Johri BN (2009) Interactions of Bacillus spp. and plants-with special reference to induced systemic resistance (ISR). Microbiol Res 164:493–513

Dakora FD, Phillips DA (2002) Root exudates as mediators of mineral acquisition in low-nutrient environments. Plant Soil 245:35–47

Danish S, Zafar-ul-Hye M, Mohsin F, Hussain M (2020) ACC-deaminase producing plant growth promoting rhizobacteria and biochar mitigate adverse effects of drought stress on maize growth. PLoS One 15(4):e0230615

Delshadi S, Ebrahimi M, Shirmohammadi E (2017) Influence of plant-growth-promoting bacteria on germination, growth and nutrients' uptake of *Onobrychis sativa* L. under drought stress. J Plant Interact 12(1):200–208

Dessaux Y, clémentC G, Faure D. 2016. Engineering the rhizosphere. Trend Plant Sci 21(3):266–278

Dobbeleare S, Croonenborghs A, Thys A, Ptacek D, Okon Y, Vanderleyden J (2002) Effects of inoculation with wild type *Azospirillum brasilense* and *A. irakense* strains on development and nitrogen uptake of spring wheat and grain maize. Biol Fertil Soils 36:284–297

Doornbos RF, van Loon LC, Bakker PAHM (2012) Impact of root exudates and plant defense signaling on bacterial communities in the rhizosphere a review. Agron Sustain Dev 32:227–243

Dotaniya ML, Meena VD (2015) Rhizosphere effect on nutrient availability in soil and its uptake by plants: a review. Proc Natl Acad Sci India Sect B Biol Sci 85:1–12

Esitken A, Yildiz HE, Ercisli S, Figen Donmez M, Turan M, Gunes A (2010) Effects of plant growth promoting bacteria (PGPB) on yield, growth and nutrient contents of organically grown strawberry. Sci Hortic 124:62–66

Farrar J, Hawes M, Jones D, Lindow S (2003) How roots control the flux of carbon to the rhizosphere. Ecology 84:827–833

Franssen HJ, Vijn I, Yang WC, Bisseling T (1992) Developmental aspects of the Rhizobium-legume symbiosis. Plant Mol Biol 19:89–107

Goteti PK, Desai S, Emmanuel LDA, Taduri M, Sultana U (2014) Phosphate solubilization potential of fluorescent *Pseudomonas* spp. isolated from diverse agro-ecosystems of India. Int J Soil Sci 9:101–110

Goteti PK, Emmanuel LD, Desai S, Shaik MH (2013) Prospective zinc solubilising bacteria for enhanced nutrient uptake and growth promotion in Maize (*Zea mays* L.). Int J Microbiol 86:96–97

Gupta G, Parihar SS, Ahirwar NK, Snehi SK, Singh V (2015) Plant Growth Promoting Rhizobacteria (PGPR): current and future prospects for development of sustainable agriculture. J Microb Biochem Technol 7:096–102

Hassan S, Mathesius U (2012) The role of flavonoids in root rhizosphere signalling: opportunities and challenges for improving plant microbe interactions. J Exp Bot 63:3429–3444

Hayat S, Faraz A, Faizan M (2017) Root exudates: composition and impact on plant–microbe interaction. In: Ahmas I, Husai FM (eds) Biofilms in plant and soil health, pp 179–193

Igiehon NO, Babalola OO (2018) Rhizosphere microbiome modulators: contributions of nitrogen fixing bacteria towards sustainable agriculture. Int J Environ Res Public Health 15(4):574

Kakar KU, Nawaz Z, Cui Z, Almoneafy AA, Ullah R, Shu QY (2018) Rhizosphere-associated *Alcaligenes* and *Bacillus* strains that induce resistance against blast and sheath blight diseases, enhance plant growth and improve mineral content in rice. J Appl Microbiol 124(3):779–796

Karnwal A (2021) Pseudomonas spp., a zinc-solubilizing vermicompost bacteria with plant growth-promoting activity moderates zinc biofortification in tomato. Int J Veg Sci 27(4):398–412

Khanna K, Jamwal VL, Gandhi SG et al (2019) Metal resistant PGPR lowered Cd uptake and expression of metal transporter genes with improved growth and photosynthetic pigments in *Lycopersicon esculentum* under metal toxicity. Sci Rep 9:5855

Korenblum E, Dong Y, Szymanski J, Panda S, Jozwiak A, Massalha H, Meir S, Rogachev I, Aharoni A (2020) Rhizosphere microbiome mediates systemic root metabolite exudation by root-to-root signaling. Proc Natl Acad Sci 117(7):3874–3883

Kumar A, Maurya BR, Raghuwanshi R (2014) Isolation and characterization of PGPR and their effect on growth, yield and nutrient content in wheat (*Triticum aestivum* L.). Biocatal Agri Biotechnol 3:121–128

Kumar P, Dubey RC, Maheshwari DK (2012) Bacillus strains isolated from rhizosphere showed plant growth promoting and antagonistic activity against phytopathogens. Microbiol Res 167:493–499

Lazcano C, Boyd E, Holmes G et al (2021) The rhizosphere microbiome plays a role in the resistance to soil-borne pathogens and nutrient uptake of strawberry cultivars under field conditions. Sci Rep 11:3188

Leontidou K, Genitsaris S, Papadopoulou A et al (2020) Plant growth promoting rhizobacteria isolated from halophytes and drought-tolerant plants: genomic characterisation and exploration of phyto-beneficial traits. Sci Rep 10:14857

Lu P, Yang T, Li L, Zhao B, Liu J (2020) Response of oat morphologies, root exudates, and rhizosphere fungal communities to amendments in a saline-alkaline environment. PLoS One 15(12):e0243301

Meena NK, Tara N, Saharan BS (2018) Review on PGPR: an alternative for chemical fertilizers to promote growth in aloe vera plants. Int J Curr Microbiol App Sci 7(03):3546–3551

Mendes R, Garbeva P, Raaijmakers JM (2013) The rhizosphere microbiome: significance of plant beneficial, plant pathogenic, and human pathogenic microorganisms. FEMS Microbiol Rev 37:634–663

Mhlongo MI, Piater LA, Madala NE, Labuschagne N, Dubery IA (2018) The chemistry of plant-microbe interactions in the rhizosphere and the potential for metabolomics to reveal signaling related to defense priming and induced systemic resistance. Front Plant Sci 9:112

Mounde L, Boh M, Cotter M, Rasche F (2015) Potential of rhizobacteria for promoting sorghum growth and suppressing *Striga hermonthica* development. J Plant Dis Prot 122(2):100–106

Nannipieri P, Judith A-J, Ceccherini MT, Pietramellara G, Renella G, Schloter M (2020) Beyond microbial diversity for predicting soil functions: a mini review. Pedosphere 30:5–17

Nannipieri P, Ascher-Jenull J, Ceccherini MT, Landi L, Pietramellara G, Renella G (2017) Microbial diversity and soil functions. Eur J Soil Sci 68:12–26

Naseem H, Bano A (2014) Role of plant growth-promoting rhizobacteria and their exopolysaccharide in drought tolerance of maize. J Plant Inter 9(1):689–701

Nautiyal CS, Srivastava S, Chauhan PS, Seem K, Mishra A, Sopory SK (2013) Plant growth-promoting bacteria *Bacillus amyloliquefaciens* NBRISN13 modulates gene expression profile of leaf and rhizosphere community in rice during salt stress. Plant Physiol Biochem 66:1–9

Nautiyal CS, Srivastava S, Chauhan PS (2008) Rhizosphere colonization: molecular determinants from plant-microbe coexistence perspective. In: Nautiyal CS, Dion P (eds) Molecular mechanisms of plant and microbe coexistence. Soil Biology, vol 15. Springer, Berlin, Heidelberg

Nazari M, Riebeling S, Banfield CC et al (2020) Mucilage polysaccharide composition and exudation in maize from contrasting climatic regions. Front Plant Sci 11:587610

Newman EI (1985) The rhizosphere: carbon sources and microbial populations. In Fitter AH (ed) Ecological interactions in soil. Blackwell Scientific Publications, Oxford, p 107

Nie L, Shah S, Rashid A, Burd GI, Dixon DG, Glick BR (2002) Phytoremediation of arsenate contaminated soil by transgenic canola and the plant growth-promoting bacterium *Enterobacter cloacae* CAL2. Plant Physiol Biochem 40:355–361

Panda B, Rahman H, Panda J (2016) Phosphate solubilizing bacteria from the acidic soils of Eastern Himalayan region and their antagonistic effect on fungal pathogens. Rhizosphere 2:62–71

Pandey C, Bajpai VK, Negi YK, Rather IA, Maheshwari DK (2018) Effect of plant growth promoting *Bacillus* spp. on nutritional properties of *Amaranthus hypochondriacus* grains. Saudi J Biol Sci 25(6):1066–1071

Pantigoso HA, Yuan J, He Y, Guo Q, Vollmer C, Vivanco JM (2020) Role of root exudates on assimilation of phosphorus in young and old Arabidopsis thaliana plants. PLoS One 15(6):e0234216

Pascale A, Proietti S, Pantelides IS, Stringlis IA (2020) Modulation of the root microbiome by plant molecules: the basis for targeted disease suppression and plant growth promotion. Front Plant Sci 10:1741

Paulin MM, Novinscak A, Lanteigne C, Gadkar VJ, Filion M (2017) Interaction between 2,4-Diacetylphloroglucinol- and Hydrogen Cyanide-producing *Pseudomonas brassicacearum* LBUM300 and *Clavibacter michiganensis* subsp. michiganensis in the Tomato Rhizosphere. Appl Environ Microbiol 83(13):e00073-17

Philippot L, Raaijmakers JM, Lemanceau P, Van Der Putten WH (2013) Going back to the roots: the microbial ecology of the rhizosphere. Nat Rev Microbiol 11:789–799

Phillips DA, Fox TC, King MD, Bhuvaneswari TV, Teuber LR (2004) Microbial products trigger amino acid exudation from plant roots. Plant Physiol 136:2887–2894

Pii Y, Penn A, Terzano R, Crecchio C, Mimmo T, Cesco S (2015) Plant-microorganism-soil interactions influence the Fe availability in the rhizosphere of cucumber plants. Plant Physiol Biochem 87:45–52

Rai AK, Johri SN, Kaur H, Basak N, Sundha P (2020) Salinity and Sodicity Influence Mutualistic Association of Beneficial Microorganism in Arid Soils. J soil salin water qual 12(1):15–21

Rai AK, Dinkar A, Basak N, Dixit AK, Das SK, Dev I, Sundha P, Chandra P, Kumar S (2021) Phosphorus nutrition of oats genotypes in acidic soils: exploiting responsive plant-microbe partnership. Appl Soil Ecol 167:104094

Rohrbacher F, St-Arnaud M (2016) Root exudation: the ecological driver of hydrocarbon rhizoremediation. Agronomy 6:19

Saha M, Maurya BR, Meena VS, Bahadur I, Kumar A (2016) Identification and characterization of potassium solubilizing bacteria (KSB) from Indo-Gangetic plains of India. Biocatal Agri Biotechnol 7:202–209

Sasse J, Martinoia E, Northen T (2018) Feed your friends: do plant exudates shape the root microbiome? Trends Plant Sci 23:25–41

Sharma SB, Sayyed RZ, Trivedi MH, Gobi TA (2013) Phosphate solubilizing microbes: sustainable approach for managing phosphorus deficiency in agricultural soils. Springerplus 2:587

Sheng XF, He LY (2006) Solubilization of potassium-bearing minerals by a wild-type strain of *Bacillus edaphicus* and its mutants and increased potassium uptake by wheat. Can J Microbiol 52:66–72

Singh RP, Jha PN (2016) The multifarious PGPR *Serratia marcescens* CDP-13 augments induced systemic resistance and enhanced salinity tolerance of wheat (*Triticum aestivum* L.). PLoS One 11:e0155026

Ström L, Owen AG, Godbold DL, Jones DL (2002) Organic acid mediated P mobilization in the rhizosphere and uptake by maize roots. Soil Biol Biochem 34:703–710

Sun F, Qiaojing O, Wang N, Guo ZX, Ou Y, Li N, Peng C(2020) Isolation and identification of potassium-solubilizing bacteria from *Mikania micrantha* rhizospheric soil and their effect on *M. micrantha* plants. Glob Ecol Conserv 23:e01141

Swain M, Samir KN, Ray R (2007) Indole-3-acetic acid production and effect on sprouting of Yam (*Dioscorea rotundata* L.) minisetts by *Bacillus subtilis* isolated from culturable cowdung microflora. Polish J Microbiol 56:103–110

Tahir M, Ahmad I, Shahid M, Shah GM, Farooq ABU, Akram M et al (2019) Regulation of antioxidant production, ion uptake and productivity in potato (*Solanum tuberosum* L.) plant inoculated with growth promoting salt tolerant *Bacillus* strains. Ecotox Environ Safe 178:33–42

Tak HI, Ahmad F, Babalola OO (2013) Advances in the application of plant growth-promoting rhizobacteria in phytoremediation of heavy metals. Rev Environ Contam Toxicol 223:33–52

Terpolilli JJ, Hood GA, Poole PS (2012) What determines the efficiency of $N_{(2)}$-fixing Rhizobium-legume symbioses? Adv Microb Physiol 60:325–389

Tiwari S, Lata C (2018) Heavy metal stress, signaling, and tolerance due to plant-associated microbes: an overview. Front Plant Sci 9:452

Tiwari S, Lata C, Chauhan PS, Nautiyal CS (2016) *Pseudomonas putida* attunes morphophysiological, biochemical and molecular responses in *Cicer arietinum* L. during drought stress and recovery. Plant Physiol Biochem 99:108–117

Trivedi P, Leach JE, Tringe SG, Sa T, Singh BK (2020) Plant–microbiome interactions: from community assembly to plant health. Nat Rev Microbiol 18(11):607–621

Vishwakarma K, Kumar N, Shandilya C, Mohapatra S, Bhayana S, Varma A (2020) Revisiting plant-microbe interactions and microbial consortia application for enhancing sustainable agriculture: a review. Front Microbiol 11:560406

Walia A, Mehta P, Chauhan A, Shirkot CK (2014) Effect of *Bacillus subtilis* strain CKT1 as inoculum on growth of tomato seedlings under net house conditions. Proc Natl Acad Sci 84(1):145–155

Xiao AW, Li Z, Li WC, Ye Z (2020) The effect of plant growth-promoting rhizobacteria (PGPR) on arsenic accumulation and the growth of rice plants (*Oryza sativa* L.). Chemosphere 242:125136

Zakikhani H, Ardakani MR, Rejali F, Gholamhoseini M, Joghan AK, Dolatabadian A (2012) Influence of diazotrophic bacteria on antioxidant enzymes and some biochemical characteristics of soybean subjected to water stress. J Integr Agric 11(11):1828–1835

Chapter 27
Strategies for Heavy Metals Remediation from Contaminated Soils and Future Perspectives

Md. Saiful Islam, Tapos Kormoker, Rahat Khan, Ram Proshad, Md. Humayun Kabir, and Abubakr M. Idris

Abstract Soil is a dynamic life-supporting component of this *Planet Earth* but its contamination with toxic heavy metals (HMs) is omnipresent throughout the planet. Abundances of these HMs in soil have augmented considerably in last 2–3 decades due to rapid industrialization, agricultural practices (fertilizers and pesticides application), and other anthropogenic activities, which causing environmental, ecological and health risks. Consequently, their remediation approaches from the environmental components are critical. Among the several procedures for metals remediation, organic residues with the plant-microbes (phyto-remediation) can simultaneously increase the fertility of soil along with the bio-remediation, which in turn is

Md. S. Islam
Department of Soil Science, Patuakhali Science and Technology University, Dumki, Patuakhali 8602, Bangladesh

T. Kormoker
Department of Emergency Management, Patuakhali Science and Technology University, Dumki, Patuakhali 8602, Bangladesh

R. Khan (✉)
Institute of Nuclear Science and Technology, Bangladesh Atomic Energy Commission, Savar, Dhaka 1349, Bangladesh
e-mail: rahatkhan.baec@gmail.com

R. Proshad
Key Laboratory of Mountain Surface Processes and Ecological Regulation, Institute of Mountain Hazards and Environment, Chinese Academy of Sciences, Chengdu 610041, Sichuan, China

University of Chinese Academy of Sciences, Beijing 100049, China

Md. H. Kabir
Department of Environmental Science and Resource Management, Mawlana Bhashani Science and Technology University, Tangail, Bangladesh

A. M. Idris (✉)
Department of Chemistry, College of Science, King Khalid University, Abha 62529, Saudi Arabia
e-mail: abubakridris@hotmail.com

Research Center for Advanced Materials Science (RCAMS), King Khalid University, Abha 62529, Saudi Arabia

P. K. Shit et al. (eds.), *Soil Health and Environmental Sustainability*,
Environmental Science and Engineering,
https://doi.org/10.1007/978-3-031-09270-1_27

thought as one of the lucrative and cost-effective approaches of HM's remediation from soil. Efficacy of phyto-remediation can be improved by simultaneous participations of plant-growth-promoting bacteria which can convert HMs into soluble and bio-available forms by the activities of siderophores, redox processes, biosurfactants, organic acids, and biomethylation. This work highlights the recent applications and advancements made hitherto to understand the molecular and biochemical mechanisms of metal-microbe-plant interactions with organic residues along with their functions in major processes belong to the phyto-remediation, for instance heavy metal detoxification, transformation, mobilization, distribution, and immobilization.

Keywords Heavy metals · Plant-microbes' interaction · Phytoremediation

27.1 Introduction

The enhancements of HMs in soil possess incredible concern as a result of their diligence nature, extensive biological half-lives and harmfulness to the environment (Bhuyan et al. 2017; Islam et al. 2015a, 2018, 2021a, b, c; Kormoker et al. 2019). Pollution of soils by various poisonous elements has been considered as a critical environmental issue in both developing and developed nations all through the globe (Islam et al. 2015b; Sun et al. 2010; Proshad et al. 2018; Khan et al. 2015, 2017, 2018, 2019a, b, 2020, 2021). Both natural as well as anthropogenic exercises, for example, mining and industrial garbage removal, use of sewage ooze, purifying of minerals, use of fertilizers and pesticides squander water, raw materials, and metropolitan soil squanders are the significant wellsprings of substantial metals in soil environment (Fig. 27.1) (Islam et al. 2014; Khan et al. 2008; Peng et al. 2018; Habib et al. 2020, 2019a, b; Habib and Khan 2021; Begum et al. 2021). However, extreme enrichment of HMs in soils from these cradles expanded heavy metal take-up by food yields and vegetables, which consecutively may prompt severe health risks to human beings (Xionget al. 2016; Rajkumar et al. 2009; Khalid et al. 2017).

The utmost well-known potentially toxic elements are chromium (Cr), copper (Cu), mercury (Hg), lead (Pb), manganese (Mn), zinc (Zn), cadmium (Cd), aluminum (Al), arsenic (As) and antimony (Sb) (Khalid et al. 2017; Duruibe et al. 2007; Rahman et al. 2022). Among these potentially toxic elements, Pb, As, Hg, and Cd are encompassed in the topmost 20 Hazardous Substances of the USEPA (United States Environmental Protection Agency) and ATSDR (Agency for Toxic Substances and Disease Registry 2012). Lead and Cd were categorized as group-B2 human carcinogens (Environmental protection Agency 2009). International Agency for Research on Cancer (IARC-WHO) has employed Cd and Cd-complexes in group-1 human's carcinogens and categorized inorganic Pb-complexes in group-2A human's carcinogens. The focal health hazards connected with HMs take account of cardiovascular illness, cognitive impairment, chronic anemia (Iqbal 2012), damage of kidneys, cancer (Wuana and Okieimen 2011), brain, nervous system, skin, bones and teeth (Järup 2003). Subsequently, these are precarious to diminish these health risks from

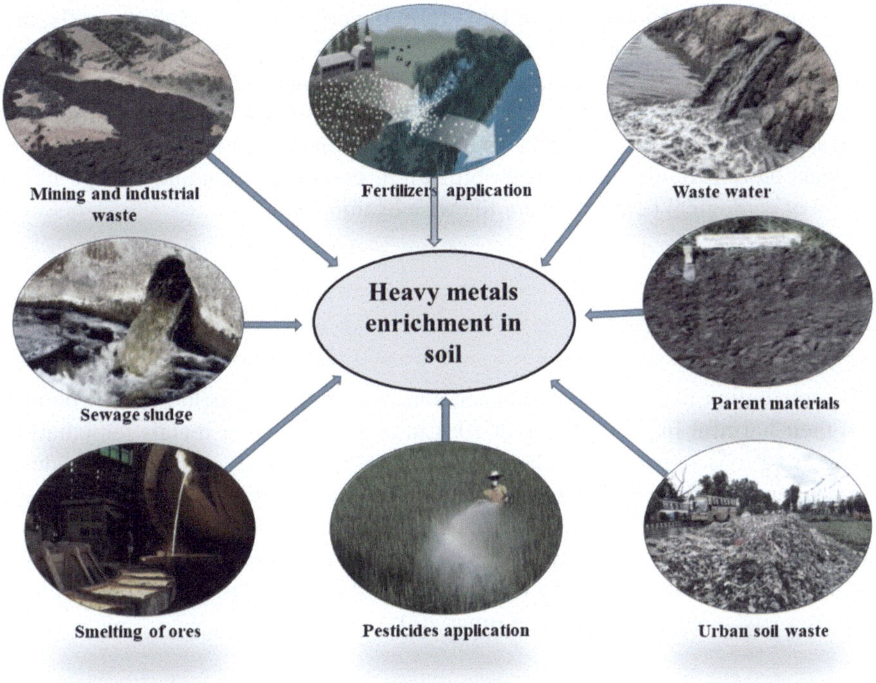

Fig. 27.1 Heavy metals enrichment in soil from different sources

substantial metals exposure, which are conceivable through the elimination of HMs from the environmental components.

In such manner, this review examines the most secure method of HMs remediation, i.e., phyto-remediation and other bio-remediation from the contaminated soil environment. As an option in contradiction of the chemical and physical remediation techniques, utilization of plants and microorganisms has been an auspicious way to deal with remediate heavy metals from defiled soils by extraction (phyto-extraction), accelerate plant biomass, increment or abatement metal accessibility in soil and encourage translocation ability of heavy metals through (bioaccumulation) soil-root-shoot tissues pathway (Glick 2010; Ma et al. 2011a; Rajkumar et al. 2012; Khalid et al. 2017; Lebeau et al. 2008; Liu et al. 2018). For improved execution of phyto-remediation, the interactions of microorganisms and plant root in rhizosphere are perceived as basic part of plant development through the creation of different metabolites (Segura et al. 2009; Badri et al. 2009). Studies focus on bioremediation either natural or inorganic poisons exclusively do not dive into the connection between the two kinds of toxins and the organisms, or the materials those are engaged with the applied strategies (Komarek et al. 2013; Chirakkara et al. 2016). In fact, a paucity of data on bio-remediation of natural poisons in co-polluted environment exist. Nonetheless, the techniques fundamental plant–microbe-metal collaborations stay slippery for the remediation of HMs from the polluted soils. Subsequently, this article endeavors to review the new advances on the sources and poisonous impact of substantial metals and to comprehend the biochemical systems for metals remediation utilizing metals–plant–microbe collaborations. Moreover, new methodologies, for example, role of plant–microbe-metal associations to accomplish the objective of metals bioremediation through mobilization, immobilization and change in the method of ecological benevolent are likewise examined.

27.2 Methodologies

HMs remediation from the contaminated soil is essential to shield the environment from their harmful impacts (Glick 2010; Liu et al. 2018). For metals remediation, a range of physicochemical techniques have been adopted, which is considered as a challenging task relating to charge, generation of hazardous by-products and technical complexity (Sheoran et al. 2011; Khalid et al. 2017). Conversely, biological techniques possibly elucidate these downsides of physicochemical remediation methods since they are easy to operate and don't have any chance to produce secondary toxic waste (Doble and Kumar 2005).

27.2.1 Physicochemical Approaches

The physicochemical methodologies (landfill, thermal, leaching, electro-reclamation and removal) have long been embraced for remediation of HMs (Sheoran et al. 2011; Barcelo and Poschenrieder 2003). Physico-chemical methodologies are quick however insufficient, expensive and occasionally may cause unfavorable consequences for soil inherent characteristics, and make secondary contamination (Glick 2010; Doble and Kumar 2005; Ali et al. 2013). Physical treatment is utilized to isolate the tainted part, normally the fine grains, from the remaining soil-matrix. The most well-known strategy for physical parting in soil decontamination utilizes rotational attrition scrubbers to separate the defiled soil portions. However, chemical remediation processing mostly involved in dissolving toxins from the most polluted portion of the contaminated soil.

27.2.2 Biological Approaches

Biological remediation process abuses normal natural cycles that permit some specific microorganisms and plants to help the remediation of HMs. These cycles happen by an assortment of techniques which includes adsorption, methylation, and redox reactions (Singh et al. 2009). Biological processes of HMs remediation methods comprise of bio-remediation, phyto-remediation, bioleaching, biostimulation, bioventing, bioreactors, bioaugmentation, land forming, and composting etc. Out of these tactics, phyto-remediation and bio-remediation are highly convenient procedures (Beskoski et al. 2011) for heavy metals remediation. Phytoremediation is one of the significant biological strategies that can be enhanced with mutual implementations of microorganism and plants (Hadi and Bano 2010; Chen et al. 2008). During last two decades, metal lenient and hyper-aggregating plant with the connections of microorganisms have acquired significant consideration because of their accumulating capability of HMs from polluted soils (bioaccumulation) and are ensuing the impacts on metals' mobilization as well as improving plant-growth (Glick 2010; Jing et al. 2007). Among various organisms, symbiotic soil microorganisms can emphatically impact plants growth as well as root improvement, which expands plants' resilience to different ecological anxieties, for example, saltiness, temperatures, heavy metals and other harmful synthetic substances (Hadi and Bano 2010; Glick 2004; Gamalero et al. 2004). Thus, larger & better accumulating plants possess higher capability to remediate inorganic as well as organic pollutants from the environment.

Heavy metals having generally high density are poisonous to living life forms at low focus (Iram et al. 2013). Different plants and microorganisms are typically cast-off for the exclusion of HMs by producing host-microbe's interaction mechanisms (Table 27.1). The involvement of microorganisms through this process is to reduce heavy metals concentration and its importance of biodiversity is increased considerably for tidy-up the metal tainted soils. All the metals are poisonous, however a portion of these are valuable at low focus where metal poison levels cause genuine mortality and morbidity. Moreover, bioavailability of heavy metals can be enhanced by expansion of natural residues to the dirts, for example, fertilizer, manure, biosolids and expands soil ripeness (Jin et al. 2011). Bioremediation is not successful just for the debasement of heavy metals yet additionally it tends to be utilized to remove undesirable substances from the ambient environmental components (e.g., air, water, soil, industrial wastes etc.). Despite the fact that many designed cycles have been produced for applying bio-remediation processes, still the economical treatments of contaminated soils have remained a tricky objective (Zeyaullah et al. 2009).

27.3 Results and Discussion

27.3.1 Phyto-Remediation of Heavy Metals

Plants can likewise be utilized to tidy up polluted soil, air or water; this is termed as phyto-remediation. There are far reaching reviews on standards and application achievability of phyto-remediation of metal from polluted soil (Chaney and Baklanov

Table 27.1 Lists of microorganisms enhanced phytoremediation of heavy metals from contaminated soil environment

Microorganisms	Host plant	Metals	Metal accumulated part	Medium	Plant growth promoting features	Mechanisms	References
Pseudomonas fluorescens G10 and *Microbacterium* sp. G16	*Brassica napus*	Pb, Cd, Zn, Cu and Ni	Roots and shoots	Soil	Production of IAA, siderophores, ACC deaminase,	Increased watersoluble Pb in solution and Pb added soil; Increased biomass production and total Pb uptake	Sheng et al. (2008a)
Enterobactorcloacae CAL2	*B. napus*	As	Roots and shoots	Promix 'BX'	IAA, ACC deaminase, siderophores, antibiotics	Increased biomass production and total As uptake	Nie et al. (2002)
Cupriavidustaiwa-nensis	*Mimosapudica*	Pb, Cu and Cd	Roots (77–98%), shoots (2–23%)	Solution	Biodegradation, biosorption, release of extra cellular products	Produced nodules andincreased metal-binding sites	Chen et al. (2008)
Azotobac-torchroococum and Rhizobium leguminosarum	*Zea mays* L	Pb	Roots, stem and leaves	Soil	IAA, decreased soil pH	Increased growth and dry biomass and total Pb uptake	Hadi and Bano (2010)
Bacillus thuringiensis GDB-1	*Alnus firma*	As, Cu, Cd, Ni, Pb and Zn	Roots and shoots	Minetailing	Production of IAA, siderophores, ACCD and solubilization of P	Bioremoval of Pb, Zn, As, Cd, Cu and Ni in metal amended and mine tailing extract medium; increased biomass, chlorophyll content, nodule number and metal (As, Cu, Pb, Ni, and Zn) accumulation in *A. firma*	Babu et al. (2013)
B. pumilus E2S2, *Bacillus* sp. E1S2, *Bacillus*sp. E4S1, *Achromobacter* sp. E4L5 and *Stenotrophomonas* sp. E1L	*Sedum plumbizincicola*	Cd, Pb and Zn	Roots and shoots	Soil	Production of IAA, siderophores, ACCD and solubilization of P	Bacterial inoculation increased water extractable Cd and Zn contents in soil; improved plant growth and metal uptake	Ma et al. (2015a)

(continued)

Table 27.1 (continued)

Microorganisms	Host plant	Metals	Metal accumulated part	Medium	Plant growth promoting features	Mechanisms	References
Bacillus sp. MN3-4	*Alnus firma* and *B. napus*	Pb, Cd, Zn, Ni and Cu	Roots and shoots	Soil	Production of IAA and siderophores	Exhibited bioremoval of Pb; increased root elongation of *B. napus* seedlings; reduced metal phytotoxicity and increase Pb accumulation in *A. firma*	Shin et al. (2012)
Pseudomonas tolaasii ACC23, *P. Fluorescens* ACC9, *Mycobacterium* sp. ACC14	*Brassica napus*	Cd	Roots and shoots	Promix 'BX'	ACC deaminase, siderophores and IAA	Increased biomass and Cd contents in above-ground tissues	Dell'Amicoet al. (2008)
P. fluorescens G10, *Microbacterium* G16	*Brassica napus*	Pb, Cd, Zn, Cu and Ni	Roots and shoots	Soil	Production of IAA, siderophores, ACCD	Increased water-soluble Pb in solution and Pb-added soil; increased biomass production and total Pb uptake	Sheng et al. (2008b)
Ralstonia sp. J1-22–2, *Pantoea agglomerans* Jp3-3, *Pseudomonas thivervalensis* Y1-3–9	*B. napus*	Cu, Pb, Cd and Ni	Roots and shoots	Promix 'BX' and soil	Production of IAA, siderophores, ACCD and solubilization of P	Increased the biomass of rape and increased Cu contents in above-ground tissues	Zhang et al. (2011b)
Pseudomonas sp. LK9	*Solanum nigrum*	Cd, Zn and Cu	Roots, stem and leaves	Soil	Biosurfactants, siderophores, organicacids	Improved soil Fe, P and heavy metal availability, shoot dry biomass and uptake of Cd, Zn and Cu	Chen et al. (2014)
Pseudomonas sp. PsM6, P.*Jessenii*PjM15	*Ricinus communis*	Zn	Roots, stem and leaves	Soil	Biosorption, mobilization, ACC deaminas, IAA, siderophores	Increased the biomass and phytoremediation of Zn	Rajkumar and Freitas (2008)

(continued)

Table 27.1 (continued)

Microorganisms	Host plant	Metals	Metal accumulated part	Medium	Plant growth promoting features	Mechanisms	References
Rahnella sp. JN6	*Brassica napus* and *Polygonumpubescens*	Cd, Pb and Zn	Roots and shoots	Soil	IAA, ACC deaminase, siderophores, phosphate solubilization	Showed high Cd, Pb, Zn tolerance and mobilization; promoted plant growth and Cd, Pb, Zn uptake by rape; high level of colonization in tissue interior of rapes	He et al. (2013a, 2013b)
Bacilus subtilis, B. cereus, Flavobacterium sp. And *Pseudomonas aeroginosa*	*Orycoprhagmusvio-laceus*	Zn	Roots and shoots	Soil	ACC deaminase, IAA, siderophores	Increased the biomass and accumulation of Zn	He et al. (2010)
Pseudomonas veronii, P. fluorescens VI8L1, *Bacillus pumilus* VI8L2, *P. fluorescens* II8L4,*P. fluorescens* VI8R2, *Acinetobacter calcoaceticus* II2R3	*Sedum alfredii*	Zn and Cd	Roots and shoots	Soil	Production of IAA, siderophores, Fixation of nitrogen, solubilization of $ZnCO_3$ and $Zn_3(PO_4)_2$	Mobilized Zn in soil, thus increased soil Zn bioavailability; improve growth and Zn accumulation by *S. alfredii*	Long et al. (2011, 2013)
Burkholderia sp. J62	*Lycopersicumescu-lentum, Z. mays*	Pb and Cd	Roots and shoots	Soil	Phosphatesolubilization, IAA, siderophores, ACC deaminase	Biosorption of Pb and Cd, promoted plant growth and reduced accumulation of Pb and Cd in roots and shoots of tomato and maize plants	Jiang et al. (2008)
Methylobacterium oryzae CBMB20, *Burkholderia* sp. CBMB40	*Lycopersicumesculentu*	Ni and Cd	Roots and shoots	Soil	Not applicable	Biosorption considerable amount of Ni and Cd, thus reduced the metal toxicity; promoted plant growth and reduced accumulation of Ni and Cd in roots and shoots of tomato plants	Madhaiyan et al. (2007)

(continued)

Table 27.1 (continued)

Microorganisms	Host plant	Metals	Metal accumulated part	Medium	Plant growth promoting features	Mechanisms	References
Staphylococcus arlettae NBRIEAG-6	*Brassica juncea*	As	Roots and shoots	Soil	IAA, siderophores, ACC deaminase	Increased biomass and As contents in above-ground tissues	Srivastavaet al. (2013)
Staphylococcus, Curtobacterium, Bacillus, Pseudomonas, Microbacterium, Arthrobater, Leifsonia, Paenibacillus	*Alyssumbertolonii*	Ni, Co, Cr, Cu and Zn	Roots and shoots	Soil	Production of siderophores	Had an ability to colonize plant tissues	Barzanti et al. (2007)
Microbacterium sp. NCr-8, Arthrobactersp. NCr-1, Bacillus sp. NCr-5, Bacillussp. NCr-9 and Kocuria sp. NCr-3	*Noccaeacaerulescens, Thlaspi perfoliatum*	Ni	Roots and shoots	Soil	Production of IAA, siderophores and ACCD	Enhanced growth and Ni translocation in plants	Visioli et al. (2014)
Achromobacter xylosoxidans Ax10	*Brassica juncea*	Cu	Roots and shoots	Soil	ACC deaminase, IAA, solubilization of P	Increased the plant biomass and accumulation of Cu	Ma et al. (2009a)
Micrococcus sp. MU1and *Klebsiella*sp. BAM1	*Helianthus annus*	Cd	Roots and shoots	Soil	IAA, ACC deaminase	Increased the plant biomass and accumulation of Cd through phytoextraction mechanism	Prapagdeeet al. (2013)
Psychrobacter sp. SRA1 and SRA2, *Bacillus cereus* SRA10	*Brassica juncea*	Ni	Roots and shoots	Soil	ACC deaminase, IAA, Psolubilization, Nimobilization	Exhibited significant levels of siderophores production and enhanced Ni accumulation in plant tissues	Ma et al. (2009b)

(continued)

Table 27.1 (continued)

Microorganisms	Host plant	Metals	Metal accumulated part	Medium	Plant growth promoting features	Mechanisms	References
Pseudomonas sp. SRI2, *Psychrobacter* sp. SRS8and*Bacillus* sp. SN9	*Brassica juncea* and *Brassica oxyrrhina*	Ni	Roots and shoots	Soil	ACC deaminase, IAA,P solubilisation, Nimobilization	Exhibited significant levels of siderophores production and enhanced Ni accumulation in plant tissues	Ma et al. (2011a)
Pseudomonas sp. A3R3	*Alyssumserpyllifolium*	Ni	Roots and shoots	Soil	Production of IAA, siderophores, ACCD and solubilization of P; Excretedcellulase and pectinase	Increased the biomass of *B. juncea* and Ni content in A. *serpyllifolium*; showed high level of colonization in tissue interior of both plant species	Ma et al. (2011a)
Bacillus subtilis, B. cereus, B. megaterium, and *Pseudomonas aeruginosa*	*Orychophragmusvio-laceus*	Cd	Roots and shoots	Soil	Production of IAA, siderophores, fixation of nitrogen, solubilization of P	Improve plant growth and rootelongation	Liang et al. (2014)
Paenibacillus macerans NBRFT5, *Bacillus endophyticus* NBRFT4, B. *pumilus* NBRFT9	*B. Juncea*	Cu, Ni and Zn	Roots, stem and leaves	Mixture of fly ash and pressmud	Produce siderophores, produce protons, organic acids and enzymes which enhance the metal mobilization and boost the phytoextraction process	Siderophores, organicacids, protons and other non-specified enzymes	Tiwari et al. (2012)
Rhizobium leguminozarum	*B. Juncea*	Zn	Above ground and Belowground	Soil	Enhanced plant growth and Zn phytoextraction	Metalchelation	Adediran et al. (2015)
Azotobacter-chroococcum HKN-5, *Bacillus megaterium* HKP-1 and *Bacillus mucilaginosun* HKK-1	*B. Juncea*	Cu, Cd, Pb and Zn	Shoots	Soil	Produced a much larger above-ground biomass, solubilize phosphate and potassium	IAA, Gibberellins	Wu et al. (2006)

(continued)

Table 27.1 (continued)

Microorganisms	Host plant	Metals	Metal accumulated part	Medium	Plant growth promoting features	Mechanisms	References
Pantoea agglomerans Jp3-3, and *Pseudomonas thivervalensis* Y1-3–9	*B. Juncea*	Cu, Pb, Cd and Ni	Roots and shoots	Quartz sand	Production of IAA, siderophores, ACCD and solubilization of P	Increased the biomass of rape and increased Cu contents in above-ground tissues	Zhanget al. (2011a,b)
Enterobacter sp. JYX7 and *Klebsiella*sp. JYX10	*Polygonumpubescens*	Cd, Pb and Zn	Roots and shoots	Soil	IAA, siderophores, ACC deaminase, phosphatesolubilizaiton	Showed high Cd, Pb and Zn tolerance and mobilization; promoted plant growth and Cd, Pb and Zn uptake by rape	Jing et al. (2014)
Rahnella sp.	*Amaranthushypochondriacus, A. Mangostanus* and *S. nigrum*	Cd	Roots and shoots	Soil	IAA, siderophores, ACC deaminase, phosphate solubilizaiton	Improved plant growth, biomass and metal uptake by Cd-hyper accumulators	Yuan et al. (2013)

2017; Sarwar et al. 2017; Pinto et al. 2015). Generally, phyto-remediation is characterized into two expansive strategies: (1) Phyto-extraction, in which substantial HMs are consumed by the plants from defiled soils followed by aggregation of those HMs in leaves and/or shoots, and (2) Phyto-stabilization, in which HMs are immobilized in soils by plant-roots (Glick 2003; Chandra et al. 2017). This innovation has been getting thoughtfulness to the researchers because of its innovative, cost-effective substitute to the more settled treatment strategies utilized at profoundly tainted sites. The utilizations of phyto-remediation innovation can likewise be arranged dependent on the accompanying mechanisms: incorporate extraction of heavy metals from soil; convergence of substantial metals in plants' tissue; debasement of metals by different abiotic and/or biotic measures; immobilization of metals in the root zone; and control of erosion, runoff, and penetration by vegetative covers (Chaney and Baklanov 2017; Macek et al. 2000). The metal (loid)s like Hg, As, & Se might be released in gaseous species by metal collector plants (e.g., *Astragalus racemosus*) into the ambient atmosphere (termed as phytovolatilization). So far 721 species of metal hyper-accumulators plants have been recognized (Reeves et al. 2017).

For heavy metals remediation from contaminated soil, several species of microbes and plants are being tested (Kcil et al. 2015; Chiang et al. 2006; Abhilash et al. 2012, 2013; Glick 2010). Nonetheless, the efficiency of every phyto-remediation measures largely be influenced by the following aspects: (i) intrinsic nature of plants such as easy propagation, production of phytoproducts, degradation and tolerance potential, (ii) microflora present in soil with plant growth promotion potential and ability to thrive in dirt, and (iii) physicochemical belongings such as water solubility, long range transport potential, persistence to the environment and phyto-accumulation of heavy metals itself (Fig. 27.2). As indicated by USEPA (2004), ~67% of contaminated sites are sullied with HMs. Thus, their remediation from defiled soils is a major issue of worldwide concern (Lesley and Colette 2013). Due to the solubility of HMs (in water or acid), they can effectively move through the pedosphere and be incorporated with Fe–Mn oxyhydroxides and organic fractions of soil. In some cases, heavy metals are impervious to the cycle of microbial-degradation (Bolan et al. 2014) and generally remediated by the assistance of hyper-accumulating plants (Peer et al. 2005). Hence, it is basic to recognize new up-and-comer types of plants and microorganisms that make due in metals tainted soil and to create innovative ways that can lessen the metals harmfulness to the remediator species.

27.3.2 Bioremediation Using Organic Residues and Microorganisms

Application of organic residues with microbial populations has been received as quite possibly the most financially savvy advances to balance out heavy metals (Hamid et al. 2019; Lu et al. 2013; Yu et al. 2011; Zhang et al. 2013). Organic residues, for instance, compost and microbes together stabilize contaminated solid waste by means

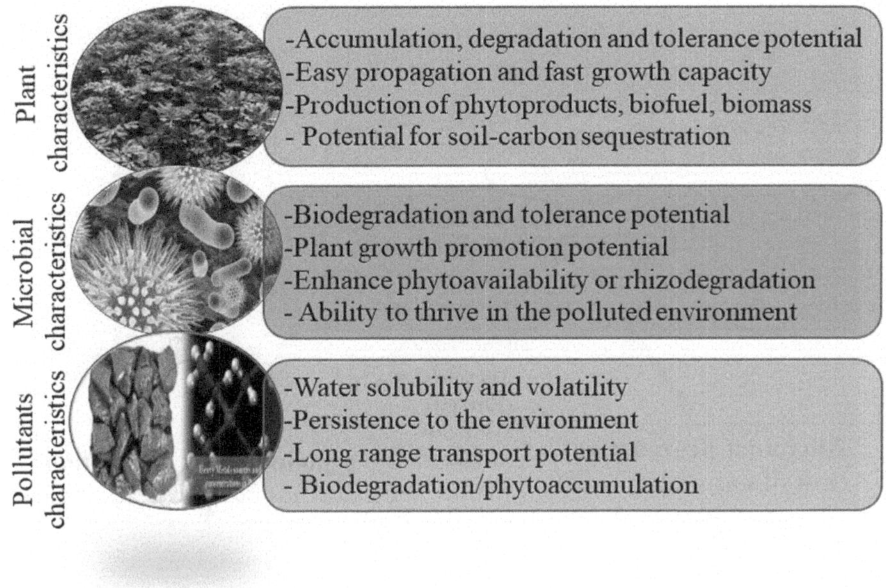

Fig. 27.2 A key role in the phytoremediation and bioremediation of heavy metals contaminated soils by plant–microbe-pollutant interaction mechanisms

of decomposition of biodegradable elements. The application of organic residues in HMs-defiled soil can cause the variety of microbial populace by altering soil pH, diminishing heavy metals dissolvability and expanding microbial biomass (Alburquerque et al. 2011). Over the past few years, enormous quantities of investigations have been conducted to validate the effectiveness of organic residues with microbes for metals remediation from soils (Megharaj et al. 2011; Fuente et al. 2011; Tandy et al. 2009). As bioremediation approaches, the combination of compost and microbes are different, as well as straight composting, bioaugmentation, integration of bulking agent and surfactant utilization (Fig. 27.3).

As the initial composting materials, organic wastes from different sources were often selected for metals bioremediation. Organic residues increase available P and total N and decreasing the heavy metals toxicity (Fernandez et al. 2009; Pedra et al. 2007). In composts, fungi and bacteria have been considered for debasement of HMs in soil which mainly relies on following mechanisms: adsorption of heavy metals using organic matters and debasement of HMs through microbes (Puglisi et al. 2007) (Fig. 27.3). Organic residues are an important source of nutrient elements for indigenous microbes which finally detoxify heavy metals through the process of decomposition and biotransformation (Namkoong et al. 2002). HMs in soil and/or compost mixtures, reviewing microbial aspects can contribute to widen the insight of biodegradation and microbial population measures, empowering one

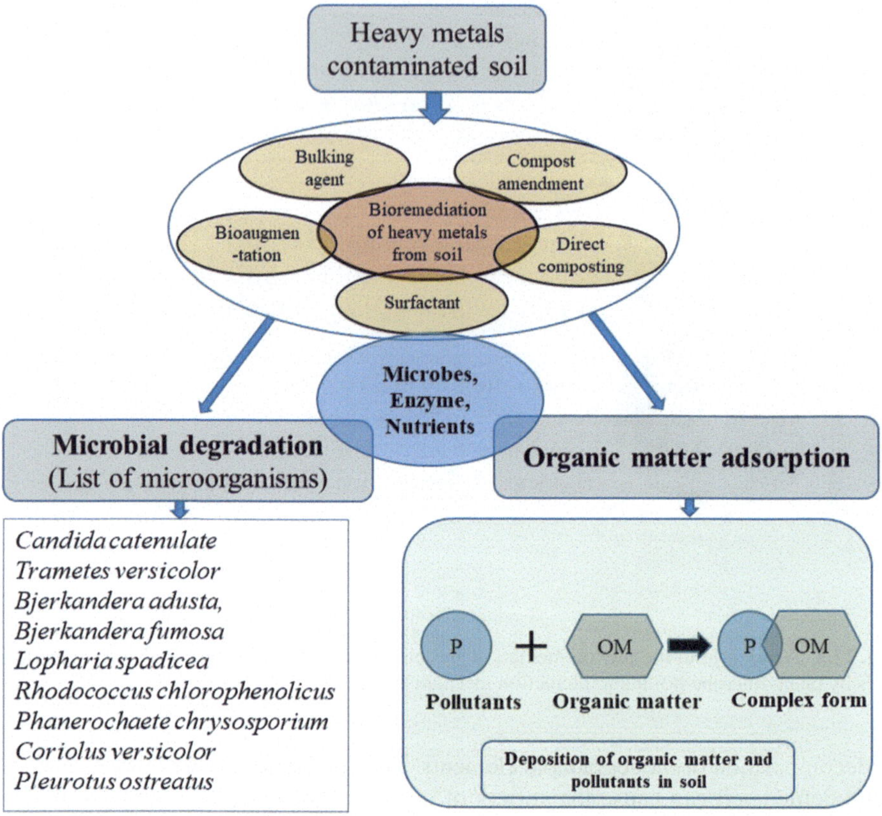

Fig. 27.3 Main strategies and mechanisms for bioremediation of contaminated soils by composting and microbes

to choose possibly appropriate microorganisms for metals bio-remediation. Utilization of organic-residues as a remediation technique requires a comprehension of microorganisms engaged with metals biodegradation and biotransformation in soils.

27.3.3 Plant-Bacteria-Metal Interaction for Phytoremediation

The resistance capacity of heavy metals in plants & microorganisms is a vital principle for metals' amassing by plant and organism-based phyto-remediation. This remediation approach has been all around exhibited that intrinsic capacity of entophytic-bacteria can support the host-plants to adapt transformation to the horrible environment & improve phytoremediation effectiveness by advancing growth, mitigating metal pressure, diminishing metals' phyto-toxicity and mobility among the different portions of plants' physiology (Ma et al. 2011a, b). When plants are subjected to the metals contaminated soil, the pressure prompts plants' inter-linked molecular and physiological mechanisms in adjusting to the unpleasant environment.

As the process of symbiosis, plant roots provide energy and supplements to organisms and consequently, microorganisms invigorate exudation from plant-roots. In this co-transformative cycle, plants & organisms coincide or compete their endurance in the altering environment. Generally, the endophytic bacterial cell advance phytoremediation cycle of HMs from polluted soil through the enhancement of plants to metal resilience and development, as along with the modification of metals aggregation in plants' physiology (Fig. 27.4, Table 27.1). For phytoremediation using plant-bacteria associated mechanisms, a substantial number of plants have been examined; however, in the field condition, numerous hyper-aggregating plants do not deliver sufficient biomass to make the cycle proficient. Therefore, the act of encouraging phytoremediation with plant growth promoting (PGP) microbes is ideal decision for metals remediation. Significant scopes for PGP microscopic organisms have been perceived which can effectively help in phyto-remediation processes (Glick 2010), e.g., metals-resilient and PGP-microbes were segregated from *Polygonum pubescens* grownup in metal-tainted soils (Jing et al. 2014). The sequestered strains were recognized as *Klebsiella* sp. and *Enterobacter* sp. which was inoculated in to *Brassica napus* for HMs amassing. These bacteria enhanced plants' growth, and Zn, Cd, and Pb build up by *B. napus*. Similar plant was also investigated where *Mycobacterium* sp. ACC14, *Pseudomonas tolaasii* ACC23 and *P. Fluorescens* ACC9 were used, and documented boosted Cd-uptake from contaminated soil (Dell'Amico et al. 2008). Effective case studies of HMs expedited by PGP bacteria are enumerated in Table 27.1.

27.3.4 Plant-Growth Promotion Mechanism

Plants are recurrently exposed to diverse environmental stresses, e.g., salinity, temperature and HMs, which change plant biochemistry and physiology (Hossain et al. 2012). Nonetheless, certain plants accompanied bacteria which have been observed to assist the host-plant to control the biotic and abiotic stresses (Rajkumar et al. 2013; Glick 2010). Even under stress conditions, many PGP bacteria are skilled in generating phytohormones such as IAA, Cytokinin and gibberellins and response to metal stress (Glick 2012). Specific bacteria produce ACC deaminase (biosynthetic precursor for ethylene) those hydrolyze ACC to α-ketobutyrate and ammonia (Ullah et al. 2015), which act as an important nitrogen source for plants growth. Therefore, beneficial endophytic microbes those stimulate plant development and phytoremediation capacity have presently received additional consideration to the researchers (Ma et al. 2015a; Zhu et al. 2014; Taghavi et al. 2009).

The remediation procedures, e.g., phytohormones generation, alleviation of metals' availability & toxicity, N_2-fixation, bio-chelation & siderophoretion, and dissolving potassium or phosphate by plant-growth-promoting-endophytes (PGPE) have extensively been suggested for microbe-aided phytoremediation of HMs from

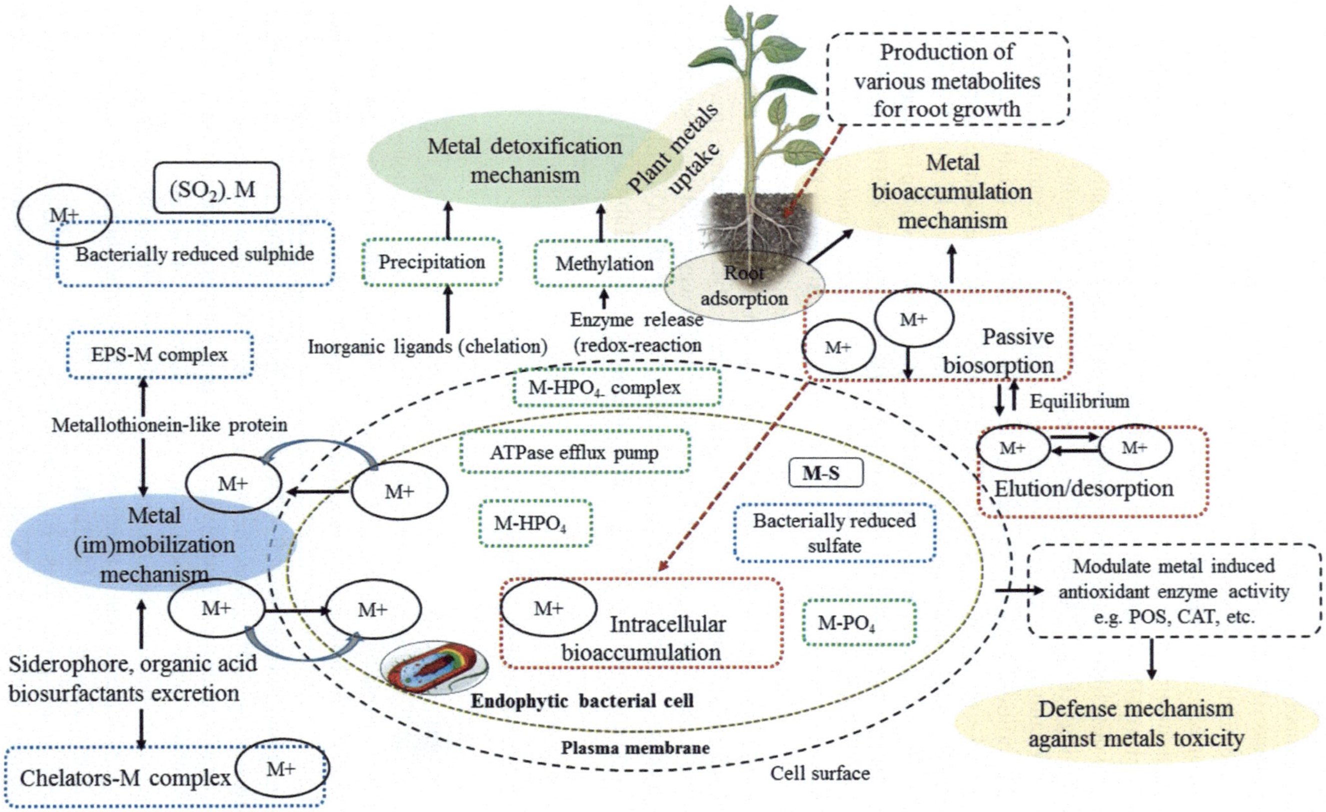

Fig. 27.4 Schematic overview of mechanisms of plant-microbe-metal interactions. M, metal; EPS-M, extracellular polymeric substances-metal; ATP, adenylpyrophosphatase

the polluted environmental components (Pereira and Castro 2014; Rajkumar et al. 2012; Ma et al. 2011a, 2015a, b, c; Ahemad and Kibret 2014; Fig. 27.4). As symbiotic association, endophytic bacteria are proficient in articulating nitrogenase inhabit through appropriately supplying fixed atmospheric N_2 to their host-plants to survive easier in nitrogen-poor soil condition (Montanez et al. 2012). Endophytic diazotrophs are well-known to provide more benefits than rhizosphere diazotrophs because of their advantageous host-endophyte allelopathies consequence. For instance, the endophytic genera *Rahnella, Acinetobacter, Burkholderia,* and *Sphingomonas* (isolated from *Salix sitchensis* and *Populus trichocarpa*) were competent to fix atmospheric N_2, thus delivered plentiful nitrogen to the plants, which in turn boosted up the plants' growth under nitrogen-poor situation (Doty et al. 2009). Gupta et al. (2013) described that N_2-fixing endophytic bacteria can also upsurge nitrogen amassing and N_2-fixation proportion in plant.

Phosphorus is a foremost vital macronutrient for biological development and growth of plant and acting a critical role in several enzymatic actions such as glucose transportation, root development and other physiological developments (Ahemad 2015; Fernandez et al. 2007). Soluble Phosphorus is frequently the limiting nutrient mineral for plant biomass creation in usual ecosystems which is only uptaken as soluble dibasic (HPO_4^{2-}) and/or monobasic ($H_2PO_4^{-}$) forms (Glass 1989). However, uptake of soluble P can be interfered by higher levels of HMs in soil, which in turn retard the botanical growth (Zaidi et al. 2006). In soil, more than 75% of applied phosphorus is unreachable for plant sowing to the development of complex forms (Ezawa et al. 2002). However, certain endophytic bacteria under metal-stress condition can dissolve precipitated phosphates by chelation (i.e., PO_4^{3-}), acidification, releasing organic acids and ion-exchanging (Nautiyal et al. 2000), or by mineralizing organo-phosphorus through extracellular acid-phosphatase secretion (van der Hiejden et al. 2008), which can augment the P accessibility. Generally, the endophytic bacteria integrate soluble-P and inhibit P from fixation and/or adsorption in soil (Khan and Joergensen 2009). Therefore, endophytic bacteria can work as a sink for accessible-P by fast mobilization action of microbial phosphorus under phosphate-limiting situation.

Iron is considered as one of the essential components for life; but maximum Fe in soil present as insoluble ferric ion (Fe^{3+}) form, e.g., hydroxides, oxides, carbonates and phosphates. Fe-obtainability may be reformed by microbial production of siderophores those can solubilize Fe under iron-shortage situations and make accessible for the plants (Chen et al. 1998). Higher order botanical sepsis can uptake Fe from plant-microbial-Fe-siderophores complexes via root facilitated chelating degradation and solubilization of inaccessible Fe by liberating phyto-siderophores (Rajkumar et al. 2009). Crucial phytohormones and vitamins can also be delivered to the plants by the endophytic bacteria for augmenting nutrients uptake and metabolic processes (Shi et al. 2009). Predominantly, phytohormones' role in defending botanical species against metals' pressure has revealed that endophytic colonization frequently upsurges nutrient uptake by means of their positive effects on root development dynamics and plant biomass production (Shi et al. 2009; Gravel et al. 2007). According to Phetcharat and Duangpaeng (2012), HMs-stress mitigation by endophytic bacteria is the outcome of nutritional amalgamation & biochemical welfares.

Harish et al. (2008) established that numerous PGPE can reduce the HM-stress effects on plants by confining phyto-pathogen either by inducing plants' resistance against pathogens attack or by biological control of pathogens. Endophytic bacteria as natural bio-controlling agent(s) can develop several inexpensive benefits over PGPR (Rajkumar et al. 2009). Furthermore, in some specific cases several hydrolytic enzymes (e.g., proteases, glucanases, and chitinases), siderophores, antimicrobial volatile organic compounds, and antibiotics can be produced by endophytic bacteria those may efficiently bounds the phytopathogen initiated plant diseases (Sheoran et al. 2015). For example (as demonstrated by Aravind et al. 2010), endophytic bacterial strain *Curtobacterium luteum* TC 10 and *Bacillus megaterium* BP 17 synthesized antibiotics which can efficiently suppress the burrowing nematode (*Radopholus similis*). *Bacillus mojavensis* strains also produce bio-surfactants C-15 surfact in, which is competent to regulate the maize mycotoxic fungus (*Fusarium verticillioides*) (Bacon et al. 2012). Furthermore, endophytic bacteria are effective as pathogens competitor for bioavailable nutrients and colonization niches for plant growth which in turn minimize the hostile environmental effects and ultimately stimulate productivity of plants (Alvin et al. 2014).

27.3.5 Alteration of Plant Metal Uptake Mechanism

27.3.5.1 Metal Stress Amelioration

For effective phytoremediation, the metal phytotoxicity is a serious issue upsetting the metal stress improvement (Shin et al. 2012). A significant number of bacteria-mediated procedures have been convoluted to alleviate the metal-stress either by deliberating plants' tolerance to heavy metals, or by relieving metal poisonousness (Ma et al. 2015a; Rajkumar et al. 2009). It has been reported that endophytic-bacteria can lessen metal-phytotoxicity by means of intracellular sequestration & accumulation (Shin et al. 2012), extracellular precipitation (Babu et al. 2015), adsorption or desorption of metal ions (Luo et al. 2011b; Guo et al. 2010) and bio-transformation of metals' ions into less- or nontoxic-forms (Zhu et al. 2014) (Fig. 27.4). Biotic or abiotic stress can be lessened by the genes encoding antibiotic or metal resistant protein(s). Shin et al. (2012) reported that endophytic bacterial-strain *Bacillus* sp. MN3-4 possesses advanced metal resistant mechanism, e.g., active export through P-type ATPase efflux pump, which can transfer metals' ions across the membranes against the concentration gradient by using the energy liberated by ATP-hydrolysis. Some cases endophytic bacteria also play a vital role in altering the functional and phenotypic characteristics of their host-plants (Liet al. 2011). These sorts of bacteria can adjust the action of plant antioxidant enzymes (e.g., glutathione peroxidase, POS, ascorbate peroxidase, SOD, CAT) as along with the lipid peroxidation, which antagonized plants' defense mechanisms (Fig. 27.4) (Zhang et al. 2010; Wan et al. 2012).

Literature works have focused on the role of endophytic bacteria for accumulation and tolerance capability of HMs in non-accumulator and hyper-accumulator plants for metals remediation. Nevertheless, it has not yet been discussed whether plants growing in metal contaminated soil can modify the survival potential or colonization of specific metal-resistant and/or beneficial microorganisms. Therefore, it is precarious to investigate the circulation diversity, and action of plant growth endorsing endophytic bacteria accompanying with numerous hyper-accumulator plants for heavy metals phytoremediation approaches in soils.

27.3.5.2 Bio-accumulation and Bio-sorption of HMs

Plant-microbes-interaction processes have been recognized to provide metal-resistance in plant body by means of bio-sorption (toxic metals remains in non-living cell) mechanisms or bio-accumulation (toxic metals in the biomass of living cells) (Rajkumar et al. 2012; Ma et al. 2011a; Fig. 27.4). Bio-accumulation practice involves two stages: metabolism-independent passive bio-sorption (e.g., metal ion exchange, chemical and physical adsorption, chelation, micro-precipitation, and surface complexation), and metabolism-dependent active bio-accumulation (e.g., transportation of metals' ions through microbial cells which includes endocytosis, carrier-mediated ion pumps and complex permeation) (Chojnacka 2010). It was revealed that bioaccumulation of heavy metals can be accounted for both abridged poisonousness and uptake of HMs in plants (Mishra and Malik 2013; Deng and Wang 2012; Ma et al. 2011a). Velásquez and Dussan (2009) conducted research on metal(oid)s like Fe, As, Hg, and Co, for bio-accumulation and bio-sorption in dead cells and living biomass of diverse *Bacillus sphaericus* strains and observed that the bio-accumulation and bio-sorption procedures accomplished by living-cells of the two most tolerant-strains were identical.

Bio-sorption in surface molecules (e.g., S-layer proteins) involves in capturing metals' ions either in dead or living cells, while bio-accumulation through helper proteins are involved for reduction of heavy metals through enzymatic process(es) (Elangovan et al. 2005) and/or essential nutrient elements (e.g., S and P) (Suarez and Reyes 2002). Once living cells exposed to heavy metals then metals can be accumulated, precipitated, sequestered, and bound within the intracellular organelles or transferred to the specific structures, depending upon the concerned element and organism (Ma et al. 2011a). Ma et al. (2015b) documented that metal resistant *Bacillus* sp. SC2b was proficient in adsorbing substantial amounts of Pb, Zn and Cd, and bacterial inoculation amended metal poisonousness by the way of biosorption, consequently unveiling a defensive consequence on host plant development. From a phyto-stabilization perspective, metals' bio-sorptions by microbial inoculants are hence invoking a precise attention. For example, Madhaiyan et al. (2007) also demonstrated that the inoculation with endophytic bacteria, *Burkholderia* sp. and *Magnaporthe oryzae* improved plants' growth but abridged the accumulation of Cd and Ni in shoots and roots of tomato plants. This outcome was as a result of the enlarged biosorption and bioaccumulation capability of heavy metal by bacterial strains.

27.3.5.3 Bioavailability and Translocation of Heavy Metals

Translocations of HMs from soils-to-plants predominantly are subjected to the bioavailability of heavy metals in soil (Glick 2010). However, the bioavailability of HMs can be exaggerated by numerous reasons, e.g., soil pH, particle size of soil, nutrient elements, organic matter content, presence of other ions and redox potential (Lebeau et al. 2008). Several investigations proposed that endophytic bacteria having a metal sequestration/ resistance pathway can lessen the bioavailability, phytotoxicity and translocation of HMs (Luo et al. 2011a) and upsurge phytoavailability by the acidification of soils, phosphate solubilization, releasing of metal-chelating agents (e.g., organic acids, siderophores, and biosurfactants), and redox activity (Ma et al. 2011a; Visioli et al. 2014; Luo et al. 2012;) (Fig. 27.4). Chen et al., (2014) showed that bacterial organic-acids frequently results in improved HMs uptake and phytoremediation. For instance, translocation of Cd (in *Brassica juncea*: Salt et al. 1995) and Ni (in *Alyssum lesbiacum*: Solanki and Dhankhar 2011) are conducted by organic/amino acids in hyper-accumulating plant. As most HMs can only be transported by forming complexes with organic-compounds (Maser et al. 2001), where several types of organic chelating agents (released by microorganisms) can alter the existing distribution and chemical-forms of HMs by the combination of metals and microorganisms in plants, subsequently enabling the transportation of HMs from roots to shoots and therefore educating phytoextraction efficacy of heavy metals (Sheng et al. 2008b).

HMs removal through the phyto bacterial bioavailability is more proficient compared to the phytoremediation alone. Degree of HMs bioavailability depends on the oxidation states of HMs, bacterial species, and soil pH. Most of the cases presented in Table 27.1 include phosphate dissolving-bacteria and those that generated siderophores, IAA and ACC-deaminase and to augment plant-growth, convert HMs into bio-available and soluble forms and finally remediate heavy metals from the contaminated soils. Consecutively, plants uptake HMs more readily as the metals are present in soluble and bio-available forms. Hence, depending on the mentioned cases, it is clear that PGP-bacteria have now been treated as biological tools to increase the effectiveness of phyto-remediation of HMs contaminated soil.

27.4 Conclusions and Future Perspective

Potential risks from soils polluted with HMs to ambient ecosystem and human-health have appealed prodigious consideration to the world scientists. Over the last few years, numerous approaches have been developed to lessen the risks raised from heavy metals contamination. Though the remediation of soils polluted with heavy metals is problematic and more challenging under varying climatic condition, contemporary advances in biotechnology has unlocked a new prospect regarding the applicability of favorable endophytic bacteria with organic residues for augmenting the biological control and plant growth as along with the metal phytoremediation. At the present time, biological remediation procedure (organic residues-plant-microbes) is considered to be one of the most effective systems for waste disposal, fertility & organic-matter content enrichment besides the bioremediation to the natural environment. Bioavailabilities of heavy metals in soils are estimated by its' accessibility for utilization, adsorption, and toxicity. Searching an appropriate methodology for investigating the biosorption, bioavailability, and bioaccumulation of heavy metals in soils is potentially significant before choosing a remediation approach for metals remediation from soils.

Suitable methodologies should be searched to elucidate the poor adaptiveness (bioaugmentation for heavy metals remediation) of exogenous microorganisms in contaminated soils. A profound understanding of microbial dynamics of communities and their lifestyle found in pedosphere is thus required to augment the effects of organic residues-plant-microbes for heavy metals remediation from contaminated soils. An exhaustive investigation of enzymatic-aspects in contaminated soils remediated by various biological techniques is still necessary. However, this fact has not yet been discussed whether plants growing in HMs contaminated soils change the survival/colonization potency of specific beneficial and/or metal resistant microorganisms. Therefore, it is decisive to investigate the activity, diversity, and distribution of endophytic microbial communities allied with various hyper-accumulating plants in phytoremediation studies. Moreover, the present study suggested earlier, plant–microbe-metals interactions and microbial-assisted phytoremediation is very essential for heavy metals remediation from contaminated soils in changing climatic conditions. Due to the complexities encountered in the extent and types of HMs pollution, the legal and social issues related to the most polluted sites, this study also emphasized that the interdisciplinary approaches such as ecology, engineering, microbiology, chemistry and geology are essential for bioremediation.

Many plant-associated microorganisms possess the capacity to confer plants development and growth by increasing diseases and abiotic stresses resistance. Nevertheless, they are frequently unable to confirm these advantageous outcomes when applied in the field condition owing to the inadequate effects of colonization. Though considerable advances have been demonstrated to understand the roles of plant-associated microorganisms in metal immobilization/mobilization process and in the implementation of these processes in HMs phytoremediation (Joshi and Juwarkar 2009; Sheng et al. 2008b; Braud et al. 2009; Shi et al. 2011; Kuffner et al. 2010; Li et al. 2010), additional advances are expected. Through improved understanding of those interdisciplinary approaches, the prospect is successfully stimulating and exploiting microbial metabolism for environmental purpose. Despite its limitations (difficulties in supplying the microorganisms with stimulating materials, problems with ensuring sufficient contacts between the pollutants & the microorganisms, and inadequate understanding of how microorganisms behave in the field), the future of bio-remediation technology appears bright due to the advancements of various disciples those shape bioremediation of HMs from contaminated environments.

Acknowledgements The authors express their gratitude to Research Center of Advanced Materials, King Khalid University, Saudi Arabia, for support (award number KKU/RCAMS/22).

Conflict of Interest The authors declare that they have no known competing financial interests or personal relationships that could have appeared to influence the work reported in this paper.

References

Abhilash PC, Dubey RK, Tripathi V, Srivastava P, Verma JP, Singh HB (2013) Adaptive soil management. Curr Sci 104:1275–1276

Abhilash PC, Powell J, Singh HB, Singh B (2012) Plant-microbe interactions: novel applications for exploitation in multipurpose remediation technologies. Trends Biotechnol 30:416–420

Adediran GA, Ngwenya BT, Mosselmans JFW, Heal KV, Harvie BA (2015) Mechanism behind bacteria induced plant growth promotion and Zn accumulation in *Brassica juncea*. J Hazard Mater 283:490–499

Ahemad M (2015) Phosphate-solubilizing bacteria-assisted phytoremediation of metalliferous soils: a review. 3 Biotechnology 5:111–121

Ahemad M, Kibret M (2014) Mechanisms and applications of plant growth promoting rhizobacteria: current perspective. J King Saud UnivSci 26:1–20

Alburquerque JA, de la Fuente C, Bernal MP (2011) Improvement of soil quality after "alperujo" compost application to two contaminated soils characterized by differing heavy metal solubility. J Environ Manag 92:733–741

Ali H, Khan E, Sajad MA (2013) Phytoremediation of heavy metals—concepts and applications. Chemosphere 91:869–881

Alvin A, Miller KI, Neilan BA (2014) Exploring the potential of endophytes from medicinal plants as sources of antimycobacterial compounds. Microbiol Res 169:483–495

Aravind R, Eapen SJ, Kumar A, Ramana KV (2010) Screening of endophytic bacteria and evaluation of selected isolates for suppression of burrowing nematode (*Radopholussimilis* Thorne) using three varieties of black pepper (*Piper nigrum* L.). Crop Prot 29:318–324

ATSDR (2012) Agency for Toxic Substance and Disease Registry, U.S. toxicological profile for cadmium. Department of Health and Humans Services, Public Health Service, Centers for Disease Control, Atlanta, Georgia, USA

Babu AG, Kim JD, Oh BT (2013) Enhancement of heavy metal phytoremediation by *Alnus firma* with endophytic *Bacillus thuringiensis* GDB-1. J Hazard Mater 250:477–483

Babu AG, Shea PJ, Sudhakar D, Jung IB, Oh BT (2015) Potential use of *Pseudomonas koreensis* AGB-1 in association with *Miscanthussinensis* to remediate heavy metal(loid)-contaminated mining site soil. J Environ Manage 151:160–166

Bacon CW, Hinton DM, Mitchell TR, Snook ME, Olubajo B (2012) Characterization of endophytic strains of *Bacillus mojavensis* and their production of surfactin isomers. Biol Control 62:1–9

Badri DV, Weir TL, vanderLelie D, Vivanco JM (2009) Rhizosphere chemical dialogues: plant-microbe interactions. Curr Opin Biotechnol 20:642–650

Barcelo J, Poschenrieder C (2003) Phytoremediation: principles and perspectives. Contrib Sci 2:333–344

Barzanti R, Ozino F, Bazzicalupo M, Gabbrielli R, Galardi F, Gonnelli C, Mengoni A (2007) Isolation and characterization of endophytic bacteria from the nickel hyperaccumulator plant *Alyssum bertolonii*. Microb Ecol 53:306–316
Begum M, Khan R, Hossain SM, Al Mamun SMMA (2021) Redistributions of NORMs in and around a gas-field (Shabazpur, Bangladesh): radiological risks assessment. J Radioanal Nucl Chem. https://doi.org/10.1007/s10967-021-08107-x
Beskoski VP, Gojgic-Cvijovic G, Milic J, Ilic M, Miletic S, Solevic T, Vrvic MM (2011) Ex situ bioremediation of a soil contaminated by mazut (heavy residual fuel oil)—a field experiment. Chemosphere 83:34–40
Bhuyan MS, Bakar MA, Akhtar A, Hossain MB, Ali MM, Islam MS (2017) Heavy metal contamination in surface water and sediment of the Meghna River, Bangladesh. Environ Nanotech Monit Manag 8:273–279
Bolan N, Kunhikrishnan A, Thangarajan R, Kumpiene J, Park J, Makino T, Kirkham MB, Scheckel K (2014) Remediation of heavy metal(loid)s contaminated soils—to mobilize or to immobilize? J Hazard Mater 266:141–166
Braud A, Jézéquel K, Bazot S, Lebeau T (2009) Enhanced phytoextraction of an agricultural Cr, Hg- and Pb-contaminated soil by bioaugmentation with siderophore producing bacteria. Chemosphere 74:280–286
Chandra R, Dubey NK, Kumar V (2017) Phytoremediation of environmental pollutants. CRC Press, Boca Raton, FL
Chen L, Luo SL, Li XJ, Wan Y, Chen JL, Liu CB (2014) Interaction of Cd hyperaccumulator*Solanumnigrum* L. and functional endophyte *Pseudomonas* sp. Lk9 on soil heavy metals uptake. Soil BiolBiochem 68:300–308
Chen LM, Dick WA, Streeter JG, Hoitink HAJ (1998) Fe chelates from compost microorganisms improve Fe nutrition of soybean and oat. Plant Soil 200:139–147
Chen WM, Wu CH, James EK, Chang JS (2008) Metal biosorption capability of *Cupriavidustaiwanensis* and its effects on heavy metal removal by nodulated*Mimosa pudica*. J Hazard Mater 151:364–371
Chaney RL, Baklanov IA (2017) Phytoremediation and phytomining: status and promise. In: Cuypers A, Vangronsveld J (eds) Advances in botanical research: phytoremediation. Academic Press, Cambridge, MA, pp 189–221
Chiang PN, Wang MKK, Chiu CY, Chou SY (2006) Effects of cadmium amendments on low-molecular-weight organic acid exudates in rhizosphere soils of tobacco and sunflower. Environ Toxicol 21:479–488
Chirakkara RA, Cameselle C, Reddy KR (2016) Assessing the applicability of phytoremediation of soils with mixed organic and heavy metal contaminants. Rev Environ Sci Biotechnol 15:299–326
Chojnacka K (2010) Biosorption and bioaccumulation–the prospects for practical applications. Environ Int 36:299–307
Dell'Amico E, Cavalca L, Andreoni V (2008) Improvement of *Brassica napus* growth under cadmium stress by cadmium-resistant rhizobacteria. Soil Biol Biochem 40:74–84
Deng X, Wang P (2012) Isolation of marine bacteria highly resistant to mercury and their bioaccumulation process. Bioresour Technol 121:342–347
Doble M, Kumar A (2005) Biotreatment of industrial effluents. Elsevier Butterworth-Heinemann
Doty SL, Oakley B, Xin G, Kang JW, Singleton G, Khan Z, Vajzovic A, Staley JT (2009) Diazotrophicendophytes of native black cottonwood and willow. Symbiosis 47:23–33
Duruibe JO, Ogwuegbuand MOC, Egwurugwu JN (2007) Heavy metal pollution and human biotoxic effects. Int J Phys Sci 2:112–118
Elangovan R, Avispa S, Rohit B, Ligy P, Chandraraj K (2005) Reduction of Cr (VI) by a *Bacillus* sp. BiotechnolLett 28:247–252
EPA (2009) Risk assessment guidance for superfund, vol. I: human health evaluation manual (part F, supplemental guidance for inhalation risk assessment). EPA-540-R-070-002
Ezawa T, Smith SE, Smith FA (2002) P metabolism and transport in AM fungi. Plant Soil 244:221–230

Fernandez JM, Plaza C, Hernandez D, Polo A (2007) Carbon mineralization in soil amended with thermally-dried and composted sewage sludges. Geoderma 137:497–503

Fernandez JM, Senesi N, Plaza C, Brunetti G, Polo A (2009) Effects of composted and thermally dried sewage sludges on soil and soil humic acid properties. Pedosphere 19:281–291

Fuente CDL, Clemente R, Martinez-Alcala I, Tortosa G, Bernal MP (2011) Impact of fresh and composted solid olive husk and their water-soluble fractions on soil heavy metal fractionation; microbial biomass and plant uptake. J Hazard Mater 186:1283–1289

Gamalero E, Trotta A, Massa N, Copetta A, Martinotti MG, Berta G (2004) Impact of two fluorescent pseudomonads and an arbuscular mycorrhizal fungus on tomato plant growth root architecture, and P acquisition. Mycorrhiza 14:185–192

Glass ADM (1989) Plant nutrition: an introduction to current concepts. Jones and Bartlett Publishers, Boston, p 234

Glick BR (2003) Phytoremediation: synergistic use of plants and bacteria to clean up the Environment. Biotechnol 21:383–393

Glick BR (2004) Bacterial ACC deaminase and the alleviation of plant stress. Adv Appl Microbiol 56:291–312

Glick BR (2010) Using soil bacteria to facilitate phytoremediation. Biotechnol Adv 28:367–374

Glick BR (2012) Plant growth-promoting bacteria: mechanisms and applications. Scientifica

Gravel V, Antoun H, Tweddell RJ (2007) Growth stimulation and fruit yield improvement of greenhouse tomato plants by inoculation with *Pseudomonas putida* or *Trichodermaatroviride*: possible role of indole acetic acid (IAA). Soil Biol Biochem 39:1968–1977

Guo HJ, Luo SL, Chen L, Xiao X, Xi Q, Wei WZ, Zeng G, Liu C, Wan Y, Chen J, He Y (2010) Bioremediation of heavy metals by growing hyperaccumulator endophytic bacterium *Bacillus* sp. L14. Bioresour Technol 101:8599–8605

Gupta G, Panwar J, Jha PN (2013) Natural occurrence of Pseudomonas aeruginosa, a dominant cultivable diazotrophicendophytic bacterium colonizing *Pennisetumglaucum* (L.) R. Br. Appl Soil Ecol 64:252–261

Habib MA, Khan R (2021) Environmental impacts of coal-mining and coal-fired power-plant activities in a developing country with global context. Spatial modeling and assessment of environmental contaminants (Chapter 24). Environmental challenges and solutions. Springer Nature Switzerland AG. https://doi.org/10.1007/978-3-030-63422-3_24

Habib MA, Islam ARMT, Bodrud-Doza M, Mukta FA, Khan R, Siddique MAB, Phoungthong K, Techato K (2020) Simultaneous appraisals of pathway and probable health risk associated with trace metals contamination in groundwater from Barapukuria coal basin, Bangladesh. Chemosphere 242. https://doi.org/10.1016/j.chemosphere.2019.125183

Habib MA, Basuki T, Miyashita S, Bekelesi W, Nakashima S, Techato K, Khan R, Majlis ABK, Phoungthong K (2019a) Assessment of natural radioactivity in coals and coal combustion residues from a coal-based thermoelectric plant in Bangladesh: Implications for radiological health hazards. Environ Monit Assess 191:27. https://doi.org/10.1007/s10661-018-7160-y

Habib MA, Basuki T, Miyashita S, Bekelesi W, Nakashima S, Phoungthong K, Khan R, Rashid MB, Islam ARMT, Techato K (2019b) Distribution of naturally occurring radionuclides in soil around a coal-based power plant and their potential radiological risk assessment. Radiochim Acta 107(3):243–259

Hadi F, Bano A (2010) Effect of diazotrophs (Rhizobium and Azobactor) on growth of maize (*Zea mays* L.) and accumulation of Lead (Pb) in different plant parts. Pak J Bot 42:4363–4370

Hamid Y, Tang L, Sohail MI, Cao X, Hussain B, Aziz MZ, Usman M, He ZL, Yang X (2019) An explanation of soil amendments to reduce cadmium phytoavailability and transfer to food chain. Sci Total Environ 660:80–96

Harish S, Kavino M, Kumar N, Saravanakumara D, Soorianathasundaramb K, Samiyappana R (2008) Biohardening with plant growth promoting rhizosphere and endophytic bacteria induces systemic resistance against banana bunchy top virus. Appl Soil Ecol 39:187–200

He CQ, Tan GE, Liang X, Du W, Chen YL, Zhi GY, Zhu Y (2010) Effect of Zn tolerant bacterial strains on growth and Zn accumulation in *Orychophragmusviolaceus*. Appl Soil Ecol 44:1–5

He H, Ye Z, Yang D, Yan J, Xiao L, Zhong T, Yuan M, Cai X, Fang Z, Jing Y (2013a) Characterization of endophytic Rahnella sp. JN6 from *Polygonumpubescens* and its potential in promoting growth and Cd Pb, Zn Uptake by Brassica Napus. Chemosphere 90:1960–1965

He J, Ma C, Ma Y, Li H, Kang J, Liu T, Polle A, Peng C, Luo Z (2013b) Cadmium tolerance in six poplar species. Environ Sci Pollut Res 20:163–174

Hossain Z, Nouri MZ, Komatsu S (2012) Plant cell organelle proteomics in response to abiotic stress. J Proteome Res 11:37–48

Iqbal MP (2012) Lead pollution—a risk factor for cardiovascular disease in Asian developing countries. Pak J Pharm Sci 25:289–294

Iram S, Uzma G, Rukh S, Ara T (2013) Bioremediation of heavy metals using isolates of filamentous fungus *Aspergillus fumigatus* collected from polluted soil of Kasur, Pakistan. Int Res J Biol Sci 2:66–73

Islam MS, Ahmed MK, Al-Mamun MH, Hoque MF (2014) Preliminary assessment of heavy metal contamination in surface sediments from a river in Bangladesh. Environ Earth Sci 73:1837–1848

Islam MS, Ahmed MK, Al-Mamun MH, Masunaga S (2015a) The concentration, source and potential human health risk of heavy metals in the commonly consumed foods in Bangladesh. Ecotox Environ Saf 122:462–469

Islam MS, Ahmed MK, Al-Mamun MH, Masunaga S (2015b) Potential ecological risk of hazardous elements in different land-use urban soils of Bangladesh. Sci Total Environ 512–513:94–102

Islam MS, Kormoker T, Idris AM, Proshad R, Kabir MH (2021a) Plant-microbe-metal interactions for heavy metal bioremediation: a review. Crop Pasture Sci. https://doi.org/10.1071/CP21322

Islam MS, Kormoker T, Mazumder M, Anika SE, Islam MT, Hemy DH, Mimi US, Proshad R, Idris AM, Kabir MH (2021b) Trace elements concentration in soil and plant within the vicinity of abandoned tanning sites in Bangladesh: an integrated chemometric approach for health risk assessment. Toxin Rev. https://doi.org/10.1080/15569543.2021.1925919

Islam MS, Idris AM, Islam ARMT, Phoungthong K, Ali MM, Kabir MH (2021c) Geochemical variation and contamination level of potentially toxic elements in land-uses urban soils. Int J Environ Anal Chem. https://doi.org/10.1080/03067319.2021c.1977286

Islam MS, Proshad R, Ahmed S (2018) Ecological risk of heavy metals in sediment of an urban river in Bangladesh. Human Ecol Risk Assess an Int J 24:699–720

Järup L (2003) Hazards of heavy metal contamination. Br Med Bull 68:167–182

Jiang C, Sheng X, Qian M, Wang Q (2008) Isolation and characterization of a heavy metal-resistant *Burkholderia* sp. from heavy metal-contaminated paddy field soil and its potential in promoting plant growth and heavy metal accumulation in metal-polluted soil. Chemosphere 72:157–164

Jin HP, Dane L, Periyasamy P, Girish C, Nanthi B, Jae-Woo C (2011) Role of organic amendments on enhanced bioremediation of heavy metal(loid) contaminated soils. J Hazard Mater 185:549–574

Jing Y, He Z, Yang X (2007) Role of soil rhizobacteria in phytoremediation of heavy metal contaminated soils. J Zhejiang Univ Sci 8:192–207

Jing YX, Yan JL, He HD, Yang DJ, Xiao L, Zhong T, Yuan M, Cai XD, Li SB (2014) Characterization of bacteria in the rhizosphere soils of *Polygonumpubescens* and their potential in promoting growth and Cd Pb, Zn uptake by *Brassica napus*. Int J Phytorem 16:321–333

Joshi PM, Juwarkar AA (2009) In vivo studies to elucidate the role of extracellular polymeric substances from Azotobacter in immobilization of heavy metals. Environ Sci Technol 43:5884–5909

Kcil A, Erust C, Ozdemiroglu S, Fonti V, Beolchini F (2015) A review of approaches and techniques used in aquatic contaminated sediments: metal removal and stabilization by chemical and biotechnological processes. J Cleaner Prod 86:24–36

Khalid S, Shahid M, Niazi NK, Murtaza B, Bibi I, Dumat C (2017) A comparison of technologies for remediation of heavy metal contaminated soils. J Geochem Explor 182:247–268

Khan KS, Joergensen RG (2009) Changes in microbial biomass and P fractions in biogenic household waste compost amended with inorganic P fertilizers. Bioresour Technol 100:303–309

Khan S, Cao Q, Zheng YM, Huang YZ, Zhu YG (2008) Health risks of heavy metals in contaminated soils and food crops irrigated with wastewater in Beijing, China. Environ Pollut 152:686–692

Khan R, Islam HMT, Islam ARMT (2021) Mechanism of elevated radioactivity in a freshwater basin: radiochemical characterization, provenance and associated hazards. Chemosphere. https://doi.org/10.1016/j.chemosphere.2020.128459

Khan R, Islam MS, Tareq, ARM, Naher K, Islam ARMT, Habib MA, Siddique MAB, Islam MA, Das S, Rashid MB, Ullah AKMA, Miah MMH, Masrura SU, Bodrud-Doza M, Sarker MR, Badruzzaman ABM (2020) Elemental and polycyclic aromatic hydrocarbons distributions in the sediments of an urban river: influence of anthropogenic runoffs. Environ Nanotech Monit Manag 14:100318. https://doi.org/10.1016/j.enmm.2020.100318

Khan R, Das S, Kabir S, Habib MA, Naher K, Islam MA, Tamim U, Rahman AKMR, Deb AK, Hossain SM (2019a) Evaluation of the elemental distribution in soil samples collected from ship-breaking areas and an adjacent island. J Environ Chem Eng. https://doi.org/10.1016/j.jece.2019a.103189

Khan R, Parvez MS, Jolly YN, Haydar MA, Alam MF, Khatun MA, Sarker MMR, Habib MA, Tamim U, Das S, Sultana S, Islam MA, Naher K, Paul D, Akter S, Khan MHR, Nahid F, Huque R, Rajib M, Hossain SM (2019b) Elemental abundances, natural radioactivity and physicochemical records of a southern part of Bangladesh: implication for assessing the environmental geochemistry. Environ Nanotech Monit Manag. https://doi.org/10.1016/j.enmm.2019b.100225

Khan R, Parvez MS, Tamim U, Das S, Islam MA, Naher K, Khan MHR, Nahid F, Hossain SM (2018) Assessment of rare earth elements, Th and U profile of a site for a potential coal based power plant by instrumental neutron activation analysis. Radiochim Acta 106(6):515–524

Khan R, Rouf MA, Das S, Tamim U, Naher K, Podder J, Hossain SM (2017) Spatial and multi-layered assessment of heavy metals in the sand of Cox's-Bazar beach of Bangladesh. Reg Stud Mar Sci 16:171–180

Khan R, Yokozuka Y, Terai S, Shirai N, Ebihara M (2015) Accurate determination of Zn in geological and cosmochemical rock samples by isotope dilution inductively coupled plasma mass spectrometry. J Anal At Spectrom 30:506–514. https://doi.org/10.1039/c4ja00344f

Kormoker T, Proshad R, Islam S, Ahmed S, Chandra K, Udding M, Rahman M (2019) Toxic metals in agricultural soils near the industrial areas of Bangladesh: ecological and human health risk assessment. Toxin Rev. https://doi.org/10.1080/15569543.2019.1650777

Komarek M, Vanek A, Ettler V (2013) Chemical stabilization of metals and arsenic in contaminated soils using oxides–a review. Environ Pollut 172:9–22

Kuffner M, De Maria S, Puschenreiter M, Fallmann K, Wieshammer G, Gorfer M et al (2010) Culturable bacteria from Zn- and Cd accumulating Salix caprea with differential effects on plant growth and heavy metal availability. J Appl Microbiol 108:1471–1484

Lebeau T, Braud A, Jezequel K (2008) Performance of bioaugmentation-assisted phytoextraction applied to metal contaminated soils: a review. Environ Pollut 153:497–522

Lesley CB, Colette D (2013) The potential use of phytoremediation for sites with mixed organic and inorganic contamination. Crit Rev Environ Sci Technol 43:217–259

Li T, Liu MJ, Zhang XT, Zhang HB, Sha T, Zhao ZW (2011) Improved tolerance of maize (*Zea mays* L.) to heavy metals by colonization of a dark septateendophyte(DSE) *Exophialapisciphila*. Sci Total Environ 409(6):1069–1074

Li WC, Ye ZH, Wong MH (2010) Metal mobilization and production of short-chain organic acids by rhizosphere bacteria associated with a Cd/Zn hyperaccumulating plant *Sedum alfredii*. Plant Soil 326:453–467

Liang X, He CQ, Ni G, Tang GE, Chen XP, Lei YR (2014) Growth and Cd accumulation of *Orychophragmusviolaceus* as affected by inoculation of Cd tolerant bacterial strains. Pedosphere 24:322–329

Liu L, Li W, Song W, Guo M (2018) Remediation techniques for heavy metal-contaminated soils: principles and applicability. Sci Total Environ 633:216–219

Long XX, Chen XM, Chen YG, Woon-Chung WJ, Wei ZB, Wu QT (2011) Isolation and characterization endophytic bacteria from hyperaccumulator *Sedum alfredii* Hance and their potential to promote phytoextraction of zinc polluted soil. World J Microbiol Biotechnol 27:1197–1207

Long XX, Chen XM, Wong JWC, Wei ZB, Wu QT (2013) Feasibility of enhanced phytoextraction of Zn contaminated soil with Zn mobilizing and plant growth promoting endophytic bacteria. Trans Nonf Met Soc China 23:2389–2396

Lu LH, Zeng GM, Fan CZ, Ren XJ, Wang C, Zhao QZhang J, Chen M, Chen A, Jian M (2013) Characterization of a laccase-like multicopper oxidase from newly isolated *Streptomyces*sp C1 in agricultural waste compost and enzymatic decolorization of azo dyes. Biochem Eng J 72:70–76

Luo S, Xu T, Chen L, Chen J, Rao C, Xiao X, Wan Y, Zeng G, Long F, Liu C, Liu Y (2012) Endophyte-assisted promotion of biomass production and metal uptake of energy crop sweet sorghum by plant-growth-promoting endophyte *Bacillus* sp. SLS18. Appl Microbiol Biotechnol 93:1745–1753

Luo SL, Chen L, Chen JL, Xiao X, Xu TY, Wan Y, Rao C, Liu CB, Liu YT, Lai C, Zeng GM (2011a) Analysis and characterization of cultivable heavy metal resistant bacterial endophytes isolated from Cd-hyperaccumulator *Solanumnigrum* L. and their potential use for phytoremediation. Chemosphere 85:1130–1138

Luo SL, Wan Y, Xiao X, Guo H, Chen L, Xi Q, Zeng G, Liu C, Chen J (2011b) Isolation and characterization of endophytic bacterium LRE07 from cadmium hyperaccumulator*Solanumnigrum* L. and its potential for remediation. Appl Microbiol Biotechnol 89:1637–1644

Ma Y, Rajkumar M, Freitas H (2009a) Inoculation of plant growth promoting bacterium *Achromobacterxylosoxidans* strain Ax10 for the improvement of copper phytoextraction by *Brassica juncea*. J Environ Manage 90:831–837

Ma Y, Rajkumar M, Freitas H (2009b) Improvement of plant growth and nickel uptake by nickel resistant-plant growth promoting bacteria. J Hazard Mater 166:1154–1161

Ma Y, Prasad MNV, Rajkumar M, Freitas H (2011a) Plant growth promoting rhizobacteria and endophytes accelerate phytoremediation of metalliferous soils. Biotechnol Adv 29:248–258

Ma Y, Rajkumar M, Luo YM, Freitas H (2011b) Inoculation of endophytic bacteria on host and non-host plants e effects on plant growth and Ni uptake. J Hazard Mater 195:230–237

Ma Y, Oliveira RS, Nai FJ, Rajkumar M, Luo YM, Rocha I, Freitas H (2015a) The hyperaccumulator *Sedum plumbizincicola* harbors metal-resistant endophytic bacteria that improve its phytoextraction capacity in multi-metal contaminated soil. J Environ Manage 156:62–69

Ma Y, Oliveira RS, Wu L, Luo Y, Rajkumar M, Rocha I, Freitas H (2015b) Inoculation with metal-mobilizing plant-growth-promoting rhizobacterium *Bacillus* sp. SC2b and its role in rhizoremediation. J Toxicol Environ Health A 78:931–944

Ma Y, Rajkumar M, Rocha I, Oliveira RS, Freitas H (2015c) Serpentine bacteria influence metal translocation and bioconcentration of *Brassica juncea* and *Ricinuscommunis* grown in multi-metal polluted soils. Front Plant Sci 5:757

Macek T, Mackova M, Kas J (2000) Exploitation of plants for the removal of organics in environmental remediation. Biotechnol 18:23–34

Madhaiyan M, Poonguzhali S, Sa T (2007) Metal tolerating methylotrophic bacteria reduces nickel and cadmium toxicity and promotes plant growth of tomato (*Lycopersiconesculentum* L.). Chemosphere 69:220–228

Maser P, Thomine S, Schroeder JI, Ward JM, Hirschi K, Sze H, Talke IN, Amtmann A, Maathuis FJ, Sanders D, Harper JF, Tchieu J, Gribskov M, Persans MW, Salt DE, Kim SA, Guerinot ML (2001) Phylogenetic relationships within cation transporter families of *Arabidopsis*. Plant Physiol 126:1646–1667

Megharaj M, Ramakrishnan B, Venkateswarlu K, Sethunathan N, Naidu R (2011) Bioremediation approaches for organic pollutants: a critical perspective. Environ Int 37:1362–1375

Mishra A, Malik A (2013) Recent advances in microbial metal bioaccumulation. Crit Rev Environ Sci Technol 43:1162–1222

Montañez A, Blanco AR, Barlocco C, Beracochea M, Sicardi M (2012) Characterization of cultivable putative endophytic plant growth promoting bacteria associated with maize cultivars (*Zea mays* L.) and their inoculation effects in vitro. Appl Soil Ecol 58:21–28

Namkoong W, Hwang EY, Park JS, Choi JY (2002) Bioremediation of diesel contaminated soil with composting. Environ Pollut 119:23–31

Nautiyal CS, Bhadauria S, Kumar P, Lal H, Mondal R, Verma D (2000) Stress induced phosphate solubilization in bacteria isolated from alkaline soils. FEMS Microbiol Lett 182:291–296

Nie L, Shah S, Burd GI, Dixon DG, Glick BR (2002) Phytoremediation of arsenate contaminated soil by transgenic canola and the plant growth promoting bacterium *Enterobacter* cloacae CAL2. Plant Physiol Biochem 40:355–361

Pedra F, Polo A, Ribeiro A, Domingues H (2007) Effects of municipal solid waste compost and sewage sludge on mineralization of soil organic matter. Soil Biol Biochem 39:1375–1382

Peer WA, Baxter IR, Richards EL, Freeman JL, Murphy AS (2005) Phytoremediation and hyperaccumulator plants. Topics in current genetics. In: Tamás MJ, Martinoia E (eds) Mol Biol Metal Homeos Detox 14:299–340

Peng W, Li X, Song J, Jiang W, Liu Y, Fan W (2018) Bioremediation of cadmium- and zinc-contaminated soil using *Rhodobactersphaeroides*. Chemosphere 197:33–41

Pereira SIA, Castro PML (2014) Diversity and characterization of culturable bacterial endophytes from *Zea mays* and their potential as plant growth promoting agents in metal-degraded soils. Environ Sci Pollut Res 21:14110–14123

Phetcharat P, Duangpaeng A (2012) Screening of endophytic bacteria from organic rice tissue for indole acetic acid production. Procedia Eng 32:177–183

Pinto AP, de Varennes A, Fonseca R, Teixeira DM (2015) Phytoremediation of soils contaminated with heavy metals: techniques and strategies. In: Ansari AA, Gill SS, Gill R, Lanza GR, Newman L (eds) Phytoremediation: management of environmental contaminants, vol 1. Springer International Publishing. Cham, Switzerland, pp 133–155

Prapagdee B, Chanprasert M, Mongkolsuk S (2013) Bioaugmentation with cadmium-resistant plant growth-promoting rhizobacteria to assist cadmium phytoextraction by Helianthus annuus. Chemosphere 92:659–666

Proshad R, Islam MS, Kormoker T (2018) Assessment of heavy metals with ecological risk of soils in the industrial vicinity of Tangail district, Bangladesh. Int J Adv Geosci 6:108–116

Puglisi E, Cappa F, Fragoulis G, Trevisan M, Del Re AA (2007) Bioavailability and degradation of phenanthrene in compost amended soils. Chemosphere 67:548–556

Rahman MA, Siddique MAB, Khan R, Reza AHMS, Khan AHAN, Akbor MA, Islam MS, Hasan AB, Hasan MI, Elius IB (2022) Mechanism of arsenic enrichment and mobilization in groundwater from southeastern Bangladesh: water quality and preliminary health risks assessment. Chemosphere. https://doi.org/10.1016/j.chemosphere.2022.133556

Rajkumar M, Ae N, Freitas H (2009) Endophytic bacteria and their potential to enhance heavy metal phytoextraction. Chemosphere 77:153–160

Rajkumar M, Freitas H (2008) Influence of metal resistant-plant growth-promoting bacteria on the growth of *Ricinuscommunis* in soil contaminated with heavy metals. Chemosphere 71:834–842

Rajkumar M, Prasad MNV, Sandhya S, Freitas H (2013) Climate change driven plant-metal-microbe interactions. Environ Int 53:74–86

Rajkumar M, Sandhya S, Prasad MNV, Freitas H (2012) Perspectives of plant-associated microbes in heavy metal phytoremediation. Biotechnol Adv 30:1562–1574

Reeves RD, Baker AJM, Jaffré T, Erskine PD, Echevarria G, van der Ent A (2017) A global database for plants that hyperaccumulate metal and metalloid trace elements. New Phytol. https://doi.org/10.1111/nph.14907

Salt DE, Prince RC, Pickering IJ, Raskin I (1995) Mechanisms of cadmium mobility and accumulation in Indian mustard. Plant Physiol 109:1427–1433

Sarwar N, Imran M, Shaheen MR, Ishaque W, Kamran MA, Matloob A, Rehim A, Hussain S (2017) Phytoremediation strategies for soils contaminated with heavymetals: modifications and future perspectives. Chemosphere 171:710–721

Segura A, Rodríguez-Conde S, Ramos C, Ramos JL (2009) Bacterial responses and interactions with plants during rhizoremediation. Microb Biotechnol 2:452–464

Sheng XF, He LY, Wang QY, Ye HS, Jiang C (2008a) Effects of inoculation of biosurfactants producing *Bacillus* sp. J119 on plant growth and cadmium uptake in a cadmium amended soil. J Hazard Mater 155:17–22

Sheng XF, Xia JJ, Jiang CY, He LY, Qian M (2008b) Characterization of heavy metal-resistant endophytic bacteria from rape *Brassica napus* roots and their potential in promoting the growth and lead accumulation of rape. Environ Pollut 156:1164–1170

Sheoran N, ValiyaNadakkakath A, Munjal V, Kundu A, Subaharan K, Venugopal V, Rajamma S, Eapen SJ, Kumar A (2015) Genetic analysis of plant endophytic *Pseudomonas putida* BP25 and chemo-profiling of its antimicrobial volatile organic compounds. Microbiol Res 173:66–78

Sheoran V, Sheoran A, Poonia P (2011) Role of hyper accumulators in phytoextraction of metals from contaminated mining sites: a review. Crit Rev Environ Sci Technol 41:168–214

Shi JY, Lin HR, Yuan XF, Chen XC, Shen CF, Chen YX (2011) Enhancement of copper availability and microbial community changes in rice rhizospheres affected by sulfur. Molecules 16:1409–1417

Shi YW, Lou K, Li C (2009) Promotion of plant growth by phytohormone producing endophytic microbes of sugar beet. Biol Fertil Soils 45:645–653

Shin M, Shim J, You Y, Myung H, Bang KS, Cho M, Kamala-Kannan S, Oh BT (2012) Characterization of lead resistant endophytic Bacillus sp. MN3-4 and its potential for promoting lead accumulation in metal hyperaccumulator *Alnus firma*. J Hazard Mater 199–200:314–320

Singh A, Kuhad RC, Ward OP (2009) Biological remediation of soil: an overview of global market and available technologies. Advances in applied bioremediation. Springer, Berlin, Heidelberg

Solanki R, Dhankhar R (2011) Biochemical changes and adaptive strategies of plants under heavy metal stress. Biologia 66:195–204

Srivastava S, Verma PC, Chaudhary V, Singh N, Abhilash PC, Kumar KV, Sharma N, Singh N (2013) Inoculation of arsenic-resistant *Staphylococcus arlettae* on growth and arsenic uptake in *Brassica juncea* (L.) Czern. Var. R-46. J Hazard Mater 262:1039–1047

Suarez P, Reyes R (2002) Heavy metal incorporation in bacteria and its environmental significance. Interciencia 27:160–172

Sun YB, Zhou QX, Xie XK, Liu R (2010) Spatial, sources and risk assessment of heavy metal contamination of urban soils in typical regions of Shenyang, China. J Hazard Mater 174:455–462

Taghavi S, Garafola C, Monchy S, Newman L, Hoffma A, Weyens N, Barac T, Vangronsveld J, van der Lelie D (2009) Genome survey and characterization of endophytic bacteria exhibiting a beneficial effect on growth and development of poplar. Appl Environ Microbiol 75:748–757

Tandy S, Healey JR, Nason MA, Williamson JC, Jones DL (2009) Remediation of metal polluted mine soil with compost: co-composting versus incorporation. Environ Pollut 157:690–697

Tiwari S, Singh SN, Garg SK (2012) Stimulated phytoextraction of metals from fly ash by microbial interventions. Environ Technol 33:2405–2413

Ullah A, Mushtaq H, Ali H, Munis MFH, Javed MT, Chaudhary HJ (2015) Diazotrophs-assisted phytoremediation of heavy metals: a novel approach. Environ Sci Pollut Res 22:2505–2514

USEPA (2004) Cleaning Up the nation's waste sites: markets and technology trends. Office of solid waste and emergency response, Washington DC, pp 3–9

van der Hiejden MGA, Bardgett RD, van Straalen NM (2008) The unseen majority: soil microbes as drivers of plant diversity and productivity in terrestrial ecosystems. Ecol Lett 11:296–310

Velásquez L, Dussan J (2009) Biosorption and bioaccumulation of heavy metals on dead and living biomass of *Bacillus sphaericus*. J Hazard Mater 167:713–716

Visioli G, D'Egidio S, Vamerali T, Mattarozzi M, Sanangelantoni AM (2014) Culturable endophytic bacteria enhance Ni translocation in the hyperaccumulator *Noccaeacaerulescens*. Chemosphere 117:538–544

Wan Y, Luo S, Chen J, Xiao X, Chen L, Zeng G, Liu C, He Y (2012) Effect of endophyte-infection on growth parameters and Cd-induced phytotoxicity of Cd-hyperaccumulator *Solanumnigrum* L. Chemosphere 89:743–750

Wu S, Cheung K, Luo Y, Wong M (2006) Effects of inoculation of plant growth-promoting rhizobacteria on metal uptake by *Brassica juncea*. Environ Pollut 140:124–135

Wuana RA, Okieimen FE (2011) Heavy metals in contaminated soil: a review of sources, chemistry, risks and best available strategies for bioremediation. ISRN Ecol 20

Xiong T, Austruy A, Pierart A, Shahid M (2016) Kinetic study of phytotoxicity induced by foliar lead uptake for vegetables exposed to fine particles and implications for sustainable urban agriculture. J Environ Sci 1–12

Yu Z, Zeng GM, Chen YN, Zhang JC, Yu Y, Li H, Liu ZF, Tang L (2011) Effects of inoculation with *Phanerochaete chrysosporium* on remediation of pentachlorophenol-contaminated soil waste by composting. Process Biochem 46:1285–1291

Yuan M, He H Xiao L, Zhong T, Liu H, Li S, Deng P, Ye Z, Jing Y (2013) Enhancement of Cd phytoextraction by two *Amaranthus* species with endophytic *Rahnella* sp. JN27. Chemosphere 103:99–104

Zaidi S, Usmani S, Singh BR, Musarrat J (2006) Significance of Bacillus subtilis strain SJ-101 as a bioinoculant for concurrent plant growth promotion and nickel accumulation in *Brassica juncea*. Chemosphere 64:991–1007

Zeyaullah M, Atif M, Islam B, Azza S, Abdelkafel SP, ElSaady MA, Ali A (2009) Bioremediation: a tool for environmental cleaning. Afr J Microbio Res 3:310–314

Zhang J, Zeng G, Chen Y, Yu M, Huang H, Fan C, Zhu Y, Li H, Liu Z, Chen M, Jiang M (2013) Impact of Phanerochaete chrysosporium inoculation on indigenous bacterial communities during agricultural waste composting. Appl Microbiol Biotechnol 97:3159–3169

Zhang X, Xia H, Li ZA, Zhuang P, Gao B (2011a) Identification of a new potential Cd-hyperaccumulator*Solanumphoteinocarpum* by soil seed bank-metal concentration gradient method. J Hazard Mater 189:414–419

Zhang XX, Li CJ, Nan ZB (2010) Effects of cadmium stress on growth and antioxidative systems in *Achnatheruminebrians* symbiotic with *Neotyphodiumgansuense*. J Hazard Mater 175:703–709

Zhang Y, He L, Chen Z, Wang Q, Qian M, Sheng X (2011b) Characterization of ACC deaminase-producing endophytic bacteria isolated from copper-tolerant plants and their potential in promoting the growth and copper accumulation of *Brassica napus*. Chemosphere 83:57–62

Zhu LJ, Guan DX, Luo J, Rathinasababapathi B, Ma LQ (2014) Characterization of arsenic-resistant endophytic bacteria from hyperaccumulators*Pterisvittata* and *Pterismultifida*. Chemosphere 113:9–16

Chapter 28
Phytoremediation of Arsenic Polluted Soil by *Brassica Nigra* L.

Soumik Chatterjee, Sutapa Deb, and Sabyasachi Chatterjee

Abstract Pollution caused by heavy metals is a prime environmental issue around the world. Among many heavy metals, arsenic (As) is one which has more toxic effect on humans, plants, animals and microorganisms. Heavy metals can be removed from contaminated soil by the aid of plants through phytoremediation process. The main objective of this study is to utilize phytoremediation of arsenic (As) from arsenic contaminated soil with the use of *Brassica nigra* L., belonging to the family Brassicaceae. This study was planned to examine the arsenic resistance capacity of *Brassica nigra* L. *Brassica nigra* L. seeds were germinated among three different concentrations of arsenic trioxide (As_2O_3) treated soil. It has been noticed that roots lengths, numbers, and growth of *Brassica nigra* L. were gradually reduced up to 100 ppm arsenic concentrations. The FESEM-EDAX was done to quantify the arsenic concentration and detection of distribution in roots, stems and leaves tissues after 40 days of phytoremediation. As a result; 0.00, 0.37, 0.61 and 0.54 atomic % of arsenic contents were found in control, 20 ppm, 50 ppm and 100 ppm arsenic treated roots respectively. Whereas, 0.00, 0.61, 0.93 and 0.93 atomic % of arsenic contents were present in control, 20 ppm, 50 ppm and 100 ppm arsenic treated stems respectively. Again, 0.00, 0.61, 0.61 and 0.54 atomic % of arsenic contents were present in control, 20 ppm, 50 ppm and 100 ppm arsenic treated leaves of *Brassica nigra* L. respectively. The result showed that *Brassica nigra* L. has the ability to uptake arsenic up to leaves. In this research, it has been noticed that *Brassica nigra* L. has higher accumulation capacities of arsenic and it can clean up contaminated soil.

Keywords Arsenic treated soil · *Brassica nigra* L. seeds · Phytormediation · FESEM-EDAX study

---

S. Chatterjee · S. Chatterjee (✉)
Department of Botany, Ramananda College, Bishnupur, Bankura, West Bengal 722122, India
e-mail: schatterjeebiotech@gmail.com

S. Deb
Department of Civil Engineering, Indian Institute of Technology, Kharagpur, West Bengal, India

P. K. Shit et al. (eds.), *Soil Health and Environmental Sustainability*,
Environmental Science and Engineering,
https://doi.org/10.1007/978-3-031-09270-1_28

28.1 Introduction

The basic element of environment is soil which constitutes ecosystem and it has an important basis for human beings. Different kinds of industry releases waste-water, live-stock manures, composts, municipal sewage sludge which are the sources of heavy metals that contaminate the soil (Abedin and Meharg 2002). Contamination of soil by heavy metals is one of the environmental issues that affect adversely the soil ecosystem. Heavy metals polluted soil has great effect on plants and human health. Heavy metals, especially arsenic is major environmental pollutant that has great effect on environment. Soil contamination by heavy metals start from centuries but it increased remarkably from last few decades due to technological advances and industrialization (Alam et al. 2001). Arsenic is highly toxic heavy metal which is easily found in traces in water sources, air and soil medium (Kang et al. 1996). Soil contamination by arsenic (As), arises through waste water irrigation and solid waste disposal from industrial activities. Heavy metals concentration in living beings can generally to be introduced through food chain (Ayodele and Abubakkar 2001). The degree of concentration influences by the amount of heavy metals and the activities taking place in a specific area. At the present time, government specifies arsenic polluted areas to avoid for continuous use but remediation of soil and reuse appears to be slow (Majid and Argue 2001). Heavy metals exhibit toxic effects towards biological soil processes and can change soil properties (Giller et al. 1998). As a result, there is a scarcity of land resources, reduction of local taxation, psychological pressure and the impairment of housing prices. Thereby, remediation of polluted soil is very essential. Many industries like iron, patrol, oil, metallurgic, ammunition, pipe and cable sheeting release arsenic as waste product which contaminates soil. Inorganic contaminants such as heavy metals creates different problem than organic contaminants (Ding et al. 2009). Heavy metals generate oxidative stress condition in the cell at high concentration and generate toxic effect on plants and animals (Zhou 2003). Heavy metals contaminate the soil strata and it is not suitable for plant growth and adversely affects the biodiversity. So, release of pollutants in the soil medium should be minimized. Therefore, several controlling steps have been executed, but those are not adequate for inspecting the contamination in the polluted soil. These traditional techniques for remediation of heavy metals from soil are not cost-effective which may cause secondary pollution of soil (Eapen and Souza 2011). Phytoremediation is one of the newly evolving technologies in field of science to remediate the heavy metals from the polluted soil, air and water sources. It may be well-defined by the usage of green plants to eliminate polluted substances from the environment (Ibeto and Okoye 2010). Phytoremediation is very useful and it can provide a comparatively cheaper, long lasting solution for remediation of contaminants from polluted soil. Phytoremediation is less disruptive and more eco-friendly technique. In this present study, usually a selected tropical plant which may have the ability to remediate arsenic from the soil was chosen. The aim of this study was to evaluate the ability of *Brassica nigra* L. to remediate arsenic (As) from the arsenic contaminated soil.

28.2 Materials and Methods

28.2.1 Collection and Processing of the Soil Samples

Soil for this experiment was collected from Durgapur industrial area (depth of 0–20 cm) in a clean sterilized plastic bag. Then soil collected was air-dried under sun bath and then stored for the further study. Various physical and chemical properties of soil were tested by the use of PUSA STFR METER in the laboratory.

28.2.2 In Vivo Culture and Treatment of Arsenic on Brassica Nigra *L. (Mustard)*

The arsenic (As) was added to the soil as arsenic trioxide (As_2O_3). Stock solution (500 ml of 200 ppm concentration) was prepared by use of arsenic trioxide. Three different concentrations of solution were prepared (20, 50, and 100 ppm) from stock solution and transferred in three different pots except the pot which was marked as control. Each pot contains 300 gm of soil and was contaminated with 20 ml of different concentration of arsenic trioxide solution (Carbonell-Barrachina et al. 1995). This experiment was done as a pot experiment at the Botany Department of Ramananda College laboratory. Healthy mustard seeds were used for this study. Mustard seeds (8 seeds per pot) were grown in each different pot for this study. Daily monitoring of the pots were done to observe seeds germination and plants were irrigated with tap water daily. After 40 days each plant of different pots were used for further experiments.

28.2.3 Field Emission Scanning Electron Microscopy (FESEM) and Energy-Dispersive X-Ray Spectroscopy (EDAX) Study

After 40 days of planting, all plants were harvested from pots and separated into four parts, namely seeds, roots, stems, and leaves. All harvested plants were washed by water to remove soil. Primary fixation of leaves, stems and roots was done. The each part sections of *Brassica nigra* L. (with and without treatment of As) immersing in 4% glutaraldehyde and in 0.2 M phosphate buffer (pH 7.2) at normal room temperature (27 ± 2 °C). After that, each plant samples were rinsed 4 times for 15–20 min with or without aldehyde fixatives (Nagpal and Grover 1994). Each plant samples were dehydrated with ethanol. Then samples were mounted and gold coated in a sputter coater machine to prevent the charge buildup due to electrons absorbed by the specimens. A 5-kV accelerating voltage was applied in order to achieve the desired magnification. FESEM using a Carl Zeiss SUPRA 40 machine (Carl Zeiss SMT GmbH, Oberkochen, Germany) attached with energy-dispersive x-ray spectroscopy (EDAX) was used for detections of Arsenic in each plant parts.

28.3 Results and Discussion

Physicochemical properties of uncontaminated soil were shown in Table 28.1. The low pH (5.13) of this soil indicates that soil was highly acidic in nature. Table 28.1 showed the physicochemical properties of soil after 40 days of phytoremediation study. After this study pH level of the contaminated soil (6.07) was increased than the uncontaminated soil pH (5.13). In this study, uncontaminated and contaminated soil has been classified as before and after harvesting the plants. This change of pH may be due to the presence of arsenic in the soil, similar trends of result were found by Tokalio et al. (2006). Soil pH plays a very vital role in the sorption of heavy metals and controls the solubility and hydrolysis of metal hydroxides states, this study found to be similar done by Jung and Thornton in 1996. Soil pH also controls ion-pair formation and also influence solubility of organic matter as well as surface charge of different organic matter (Fe, Mn, and Al-oxides). These information clearly indicated that heavy metals uptake by plants is influenced by soil pH, organic matter, and ion exchange (cation) ability and also on plant species as per the study of Rosselli et al. (2003). Before phytoremediation study, it is to be noticed that the uncontaminated soil had higher organic carbon (1.36%) than the contaminated soil (1.25%) (Reported in Table 28.1). There was a slight decrease of nitrogen content in contaminated soil (107 kg/ha) than the normal soil nitrogen content (118.7 kg/ha) while phosphorous content was higher (50 kg/ha) in contaminated soil than the uncontaminated soil (43.22 kg/ha). The uncontaminated soil showed that the seeds

Table 28.1 Physicochemical properties of soil before and after harvesting the plants (*Brassica nigra* L.)

Properties of soil	Test value of uncontaminated soil	Test value of contaminated soil
pH	5.13	6.07
Organic Carbon (%)	1.36	1.25
Nitrogen (kg/ha)	118.70	107.00
Phosphorus (kg/ha)	43.22	50.00
Potassium (kg/ha)	475.50	390.07
Sulphur (mg/kg)	83.80	70.06
Zinc (mg/kg)	42.80	30.50
Boron (mg/kg)	1.56	1.17
Iron (mg/kg)	31.90	20.08

had potassium content of 475.5 kg/ha before sowing and after sowing, seed had potassium content 390.07 kg/ha. Uncontaminated soil had more quantity of sulphur, zinc, boron and iron contents respectively as compared to the contaminated soil after harvesting the plants. This increase or decrease of soil properties might be due to the arsenic added to the soil and phytoremediation by plant. Ryser and Sauder in 2006 found similar results by use of *Hieracium piloselloides* which belongs to Asteraceae family.

Arsenic treatment had significant effects on germination of *Brassica nigra* L. as compared to control plants (Figs. 28.1, 28.2 and Table 28.2). Arsenic had also significant effect on growth of roots, shoot and seedling of *Brassica nigra* L. at different arsenic concentrations as compared to control sample which is incorporated with the study of Abedin et al. (2002). The results indicated that the roots were highly affected by all the concentrations of Arsenic (upto 100 ppm) as compared stems of *Brassica nigra* L. The germination rate of *Brassica nigra* L. were gradually reduced with the increasing arsenic concentration in pots (Table 28.2, Fig. 28.1). In control condition, all seeds (40) were germinated but in 100 ppm concentration of arsenic only 17 seeds were germinated. Arsenic inhibited the germination rate of *Brassica nigra* L. gradually upto 100 ppm concentration. After 40 days of *Brassica nigra* L. growth the height was measured for control, 20 ppm, 50 ppm and 100 ppm of

Fig. 28.1 Cultivation of *Brassica nogra* L. seeds in different pots with varying concentrations of arsenic

Fig. 28.2 Germination rate of *Brassica nigra* L. with different arsenic concentrations

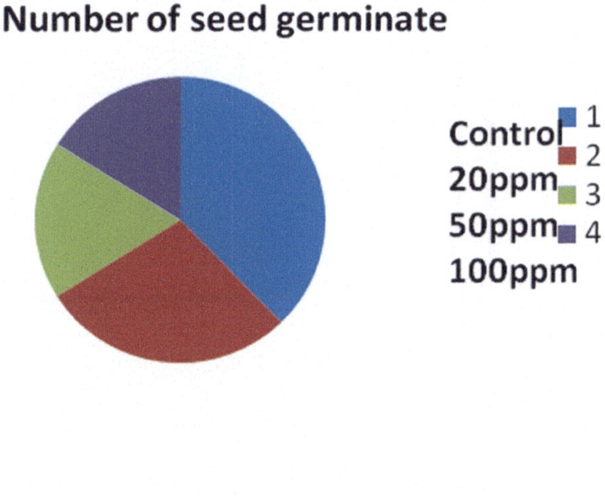

Table 28.2 Germination rate of *Brassica nogra* L. with different arsenic concentrations

Sl. no	Heavy metal concentration (ppm) As_2O_3	Number of seeds	Time (days)	Number of seeds germinate	Germination rate (%)
1	Control	40	10	40	100.0
2	20	40	10	30	75.0
3	50	40	10	19	47.5
4	100	40	10	17	42.5

arsenic condition and found to be 20.5 cm, 19.0 cm, 17.0 cm and 15.4 cm respectively (Fig. 28.3). Lateral roots were gradually decreased and roots length totally inhibited in 100 ppm concentration of arsenic (Figs. 28.4, 28.5, 28.6 and Table 28.3). The *Brassica nigra* L. flowers and stamens structure were unchanged but flowers size were decreased (Fig. 28.8).

Many black spots and leaf tissue damages were appeared after 30 days in *Brassica nigra* L. and also chlorosis disease was observed after 25 days on *Brassica nigra*

Fig. 28.3 Growth rate of *Brassica nigra* L. was gradually decreased with the increasing arsenic concentrations after 20 and 40 days

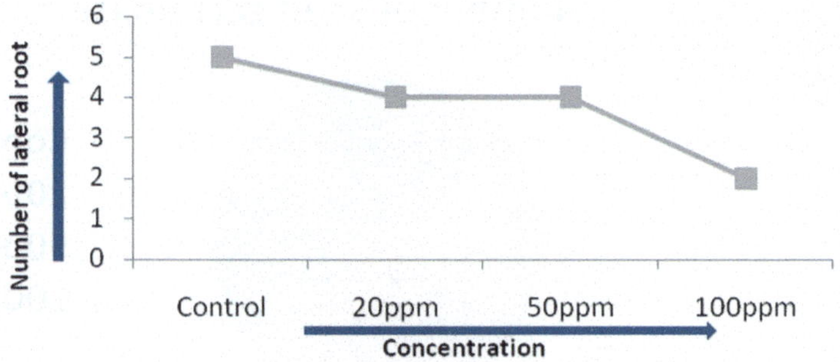

Fig. 28.4 Number of Lateral roots of *Brassica nigra* L.

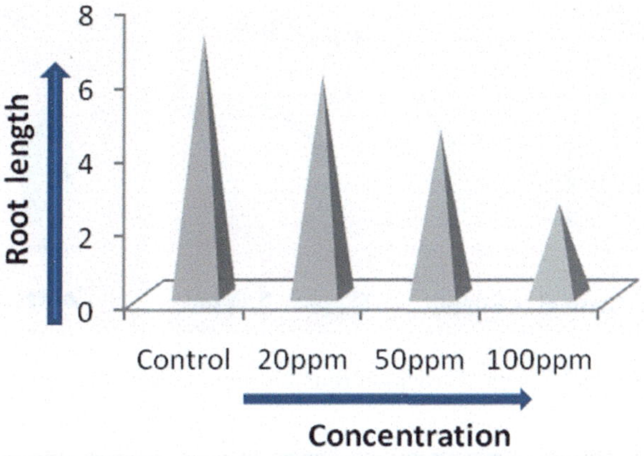

Fig. 28.5 Root lengths of *Brassica nigra* L.

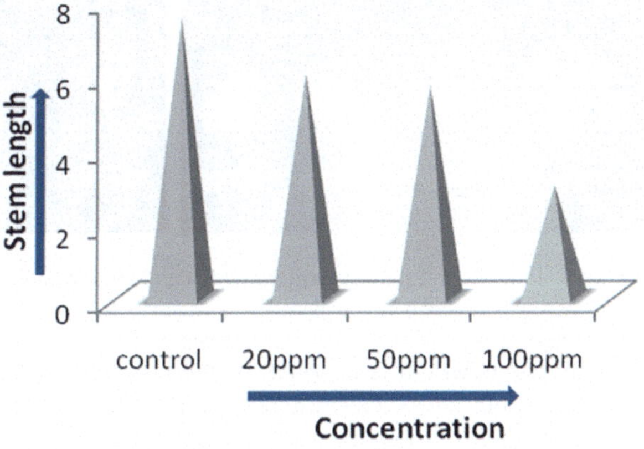

Fig. 28.6 Stem lengths of *Brassica nigra* L.

Table 28.3 Roots and stem lengths of *Brassica nigra* L. at different arsenic concentrations

SL. no	Heavy metal concentration (ppm) As_2O_3	Time (days)	Number of lateral roots	Root length (cm)	Stem length (cm)
1	Control	20	5	7.1	7.5
2	20	20	4	6.0	6.0
3	50	20	4	4.5	5.7
4	100	20	2	2.5	3.0

L. (Figs. 28.7 and 28.8). These results indicated that this plant might uptake and stored more amount of arsenic in leaves. This study also proved that the application of various concentrations of arsenic contributes the decreased seedling growth in *Brassica nigra* L. (Fig. 28.3). There was significant reduction in seed germination rate of *Brassica nigra* L. which might be due to its resistance to tolerate arsenic at all

Fig. 28.7 Tissue damages of *Brassica nigra* L. leaves after 25 days of phytoremediation

Fig. 28.8 Flowers and stamens structure of arsenic treated and untreated *Brassica nigra* L.

concentrations which is in some extent, reported by Jahan et al. (2003) (Figs. 28.1, 28.2 and Table 28.2). Germination and establishment is critical stage in plant life cycle and can be affected in the presence of arsenic but this plant was capable to grow by tolerating a certain concentration of arsenic (Fig. 28.3). It was observed that among heavy metals, arsenic is more toxic at higher concentrations. In the present study, the toxicity of arsenic (at 20, 50 and 100 ppm concentrations) was found in seedling growth and yield of *Brassica nigra* L. were affected but not severely (Fig. 28.3). It supports phytoremediation activity which is corroborated with the findings of Rahman et al. (2004). The response of *Brassica nigra* L. seedlings at 100 ppm concentration helped in understanding the tolerance limit at Arsenic stress.

The FESEM-EDAX analyzer produced a spectrum of the Arsenic present in focus areas of the arsenic treated or untreated samples that allowing detectable elements to be quantized or mapped (Figs. 28.9, 28.10 and 28.11) after 40 days of phytoremediation, followed by the same method of Singh et al. (2004) in case of corn and potato. Arsenic was detected in root internal tissues of different concentrations of arsenic treated *Brassica nigra* L. As a result 0.00, 0.37, 0.61 and 0.54 atomic % of arsenic were found to present in control, 20 ppm, 50 ppm and 100 ppm arsenic treated roots respectively (Table 28.4). So arsenic might have transported to the roots of *Brassica nigra* L. It was also found that 0.00,0.61, 0.93 and 0.93 atomic % of arsenic were present in control, 20 ppm, 50 ppm and 100 ppm arsenic treated stems of *Brassica*

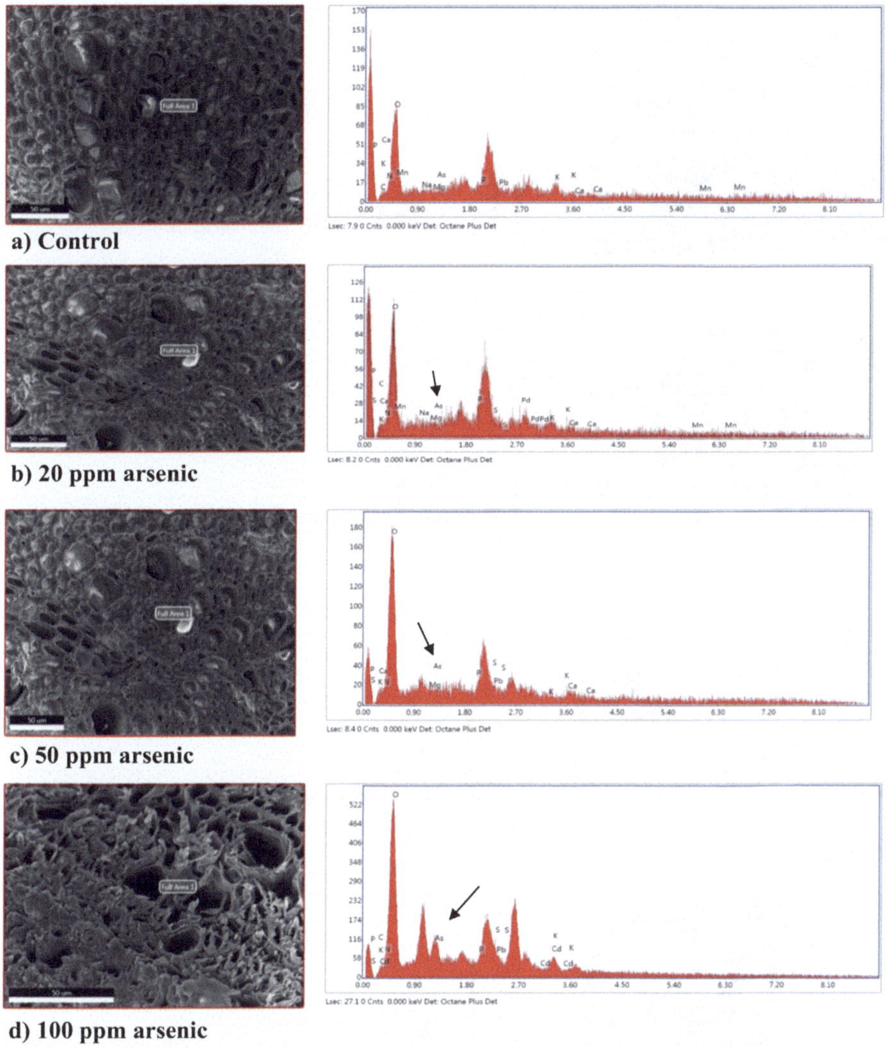

Fig. 28.9 Surface morphology through FESEM micrographs of *Brassica nigra* L. roots with respect to **a** control, **b** 20 ppm, **c** 50 ppm, and **d** 100 ppm of arsenic and corresponding (EDAX) analysis

nigra L. respectively (Table 28.4) which indicated that arsenic was transported to the stem portions. Again, 0.00, 0.61, 0.61 and 0.54 atomic % of arsenic were present in control, 20 ppm, 50 ppm and 100 ppm arsenic treated leaves of *Brassica nigra* L. respectively (Table 28.4). So it is revealed that *Brassica nigra* L. had the ability to uptake arsenic up to the leaves and this plant has well established phytoremediation capability.

The presence of arsenic content in the leaves indicated availability of arsenic in soils as plants thought to absorb arsenic, this study found to be similar done by

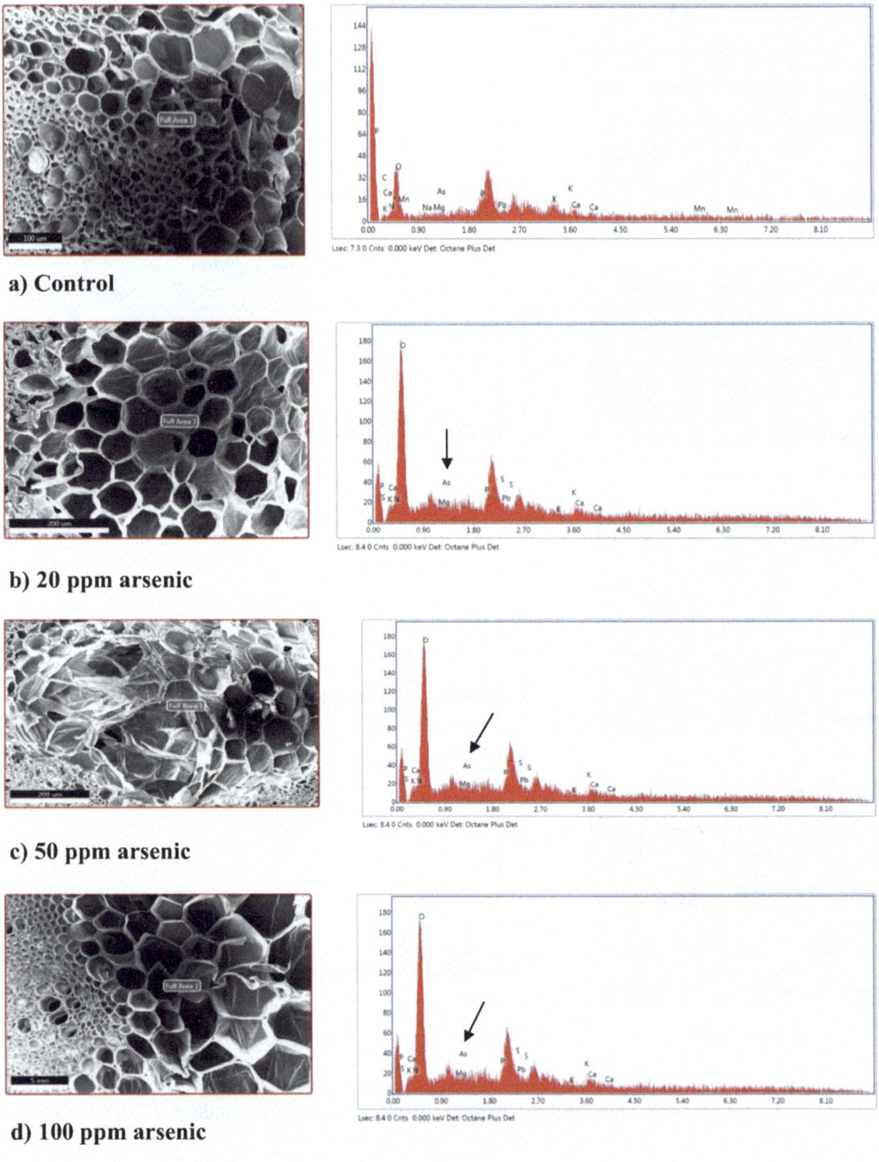

Fig. 28.10 Surface morphology through FESEM micrographs of *Brassica nigra* L. stems with respect to **a** control, **b** 20 ppm, **c** 50 ppm, and **d** 100 ppm of arsenic contents and corresponding (EDAX) analysis

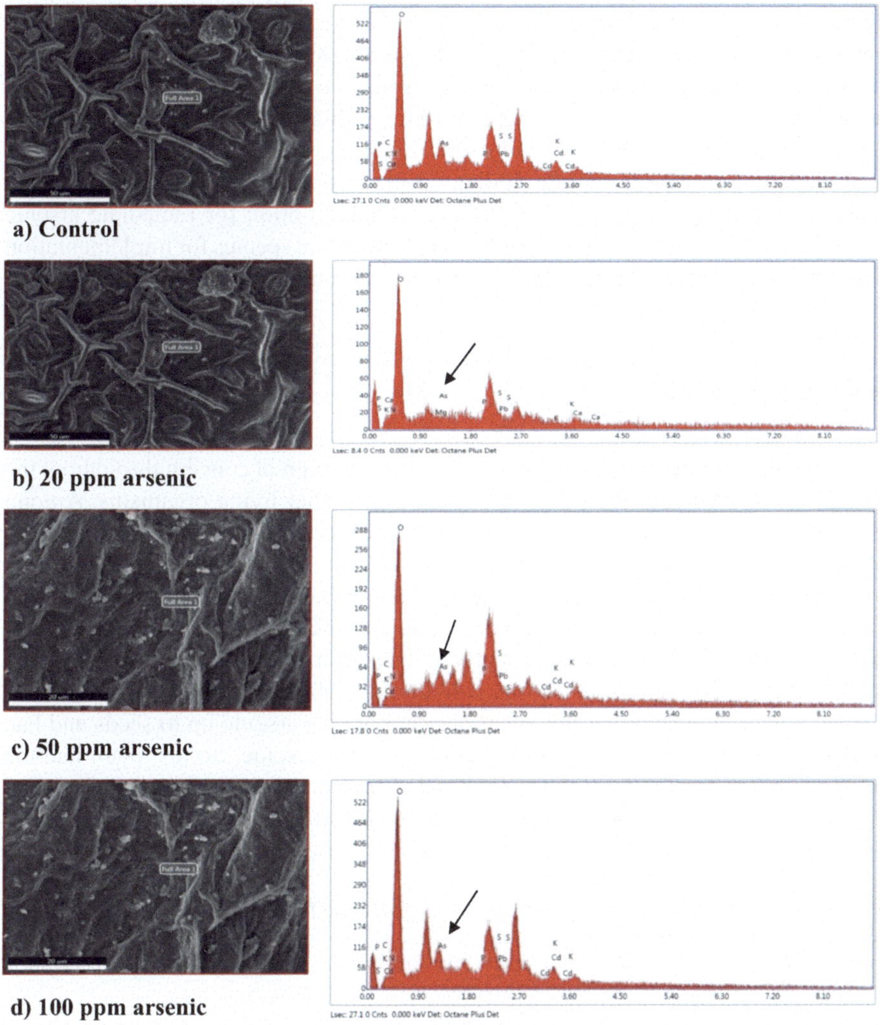

Fig. 28.11 Surface morphology through FESEM micrographs of *Brassica nigra* L. leaves with respect to **a** control, **b** 20 ppm, **c** 50 ppm, and **d** 100 ppm of arsenic contents and corresponding (EDAX) analysis

Table 28.4 Energy-dispersive X-ray spectroscopy (EDAX) study of *Brassica nigra* L.

Plants parts of *Brassica nigra* L.	Atomic % of as in control	Atomic % of as in 20 ppm	Atomic % of as in 50 ppm	Atomic % of as in 100 ppm
Roots	0.00	0.37	0.61	0.54
Stems	0.00	0.62	0.93	0.93
Leaves	0.00	0.61	0.61	0.54

Chaudhary et al. (2000). Those plants were survived in high concentration of arsenic contaminated soils produce fruits and seeds without any abnormalities. As a result *Brassica nigra* L. could not uptake arsenic up to seeds of fruits. The *Brassica nigra* L. had ability to accumulated Arsenic and may be specifically used for phytoextraction of arsenic from arsenic contaminated soils. This study was found to be similar done by Srivastava et al. (2006). These findings confirmed that phytoremediation is one of the biological process that provide a sustainable option for remediate arsenic contaminated soils. It is very crucial to select proper plant species for implementation of phytoremediation techniques for specific target metals.

28.4 Conclusion

Environmental pollution by heavy metals has become a great concern throughout the world because of their effects on public health and other living organisms. Among heavy metals arsenic is very toxic for environment. It is well established that phytoremediation is one of the best technology to remediate soil and restore balance to a stressed environment. This study demonstrated that *Brassica nigra* L. had potential to remediate arsenic from contaminated soil. This plant generally uptook arsenic up to leaves after 40 days which clearly reflects that this plant had the ability to clean up the contaminated soils through phytoremediation. Fruits and seeds were not affected which indicated that this plant may not uptake arsenic up to seeds and had resistance power to tolerate certain concentrations of arsenic. So this plant had the ability of absorb arsenic from the contaminated soils and restore the soil ecosystem. The study of ecosystem of rhizospheric region of *Brassica nigra* L. in presence of varying concentrations of arsenic will be our future area of investigation.

Acknowledgements The authors gratefully acknowledge the financial assistance received from Ramananda College, Bishnupur, Bankura to carry out this research work. The authors would like to extend their sincere gratitude to the Central Research Facility of Indian Institute of Technology, Kharagpur for conducting the FESEM-EDAX study.

References

Abedin MJ, Meharg AA (2002) Relative toxicity of arsenite and arsenate on germination and early seedling growth of rice (*Oryza sativa* L.). Plant Soil 24(3):57–66

Abedin MJ, Cottep-Howells J, Meharg AA (2002) Arsenic uptake and accumulation in rice (*Oryza sativa* L.) irrigated with contaminated water. Plant Soil 24(2):311–319

Alam GM, Tokunaga S, Maekawa T (2001) Extraction of arsenic in a synthetic arsenic contaminated soil using phosphate. Chemosphere 43(8):1035–1041

Ayodele JT, Abubakkar MB (2001) Trace metal levels in Tiga lake, Kano, Nigeria. Trop Environ Resour 3(1):230–237

Carbonell-Barrachina AA, Burlo-Carbonell F, Mataix-Beneyto J (1995) Arsenic uptake distribution and accumulation in tomato plants and effect of arsenic on plant growth and yield. Plant Nutri 18:1237–1250

Chaudhary UK, Biswas BK, Roy Choudhury T (2000) Ground water arsenic contamination in Bangladesh and West Bengal, India. Environ Health Perspect 10(8):388–397

Cox MS, Bell PF, Kovar JL (1996) Different tolerance of canola to arsenic when grown hydroponically or in soil. J Plant Nutri 19(1):1599–1610

Ding ZH, Hu X, Yin DQ (2009) Application of chelants in remediation of heavy metals-contaminated soil. Ecol Environ Sci 18(2):777–782

Eapen S, Souza SF (2011) Prospects of genetic engineering of plants for phytoremediation of toxic metals. Biotechnol Adv 23(97):114–115

Giller KE, Witter E, Mcgrath SP (1998) Toxicity of heavy metals to microorganisms and microbial processes in agricultural soils: a review. Soil Biol Biochem 30(10):1389–1414

Ibeto CN, Okoye CO (2010) High levels of heavy metals in blood of urban population in Nigeria. Res J Environ Sci 4(4):371–382

Jung MC, Thornton I (1996) Heavy metal contamination of soils and plants in the vicinity of a lead-zinc mine, Korea. Appl Geochem 11(1):53–59

Kang LJ, Li XD, Liu JH, Zhang XY (1996) The effect of arsenic on the growth of rice and residues in a loam paddy soil. J Jilin Agric Univ 18(1):58–61

Majid A, Argue S (2001) Remediation of heavy metal contaminated solid wastes using agglomeration techniques. Miner Eng 14(11):1513–1525

Marin AR, Masscheleyn PH, Patrick WH (1992) The influence of chemical form and concentration of arsenic on rice growth and tissue arsenic concentration. Plant Soil 13(9):175–183

Nagpal A, Grover IS (1994) Genotoxic evaluation of systemic pesticides in *Allium cepa* L. mitotic effects. Nucleus 37:99–105

Rahman MA, Rahman MM, Miah MA, Khaled HM (2004) Influence of soil arsenic concentrations in rice (*Oryza sativa* L.). J Sub-trop Agric Res Dev 2(1):24–31

Rosselli W, Keller C, Boschi K (2003) Phytoextraction capacity of trees growing on a metal contaminated soil. Plant Soil 25(2):265–272

Ryser P, Sauder WR (2006) Effects of heavy-metal contaminated soil on growth, phenology and biomass turnover of *Hieracium piloselloides*. Environ Pollut 14(1):52–61

Singh J, Kaur L, Singh N (2004) Effect of acetylation on some properties of corn and potato starches. Starch/Starke 56(1):586–601

Srivastava M, Ma LQ, Santos JA (2006) Three new arsenic hyper-accumulating ferns. Sci Total Environ 64(1):24–31

Tokalio S, Kartal S, Ultekin AG (2006) Investigation of heavy-metal uptake by vegetables growing in contaminated soils using the modified BCR sequential extraction method. Int J Environ Anal Chem 86(6):417–430

Zhou QX (2003) Interaction between heavy metals and nitrogen fertilizers applied in soil-vegetable systems. Bull Environ Contam Toxicol 71(11):338–344

Chapter 29
π-π Interaction: Defining the Role and Relevance in Environmental Detoxification of Heavy Metals from Soil

Varun Dhiman and Deepak Pant

Abstract Heavy metals contamination of the soil due to unreasonable hazardous waste disposal is a serious environmental concern. The entry of organic pollutants/heavy metals in the terrestrial and other environmental spheres attracted the attention of the global scientific community which in turn led to the development of numerous remediation pathways for their effective treatment. Environmental detoxification of heavy metals involves the various processes that limit the availability and toxicity of metal and protect biological targets from them. Mechanistically, it can be described with reference to the chemical skeletal of pollutant/metal (M) and nearby host (H) moiety through complexation. The most important heavy metal detoxification method involves complexation initiated by π-cloud extension. The π-π interaction regarded as powerful noncovalent intermolecular interactions that direct the supramolecular architectures and synthesis of organic molecules for bioremediation purposes. Conjugated π-systems in many ways affect the physical properties of toxic substances involving the application of coordination chemistry. These interactions govern atrazine uptake upto 84–95% and enhance leaching efficiency. However, controlling metal–organic complex motifs through coordination bonds remains the center of attention while role of π-π interactions has been ignored. Revealing the contribution of π-π interactions, current chapter aims to provide a general overview of π-π interactions and their role in remediation of organic pollutants/heavy metals present in soil and other environmental spheres.

Keywords Complexation · Environment · Heavy metals · Ligands · Pollutants · Remediation · Soil Health · Sustainability · π-π interactions

V. Dhiman (✉) · D. Pant
Department of Environmental Sciences, Central University of Himachal Pradesh, Dharamshala 176215, India
e-mail: varundhiman79@gmail.com

P. K. Shit et al. (eds.), *Soil Health and Environmental Sustainability*,
Environmental Science and Engineering,
https://doi.org/10.1007/978-3-031-09270-1_29

29.1 Introduction

The continuous release of pollutants in the form of heavy metals and their complexes in the natural soil environment and other environmental spheres due to unsustainable growth and excessive industrialization causes serious toxic hazards to the existed flora and fauna (Pant et al. 2018; Zhang and Wang 2020; Dhiman et al. 2020). Soil health is highly influenced by numerous chemicals released from agricultural runoff and non-scientific disposal practices (Sharma et al. 2018; Meena et al. 2020). Consequently, exceeds the pollutants/ metal concentration than the maximum permissible limits which in turn causes potential environmental risk. Heavy metals (e.g. Cd, Cr, Cu, Ni, Zn, As, Hg), PAH, pharmaceuticals, pesticides, herbicides, weedicides, etc. are some of the major contributors to soil pollution (Li et al. 2020; Yuvaraj and Mahendran 2020; Xu et al. 2021). These pollutants deteriorate the physiochemical properties of the soil.

Inspired from the extreme aromaticity of the organic pollutants and their derivatives paves the way for the development of a supramolecular structure for the adsorption and decontamination of environmental pollutants using π cloud extension (Chen et al. 2018; Bettini et al. 2020). π-π interactions are generally non-covalent interactions of aromatic moieties, enhancing the π cloud and could play a crucial role in understanding the chemistry behind the complexations (Neel et al. 2017; Lin et al. 2020). The π-orbitals are the major systems in π-π stacking interactions (Fu et al. 2016). Further, magical materials such as biochar or different pyrogenic carbonaceous materials have been extensively modified using π-π interactions (Xiao and Pignatello 2015; Pignatello et al. 2017). It has been observed that the organic pollutants of environmental concern are usually known to contain positively charged amino groups at environmental Ph (Ambaye et al. 2020). These positively charged aromatic amines can act as π acceptor ligands in inducing π-π interactions within the PCM for the development of adsorbents for the pollutants decontamination from the soil (Xiao and Pignatello 2015). Charge polarization of the ring quadrupole of organic contaminants has an important role in causing π-π interactions (Dougherty 2013; Aliakbar Tehrani and Kim 2016). The amines associated within the natural soil environment are highly capable in π-π electron donor–acceptor interactions (Zhu et al. 2004). This capability can modify the behavior of decontamination adsorbents through charge-quadrupole interactions with the highly electron-rich surface of organic contaminants.

29.2 Soil Environment

Organic matter, water, air, minerals, and living fauna constitute the soil which is the topmost layer of the earth's crust and is generally referred to as the skin of the earth (Kutílek and Nielsen 2015; Bhattacharyya and Pal 2015). The complex interactions of these constituents resulted in the development of soil/terrestrial environment having a dynamic origin (Wagenet and Hutson 1997; Mandal 2016; Al-Kaisi et al. 2017). Pedogenesis (soil formation) occurs because of the complex phenomenon of rock weathering which involves various physical and chemical processes (Lavelle and Spain 2001). Soil environment provides a base for life sustainability (Hillel 2007) and to perform its life-sustaining functions, it requires a healthy texture but a series of anthropogenic degradation and toxication processes interfere in its normal functioning by deteriorating the soil structure (Carter et al. 1997; Lal 2012; Zhao and Hou 2019). These in turn cause productivity loss and low soil utility in terms of material use (Garbuio et al. 2012). The soil environment particularly facing the problem of erosion (Borrelli et al. 2020), low organics (Obalum et al. 2017), biodiversity loss of residing flora and fauna (Geisen et al. 2019), salinization (Singh 2021), sealing (Artmann 2014), compaction (Shah et al. 2017) and diffused contaminants that highly alter the soil quality which takes the shape of serious and widespread environmental problem across the globe (Sethi and Gupta 2020).

29.3 Major Soil Pollutants

Several pollutants seriously affect the soil environment. PAHs, antibiotics, heavy metals, and hydrophobic organic pollutants are some of the major soil pollutants (Fig. 29.1) that have a significant impact on soil. These are discussed below:

PAHs

These are ubiquitous organic environmental pollutants generated from incomplete combustion of wood, oil, coal, petrol, etc. (Abdel-Shafy and Mansour 2016). They sorbet to aerosols when in the gas phase in the ambient air (Abdel-Shafy and Mansour 2016; Hussain et al. 2018). They are highly toxic and cause mutations and carcinogenicity in the exposed animals (Rengarajan et al. 2015). They entered the terrestrial environment through dry or wet deposition processes (Zhang et al. 2015a). On their deposition to the soil, these start bonding with the soil particles and mobilize themselves in the subsurface of the soil environment. Due to their higher toxicity, PAHs must be removed from the soil. PAH sorption is one of the efficient methods from its removal from the terrestrial environment (Karaca et al. 2016). The cation-π interactions further enhance the sorption behavior of the carbonaceous materials (Qu et al. 2008). This is discussed in further sections of the chapter.

Antibiotics

These are highly complex and environmental persistent molecules having diverse functional groups in their chemical structures (Cycoń et al. 2019). Antibiotics production increases at an annual rate of 100,000–200,000 tons per year worldwide (Van Boeckel et al. 2015). Their continuous release in the soil environment through municipal wastewater, untreated sewage, urine, and feces in the form of active pharmaceutical ingredients causes serious environmental concerns. Cation-π interactions

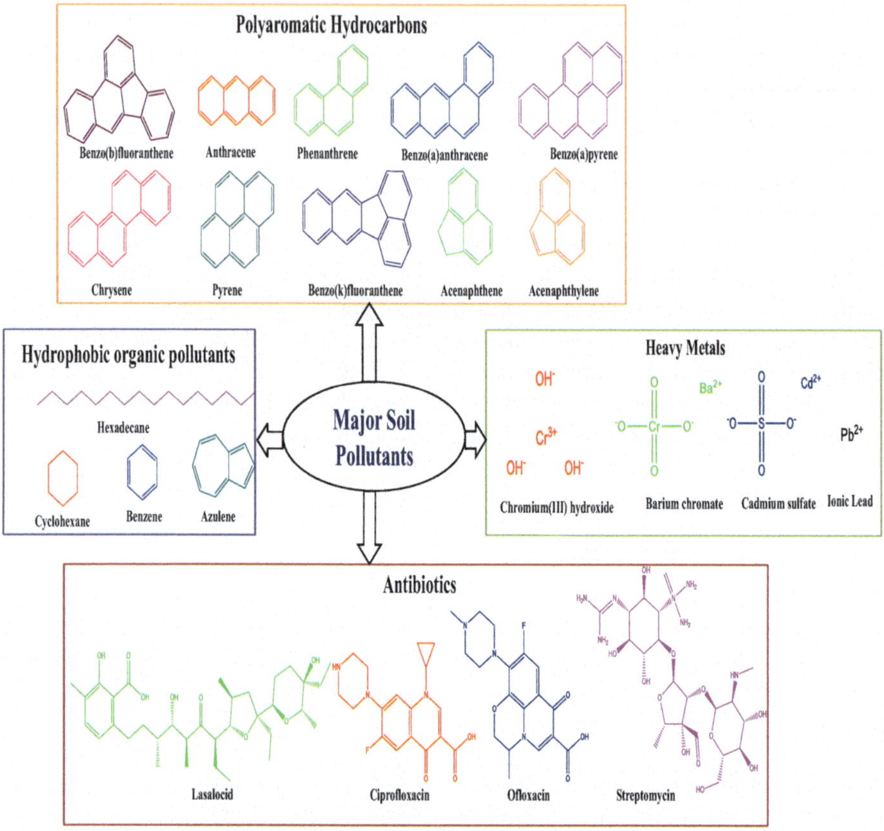

Fig. 29.1 Major pollutants of soil environment (Havugimana et al. 2015; Cycoń et al. 2019; Patel et al. 2020)

help in their sorption by determining drug-receptor interactions and macromolecular structures (Zhao et al. 2017).

Heavy metals

Continuous toxic emissions from industries, mine tailings, fertilizers, paints, leaded gasoline, metal scrap, coal burning, atmospheric deposition, petro-products, paints, etc. contaminated the soil environment (Dhiman 2020; Dhiman et al. 2020; Dhiman and Pant 2021). Heavy metals constitute the major part of these above-mentioned products. Lead, zinc, chromium, copper, mercury, cadmium are some of the majorly identified inorganic soil pollutants (Abioye 2011; Wuana and Okieimen 2011).

Hydrophobic organic pollutants

Apart from PAH, other hydrophobic organo-xenobiotics such as petro-hydrocarbons and PCBs cause soil and groundwater contamination (Zhang et al. 2013; Srivastava et al. 2019; Truskewycz et al. 2019). Due to their higher toxicological impacts, these pollutants cause serious environmental hazards. For the efficient and effective removal of target pollutants, selective adsorption techniques enhanced by π-π interactions have been used.

29.4 Soil-Pollutants Interactions and Role of Complexation

Soil-pollutants interact using physical, chemical, and biological mechanisms (Mirsal 2004). In physical interaction, the particular pollutant is adsorbed on the granular or intergranular space of the soil particle (Ye et al. 2017). The study of the mechanism behind the soil-pollutants interactions in the environmental sphere helps in the development of efficient soil remediation practices. The study of complexation during these interactions is one of the important aspects of soil remediation. Organic pollutants make complexes with the help of chemical reactions which further enhance their biotoxicity by contributing synergistic effects on chemical mechanisms (Kawaguchi and Kyuma 1959; Biswas et al. 2018). Soil is known to contain a simultaneous occurrence of potentially toxic elements with numerous pollutants (Ye et al. 2017). Their chemical interactions results in the complexation that limits the bioavailability of pollutant. For example, when dissolved negatively charged 2,4-DCP combined with the chromium ions (aqueous solution) at a pH ranges from 7.2 to 7.4, a neutral complex has been formed and it has been observed that the mobility of both the reactants and products seems to be decreased (Fig. 29.2).

Metal cations show a higher level of affection toward organic pollutants with higher electron density which enhances the cation-π bonding (Biswas et al. 2018). The whole phenomenon is described as the “Salting in” effect (Chiu and Dural 1997). Different bidentate and multidentate organic ligands and potentially toxic elements from the complexes through self-assembly involving coordination bonding

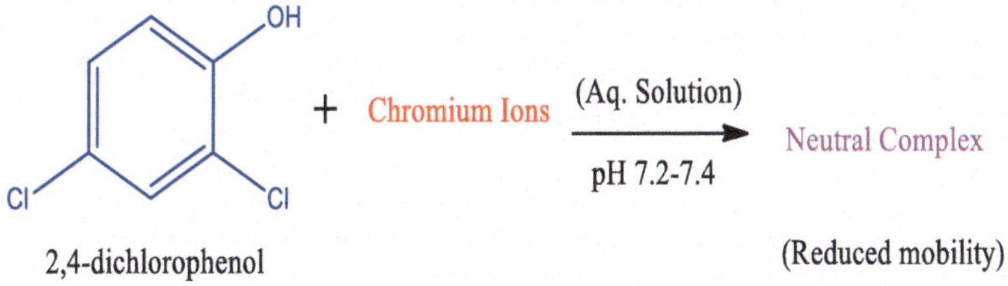

Fig. 29.2 The reaction mechanism of 2,4-DCP with chromium ions results in the formation of the neutral complex with reduced mobility (Ye et al. 2017)

that in turn changes the existing structure of the pollutant in the terrestrial environment (Torri and Corrêa 2012; Chen et al. 2020). Soil pollutants and the metal ions shows enhanced cation-π bonding between their aromatic rings which enhances the adsorption of these pollutants from the soil. The naturally existing soil pollutants have different adsorption behavior thus influencing their retention and releasing time in the terrestrial environment (Huang et al. 2020). The adsorption is mainly regulated by surface modifications, competition on adsorption sites, ion exchange, co-precipitation, and binding forces of functional groups. The factors involved in the regulation of adsorption of soil contaminants in the terrestrial environment are represented in Fig. 29.3.

Consequently, the role of the complexation mechanism is an essential factor for soil remediation. It further enhances the bioavailability and solubility of pollutants.

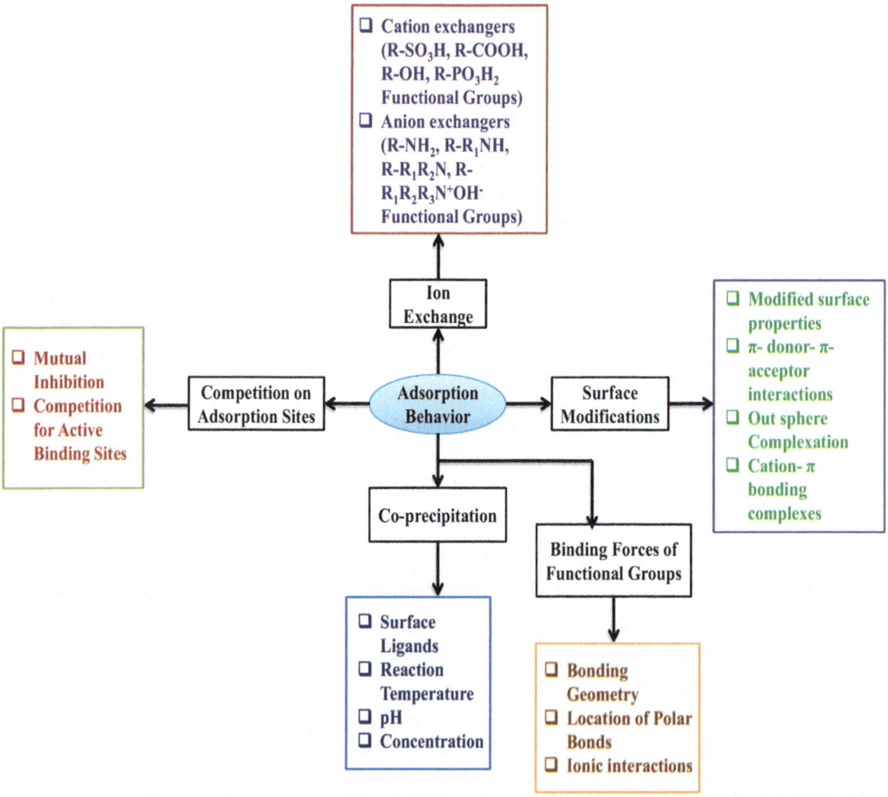

Fig. 29.3 Regulation factors of the soil contaminants adsorption in the terrestrial environment (Jørgensen 1989; Ye et al. 2017)

29.5 Cation-π Interaction and Host–Guest Complexation

The highly polarized ionic species electrostatically shows strong binding with a neutral molecule which in turn causes cation-π interactions with the neutral π system (Ma and Dougherty 1997). The different studies reveal the preferential interaction of K^+ ions with the benzene which was further validated by analyzing the molecular plane of benzene and K^+ ion interactions (Ferretti et al. 2020). Furthermore, studies provide a quantitative analysis of cation-π interaction strength. The study by Dougherty 2013on cation-π interactions explores molecular recognition in the different host–guest models. It has been observed that K^+ ions are highly potent and specific for aromatic complexes (Ma and Dougherty 1997). Different techniques (Table 29.1) have been employed for understanding the intrinsic binding mode of cation-π complexes.

29.6 Sorption Behavior and π-π Interactions in Soil Remediation

The π-π interactions are highly potent and powerful interactions that can be useful in the development of self-assembled supramolecular engineered materials for soil remediation (Deng et al. 2020). The donor–acceptor π-π interactions influence the sorption behavior of stacked supramolecular motifs (Zhu et al. 2004). This interaction causes π cloud extension and hence enhances the surface area and sorption capacity for organic pollutants in the soil. The sorption behavior of organic pollutants in the natural environment determines their fate by analyzing its mobilization, bioavailability, and biodegradation mechanism (Zhang et al. 2015b). The soil contaminants are retained by soil-based organic matter and interactions have occurred between aromatic rings, functional groups, amorphous and microporous structures which generally influences the sorption affinity (Weber et al. 2002; Huang et al. 2003; Gunasekara and Xing 2003; Kang and Xing 2005; Ran et al. 2007). For example, PAH sorption was highly facilitated by strong, non-covalent aromatic π donors and cation interactions. With this facilitation, PAH sorption occurs at the mineral surface (Zhu et al. 2004; Qu et al. 2008). The cation π interaction strength of PAH is mainly regulated by co-existing soil pollutants, delocalized π electrons, and exchangeable cations (Zhu et al. 2004; Qu et al. 2008; Zhang et al. 2011; Vasudevan et al. 2013). Similarly, in the case of lead detoxification from the soil environment, cation π bonding has been observed between sorbed phenanthrene aromatic ring and associated lead molecule (Zhu et al. 2004; Zhang et al. 2011). The large ionic radius of lead rendered the cation π interaction which makes lead more liable for strong π interactions (Zhang et al. 2015b).

Similarly, biochar has been modified as a new functional material for soil remediation purposes. It efficiently removes soil pollutants and proved beneficial in soil

Table 29.1 Different techniques are employed for understanding the intrinsic binding mode of cation-π complexes

Techniques	Complexes and interactions	Uses	References
CID	Ag^+ with $C_9H_{11}NO_2$ mono and dimer	It helps in understanding the dissociation reaction	Shoeib et al. (2002)
	Mono and Bis-Benzene complexes	Sequential bond dissociation energy	Armentrout and Rodgers (2000)
FT-ICR	Complexes of different atomic ions with coronene complexes of Na^+ ion	Analyze formation of metal ions-π complexes	Pozniak and Dunbar (1997)
Threshold CID	Metal ions-π complexes	Measure affinity and strength of cation-π interaction for alkali metal ions complexed with vast π system	munugama and Rodgers (A2000, 2002), Ruan and Rodgers (2004), Hallowita et al. (2008, 2009)
ESI–MS	Hydrated divalent alkaline earth metals and benzene interaction	Non-covalent complex with phosphorylated residues	Rodriguez-Cruz and Williams (2001)
IRPD	Alkali metal ions and crown ether complexes	Determination of neutral vibrations of crown ether when forming complexes with alkali metal ions	Rodriguez et al. (2010)
IRMPD	Di and Tri-peptides complexed with alkali metal ions	Characterization of the behavior of small and large alkali metal ions on complexation with Phe ligands via cation-π interactions	Dunbar et al. (2009, 2010, 2011)
UVRR	Cation-π interactions between metal ions and pyrrole groups of diazacrown ether	It analyzes the cation-π interactions between the numerous protein structures	Schlamadinger et al. (2011)

remediation (Yang et al. 2019). π cloud extension helps in the development of modified biochar having the superior capability to detoxify the soil. The delocalized π electrons help in the surface complexation of different heavy metals like copper, zinc, and cadmium on modified biochar (Xu et al. 2013). Likewise, the π electron cloud lessen the vacancies on the biochar surface area and promotes lead adsorption (Yu et al. 2016). π-π interactions also promote the sorption of sulfonamide species (Ahmed et al. 2017).

The π-π interactions improve the sorption capacity of the newly developed sorbent material. They governs the metal organic frameworks by enhancing pore size which results in higher uptake of commonly used herbicide, atrazine up to 84–95% which is attributed to the enhanced π-π interactions (Akpinar et al. 2019).

Antibiotic such as oxytetracycline was also found to be absorbed by biochar. This adsorption mechanism is also mediated by π-π interactions which further involve metal bridging, cation exchange, and surface complexation (Jia et al. 2013). Different carbon-based materials help in antibiotic removal from the terrestrial environment as they are highly useful in agricultural soil amendments. The aromaticity of antibiotics is responsible for their adsorption. The fluorescence experiments using a confocal laser scanning microscope are used for observing π-π interactions. Studies reveal that as the number of π rings increases, the adsorption rate is also increased (Peng et al. 2016).

Aromatic π systems are responsible for higher adsorption of organic matter and fire derived black carbon from the soil. π donor-acceptor interactions enhances adsorption energy ranges from 4 to 167 kJ mol^{-1} which shows a substantial sorptive potential for aromatic compounds present in the terrestrial sphere (Akpinar et al. 2019).

Recently a study confirms the role of cation-π interactions in enhancing the leaching efficiency. It has been observed that leaching efficiency of hydroxamate siderophore produced by *Pseudomonas fluorescens HMP01* is higher for heavy metals and PAH molecules present in the soil. The cation-π interactions and coordination causes 90.2 mg/kg of Phenanthrene uptake in the contaminated soil (Yi et al. 2022). The above discussion proves the role of π-π interactions in soil remediation.

Conclusion

The current chapter illustrates the universality of π-π interactions in heavy metals bioremediation from the soil environment. The chapter further reveals that π-π interactions are of exceptional importance in soil remediation. Modern studies report the role of π-π interactions among heavy metals and soil environment. The sorption of π system compounds on the soil surface, minerals using carbonaceous biochar, and other substances have been observed. It has been further revealed that the underlying mechanisms of heavy metals removal from polluted soil under the influence of co-existing factors are to be further explored.

Acknowledgements The authors are grateful to the Central University of Himachal Pradesh for providing the necessary facilities for this work.

Authors Contributions Varun Dhiman: Planning, conceptualization, manuscript writing, data arrangement; Deepak Pant: Editing, feedback, and supervision.

Funding There is no external funding for this research.

Competing Interests The authors declare that they have no known competing financial interests or personal relationships that could have appeared to influence the work reported in this paper.

Availability of Data and Materials Data sharing applies to this article.

References

Abdel-Shafy HI, Mansour MSM (2016) A review on polycyclic aromatic hydrocarbons: source, environmental impact, effect on human health and remediation. Egypt J Pet 25:107–123. https://doi.org/10.1016/j.ejpe.2015.03.011

Abioye OP (2011) Biological remediation of hydrocarbon and heavy metals contaminated soil. Soil Contam 7:127–142

Ahmed MB, Zhou JL, Ngo HH et al (2017) Single and competitive sorption properties and mechanism of functionalized biochar for removing sulfonamide antibiotics from water. Chem Eng J 311:348–358. https://doi.org/10.1016/j.cej.2016.11.106

Al-Kaisi MM, Lal R, Olson KR, Lowery B (2017) Chapter 1—Fundamentals and functions of soil environment. In: Al-Kaisi MM, Lowery BBT-SH and I of A (eds). Academic Press, pp 1–23

Akpinar I, Drout RJ, Islamoglu T, Kato S, Lyu J, Farha OK (2019) Exploiting π–π interactions to design an efficient sorbent for atrazine removal from water. ACS Appl Mater Interfaces 11(6):6097–6103

Aliakbar Tehrani Z, Kim KS (2016) Functional molecules and materials by π-Interaction based quantum theoretical design. Int J Quantum Chem 116:622–633. https://doi.org/10.1002/qua.25109

Ambaye TG, Vaccari M, van Hullebusch ED et al (2020) Mechanisms and adsorption capacities of biochar for the removal of organic and inorganic pollutants from industrial wastewater. Int J Environ Sci Technol. https://doi.org/10.1007/s13762-020-03060-w

Amunugama R, Rodgers MT (2000) Absolute alkali metal ion binding affinities of several azines determined by threshold collision-induced dissociation and ab initio theory. Int J Mass Spectrom 195:439–457

Amunugama R, Rodgers MT (2002) Influence of substituents on cation-π interactions. 1. Absolute binding energies of alkali metal cation–toluene complexes determined by threshold collision-induced dissociation and theoretical studies. J Phys Chem A 106:5529–5539

Armentrout PB, Rodgers MT (2000) An absolute sodium cation affinity scale: threshold collision-induced dissociation experiments and ab initio theory. J Phys Chem A 104:2238–2247. https://doi.org/10.1021/jp991716n

Artmann M (2014) Assessment of soil sealing management responses, strategies, and targets toward ecologically sustainable urban land use management. Ambio 43:530–541. https://doi.org/10.1007/s13280-014-0511-1

Bettini S, Valli L, Giancane G (2020) Applications of photoinduced phenomena in supramolecularly arranged phthalocyanine derivatives: a perspective. Mol 25

Bhattacharyya T, Pal D (2015) The soil : a natural resource. In: Soil science: an introduction. Indian Society of Soil Science, pp 39–56

Biswas B, Qi F, Biswas JK, et al (2018) The fate of chemical pollutants with soil properties and processes in the climate change paradigm—a review. Soil Syst 2

Borrelli P, Robinson DA, Panagos P, et al (2020) Land use and climate change impacts on global soil erosion by water (2015–2070). Proc Natl Acad Sci 117:21994–22001. https://doi.org/10.1073/pnas.2001403117

Carter MR, Gregorich EG, Anderson DW et al (1997) Chapter 1 Concepts of soil quality and their significance. In: Gregorich EG, Carter MRBT-D in SS (eds) Soil quality for crop production and ecosystem health. Elsevier, pp 1–19

Chen L, Zhang X, Cheng X et al (2020) The function of metal–organic frameworks in the application of MOF-based composites. Nanoscale Adv 2:2628–2647. https://doi.org/10.1039/D0NA00184H

Chen T, Li M, Liu J (2018) π–π stacking interaction: a nondestructive and facile means in material engineering for bioapplications. Cryst Growth Des 18:2765–2783. https://doi.org/10.1021/acs.cgd.7b01503

Chiu S-L, Dural NH (1997) Feasibility study on salting-in of organic pollutants to enhance the effectiveness of ex-situ soil washing. Ind Eng Chem Res 36:2435–2439. https://doi.org/10.1021/ie960571t

Cycoń M, Mrozik A, Piotrowska-Seget Z (2019) Antibiotics in the soil environment—degradation and their impact on microbial activity and diversity. Front Microbiol 10:338. https://doi.org/10.3389/fmicb.2019.00338

Deng J-H, Luo J, Mao Y-L, et al (2020) π-π stacking interactions: non-negligible forces for stabilizing porous supramolecular frameworks. Sci Adv 6:eaax9976. https://doi.org/10.1126/sciadv.aax9976

Dhiman V (2020) Preliminary toxicity assessment of chromium (Cr) and lead (Pb) on terrestrial snail (Helix aspersa). Arch Agric Environ Sci 5:67–72. https://doi.org/10.26832/24566632.2020.0501010

Dhiman V, Pant D (2021) Environmental biomonitoring by snails. Biomarkers 26:221–239. https://doi.org/10.1080/1354750X.2020.1871514

Dhiman V, Pant D, Kumari S, Kumar S (2020) Recent advances in novel remediation processes towards heavy metals removal from wastewaters. In: Advances in environmental pollution management: wastewater impacts and treatment technologies. Agro Environ Media, pp 77–99

Dougherty DA (2013) The cation-π interaction. Acc Chem Res 46:885–893. https://doi.org/10.1021/ar300265y

Dunbar RC, Steill JD, Polfer NC, Oomens J (2009) Peptide length, steric effects, and ion solvation govern Zwitterion stabilization in barium-chelated Di- and tripeptides. J Phys Chem B 113:10552–10554. https://doi.org/10.1021/jp905060n

Dunbar RC, Steill JD, Oomens J (2010) Cationized phenylalanine conformations characterized by IRMPD and computation for singly and doubly charged ions. Phys Chem Chem Phys 12:13383–13393. https://doi.org/10.1039/C0CP00784F

Dunbar RC, Steill JD, Oomens J (2011) Encapsulation of metal cations by the PhePhe ligand: a cation−π ion cage. J Am Chem Soc 133:9376–9386. https://doi.org/10.1021/ja200219q

Ferretti A, d'Ischia M, Prampolini G (2020) Benchmarking cation−π interactions: assessment of density functional theory and Möller-Plesset second-order perturbation theory calculations with optimized basis sets (mp2mod) for complexes of benzene, phenol, and catechol with Na^+, K^+, Rb^+, and Cs^+. J Phys Chem A 124:3445–3459. https://doi.org/10.1021/acs.jpca.0c02090

Fu X-X, Li J-F, Zhang R-Q (2016) Strong orbital interaction in pi-pi stacking system. arXiv Prepr arXiv

Garbuio FJ, Howard JL, dos Santos LM (2012) Impact of human activities on soil contamination. Appl Environ Soil Sci 2012:619548. https://doi.org/10.1155/2012/619548

Geisen S, Wall DH, van der Putten WH (2019) Challenges and opportunities for soil biodiversity in the Anthropocene. Curr Biol 29:R1036–R1044. https://doi.org/10.1016/j.cub.2019.08.007

Gunasekara AS, Xing B (2003) Sorption and desorption of naphthalene by soil organic matter. J Environ Qual 32:240–246. https://doi.org/10.2134/jeq2003.2400

Hallowita N, Carl DR, Armentrout PB, Rodgers MT (2008) Dipole effects on cation−π interactions: absolute bond dissociation energies of complexes of alkali metal cations to N-methylaniline and N, N-dimethylaniline. J Phys Chem A 112:7996–8008

Hallowita N, Udonkang E, Ruan C et al (2009) Inductive effects on cation–π interactions: structures and bond dissociation energies of alkali metal cation–halobenzene complexes. Int J Mass Spectrom 283:35–47

Havugimana E, Bhople BS, Kumar A et al (2015) Soil pollution—major sources and types of soil pollutants. Environ Sci Eng 11:53–86

Hillel D (2007) Soil in the environment: crucible of terrestrial life. Elsevier

Huang W, Peng P, Yu Z, Fu J (2003) Effects of organic matter heterogeneity on sorption and desorption of organic contaminants by soils and sediments. Appl Geochem 18:955–972. https://doi.org/10.1016/S0883-2927(02)00205-6

Huang B, Yuan Z, Li D et al (2020) Effects of soil particle size on the adsorption, distribution, and migration behaviors of heavy metal(loid)s in soil: a review. Environ Sci Process Impacts 22:1596–1615. https://doi.org/10.1039/D0EM00189A

Hussain K, Hoque RR, Balachandran S et al (2018) Monitoring and risk analysis of PAHs in the environment BT—handbook of environmental materials management. In: Hussain CM (ed). Springer International Publishing, Cham, pp 1–35
Jia M, Wang F, Bian Y et al (2013) Effects of pH and metal ions on oxytetracycline sorption to maize-straw-derived biochar. Bioresour Technol 136:87–93. https://doi.org/10.1016/j.biortech.2013.02.098
Jørgensen SE (1989) Chapter 4—Adsorption and ion exchange. In: Jørgensen SE, Gromiec MJBT-D in EM (eds) Mathematical submodels in water quality systems. Elsevier, pp 65–81
Kang S, Xing B (2005) Phenanthrene sorption to sequentially extracted soil humic acids and humins. Environ Sci Technol 39:134–140. https://doi.org/10.1021/es0490828
Karaca G, Baskaya HS, Tasdemir Y (2016) Removal of polycyclic aromatic hydrocarbons (PAHs) from inorganic clay mineral: bentonite. Environ Sci Pollut Res 23:242–252. https://doi.org/10.1007/s11356-015-5676-z
Kawaguchi K, Kyuma K (1959) On the complex formation between soil humus and polyvalent cations. Soil Sci Plant Nutr 5:54–63. https://doi.org/10.1080/00380768.1959.10430895
Kutílek M, Nielsen DR (2015) Soil is the skin of the planet earth. In: Soil. Springer, pp 13–19
Lal R (2012) Climate change and soil degradation mitigation by sustainable management of soils and other natural resources. Agric Res 1:199–212. https://doi.org/10.1007/s40003-012-0031-9
Lavelle P, Spain AV (eds) (2001) Soil formation BT—soil ecology. Springer, Netherlands, Dordrecht, pp 143–200
Li Y, Liu M, Li R et al (2020) Polycyclic aromatic hydrocarbons in the soils of the Yangtze River delta urban agglomeration, China: influence of land cover types and urbanization. Sci Total Environ 715:137011. https://doi.org/10.1016/j.scitotenv.2020.137011
Lin C, Skufca J, Partch RE (2020) New insights into prediction of weak π–π complex association through proton-nuclear magnetic resonance analysis. BMC Chem 14:66. https://doi.org/10.1186/s13065-020-00718-x
Ma JC, Dougherty DA (1997) The cation−π interaction. Chem Rev 97:1303–1324. https://doi.org/10.1021/cr9603744
Mandal UK (2016) Soil physical and chemical properties in relation to conservation of natural resources
Meena RS, Kumar S, Datta R et al (2020) Impact of agrochemicals on soil microbiota and management: a review. L 9
Mirsal IA (2004) Pollution mechanisms and soil—pollutants interaction BT—soil pollution: origin, monitoring & remediation. In: Mirsal IA (ed) Soil pollution. Springer, Berlin Heidelberg, Berlin, Heidelberg, pp 111–139
Neel AJ, Hilton MJ, Sigman MS, Toste FD (2017) Exploiting non-covalent π interactions for catalyst design. Nature 543:637–646. https://doi.org/10.1038/nature21701
Obalum SE, Chibuike GU, Peth S, Ouyang Y (2017) Soil organic matter as sole indicator of soil degradation. Environ Monit Assess 189:176. https://doi.org/10.1007/s10661-017-5881-y
Pant D, Giri A, Dhiman V (2018) Bioremediation techniques for E-waste management BT—waste bioremediation. In: Varjani SJ, Gnansounou E, Gurunathan B et al (eds) Waste bioremediation. Springer Singapore, Singapore, pp 105–125
Patel AB, Shaikh S, Jain KR et al (2020) Polycyclic aromatic hydrocarbons: sources, toxicity, and remediation approaches. Front Microbiol 11:2675
Peng B, Chen L, Que C et al (2016) Adsorption of antibiotics on graphene and biochar in aqueous solutions induced by π-π interactions. Sci Rep 6:31920. https://doi.org/10.1038/srep31920
Pignatello JJ, Mitch WA, Xu W (2017) Activity and reactivity of pyrogenic carbonaceous matter toward organic compounds. Environ Sci Technol 51:8893–8908. https://doi.org/10.1021/acs.est.7b01088
Pozniak BP, Dunbar RC (1997) Monomer and dimer complexes of coronene with atomic ions. J Am Chem Soc 119:10439–10445. https://doi.org/10.1021/ja9716259

Qu X, Liu P, Zhu D (2008) Enhanced sorption of polycyclic aromatic hydrocarbons to tetra-alkyl ammonium modified smectites via cation−π interactions. Environ Sci Technol 42:1109–1116. https://doi.org/10.1021/es071613f

Ran Y, Sun K, Yang Y et al (2007) Strong sorption of phenanthrene by condensed organic matter in soils and sediments. Environ Sci Technol 41:3952–3958. https://doi.org/10.1021/es062928i

Rengarajan T, Rajendran P, Nandakumar N et al (2015) Exposure to polycyclic aromatic hydrocarbons with special focus on cancer. Asian Pac J Trop Biomed 5:182–189. https://doi.org/10.1016/S2221-1691(15)30003-4

Rodriguez-Cruz SE, Williams ER (2001) Gas-phase reactions of hydrated alkaline earth metal ions, $M^{2+}(H_2O)n(M = Mg, Ca, Sr, Ba$ and $n = 4–7)$, with benzene. J Am Soc Mass Spectrom 12:250–257. https://doi.org/10.1016/S1044-0305(00)00224-5

Rodriguez JD, Kim D, Tarakeshwar P, Lisy JM (2010) Exploring gas-phase ion−ionophore interactions: infrared spectroscopy of argon-tagged alkali ion-crown ether complexes. J Phys Chem A 114:1514–1520. https://doi.org/10.1021/jp907838r

Ruan C, Rodgers MT (2004) Cation−π interactions: structures and energetics of complexation of Na^+ and K^+ with the aromatic amino acids, phenylalanine, tyrosine, and tryptophan. J Am Chem Soc 126:14600–14610

Schlamadinger DE, Daschbach MM, Gokel GW, Kim JE (2011) UV resonance Raman study of cation−π interactions in an indole crown ether. J Raman Spectrosc 42:633–638. https://doi.org/10.1002/jrs.2781

Sethi S, Gupta P (2020) Soil contamination: a menace to life. In: Soil contamination. IntechOpen

Shah AN, Tanveer M, Shahzad B et al (2017) Soil compaction effects on soil health and crop productivity: an overview. Environ Sci Pollut Res 24:10056–10067. https://doi.org/10.1007/s11356-017-8421-y

Sharma A, Gupta AK, Ganguly R (2018) Impact of open dumping of municipal solid waste on soil properties in mountainous region. J Rock Mech Geotech Eng 10:725–739. https://doi.org/10.1016/j.jrmge.2017.12.009

Shoeib T, Cunje A, Hopkinson AC, Siu KWM (2002) Gas-phase fragmentation of the Ag^+-phenylalanine complex: cation-π interactions and radical cation formation. J Am Soc Mass Spectrom 13:408–416. https://doi.org/10.1016/S1044-0305(02)00353-7

Singh A (2021) Soil salinization management for sustainable development: a review. J Environ Manag 277:111383. https://doi.org/10.1016/j.jenvman.2020.111383

Srivastava M, Srivastava A, Yadav A, Rawat V (2019) Source and control of hydrocarbon pollution. In: Hydrocarbon pollution and its effect on the environment. IntechOpen

Torri SI, Corrêa RS (2012) Downward movement of potentially toxic elements in biosolids amended soils. Appl Environ Soil Sci 2012:145724. https://doi.org/10.1155/2012/145724

Truskewycz A, Gundry TD, Khudur LS et al (2019) Petroleum hydrocarbon contamination in terrestrial ecosystems-fate and microbial responses. Molecules 24:3400. https://doi.org/10.3390/molecules24183400

Van Boeckel TP, Brower C, Gilbert M et al (2015) Global trends in antimicrobial use in food animals. Proc Natl Acad Sci U S A 112:5649–5654. https://doi.org/10.1073/pnas.1503141112

Vasudevan D, Arey TA, Dickstein DR et al (2013) Nonlinearity of cationic aromatic amine sorption to aluminosilicates and soils: role of intermolecular cation−π interactions. Environ Sci Technol 47:14119–14127. https://doi.org/10.1021/es403389a

Wagenet RJ, Hutson JL (1997) Soil quality and its dependence on dynamic physical processes. J Environ Qual 26:41–48. https://doi.org/10.2134/jeq1997.00472425002600010007x

Weber WJ, Kim SH, Johnson MD (2002) Distributed reactivity model for sorption by soils and sediments. 15. High-concentration co-contaminant effects on phenanthrene sorption and desorption. Environ Sci Technol 36:3625–3634. https://doi.org/10.1021/es020557+

Wuana RA, Okieimen FE (2011) Heavy metals in contaminated soils: a review of sources, chemistry, risks and best available strategies for remediation. Isrn Ecol 2011. https://doi.org/10.5402/2011/402647

Xiao F, Pignatello J (2015) π + −π interactions between (hetero)aromatic amine cations and the graphitic surfaces of pyrogenic carbonaceous materials. Environ Sci Technol 49. https://doi.org/10.1021/es5043029

Xu X, Cao X, Zhao L et al (2013) Removal of Cu, Zn, and Cd from aqueous solutions by the dairy manure-derived biochar. Environ Sci Pollut Res 20:358–368. https://doi.org/10.1007/s11356-012-0873-5

Xu Y, Yu X, Xu B et al (2021) Sorption of pharmaceuticals and personal care products on soil and soil components: influencing factors and mechanisms. Sci Total Environ 753:141891. https://doi.org/10.1016/j.scitotenv.2020.141891

Yang X, Zhang S, Ju M, Liu L (2019) Preparation and modification of biochar materials and their application in soil remediation. Appl Sci 9

Ye S, Zeng G, Wu H et al (2017) Co-occurrence and interactions of pollutants, and their impacts on soil remediation—a review. Crit Rev Environ Sci Technol 47:1528–1553. https://doi.org/10.1080/10643389.2017.1386951

Yi S, Li F, Wu C, Wei M, Tian J, Ge F (2022) Synergistic leaching of heavy metal-polycyclic aromatic hydrocarbon in co-contaminated soil by hydroxamate siderophore: role of cation-π and chelation. J Hazard Mater 424:127514

Yu H, Liu J, Shen J et al (2016) Preparation of MnOx-loaded biochar for Pb^{2+} removal: adsorption performance and possible mechanism. J Taiwan Inst Chem Eng 66:313–320. https://doi.org/10.1016/j.jtice.2016.07.010

Yuvaraj M, Mahendran PP (2020) Soil pollution causes and mitigation measures. Biot Res Today 2:550–552

Zhang Q, Wang C (2020) Natural and human factors affect the distribution of soil heavy metal pollution: a review. Water Air Soil Pollut 231:1–13. https://doi.org/10.1007/s11270-020-04728-2

Zhang W, Zhuang L, Yuan Y et al (2011) Enhancement of phenanthrene adsorption on a clayey soil and clay minerals by coexisting lead or cadmium. Chemosphere 83:302–310. https://doi.org/10.1016/j.chemosphere.2010.12.056

Zhang X, Li F, Liu T et al (2013) the influence of polychlorinated biphenyls contamination on soil protein expression. ISRN Soil Sci 2013:126391. https://doi.org/10.1155/2013/126391

Zhang L, Cheng I, Wu Z et al (2015a) Dry deposition of polycyclic aromatic compounds to various land covers in the Athabasca oil sands region. J Adv Model Earth Syst 7:1339–1350. https://doi.org/10.1002/2015MS000473

Zhang W, Zheng J, Zheng P et al (2015b) The roles of humic substances in the interactions of phenanthrene and heavy metals on the bentonite surface. J Soils Sediments 15:1463–1472. https://doi.org/10.1007/s11368-015-1112-8

Zhao L, Hou R (2019) Human causes of soil loss in rural karst environments: a case study of Guizhou, China. Sci Rep 9:3225. https://doi.org/10.1038/s41598-018-35808-3

Zhao Q, Zhang S, Zhang X et al (2017) Cation–Pi interaction: a key force for sorption of fluoroquinolone antibiotics on pyrogenic carbonaceous materials. Environ Sci Technol 51:13659–13667. https://doi.org/10.1021/acs.est.7b02317

Zhu D, Hyun S, Pignatello JJ, Lee LS (2004) Evidence for π−π electron donor−acceptor interactions between π-donor aromatic compounds and π-acceptor sites in soil organic matter through pH effects on sorption. Environ Sci Technol 38:4361–4368. https://doi.org/10.1021/es035379e

Chapter 30
Assessment of Ecological and Human Health Risk of Soil Heavy Metals Pollution: Study from Chotanagpur Plateau Region, India

Baisakhi Chakraborty, Sambhunath Roy, Biswajit Bera, Partha Pratim Adhikary, Debashish Sengupta, and Pravat Kumar Shit

Abstract Soil toxic metals pollution has been significantly increased during the last three decades mainly due to intensive agricultural practices, and unplanned rapid development actives. The present study, soil heavy metal pollution load, its ecological risk and impact on human health has been analyzed of Chotanagpur plateau fringe region, India. A total of 96 soil samples have collected both topsoil (0–20 cm) and subsoil (20–50 cm) from different land-use practices at random. The soil samples were analyzed and assessed of heavy metals (HM) pollution load such as Cr, Fe, Mn, Ni, Zn, Cu, As, Sr, Pb and Zr. Mean values of Cr, Mn, Fe, Ni, Zn, and As in all land use practices of topsoil were found 499.67, 1711.75, 4061.81, 459.19, 470.58 and 18.88 respectively. Mean values of Cr, Mn, Fe, Ni, Zn, and As in subsoil were found 314.56, 868.5, 2287.33, 231.73, 279.35 and 12.23 respectively. This pollution load is 3–5 times higher than the world's normal standard guidelines. The results showed that topsoil of industrial and semi urban areas was mostly polluted than the agricultural field. Soil pollution mainly occurred by iron, manganese, copper, zircon contamination in this study region. Ecological risk (RI) was identified as moderate level at topsoil of industrial and township regions. Human health risk

B. Chakraborty · S. Roy · P. K. Shit (✉)
PG Department of Geography, Raja N. L. Khan Women's College (Autonomous), Gope Palace, Midnapore, West Bengal 721102, India
e-mail: pravatgeo2007@gmail.com

B. Bera
Department of Geography, Sidho Kanho Birsha University, University Campus Road, Ranchi-Purulia Rd, Purulia, West Bengal 723104, India
e-mail: biswajitbera007@gmail.com

P. P. Adhikary
ICAR Indian Institute of Water Management, Bhubaneswar, Odisha 751023, India
e-mail: ppadhikary@gmail.com

D. Sengupta
Department of Geology and Geophysics, Indian Institute of Technology (IIT), Kharagpur, West Bengal 721302, India
e-mail: dsgg@gg.iitkgp.ac.in

P. K. Shit et al. (eds.), *Soil Health and Environmental Sustainability*,
Environmental Science and Engineering,
https://doi.org/10.1007/978-3-031-09270-1_30

analysis indicated children of those highly contaminated regions are vulnerable to the non-carcinogenic types of diseases by dermal contaminates. Our findings may be valuable to assist the understanding of soil pollution concerning public health in different land-use practices.

Keywords Anthropogenic activities · Ecological risk · Health hazards · Heavy metals · GIS analysis · Sustainable management

30.1 Introduction

Now a day, heavy metals (HM) pollution in soil is one of the important issues and changes of environmental deterioration worldwide. Metals, those densities are >5 g/cm^3 are referred as HM (Yang et al. 2018). Aluminium, cadmium, iron, lead, zinc, manganese, nickel, selenium are such metals those can't be degraded or destroyed easily in environment (Selvi et al. 2019). In most of the countries, uncontrolled development of urban areas, industries, automobile services, agricultural practices etc. noticeably help to increase load of HM in soil. Many researchers reported that metal processing, power plants, chemical factories, urban sewage, automobile emission and usages of fertilizers, pesticides in crop fields are the main supplier of HM in the soils of developing regions of the world (Bilos et al. 2001; Chen et al. 2014; Zhang et al. 2015; Taiwo et al. 2017; Ali et al. 2019).

Lancet Commission (2017) reported that soil pollution is a raising global issues and most vulnerable to health risk and well-being (Landrigan et al. 2017). WHO (2017) estimated globally, about twenty-four percent of population is contaminated by soil-transmitted helminths (parasitic worms), affecting primarily poorest peoples without sanitary measures (FAO and UNEP (2021). HM from topsoil leached to subsurface soil with percolating water and contaminated it in various dumping areas of mines and industrial region (Dang et al. 2002; De and Mitra 2004; Chakraborty et al. 2021). It adversely affects the natural fertility of soil as well human health (Sun et al. 2019; Wang et al. 2019; Adimalla 2020). HM in soil dust particles can enter to human body via three exposure pathways as ingestion, inhalation and dermal contact and also indirectly by transfer of pollutant into the food chain (Verma et al. 2019; Jiang et al. 2020; Chakraborty et al. 2021). Consequently, intake of those HM in a long term can cause serious health related issues to human body and leads to health hazards in many effected regions (Taiwo et al. 2017). Major respiratory diseases, cardiovascular diseases, birth defects, malfunction of central nervous system could be result from continuous intake of HM with soil dust for long time period (Greening 2011; Adimalla and Wang 2018; Yang et al. 2019). Many other chronic diseases like insomnia, memory loss, gastro-intestinal disorder, lung cancer and even death happened sometimes due to heavy metal intake through any of these three pathways (Jiang et al. 2017; Ali et al. 2019; Luo et al. 2019).

Many scholars reported that, soil heavy metal pollution is often related to human health risk (Li et al. 2014; Pan et al. 2018; Yang et al. 2018; Kashyap et al. 2019;

Jiang et al. 2020). Therefore, it is very necessary to evaluate pollution load to the soil and potential health risk to human by HM at any developing region. Previously, many scientific study were conducted on the assessment of pollution load of HM in soil using different indexing methods (Das and Chakrapani 2011; Ameh 2013; Sahoo et al. 2016; Li et al. 2017; Zang et al. 2017; Qu et al. 2018; Chakraborty et al. 2021). Human health risks were assessed in various part of China for non-carcinogenic and carcinogenic type of diseases (Zang et al. 2017; Yang et al. 2018). In India, health risk to heavy metal consumption of soil dust were analysed in a developing region of Telangana state (Adimalla et al. 2020). It has been reported that nearly 32 important cities in India are highly contaminated by HM in soil and vulnerable to health risk potentiality of their citizen (Adimalla et al. 2020b). All above studies clearly depict developing region are more prone to soil pollution related with HM and associated health hazards.

However, Chotanagpur plateau region (India) is well known for its vast industrial development with advantages of huge mineral resources and fertile soil of Damodar river basin. Therefore, many small and big industries with allied activities, urban sectors, and agricultural activities are developed in this region. During last decades non-carcinogenic health risk are affected this region due to soil–water pollution (Bera and Ghosh 2019; Bera et al. 2021; De et al. 2021). This work provides for the first time spatial information on soil toxic metals pollution level, concerning health risk in the tosoil and subsoil. Therefore, in this study the main objectives are (i) to ground level assessment of pollution load of HM in different depth of soil, (ii) to evaluate ecological and human health risk (non-carcinogenic type), and (iii) to suggests possible remediation measures for sustainable development.

30.2 Materials and Methods

30.2.1 Study Area

Santuri block of Purulia district of West Bengal is one of the important administrative regions under Chotanagpur physiographic division. From last few decades this block is developing with settlements, agriculture, industrial activities. Santuri block is situated in the north-east location of Purulia district. Geographically it is located between 23° 27′ 43″ N to 23° 39′ 35″ N and 86° 45′ 50″ E to 86° 54′ 33″ E with an area of 179.69 km^2 (Fig. 30.1). Geomorphological point of view this region is an undulating plateau fringe with granite-gneiss, mica-schist geological formation (Bera and Ghosh 2019). This block is faced scorching heat at summer season with 45 °C average temperature. In winter season, dry climatic condition with 4 to 5 °C average temperature is experienced here. This block is surrounded by vast coalfield region of Raniganj, Neturia, Bokaro formations. Availability of rail routes and roadways promotes semi urban areas with dense settlement in this region.

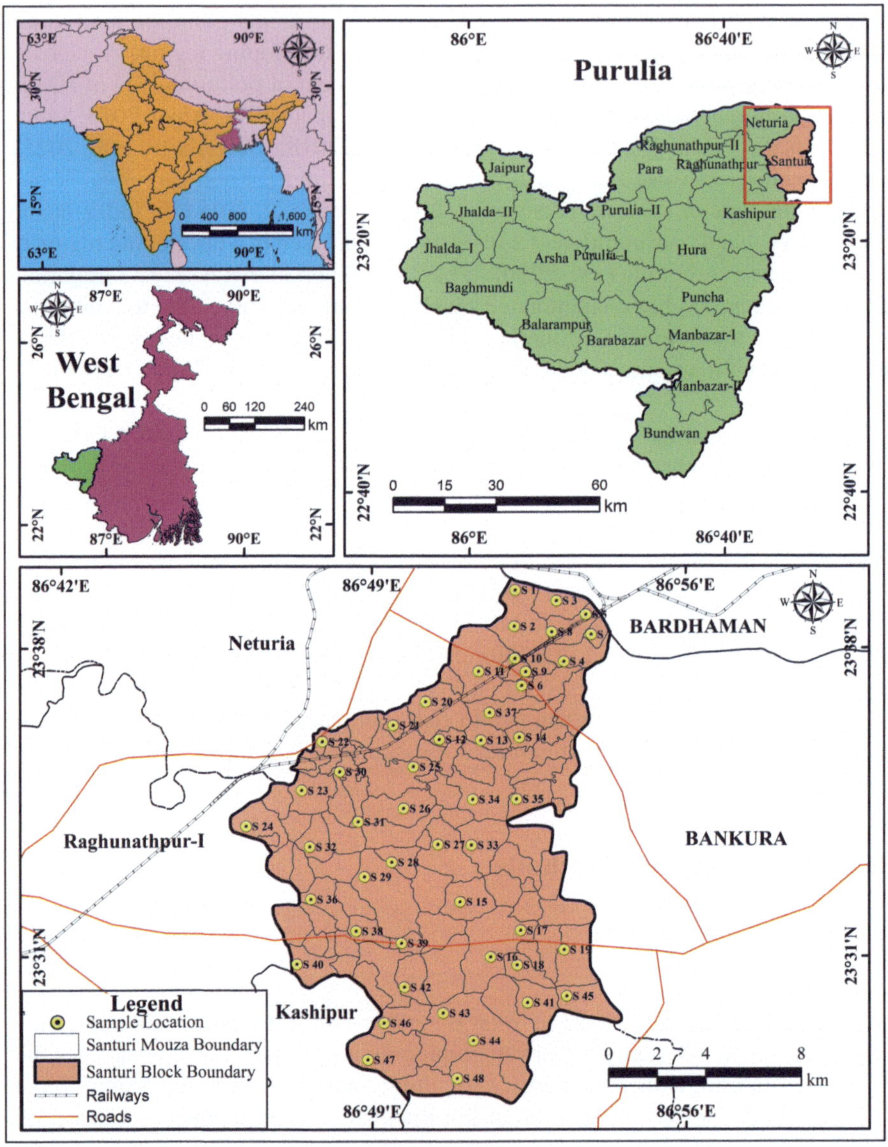

Fig. 30.1 Location map of the study area and sampling sites

30.2.2 Collection of Soil Samples and Its Procedures

Sample soils were collected from 0 to 20 cm depth of topsoil and 20 to 50 cm depth of subsoil at randomly selected 48 different land use areas of the Santuri block. A total of 96 soil samples were collected between November and December, 2019. Handheld

GPS (Garman e Trex 30 GPS) was used to record each sampling location. Wooden shovel was used to collect all samples and preserved in pre cleaned polyethylene bags with self-lock system (Chakraborty et al. 2021). Three sub samples from each sampling site were gained and mixed to get a bulk contain of 1 kg. All samples were identified by their specific code labelled to them. One soil sample from nearby uncontaminated area was collected to determine regional(local) background value of each heavy metals (HM) of the soil. After carried properly to laboratory, Sample soils were dried for evaporate moisture at normal temperature and after that it were grinded. A swing grinding mill (US standards) was used for sieved soil dust of 200 mesh size. Aluminium cups contained with boric acid were used to spread soil sample of 1 g over it and amount of 20 tons pressured was applied to get pallets from the soil samples. HM from prepared soil pallets were measured by using X-ray fluorescence spectrometry (XRF) in laboratory. Philips MagiXPRO-PW2440 XRF with fully automated microprocessor and 4 KW X-ray generator was used to obtain HM quantity in each sample. HM such as Cr, Fe, Mn, Ni, Zn, Cu, As, Sr, Pb and Zr were measured from this method (Adimalla et al. 2020). World normal value of each heavy metal was used as their standard limit of concentration in the soil.

30.2.3 Quantification of Soil Heavy Metals (HM) Pollution

To quantify the pollution load of HM in soil, base line data is most important. In this study, we used world average values and standard regional (local) background values for analysis HM pollution level (Jiang et al. 1996; Rubio et al. 2020). In this study, regional (local) background values of HM were determined from nearby uncontaminated soil sample (Chakraborty et al. 2021).

30.2.3.1 Contamination Factor (CF)

Contamination factor (CF) determines the contamination load of each metal in the soil of an area. It is measured by following the equation below (Cabrera et al. 1999)

$$CF_i = \frac{C_{metal(sample)}}{C_{metal(background)}}$$

Values of CF has been classified in 4 categories i.e. CF < 1 indicates low contamination of HM to the soil, CF = >1 to <3 indicates moderate contamination, CF = >3 to <6 indicates considerable contamination and CF > 6 indicates very high contamination (Muller 1969).

30.2.3.2 Pollution Load Index (PLI)

This index method is used for determining the level of pollution load contributed by all considered HM at different sites (Tomlinson et al. 1980). PLI helps to delineate contamination sites of an area caused by pollutants and taking necessary management plan to remediate. Calculation of PLI was conducted by following equation:

$$PLI = (CF_1 \times CF_2 \times CF_3 \times \cdots CF_n)^{1/n}$$

where, CF indicates contamination factor and n indicates number of contaminants. PLI can be divided into two categories as PLI $\leq$ 1 (pollutants background level are present), and PLI $\geq$ 1 (soil quality deterioration or pollutants exceeds their background level).

30.2.3.3 *Ecological Risk Factor (E_r)*

Ecological risk factor provides toxicity response of any single heavy metal contaminant to its own. It has been calculated by following formula (Hakanson 1980).

$$ER_i = T_r \times CF_i$$

where, T_r means toxicity response of HM. CF means contamination factor of HM. In this study six HM were selected according to their toxic factors as follows Zn = 1 < Cr = 2 < Ni = Cu = Pb = 5 < As = 10. E_r can be divided into four categories i.e. E_r < 40 (practically uncontaminated), E_r = 40–80 (moderately contaminated), E_r = 80–160 (heavily contaminated), E_r = 160–320 (extremely contaminated).

30.2.3.4 Potential Ecological Risk Index (RI)

Potential ecological risk index is a modified index method to assess degree of sensitivity towards environment in respect of soil HM. It was first applied by Hakanson, 1980 to study the contamination of HM in coastal sediments. It can be calculated through following formula.

$$RI = \sum_{i=1}^{n=6} ER_i$$

where, RI is the sum of risk factor, ER_i denotes ecological risk factor of individual heavy metal. Value of RI can be classified into four groups as RI < 150 (practically uncontaminated), RI = 150–300 (moderately contaminated), RI = 300–600 (heavily contaminated) and RI > 600 (extremely contaminated).

30.2.3.5 Human Health Risk Assessment (HHR)

The bare part of earth surface or topsoil is highly vulnerable to HM pollution (Wang et al. 2020). Industrial smoke dusts, automobile emission, mining extraction, spoil dumping, domestic sewage, fertiliser, pesticides all practices contained with HM contaminated top layer of soil surface at first. Maximum crop plants spread their roots up to top layer (0–20 cm) and absorbed nutrients along with metal contains (Yang et al. 2018). These metals can be entered to human body through intake food chain, or by respiratory system or by direct skin contact with particles of road dust. Thus, pollution on topsoil acts as important issue of human health risk of any region.

In this study, HHR has been considered for contamination of HM on topsoil for three exposure pathways such as ingestion, inhalation and dermal. Health risk was determined by suggested method of USEPA (US Environment Protection Agency) for the assessment of non-carcinogenic diseases of adults (male and female) and children. Health risks of non-carcinogenic diseases were calculated using the formulas below:

$$ADD_{ingestion} = \frac{C_{soil} \times IngR \times EF \times ED \times F}{BW \times ET} \times 10^{-6}$$

$$ADD_{inhalation} = \frac{C_{soil} \times InhR \times EF \times ED \times F}{BW \times ET \times PEF}$$

$$ADD_{dermal} = \frac{C_{soil} \times EF \times ED \times F \times SA \times AF}{BW \times AT} \times 10^{-6}$$

$$HQ = \frac{ADD}{RfD}$$

$$HI = \sum_{i=1}^{n} HQ$$

where, $ADD_{ingestion}$, $ADD_{inhalation}$ and ADD_{dermal} indicates daily intake of HM dust through three exposure pathways (mg/kg/day). C_{soil} is HM concentration in soil (mg/kg), IR_{ing}, IR_{inh} are the ingestion and inhalation rate of metal i.e. $IR_{ing} = 100$ and 200 mg/kg, $IR_{inh} = 20$ and 5 mg/kg/day for adult and children, respectively. EF is exposure frequency (250 days/year). ED is exposure duration (30 years for adult and 12 years for children). BW is average body weight (60 kg for adult and 15 kg for children). ET is mean exposure time i.e. 10,950 days for adult and 4380 days for children. F is fraction of time spent at open area in a day (6.94%). CF is conversion factor i.e. 10^{-6} kg/mg. SA is exposed skin surface area (4350 cm^2/day). AF is adherence factor (0.07 mg/cm^2). RfD is reference dose of each HM (mg/kg/day) suggested by USEPA. Values of HI > 1 indicates high possibility of health risk and HI < 1 indicates no possible health hazard of non-carcinogenic type in the study area.

30.2.3.6 Statistical and Spatial Analysis

Statistical analysis of the data related to HM for determination of general tendency of their concentration, descriptive statistics and correlation analysis (0.05% level of significance) have been performed using SPSS 16 software. Spatial analysis for identification of regional variation of metal pollution load has been derived through inverse distance weightage (IDW) method using Arc GIS 10.4 software.

30.3 Results

30.3.1 Distribution of Soil Heavy Metals (HM)

Topsoil: Descriptive statistics of ten (10) HM on topsoil (0–20 cm) of Santuri block is presented in Table 30.1. Mean values of Cr, Mn, Fe, Ni, Zn, and As in all land use practices of topsoil were found 499.67, 1711.75, 4061.81, 459.19, 470.58 and 18.88 respectively. Mean concentration of HM indicated its descending order as Fe > Mn > Zr > Cr > Zn > Ni > Cu > Sr > Pb > As. Higher concentration of iron in topsoil indicated its natural sources i.e. weathering of parent rocks and also industrial effluences in different sites of the study area. Abundant concentration of each metal was found at industrial (22.92% sample sites) and semi-urban (31.25% sample sites) areas of this region. Comparatively lower concentration of HM was found in agricultural region due to mixing of fertilisers and pesticides in its soil of top layer. Though, average value of all metals exceeds their world normal value in a high magnitude. Correlation matrix (Fig. 30.2a) showed highly positive correlation of HM with each other in this developing region. Spatial mapping indicated northern part of Santuri block was most enriched with toxic heavy metal contamination in its topsoil. Metal concentration has been classified into four groups as 'very high', 'high', 'moderate', 'low' and presented in Fig. 30.3.

Subsoil: Table 30.1 illustrated descriptive statistics of ten (10) HM of 48 subsurface soil samples. Mean values of Cr, Mn, Fe, Ni, Zn, and As in subsoil were found 314.56, 868.5, 2287.33, 231.73, 279.35 and 12.23 respectively. Mean concentration of these metals can be arranged as Fe > Mn > Zr > Cr > Zn > Ni > Cu > Sr > Pb > As. Leaching of HM with percolating water increased metal load to the subsurface soil of industrial and semi township areas. High concentration of each metal indicated subsurface soils of industrial and township areas are heavily affected by leached pollutant from topsoil. All HM except Sr crossed world normal value of their concentration in soil. Correlation coefficient matrix of subsurface metals indicated high positive correlation ($r > 0.80$) among each other (Fig. 30.2b). Spatial mapping showed subsurface soil of northern and south eastern parts were moderate to highly polluted than other region of Santuri block (Fig. 30.3).

Table 30.1 Descriptive statistics of soil heavy metals in top soil (0–20 cm) and subsurface soil (20–50 cm)

	Parameters	Cr	Mn	Fe	Ni	Cu	Zn	As	Pb	Sr	Zr
Ref. value	World normal	70.9	571	3800	23	28.2	67.8	11.4	28.4	200	307
Top soil	Mean	499.67	1711.75	4061.81	459.19	415.65	470.58	18.88	137.27	282.42	585.17
	Standard deviation	221.62	1520.8	3097.53	216.2	191.36	246.52	8.87	111.26	175.06	232.22
	Minimum	231	328	809	68	123	156	7	32	110	325
	Maximum	964	4967	9987	895	798	998	35	386	695	1159
	Background value	123	180	2590	85	55	85	8	44	55	120
Subsurface soil	Mean	314.56	868.5	2287.33	231.73	206.1	279.35	12.23	69.73	171.15	318.58
	Standard deviation	244.59	970.68	2238.97	220.96	136.96	196.09	7.3	54.28	136.92	154.1
	Minimum	45	138	409	15	22	55	5	16	56	165
	Maximum	837	2650	6998	765	487	695	29	199	586	690
	Background value	121	176	2585	85	55	85	7	42	50	118

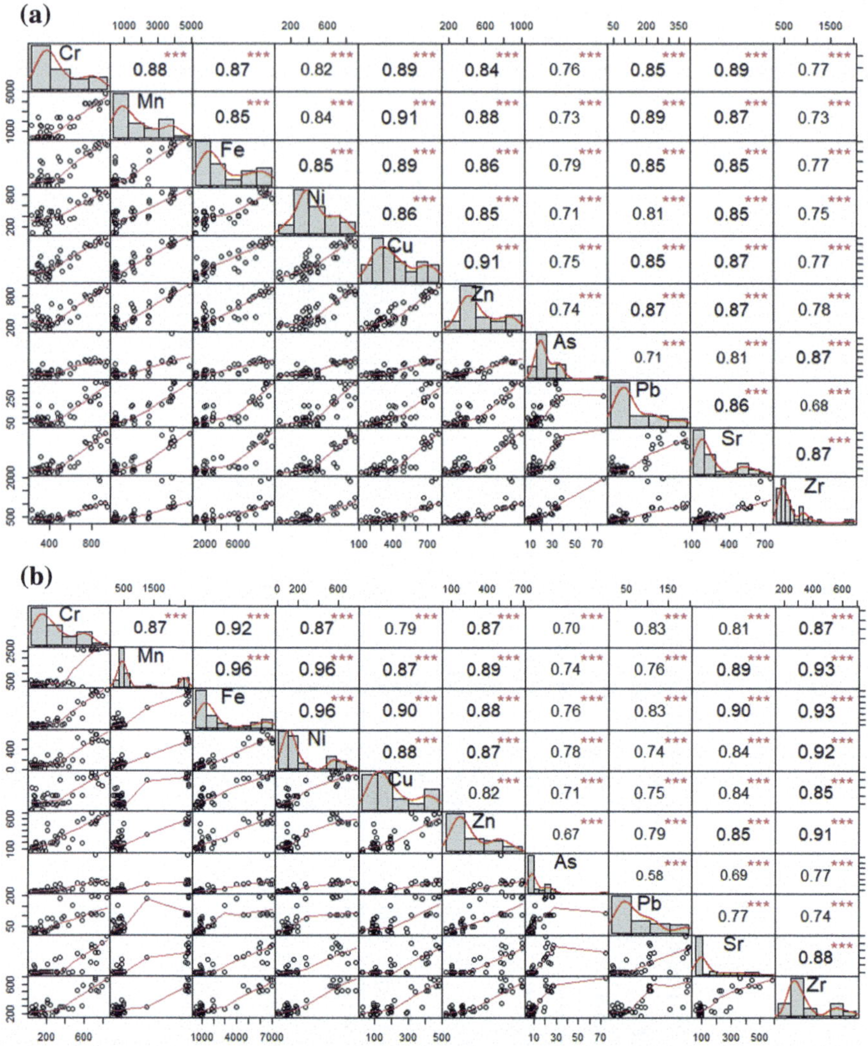

Fig. 30.2 **a** Pearson correlation matrix of topsoil (0–20 cm) heavy metals *** significant at $p < 0.01$; ** significant at $p < 0.05$; * significant at $p < 0.1$. **b** Pearson correlation matrix of sub-soil (20–50 cm) heavy metals. *** significant at $p < 0.01$; ** significant at $p < 0.05$; * significant at $p < 0.1$

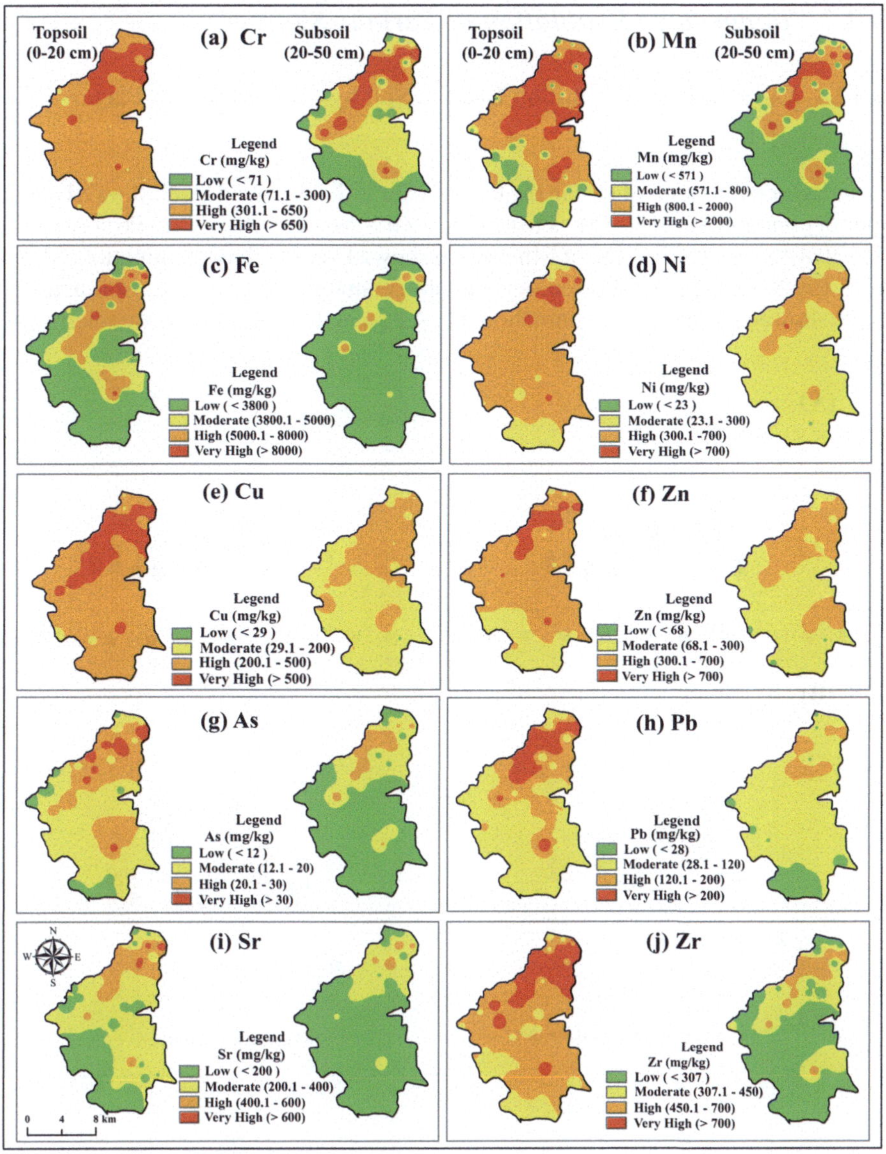

Fig. 30.3 Spatial distribution of soil heavy metals in Santuri block

30.3.2 Importance Evaluation of Various Sources

30.3.2.1 Assessment of Pollution Load in Soil by Contamination Factor (CF)

Topsoil: Contamination of HM in the topsoil over its background level, showed mean order of contamination of HM as Mn > Cu > Zn > Ni > Zr > Sr > Cr > Pb > As > Fe (Fig. 30.4A). Mean CF value of Mn and Cu showed very high contamination. Mean CF of Cr, Ni, Zn, Sr, Zr showed considerable contamination to its topsoil and Fe, As, Pb showed moderate contamination. Fe, As generated naturally by weathering of

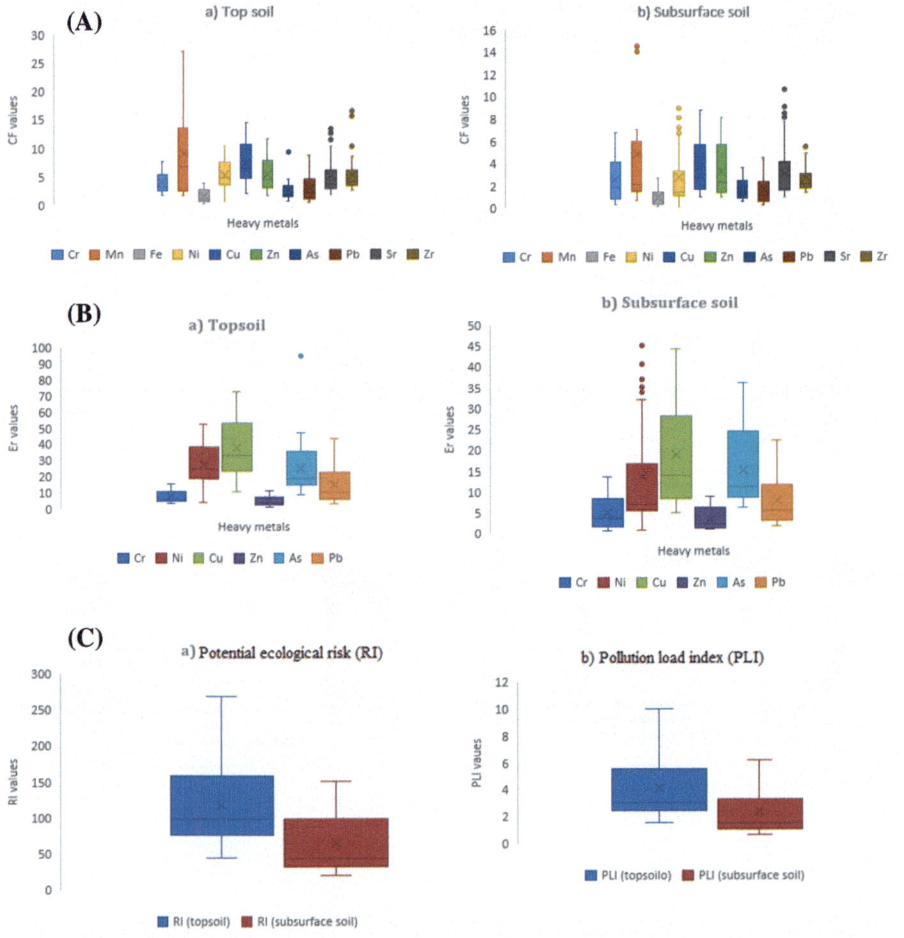

Fig. 30.4 Box and whisker plots indicating average, maximum, minimum values of each heavy metals **A** contamination factors; **B** ecological risk factor and **C** Risk Index **a** potential ecological risk (RI); **b** PLI values

parent materials in this plateau region. Therefore, background values of these metals from uncontaminated sites were also found higher than other HM in this region. It reduces values of contamination load of those metals in this study area.

Subsoil: Contamination factor of HM to its subsurface soil can be arranged by their mean values as Mn > Cu > Zn > Sr > Ni > Zr > Cr > Pb > As > Fe (Fig. 30.4A). Mean CF of Mn, Cu, Zn, Sr showed considerable contamination to its subsurface soil. Mean CF of Cr, Ni, As, Pb, Zr illustrated moderate contamination and Fe showed low contamination to the sub-soil.

30.3.2.2 Potential Ecological Risk Assessment (RI)

Ecological risk of six HM has been presented in Fig. 30.4B. In the topsoil, average value of E_r of six metals suggested its order of risk potentiality as Cu > Ni > As > Pb > Zn > Cr. Average E_r values of all these HM indicated practically uncontaminated ecological quality of topsoil of the study area ($E_r < 40$). Potential ecological risk factor by these six metals at 48 locations indicated its range from 44.71 to 269.09 with 120.04 as mean value). Mean value of topsoil RI represented potentially uncontaminated to ecological risk but RI value of 25% sample sites indicated moderate contamination to its soil (RI = 150–300). These sites are influenced by industrial and urban activities mainly. Other 75% sample sites showed un-contamination to potential ecological risk (RI < 150) of the study area.

Sensitivity of toxic metals to biological environment of subsurface soil indicated average value of Er of selected six metals can be arranged as Cu > As > Ni > Pb > Cr > Zn (Fig. 30.4B) and indicated all toxic metals of subsurface layer were potentially uncontaminated to ecological quality ($E_r < 40$). RI values of 48 subsurface soil sample showed its range from 26.03 to 151.69 with 65.21 as mean value (). 2.08% samples were moderately uncontaminated in nature. Rest 97.92% soil samples of subsurface layer were practically uncontaminated to ecological risk.

30.3.2.3 Overall Pollution Load Assessment by Pollution Load Index (PLI)

Pollution level by all selected HM at top soils of 48 sample sites showed its range from 1.57 to 10.03 with 4.21 as mean value (Fig. 30.4C.b). All PLI values indicated deterioration of soil quality (PLI > 1) due to mixing of HM from industries, automobile emission, domestic sewage and agricultural field of this developing region. Highest PLI value was found from S6 sample site. In this site, sponge iron industries have been developed and generate huge metal dusts from it. Lowest PLI value was found from S1 sample site, where low agricultural practice helps to low mixing of HM in the topsoil. Values of PLI were categorized into four groups for clear understanding of spatial distribution of pollution load in this study area, such as high pollution (PLI > 6), moderate pollution (PLI = 3–6), low pollution (PLI = 1–3) and no pollution (PLI < 1) to the soil (Table 30.2).

Table 30.2 Soil pollution load index (PLI)

Level of pollution	Top soil		Subsoil		Land use practices
	No of samples	Percentage (%)	No of samples	Percentage (%)	
No pollution (<1)	2	4.16	06	12.50	Low agricultural practice
Low pollution (1–3)	20	41.67	30	62.50	High agricultural practice
Moderate pollution (3–6)	15	31.25	11	22.92	Domestic influences with agricultural dominance and semi-urban areas
High pollution (>6)	11	22.92	01	2.08	Industrially influenced areas

Pollution load of subsurface soil of the study area showed its range from 0.72 to 6.21 with 2.34 mean value (Fig. 30.4C.b). Highest PLI of subsurface soil was found from S6 location. This site is adjacent to the iron industries and huge spoil dumping of minerals, brings metal load to its soil and deeply affects its subsurface soil quality by leaching process. Mean PLI of subsurface soil indicated high pollution load to the soil (PLI > 1). Sample sites S1, S2, S42, S44, S45 and S48 showed their PLI value < 1, denoted no pollution to its underlying soil. Spatial distribution of pollution load index showed that northern and small part of eastern side of this block was moderately polluted (PLI = 3–6) by HM.

30.3.2.4 Spatial Distribution of PLI and RI

Spatial zonation mapping on Arc GIS indicated northern part of this block is highly polluted by metal load due to industrial agglomeration in its topsoil (Fig. 30.5a). In the subsurface soil, All other sites covering mainly the agricultural region indicated low pollution (PLI = 1–3) (Fig. 30.5). Spatial map of RI on topsoil indicated northern and few parts of central area has higher potentiality to ecological risk by soil HM (Fig. 30.5b). RI of subsurface soil suggested there are very less ecological risk in most all over the study area (Fig. 30.5b).

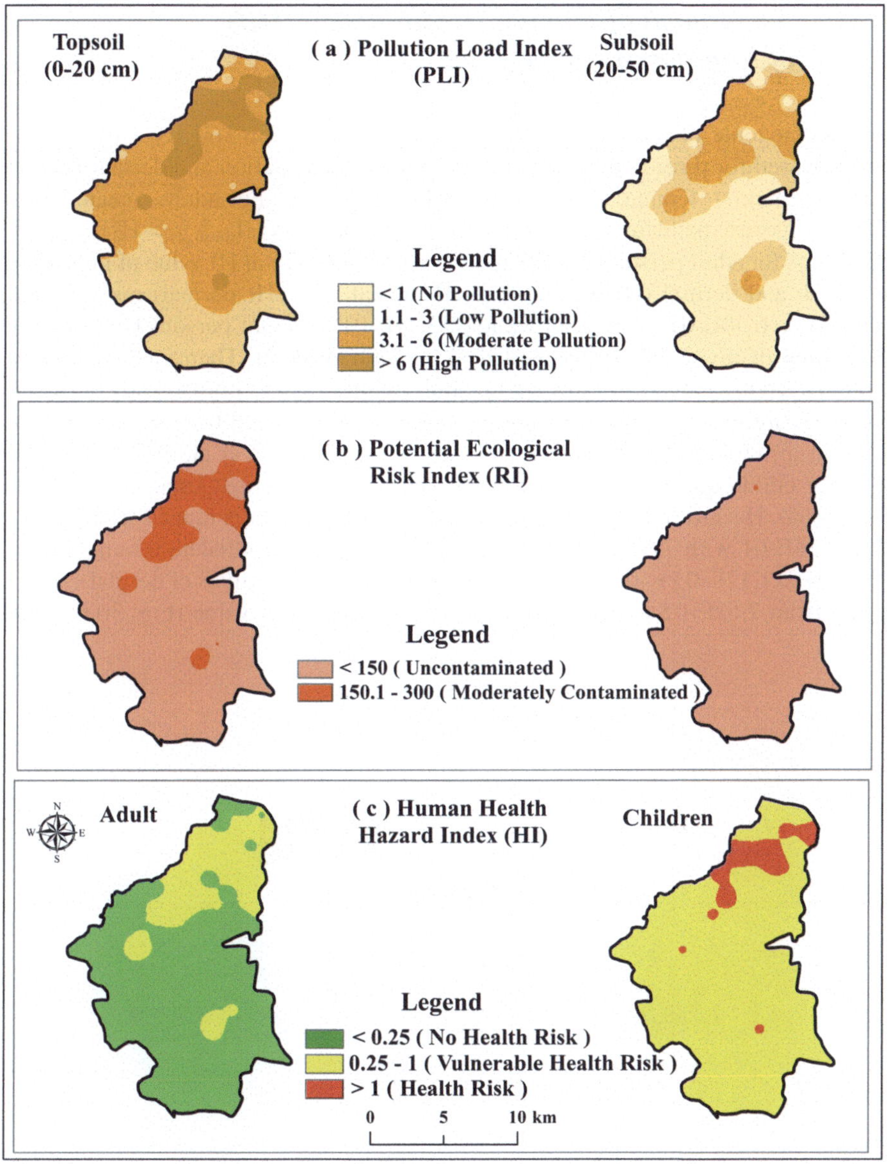

Fig. 30.5 Spatial distribution of pollution load of heavy metals **a** for topsoil (left) and subsurface soil (right); **b** Potential ecological risk index for topsoil (left) and subsurface soil (right); and **c** non-carcinogenic health hazard risk of human body

30.3.3 Assessment of Human Health Risk (HHR) of Non-carcinogenic Type

Non-carcinogenic health risk of HM consumption from top soil layer (0–20 cm) were assessed for three exposure route i.e. ingestion, inhalation and dermal contact by adults (male + female) and children of the study area. For adults, mean values of three exposure pathways showed its decreasing order as $HI_{(dermal)} > HI_{(ingestion)} > HI_{(inhalation)}$ for adult persons as well as for children also. Total HI value of ingestion, inhalation and dermal (HI ingestion + HI inhalation + HI dermal) ranged from 1.16E-01 to 6.40E-01 with mean value of 2.31E-01 for adult persons (Fig. 30.6a). No HI value of any sample site exceeded >1 for adult residents. Therefore, no obvious health related hazard can be expected for adults of this region. Similar study on health risk (non-carcinogenic) also showed no obvious health hazard for soil heavy metal intake in different study areas (Taiwo et al. 2017; Adimalla et al. 2020). Total HI value for children ranged from 3.30E-01 to 1.62 with average value of 6.70E-01 (Fig. 30.6a). Hazard Index (HI) values of adults for ingestion is ranged from 4.03E-02 to 2.24E-01 with 7.84E-02 as mean value. HI values of inhalation ranged from 2.26E-04 to 1.12E-03 with 4.62E-04 as mean value, and HI values of dermal contact ranged from 7.62E-02 to 4.14E-01 with 1.52E-01 as mean value (Fig. 30.6b). On

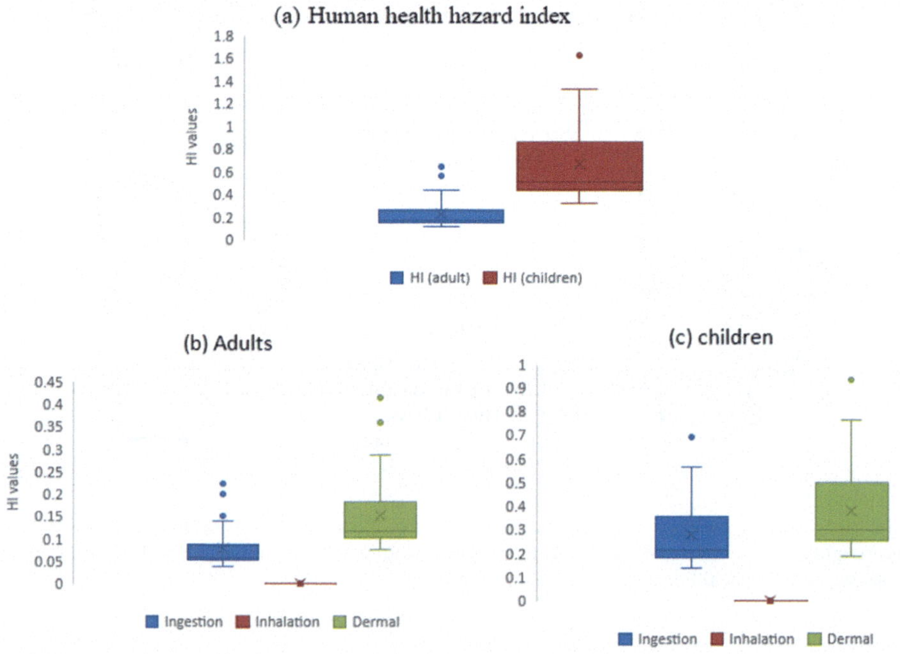

Fig. 30.6 Box and whisker plot showed mean, maximum, minimum values of **a** Human health hazard index and HI values of three exposure pathways for **b** adults and **c** children in the study area

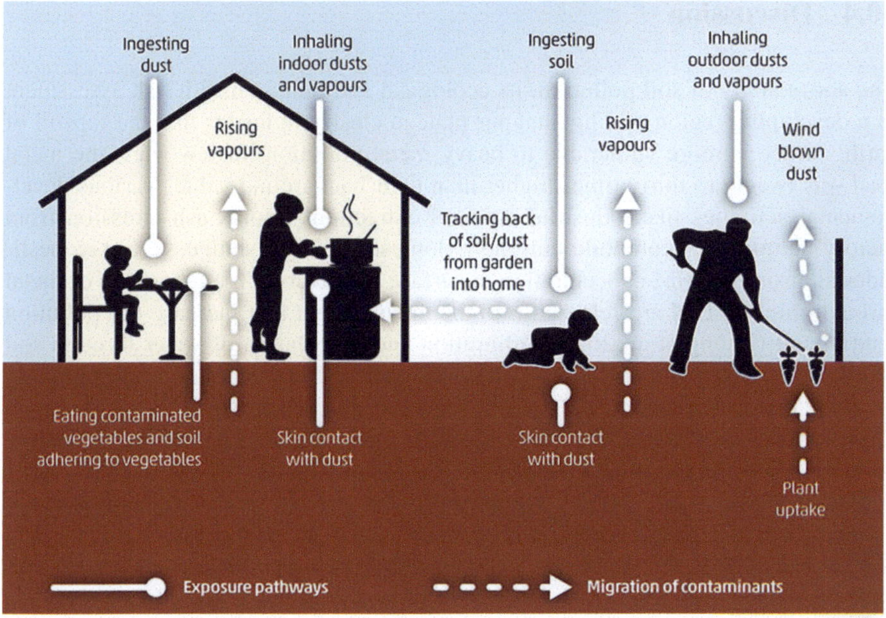

Fig. 30.7 Main exposure pathways to soil pollution. Source: adapted from Environment Agency of Great Britain, 2008 (http://www.fao.org/3/cb4894en/online/src/html/chapter-04-3.html)

the other hand, for children, HI values of ingestion ranged from 1.42E-01 to 6.91E-01 with 2.83E-01 as mean value. HI values of inhalation ranged from 2.26E-04 to 1.12E-03 with 4.66E-04 as mean value, and HI values of dermal ranged from 1.87E-01 to 9.34E-01 with 3.86E-01 as mean value (Fig. 30.6c). Average value of total HI indicates children are more prone to health related hazard by intake soil dust than adult persons. Children have more probability to non-carcinogenic health risk because of their physical activities such as plying with soil dust, ingestion of dust by hands and higher respiration rates (Fig. 30.7) (Jiang et al. 2017). Industrial and semi urban sites (20.83% samples) indicated HI > 1 for children residents. This indicates, there are obvious health related hazards (non-carcinogenic) for children population in Santuri block. Hazard Index values were classified into three categories in this study for identification of potential health risk zone as 'no possible health risk zone' (HI < 0.25), 'vulnerable health risk zone' (HI = 0.25–1), and 'health risk zone' (HI > 1). Spatial mapping using IDW on Arc GIS shows northern part of this block is much sensitive to potential health risk for adult and children population both (Fig. 30.5c).

30.4 Discussion

The above study of soil pollution, its ecological and human health risk assessment in a developing region of Chotanagpur plateau clearly indicates that the topsoil of earth surface is more vulnerable to heavy metal pollution. In few sites, the metal load was twenty to thirty times higher than their background value. Various developmental activities like industrial, mineral extraction, smoke, ash emission from factory chimneys, automobile emission, domestic sewage sludge, fertiliser, pesticides all promotes HM concentration in surface soil. During the extraction of metal through mining, HM are released into the environment and increase the pollution load in soil (Zhong et al. 2020). Infiltration and percolation of water through soil profile helps to leaching of HM to subsurface soil and contaminate it. In this study area, the concentration of each heavy metal exceeded their world normal value in topsoil as well as subsurface soil also except Sr for subsurface layer. Iron, manganese and zircon were most abundant metals in this region at both the soil layers.

Pollution load assessment by contamination factor indicates high manganese and copper load in the topsoil but this load decreases slowly in the subsurface soil. Overall assessment of pollution load in the surface soil by PLI suggested that all sample sites were exceeded threshold value (PLI > 1) of pollution load. At subsurface layer, 12.5% sample sites indicated no pollution (PLI < 1). Potential ecological risk (RI) of top soil showed that 25% of sample sites were moderately contaminated. On the other hand, 97.92% soil samples of subsurface layer showed practically uncontaminated to ecological environment. Human health risk of non-carcinogenic type via ingestion, inhalation and dermal contact of adult and children showed children were highly vulnerable to health hazard than adult population of the study area. Study from many parts of the world found that children are more vulnerable to health risk than adult (Zang et al. 2017; Yang et al. 2018; Adimalla et al. 2020). This study indicates that industrial regions are more polluted with heavy metal followed by semi urban areas. In Santuri block, unscientific establishment of sponge iron industries, cement factories and chemical factories produced huge amount of metal pollution to its surrounding environment. But, agricultural fields carried comparatively lower pollution load because of low usages of chemical fertilisers for crop production. The studies on HM pollution load on industrial and agricultural soil in other parts of the world indicate that mining and industrial sectors contributes more pollution by metal loads than crop fields (Liu et al. 2016; Fan et al. 2017; Yang et al. 2018).

30.5 Possible Remediation Strategies to Control Ecological and Human Health Risk from Soil Heavy Metal Pollution

Heavy metal contamination on pedosphere seriously makes negative impression on environmental quality. Though, limited quantity of metals presents in soil particles helps to promote soil productivity and ionic activity and increases plant growth. But, uncontrolled developmental projects lead to limitless metal dust release on open environment (2017) (Ali et al. 2019). In the present study industrial and semi urban areas inflicted significant pollution load to its topsoil and subsurface soil also. Ecological risk indices affirmed that industrial and semi urban sectors of this developing area are moderately contaminated to its topsoil ecological health. This can be appeared as a major environmental issue in near future, if there should not take any remedial measured to check it out (Chakraborty et al. 2021). High concentration of toxic metals in topsoil often creates health hazard of human body. Toxic metals can be entered through food chain, respiration or dermal contact. Higher quantity of HM in topsoil infiltrates with percolating water and entered to subsurface zone. HM of subsurface zone slowly mixed with groundwater and contaminate it also (Adhikary et al. 2011). Therefore, heavy metals (HM) can seriously impose to health hazard in this area. On this concern, scientific and affordable remediation measures are highly needed for holistic development of economic, ecologic and human life. Here, some possible remedial strategies are proposed to control soil heavy metal pollution.

1. To control metal dust emission on open air, industries must have to be installed high chimney machine with filtration facilities.
2. Metal dumping near industries should be controlled by restricting their unloading here and there without any particular dumping area. This could help to low mixing of heavy metals (HM) to soil directly or by surface runoff.
3. Metallurgical industries based with iron, coal etc. must follow proper waste management strategies of their solid or liquid effluents. This could help to low mixing of iron, manganese, chromium, and lead directly into soil.
4. Long term deposition of metal dumps, domestic solid wastes, and industrial waste should be prohibited or strictly restricted. It is because; materials with high metal contain leaches to subsurface soil and contaminate its environment as well groundwater in a long term period.
5. Electric burner should be installed by local governing bodies, though this technique is costlier one. Therefore, cheap measures such as cementation of base ground of waste disposal areas, quick utilisation of metal stocks by industries should be applied for restricting the heavy metal contamination.
6. Old tools of automobile vehicles must be changed to new tools and this practice should be encouraged by governmental authorities to their citizen to avoid high smoke and metal emission from old automobile machines.
7. Using of bio-fertilisers, instead of chemical products should be encouraged by farmers to control high mixing of toxic metals to field soil.
8. Children are more vulnerable to health risk by metal pollution of soil. Therefore, adequate protection measures and school level learning of hygiene maintenance should be followed by guardians and teachers.
9. Afforestation at road side areas of urban sectors, park, school or bare land should be encouraged by local authorities to control road dust and metal leaching to subsurface soil.

10. Soil sample analysis of contaminated area in a regular time period basis should be maintained to avoid ecological or health hazard properly.
11. Most of all, public awareness is very important to prohibit metal contamination, as because it is such a critical issue to be controlled by any single authorities. Therefore, self-awareness and honest effort to protect environment should be always helpful to reduce any kind of anthropogenic hazard caused by HM.
12. Using of modern technologies as GIS techniques should be helpful to identify spatial concentration of heavy metal load to soil and associated risk potential areas for taking further management strategies easily. In this study, GIS analysis significantly helped to indicate spatial zonation of HM pollution, its load to environment and potential human health risk of local residence.

30.6 Conclusion

Economic development is undoubtedly necessary for any nation's prosperity. But unscientific mining, industrial practice, waste disposal, land dumping, automobile usages etc. helps to increase heavy metals (HM) concentration in soil. In this study soil HM of Santuri block was analysed for identifying pollution load, ecological and human health risk. Analysis showed iron is most dominant metal found in the soil. It is naturally occurred by weathering of base rocks and also supplied by industrial activities in this area. Top soils of industrial and semi urban regions were highly polluted by heavy metal load. Comparatively, agricultural topsoil is less contaminated by metal load. Low productivity and moderate usages of fertilisers helped to low mixing of metals to its soil. Subsurface soil of industrial and semi urban sites also indicated moderate to low pollution by metal concentration. Assessment of potential ecological risk factor also depicts industrial and semi townships are moderately contaminated to ecological quality of its topsoil. Health hazard indicated children residence of industrial and township areas are prone to obvious health risk of non-carcinogenic type mainly by dermal contact in this region. Therefore, possible remediation strategies must be practiced and implemented by governing bodies, general public and entrepreneurs. GIS spatial analysis and cost efficient scientific remediation measures may be valuable to assist the understanding of soil pollution concerning public health in different land use practices.

Data Availability The datasets used and analyzed during the current study are available from the corresponding author on reasonable request.

Ethic Declarations

Competing Interests The authors declare that they have no competing interests.

Consent for Publication Not applicable.

Ethics Approval and Consent to Participate Not applicable.

References

Adhikary PP, Dash CJ, Bej R, Chandrasekharan H (2011) Indicator and probability kriging methods for delineating Cu, Fe, and Mn contamination in groundwater of Najafgarh Block, Delhi, India. Environ Monit Assess 176(1–4):663–676

Adimalla N (2020) heavy metals contamination in urban surface soils of Medak province, India, and its risk assessment and spatial distribution. Environ Geochem Health 42(1):59–75. https://doi.org/10.1007/s10653-019-00270-1

Adimalla N (2020b) Heavy metals pollution assessment and its associated human health risk evaluation of urban soils from Indian cites. Environ Geochem Health. https://doi.org/10.1007/s10653-019-00324-4

Adimalla N, Wang H (2018) Distribution, contamination, and health risk assessment of heavy metals in surface soils from northern Telangana, India. Arab J Geosci 11(21):684. https://doi.org/10.1007/s12517-018-4028-y

Adimalla N, Chen J, Qian H (2020) Spatial characteristics of heavy metal contamination and potential human health risk assessment of urban soils: a case study from an urban region of South India. Ecotoxicol Environ Saf 194:110406

Ali L, Rashid A, Khattak SA, Zeb M, Jehan S (2019) Geochemical control of potential toxic elements (PTEs), associated risk exposure and source apportionment of agricultural soil in Southern Chitral, Pakistan. Microchem J 147:516–523

Ameh EG (2013) Multivariate statistical analysis and enrichment of heavy metal contamination of soil around Okaba coalmines. Am Eurasian J Agron 6:09e18

Bera B, Bhattacharjee S, Chamling M et al (2021) Fluoride hazard and risk enumeration of hard rock unconfined aquifers in the extended part of Chhota Nagpur Gneissic Complex. J Geol Soc India 97:199–209. https://doi.org/10.1007/s12594-021-1651-0

Bera B, Ghosh A (2019) Fluoride dynamics in hydrogeological diversity and fluoride contamination index mapping: a correlation study of North Singbhum Craton, India. Arab J Geosci 802. https://doi.org/10.1007/s12517-019-4994-8

Bilos C, Colombo JC, Skorupka CN, Rodriguez Presa MJ (2001) Sources, distribution and variability of airborne trace metals in La Plata City area, Argentina. Environ Pollut 111:149–158

Cabrera F, Clemente L, Barrientos ED, Lopez R, Murillo JM (1999) Heavy metal pollution of soils affected by the Guadiamar toxic flood. Sci Total Environ 242:117e129

Chakraborty B, Bera B, Roy S, Adhikary PP, Sengupta D, Shit PK (2021) Assessment of non-carcinogenic health risk of heavy metal pollution: evidences from coal mining region of eastern India. Environ Sci Pollut Res. https://doi.org/10.1007/s11356-021-14012-3

Chen K, Huang L, Yan BZ, Li HB, Sun H, Bi J (2014) Effect of lead pollution control on environmental and childhood blood lead level in Nantong, China: an interventional study. Environ Sci Technol 48:12930

Dang Z, Liu C, Haigh MJ (2002) Mobility of heavy metals associated with the natural weathering of coal mine spoils. Environ Pollut 118:519e526

Das SK, Chakrapani GJ (2011) Assessment of trace metal toxicity in soils of Raniganj Coalfield, India. Environ Monit Assess 177:63e71

De A, Mridha D, Ray I, Joardar M, Das A, Chowdhury NR, Roychowdhury T (2021) Fluoride exposure and probabilistic health risk assessment through different agricultural food crops from fluoride endemic Bankura and Purulia districts of West Bengal, India. Front Environ Sci 9:713148. https://doi.org/10.3389/fenvs.2021.713148

De S, Mitra AK (2004) Mobilisation of heavy metals from mine spoils in a part of Raniganj Coalfield, India: causes and effects. Environ Geosci 11:65e76

Fan Y, Zhu TP, Li MT, He JY, Huang RX (2017) Heavy metal contamination in soil and brown rice and human health risk assessment near three mining areas in central China. J Health Eng 2017:1–9

FAO and UNEP (2021) Global assessment of soil pollution: report. Rome. https://doi.org/10.4060/cb4894en

Greening T (2011) Quantifying the impacts of vehicle-generated dust: a comprehensive approach. Transport Research Support Program. World Bank, Washington

Hakanson L (1980) Ecological risk index for aquatic pollution control, a sedimentological approach. Water Res 14:975e1001

http://www.fao.org/3/cb4894en/online/src/html/chapter-04-3.html

Jiang Y, Chao S, Liu J, Yang Y, Chen Y, Zhang A, Cao H (2017) Source apportionment and health risk assessment of heavy metals in soil for a township in Jiangsu Province, China. Chemosphere 168:1658–1668

Jiang DZ, Teng EJ, Liu YL (1996) The contribution of difference on the element background values in soils and the analysis of variance of single factor on soil groups. Environ Monit China 2:21e24

Jiang, Jiang HH, Cai LM, Wen HH, Hu GC, Chen LG, Luo J (2020) An integrated approach to quantifying ecological and human health risks from different sources of soil heavy metals. Sci Total Environ 701:134466

Kashyap R, Sharma R, Uniyal SK (2019) Distribution of heavy metals in habitation land-use soils with high ecological risk in urban and peri-urban areas. Int J Environ Sci Technol 16:8093–8106

Landrigan PJ, Fuller R, Acosta NJR, Adeyi O et al. (2017) The Lancet Commission on pollution and health. The Lancet 391 (10119):462–512

Lancet Commission (2017) Dementia prevention, intervention, and care, 16; 390 (10113):2673–2734

Li Z, Ma Z, van der Kuijp TJ, Yuan Z, Huang L (2014) A review of soil heavy metal pollution from mines in China: pollution and health risk assessment. Sci Total Environ 468–469:843–853

Li C, Li F, Wu Z, Cheng J (2017) Exploring spatially varying and scale-dependent relationships between soil contamination and landscape patterns using geographically weighted regression. Appl Geogr 82:101–114

Liu G, Wang J, Zhang E, Hou J, Liu X (2016) Heavy metal speciation and risk assessment in dry land and paddy soils near mining areas at Southern China. Environ Sci Pollut Res Int 23:8709–8720

Luo X, Ren B, Hursthouse AS, Jiang F, Deng R-J (2019) Potentially toxic elements (PTEs) in crops, soil, and water near Xiangtan manganese mine, China: potential risk to health in the foodchain. Environ Geochem Health. https://doi.org/10.1007/s10653-019-00454-9

Muller G (1969) Index of geoaccumulation in sediments of the Rhine river. GeoJournal 2:108–118

Pan L, Wang Y, Ma J, Hu Y, Su B, Fang G, Wang L, Xiang B (2018) A review of heavy metal pollution levels and health risk assessment of urban soils in Chinese cities. Environ Sci Pollut Res 25(2):1055–1069

Qu M, Wang Y, Huang B, Zhao Y (2018) Source apportionment of soil heavy metals using robust absolute principal component scores-robust geographically weighted regression (RAPCS-RGWR) receptor model. Sci Total Environ 626:203–210

Rubio B, Nombela MA, Vilas F (2000) Geochemistry of major and trace elements in sediments of the Ria de Vigo (NW Spain): an assessment of metal pollution. Marine Pollution Bulletin 40: 968–980

Sahoo PK, Equeenuddin SM, Powell MA (2016) Trace elements in soils around coal mines: current scenario, impact and available techniques for management. Curr Pollut Rep 2:1e14. https://doi.org/10.1007/s40726-016-0025-5

Selvi A, Rajasekar A, Theerthagiri J, Ananthaselvam A, Sathishkumar K, Madhavan J, Rahman PKSM (2019) Integrated remediation processes toward heavy metal removal/recovery from various environments—a review. Front Environ Sci 7. https://doi.org/10.3389/fenvs.2019.00066

Sun L, Guo D, Liu K, Meng H, Zheng Y, Yuan F, Zhu G (2019) Levels, sources, and spatial distribution of heavy metals in soils from a typical coal industrial city of Tangshan, China. CATENA 175:101–109

Taiwo AM, Awomeso JA, Taiwo OT et al (2017) Assessment of health risks associated with road dust in major traffic hotspots in Abeokuta metropolis, Ogun state, southwestern Nigeria. Stoch Environ Res Risk Assess 31:431–447

Tomlinson DL, Wilson JG, Harris CR, Jeffrey DW (1980) Problems in the assessments of heavy-metal levels in estuaries and formation of a pollution index. Helgol Meeresunters 33:566e575

Verma A, Kumar R, Yadav S (2019) Distribution, pollution levels, toxicity, and health risk assessment of metals in surface dust from Bhiwadi industrial area in North India. Hum Ecol Risk Assess 1–21

Wang F, Zhao W, Chen Y (2019) Spatial variations of soil heavy metal potential ecological risks in typical moso bamboo forests of Southeast China. Bull Environ Contam Toxicol 102(2):224–230

Wang Z, Xiao J, Wang L, Liang T, Gao Q, Guan Y, Rinklebe J (2020) Elucidating the differentiation of soil heavy metals under different land uses with geographically weighted regression and self-organizing map. Environ Pollut 260:114065

World Health Organization (2017) Guidelines for drinking water quality: fourth edition incorporating the first addendum. Geneva, Switzerland

Yang Q, Li Z, Lu X, Duan Q, Huang L, Bi J (2018) A review of soil heavy metal pollution from industrial and agricultural regions in China: pollution and risk assessment. Sci Total Environ 642:690–700

Yang W, Wang D, Wang M, Zhou F, Huang J, Xue M, Dinh QT, Liang D (2019) Heavy metals and associated health risk of wheat grain in a traditional cultivation area of Baoji, Shaanxi, China. Environ Monit Assess 191(7):428

Zang F, Wang S, Nan Z, Ma J, Zhang Q, Chen Y, Li Y (2017) Accumulation, spatio-temporal distribution, and risk assessment of HM in the soil-corn system around a polymetallic mining area from the Loess Plateau, northwest China. Geoderma 306:188–196

Zhang Y, Lu W, Yang Q (2015) The impacts of mining exploitation on the environment in the Changchun—Jilin Tumen economic area, Northeast China. Nat Hazards 76(2):1019–1038

Zhong X, Chen Z, Li Y, Ding K, Liu W, Liu Y, Yuan Y, Zhang M, Baker AJM, Yang W, Fei Y, Wang Y, Chao Y, Qiu R (2020) Factors influencing heavy metal availability and risk assessment of soils at typical metal mines in Eastern China. J Hazard Mater. https://doi.org/10.1016/j.jhazmat.2020.123289

Chapter 31
Bioremediation Approaches for Curbing the Potential of Toxic Element for Sustainable Agriculture

Supriya Pandey, Pooja Thathola, Dinesh Chandola, Sumit Rai, and Ashish Rai

Abstract With increasing urbanization, industrialization and adaptation of modern agricultural practices, soil contamination has become a prime concern. Soil health is crucial for the health of environment, ecosystems, world economy and for human population as well, but due to various natural and anthropogenic activities soil health gets depleted that ultimately affect plant and human health. Various techniques have been introduced that involved in transformation and biodegradation of organic and inorganic contaminants in soils with plant based and microorganisms-based remediation techniques that takes place in plants as well as in soil. Ability of plant to uptake contaminants is well studied but plants in combination with microorganisms have been found to increase the ability to degrade many compounds. Thus, this review is a scattered literature that highlights the risks associated with organic and inorganic contaminants to soil, plant and environment as a whole and possible bioremediation techniques have also been presented.

Keywords Bioremediation · Phytoremediation · Toxic elements · Sustainable Agriculture

31.1 Introduction

Increasing world population demands large production of consumer goods thus, need of rapid growth in industrial sector increasing day by day. With the increasing growth in industrial sector, and anthropogenic induced climate change (Kanwal et al. 2019) increases chances of releasing high number of toxic compounds that affecting human life and environment (Singh et al. 2020; Kumar et al. 2018). With

S. Pandey · P. Thathola · D. Chandola · S. Rai (✉)
GB Pant National Institute of Himalayan Environment, Kosi-Katarmal, Almora, Uttarakhand 263643, India
e-mail: sumitssac101@gmail.com; rai.sumit@gbpihed.nic.in

A. Rai
Krishi Vigyan Kendra, Rajendra Prasad Central Agricultural University, Parsauni East Champaran, Dr., Pusa 848125, Bihar, India

P. K. Shit et al. (eds.), *Soil Health and Environmental Sustainability*, Environmental Science and Engineering,
https://doi.org/10.1007/978-3-031-09270-1_31

the advancement in time among the inorganic pollutants, heavy metals are the prime topic of research among researches and menace among ecologists; they are continuously engaged in findings solution of these problems that can be accommodative in reducing toxic effect of various pollutants affecting human health, ecosystem and whole environment.

Modern agricultural practices involve the application of agrochemicals and inorganic fertilizer that are the cause of degradation of ecosystem and environment (Malik et al. 2017). And these environmental pollutants become toxic in nature and cause negative effect on human health and environment. These contaminants can be naturally synthesized or can be of synthetic origin. As synthetic contaminants like pesticides, dyes are synthetically derived thus they are new to any type of ecosystem, therefore their degradation is not possible for the nature's self-cleaning system, as it is quite slow and insufficient for elimination of synthetic pollutants (Rai et al. 2014, 2016).

On the other hand heavy metals are the most hazardous pollutants of the environment, and are the most common toxic pollutants of soil viz. arsenic, cadmium, chromium, copper, zinc and mercury. And major sources of heavy metals into the agricultural soil systems are application of sewage sludge, organic waste manure, industrial byproducts, and irrigation with waste water (Khan et al. 2013; Tóth et al. 2016; Srivastava et al. 2016; Sharma et al. 2017; Woldetsadik et al. 2017). Heavy metals constitute a group of inorganic chemical hazards, most commonly found at heavy metals contaminated sites are Pb, Cr, As, Zn, Cd, Cu, Hg and Ni). Some heavy metals like Co, Cu, Fe, Mn, Mo, Ni, V, and Zn are required in minute quantities by organisms. However, excessive amounts of these elements can become harmful to organisms and environment as well. In the industrialized world Soil contamination by heavy metals is one of the most important anticipation (Hinojosa et al. 2004). Automobile exhaust, coal burning, erosion of metal structures, and refuse incineration are the major sources of accumulation of heavy metals in agricultural soil. It is reported that in agriculture, the utilization of fertilizers, manures, and pesticides have contributed to the accumulation of heavy metals in soils (Senesil et al. 1999). Developing remediation processes for toxic environmental contaminants gives us not only an economically viable opportunity for management of environmental pollutants but also a suitable alternative to costly physicochemical processes (Chanwala et al. 2019).

31.2 Toxic Elements in Agricultural Soil

We all depend on natural resources for our basic requirement like food, fodder, fiber and fuel, thus soil quality is an immediate concern among us. But various anthropogenic activities laid intense pressure on land resources via industries and various other activities as a result soil gets polluted and its quality deteriorates. Broadly soil pollutants are divided into two categories that are shown in Fig. 31.1. Soil is contaminated with various organic pollutants from various sources like industrial waste

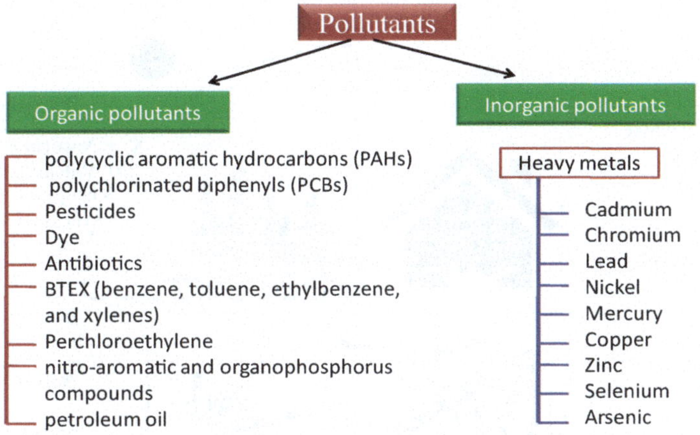

Fig. 31.1 Categories of pollutants

discharges containing toxic pollutants, improper disposal of solid waste, and accidental spillage of contaminants during transportation or handling of hazardous materials and indiscriminate use of pesticide and fertilizers. Organic pollutants can be both naturally occurring and synthesized compounds and may pollute agro-ecosystem either deliberately through crop management activities or accidentally through entry of industrial/urban wastes. Their behavior and fate in the soil depend mainly on their chemical structure, which determines interactions with mineral particulates, organic matter, water, gas and biota of the soil (Gurjar et al. 2017; Saxena and Rai 2020).

From thousands of years various types of compounds have been used by humans for various purposes. But their excessive utilization cause soil pollution like Inorganic pollutants (Heavy metals) by accumulating in the soil and cause soil contamination via emission from speedily spread-out industrial sectors, disposal of heavy metal wastes, paints, land uncontrolled application of fertilizers in fields, animal manures, sewage sludge, pesticides, coal combustion residues, spillage of petrochemicals, and atmospheric deposition (Khan et al. 2008; Zhang et al. 2010). A large number of biotic and abiotic byproducts affects agroecosystem, among them heavy metals, fertilizers, pesticides, and sewage sludge are the most common (Alloway 2013).

31.2.1 Inorganic Toxic Elements

Due to advancement in agriculture technology and rapid development of industries, without even realizing their impact on environment and in ecosystem, utilization of inorganic elements by natural and anthropogenic source is most common (Fig. 31.2). Inorganic elements (heavy metals) are considered as a part of soil. When soil gets highly concentrated or loaded with these elements, they negativity affects plant growth and yield by polluting soil and considered as toxic soil contaminants. The

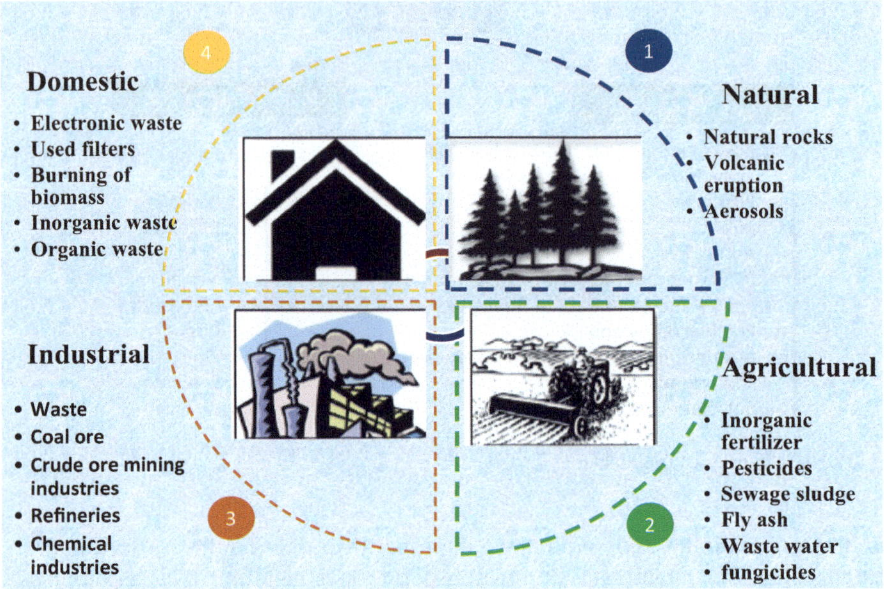

Fig. 31.2 Overview of sources of heavy metal pollution

low availability of macro-nutrients in plants and soil acidity are the main problems associated with the heavy metals toxicity.

31.2.1.1 Cadmium

Accumulation of Cd in soil is a worldwide problem which is mainly affected by soil pH and content of organic matter. As pH of soil decreases, Cd bioavailability increases, thus cause imperfection in soil properties. Study conducted by Liao et al. (2005) reported toxic effect of Cadmium by causing inhibitory effects on soil microbial activities, microbial growth, and microbial metabolic processes on paddy soil. Combined negative effect of Cd and salinity by reduction of microbial respiration and content of microbial biomass in soil was reported by Raiesi and Sadeghi (2019). An (2004) argued that Cd is highly mobile in the soil and resulted in high toxicity that affected the essential microorganisms, inhibited microbial activities, and absorbed the organic matter present in the soil, as well as physicochemical characteristics has also changed. Although worldwide natural soil content of Cd is reported to vary in the range between 0.06 and 1.1 $\mu g\ g^{-1}$ whereas its concentration exceeding 0.5 $\mu g\ g^{-1}$ in soils contaminated from anthropogenic sources. At a given soil concentration Plant species vary greatly in their ability to translocate Cd like: lettuce > spinach, tomato, carrot > rape, cabbage, radish > maize, mustard > sunflowers, bean, pea, cucumber > wheat, oat (Saha et al. 2017).

31.2.1.2 Lead

It is suggested that very high concentrations of Pb in the soil are necessary to exhibit toxicity because plants transfer little amount of Pb^{+2} from soil to aboveground parts then enriching it at the root surface and immobilize it in the cell wall of the root (Saha et al. 2017). Pb is categorised as hazardous heavy metal pollutant due to its high toxicity (Qi et al. 2018). The effects of Pb on soil cause reduction in soil nutrients, microbial diversity, and soil fertility (Dotaniya et al. 2020). Various studies have been done on effect of Pb on agricultural soil suggested that Absorption and retention of Pb cause serious impact on soil properties (Vega et al. 2010). Inhibition of enzymes' activities has also been reported by Pb exposure for long time. Lead is regarded as a useful and toxic metal for plant growth at the same time (Uzu et al. 2009) but are not essential for plant growth (Diaconu et al. 2020). Even at low concentration Pb can be highly toxic to plants, which inhibit plant growth, yield and productivity (Ashraf et al. 2017).

A study on soybean crops done by Hamid et al. (2010) reported the toxic effect of Pb on crop growth, decrease in the chlorophyll content in the plant. Inhibition in seed germination and decreased the protein content was reported by Kushwaha et al. (2018)

31.2.1.3 Copper

Cu is an important micronutrient for plants and essential element for soil. Copper availability is higher in acidic soil than in alkaline and organic matter (Brun et al. 2003). Cu based fungicides are the main cause of high accumulation rate of Cu in agricultural soil. A study suggests that the range of Cu concentration in agricultural soil is between 5 and 30 mg Kg^{-1} (Brun et al. 2003). Cu is an essential element for plant nutrition and seed production. However, at high concentrations, Cu is considered a very toxic metal (Chiou and Hsu 2019). Decrease in crop yield, biosynthesis of chlorophyll, and plant productivity by modification of photosynthesis and nutrients due to Cu toxicity has been recorded by Adrees et al. (2015). Aly and Mohamed (2012) also reported the negative effects of the high concentration of Cu on maize plants.

31.2.1.4 Zinc

The Zn toxicity in soil has a notable relationship with the soil enzyme's active sites, as Zn replaces certain cations that are crucial for cell performance (Łukowski and Dec 2018). Barman et al. (2018) reported negative effect on soil characteristics such as pH, the content of organic matter, bicarbonate content, and impedes the role of Mg and Fe in the soil due to Zn deficiency. Zn is a crucial micronutrient for plants (Song et al. 2019). Accumulation of Zn causes severe damages in plant roots and shoots. Hammerschmitt et al. (2020) demonstrated Zn toxicity on the young peach tree and shows that accumulation of Zn in the root system prevented the elements from transportation to the leaves. Decrease in the root length and the photosynthesis rate was reported by Song et al. (2019) due to exposure of Zn. Due to effect of Zn hyperaccumulation in plants all the physiological and biochemical mechanisms in plants were affected, indicating the harmful effect of Zn precipitation in plants (Balafrej et al. 2020).

31.2.2 Organic Toxic Element

31.2.2.1 Fertilizer

Fertilizers are one of leading source of Heavy Metals Accumulation in Agricultural Soil and Plant. They improve soil fertility by supplying nutrients that enhance plant growth and productivity and increase organic matter in soil as well (Meng et al. 2020). Fertilizers, contain organic and inorganic elements that are responsible for producing heavy metals in the soil. Excessive use of fertilizers for a long time results in heavy metals accumulation in agricultural soils that reduces soil fertility, and consequently decreases plant growth and productivity (Ai et al. 2020). After heavy metal contamination it is extremely challenging to recover the soil environment to its original state. Cu, Zn, and Cd have a higher accumulation potential in agricultural soil due to the long-term use of fertilizers (Qin et al. 2020) Therefore, they enter into the food chain and reach animals and humans (Liu et al. 2020). Regularly large quality of fertilizers has been introduced into soils in intensive farming systems to provide adequate NPK for plant growth. The compounds used to supply these elements contain trace amounts of heavy metals (e.g., Cd and Pb) as impurities, which, after continued fertilizer, application may significantly increase their content in the soil Metals, such as Cd and Pb, have no known such physiological activity on plants.

31.2.3 Pesticides

Agricultural fields are the largest consumer of the global production of pesticides. Abhilash and Singh (2009) reported that, an average of about 2 Mt. of pesticides are consumed every year throughout the world; and out of these 24% is consumed in USA, 45% in Europe and 25% in rest of the world. Pesticides can be generated naturally or synthetically and widely used for controlling harmful weeds (herbicides), fungi (fungicides), bacteria (bactericides), and insect infestations (insecticides) in the agricultural field (Khalek et al. 2018), and also to prevent rodenticides, molluscicides, and nematicides. At global level, the consumption of herbicide in India is the highest especially in tropical countries where climatic conditions are predominantly warm humid; therefore utilization of insecticides is more.

Due to longer persistence in the environment, some of the pesticides were banned several decades before. However, because they are cost effective, easily available, and display a wide spectrum of bioactivity some of these pesticides are still preferred by the small farmers (Devi and Raha 2013). Low solubility and high structural stability of the pesticide limit their degradation in soil via chemical and biochemical processes by plants and microbes. Most of the hydrophobic pesticides are adsorbed to soil surface or to organic matter and get sequestrated into tiny pores of soil matrix, becoming less bioavailable. Soil microorganism mainly degrades pesticides and converts them into its less toxic form but many times, it was found that formed end products become more toxic than the original pesticide. Some of the factors that influence their persistency are microbial diversity, rainfall, soil temperature, exposure to sunlight, application rate as well as their solubility and mobility in soil (Saha et al. 2017).

31.2.4 Dye Pollutants

Rapid increase in human population and agricultural activities has aggravated the problem of environmental contamination globally. Chanwala et al. 2019; Yang et al. (2019a, b) reported that rapid increase in textile, petroleum-based industries cause accumulation of large amount of unwanted dyes and organic contaminants into the environment that affecting human health and environment. Currently, dye degradation is a challenging problem worldwide. It has been reported that many azo dyes and their products after degradation such as aromatic amine are potentially toxic and carcinogenicin nature (Khalid et al. 2008; Dafale et al. 2008). Under natural conditions many of dyes are resistant to degradation and remediation and through conventional remediation methods (Tahir et al. 2016). Thus, advancement in remediation techniques is required. As a result of irrigation with effluent from dye industries, considerable accumulation of total organic dyes in cultivated soil had been observed and such accumulation has also been transported in plant tissue (Chandanshive et al. 2018). Evidences of plant uptake of dye compounds by plants were provided by Uera et al. (2007), Muthunarayanan et al. (2011).

31.2.5 Antibiotics

As number of pharmaceutical prescriptions raising for the aging population it results in a higher discharge of the medicaments and their metabolites in sewage water. Subsequently, the presence of pharmaceuticals in surface water has increased in recent years and it has been accepted that this tendency will continue in the near future. A review indicated absorption of antibiotics by crops from contaminated soils and detected concentrations in different plant parts ranged between 0.9 and 6.1 mg/kg (Du and Liu 2012).

31.3 Impact of Inorganic Pollutants on Soil and Plants

31.3.1 Soil

Soils are the major source for heavy metals released into the environment by various anthropogenic activities. Due to adverse effect on microbial communities and plant quality, yield, heavy metals especially Cu, Ni, Cd, Zn, Cr, and Pb (Hinojosa et al. 2004) are considered as one of leading rootage of soil pollution. Inauspicious outcome of heavy metals on soil biological (Friedlová 2010) and biochemical properties i.e. organic matter, clay contents and pH are well referenced. Number, diversity, and activities of soil microorganisms have been affected by accumulation of heavy metals. Magnuson et al. (2001) reported that soil aeration, microbial activity, and mineral composition influence heavy metal availability in soils. Biodegradation of organic contaminants has been severely inhibited by the presence of toxic metals in the soil. The effect of heavy metal pollutants on soil and plant health and how they become contaminants in the environment is depicted in Figs. 31.3 and 31.4.

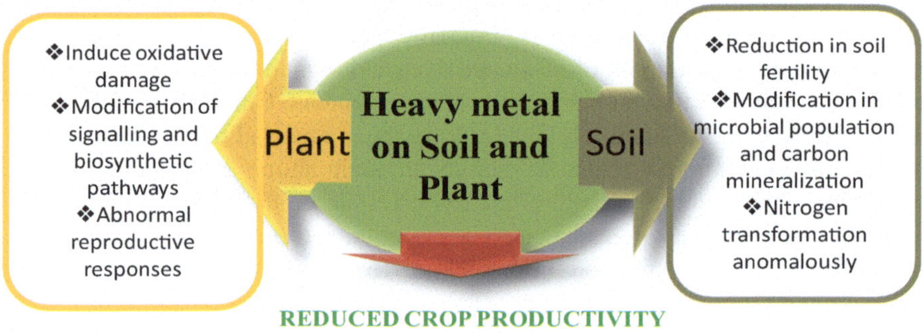

Fig. 31.3 Effect of heavy metal pollutants on soil and plant health

Fig. 31.4 Possible reasons why heavy metal become contaminants in the soil environment (adapted from D'Amore et al. 2005)

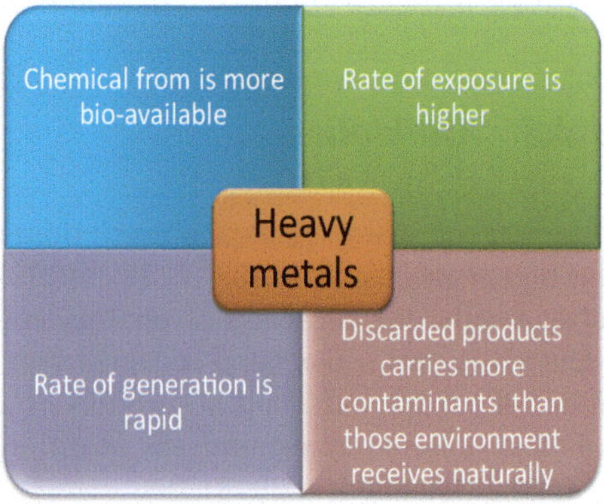

31.3.2 *Plants*

Plant uptake those heavy metals that are present in soluble form in the soil and can be easily solubilize by root exudates. Though plants need presence of heavy metal (Co, Cu, Fe, Mn, Mo, Ni and Zn) for their proper growth, functioning and metabolism but excess amount of these heavy metals can be toxic to plants (Garrido et al. 2005; Rascio and Navari-Izzo 2011), they directly or indirectly affect plant growth and organisms those are depend on plants. Some of heavy metals such as As, Cd, Hg, Pb or Se are not essential for plants growth, since they do not perform any known physiological function in plants. As industrialization increase exposure of plants to various biotic and abiotic stresses like heat, cold, drought, high light intensity, UV radiations, heavy metals, and pollutants such as O_3 and SO_2 (Dezhban et al. 2015; Kumar et al. 2016) increases. Consequently, high level of reactive oxygen species (ROS) like singlet oxygen ($1/2O_2$), hydroxylradical (HO•), superoxide radical (O•–2), and hydrogen peroxide (H_2O_2) are produced (Wang et al. 2015).

31.4 Bioremediation Strategies for Sustainable Agriculture

31.4.1 *Plant Mediated Remediation of Heavy Metal Polluted Soil*

Phytoremediation is a plant mediated process for cleaning up of contaminated sites and for stabilization of contaminated soils and ground water (Albeto and Sigua 2013). There are certain plants which are capable of metabolizing toxic compounds from their environment therefore they are used in contaminated sites for remediation of pollutants. Phytoremediation has become the emerging concept in the last few decades where issue of removal of soil pollution has gain great interest among the researchers (Thathola et al. 2019).

Everyday new chemical compounds have been continuously produced. In last few decades production and consumption of harmful supplements have been increased substantially, in developed countries its production also increases. Their persistence may affect the ecosystem, including agricultural product, water quality as well as soil microorganisms and human health (Fig. 31.5). Methods used in the remediation process are very costly and also can be destructive for the ecosystem present in soil (Hooda 2007) so, an effective approach for the removal should be applied which can be environment friendly, cost effective and easy for establishment and reestablishment of crop in contaminated soil. Thus, to ensure more effective cleanup of contaminated soil phytoremediation techniques can be employed in which plants are used to remove, pollutants or convert them into less toxic form in effective and cost-effective manner, this method has received great attention in last decades (Glick 2010).

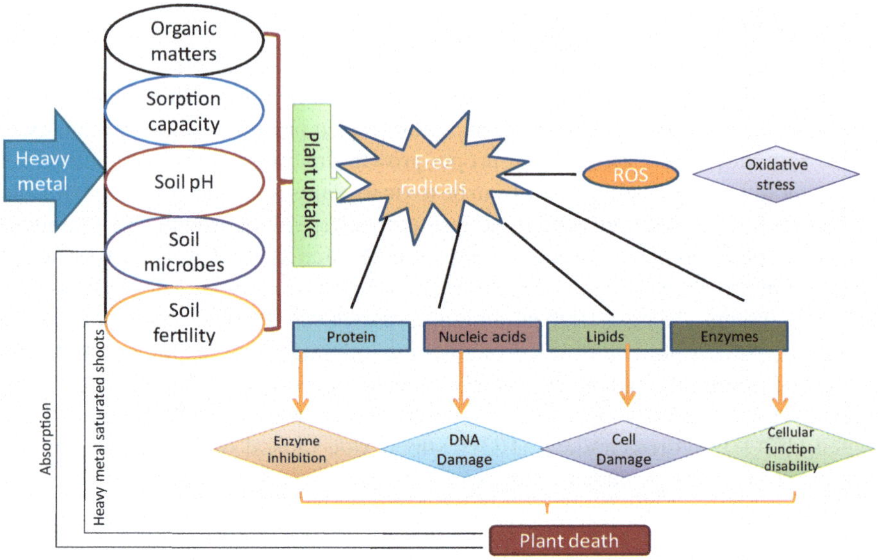

Fig. 31.5 Mechanism of action and pathway of heavy metals toxicity in soil and plant (adapted from Alengebawy et al. 2021)

Remediation methods currently applicable for removal of contamination from contaminated soil are expensive and cause damage to ecosystem and do not take place in sustainable way. Therefore, natural remediation techniques such as phytoremediation have been developed to provide more ecofriendly and cost-effective cleanup of contaminated sites. This technique has been used to treat wide variety of chemicals including metals etc. But some of the major limitations of this technology include phytotoxicity, slow degradation, and limited contamination uptake. Thus to overcome this, plant associated bacteria such as endophytes have been exploited for improving phytoremediation efficiency of plants for several pollutants Almost all pants in their natural state are colonized by endophytes, that form beneficial associations with their hosts, they offers various benefits to their hosts such as enhancing plant growth through phytohormone production, resistance to environmental stresses, supplying biologically fixed nitrogen and producing important compounds like medicinal, agricultural and industrial. Enhancement in phytoremediation by using endophytes has been shown to improve uptake and degradation of several toxins from plants.

Among all the treatment technology, phytoremediation is the low cost and effective green technology for the removal of these contaminants from the environments with no harmful effect. To ensure more effective clean-up polluted soils, combination of microorganisms and plants is an effective approach for bioremediation process. In the era of advancement soil pollution become a serious concern for environment flora and fauna, to decrease the effect of pollution plant-based remediation technique act as an advance approach in which removal of contamination from soil is possible where plants and their associated microflora are used to eliminate the contaminants from the soil (Table 31.1). As this method is achieved via natural processes, it comes under environment friendly approach, and is an economically remediation techniques. For the remediation of contaminated soil use of vegetation is a promising, cost- effective approach. Plants convert organic contaminants to fewer toxic metabolites; they also stimulate the degradation of organic compounds in rhizosphere by release of root exudates and enzymes (Karthikeyan and Kulakow 2003). Phytoremediation of contaminated soils can be achieved via different mechanisms. These mechanisms have been discussed hereunder.

31.4.1.1 Phytostablisation

Process of phytostabilisation involves the use of metal tolerant plant species to immobilize heavy metals and to decrease their bioavailability by this process it prevents the migration of heavy metals into the ecosystem and reduce the chance of their transfer in food chain. Contaminants stabilize in the roots of plants or within the rhizosphere. Contaminants are absorbed and accumulated in the roots and absorbed onto the roots, or precipitated in the rhizosphere. It not only prevents migration of contaminants into the groundwater or air, but also reduces the bioavailability of the contaminant (Pandey and Bagga 2013).

31.4.1.2 Phytoextraction

This process uses ability of plant to accumulate contaminants from the soil, plant root uptake metal contaminants and translocate it to their above soil tissues. Phytoextraction is an important process for sites having more than one type of metal contaminants. Phytoextraction involves repeated cropping of plant until the metal concentration in the soil has reached the acceptable (targeted) level (Pandey and Bagga 2013).

31.4.1.3 Rhizofiltration

In an aquatic environment the toxic metals and other contaminants are remediated using plant biomass by a phytoremediation approach, known as rhizofiltration (Kumar et al. 2019). In this method plants are grown directly in contaminated soil and plant roots absorb contaminants from soil. Remediation of contaminated ground water is concerned with this process. Plants either adsorb contaminants in root surface or absorbed contaminants by their roots and absorb contaminating metals and concentrate them in roots and shoots. Agricultural runoff and industrial discharge can be treated by rhizofiltration (Yadav et al. 2011).

Table 31.1 Role of plant in bioremediation of heavy metal polluted soil

S. No.	Plant	Contaminants	Remarks	References
1.	*Cannabis sativa L.*	Metalaxyl-M and metribuzin, (pesticides) and BPA, 17β-estradiol (E2), and 4-tert-octylphenol (OP) (Endocrine disturbing compounds)	On average, 12, 11, 10, 9, and 14% removal of metalaxyl-M, metribuzin, BPA, E2, and OP	Loffredo et al. (2021)
2.	*Galium mollugo* and *Stellaria holostea*	Zn and Cd	*G. mollugo* and *S. holostea* had a hyperaccumulator behavior for Cd and Zn	Antoniadis et al. (2021)
3.	*Leersia oryzoides (rice-cut grass)*	Arsenic	Removal of As during periodic mowing over a growing season by phytoextraction technique	Ampiah-Bonney and Lanza (2007)
4.	*Scirpus littoralis*	Pb, Zn, Ni, Mn, Cu	*Scirpus littoralis* accumulated Mn, Ni, Cu, Zn and Pb upto a below ground organs in 90 days	Bhattacharya et al. (2006)
5.	*Calotropis procera*	Arsenic	Efficient in remediating significant quantities of As after 15 and 30 d when exposed to a range of concentrations	Singh and Fulzele (2021)
6.	*Pteris vittata*	Arsenic	accumulate Cr in its sporophytic and gametophytic biomass which is a new finding adding to its ability for hyperaccumulation	Kalve et al. (2011)
7.	*Corrigiola telephiifolia*	As, Pb	C. telephiifolia could be considered a Pb accumulator and an As hyperaccumulator plant	García-Salgado et al. (2012)
8.	*Pteridium Aquilinum, Corrigiola Telephiifolia and Cyperus Exaltatus*	As	Intense accumulative capacity for As by *Pteridium Aquilinum, Corrigiola Telephiifolia and Cyperus Exaltatus* for cleanup and restoration of As-contaminated soils	Onyia et al. (2020)

(continued)

Table 31.1 (continued)

S. No.	Plant	Contaminants	Remarks	References
9.	*Turnip Landraces*	Cd	Showed Cd accumulation capacity but it was not advised to consume turnips cultivated in an environment that exceeds safe Cd levels	Li et al. (2016)
10.	*Spartina alterniflora*	Cu	Accumulation of metal in leaves, rhizomes and fine roots, the highest Cu accumulations were detected under 800 mg kg^{-1} Cu. The highest Cu accumulation in stem was revealed under 200 mg kg^{-1} Cu	Chai et al. (2014)
11.	*Moringa oleifera*	Cd	Removal of Cd through leaf extraction	Howladar (2014)
12.	*Lavandula vera L.*	Cd, Pb, and Zn	*Lavandula vera L.* was found to be hyperaccumulators of lead and the accumulators of cadmium and zinc,	Angelova et al. (2015)
13.	*Portulaca oleracea L.*	Cr	Cr accumulation (150–190 mg/kg dry weight) in harvestable parts of *Portulaca*	Kale et al. (2015)
14.	*Atriplex Halimus, Medicago Lupulina and Portulaca Oleracea*	Pb, Ni, and Zn	Plant metal uptake efficiency *A. halimus* > *M. lupulina* > *P. oleracea*, *A. halimus* and *M. lupulina* could be successfully used in phytoremediation, and in phytostabilization, in particular	Amer et al. (2013)
15.	*Zygosaccharomyces rouxii*	Cd, Zn, Cu, and Pb	Extraction and exclusion of heavy metals	Li et al. (2013)
16.	*Halimione portulacoides*	Zn	*H. portulacoides* cuttings can be useful in the restoration of metal-polluted soils	Andrades-Moreno et al. (2013)
17.	*Suaeda salsa*	Pb and Zn	Accumulation of metals in roots of *Suaeda salsa*	Wu et al. (2013)
18.	*Salicornia ramosissima*	Cd	Accumulation of metal in roots and Cd bioaccumulation decreased with the increase of salinity and Cd concentration	Pedro et al. (2013)

(continued)

Table 31.1 (continued)

S. No.	Plant	Contaminants	Remarks	References
19.	*Arthrocnemum macrostachyum*	Cd	Accumulation of metal in roots and *A. macrostachyum* demonstrated hypertolerance to cadmium stress	Redondo-Gómez et al. (2010)
20.	*Commelina communis*	Cu	Accumulation of metal in roots, copper influx of hyperaccumulator roots was higher than that of nonaccumulator roots	Wang and Zhong, (2011)
21.	*Arabidopsis thaliana*	Cd	Low concentration of salt alleviates Cd-induced growth inhibition and increases Cd accumulation in *Arabidopsis thaliana*	Xu et al. (2010)
22.	*Ocimum tenuiflorum L., Ocimum gratissimum L., and Ocimum basilicum L.*	As	Plants accumulated high amount of As ($\mu g\ g^{-1}$ dry weight) (662 in *O.tenuiflorum*, 764 in *O. basilicum* and 831 in *O. gratissimum* at 100 μM As(III) after 10 days with the order of accumulation being roots > stem > leaves	Siddiqui et al. (2013)

31.4.1.4 Phytovolatilisation

In this process of phytoremediation plants release contaminants present in soil into the environment in volatile form. Uptake and transpiration of contaminants present in soil by a plant and its release to the atmosphere by various process applied by plants i.e. contaminants uptake, plant metabolism and plant transpiration. By this process contaminants could be transformed to less-toxic forms, such as elemental mercury and dimethyl selenite gas (Pandey and Bagga 2013).

31.4.1.5 Phytodegradation

This process involves degradation of organic contaminants by release of enzymes from roots or through metabolic activities within plant tissue. In phytodegradation process organic contaminants are taken up by roots and metabolized in plant tissues to less toxic substance (Greipsson 2011). By metabolic process driven by plants they hydrolyse organic compounds into smaller units i.e. absorbed by plants. It can be a promising technology to remediate soil, sediments, sludges, and groundwater. Surface water can also be remediated using Phytodegradation (Pandey and Bagga 2013).

31.4.2 Beneficial Interaction Between Micro-plant: Toxic Element Remediation

31.4.2.1 Endophytes

Plants-microbes interaction is beneficial for plant as well for microorganisms, as the plant provides the habitat as well as nutrients to the associated rhizosphere and endophytic bacteria and in return, the bacteria enhance the stress tolerance of the plant thus improving plant growth and detoxify the plant environment by degrading the pollutant. However, the concept of using endophytic bacteria to improve phytoremediation efficiency has been proposed relatively recently. Endophytic bacteria equipped with pollutant degradation pathways and metabolic activities can diminish both phytotoxicity and evapotranspiration of volatile organic compounds. As endophytic bacteria colonize the plant interior, they can interact more closely with their host plant as compared to rhizobacteria (Fig. 31.6).

Bacterial endophytes that assist the process of Phytoremediation has highly recommended for cleaning up of contaminated polluted soil, endophytic bacteria alleviate toxicity in plant due to soil contamination by their own resistance system and also facilitate plant growth under stress condition (Ma et al. 2016). For tolerating effect of contaminants present in soil, microbes develop mechanism: They involve in efflux, complexation or reduction of contaminants and use them as terminal electron acceptor in anaerobic respiration. Plant growth is adversely affected by various organic contaminants that become a serious global environmental problem day by day. It also alters the composition and activity of soil microbial communities (Li et al. 2017).

Endophytes plays a crucial role in enhancing the process of phytoremediation, this process has been applied for uptake and degradation of several toxins (Fig. 31.7). Phytoremediation method can be used as an alternative method of remediation. But

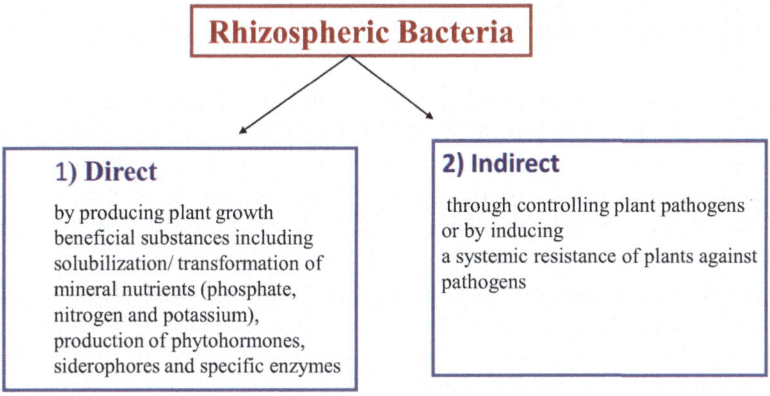

Fig. 31.6 Involvement of rhizospheric bacteria in the improvement of plant growth in polluted soil

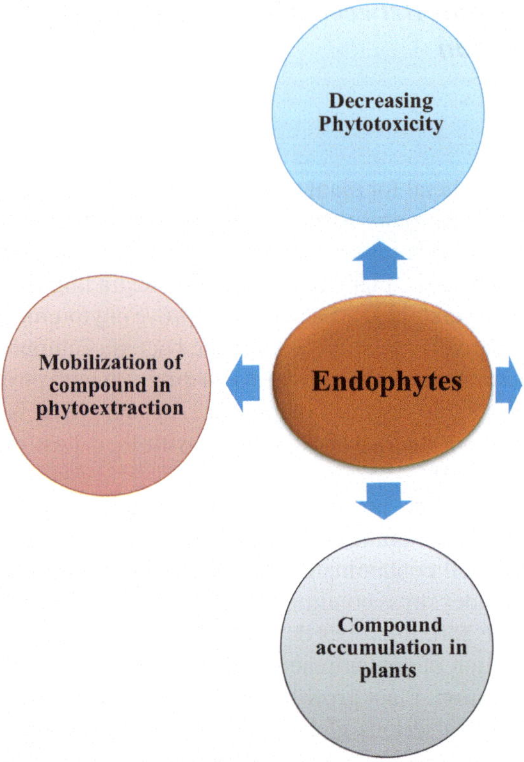

Fig. 31.7 Role of endophytes in phytoremediation

this process is of great challenge due to its drawbacks like if anaerobic microbes are used; there are chances/potential in production of more harmful by product. For example, in the bioremediation of Trichloroethylene more toxic by product like vinyl chloride and cis-dichloroethylene can be produces that remain in the environment for long time.

Rhizospheric microorganisms directly or indirectly enhance plant growth hence they are considered as plant growth promoting organisms. Plant and microorganisms are having symbiotic relationship between them in which plant roots provide growth for microorganisms, and microorganism act as biocatalysts to eliminate contaminants. The interaction of plants and rhizobial microbes can be beneficial for the removal of harmful pollutants from the soil. (Kumar et al. 2019). Microorganisms play an important role in Phytoremediation process they enzymatically convert pollutants (hazardous products) to nonhazardous products. Butdetoxification process for removal of pollutants will proceed efficient only when microorganism get favorable conditions for their growth and activity.

Soil microorganisms play important role in several nutrient transformation processes in soil like nitrogen fixation, nitrification, ammonification, phosphate solubilisation etc. As antibiotics are meant to kill mainly microorganisms, their entry is expected to have harmful effects on agriculturally important soil microorganisms. Repeated contamination through livestock manure application and wastewater irrigation during cropping can accumulate antibiotics in the soil, which may reach beyond the level of the threshold inhibitory concentrations for agriculturally important microorganisms in the soil ecosystem (Table 31.2).

Phytoremediation has been used to treat variety of chemicals like metals, organic excess nutrients and radionuclides. The increase in heavy metal pollution in the agro ecosystem has been become a serious problem worldwide (Fig. 31.8). These metals do not have the capacity to decay in the nature on their own and remain in the nature for long time and become toxic to plants animals and human beings. Anthropogenic activities like removal of heavy metals from industries, waste incineration in the agricultural area is mostly responsible for this problem. To overcome these problem endophytes assisted phytoremediation method can be used.

Anthropogenic activities have aggravated the problem of environmental contamination globally (Yang et al. 2019a, b). The currently used physicochemical methods have limitation requiring the search for environment-friendly options for contaminant removal from a system. Application of microbes, especially diverse bacterial species, is attractive because of their fast growth and easy adaptation even under harsh environmental conditions (Kumar et al. 2020). Many of the xenobiotic compounds are either nonbiodegradable or slowly degradable by microbial communities. Because the indigenous bacteria are less efficient in degrading the pollutants, designing of engineered bacterial systems for enhanced degradation of hazardous contaminants could serve as a suitable approach for mitigating the impacts of pollutants on environment and human health. However, the release of transgenic bacteria for field application to clean the noxious contaminants is a matter of controversy. The release of engineered bacteria may affect the natural microbial diversity, and the genes for catabolism of a particular substance may be harbored by pathogenic microbes. Lots of research has been performed to detect the presence of transgenic bacteria under environmental conditions along with the development of suicidal genetically engineered microbes to restrict their transfer into native microorganisms (Sarma and Prasad 2019). Hence, the use of genetically engineered bacteria could serve as a viable option for management of contaminated sites (Table 31.3).

31.4.2.2 PGPR

Plant growth-promoting rhizobacteria (PGPR) are those bacteria that promote plant growth by colonizing the plant root. PGPR assists the plants to uptake nutrients from the environment or preventing plant diseases (Zhuang et al. 2007), but using PGPR in phytoremediation is a new and promising approach to remove contaminants in the environment. It is seen that using plants alone for remediation confronts many limitations. Recently, the application of PGPR in association with plants has been

Table 31.2 Role of microorganisms in bioremediation

S. No.	Organic pollutants	Microbes involved	Remarks	References
1.	Remazol red	*Lysinibacillus sp.*	87 and 72% decolorization with 69% and 62% COD reduction within 48 and 96 h,	Saratale et al. (2013)
2.	Congo red	*Brevibacillus parabrevis*	Removal of 95.71% of dye sample	Talha et al. (2018)
3.	Reactive red 31	*Aspergillus bombycis*	Decolorization potential, chemical oxygen demand (COD) and total organic carbon reduction (TOC) was 99.02, 94.19, and 83.97%, respectively, for 20 mg/L of dye concentration at 12 h	Khan and Fulekar (2017)
4.	Triphenylmethane dyes; crystal violet Cotton blue	*Coriolopsis sp.*	Decolourization activities by filamentous biofilm for CB (79.6%) and CV (85.1%), as compared to free-mycelium forms 72.6 and 58.3%, for CB and CV, respectively	Munck et al. (2018)
5.	Scarlet RR	*Peyronellaea prosopidis*	68, 88 and 91% reduction was recorded in the biological oxygen demand, chemical oxygen demand and color intensity of the textile industry effluent	Bankole et al. (2018)
6.	Indigo blue	*Cyanobacterium Phormidium*	Degrade the dyes present in a textile effluent; therefore, can be used in a tertiary treatment of effluents with recalcitrant compounds	Dellamatrice et al. (2017)
7.	Reactive green 19, reactive blue 160	*Enterobacter cancerogenus*	Disclose decolorized metabolites-accumulation as MFC strategy for decolorization	Chen et al. (2016)
8.	Heavy metals (Cu, Zn)	*B. thuringiensis* A1-3, *P. aeruginosa* A-33, *B. cereus* A1-5, and *B. anthracis* A1-7	The maximum biosorption efficiency was noted at Cu 92.7% and Zn 90.3% by bacterial consortium	Anusha et al. (2021)

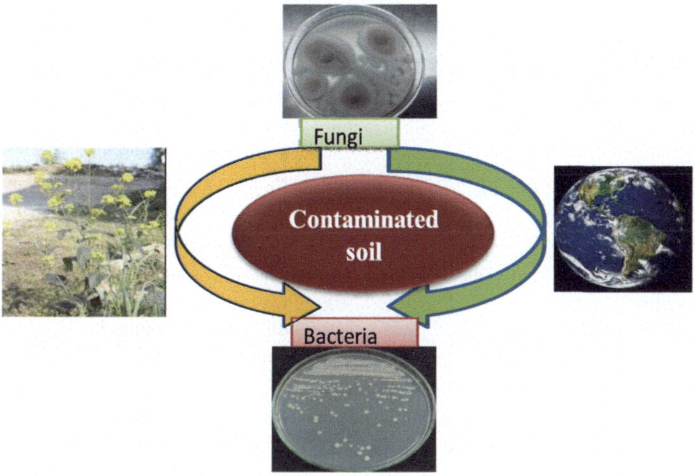

Fig. 31.8 The process of waste bioremediation

extended to remediate contaminated soils. PGPR are heterogenous group of bacteria reside in both rhizospheric and endophytic regions of plants that are directly or indirectly involved in growth promotion. Plant growth promoting microbes are well studied for their biodegradation activity against various pollutants (Shaheen and Sundari 2013; Pratibha et al. 2015). Some reported PGPR for their bioremediation activity have been discussed in Table 31.4.

31.5 Conclusion

Soil is one of the precious natural resource; it is more than the basis for development. But as the urbanization, industrialization and modern agricultural practices increases soil health gets depleted. Soil is a precious nonrenewable commodity that is continuously threatened by destructive anthropogenic activities. Intensification of agriculture land use and adaptation of modern agriculture techniques have led to the accumulation of organic and inorganic contaminants in the soil that are increasing day by day at their alarming level, which not only affect soil and plant health but directly or indirectly adversely affect organisms depend on them. Time to time researches introduce various techniques for the better management of contaminants like Plant and microbes mediated remediation techniques but, in spite of introducing cost effective and environment friendly technologies, some are not commercially available or they are far away from the reach of people may be due to inadequate awareness of their advantages, although further execution should be pursued to make these technologies more effective.

Table 31.3 Degradation of organic contaminants by genetically engineered microorganisms

S. No.	GMMO's	Contaminants	Observation/Remarks	References
1.	*Pseudomonas fluorescens*	Hexahydro-1,3, 5-trinitro-1,3,5-triazine (RDX)	Pollutant degradation was high in presence of a-aminolevulinic acid	Lorenz et al. (2013)
2.	*Rhodococcus erythropolis strains*	Phenol	Engineered cells were 50% much effective in phenol degradation	ZÝdkovß et al. (2013)
3.	*Cupriavidus necator* JMP134-ONP	Nitrophenol	The transgenic bacteria was able to degrade different nitrophenols simultaneously	Hu et al. (2014)
4.	*Pseudomonas putida* KTUe	Organophosphates, pyrethroids, and carbamates	The modified bacteria was able to degrade 50 ppm of selected pesticides within 30 h	Gong et al. (2018)
5.	*Escherichia coli* BL21AI-GOS	Organophosphates	The bacteria could survive only in the presence of contaminant and commended suicide in the absence of contaminant, the engineered microbe is safe for environmental applications	Li and Wu (2009)
6.	*Escherichia coli* JM109	C.I. Direct Blue 71	Dye content removal from 150 to 27.4 ppm within 12 h	Jin et al. (2009)
7.	*Cupriavidus necator* RW112	Monochlorobenzoates and 3,5-dichlorobenzoate	First description of aerobic utilization of Aroclor mixtures	Wittich and Wolff (2007)

Table 31.4 Role of endophytes and PGPR in bioremediation

S. No.	Microorganisms	Category	Plants	Contaminants	Observation/Remarks	References
1.	*Ochrobactrum sp., Bacillus sp.*	Endophytes	Rice *(Oryza sativa L.),*	Cd Pb	Toxic effect of metal was reduced in presence of these bacteria and Germination percentage, relative root elongation (RRE), amylase, and protease activities were increased	Pandey et al. (2013)
2.	*Pseudomonas sp.; Pseudomonas fluorescence; Bacillus cereus*	PGPR	*Zea mays*	Heavy metal Pb, Cd, Ni	Ag-nano particles augmented the PGPR-induced increase in root area and root length	Khan and Bano (2016)
3.	*Penicillium funiculosum LHL06*	Endophytes	Soybean/Cucumber plants	Heavy metals Cu	Cu metal-removal potential	Khan and Lee (2013)
4.	*Bacillus licheniformis NCCP-59*	PGPR	*Oryza sativa*	Heavy metal Ni	Improved seed germination under Ni stress and safeguarded against toxicity	Jamil et al. (2014)
5.	*Achromobacter xylosoxidans Ax10*	PGPR	*Brassica juncea*	Cu	Enhance Cu uptake; increase root, shoot length and fresh, dry weights	Ma et al. (2009)
6.	*Methylobacterium,*	Endophytes	Hybrid Poplar trees (*Populus deltoids x nigra*),		Biodegradation of numerous nitro-aromatic compounds such as 2,4,6-trinitrotoluene	Van Aken et al. (2004)
7.	*Pseudomonas fluorescens G10; Microbacterium sp. G16 (EN)*	PGPR	*Brassica napus*	Heavy metal Pb	Increase Pb uptake in shoots, increase root length, shoot and dry weight	Sheng et al. (2008)

(continued)

Table 31.4 (continued)

S. No.	Microorganisms	Category	Plants	Contaminants	Observation/Remarks	References
8.	*Pseudomonas*	Endophytes	Pea		Degradation of organochlorine herbicide, 2,4-dichlorophenoxyacetic acid (2,4-D)	Germaine et al. (2006)
9.	*Ochrobactrumhaematophilum ZR 3–5, Acidovorax oryzae ZS 1–7, Frigoribacterium faeni ZS 3–5 and Pantoea allii ZS 3–6*	Endophytes	*Zea mays*	Heavy metals	Increase root elongation and biomass of maize seedlings grown in soil contaminated with Cd and Zn	Pereira and Castro (2014)
10.	*Pseudomonas putida*	PGPR	Zea mays	Cd, Ni	PGPR exhibited significant increases in Na, K, Ca, Fe and Zn concomitant with significant decreases in Cd and Ni contents were recorded in the rhizosphere soil of maize inoculated with *Pseudomonas putida*	Asadullah et al. (2021)
11.	*Enterobacter sp.*	Endophytes	Wheat (*Triticum aestivum*) and peanut (*Arachis hypogaea*)	Mn^{2+}, Ni^{2+}, Zn^{2+}, Cu^{2+}, CO^{2+} and Fe^{2+}	Potential of the isolate as an efficient bio-inoculant in bioremediation	Sayyed et al. (2015)

Acknowledgements We gratefully acknowledge Director, GBPNIHE, Almora and Head, CEA&CC, GBPNIHE, Almora for their kind support and guidance during the manuscript preparation.

References

Abdel Khalek STT, Mostafa ZK, Hassan HA, El-Bar A, Marah M, Abu El-Hassan GM (2018) A new list to the Entomofauna associated with Faba Bean, Vicia faba L. (Fabales: Fabaceae) grown in El-Kharga Oasis, new valley governorate, Egypt. Egypt Acad J Biol Sci A Entomol 11(2):95–100. https://doi.org/10.21608/eajb.2018.11901

Abhilash PC, Singh N (2009) Pesticide use and application: an Indian scenario. J Hazard Mater 165(1–3):1–12. https://doi.org/10.1016/j.jhazmat.2008.10.061

Adrees M, Ali S, Rizwan M, Ibrahim M, Abbas F, Farid M, Zia-ur-Rehman M, Irshad MK, Bharwana SA (2015) The effect of excess copper on growth and physiology of important food crops: a review. Environ Sci Pollut Res 22(11):8148–8162

Ai P, Jin K, Alengebawy A, Elsayed M, Meng L, Chen M, Ran Y (2020) Effect of application of different biogas fertilizer on eggplant production: analysis of fertilizer value and risk assessment. Environ Technol Innov 19:101019

Alengebawy A, Abdelkhalek ST, Qureshi SR, Wang MQ (2021) Heavy metals and pesticides toxicity in agricultural soil and plants: ecological risks and human health implications. Toxics 9(3):42

Alloway BJ (2013) Sources of heavy metals and metalloids in soils. In: Heavy metals in soils. Springer, Dordrecht, pp 11–50

Aly AA, Mohamed AA (2012) The impact of copper ion on growth, thiol compounds and lipid peroxidation in two maize cultivars ('Zea mays' L.) grown 'in vitro'. Aust J Crop Sci 6(3):541–549

Amer N, Chami ZA, Bitar LA, Mondelli D, Dumontet S (2013) Evaluation of Atriplex halimus, Medicago lupulina and Portulaca oleracea for phytoremediation of Ni, Pb, and Zn. Int J Phytorem 15(5):498–512. https://doi.org/10.1080/15226514.2012.716102

Ampiah-Bonney RJ, Lanza GR (2007) Phytoextraction of arsenic from soil by *Leersia Oryzoides*. Int J Phytoremediation 9(1):31–40. https://doi.org/10.1080/15226510601139383

An YJ (2004) Soil ecotoxicity assessment using cadmium sensitive plants. Environ Pollut 127(1):21–26. https://doi.org/10.1016/S0269-7491(03)00263-X

Andrades-Moreno L, Cambrollé J, Figueroa ME, Mateos-Naranjo E (2013) Growth and survival of Halimione portulacoides stem cuttings in heavy metal contaminated soils. Mar Pollut Bullet 75(1–2):28–32. https://doi.org/10.1016/j.marpolbul.2013.08.015

Angelova VR, Grekov DF, Kisyov VK, Ivanov KI (2015) Potential of lavender (Lavandula vera L.) for phytoremediation of soils contaminated with heavy metals. Int J Biol Biomol Agric Food Biotechnol Eng 9:465–472

Antoniadis V, Shaheen SM, Stärk HJ, Wennrich R, Levizou E, Merbach I, Rinklebe J (2021) Phytoremediation potential of twelve wild plant species for toxic elements in a contaminated soil. Environ Int 146:106233. https://doi.org/10.1016/j.envint.2020.106233

Anusha P, Natarajan D, Narayanan M (2021) Heavy metal removal competence of individual and bacterial consortium, evolved from metal contaminated soil. Mater Today Proc. https://doi.org/10.1016/j.matpr.2021.01.566

Asadullah AB, Javed H (2021) PGPR assisted bioremediation of heavy metals and nutrient accumulation in zea mays under saline sodic soil. Pak J Bot 53(1):31–38

Ashraf U, Kanu AS, Deng Q, Mo Z, Pan S, Tian H, Tang X (2017) Lead (Pb) toxicity; physio-biochemical mechanisms, grain yield, quality, and Pb distribution proportions in scented rice. Front Plant Sci 8:259. https://doi.org/10.3389/fpls.2017.00259

Balafrej H, Bogusz D, Triqui ZEA, Guedira A, Bendaou N, Smouni A, Fahr M (2020) Zinc hyperaccumulation in plants: a review. Plants 9(5):562

Bankole PO, Adekunle AA, Obidi OF, Chandanshive VV, Govindwar SP (2018) Biodegradation and detoxification of Scarlet RR dye by a newly isolated filamentous fungus, Peyronellaea Prosopidis. Sustain Environ Res 28(5):214–222

Barman H, Das SK, Roy A (2018) Zinc in soil environment for plant health and management strategy. Univ J Agric Res 6(5):149–154

Bhattacharya T, Banerjee DK, Gopal B (2006) Heavy metal uptake by *Scirpus littoralis* Schrad. from fly ash dosed and metal spiked soils. Environ Monitor Assessment 121(1–3):363–380

Brun LA, Le Corff J, Maillet J (2003) Effects of elevated soil copper on phenology, growth and reproduction of five ruderal plant species. Environ Pollut 122(3):361–368. https://doi.org/10.1016/S0269-7491(02)00312-3

Chai M, Shi F, Li R, Qiu G, Liu F, Liu L (2014) Growth and physiological responses to copper stress in a halophyte Spartina alterniflora (Poaceae). Acta Physiol Plant 36(3):745–754

Chandanshive VV, Kadam SK, Khandare RV, Kurade MB, Jeon BH, Jadhav JP, Govindwar SP (2018) In situ phytoremediation of dyes from textile wastewater using garden ornamental plants, effect on soil quality and plant growth. Chemosphere 210:968–976

Chanwala J, Kaushik G, Dar MA, Upadhyay S, Agrawal A (2019) Process optimization and enhanced decolorization of textile effluent by Planococcus sp. isolated from textile sludge. Environ Technol Innov 13:122–129. https://doi.org/10.1016/j.eti.2018.11.008

Chen BY, Ma CM, Han K, Yueh PL, Qin LJ, Hsueh CC (2016) Influence of textile dye and decolorized metabolites on microbial fuel cell-assisted bioremediation. Biores Technol 200:1033–1038. https://doi.org/10.1016/j.biortech.2015.10.011

Chiou WY, Hsu FC (2019) Copper toxicity and prediction models of copper content in leafy vegetables. Sustainability 11(22):6215. https://doi.org/10.3390/su11226215

Dafale N, Rao NN, Meshram SU, Wate SR (2008) Decolorization of azo dyes and simulated dye bath wastewater using acclimatized microbial consortium–biostimulation and halo tolerance. Biores Technol 99(7):2552–2558. https://doi.org/10.1016/j.biortech.2007.04.044

D'amore JJ, Al-Abed SR, Scheckel KG, Ryan JA (2005) Methods for speciation of metals in soils: a review. J Environ Qual 34(5):1707–1745

Dellamatrice PM, Silva-Stenico ME, Moraes LABD, Fiore MF, Monteiro RTR (2017) Degradation of textile dyes by cyanobacteria. Brazilian J Microbiol 48:25–31. https://doi.org/10.1016/j.bjm.2016.09.012

Devi NL, Raha P (2013) Contamination of Organochlorine Pesticides (OCPs) in India. Bull Environ Sci Res 2:9–14

Dezhban A, Shirvany A, Attarod P, Delshad M, Matinizadeh M, Khoshnevis M (2015) Cadmium and lead effects on chlorophyll fluorescence, chlorophyll pigments and proline of Robinia pseudoacacia. J For Res 26(2):323–329. https://doi.org/10.1007/s11676-015-0045-9

Diaconu M, Pavel LV, Hlihor RM, Rosca M, Fertu DI, Lenz M, Corvini PX, Gavrilescu M (2020) Characterization of heavy metal toxicity in some plants and microorganisms—a preliminary approach for environmental bioremediation. New Biotechnol 56:130–139. https://doi.org/10.1016/j.nbt.2020.01.003

Dotaniya ML, Dotaniya CK, Solanki P, Meena VD, Doutaniya RK (2020) Lead contamination and its dynamics in soil–plant system. In: Gupta D, Chatterjee S, Walther C (eds) Lead in plants and the environment. Radionuclides and heavy metals in the environment. Springer

Du L, Liu W (2012) Occurrence, fate, and ecotoxicity of antibiotics in agro-ecosystems. A review. Agron Sustain Dev 32(2):309–327

Friedlova M (2010) The influence of heavy metals on soil biological and chemical properties. Soil Water Res 5(1):21–27. https://doi.org/10.17221/11/2009-SWR

García-Salgado S, García-Casillas D, Quijano-Nieto MA, Bonilla-Simón MM (2012) Arsenic and heavy metal uptake and accumulation in native plant species from soils polluted by mining activities. Water Air Soil Pollut 223(2):559–572

Garrido S, Del Campo GM, Esteller MV, Vaca R, Lugo J (2005) Heavy metals in soil treated with sewage sludge composting, their effect on yield and uptake of broad bean seeds (Vicia faba L.). Water Air Soil Pollut 166(1):303–319

Germaine KJ, Liu X, Cabellos GG, Hogan JP, Ryan D, Dowling DN (2006) Bacterial endophyte-enhanced phytoremediation of the organochlorine herbicide 2, 4-dichlorophenoxyacetic acid. FEMS Microbiol Ecol 57(2):302–310

Glick BR (2010) Using soil bacteria to facilitate phytoremediation. Biotechnol Adv 28(3):367–374

Gong T, Xu X, Dang Y, Kong A, Wu Y, Liang P, Wang S, Yu H, Xu P, Yang C (2018) An engineered Pseudomonas putida can simultaneously degrade organophosphates, pyrethroids and carbamates. Sci Total Environ 628:1258–1265. https://doi.org/10.1016/j.scitotenv.2018.02.143

Greipsson S (2011) Phytoremediation. Nat Educ Knowledge 3(10):7

Gurjar OP, Meena R, Latare AM, Rai S, Kant S, Kumar A, Kumar A (2017) Effects of sewage wastewater irrigation compare to ground water irrigation on soil physico-chemical properties. Int J Chem Stud 5(6):265–267

Hamid N, Bukhari N, Jawaid F (2010) Physiological responses of Phaseolus vulgaris to different lead concentrations. Pak J Bot 42(1):239–246

Hammerschmitt RK, Tiecher TL, Facco DB, Silva LO, Schwalbert R, Drescher GL, Trentin E, Somavilla LM, Kulmann MS, Silva IC, Tarouco CP (2020) Copper and zinc distribution and toxicity in 'Jade'/'Genovesa' young peach tree. Scientia Horticulturae 259:108763. https://doi.org/10.1016/j.scienta.2019.108763

Hinojosa MB, Carreira JA, García-Ruíz R, Dick RP (2004) Soil moisture pre-treatment effects on enzyme activities as indicators of heavy metal-contaminated and reclaimed soils. Soil Biol Biochem 36(10):1559–1568. https://doi.org/10.1016/j.soilbio.2004.07.003

Hooda V (2007) Phytoremediation of toxic metals from soil and waste water. J Environ Biol 28(2):367

Howladar SM (2014) A novel Moringa oleifera leaf extract can mitigate the stress effects of salinity and cadmium in bean (Phaseolus vulgaris L.) plants. Ecotoxicol Environ Saf 100:69–75. https://doi.org/10.1016/j.ecoenv.2013.11.022

Hu F, Jiang X, Zhang JJ, Zhou NY (2014) Construction of an engineered strain capable of degrading two isomeric nitrophenols via a sacB-and gfp-based markerless integration system. Appl Microbiol Biotechnol 98(10):4749–4756

Jamil M, Zeb S, Anees M, Roohi A, Ahmed I, ur Rehman S, Rha ES (2014) Role of Bacillus licheniformis in phytoremediation of nickel contaminated soil cultivated with rice. Int J Phytoremediat 16(6):554–571.https://doi.org/10.1080/15226514.2013.798621

Jin R, Yang H, Zhang A, Wang J, Liu G (2009) Bioaugmentation on decolorization of CI Direct Blue 71 by using genetically engineered strain Escherichia coli JM109 (pGEX-AZR). J Hazard Mater 163(2–3):1123–1128. https://doi.org/10.1016/j.jhazmat.2008.07.067

Kale RA, Lokhande VH, Ade AB (2015) Investigation of chromium phytoremediation and tolerance capacity of a weed, Portulaca oleracea L. in a hydroponic system. Water Environ J 29(2):236–242. https://doi.org/10.1111/wej.12106

Kalve S, Sarangi BK, Pandey RA, Chakrabarti T (2011) Arsenic and chromium hyperaccumulation by an ecotype of Pteris vittata–prospective for phytoextraction from contaminated water and soil. Curr Sci 888–894

Kanwal MS, Mukherjee S, Joshi R, Rai S (2019) Impact assessment of changing environmental and socio-economical factors on crop yields of central Himalaya with emphasis to climate change. Environ Ecol 37(1B):324–332

Karthikeyan R, Kulakow PA (2003) Soil plant microbe interactions in phytoremediation. In: Phytoremediation. Springer, Berlin, Heidelberg, pp 52–74

Khalid A, Arshad M, Crowley DE (2008) Decolorization of azo dyes by Shewanella sp. under saline conditions. Appl Microbiol Biotechnol 79(6):1053–1059

Khan AL, Lee IJ (2013) Endophytic Penicillium funiculosum LHL06 secretes gibberellin that reprograms Glycine max L. growth during copper stress. BMC Plant Biol 13(1):1–14

Khan MU, Malik RN, Muhammad S (2013) Human health risk from heavy metal via food crops consumption with wastewater irrigation practices in Pakistan. Chemosphere 93(10):2230–2238. https://doi.org/10.1016/j.chemosphere.2013.07.067

Khan N, Bano A (2016) Role of plant growth promoting rhizobacteria and Ag-nano particle in the bioremediation of heavy metals and maize growth under municipal wastewater irrigation. Int J Phytorem 18(3):211–221

Khan R, Fulekar MH (2017) Mineralization of a sulfonated textile dye reactive red 31 from simulated wastewater using pellets of Aspergillus bombycis. Bioresour Bioprocess 4(1):1–11

Khan S, Cao Q, Zheng YM, Huang YZ, Zhu YG (2008) Health risks of heavy metals in contaminated soils and food crops irrigated with wastewater in Beijing, China. Environ Pollut 152(3):686–692. https://doi.org/10.1016/j.envpol.2007.06.056

Kumar M, Rai AK, Rai S, Rani P, Anjum M (2018) Effect of integrated nutrient management on physico-chemical soil properties under rice crop in hot sub humid ecoregion of middle Gangetic plains of India. J Pure Appl Microbiol 10(4):3051–3056

Kumar A, Kumar A, Singh R, Singh R, Pandey S, Rai A, Singh VK, Rahul B (2020) Genetically engineered bacteria for the degradation of dye and other organic compounds. Abatement Environ Pollut 331–350. https://doi.org/10.1016/b978-0-12-818095-2.00016-3

Kumar D, Singh DP, Barman SC, Kumar N (2016) Heavy metal and their regulation in plant system: an overview. Plant Responses Xenobiot 19–38. sci.2010.08.016

Kumar N, Jeena N, Gangola S, Singh H (2019) Phytoremediation facilitating enzymes: an enzymatic approach for enhancing remediation process. In: Smart bioremediation technologies. Academic Press, pp 289–306

Kushwaha A, Hans N, Kumar S, Rani R (2018) A critical review on speciation, mobilization and toxicity of lead in soil-microbe-plant system and bioremediation strategies. Ecotoxicol Environ Safety 147:1035–1045. https://doi.org/10.1016/j.ecoenv.2017.09.049

Li C, Xu Y, Jiang W, Dong X, Wang D, Liu B (2013) Effect of NaCl on the heavy metal tolerance and bioaccumulation of Zygosaccharomyces rouxii and Saccharomyces cerevisiae. Biores Technol 143:46–52. https://doi.org/10.1016/j.biortech.2013.05.114

Li Q, Wu YJ (2009) A fluorescent, genetically engineered microorganism that degrades organophosphates and commits suicide when required. Appl Microbiol Biotechnol 82(4):749–756

Li X, Meng D, Li J, Yin H, Liu H, Liu X, Cheng C, Xiao Y, Liu Z, Yan M (2017) Response of soil microbial communities and microbial interactions to long-term heavy metal contamination. Environ Pollut 231:908–917

Li X, Zhang X, Yang Y, Li B, Wu Y, Sun H, Yang Y (2016) Cadmium accumulation characteristics in turnip landraces from China and assessment of their phytoremediation potential for contaminated soils. Front Plant Sci 7:1862. https://doi.org/10.3389/fpls.2016.01862

Liao M, Luo YK, Zhao XM, Huang CY (2005) Toxicity of cadmium to soil microbial biomass and its activity: effect of incubation time on Cd ecological dose in a paddy soil. J Zhejiang Univ Sci B 6(5):324

Liu YM, Liu DY, Zhang W, Chen XX, Zhao QY, Chen XP, Zou CQ (2020) Health risk assessment of heavy metals (Zn, Cu, Cd, Pb, As and Cr) in wheat grain receiving repeated Zn fertilizers. Environ Pollut 257:113581. https://doi.org/10.1016/j.envpol.2019.113581

Loffredo E, Picca G, Parlavecchia M (2021) Single and combined use of Cannabis sativa L. and carbon-rich materials for the removal of pesticides and endocrine-disrupting chemicals from water and soil. Environ Sci Pollut Res 28(3):3601–3616

Lorenz A, Rylott EL, Strand SE, Bruce NC (2013) Towards engineering degradation of the explosive pollutant hexahydro-1, 3, 5-trinitro-1, 3, 5-triazine in the rhizosphere. FEMS Microbiol Lett 340(1):49–54. https://doi.org/10.1111/1574-6968.12072

Łukowski A, Dec D (2018) Influence of Zn, Cd, and Cu fractions on enzymatic activity of arable soils. Environ Monit Assess 190(5):1–12

Ma Y, Rajkumar M, Freitas H (2009) Inoculation of plant growth promoting bacterium Achromobacter xylosoxidans strain Ax10 for the improvement of copper phytoextraction by Brassica juncea. J Environ Manage 90(2):831–837

Ma Y, Rajkumar M, Zhang C, Freitas H (2016) Inoculation of Brassica oxyrrhina with plant growth promoting bacteria for the improvement of heavy metal phytoremediation under drought conditions. J Hazard Mater 320:36–44. https://doi.org/10.1016/j.jhazmat.2016.08.009

Magnuson ML, Kelty CA, Kelty KC (2001) Trace metal loading on water-borne soil and dust particles characterized through the use of Split-flow thin-cell fractionation. Anal Chem 73(14):3492–3496. https://doi.org/10.1021/ac0015321

Malik Z, Ahmad M, Abassi GH, Dawood M, Hussain A, Jamil M (2017) Agrochemicals and soil microbes: interaction for soil health. In: Xenobiotics in the soil environment. Springer, Cham, pp 139–152

Meng L, Alengebawy A, Ai P, Jin K, Chen M, Pan Y (2020) Techno-economic assessment of three modes of large-scale crop residue utilization projects in china. Energies 13(14):3729. https://doi.org/10.3390/en13143729

Munck C, Thierry E, Gräßle S, Chen SH, Ting ASY (2018) Biofilm formation of filamentous fungi Coriolopsis sp. on simple muslin cloth to enhance removal of triphenylmethane dyes. J Environ Manage 214:261–266

Muthunarayanan V, Santhiya M, Swabna V, Geetha A (2011) Phytodegradation of textile dyes by water hyacinth (Eichhornia crassipes) from aqueous dye solutions. Int J Environ Sci 1(7):1702–1717

Onyia PC, Ozoko DC, Ifediegwu SI (2020) Phytoremediation of arsenic-contaminated soils by arsenic hyperaccumulating plants in selected areas of Enugu State, Southeastern, Nigeria. Geol Ecol Landscapes 1–12. https://doi.org/10.1080/24749508.2020.1809058

Pandey S, Ghosh PK, Ghosh S, De TK, Maiti TK (2013) Role of heavy metal resistant Ochrobactrum sp. and Bacillus spp. strains in bioremediation of a rice cultivar and their PGPR like activities. J Microbiol 51(1):11–17

Pandey SS, Bagga D (2013) Phytoremediation an alternative. Int J Environ Eng Manage 4(5):483–488

Pedro CA, Santos MS, Ferreira SM, Gonçalves SC (2013) The influence of cadmium contamination and salinity on the survival, growth and phytoremediation capacity of the saltmarsh plant Salicornia ramosissima. Mar Environ Res 92:197–205. https://doi.org/10.1016/j.marenvres.2013.09.018

Pereira SIA, Castro PML (2014) Diversity and characterization of culturable bacterial endophytes from Zea mays and their potential as plant growth-promoting agents in metal-degraded soils. Environ Sci Pollut Res 21(24):14110–14123

Pratibha C, Teja T, Krishna PM (2015) Effect of chemical treatments on the germination and subsequent seedlings growth of papaya (Carica papaya L.) seeds cv. Pusa Nanha. J Agric Eng Food Technol 2(3):189–191

Qi X, Xu X, Zhong C, Jiang T, Wei W, Song X (2018) Removal of cadmium and lead from contaminated soils using sophorolipids from fermentation culture of Starmerella bombicola CGMCC 1576 fermentation. Int J Environ Res Public Health 15(11):2334. https://doi.org/10.3390/ijerph15112334

Qin G, Niu Z, Yu J, Li Z, Ma JY, Xiang P (2020) Soil heavy metal pollution and food safety in China: effects, sources and removing technology. Chemosphere 129205

Rai AK, Rakshit A, Rai S, Parihar M, Seth V (2016) Factors responsible for phosphorus uptake efficiencies of crop species in hot sub humid eco-region of Middle Gangetic Plains of India. J Pure Appl Microbiol 10(2):1303–1310

Rai S, Rani P, Kumar M, Rai AK, Shahi SK (2014) Effect of integrated use of vermicompost, FYM, PSB and azotobactor on physicochemical properties of soil under onion crop. Environ Ecol 32:1797–1803

Raiesi F, Sadeghi E (2019) Interactive effect of salinity and cadmium toxicity on soil microbial properties and enzyme activities. Ecotoxicol Environ Safety 168:221–229. https://doi.org/10.1016/j.ecoenv.2018.10.079

Rascio N, Navari-Izzo F (2011) Heavy metal hyperaccumulating plants: how and why do they do it? And what makes them so interesting? Plant Sci 180(2):169–181

Redondo-Gómez S, Mateos-Naranjo E, Andrades-Moreno L (2010) Accumulation and tolerance characteristics of cadmium in a halophytic Cd-hyperaccumulator, Arthrocnemum macrostachyum. J Hazard Mater 184(1–3):299–307. https://doi.org/10.1016/j.jhazmat.2010.08.036

Saha JK, Selladurai R, Coumar MV, Dotaniya ML, Kundu S, Patra AK (2017) Major inorganic pollutants affecting soil and crop quality. In: Soil pollution-an emerging threat to agriculture. Springer, Singapore, pp 75–104

Saratale RG, Gandhi SS, Purankar MV, Kurade MB, Govindwar SP, Oh SE, Saratale GD (2013) Decolorization and detoxification of sulfonated azo dye CI Remazol Red and textile effluent by isolated Lysinibacillus sp. RGS. J Biosci Bioeng 115(6):658–667. https://doi.org/10.1016/j.jbiosc.2012.12.009

Sarma H, Prasad MNV (2019) Metabolic engineering of rhizobacteria associated with plants for remediation of toxic metals and metalloids. In: Prasad MNV (ed) Transgenic plant technology for remediation of toxic metals and metalloids. Academic Press, pp 299e318

Saxena S, Rai S (2020) Okara: a low-cost adsorbent for textile waste water treatment. Res Biotica 2(2):26–29

Sayyed RZ, Patel PR, Shaikh SS (2015) Plant growth promotion and root colonization by EPS producing Enterobacter sp. RZS5 under heavy metal contaminated soil. Indian J Exp Biol 53(2). http://nopr.niscair.res.in/handle/123456789/30443

Senesil GS, Baldassarre G, Senesi N, Radina B (1999) Trace element inputs into soils by anthropogenic activities and implications for human health. Chemosphere 39(2):343–377

Shaheen S, Sundari K (2013) Exploring the applicability of PGPR to remediate residual organophosphate and carbamate pesticides used in agriculture fields. Int J Agric Food Sci Technol 4(10):947–954

Sharma B, Sarkar A, Singh P, Singh RP (2017) Agricultural utilization of biosolids: a review on potential effects on soil and plant grown. Waste Manage 64:117–132

Sheng XF, Xia JJ, Jiang CY, He LY, Qian M (2008) Characterization of heavy metal-resistant endophytic bacteria from rape (Brassica napus) roots and their potential in promoting the growth and lead accumulation of rape. Environ Pollut 156(3):1164–1170. https://doi.org/10.1016/j.envpol.2008.04.007

Siddiqui F, Krishna SK, Tandon PK, Srivastava S (2013) Arsenic accumulation in Ocimum spp. and its effect on growth and oil constituents. Acta Physiologiae Plantarum 35(4):1071–1079

Singh AP, Singh SK, Rai S, Kumar M (2020) Soil carbon dynamics in relation to soil surface management and cropping system. In: Carbon management in tropical and sub-tropical terrestrial systems. Springer, pp 159–172

Singh S, Fulzele DP (2021) Phytoextraction of arsenic using a weed plant Calotropis procera from contaminated water and soil: growth and biochemical response. Int J Phytorem 1–9. https://doi.org/10.1080/15226514.2021.1895717

Song C, Yan Y, Rosado A, Zhang Z, Castellarin SD (2019) ABA alleviates uptake and accumulation of zinc in grapevine (Vitis vinifera l.) by inducing expression of ZIP and detoxification-related genes. Front Plant Sci 10:872

Srivastava V, de Araujo ASF, Vaish B, Bartelt-Hunt S, Singh P, Singh RP (2016) Biological response of using municipal solid waste compost in agriculture as fertilizer supplement. Rev Environ Sci Bio/Technol 15:677–696. https://doi.org/10.1007/s11157-016-9407-9

Tahir U, Yasmin A, Khan UH (2016) Phytoremediation: potential flora for synthetic dyestuff metabolism. J King Saud Univ-Sci 28(2):119–130

Talha MA, Goswami M, Giri BS, Sharma A, Rai BN, Singh RS (2018) Bioremediation of Congo red dye in immobilized batch and continuous packed bed bioreactor by Brevibacillus parabrevis using coconut shell bio-char. Biores Technol 252:37–43. https://doi.org/10.1016/j.biortech.2017.12.081

Thathola P, Chandola D, Agnihotri V, Rai S (2019) Phytoremeditation: a potential tool for waste water recycling. Res Biotica 1(1):05–08

Tóth G, Hermann T, Da Silva MR, Montanarella L (2016) Heavy metals in agricultural soils of the European Union with implications for food safety. Environ Int 88:299–309. https://doi.org/10.1016/j.envint.2015.12.017

Uera RB, Paz-Alberto AM, Sigua GC (2007) Phytoremediation potentials of selected tropical plants for ethidium bromide. Environ Sci Pollut Res-Int 14(7):505–509

Uzu G, Sobanska S, Aliouane Y, Pradere P, Dumat C (2009) Study of lead phytoavailability for atmospheric industrial micronic and sub-micronic particles in relation with lead speciation. Environ Pollut 157(4):1178–1185. https://doi.org/10.1016/j.envpol.2008.09.053

Van Aken B, Peres CM, Doty SL, Yoon JM, Schnoor JL (2004) Methylobacteriumpopuli sp. nov., a novel aerobic, pink-pigmented, facultatively methylotrophic, methane-utilizing bacterium isolated from poplar trees (Populus deltoides × nigra DN34). Int J Syst Evol Microbiol 54(4):1191–1196

Vega FA, Andrade ML, Covelo EF (2010) Influence of soil properties on the sorption and retention of cadmium, copper and lead, separately and together, by 20 soil horizons: comparison of linear regression and tree regression analyses. J Hazard Mater 174(1–3):522–533. https://doi.org/10.1016/j.jhazmat.2009.09.083

Wang H, Zhong G (2011) Effect of organic ligands on accumulation of copper in hyperaccumulator and nonaccumulator Commelina communis. Biol Trace Elem Res 143(1):489–499

Wang Y, Shen H, Xu L, Zhu X, Li C, Zhang W, Xie Y, Gong Y, Liu L (2015) Transport, ultrastructural localization, and distribution of chemical forms of lead in radish (Raphanus sativus L.). Front Plant Sci 6:293. https://doi.org/10.3389/fpls.2015.00293

Wittich RM, Wolff P (2007) Growth of the genetically engineered strain Cupriavidus necator RW112 with chlorobenzoates and technical chlorobiphenyls. Microbiology (reading) 153(Pt 1):186–195. https://doi.org/10.1099/mic.0.29096-0 (PMID: 17185547)

Woldetsadik D, Drechsel P, Keraita B, Itanna F, Gebrekidan H (2017) Heavy metal accumulation and health risk assessment in wastewater-irrigated urban vegetable farming sites of Addis Ababa, Ethiopia. Int J Food Contam 4(9). https://doi.org/10.1186/s40550-017-0053-y

Wu H, Liu X, Zhao J, Yu J (2013) Regulation of metabolites, gene expression, and antioxidant enzymes to environmentally relevant lead and zinc in the halophyte Suaeda salsa. J Plant Growth Regul 32(2):353–361

Xu J, Yin H, Liu X, Li X (2010) Salt affects plant Cd-stress responses by modulating growth and Cd accumulation. Planta 231(2):449–459

Yadav BK, Siebel MA, Van Bruggen JJ (2011) Rhizofiltration of a heavy metal (lead) containing wastewater using the wetland plant *Carex pendula*. CLEAN–Soil Air Water 39(5):467–474

Yang CF, Liu SH, Su YM, Chen YR, Lin CW, Lin KL (2019a) Bioremediation capability evaluation of benzene and sulfolane contaminated groundwater: determination of bioremediation parameters. Scie Total Environ 648:811e818

Yang L, Xue J, He L, Wu L, Ma Y, Chen H, Li H, Peng P, Zhang Z (2019b) Review on ultrasound assisted persulfate degradation of organic contaminants in wastewater: influences, mechanisms and prospective. Chem Eng J 378:122146

Zhang MK, Liu ZY, Wang H (2010) Use of single extraction methods to predict bioavailability of heavy metals in polluted soils to rice. Commun Soil Sci Plant Anal 41(7):820–831. https://doi.org/10.1080/00103621003592341

Zhuang X, Chen J, Shim H, Bai Z (2007) New advances in plant growth-promoting rhizobacteria for bioremediation. Environ Int 33(3):406–413. https://doi.org/10.1016/j.envint.2006.12.005

ZÝdkovß L, Szők J, Ruckß L, Pßtek M, Nešvera J (2013) Biodegradation of phenol using recombinant plasmid-carrying Rhodococcus erythropolis strains. Int Biodeterior Biodegradation 84:179–184. https://doi.org/10.1016/j.ibiod.2012.05.017

Index

P. K. Shit et al. (eds.), *Soil Health and Environmental Sustainability*, Environmental Science and Engineering,
https://doi.org/10.1007/978-3-031-09270-1

S

T

V

W

The manufacturer's authorised representative in the EU is Springer Nature Customer Service Centre GmbH, Europaplatz 3, 69115 Heidelberg, Germany. If you have any concerns regarding our products, please contact ProductSafety@springernature.com

Printed and bound by CPI Group (UK) Ltd, Croydon, CR0 4YY
30/09/2026
02225493-0020